Peptide Materials

Peptide Materials: From Nanostructures to Applications

Edited by

CARLOS ALEMÁN

Department of Chemical Engineering ETSEIB,
Polytechnic University of Catalonia, Spain

and

ALBERTO BIANCO

Institut de Biologie Moléculaire et Cellulaire, CNRS, France

and

MARIANO VENANZI

Department of Chemical Sciences and Technologies,
University of Rome Tor Vergata, Italy

WILEY

This edition first published 2013

Registered Office
John Wiley & Sons, Ltd, The Atrium, Southern Gate, Chichester, West Sussex, PO19 8SQ, United Kingdom

For details of our global editorial offices, for customer services and for information about how to apply for permission to reuse the copyright material in this book please see our website at www.wiley.com.

Library of Congress Cataloging-in-Publication data applied for

A catalogue record for this book is available from the British Library.

ISBN: 9781119953739

Set in 10/12pt Times by SPi Publisher Services, Pondicherry, India

1 2013

Contents

Preface

In 2008 the 30th Conference of the European Peptide Society, held in Helsinki, included for the first time in the Conference program a session devoted to Peptide Materials. At the 2010 Spring Conference of the European Materials Research Society, held in Strasbourg in June 2010, a two-day symposium was dedicated to peptide-based materials. One year after, the *Journal of Peptide Science*, the official journal of the European Peptide Society, published a special issue containing selected contributions to the Symposium (J. Pept. Sci., February 2011, 17, no. 2, pages 73–168).

All these facts testify to the explosion of interest that the study of the design, synthesis, characterization and application of peptide-based materials has experienced. In the last few years, the advancement of the knowledge in this field has been really impressive on both the fundamental side (i.e. the molecular mechanisms and forces that determine the growth of nanometric architectures from basic structural peptide motifs) and applications (i.e. the design of peptide-based devices). Peptide nanowires, nanotubes, self-assembled monolayers and fibers were prepared and characterized by using smart amino acid building blocks, derivatized by well-established and creative peptide chemistry to accomplish specific functions. This has been mainly achieved by mimicking the strategies that Nature uses in the construction of complex supramolecular aggregates, essentially based on self-assembly and self-organization.

The advantage of using peptides as elements of a molecular LEGO resides in the possibility to attain, by proper selection of the single amino acid components, specific secondary structures (α- or 3_{10}-helices, β-turns, β-sheets, coiled coils) and to realize biocompatible surfaces with predetermined hydrophobic/hydrophilic properties. Amphiphilic peptides are shown to be particularly suitable to generate ordered nano- and micro-sized superstructures via hierarchic self-assembly.

Peptides are also playing a key role in the new generation of hybrid materials, as, for example, peptide–inorganic biomineralized systems and peptide/polymer conjugates. This interdisciplinary field connects the world of functional biomolecules to conventional inorganic or polymeric materials, used not only as a solid support, but also as an active medium affecting the functional and structural properties of the bio-layer. This led to the preparation of peptide adducts with quantum dots, carbon nanotubes, ion metals and nanoparticles, producing smart materials endowed with peculiar magnetic and electric properties or suitably functionalized for imaging, bioelectronics, biosensing and molecular recognition.

This book represents the first attempt to make order in this rapidly growing field, gathering knowledge and experiences from leading laboratories. The purpose of this book is not only to collect seminal contributions and diffuse knowledge in the field but also to establish some firm points on the mechanisms and general laws governing the physicochemical behaviour of these systems and to envisage future applications based on these specialized, high-tech materials.

The book is organized in four sections: Part I – Fundamentals of Peptide Materials (Chapters 1 to 4); Part II – Peptide Nanostructures (Chapters 5 to 7); Part III – Peptide Conjugates and Hybrid Materials (Chapters 8 to 11); Part IV – Applications of Peptide Materials (Chapters 12 to 15). In the first part Gil Rosenman (Tel Aviv University, Israel) introduces fundamental concepts on the physics of peptide nanostructures and discusses their potential nanotechnology applications. It is a fascinating chapter, introducing the unique physical properties of peptide nanostructures with applications in optoelectronics (piezoelectricity, second harmonic generation) and therapeutics, as well. Claudio Toniolo (University of Padova, Italy) resumes the synthetic strategies that allow peptide building blocks of well-defined conformational properties to be obtained. The rich family of possible secondary structures, i.e. helices (α–, 3_{10}–, 2.2_7– and collagen helices), turns (α–, β– and γ-turns) and extended conformations (β-pleated sheets; a fully extended structure also termed 2.0_5-helix), are introduced and discussed, in view of their utility in the assembly of 3D structures. Carlos Aleman (Technical University of Catalunya, Spain), in Chapter 3, introduces the structural and dynamical aspects of the conformational preferences of peptides and peptide-containing hybrid systems by computational methods. Concepts and examples are presented with a pedagogical approach suitable for researches interested in the application of computational tools to investigate peptide materials. In conclusion of this introductory section, Mariano Venanzi (University of Rome Tor Vergata, Italy) describes in detail the electron transfer (ET) properties of peptide scaffolds in solution and immobilized on solid substrates. Current theoretical models are introduced and the results of seminal experimental results discussed, focusing on the major achievements in the field. Special emphasis is devoted to the distance dependence of the ET efficiency and the molecular mechanisms actually governing the ET process.

The second part of the book, dedicated to peptide-based nanostructures, begins with the contribution of Shunsaku Kimura (University of Kyoto, Japan) who describes the possibility of obtaining nanostructures (peptide vesicles, dendrimers, protein cages) from the assembly of peptide building blocks. The function of peptide assemblies and their possible applications as nanocarriers for imaging and therapeutic issues is also discussed. The principles of the peptide nanostructure design are introduced by Ruth Nussinov (National Cancer Institute, USA) in Chapter 6. Protein molecules and peptide building blocks are shown to self-assemble in a variety of nanoshapes, peptide nanotubes being the simplest 3D elements in the construction of complex architectures. Sandeep Verma (Indian Institute of Technology Kanpur, India) introduces soft spherical structures obtained from amphiphilic peptides and peptide–polymer hybrids. Strikingly different designs of these peptide constructs are described for creating a stimuli responsive and efficient delivery vehicle for biomedical applications.

Gregg Dieckmann (Texas University, USA) opens the section dedicated to peptide conjugates and peptide-based materials with a contribution concerning the functionalization of carbon nanotubes (CNTs) with peptides. The unique properties associated with peptide/CNT complexes are described, focusing on the effect that the peptide secondary structure (α-helices, β-turns, cyclic peptides) exerts in the dispersion of CNTs. The potential use of peptide/nucleic acid complexes for medical applications is reviewed by Burkhard Bechinger (University of Strasbourg, France), focusing on the modification of the physicochemical properties of nucleic acids caused by incorporation of peptide segments. Multifunctional peptide sequences are introduced with the aim to enhance the delivery of peptide/nucleic acid conjugates to the nucleus and to intracellular compartments and combine the different functionalities for the design of efficient nanovectors. The contribution of Heinz-Berhanrd Kraatz (University of Toronto, Ontario, Canada) concerns the properties of ferrocene–peptide conjugates to probe the interactions with ion and molecular targets and to investigate peptide–protein interactions.

This part ends with the important issue introduced by Claire-Marie Pradier (Pierre et Marie Curie University, Paris, France), discussing the principles governing the adsorption of peptides on metal and oxide surfaces. The dependence on the chemical nature and structure of the surface as well as on the adsorption conditions is clearly highlighted, together with the intermolecular interactions that, prevailing on the surface–peptide interaction, lead to peptide self-assembled mesoscopic structures.

The final section of the book is devoted to the presentation of some selected applications of peptide-based materials that we consider of special interest and prompting real applications in near future. Hicham Fenniri (University of Alberta, Edmonton, Canada) describes the potential use of nanotubular architectures functionalized with peptides for tissue engineering and drug delivery purposes. The role of peptides for enhancing the coating of titanium implant or the incorporation in hydrogel formulation is well delineated in his contribution. A second contribution from the group of Claudio Toniolo in Padova describes an innovative application of peptides as molecular switches. In this approach the control of the conformational transition between different ordered structures, i.e. α-helix/β-pleated sheet, type-I poly-(L-pro)$_n$/ type II poly-(L-pro)$_n$, 3_{10}-helix/α-helix, is devised as the structural motif driving molecular motors. Peptide therapeutics is also the focus of the contribution of Sylviane Muller (CNRS, Institut of Molecular and Cellular Biology, Strasbourg, France), describing how the problems of noninvasive administration could be circumvented by the design of peptide drugs. It is shown that peptides can guarantee higher activity per unit mass, increased selectivity and specificity, greater stability at storage, weaker intrinsic immunogenicity, better organ or tumor penetration. Limitations on the application of peptide drugs are, however, also discussed. Back to inorganics, Takeshi Serizawa (Tokyo Institute of Technology, Japan) illustrates recent development and applications of polymer/peptide conjugates, highlighting the specificity of the interactions at the polymer/peptide interface.

The book covers fast-growing fields of research, characterized by a truly interdisciplinary approach, and requiring the contribution of synthetic peptide chemistry, physicochemical characterization, structural determination and conformational analysis, computational techniques for molecular modeling, imaging with atomic resolution for morphological studies of surfaces, biomedical expertise, medicinal chemistry, engineering for the realization of peptide-based devices. In this sense, this book is unique not only for its content but, more importantly, for its approach and inspiration. We do hope to have edited a book of intellectual stimulus for a broad readership and that graduate students and young researchers could find in this book not only a source of information but, above all, inspiration for creative research activity.

Carlos Alemán, Alberto Bianco and Mariano Venanzi
Barcelona, Strasbourg, Rome
October 2012

List of Contributors

Carlos Alemán Departament d'Enginyeria Química, E.T.S. d'Enginyers Industrials de Barcelona and Center for Research in Nano-Engineering, Universitat Politècnica de Catalunya, Spain.

Alaaeddin Alsbaiee Department of Chemistry, National Institute for Nanotechnology and University of Alberta, Canada.

Nadav Amdursky Department of Materials and Interfaces, Faculty of Chemistry, Weizmann Institute of Science, Israel.

Burkhard Bechinger Institut de Chimie, CNRS, Université de Strasbourg, France.

Samaneh Beheshti Department of Physical and Environmental Sciences and Department of Chemistry, University of Toronto, Canada.

Rachel L. Beingessner National Institute for Nanotechnology, Canada.

Peter Beker School of Electrical Engineering, Iby and Aladar Fleischman, Faculty of Engineering, Tel Aviv University, Israel.

Oscar Bertran Departament de Física Aplicada, EEI, Universitat Politècnica de Catalunya, Spain.

Idit Buch Department of Human Molecular Genetics and Biochemistry, Sackler Institute of Molecular Medicine, Sackler Faculty of Medicine, Tel Aviv University, Israel.

Jordi Casanovas Departament de Química, Escola Politècnica Superior, Universitat de Lleida, Spain.

Marco Crisma Institute of Biomolecular Chemistry, Padova Unit, CNR, Department of Chemistry, University of Padova, Italy.

Louis Danicher CNRS UPR22, Institut Charles Sadron, France.

Gregg R. Dieckmann Department of Chemistry and Alan G. MacDiarmid NanoTech Institute, The University of Texas at Dallas, USA.

Hicham Fenniri Department of Chemistry, National Institute for Nanotechnology and University of Alberta, Canada.

Fernando Formaggio Institute of Biomolecular Chemistry, Padova Unit, CNR, Department of Chemistry, University of Padova, Italy.

Yves Frère CNRS UPR22, Institut Charles Sadron, France.

Emanuela Gatto Department of Chemical Sciences and Technologies, University of Roma Tor Vergata, Italy.

Nidhi Gour Department of Chemistry, Indian Institute of Technology Kanpur, India.

Vincent Humblot Laboratoire de Réactivité de Surface, UMR CNRS 7197, Université Pierre et Marie Curie, France.

Shunsaku Kimura Department of Material Chemistry, Graduate School of Engineering, Kyoto University, Japan.

Anton S. Klimenko Department of Chemistry and Alan G. MacDiarmid NanoTech Institute, The University of Texas at Dallas, USA.

Heinz-Bernhard Kraatz Department of Physical and Environmental Sciences and Department of Chemistry, University of Toronto, Canada.

K. Vijaya Krishna Department of Chemistry, Indian Institute of Technology Kanpur, India.

Jessem Landoulsi Laboratoire de Réactivité de Surface, UMR CNRS 7197, Université Pierre et Marie Curie, France.

Sanela Martić Department of Physical and Environmental Sciences and Department of Chemistry, University of Toronto, Canada.

Alessandro Moretto Institute of Biomolecular Chemistry, Padova Unit, CNR, Department of Chemistry, University of Padova, Italy.

Sylviane Muller CNRS UPR9021, Institut de Biologie Moléculaire et Cellulaire, Immunologie et Chimie Thérapeutiques, France.

Ruth Nussinov Department of Human Molecular Genetics and Biochemistry, Sackler Institute of Molecular Medicine, Sackler Faculty of Medicine, Tel Aviv University, Israel and Center for Cancer Research Nanobiology Program, SAIC-Frederick, Inc., National Cancer Institute, NCI-Frederick, USA.

Claire-Marie Pradier Laboratoire de Réactivité de Surface, UMR CNRS 7197, Université Pierre et Marie Curie, France.

Guillermo Revilla-López Departament d'Enginyeria Química, E.T.S d'Enginyers Industrials de Barcelona, Universitat Politècnica de Catalunya, Spain.

Gil Rosenman School of Electrical Engineering, Iby and Aladar Fleischman, Faculty of Engineering, Tel Aviv University, Israel.

Toshiki Sawada Department of Organic and Polymeric Materials, Tokyo Institute of Technology, Japan.

Takeshi Serizawa Department of Organic and Polymeric Materials, Tokyo Institute of Technology, Japan.

Baljit Singh Department of Veterinary Biomedical Sciences, Western College of Veterinary Medicine, University of Saskatchewan, Canada.

Claudio Toniolo Institute of Biomolecular Chemistry, Padova Unit, CNR, Department of Chemistry, University of Padova, Italy.

Juan Torras Departament d'Enginyeria Química, EEI, Universitat Politècnica de Catalunya, Spain.

Chung-Jung Tsai Center for Cancer Research Nanobiology Program, SAIC-Frederick, Inc., National Cancer Institute, NCI-Frederick, USA.

Motoki Ueda Department of Material Chemistry, Graduate School of Engineering, Kyoto University, Japan.

Mariano Venanzi Department of Chemical Sciences and Technologies, University of Roma Tor Vergata, Italy.

Sandeep Verma Department of Chemistry and DST Unit of Excellence on Soft Nanofabrication, Indian Institute of Technology Kanpur, India.

Thomas J. Webster Division of Engineering and Institute for Molecular and Nanoscale Innovation, Brown University, USA.

David Zanuy Departament d'Enginyeria Química, E.T.S d'Enginyers Industrials de Barcelona, Universitat Politècnica de Catalunya, Spain.

Part I

Fundamentals of Peptide Materials

1

Physics of Peptide Nanostructures and Their Nanotechnology Applications

Nadav Amdursky[1], *Peter Beker*[2] *and Gil Rosenman*[2]

[1] *Department of Materials and Interfaces, Faculty of Chemistry, Weizmann Institute of Science, Israel*

[2] *School of Electrical Engineering, Iby and Aladar Fleischman, Faculty of Engineering, Tel Aviv University, Israel*

Acronyms

Aβ	Amyloid-β
AFM	Atomic force microscope
CD	Circular dichroism
DSC	Differential scanning calorimeter
ESEM	Environmental scanning electron microscope
FF	Diphenylalanine
FT-IR	Fourier transform infrared
FWHM	Full width at half maximum
HFIP	1,1,1,3,3,3-hexafluoro-2-propanol
IP	In plane
OOP	Out of plane
PFM	Piezoelectric force microscopy
PL	Photoluminescence

Peptide Materials: From Nanostructures to Applications, First Edition. Edited by Carlos Alemán, Alberto Bianco and Mariano Venanzi.

PLE	Photoluminescence excitation
PNF	Peptide nanofiber
PNT	Peptide nanotube
QC	Quantum confinement
QD	Quantum dot
QW	Quantum well
SHG	Second harmonic generation
STEM	Scanning transmission electron microscope
TGA	Thermal gravimetric analysis
ToF-SIMS	Time of flight secondary ion mass spectrometry
XPS	X-ray photoelectron spectroscopy
XRD	X-ray diffraction

1.1 Introduction to Peptide Nanotubes

The concept of nanotechnology has emerged in 1991 as a new discipline in the materials sciences, and it was believed that science has entered an era where we can control the locations of individual atoms [1] and to self-assemble nanoscale factories [2]. Nanomaterials can possess unique properties that differ from the same materials in the macroscale. They can be roughly divided into two kinds of groups, inorganic and organic materials. Today the inorganic materials are well studied and mostly used in nanotechnological devices.

Among the organic materials we can find the subgroup of (bio-)organic (which can also be called bio-inspired) materials. Bio-organic materials are fabricated from molecules that are composed of biological elements, which in some cases can be chemically synthesized. One of the main differences between inorganic and bio-organic nanomaterials is the production process. Making inorganic nanomaterial structures or devices is usually done using 'top-down' techniques, such as lithography. However, as the size of the inorganic nanomaterial decreases, it becomes more complicated and expensive to use 'top-down' techniques. On the other hand, bio-inspired nanomaterials are produced by 'bottom-up' techniques. In the 'bottom-up' approach, single biomolecules interact with one another using basic molecular recognition principles, to form a supramolecular structure. In general, noncovalent interactions, such as van der Waals, hydrophobic/hydrophilic, dipole–dipole, electrostatic, and aromatic, play a major role in the 'bottom-up' process of forming the bio-organic supramolecular nanomaterial structure from its elementary building blocks [3].

Although there is an enormous variety of structures in the biological world, the set of building blocks is relatively small. In general, we can divide the biological world building blocks into four thematic subgroups: amino acids, sugars, nucleotides, and lipid molecules [4]. These building blocks can assemble (covalently or noncovalently) into supramolecular structures: amino acids into peptide or proteins, nucleotides into DNA/RNA, lipids into membranes, and more. The use of the self-organization process of biological materials has been developed into a new branch of nanotechnology: bionanotechnology. In this branch, researchers try to use and integrate biological materials in nanotechnological platforms, such as DNA- or protein-based sensors, sophisticated lipid-based drug delivery systems, DNA tweezers, and many more (detailed reviews about bionanotechnology can be found in references [3] and [5–7]). In this chapter we will focus on self-assembled peptide nanostructures [8].

The first one to use the term peptide nanotubes (PNTs) was Ghadiri [9] in 1993. The Ghadiri group used cyclic peptide, which contains an even number of alternating D- and L-amino acids. The cyclic peptides can self-assemble to form nanocrystalline PNTs, which are in the micrometer length scale, with a diameter of 7–8 Å [9–11] (Figure 1.1a). Since the discovery in 1993, hundreds of works have been conducted in exploring the cyclic PNT properties toward antimicrobial materials [12, 13], incorporation at artificial photosystems [14], adaptors for biosensors [15–17], membrane transporters [18, 19], and more (a detailed review on cyclic peptides can be found in reference [16]).

Kimura has investigated the field of peptide engineering further with an emphasis on tubular structures composed of cyclic β-peptides, which consist of β-amino acids with an amino group bonded to the β carbon rather than the α carbon (as in the 20 metabolic biological amino acids) [20–22]. He proposed models for the self-assembly of molecular architectures on the basis of molecular dipoles, and by that method opened the avenue to a new interdisciplinary field – 'molecular dipole engineering' (Figure 1.1b). The strong and directional dipole–dipole interaction can help to arrange molecules in a specific way, either when dipole units are incorporated into the molecule or when molecules are placed in an electric field.

PNTs composed of β-amino acids exhibit a strong dipole moment along the nanotube's primer axis. Those PNTs have a strong tendency to associate together to form thick bundles, probably because the dipole–dipole interactions between the PNTs attract them to take an antiparallel orientation, canceling out the total dipole for stabilization [20]. Interestingly, by integration of cyclohexyl groups into the cyclic β-peptides, the PNTs have self-assembled into bundles with all the amide groups pointing in the same direction in the bundle.

This parallel arrangement in the bundle is highly unique. A plausible explanation is that the cyclohexyl groups fit in the spaces between the nanotubes in an interdigital manner, stabilizing the parallel orientation [21]. The strong electric field generated by the dipole can influence charge movements in molecular assemblies. Bio-organic structures with a strong dipole moment can be applicable to various fields, such as molecular electronics and medicinal chemistry [21], as well as functional nanomaterials in nanopiezotronics or nanophotronics due to observation of strong piezoelectric and second harmonic generation effects in some of the peptide nanostructures [23–25].

Another kind of peptide tubular structure (which is the main scope of this chapter) is formed from dipeptides. The first person who scaled down and showed that small dipeptides can self-assemble into ordered PNT-like crystalline structures was Görbitz [26–28] in 2001. By using only crystallographic techniques he was able to characterize the conformation packaging of over 160 dipeptides, which can self-assemble into tubular-like supramolecular structures [27, 29, 30]. Among the large variety of dipeptides that can self-assemble into supramolecular crystal structures that Görbitz has considered, we can find the diphenylalanine (FF) peptide [26].

The breakthrough for FF-based PNTs was in 2003 by the work of Reches and Gazit [31]. They discovered the formation of self-assembled FF PNTs in an aqueous solution. The inspiration for the formation of FF PNTs came from amyloid protein fibrils. Naturally self-assembled protein fibrils, which are associated with neurodegenerative diseases, have been thoroughly researched in the past century. The most common and studied disorder is Alzheimer's disease, with a defined, well-known fibrils structure made of amyloid-β (Aβ)

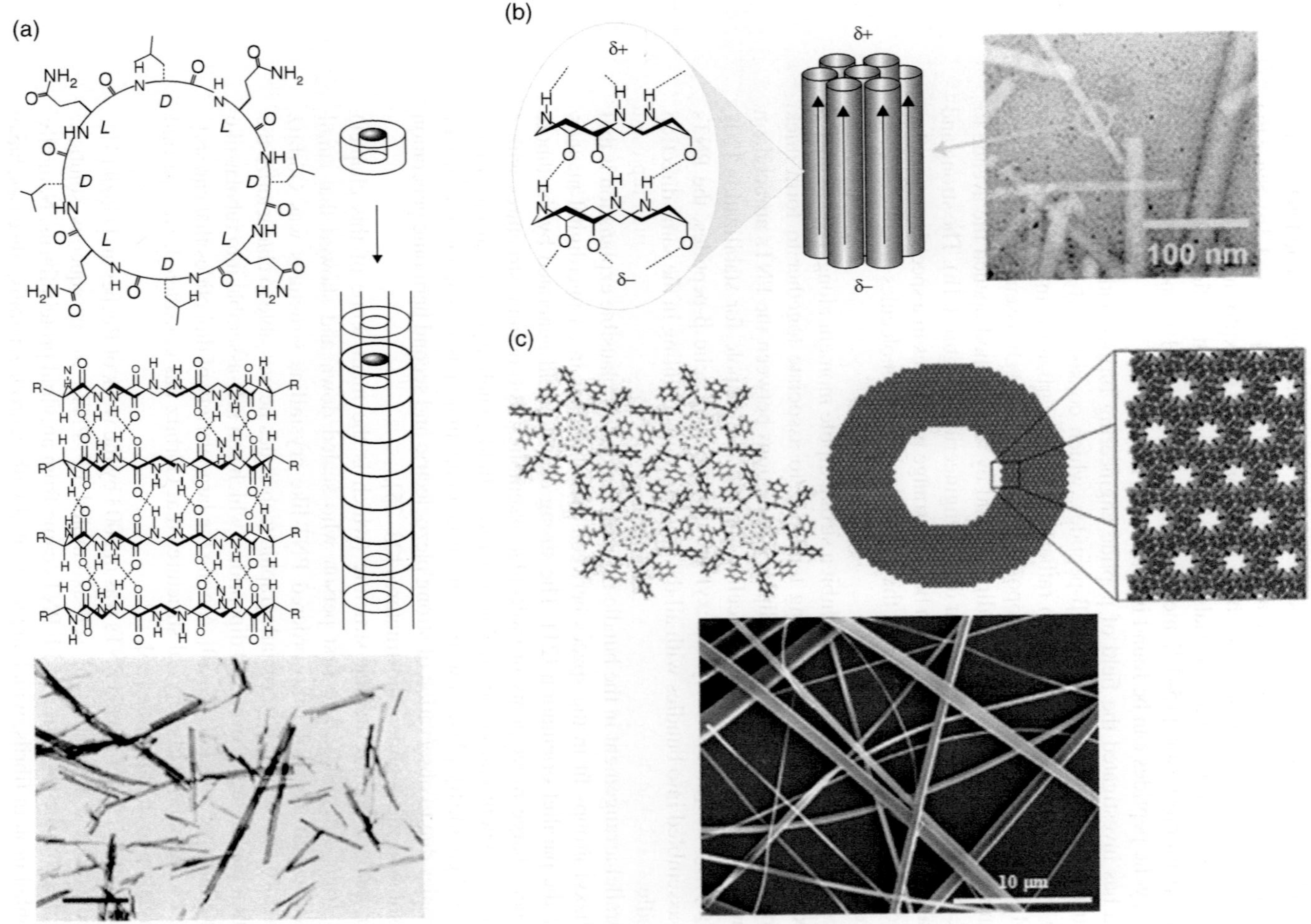

Figure 1.1 *(a) Alternating D- and L-amino acids PNT: scheme (upper part) and morphology (bottom part). (b) The concept of dipole engineering, molecular packing scheme and morphology. (c) Molecular packing (upper part) and morphology (bottom part) of FF PNTs. Reproduced with permission from references [9], [22], [40], and [45]. Copyright (1993) Nature Publishing Group, (2006) Royal Society of Chemistry, and (2010) American Chemical Society (see color plate figure)*

peptide [32–34]. A partial list of other amyloid diseases includes Parkinson disease, type II diabetes, amyloidosis, medullary carcinoma of the thyroid, and prion diseases. Apparently, there is a great significance to the presence of the aromatic residues at the self-assembly process in the amyloid fibril formation, due to aromatic π–π interactions [35]. Following the determination of the smallest core recognition motif of the Aβ protein to be the diphenylalanine element, an FF PNT was discovered [31]. The FF PNTs are long and hollow nanotubes (Figure 1.1c) and, like other biological entities, have the ability to form in mild conditions in water and are biocompatible. Following the recognition that the small FF dipeptides can self-assemble to a tubular structure, Gazit and coworkers purposed dozens of other small di-peptides, composed of natural and un-natural amino acids, which can self-assemble into peptide nanostructures [36]. The common feature of all the purposed dipeptides is the presence of an aromatic region, which seems to have a crucial role in the unique properties of the self-assembled structure, as will be described later in this chapter. This triggered a decade of enormous study on FF-based nanostructures and their applications [37–39]. The most studied FF-based nanostructure is the FF PNT, while other FF-based nanostructures, such as peptide spheres composed of *t*-butyloxy-carbonyl (Boc)-FF and peptide fibrils composed of fluorenylmethyloxycarbonyl (Fmoc)-FF, are also well studied in the literature.

The question in this context is '*Why do the FF-based nanostructures possess such exceptional properties*?' To answer this question we need to consider the basic features of FF nanotubes by referring to their intrinsic nanostructure. An FF crystal structure possesses a noncentrosymmetric hexagonal space group of $P6_1$ [26, 40]. This crystalline class should demonstrate diverse physical effects described by tensors of the odd ranks [24, 41]. As seen in Figure 1.2, the class is situated in the center of the four ellipses, which represent piezoelectricity, second harmonic generation (SHG), optical activity (optical rotation), pyroelectricity and enantiomorphism. Moreover, the space group $P6_1$ also permits the existence of electrical spontaneous polarization and therefore could demonstrate ferroelectric properties. The bottom of Figure 1.2 shows the odd-rank tensor of the space group.

Another set of physical properties of FF-PNTs are defined by their low-dimensional crystalline highly ordered subunits of the supramolecular structure. They demonstrate exceptional electron-hole quantum confinement (QC) phenomena, indicating the formation of quantum dots (QDs) and quantum wells (QWs) in these self-assembled bio-inspired nanostructures [42–45]. These effects are well known for semiconductor low-dimensional materials, but were never observed in bio-organic structures [46].

Other intriguing features are related to the morphological structure of the tubes. The FF PNT structure can be described as a two-dimensional sheet in which the intermolecular hydrogen bonding along the backbone of the dipeptides is one dimensional, which is being wrapped into a tube. In this manner the hydrophobic side chains are positioned outside the tube and the amine and carboxyl groups inside the tube, creating a hydrophilic pore [26, 40, 47] (Figure 1.1c). This unique crystallographic structure contains hydrophilic channels embedded in a hydrophobic matrix.

In this chapter we will discuss the mentioned intrinsic physical properties and application of FF PNTs. In this context we will divide the chapter into several thematic parts. The first part will discuss the unique optical properties of FF PNTs. These properties are related to the electronic structure and are defined by observed QC optical phenomena, which is due to the peptide nanoscale packing. The second part of this chapter will focus on the

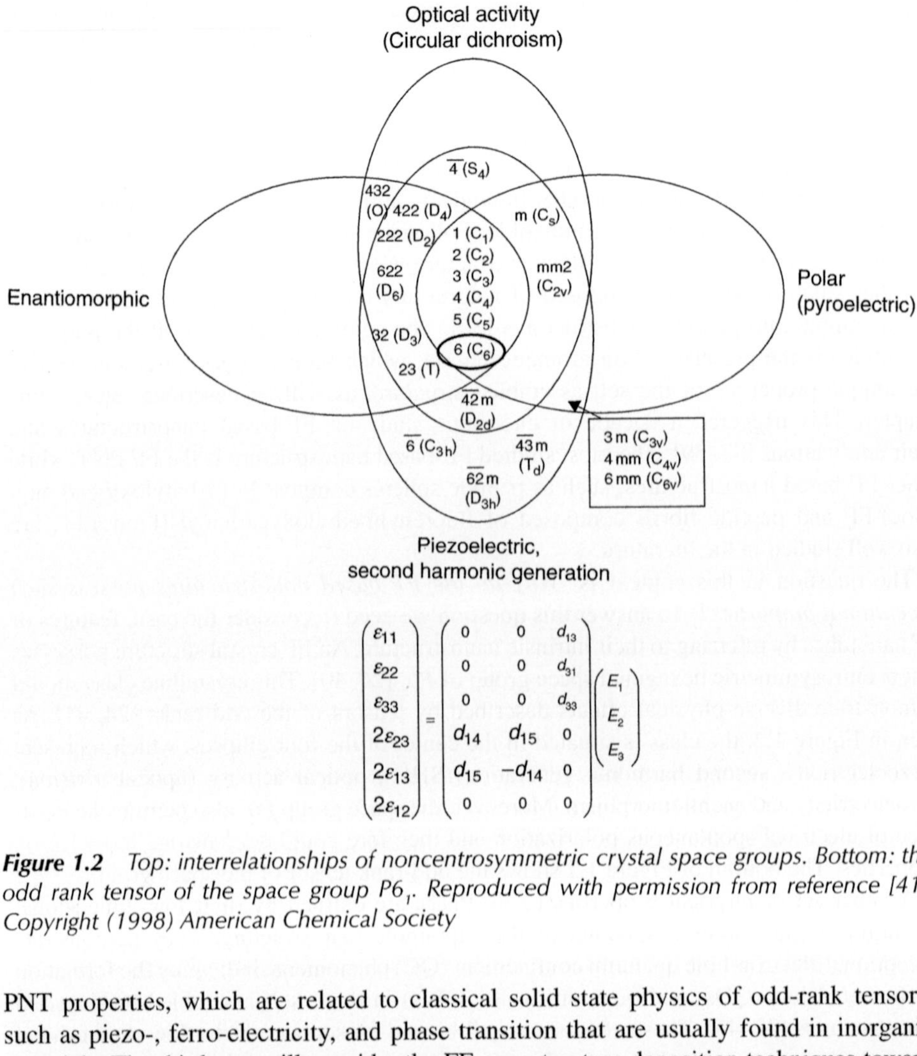

Figure 1.2 Top: interrelationships of noncentrosymmetric crystal space groups. Bottom: the odd rank tensor of the space group P6₁. Reproduced with permission from reference [41]. Copyright (1998) American Chemical Society

PNT properties, which are related to classical solid state physics of odd-rank tensors, such as piezo-, ferro-electricity, and phase transition that are usually found in inorganic materials. The third part will consider the FF nanostructure deposition techniques toward the integration of FF PNTs in nanotechnological devices, which will be discussed in the last part of this chapter.

1.2 Optical Properties and Quantum Confinement of FF-based Nanostructures

The electronic structure of a material defines its optical properties. QC electronic/optical effects are originally ascribed to specific electron density of states in the low-dimensional structures. By means of optical absorption, photoluminescence (PL) and PL excitation (PLE) it is easy to follow the electronic properties of the crystalline structure. While forming the peptide nanostructures in solution there is a critical monomer concentration,

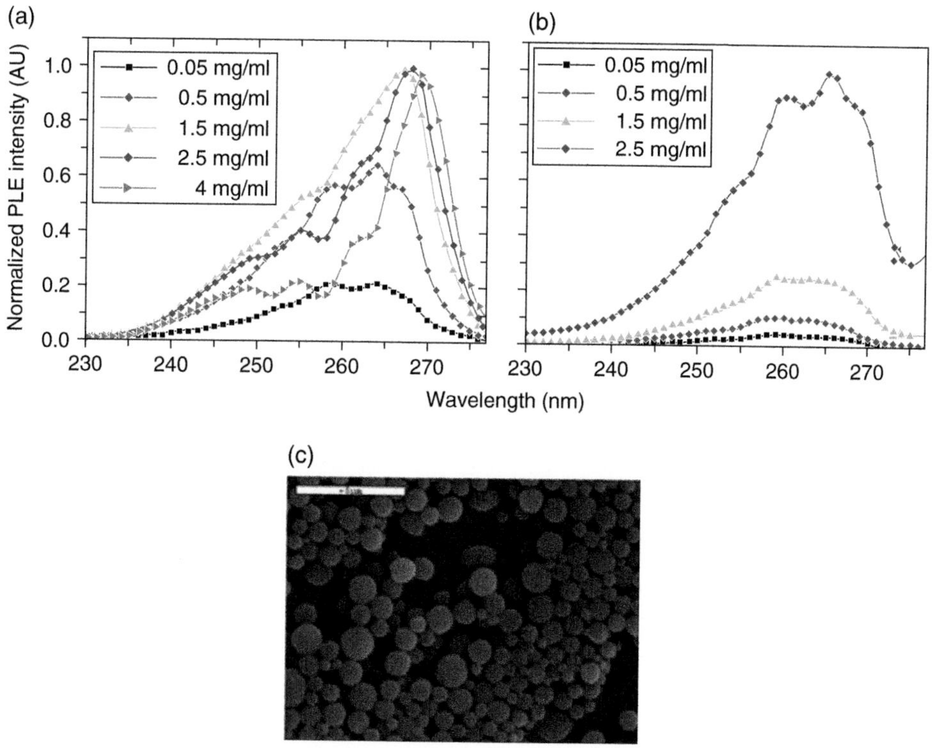

Figure 1.3 PLE spectrum of a solution of Boc-FF (a) monomers and (b) aggregates at different concentrations. (c) SEM image of Boc-FF spheres. Reproduced with permission from reference [44]. Copyright (2009) American Institute of Physics

below which the structures cannot be formed. Figure 1.3a shows the PLE spectrum of a solution of Boc-FF monomers at different concentrations [44]. Boc-FF forms spheres in solution with ethanol and water (Figure 1.3c), while the minimal sphere forming concentration of Boc-FF is ~2 mg/ml. As can be seen in the figure, the spectrum changes dramatically as the spheres are formed, from a broad shapeless excitation spectrum to a sharp excitation peak, located at 270 nm. If the solution of the Boc-FF monomers is pure water (without the addition of ethanol), the Boc-FF monomers will not form the ordered sphere structure, but rather they will aggregate into an amorphous structure. Figure 1.3b shows the PLE spectrum of the Boc-FF aggregates at different concentrations. Unlike the ordered structure, the Boc-FF aggregates do not exhibit changes in the excitation spectrum of the structures as a function of the concentration. The PLE spectrum of the Boc-FF aggregates remains broad and a shapeless peak for even high concentrations. The change in the excitation spectrum implies the formation of a new kind of exciton with different electronic properties than the original electronic properties of the monomer.

In order to follow the formation of the exciton, it is easier to use the peptide hydrogel platform, which self-assemble from Fmoc-FF molecules, since the formation of peptide spheres from Boc-FF molecules or the formation of PNTs from FF molecules is very fast. Figure 1.4a shows the PLE spectrum of the Fmoc-FF molecules during their

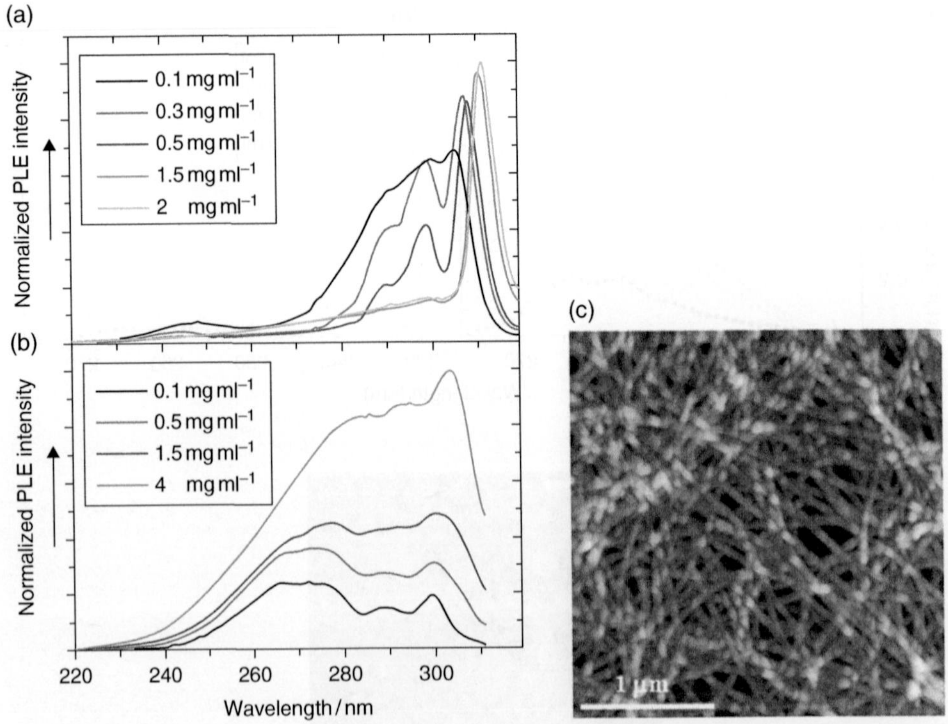

Figure 1.4 *PLE spectrum of a solution of Fmoc-FF (a) monomers and (b) aggregates at different concentrations. (c) AFM image of the Fmoc-FF fibrillar structure. Reproduced with permission from reference [42]. Copyright (2010) John Wiley & Sons, Ltd (see color plate figure)*

self-assembly process into a peptide nanofibril network (Figure 1.4c) [42]. The PLE spectrum of the Fmoc-FF molecules exhibit similar characteristics to those of the PLE spectrum of the Boc-FF molecules (Figure 1.3a). Similar to Boc-FF, Fmoc-FF also cannot form the structures in pure water, and it tends to aggregate. The PLE spectrum of the Fmoc-FF aggregates (Figure 1.4b) exhibits the same characteristics as the Boc-FF aggregates, a broad and shapeless PLE spectrum even in high concentrations. Since the self-assembly process of Fmoc-FF into the fibril network takes ~2 minutes, it is possible to follow the formation of the exciton. Figure 1.5a shows the PLE spectrum of Fmoc-FF at time intervals of 10 s. This graph shows the formation of the excitation peak and that its full width at half maximum (FWHM) decreases. Figure 1.5b shows the intensity of the peak as a function of time. The increase in the exciton peak intensity can be clearly seen at around 125 s. Figure 1.5c shows the FWHM of the exciton peak as a function of time, in which a noticeable decrease in the peak's FWHM can be observed, until it reaches a very low width of 3.75 nm.

The change in the PLE spectrum of the molecules, from a broad and shapeless excitation spectrum to a very narrow peak, is evidence of the crystalline structure formed in the peptide spheres [48, 49]. Figure 1.6a shows the location of the narrow PLE peak of the Boc-FF at a high concentration (after the spheres have formed) together with the optical

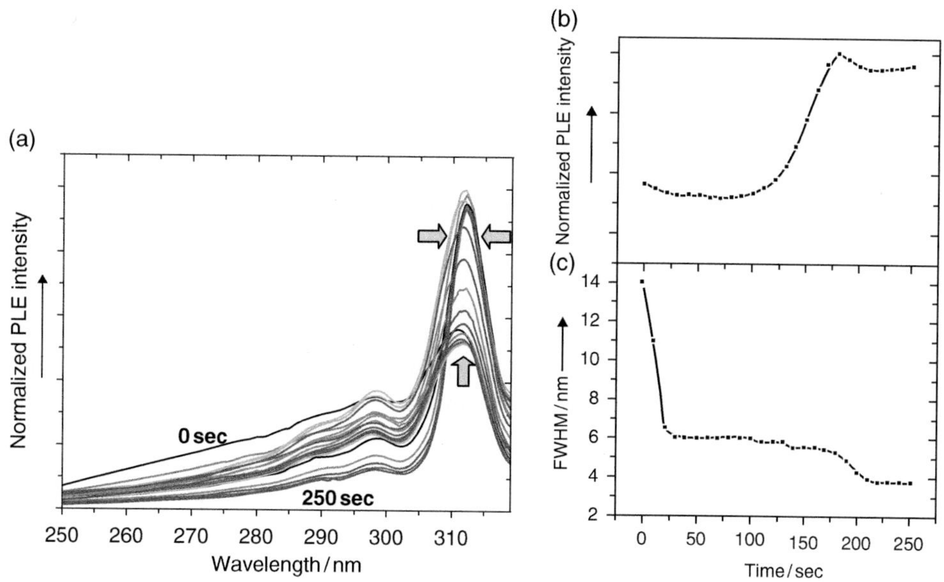

Figure 1.5 *(a) PLE spectrum of Fmoc-FF at time intervals of 10 s. (b) The intensity of the peak as a function of time. (c) FWHM of the exciton peak as a function of time. Reproduced with permission from reference [42]. Copyright (2010) John Wiley & Sons, Ltd*

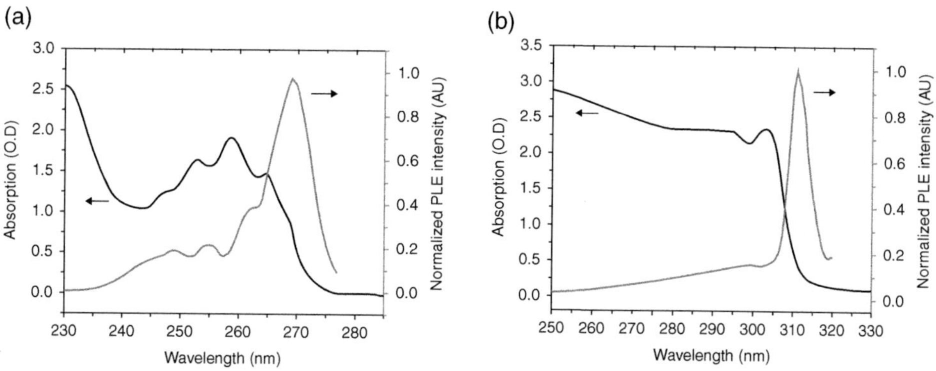

Figure 1.6 *PLE and optical absorption of the high concentrated sample of (a) Boc-FF and (b) Fmoc-FF. Reproduced with permission from references [42] and [44]. Copyright (2010) John Wiley & Sons, Ltd and (2009) American Institute of Physics*

absorption spectrum at the same monomer concentration. The narrow PLE peak is located at the red edge of the absorption spectrum. Excitons that are located at the red edge of the absorption spectrum are typical of quantum confined systems [50].

QC has been mainly ascribed up to now to inorganic structures that are characterized by the enhancement of exciton effects, when electrons and holes are tightly squeezed inside a confined region with a dimension of a dozen angstrom, corresponding to the de-Broglie wavelength of the confined carriers of charge. Due to a strong Coulomb interaction they form a stable neutral quasiparticle exciton. QC provides a dramatic increase in the exciton binding energy and oscillator strength, which may allow the observation of an exciton luminescence at room temperature and even above it [51]. There are several levels of confinement: QWs that confine in 1D (2D object), quantum wires that confine in 2D (1D object), and QDs that confine in 3D (0D object).

The spike-like optical absorption spectrum of Boc-FF (Figure 1.6a), in which the narrow PLE peak is located in its red edge, suggests that the newly formed crystalline confined structure resembles inorganic QD. The electron density of states (DOSs) of a zero dimension structure QD is described by a spike-like behavior [52] and the optical absorption spectrum of a matter should follow its electronic DOS.

Both the absorption and PLE spectra (Figure 1.6a) show multiple peaks. The optical absorption peaks are located at 265 nm (4.68 eV), 259 nm (4.79 eV), 253 nm (4.90 eV), and 248 nm (5.0 eV). The energy interval between two neighboring peaks, both for absorption and for PLE, is the same and equal to 0.10–0.11 eV. The PLE peak related to the excitation around 265–269 nm has the highest intensity. The intensity of the other PLE peaks gradually and monotonically decreases with their transition from this first and main peak. The larger the energy interval between the fundamental absorption peaks the less its intensity. Such an absorption and PLE behavior is typical for local centers where the excited electron interacts with lattice vibrations [53]. Therefore the observed spectrum may be considered as the well-known effect of a phononless exciton absorption line at 265 nm and its phonon replicas at 259 nm, 253 nm, and 248 nm. The energy interval between the resulting maxima is equal to the phonon energy $\hbar\omega_{ph}=0.10-0.11$ eV, which actively interacts with the excited exciton. The energy of the active phonons found in this study is often observed in various molecular crystals. For example, the benzene crystal, which is also related to aromatic compounds such as Boc-FF, has an identical energy to that of the active phonons $\hbar\omega_{ph}=0.12$ eV [54].

Similar to the Boc-FF sphere structure, the fibril network that is formed by Fmoc-FF molecules also exhibits the same phenomenon, in which the narrow exciton peak is located at the red edge of a spike-like absorption spectrum (Figure 1.6b). The different location of the exciton peak at the Fmoc-FF structures (315 nm) suggests that, similar to the common QD structure, the final dimension of the confined crystalline structure is different from that in Boc-FF structures.

In order to observe the nanocrystalline nature of the peptide nanostructures the FF PNT structures have been used. The FF PNTs can self-assemble and disassemble according to the solution that surrounds the structures [45]. In a ~98–99% water solution (the remaining 1–2% are hexafluoro-2-propanol) the FF monomers form the rigid PNTs. However, upon transferring the tubes to pure organic solvent, as anhydrous methanol, the PNTs disassemble into their elementary building blocks. The elementary building blocks can be seen by atomic force microscopy (Figure 1.7a) and have a small size distribution of 2.12±0.15 nm (Figures 1.7b and c). To conclude that these elementary building blocks of FF PNTs are actually the quantum confined crystalline region that can be seen by optical measurement, they need to have an identical optical signature. Figure 1.7d shows the similar PLE spectrum of the dissolved FF PNTs in comparison to the high-concentrated sample, in which the structures can form. Moreover, the elementary building blocks possess the same X-ray diffraction (XRD) pattern as regular FF PNTs (Figure 1.7e), pointing out that the building blocks have the same crystal structure as the FF PNTs.

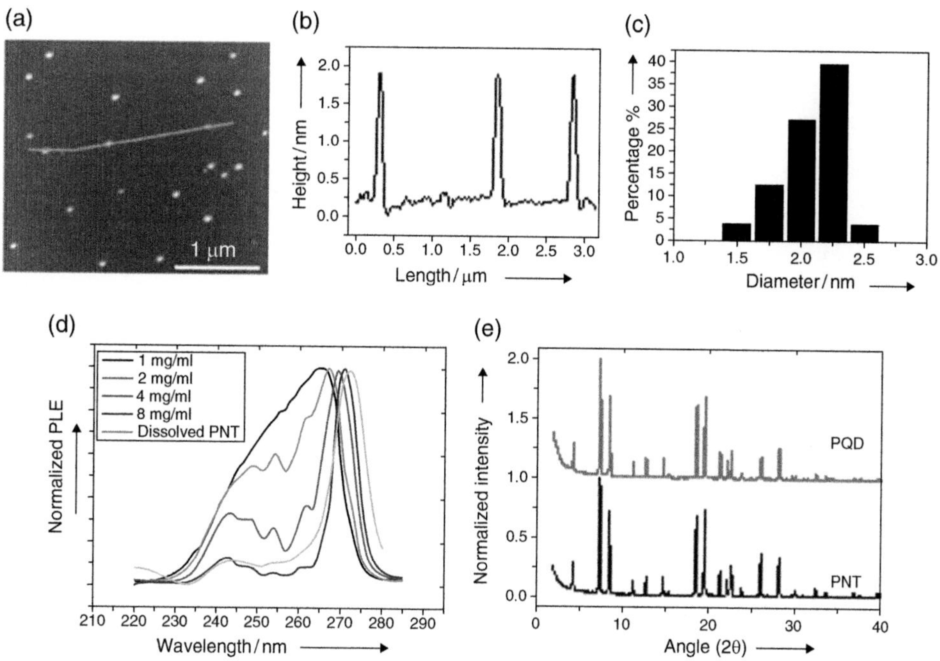

Figure 1.7 *(a) AFM images of dissolved FF PNT. (b) Cross-section of (a). (c) Size distribution of the dots. (d) PLE spectrum of the dissolved FF PNTs in comparison to the high-concentrated sample. (e) XRD patterns of before and after dissolution of FF PNTs. Reproduced with permission from references [24] and [45]. Copyright (2010) American Chemical Society and Taylor & Francis Ltd*

1.3 Odd-Tensor Related Physical Properties

The nanocrystalline structure of FF PNT possesses a $P6_1$ noncentrosymmetric space group. This odd-tensor space group permits the existence of several physical properties such as SHG, ferroelectricity, pyroelectricity, piezoelectricity, and optical activity (Figure 1.2). In the context of this section we will discuss the piezoelectric and SHG properties of FF PNTs.

Piezoelectricity is the ability of some materials to generate an electric potential in response to applied mechanical stress. The piezoelectric effect is reversible, in which materials that exhibit the direct piezoelectric effect (the production of electricity when stress is applied) also exhibit the converse piezoelectric effect (the production of stress and/or strain when an electric field is applied), which is the most common use of the piezoelectric effect.

Biological materials demonstrate pronounced piezoelectric phenomenon of electromechanical coupling, which is considered to be a universal intrinsic feature of them. The origin of this fundamental physical property is ascribed to high structural ordering of biological systems with a low symmetry configuration of elementary cells, based on their helical or chiralic dissymmetry [55]. Plants, animal, and human tissues such as wood, bones, skin, tendon, as well as elementary biological units of amino acids, which are the building blocks of proteins and peptides, reveal pronounced piezoelectric properties [56–59]. High-resolution studies using piezoelectric force microscopy (PFM) have been

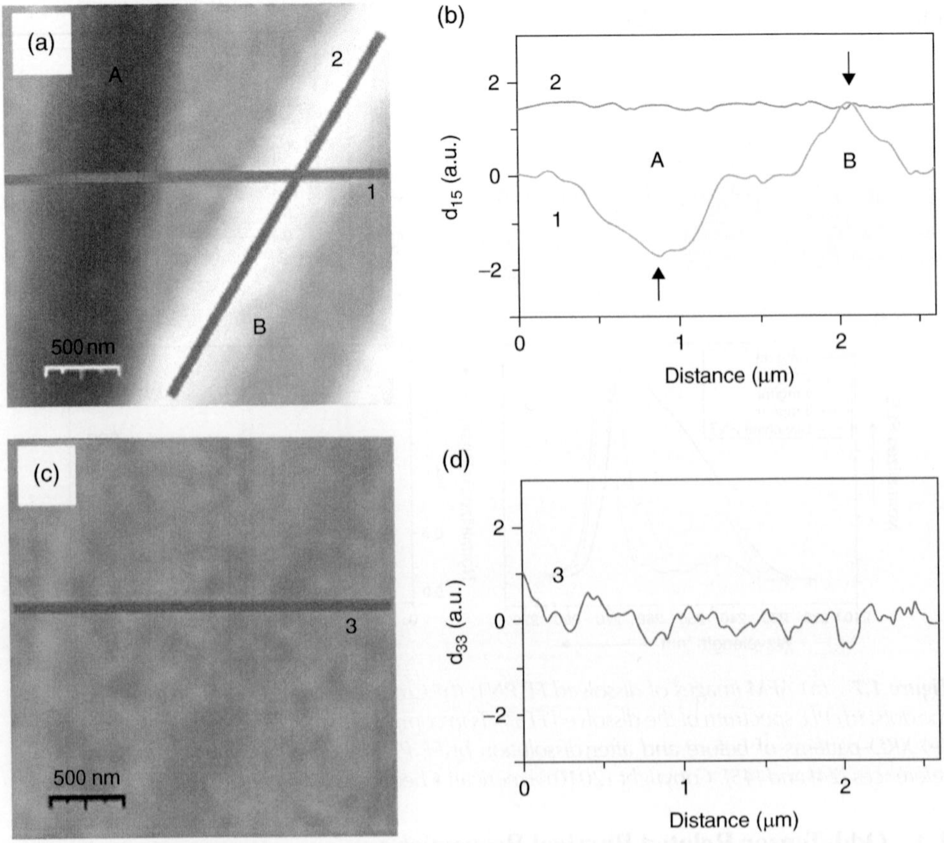

Figure 1.8 *(a) PFM IP signal of FF-PNTs. (b) Cross-section along the lines in (a). (c) PFM OOP signal of FF-PNTs. (d) Cross-section along the line in (c). Reproduced with permission from reference [25]. Copyright (2009) American Chemical Society (see color plate figure)*

applied to different biological structures including human bone, teeth, canine femoral cartilage, deer antlers, and butterfly wings [60–63]. These experiments have allowed to image bone collagen matrix in nanometer scale and to find the internal structure and orientation of protein fibrils.

In PFM measurements, the common topography acquisition is followed by a PFM regime scan. At this regime the conducting tip is scanned in contact mode while an ac voltage (V_{ac}) is applied between the tip and Au electrode. In these conditions both out-of-plane (OOP) and in-plane (IP) polarization components can be measured [64]. In the case of PFM measurements of FF PNTs there is a crucial importance to the tube orientation in respect to the atomic force microscope (AFM) tip. When the scan is performed along the tube axis, both polarizations can be detected, the IP signal due to a shear component of the tensor of piezoelectric coefficient (d_{15}), corresponding to a polarization parallel to the tube axis, and the OOP signal that reflects polarization along the tube radius. Figures 1.8a and c show both the IP and OOP signals of FF PNTs, respectively (Figures 1.8b and d are the cross-sections along the lines in Figures 1.8a and c, respectively) [25].

It can easily be seen in the figure that only the shear component, IP signal, can be observed. The sole existence of the shear component unequivocally suggests that the only polarization component existing in the PNT is along the tube axis.

In order to understand the origin of a piezoresponse, we should consider the possible d_{ij} matrix for the space group $P6_1$ [26] in accordance to the electric field of the tip surface. From the tensor of the FF PNT (Figure 1.2) it can be seen that while applying an E_3 electric field the d_{33} piezoelectric coefficient (OOP component) can be measured. This direction is the Z axis of the tube, meaning along the tube axis. Due to the limitation of the AFM system and the available deposition techniques of FF PNT, the PFM response of the standing tubes cannot be measured. However, measuring the PFM response of lying on the surface tubes is possible. While looking at the tensor again, we can see that by applying E_1 (which is the same as E_2 due to the tube symmetry) the d_{15} shear component coefficient (IP component) can be measured. The spontaneous polarization vector, located parallel to the tube axis, can be easily tracked by physically rotating the PNT sample (Figures 1.9a and b). By physically rotating the PNT sample by 180°, the spontaneous polarization vector (schematically shown in Figure 1.9c) should follow the rotation and point in the opposite direction (with respect to the AFM tip), and thus a contrast reversal in the PFM image should be observed, as seen in the figures. The component of the shear deformation is parallel to the tube axis and proportional to $1/2(d_{15}V_{ac}\cos(\alpha))$, where α is the angle between the tube axis and scanning direction and d_{15} is the shear piezocoefficient. Figure 1.9d shows the PFM signal of the FF PNT as a function of α. The data are in perfect agreement with the expected cosine function. The piezoelectric signal is linearly dependent on the applied bias; thus the observed IP PFM signal should also be linear to the applied bias. Figure 1.9e shows the PFM image of a tube at different alternating current (AC) biases, from 0V to 16V. The linear dependence of the PFM signal as a function of the applied bias (for both 'dark' and 'bright' tubes) is shown in Figure 1.9f.

The piezoelectric coefficient of FF PNTs was estimated by comparison with the well-known piezoelectric crystal of $LiNbO_3$ (LNO) [25]. Figure 1.9d demonstrates a similar angle dependence, measured under identical conditions as a 100nm thick PNT, on the Y-cut surface of LNO crystals with 3m symmetry. If the cantilever is scanned parallel to the Z direction of the Y-cut LNO, there is only one component of the shear displacement that is present when the E-field is applied along the Y direction [65]. For an arbitrary orientation of the crystal, the same equation, $1/2(d_{15}V_{ac}\cos(\beta))$, applies, where now β is the angle between the scanning direction and the Z axis. As shown in Figure 1.9c, the same dependence with the maximum value slightly shifted to the negative side is observed between FF PNTs and LNO. The observed maximum excludes any angle dependence (critical for a shear response) and allows direct comparison of the piezoelectric activities in both materials. It is seen that the effective d_{15} coefficient for a 100nm tube is about 2 times smaller than in bulk LNO [66, 67], thus giving a rough estimation of the value of a shear coefficient of ~35 pm/V for FF PNTs.

An additional odd-rank tensor related characteristic that was explored in FF PNTs is SHG [23]. SHG is a process in which photons interacting with a nonlinear material are effectively 'combined' to form new photons with twice the energy, and therefore twice the frequency and half the wavelength of the initial photons. Figure 1.10a shows the emission spectrum of FF PNTs, following optical excitation with a wavelength of $\lambda_{pump} = 800$ nm, where the pronounced narrow SHG peak is clearly seen. The SHG nature

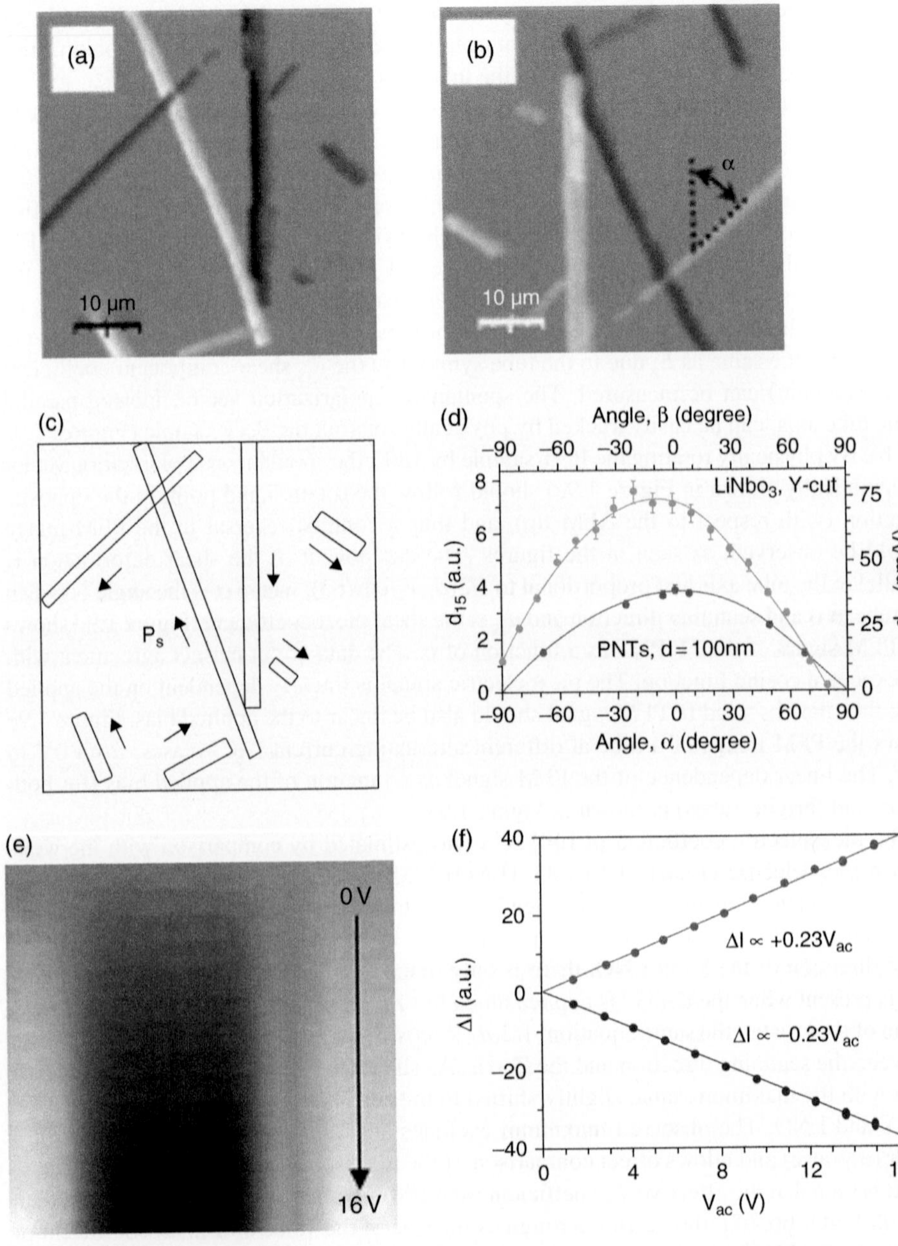

Figure 1.9 PFM IP signal of FF-PNTs (a) before and (b) after the rotation of 180°. (c) Scheme of the spontaneous polarization vector along the tubes in (a) and (b). (d) PFM IP signal of FF-PNTs as a function of α, in comparison to an LNO crystal. (e) PFM IP signal of FF-PNTs at different ac biases, from 0V to 16V. (f) PFM IP signal of FF-PNTs as a function of ac bias. Reproduced with permission from reference [25]. Copyright (2010) American Chemical Society (see color plate figure)

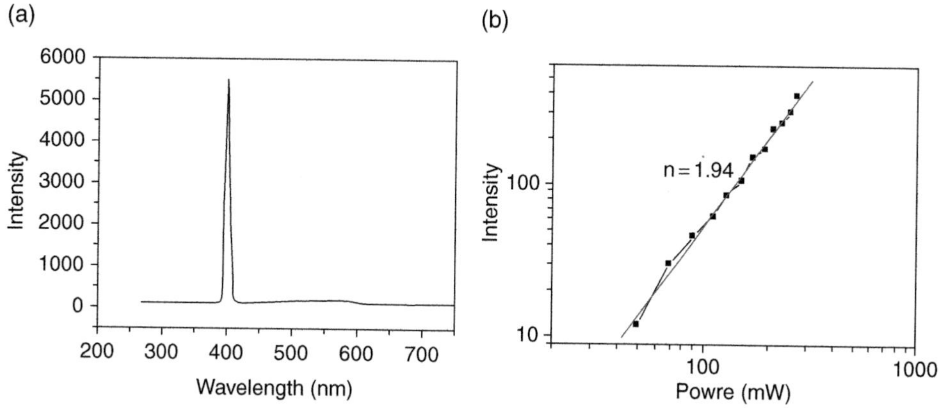

Figure 1.10 (a) Emission spectrum of FF PNTs following optical excitation with a wavelength of λ_{pump} = 800 nm. (b) Emission intensity as a function of the pump laser power. Reproduced with permission from reference [23]. Copyright (2011) American Chemical Society

of the optical response of the tubes was further validated by measuring its intensity as a function of the pump laser power (Figure 1.10b), showing a quadratic dependence, which is a characteristic of an SHG response.

1.4 Thermal Induced Phase Transition in Peptide Nanotubes

In the previous sections we discussed the diverse physical properties of the FF-based peptide structures. The peptide structures were formed in ambient conditions, at room temperature, and atmospheric pressure. In this section we will review the numerous transitions that the FF PNT undergoes during thermal induction at 150 °C. In this context, we will review the unique changes in every property aspect of the peptide structure: molecular, morphological, crystal structure, piezoelectric, SHG, wettability, optical, and even conductivity properties.

Classic structural phase transitions in organic and inorganic crystals are generally described by small atomic displacements, where each phase is characterized by a specific space group, which unambiguously defines its physical properties [68–74]. Contrary to the common models of structural phase transitions, such as ferroelectric phase transitions described by the Landau–Ginzburg–Devonshire theory [70], the phenomenon of a chemical conformational induced transition is very common in organic and biological complexes. Such a conformational transition results from variations at the molecular level [75–79] and often leads to dramatic variation of molecular crystal symmetry [80]. The molecular transformation may involve strong covalent bonds or weak noncovalent interactions like hydrogen bonds, van der Waals, hydrophobic, aromatic, and π-stacking interactions. Due to energy considerations, the covalent related transitions are usually irreversible, whereas the noncovalent related transition is usually reversible [81, 82]. One of the molecular transformations that involve covalent bonding is the irreversible formation of diketopiperazine from small peptides [83–85], as in the phase transition of FF PNTs.

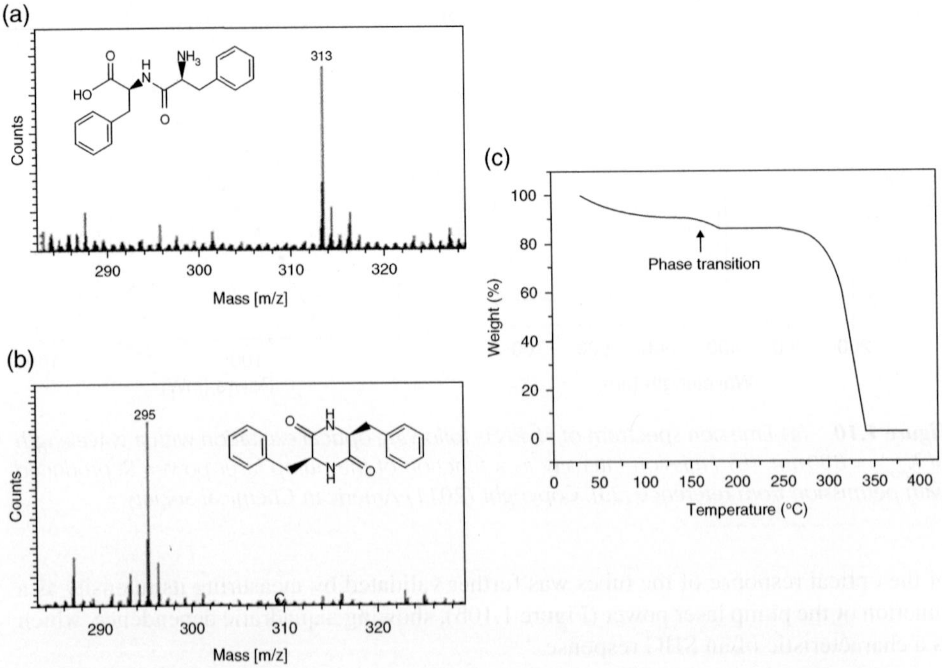

(d)

Atomic ratio	Linear structure theoretial value	Native phase experimental value	Cyclic structure theoretial value	Thermal induced phase experimental value
C : O	18 : 3 = 6 : 1	6.4 : 1	18 : 2 = 9 : 1	8.40 : 1
C : N	18 : 2 = 9 : 1	8.92 : 1	18 : 2 = 9 : 1	8.95 : 1
N : O	2 : 3 = 0.67 : 1	0.73 : 1	2 : 2 = 1 : 1	0.95 : 1

Figure 1.11 *ToF-SIMS analysis of (a) before and (b) after the phase transition. (c) TGA analysis of FF PNTs. (d) XPS ratio values before and after the phase transition. Reproduced with permission from reference [23]. Copyright (2011) American Chemical Society*

1.4.1 Changes in the Structure Properties during the Phase Transition Process

Thermal gravimetric analysis (TGA) and differential scanning calorimeter (DSC) analysis of FF PNTs show that a phase transition occurs at ~150 °C (Figure 1.11a) [23, 86, 87]. Time of flight secondary ion mass spectrometry (ToF-SIMS) (Figures 1.11b and c) and X-ray photoelectron spectroscopy (XPS) analysis (Figure 1.11d) explain the observed phase transition as a molecular transition from the native phase structures that are formed by linear-FF peptides (MW = 313 g/mol) to a thermally induced phase that is formed by cyclic-FF molecules (MW = 295 g/mol) [23]. The cyclization process is accompanied by the release of a water molecule, which corresponds to the change in the molecular weight (observed by ToF-SIMS) that leads to the observed mass loss in the TGA and DSC analysis.

The thermally induced molecular transition changes the morphological structure of the tube, from a hollow PNT to a closed low diameter peptide nanofiber (PNF) (Figure 1.12a

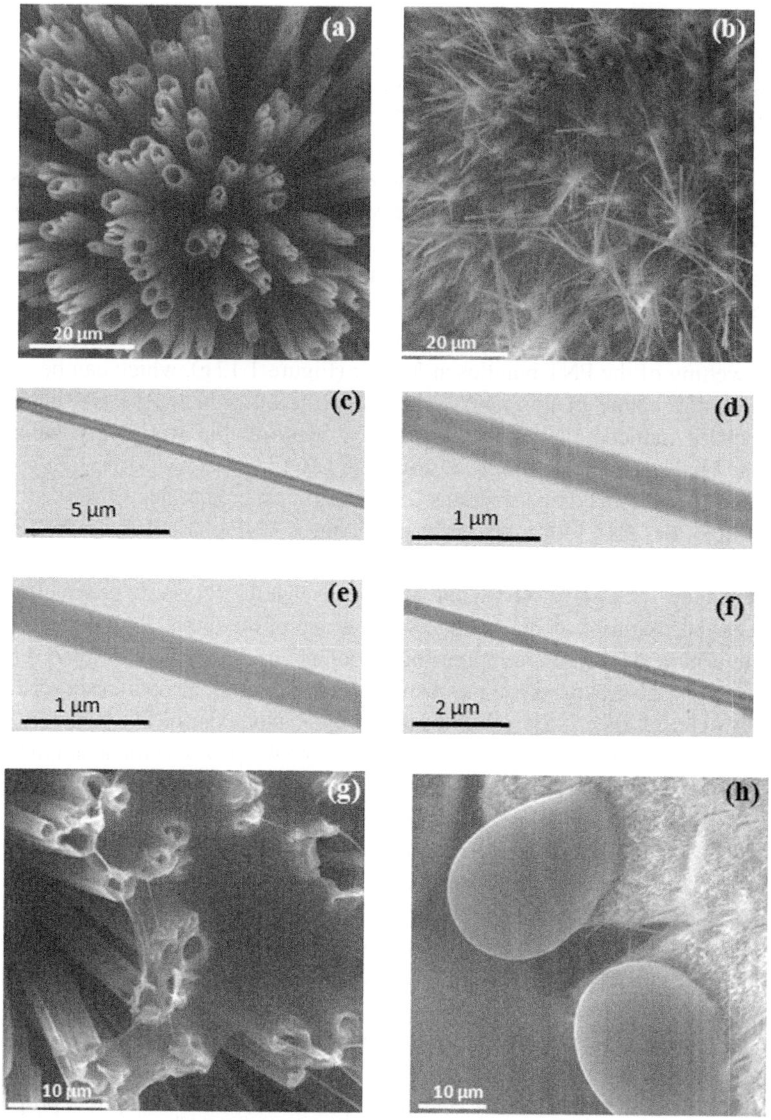

Figure 1.12 SEM images of (a) before and (b) after the phase transition. (c) to (f) The morphology dynamic of the phase transition as observed by STEM. High-resolution wettability behavior of the structures (g) before and (h) after the phase transition. Reproduced with permission from reference [23]. Copyright (2011) American Chemical Society

and b) [23, 87]. The dynamic of the central hole closing can be observed by the scanning transmission electron microscope (STEM) [23]. Figures 1.12c to f show a series of STEM images of a single PNT during heat induction by the STEM electron beam. Focusing the electron beam on the PNT induced the thermal induction, which caused a reduction of ~15% in the structure diameter (corresponding to a reduction of ~30% in the structure volume). In the final lower magnification STEM image (Figure 1.12f) the changes between the thermally induced and the native parts can be observed.

The crystal structure of the PNT in its native phase [40] (prior to the thermal induction) suggests a representation for the inner surface of the nanotubes, containing multiple hydrophilic/hydrophobic channels. In addition to the closing of the central hole, the multiple channel network is also closing. The alignment of the multiple channels causes the native phase PNT to be hydrophilic. The wetting behavior of nanoscale objects can be observed by high-resolution environmental scanning electron microscope (ESEM) measurements. In ESEM measurements of wetting behavior the sample is cooled (~2 °C) at a low pressure (~5 torr), followed by a gradual increase of the pressure until water condensation occurs. The wetting dynamics of bundles of FF PNTs at the native phase shows a rapid full wetting of the PNT bundles at 5.9 torr (Figure 1.12 g), which can be explained by the hydrophilic nature of this native phase. On the other hand, the wetting dynamic of the thermally induced PNF phase dictates a hydrophobic wettability behavior, as demonstrated by the repelling of water in Figure 1.12 h [23].

The molecular and structural transitions do not permit the existence of the native phase crystal packing of the PNT [40] in the thermally induced PNF phase. Indeed, XRD patterns show that a crystal structure phase transition has occurred during the thermal induction process [23, 86, 87] (Figure 1.13a). The XRD pattern indicates that the $P6_1$ space group of the native PNT phase has been transformed to a *Pbca* space group of the thermally induced PNF [23]. This phase transition also represents a transition from a noncentrosymmetric space group to a centrosymmetric space group. As a consequence, all of the described odd-tensor related properties, which were described in the previous section, cannot exist in the thermally induced phase. Rosenman *et al.* showed the disappearance of both the piezoelectrical and SHG activities of the structure during the phase transition process [23]. The new molecular packing of the cyclo-FF PNF has been resolved by Park *et al.* [88] and can be seen in Figure 1.13b.

As described previously, the unique nanocrystalline nature of the FF PNT native phase corresponds to its exceptional optical properties. Thus the change in the crystal packing of the structures dictates the change in the optical properties as well. The thermally induced phase no longer has the unique QD-like optical parameters of the native phase. Whereas the PNT native phase absorbs light only at wavelengths below 300 nm, the PNF thermally induced phase has new absorption levels at 305 nm and 370 nm, as can be seen by diffuse reflectance spectroscopy (Figure 1.13c) [88]. The new absorption results in an emission in the visible light (Figure 1.13d), which makes the thermally induced PNFs to fluoresce in blue under UV excitation. The new optical properties resemble the optical properties of PNF that were formed by the vapor deposition technique (see Section 5.2). Studies have shown that unlike the QD behavior of FF PNTs, the PNF have optical properties as QWs [43]. The rest of the optical properties, such as a circular dichroism (CD) pattern (Figure 1.13e) and Fourier transform infrared (FT-IR) spectroscopy (Figure 1.13f) also change during the phase transition process [87]. The new electronic properties of the thermally induced PNF phase changes the electric conductance of the structures. The native phase PNT structures are composed of nanocrystalline regions with strongly confined excitons. This causes the structure to be highly insulating. However, following the phase transition process into cyclo-FF PNFs, the new structures possess elevated electrical conductivity, as shown by current voltage measurement of a single PNF between two electrodes [88].

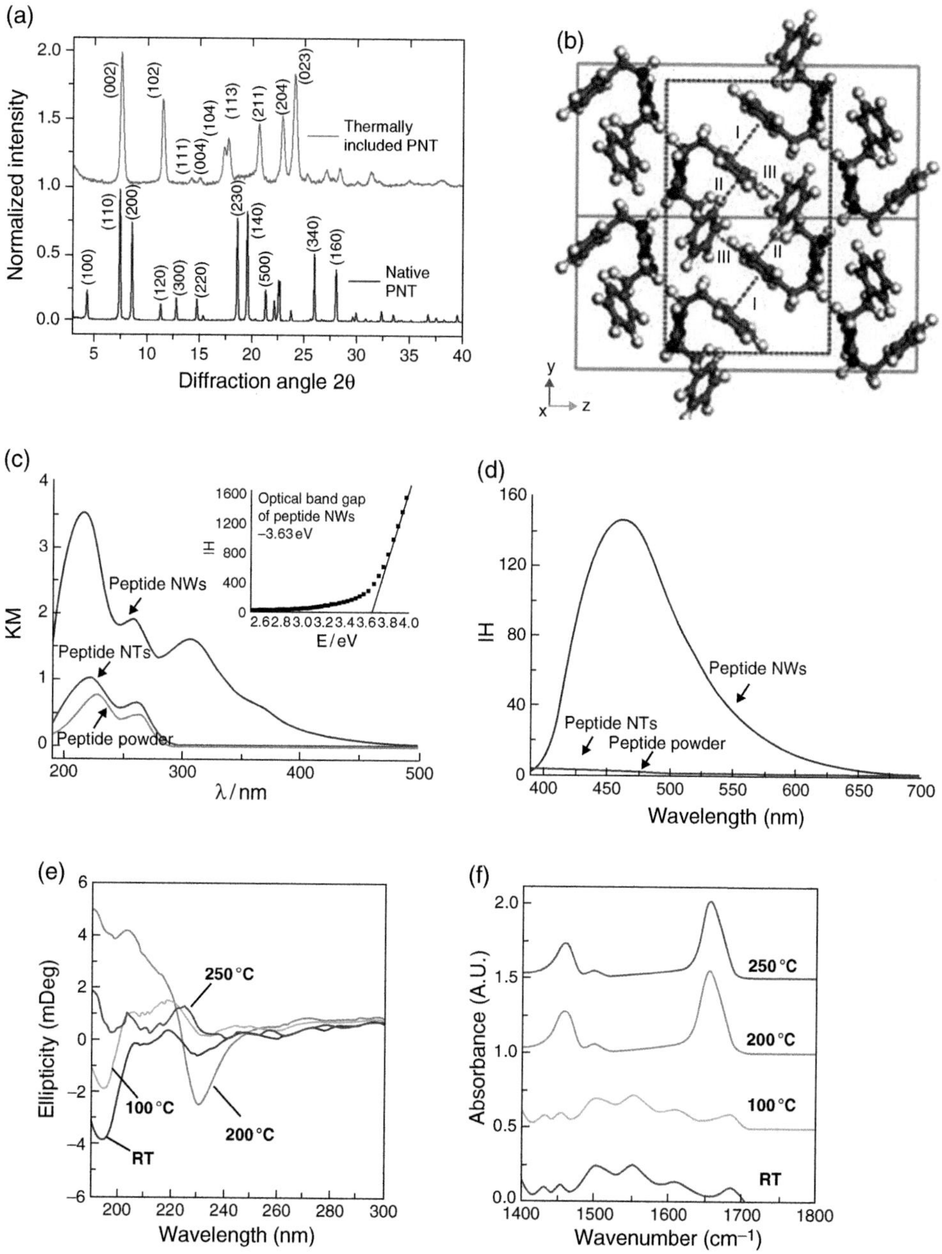

Figure 1.13 *(a) XRD pattern change during the phase transition. (b) Molecular packing of the cyclo-FF PNF. (c) Diffuse reflectance spectroscopy and (d) emission spectrum of the cyclo-FF PNF. (e) CD and (f) FT-IR pattern change during the phase transition. Reproduced with permission from references [23], [87], and [88]. Copyright (2011) American Chemical Society, and (2010) and (2011) John Wiley & Sons, Ltd*

1.4.2 Phase Transition Classification of the Thermally Induced Process

In the previous section we discussed the phase transition process in the peptide nanostructure, which demonstrates remarkable and deep transformation of all fundamental physical properties at all levels, such as molecular, electronic, optical, piezoelectric, nonlinear optical, crystalline symmetry, morphological, and more. Moreover and importantly, this phase transition is irreversible.

According to the basic physical classification, there are two main classes of phase transitions: distortive and reconstructive. The first class is characterized by invariable molecular (atomic) composition preserving stable chemical bonds, which could slightly change their length and orientation. As a result, distortive phase transition is described by small atomic displacements of 0.01–0.1 Å, allowing its reversibility. The key feature of these phase transitions is a group–subgroup relation, when a low-temperature low-symmetry phase is a subgroup of a high-temperature parent high-symmetry phase. A good example is the ferroelectric phase transition, which is accompanied by disappearance of spontaneous polarization. In the ferroelectric phase transition the ferroelectric-related properties, namely piezoelectric and SHG, cannot be observed due to centrosymmetric structure of the parent high-temperature phase.

An additional class of phase transitions is reconstructive transitions. This class of phase transition is completely different. It involves breaking a part of the chemical bonds of the initial phase as their high- and low-symmetry phases lack a group–subgroup relationship, and the transitions are strongly first order.

The discussed phase transition distinctly shows that the phase transition in FF PNT/PNF is accompanied by a profound reconstruction of the chemical covalent bonds, and creation of a new type of cyclic peptide molecules from linear ones. High- and low-temperature parent and distortive phases are not linked by group–subgroup relations. The found morphologic transformation, from hollow FF PNT to nanofibers cyclo-FF PNF, is the evidence that the phase transition passes through a profound variation of the structure, when atomic displacements are much larger than those in distortive phase transitions, and it might be characterized as a structural collapse. Such a physical picture reminds us of allotropic phase transitions. Thus we could relate the described phase transition in peptide nanostructures to reconstructive phase transitions. Similar to distortive phase transitions, this phase transition also includes the disappearance of piezoelectric, nonlinear optical properties and spontaneous polarization in the high-temperature centrosymmetric phase, as seen during the FF PNT phase transition.

1.5 Deposition Techniques of PNT

The commercial powder of FF (usually purchased from Bachem or Sigma-Aldrich) is a mixture of PNT and aggregates. In order to dissolve the powder to FF monomers, a strong organic solvent should be used, in which the most common used organic solvent in the scientific community is 1,1,1,3,3,3-hexafluoro-2-propanol (HFIP). As a consequence, the vast majority of the reported deposition processes of FF PNTs (or other structures related to FF, such as ribbons or spheres) start with dissolving the FF powder in HFIP. We will review in this section the state of the art deposition techniques of FF nanostructures that include deposition from solution or in the dry state.

Figure 1.14 SEM images of FF PNTs from HFIP solvent. Reproduced with permission from reference [89]. Copyright (2006) Nature Publishing Group

1.5.1 Wet Deposition Techniques

Most of the deposition techniques of FF-related structures are in a wet atmosphere. The first experimental procedure for forming FF PNTs in an aqueous solution followed by depositing the structures randomly on a surface [31] involved dissolving the FF powder in a low-volume solution of HFIP, in a final concentration of 100 mg/ml. FF can be dissolved in HFIP; hence the solution of FF in HFIP is clear. The next step involves diluting the concentrated solution of FF in HFIP in water to a final concentration of ~2 mg/ml. The solution of FF in water is opaque, meaning that the structures have formed and aggregated. The formed structures are very inhomogeneous regarding their diameter and length, in which their diameter can range from dozens of nanometers to a few micrometers, and their length can reach to even more than a millimeter. This experimental procedure to form PNTs in an aqueous solution is the most common one discussed in the literature. In order to deposit the structures, for further analysis, such as to image the structures by electronic microscopy, AFM, or even by optical microscopy (in cases of large tubes), for solid-state spectroscopic analysis and more, the aqueous solution can simply be drop-cast and dried on any given structure (Figure 1.1c shows a scanning electron microscopy (SEM) image of this kind of deposition technique).

The HFIP solvent has a very high vapor pressure (159 mmHg at 25 °C). As a consequence, further safety regulations should be taken while handling it. Moreover, the low-volume concentrated sample of FF in HFIP can be evaporated quickly if the HFIP-containing tubes remain open during the procedure. However, the fast evaporation process of HFIP can also be beneficial. Reches and Gazit have shown [89] that simply by depositing a small aliquot of FF in HFIP on any given substrate a vertically oriented FF PNT can be formed (Figure 1.14).

The latter deposition technique of vertically oriented FF PNT deposition from HFIP solution is very hard to control, due to the very fast evaporation process of HFIP. Park *et al.* [86, 87, 90] improved this method by controlling the deposition environment. In order to prevent the interaction of HFIP with water in the air, and thus prevent rapid PNTs forming, they deposited the FF containing HFIP solution in a dry box. In this condition the HFIP forms an amorphous thin layer (74 and 145 nm for FF concentrations of 1 and 2 mg/ml, respectively). The amorphous thin layer can be exposed to water

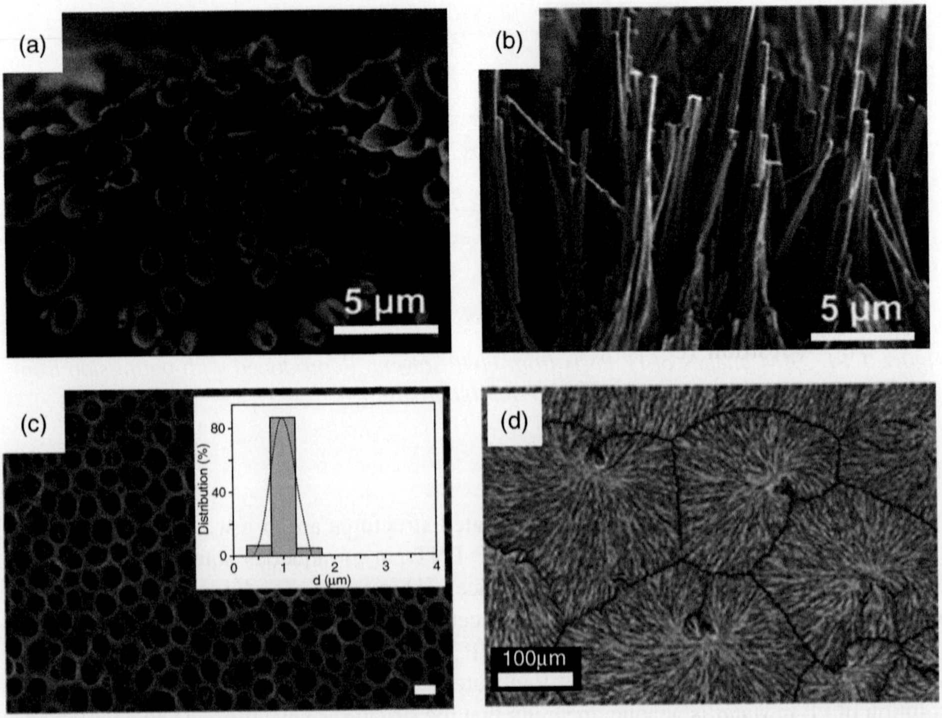

Figure 1.15 *SEM images of (a) PNTs formed by exposure of an amorphous FF layer to water vapor, (b) PNFs formed by exposure of an amorphous FF layer to aniline vapors at 150 °C, (c) an FF PNT honeycomb scaffold (scale bar is 1 μm), (d) spherulitic-like films of FF PNTs. Reproduced with permission from references [87], [91], and [92]. Copyright (2010), (2011), and (2007) John Wiley & Sons, Ltd*

vapor, which allows the formation of the tube structure [90] (Figure 1.15a). The same amorphous layer can also be exposed to aniline vapors and aging in the temperature of 150 °C to form a vertically aligned nanowire structure [86, 87] (Figure 1.15b). As previously described, the FF PNTs can undergo a phase transition process, where the linear peptide structure transforms into a cyclic diketopiperazine structure. Although Park *et al.* do not mention it in their original publication, all of their experimental data (mass spectrometry, PL spectra, XRD pattern, and morphology) suggest that the vertically aligned nanowire structures are composed of cyclic-FF.

A recently reported deposition method, which resembles the latter vapor-related deposition technique, resulted in an exceptional FF structure in the shape of a honeycomb [91]. In this deposition technique, the high concentrated FF solution in HFIP is diluted in toluene to a very low concentration solution (25 μg/ml). Small droplets of the low-concentration solution was placed on to a surface and blown with moist air (80% humidity) with a controllable air speed, in what is called the 'breath figure'. This procedure resulted in an FF honeycomb scaffold with an adjustable pore size diameter as a function of the blown air speed (Figure 1.15c).

The simple deposition technique, which involves drying an aliquot of FF in an aqueous solution on a given substrate, produces randomly oriented horizontal (lying on the surface) FF PNTs. An ordered homogenous layer of FF PNTs that will cover a desirable area cannot be achieved using this kind of procedure, no matter what the concentration of the FF in the solution is. Richter *et al.* [92, 93] showed the formation of thick spherulitic-like films of FF PNT that can cover a large area with horizontally aligned PNTs (Figure 1.15d). This deposition procedure involves dilution of FF in *N*-methyl pyrrolidone (a solvent that is known to increase the solubility of various organic materials). The diluted FF sample was then cast on to a preheated (at 60 °C) surface, which was later cooled back to room temperature.

1.5.2 Dry Deposition Technique

In the previous section we reviewed the known deposition techniques of FF PNTs from solution. The unique physical properties of FF PNTs that we described earlier promote the idea of using FF PNTs in the nano/microtechnological industry [37, 38]. However, there is no tolerance to wet solutions in some of the nano/microtechnological industry environment. In order to bypass this obstacle, we need to use a solid-state deposition technique, which does not involve the use of solutions. A solution for this obstacle can be found in the recent development large-scale bottom-up technology of PNT coatings based on physical vapor deposition (PVD) [94]. PVD is one of the basic technologies in microelectronics and its application to biomolecule deposition allows the unique structure of FF structures to be integrated in a facile way into the production line of devices. The coating technique is based on the vaporization process from a solid source in the form of atoms or molecules. The vaporized flux is transported through vacuum, a low-pressure gaseous environment or plasma towards the substrate where it condenses. In a thermal vapor evaporation process, the source material is heated to a temperature where there is an appreciable vapor pressure. During the deposition of the peptide coating, the monomer powder is placed in the sample holder, which is directly connected to the heater, allowing the evaporation of the biomaterial at a definite temperature specific for the evaporated peptide material. A substrate is placed on the substrate holder at a definite distance above the source material. The coating is then formed on the downward side of the substrate.

Except for the FF fibrillar structures, additional structures have been formed by PVD with the use of dipeptides and amino acids, such as phenylalanine (F), tryptophan (W), phenylalanine-tryptophan (FW), tyrosine (Y), dityrosine (YY), and dialanine (AA). Figure 1.16 shows side views of SEM images of the deposited FF structures. The observed SEM data demonstrate normally oriented, homogeneously distributed arrays of the nanofibrillar structure with density of $\sim 10^8\,cm^{-2}$. The powder placement needs to be heated at 150 °C for an effective evaporation process. As discussed previously, the heating causes a phase transition process and the formed structures possess the physical properties of the thermally induced phase. The structure parameters can be varied in a wide range by adjusting the deposition parameters, thus enabling the control of the average height (1–70 μm) and surface density (10^5–$10^8\,cm^{-2}$). Figure 1.16a shows a side view of a thick array of nanotubes, of ~65 μm height, whereas Figure 1.16b shows a side view of a thinner array of nanotubes, of ~5 μm height. The described method allows dense homogeneous PNF coatings of large area to be formed using biological molecules.

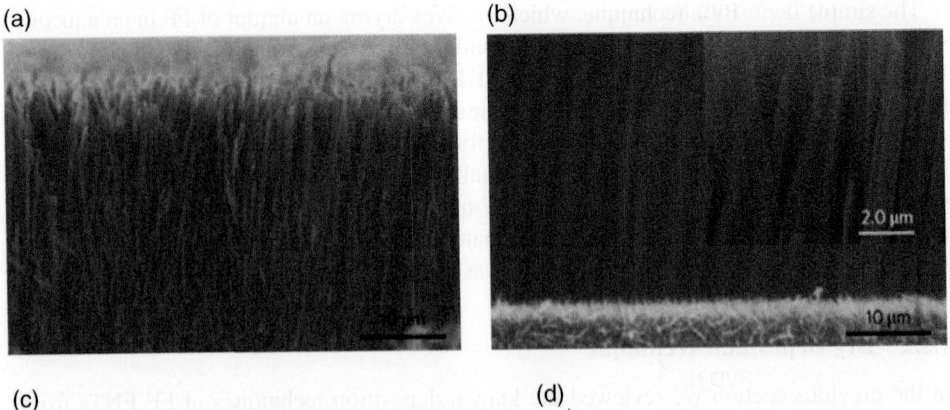

***Figure 1.16** SEM side view of (a) thick and (b) thin surfaces of PVD-formed structures. (c) ToF-SIMS and (d) QCM analysis of the PVD-formed structures. Reproduced with permission from reference [94]. Copyright (2009) Nature Publishing Group*

The chemical structure of the self-assembled PVD nanotubular structures was examined using ToF-SIMS. This method showed that PVD FF-nanotubular arrays represent structures from both sides of the phase transition, characterized by two different molecular compositions. The ToF-SIMS positive ion spectrum shows two characteristic peaks: a high peak at 295 *m/z* (mass to charge ratio), which corresponds to the cyclo-FF, and a low peak at 313 *m/z*, which corresponds to the linear-FF (Figure 1.16c).

The experimental data described above distinctly show that the PVD technique provides deposition of self-assembled nanotubular structures having different chemical/molecular conformations, namely, that a small quantity of nanotubular units are composed of linear peptide molecules while the majority of them are composed of cyclic peptides. The PVD deposition method is performed under thermal conditions in which the FF building blocks are vaporized on to a surface. According to the literature dipeptides can form cyclic analogs upon heating [95, 96]. The monomer powder placement needs to be heated above 250 °C for an effective evaporation process.

The raw PVD material used in the mentioned PVD procedure is linear FF-monomer powder. As previously discussed, the temperature where the FF-linear molecular transforms into the cyclic molecules is around 150 °C. Therefore, the first step toward controllable PVD technology is temperature control of the monomer heater, allowing the evaporation and deposition of either linear or cyclic FF-molecules for fabrication of peptide nanotubular structures of solely one type.

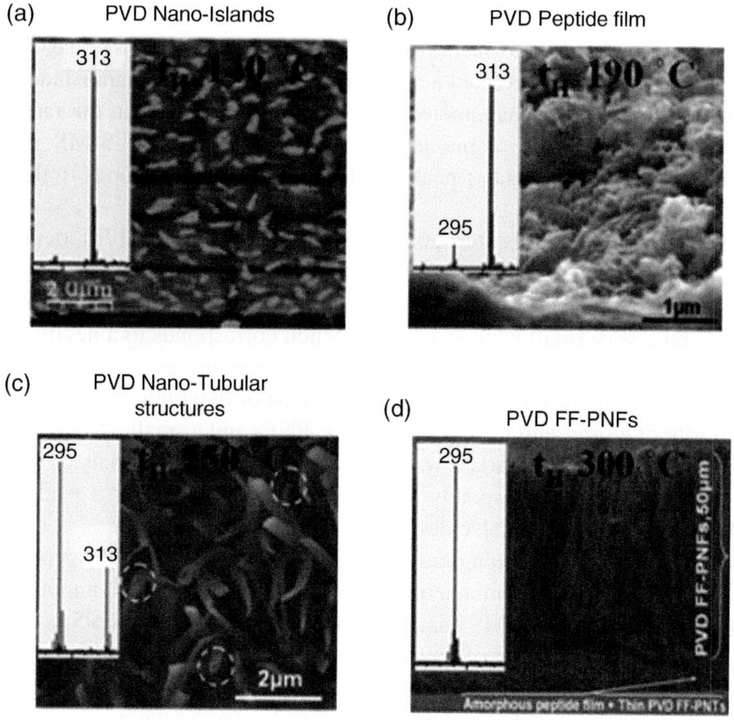

Figure 1.17 *High-resolution microscopy and ToF-SIMS spectra of different PVD surfaces, evaporated at heater temperatures of T_H ~140, 190, 250, 300 °C (a to d, respectively)*

1.5.2.1 Mechanism of PVD Deposition Technique

The deposition process of the evaporated peptide monomer versus the heater's temperature was investigated at a vacuum pressure of 10^{-6} mbar by using the piezoelectric quartz crystal microbalance (QCM) method (Figure 1.16d). Conducted analysis of the QCM thickness deposition rate showed that the PVD process occurs in two different temperature regions. The first peak of FF-peptide monomer powder sublimation and deposition is observed at T_H = 140–190 °C (T_H is the temperature of the heater). The second QCM peak is observed at T_H ~230–300 °C. It is clear that temperature distribution of the peptide monomer powder placed on the heater holder is highly inhomogeneous due to the low thermal conductivity of the peptide powder and lack of tight thermal contact between the peptide raw material and the heater's surface. However, the two separated temperature regions are the direct evidence of evaporation and deposition of two different materials. The used raw material is linear peptide monomer, which passes the cyclization process at around 150 °C. Thus, the first region of the heater temperature, T_H ~140–190 °C, is related to evaporation of the linear FF-peptide biomolecules while the second QCM peak is responsible for evaporation and deposition of the cyclic peptides, which are the product of the molecular transformation at high temperatures of the heater.

Figures 1.17a to d shows high-resolution microscopy and ToF-SIMS spectra of different PVD surfaces, evaporated at different heater temperatures, T_H ~140, 190, 250, 300 °C. The temperature of the substrate holder was kept at T_S ~90 °C, which is also lower than that of the FF-peptide molecular transformation temperature point. The AFM image of the thin coating (Figure 1.17a) clearly revealed a monolayer array of peptide nanoislands of irregular shape. They elongated in a nanometer scale along the substrate, in the range of 50 to 500 nm (with heights reaching a maximum of ~10 nm). The ToF-SIMS positive ion spectrum showed an expected major peak at 313 *m/z*, which corresponds to the linear-FF composition.

The SEM image of the next heater temperature, 190 °C (Figure 1.17b), demonstrates a nonuniform layer of the peptide coating, without any nanotubular structures that could be observed. The ToF-SIMS analysis showed a major peak at 313 *m/z*, which corresponds to the linear-FF, and a very small peak at 295 *m/z*, which corresponds to a negligible amount of cyclic-FF. However, upon increasing the heater temperature to 250 °C, a formation of PNTs was observed (Figure 1.17c). These PNTs consist of two quite dissimilar morphologies: the majority of closed-end nanofiber PVD FF-PNFs and a small part of the structures with an opened-top structure (marked with circles). The ToF-SIMS analysis showed both a peak at 295 *m/z*, which corresponds to the cyclic-FF PNFs, and a peak at 313 *m/z* corresponding to the linear-FF molecules.

The upper possible deposition temperature limit (prior to burning of the peptide powder) is at T_H ~300 °C. The cross SEM image of the surface shows an array of nanotubular structures (Figure 1.17d). The ToF-SIMS analysis showed only one peak at 295 *m/z*, which corresponds to cyclic-FF. As a result of increasing the heater temperature, the linear form of the FF monomer undergoes a chemical reaction between the N- and C-termini of the amino acid, causing a cyclization of the backbone, to form the diketopiperazine, cyclic-FF. The cyclic-FF molecules self-assemble during PVD to form an array of nanofibers, PNFs. The inhomogeneous temperature distribution within the precursor-peptide powder placed on the heater of the PVD system is the cause for the formation of the three layers in the final peptide coating (Figure 1.17d).

1.5.2.2 Integration of PVD in Industrial Processes

The main advantage of using the PVD deposition technique is integration in the common nano/microtechnological industry. Rosenman *et al.* [97] recently showed a simple strategy for a wafer scale assembly of highly ordered and vertically aligned PNF arrays on substrates, which combines top-down and bottom-up nanofabrication schemes. The presented patterned PNF arrays combine the attributes of photolithography with bottom-up self-assembly synthesis. The key solution for selective removal of the PNTs from desirable surfaces was using hydrofluoric acid (HF). Although HF is considered to be a weak acid, it has a known ability to dissolve oxides. Figure 1.18 shows the fabrication process of a PNF array. The process begins with a standard double-sided polished 100 mm diameter and 500 μm thickness single crystal silicon wafer with a thermally grown 5 μm thick silicon dioxide (SiO_2) layer on the front and back sides (Figure 1.18a). By means of standard photolithography methods and by using reactive ion etching (RIE), holes of 50 μm diameter with a 35 μm period were patterned and etched in the SiO_2 upper layer to a depth of 5 μm (Figures 1.18a to d). At the following stage, the vapor deposition technology of PNFs was applied (Figure 1.18e). As a result, the substrate surface was covered with a uniform coating of vertically aligned PNF with a height of about 5 μm. The process was accomplished

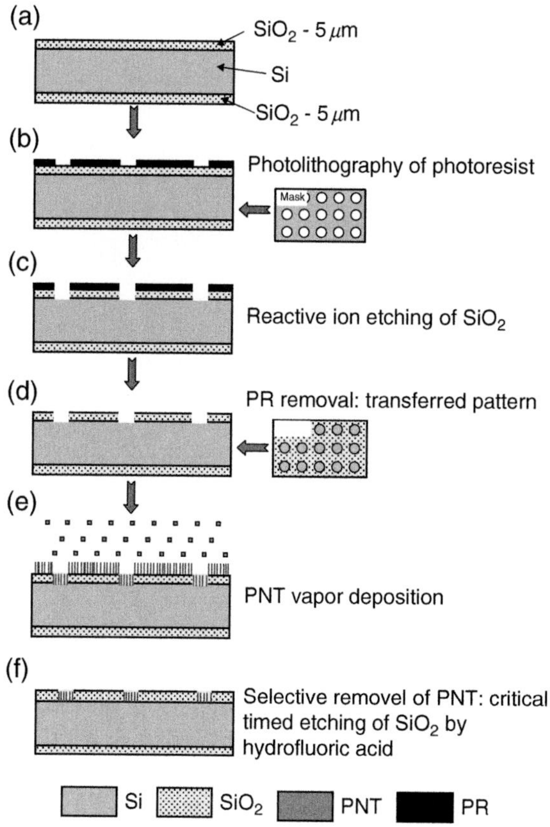

Figure 1.18 *Schematic diagram of the PNT bundles fabrication process. Reproduced with permission from reference [97]. Copyright (2010) Elsevier*

by selectively removing the PNF coating above the SiO_2 layer. Immersing the specimen in HF of 49% at 25 °C for 60 s and rinsing it in water caused partial etching of SiO_2 (about 0.7 μm), which caused the removal of the peptide coating (Figure 1.18f). The final architecture is composed from the PNF arrays within the substrate holes. Figure 1.19 shows SEM images of the substrate before (Figures 1.19a and b) and after (Figures 1.19c and d) the etching process by HF. It should be mentioned that the PNFs were not damaged and completely retained their structure in HF.

1.6 Applications of PNTs

The aforementioned diverse properties of FF PNTs have encouraged many researchers in the field to propose various application pathways for utilizing the FF PNTs. We can divide the application pathways into two thematic niches: (1) using the intrinsic properties of FF PNTs for nanotechnological applications and (2) using the tubular shape of the PNTs as a scaffold for deposition.

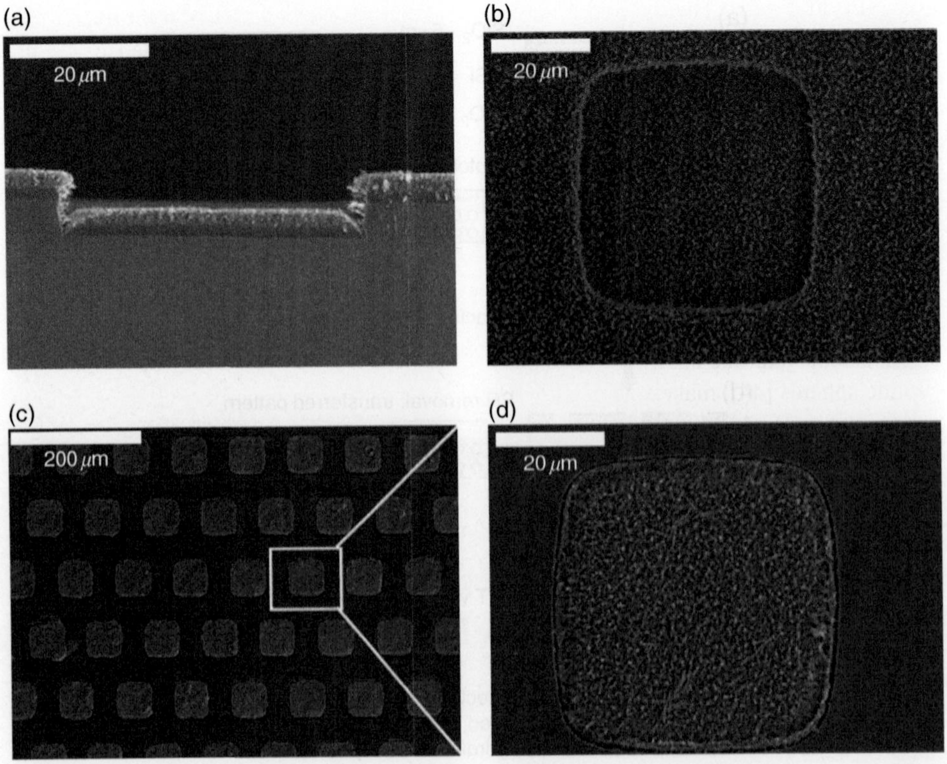

Figure 1.19 *SEM images of patterned arrays of PNT fabricated by vapor deposition technology. (a) Cross-section view of patterned substrate covered by a PNT coating. (b) Top view of the patterned substrate covered by a PNT coating. (c) Top view of PNT bundles after HF release. (d) Enlargement view of a PNT bundle image (PR-photoresist). Reproduced with permission from reference [97]. Copyright (2010) Elsevier*

1.6.1 PNTs for Nanotechnological Applications

The unique intrinsic properties of FF PNT promote the use of these structures in diverse nanotechnological applications. In the native phase of FF PNTs, which are composed of linear-FF peptides, the structure is composed of nanocrystalline units that resemble QD electronic structures [45]. These QD-like assemblies promote the use of FF structures for UV light emitting devices. Following the thermally induced phase transition process, the newly formed PNF structure changes its electronic and optical properties. The PNF structure is also the result of using the PVD deposition process [43, 94] and the aniline vapour-based deposition [86, 90]. One of the outcomes of the transformation to the PNF architecture is the formation of a new PL in the blue region, which can be transformed in blue light emitting devices.

As described previously, several deposition techniques can form films of peptide structures. During the phase transition the structures change their wettability behavior from a hydrophilic behavior of the native phase linear-FF PNT to a hydrophobic behavior of the cyclic-FF PNF [23]. The hydrophobic surface of PNF structures can be used for making hydrophobic coats for surfaces. This application was demonstrated by Park and coworkers who showed the formation of superhydrophobic surfaces, which exhibit a contact angle of 152°, by using the thermal aniline vapor deposition technique of FF [98]. This application pathway can resulted in self-cleaning surfaces.

An additional outcome of the transformation to a PNF structure can be found in enhanced electron transport properties. Park and coworkers [88] examined the current–voltage properties of the native phase PNTs in comparison to the cyclic-FF PNFs. They show that while the native phase PNTs are highly insulating, the cyclic-FF PNFs are fairly electron transport structures. This semi-conducting behavior of the structures can pave the way to using the PNF structures in nanoelectronic devices.

The nanocrystalline nature of the native PNT structures, as well as of the Boc-FF-based peptide spheres [44], makes the structures extremely rigid in comparison to other biological structures. By AFM methodology it was shown that these structures have a remarkably high Young's modulus, from 20 GPa for the FF PNT structure [99] up to hundreds of GPa for the Boc-FF peptide spheres [100]. The high rigidity of the structures promotes their use as reinforcement for composite materials. This application pathway was recently shown by reinforcement of epoxy by FF PNTs [101].

The nanocrystalline structures self-assemble into a noncentrosymmetric space group of $P6_1$. As described in previous sections, the noncentrosymmetric crystal structure resulted in a strong piezoelectric effect of the PNTs [25]. Moreover, the odd-tensor nature of the FF PNT crystal structure should also result in a ferroelectric and nonlinear optical phenomenon. The odd-tensor related properties can result in peptide-based nanopiezoelectrics, in ferroelectric devices (such as ferroelectric capacitors, switchers, etc.), and in nonlinear optical applications (such as nonlinear light emitting diodes, amplifiers, frequency converters, etc.).

The PNTs can also be used in electrochemical applications. There are two electrochemical related applications of FF PNTs. The first one is toward biosensors, where it was found that FF PNT decorated electrodes show a superior electrochemical response in comparison to bare electrodes [102, 103]. In this context, the PNT modified electrodes were examined for the response of hydrogen peroxide and NADH, and for detection of glucose by glucose oxidase, which was attached to the FF PNTs. The second electrochemical application is toward the formation of peptide-based supercapacitors. Rosenman and coworkers have shown that by using the PVD technique, the double-layer capacitance density of the PNT-modified electrode ($480\,\mu F/cm^2$) can increase by 30-fold in comparison to the bare electrode ($16\,\mu F/cm^2$) [94]. Later on they showed that the phase of the PNT is crucial for the superior capacitance properties of the supercapacitor [104, 105]. The hydrophilic nature of the native phase FF PNT permits access of the electrolyte to the electrode, while the hydrophobic cyclic-FF PNF blocks access. According to that, they were able to increase the double-layer capacitance density of the PNT-modified electrode by up to $800\,\mu F/cm^2$.

The self-assembled structure of the PNT is formed by a network of noncovalent bonds. This self-assembly mechanism enables the integration of other molecules, which will be incorporated in the final PNT structures. The possible applications of the integration approach are dependent on the incorporated molecule. Ryu *et al.* [106] showed the incorporation of photosensitizers and lanthanide ions (such as terbium and europium) in different concentrations to FF PNTs, thus creating luminescent PNTs in a wide range of emission wavelengths. In this case the formed exciton of the FF was used to excite the photosensitizers and/or lanthanide ions. A second example is the integration of photosynthetic units (a porphyrin derivative) to the self-assembly process of FF PNTs [107]. Along with the assistance of Pt nanoparticles, the formed light harvesting PNTs could mimic natural photosynthesis.

1.6.2 PNTs as a Deposition Scaffold

The tubular structure of FF PNTs can also be used only for its geometrical shape and not due to its FF-peptide related intrinsic properties. The internal hole of the PNTs was used as a template for forming silver nanowires [31]. The Ag ions in solution can enter the inner hole of the tubular structure, where they can be reduced with a reducing agent to form nanowires. The peptide structure can be easily removed by the use of proteases, thus creating hanging free silver nanowires, ~20 nm in diameter. In a later work, Gazit and coworkers have shown the metallization of the outer surface of the nanotube by gold, in addition to the inner metallization by silver, thus creating a metal–insulator (peptide)–metal coaxial wire [108].

An additional use of the inner hole of the tubular structure can result in nanofluidic devices. Sopher *et al.* [109] have used standard photolithography techniques to successfully link two PDMS channels by FF PNTs. By filling only one channel with water they showed that the FF PNTs can serve as an efficient nanochannel for water to go from one channel to another.

The outer tube geometry was recently used by Kim and coworkers as a template for TiO_2 structures [110, 111]. They used atomic layer deposition to deposit a nanoscale layer of TiO_2 on a surface of FF PNTs. Later on, the FF PNTs were removed by calcination at 400 °C, thus creating highly entangled hollow TiO_2 nanoribbons. These nanoribbons were used for creating an Li secondary battery electrode [110] and for dye-sensitized solar cells [112].

1.7 Conclusion

In this chapter we reviewed the vast progress that has been done in the last decade in understanding the unique properties of FF peptide structures and the development of several deposition techniques toward possible applications. The unique properties of FF peptide structures consists of the formation of nanocrystalline regions with confined excitons, which resemble the QC structure of QDs. The QD-like nanocrystalline structures self-assemble into the tube formation with a noncentrosymmetric space group of $P6_1$. The odd-rank tensor of this space group permits the existence of several odd-rank tensor related properties, such as piezo-, ferroelectricity and nonlinear optics. Moreover, the nanocrystalline nature of the self-assembled structure also resulted in the most rigid structure to be found in biomolecules.

The unique properties of FF PNTs have promoted the invention of new methods to deposit the structures. In addition to the common drop-cast deposition technique, we also reviewed the fast evaporation, PVD, and aniline vapor techniques. The development of the various deposition techniques paved the road to several possible applications of FF peptide structures, such as light emitting devices, piezoelectric actuators, biosensor platform, supercapacitors, reinforcement for composite materials, scaffolds for nanoscale metal wires, as nanofluidic channels, and more. Though none of the proposed applications reached the final stage of commercially available applications, the relatively easy deposition techniques of FF PNTs in addition to their cheap cost and biocompatibility are very promising factors for commercial availability of FF-based applications in the future.

References

[1] Quate, C.F. (1991) Nanotechnology – switch to atom control. Nature 352: 571.
[2] Aldhous, P. (1991) Nanotechnology – wanted – home-grown industry. Nature 352: 746.
[3] Zhang, S.G. (2003) Fabrication of novel biomaterials through molecular self-assembly. Nat. Biotechnol. 21: 1171–1178.
[4] Alberts, B., Johnson, A., Lewis, J., Raff, M., Roberts, K., Walter, P. (2002) Molecular Biology of the Cell. Garland Science, New York.
[5] de la Rica, R., Matsui, H. (2010) Applications of peptide and protein-based materials in bionanotechnology. Chem. Soc. Rev. 39: 3499–3509.
[6] Langer, R., Peppas, N.A. (2003) Advances in biomaterials, drug delivery, and bionanotechnology. Aiche J. 49: 2990–3006.
[7] Zhang, S.G., Marini, D.M., Hwang, W., Santoso, S. (2002) Design of nanostructured biological materials through self-assembly of peptides and proteins. Curr. Opin. Chem. Biol. 6: 865–871.
[8] Gao, X.Y., Matsui, H. (2005) Peptide-based nanotubes and their applications in bionanotechnology. Adv. Mater. 17: 2037–2050.
[9] Ghadiri, M.R., Granja, J.R., Milligan, R.A., McRee, D.E., Khazanovich, N. (1993) Self-assembling organic nanotubes based on a cyclic peptide architecture. Nature 366: 324–327.
[10] Ghadiri, M.R. (1995) Self-assembled nanoscale tubular ensembles. Adv. Mater. 7: 675–677.
[11] Hartgerink, J.D., Granja, J.R., Milligan, R.A., Ghadiri, M.R. (1996) Self-assembling peptide nanotubes. J. Am. Chem. Soc. 118: 43–50.
[12] Fernandez-Lopez, S., Kim, H.S., Choi, E.C., Delgado, M., Granja, J.R., Khasanov, A., Kraehenbuehl, K., Long, G., Weinberger, D.A., Wilcoxen, K.M., Ghadiri, M.R. (2001) Antibacterial agents based on the cyclic D,L-alpha-peptide architecture. Nature 412: 452–455.
[13] Fletcher, J.T., Finlay, J.A., Callow, M.E., Callow, J.A., Ghadiri, M.R. (2007) A combinatorial approach to the discovery of biocidal six-residue cyclic D,L-alpha-peptides against the bacteria methicillin-resistant *Staphylococcus aureus* (MRSA) and *E. coli* and the biofouling algae *Ulva linza* and *Navicula perminuta*. Chem. Eur. J. 13: 4008–4013.
[14] Brea, R.J., Castedo, L., Granja, J.R., Herranz, M.A., Sanchez, L., Martin, N., Seitz, W., Guldi, D.M. (2007) Electron transfer in Me-blocked heterodimeric alpha, gamma-peptide nanotubular donor–acceptor hybrids. PNAS 104: 5291–5294.
[15] Motesharei, K., Ghadiri, M.R. (1997) Diffusion-limited size-selective ion sensing based on SAM-supported peptide nanotubes. J. Am. Chem. Soc. 119: 11306–11312.
[16] Brea, R.J., Reiriz, C., Granja, J.R. (2010) Towards functional bionanomaterials based on self-assembling cyclic peptide nanotubes. Chem. Soc. Rev. 39: 1448–1456.
[17] Sanchez-Quesada, J., Ghadiri, M.R., Bayley, H., Braha, O. (2000) Cyclic peptides as molecular adapters for a pore-forming protein. J. Am. Chem. Soc. 122: 11757–11766.
[18] Ghadiri, M.R., Granja, J.R., Buehler, L.K. (1994) Artificial transmembrane ion channels from self-assembling peptide nanotubes. Nature 369: 301–304.
[19] Sanchez-Quesada, J., Kim, H.S., Ghadiri, M.R. (2001) A synthetic pore-mediated transmembrane transport of glutamic acid. Angew. Chem. Int. Ed. 113: 2571–2574.
[20] Ishihara, Y., Kimura, S. (2010) Nanofiber formation of amphiphilic cyclic tri-beta-peptide. J. Pept. Sci. 16: 110–114.

[21] Kimura, S. (2008) Molecular dipole engineering: new aspects of molecular dipoles in molecular architecture and their functions. Org. Biomol. Chem. 6: 1143–1148.
[22] Fujimura, F., Fukuda, M., Sugiyama, J., Morita, T., Kimura, S. (2006) Parallel assembly of dipolar columns composed of a stacked cyclic tri-[small beta]-peptide. Org. Biomol. Chem. 4: 1896–1901.
[23] Amdursky, N., Beker, P., Koren, I., Bank-Srour, B., Mishina, E., Semin, S., Rasing, T., Rosenberg, Y., Barkay, Z., Gazit, E., Rosenman, G. (2011) Structural transition in peptide nanotubes. Biomacromolecules 12: 1349–1354.
[24] Amdursky, N., Beker, P., Schklovsky, J., Gazit, E., Rosenman, G. (2010) Ferroelectric and related phenomena in biological and bioinspired nanostructures. Ferroelectrics 399: 107–117.
[25] Kholkin, A., Amdursky, N., Bdikin, I., Gazit, E., Rosenman, G. (2009) Strong piezoelectricity in bioinspired peptide nanotubes. ACS Nano 4: 610–614.
[26] Görbitz, C.H. (2001) Nanotube formation by hydrophobic dipeptides. Chem. Eur. J. 7: 5153–5159.
[27] Görbitz, C.H. (2002) Peptide structures. Curr. Opin. Solid State Mater. Sci. 6: 109–116.
[28] Görbitz, C.H. (2002) An exceptionally stable peptide nanotube system with flexible pores. Acta Crystallogr., Sec. B: Struct. Sci. 58: 849–854.
[29] Görbitz, C.H. (2010) Structures of dipeptides: the head-to-tail story. Acta Crystallogr., Sec. B: Struct. Sci. 66: 84–93.
[30] Görbitz, C.H. (2007) Microporous organic materials from hydrophobic dipeptides. Chem. Eur. J. 13: 1022–1031.
[31] Reches, M., Gazit, E. (2003) Casting metal nanowires within discrete self-assembled peptide nanotubes. Science 300: 625-627.
[32] Husby, G., Stenstad, T., Magnus, J.H., Sletten, K., Nordvag, B.Y., Marhaug, G. (1994) Interaction between circulating amyloid fibril protein precursors and extracellular tissue matrix components in the pathogenesis of systemic amyloidosis. Clin. Immunol. Immunopathol. 70: 2–9.
[33] Kelly, J.W. (1996) Alternative conformations of amyloidogenic proteins govern their behavior. Curr. Opin. Struct. Biol. 6: 11–17.
[34] Xing, Y.M., Higuchi, K. (2002) Amyloid fibril proteins. Mech. Ageing Dev. 123: 1625–1636.
[35] Gazit, E. (2002) A possible role for π-stacking in the self-assembly of amyloid fibrils. Faseb J. 16: 77–83.
[36] Reches, M., Gazit, E. (2006) Designed aromatic homo-dipeptides: formation of ordered nanostructures and potential nanotechnological applications. Phys. Biol. 3: S10–S19.
[37] Yan, X.H., Zhu, P.L., Li, J.B. (2010) Self-assembly and application of diphenylalanine-based nanostructures. Chem. Soc. Rev. 39: 1877–1890.
[38] Gazit, E. (2007) Self-assembled peptide nanostructures: the design of molecular building blocks and their technological utilization. Chem. Soc. Rev. 36: 1263–1269.
[39] Rosenman, G., Beker, P., Koren, I., Yevnin, M., Bank-Srour, B., Mishina, E., Semin, S. (2011) Bioinspired peptide nanotubes: deposition technology, basic physics and nanotechnology applications. J. Pept. Sci. 17: 75–87.
[40] Görbitz, C.H. (2006) The structure of nanotubes formed by diphenylalanine, the core recognition motif of Alzheimer's beta-amyloid polypeptide. Chem. Commun. 2332–2334.
[41] Halasyamani, P.S., Poeppelmeier, K.R. (1998) Noncentrosymmetric Oxides. Chem. Mater. 10: 2753–2769.
[42] Amdursky, N., Gazit, E., Rosenman, G. (2010) Quantum confinement in self-assembled bioinspired peptide hydrogels. Adv. Mater. 22: 2311–2315.
[43] Amdursky, N., Molotskii, M., Aronov, D., Adler-Abramovich, L., Gazit, E., Rosenman, G. (2009) Blue luminescence based on quantum confinement at peptide nanotubes. Nano Lett. 9: 3111–3115.
[44] Amdursky, N., Molotskii, M., Gazit, E., Rosenman, G. (2009) Self-assembled bioinspired quantum dots: optical properties. Appl. Phys. Lett. 94: 261907.
[45] Amdursky, N., Molotskii, M., Gazit, E., Rosenman, G. (2010) Elementary building blocks of self-assembled peptide nanotubes. J. Am. Chem. Soc. 132: 15632–15636.
[46] Hauser, C.A.E., Zhang, S.G. (2010) Nanotechnology: peptides as biological semiconductors. Nature 468: 516–517.

[47] Gorbitz, C.H. (2007) Microporous organic materials from hydrophobic dipeptides. Chem. Eur. J. 13: 1022–1031.

[48] Nelson, R.J. (1982) In: Excitons. North-Holland, Amsterdam.

[49] Schubert, E.F., Gobel, E.O., Horikoshi, Y., Ploog, K., Queisser, H.J. (1984) Alloy broadening in photoluminescence spectra of Alxgal-Xas. Phys. Rev. B 30: 813–820.

[50] Mi, X.W., Cao, J.C., Zhang, C. (2004) Optical absorption in terahertz-driven quantum wells. J. App. Phys. 95: 1191–1195.

[51] Singh, J. (1995) Semiconductor Optoelectronics: Physics and Technology. McGraw-Hill International Editions, Columbus.

[52] Svelto, O. (1998) Principles of Lasers. Pleneum Press, New York.

[53] Ridley, B.K. (1999) Quantum Processes in Semiconductors. Oxford University Press, Oxford.

[54] Lalov, I.J., Zhelyazkov, I. (2007) Vibronic approach in the theory of the excitonic spectra of molecular crystals. Phys. Rev. B 75: 245435.

[55] Lang, S.B. (1966) Pyroelectric effect in bone and tendon. Nature 212: 705.

[56] Bazhenov, V.A., Kytmanov, A.V. (1967) New data on symmetry of piezoelectric parameters of wood. Sov. Phys. Crystallogr. 11: 784–786.

[57] Shamos, M.H., Lavine, L.S. (1967) Piezoelectricity as a fundamental property of biological tissues. Nature 213: 267–269.

[58] Williams, W.S. (1982) Piezoelectric effects in biological-materials. Ferroelectrics 41: 359–380.

[59] Lemanov, V.V. (2000) Piezoelectric and pyroelectric properties of protein amino acids as basic materials of Soft State Physics. Ferroelectrics 238: 211–218.

[60] Gruverman, A., Wu, D., Rodriguez, B.J., Kalinin, S.V., Habelitz, S. (2007) High-resolution imaging of proteins in human teeth by scanning probe microscopy. Biochem. Biophys. Res. Commun. 352: 142–146.

[61] Halperin, C., Mutchnik, S., Agronin, A., Molotskii, M., Urenski, P., Salai, M., Rosenman, G. (2004) Piezoelectric effect in human bones studied in nanometer scale. Nano Lett. 4: 1253–1256.

[62] Kalinin, S.V., Rodriguez, B.J., Jesse, S., Thundat, T., Gruverman, A. (2005) Electromechanical imaging of biological systems with sub-10 nm resolution. Appl. Phys. Lett. 87: 053901.

[63] Rodriguez, B.J., Kalinin, S.V., Shin, J., Jesse, S., Grichko, V., Thundat, T., Baddorf, A.P., Gruverman, A. (2006) Electromechanical imaging of biomaterials by scanning probe microscopy. J. Struct. Biol. 153: 151–159.

[64] Abplanalp, M., Eng, L.M., Gunter, P. (1998) Mapping the domain distribution at ferroelectric surfaces by scanning force microscopy. Appl. Phys. A 66: S231–S234.

[65] Heider, U., Weis, O. (1993) Distortion-free, calibrated Linbo3 piezoscanner for probe microscopes with atomic-resolution. Rev. Sci. Instrum. 64: 3534–3537.

[66] Smith, R.T., Welsh, F.S. (1971) Temperature dependence of elastic, piezoelectric, and dielectric constants of lithium tantalate and lithium niobate. J. App. Phys. 42: 2219–2230.

[67] Warner, A.W., Onoe, M., Coquin, G.A. (1967) Determination of elastic and piezoelectric constants for crystals in class (3 m). J. Acoust. Soc. Am. 42: 1223–1231.

[68] Akutagawa, T., Koshinaka, H., Sato, D., Takeda, S., Noro, S.I., Takahashi, H., Kumai, R., Tokura, Y., Nakamura, T. (2009) Ferroelectricity and polarity control in solid-state flip-flop supramolecular rotators. Nat. Mater. 8: 342–347.

[69] Horiuchi, S., Tokunaga, Y., Giovannetti, G., Picozzi, S., Itoh, H., Shimano, R., Kumai, R., Tokura, Y. (2010) Above-room-temperature ferroelectricity in a single-component molecular crystal. Nature 463: 789–793.

[70] Lines, M.E., Glass, M.A. (2001) Principles and Applications of Ferroelectrics and Related Materials. Oxford University Press, USA.

[71] Spanier, J.E., Kolpak, A.M., Urban, J.J., Grinberg, I., Lian, O.Y., Yun, W.S., Rappe, A.M., Park, H. (2006) Ferroelectric phase transition in individual single-crystalline $BaTiO_3$ nanowires. Nano Lett. 6: 735–739.

[72] Capillas, C., Perez-Mato, J.M., Aroyo, M.I. (2007) Maximal symmetry transition paths for reconstructive phase transitions. J. Phys. Condens. Matter 19: 275203.

[73] Horiuchi, S., Tokura, Y. (2008) Organic ferroelectrics. Nat. Mater. 7: 357–366.

[74] Kishine, J., Luty, T., Yonemitsu, K. (2004) Ferroelectric phase transition, ionicity condensation, and multicriticality in charge-transfer organic complexes. Phys. Rev. B 69: 075115.

[75] Bombelli, F.B., Berti, D., Milani, S., Lagi, M., Barbaro, P., Karlsson, G., Brandt, A., Baglioni, P. (2008) Collective headgroup conformational transition in twisted micellar superstructures. Soft Matter 4: 1102–1113.
[76] Higuchi, T., Yabu, H., Onoue, S., Kunitake, T., Shimomura, M. (2008) Preparation of lamella-structured block-copolymer particles and their irreversible lamella-disorder phase transition. Colloids Surf. A 313: 87–90.
[77] Noe, F., Fischer, S. (2008) Transition networks for modeling the kinetics of conformational change in macromolecules. Curr. Opin. Struct. Biol. 18: 154–162.
[78] Wang, F., Wenslow, R.M., Dowling, T.M., Mueller, K.T., Santos, I., Wyvratt, J.M. (2003) Characterization of a thermally induced irreversible conformational transition of amylose tris(3,5-dimethylphenylcarbamate) chiral stationary phase in enantioseparation of dihydropyrimidinone acid by quasi-equilibrated liquid chromatography and solid-state NMR. Anal. Chem. 75: 5877–5885.
[79] Zhang, M., Tanaka, T., Ikura, M. (1995) Calcium-induced conformational transition revealed by the solution structure of Apo Calmodulin. Nat. Struct. Mol. Biol. 2: 758–767.
[80] Dunitz, J.D. (1991) Phase-transitions in molecular-crystals from a chemical viewpoint. Pure Appl. Chem. 63: 177–185.
[81] Huheey, J.E., Keiter, E.A., Keiter, R.L. (1997) Inorganic Chemistry: Principles of Structure and Reactivity. Prentice Hall, Upper Saddle River, NJ, USA.
[82] Bruice, P.Y. (2003) Organic Chemistry. Prentice Hall, Upper Saddle River, NJ, USA.
[83] Martins, M.B., Carvalho, I. (2007) Diketopiperazines: biological activity and synthesis. Tetrahedron 63: 9923–9932.
[84] Herbert, R.B. (1989) Biosynthesis of Secondary Metabolites. Springer, New York.
[85] Huang, Z.Z., Leman, L.J., Ghadiri, M.R. (2008) Biomimetic catalysis of diketopiperazine and dipeptide syntheses. Angew. Chem. Int. Ed. 47: 1758–1761.
[86] Ryu, J., Park, C.B. (2008) High-temperature self-assembly of peptides into vertically well-aligned nanowires by aniline vapor. Adv. Mater. 20: 3754–3758.
[87] Ryu, J., Park, C.B. (2010) High stability of self-assembled peptide nanowires against thermal, chemical, and proteolytic attacks. Biotechnol. Bioeng. 105: 221–230.
[88] Lee, J.S., Yoon, I., Kim, J., Ihee, H., Kim, B., Park, C.B. (2011) Self-assembly of semiconducting photoluminescent peptide nanowires in the vapor phase. Angew. Chem. Int. Ed. 50: 1164–1167.
[89] Reches, M., Gazit, E. (2006) Controlled patterning of aligned self-assembled peptide nanotubes. Nat. Nanotechnol. 1: 195–200.
[90] Ryu, J., Park, C.B. (2008) Solid-phase growth of nanostructures from amorphous peptide thin film: effect of water activity and temperature. Chem. Mater. 20: 4284–4290.
[91] Du, M.C., Zhu, P.L., Yan, X.H., Su, Y., Song, W.X., Li, J.B. (2011) Honeycomb self-assembled peptide scaffolds by the breath figure method. Chem. Eur. J. 17: 4238–4245.
[92] Hendler, N., Sidelman, N., Reches, M., Gazit, E., Rosenberg, Y., Richter, S. (2007) Formation of well-organized self-assembled films from peptide nanotubes. Adv. Mater. 19: 1485–1488.
[93] Sidelman, N., Rosenberg, Y., Richter, S. (2010) Peptide-based spherulitic films – formation and properties. J. Colloid Interface Sci. 343: 387–391.
[94] Adler-Abramovich, L., Aronov, D., Beker, P., Yevnin, M., Stempler, S., Buzhansky, L., Rosenman, G., Gazit, E. (2009) Vapor-deposited self-assembled peptide nano-array for energy storage and microfluidics devices. Nat. Nanotechnol. 4: 849–854.
[95] Li, J., Brill, T.B. (2003) Spectroscopy of hydrothermal reactions. 27. Simultaneous determination of hydrolysis rate constants of glycylglycine to glycine and glycylglycine-diketopiperazine equilibrium constants at 310–330 degrees C and 275 bar. J. Phys. Chem. A 107: 8575–8577.
[96] Balibar, C.J., Walsh, C.T. (2006) GliP, a multimodular nonribosomal peptide synthetase in *Aspergillus fumigatus*, makes the diketopiperazine scaffold of gliotoxin. Biochemistry 45: 15029–15038.
[97] Shklovsky, J., Beker, P., Amdursky, N., Gazit, E., Rosenman, G. (2010) Bioinspired peptide nanotubes: deposition technology and physical properties. Mater. Sci. Eng. B 169: 62–66.
[98] Lee, J.S., Ryu, J., Park, C.B. (2009) Bio-inspired fabrication of superhydrophobic surfaces through peptide self-assembly. Soft Matter 5: 2717–2720.

[99] Kol, N., Adler-Abramovich, L., Barlam, D., Shneck, R.Z., Gazit, E., Rousso, I. (2005) Self-assembled peptide nanotubes are uniquely rigid bioinspired supramolecular structures. Nano Lett. 5: 1343–1346.

[100] Adler-Abramovich, L., Kol, N., Yanai, I., Barlam, D., Shneck, R.Z., Gazit, E., Rousso, I. (2010) Self-ssembled organic nanostructures with metallic-like stiffness. Angew. Chem. Int. Ed. 49: 9939–9942.

[101] Even, N., Adler-Abramovich, L., Buzhansky, L., Dodiuk, H., Gazit, E. (2011) Improvement of the mechanical properties of epoxy by peptide nanotube fillers. Small 7: 1007–1011.

[102] Yemini, M., Reches, M., Rishpon, J., Gazit, E. (2005) Novel electrochemical biosensing platform using self-assembled peptide nanotubes. Nano Lett. 5: 183–186.

[103] Yemini, M., Reches, M., Gazit, E., Rishpon, J. (2005) Peptide nanotube-modified electrodes for enzyme–biosensor applications. Anal. Chem. 77: 5155–5159.

[104] Beker, P., Koren, I., Amdursky, N., Gazit, E., Rosenman, G. (2010) Bioinspired peptide nanotubes as supercapacitor electrodes. J. Mater. Sci. 45: 6374–6378.

[105] Beker, P., Rosenman, G. (2010) Bioinspired nanostructural peptide materials for supercapacitor electrodes. J. Mater. Res. 25: 1661–1666.

[106] Ryu, J., Lim, S.Y., Park, C.B. (2009) Photoluminescent peptide nanotubes. Adv. Mater. 21: 1577–1581.

[107] Kim, J.H., Lee, M., Lee, J.S., Park, C.B. (2012) Self-assembled light-harvesting peptide nanotubes for mimicking natural photosynthesis. Angew. Chem. Int. Ed. 51: 517–520.

[108] Carny, O., Shalev, D.E., Gazit, E. (2006) Fabrication of coaxial metal nanocables using a self-assembled peptide nanotube scaffold. Nano Lett. 6: 1594–1597.

[109] Sopher, N.B., Abrams, Z.R., Reches, M., Gazit, E., Hanein, Y. (2007) Integrating peptide nanotubes in micro-fabrication processes. J. Micromech. Microeng. 17: 2360–2365.

[110] Kim, S.-W., Han, T.H., Kim, J., Gwon, H., Moon, H.-S., Kang, S.-W., Kim, S.O., Kang, K. (2009) Fabrication and electrochemical characterization of TiO_2 three-dimensional nanonetwork based on peptide assembly. ACS Nano 3: 1085–1090.

[111] Han, T.H., Oh, J.K., Park, J.S., Kwon, S.-H., Kim, S.-W., Kim, S.O. (2009) Highly entangled hollow TiO_2 nanoribbons templating diphenylalanine assembly. J. Mater. Chem. 19: 3512–3516.

[112] Han, T.H., Moon, H.S., Hwang, J.O., Seok, S.I., Im, S.H., Kim, S.O. (2010) Peptide-templating dye-sensitized solar cells. Nanotechnology 21: 185601.

2

Chemistry of Peptide Materials: Synthetic Aspects and 3D Structural Studies

Fernando Formaggio, Alessandro Moretto, Marco Crisma and Claudio Toniolo

Institute of Biomolecular Chemistry, Padova Unit, CNR, Department of Chemistry, University of Padova, Italy

Acronyms

Aib	α-Aminoisobutyric acid or C^{α}-methylalanine
(αMe)Phe	C^{α}-methylphenylalanine
(αMe)Val	C^{α}-methylvaline
BOP	(Benzotriazol-1-yloxy)tris(dimethylamino)phosphonium
CIP	2-Chloro-1,3-dimethylimidazolidium phosphate
Db_zg	$C^{\alpha,\alpha}$-dibenzylglycine
Deg	$C^{\alpha,\alpha}$-diethylglycine
Dϕg	$C^{\alpha,\alpha}$-diphenylglycine
Dp_ng	$C^{\alpha,\alpha}$-di-*n*-propylglycine
EDC	1-(3-Dimethylamino)propyl-3-ethylcarbodiimide
HATU	2-(1*H*-7-aza-1,2,3-benzotriazol-1-yl)-1,1,3,3-tetramethyl uronium hexafluorophosphate
HBTU	2-(1*H*-1,2,3-benzotriazol-1-yl)-1,1,3,3-tetramethyluronium hexafluorophosphate
HOAt	7-Aza-1-hydroxy-1,2,3-benzotriazole
HOBt	1-Hydroxy-1,2,3-benzotriazole
HOSu	1-Hydroxysuccinimide
Iva	Isovaline or C^{α}-methyl-α-aminobutyric acid

Peptide Materials: From Nanostructures to Applications, First Edition. Edited by Carlos Alemán, Alberto Bianco and Mariano Venanzi.

PyBOP	(Benzotriazol-1-yloxy)tripyrrolidinophosphonium
SPPS	Solid-phase peptide synthesis
TBTU	2-(1*H*-1,2,3-benzotriazol-1-yl)-1,1,3,3-tetramethyluronium tetrafluoroborate

2.1 Introduction

This chapter deals with α-amino acids and α-peptides only, as they are currently the most extensively exploited in the materials chemistry field. However, it is worth remembering that studies and applications of β-peptides (and γ- and δ- as well, although to a limited extent) are growing significantly.

In 1907 Emil Fischer published the chemical synthesis of the 18-mer peptide H–L-Leu–$(Gly)_3$–L-Leu–$(Gly)_3$–L-Leu–$(Gly)_8$–Gly–OH, which represents a milestone in the history of peptide chemistry [1]. After Fischer's remarkable achievement, it took almost 50 years for peptide chemists to synthesize a peptide longer than an octadecapeptide [2]. Only the introduction of the carbodiimide activating reagent in 1955 [3] and the SPPS technique in 1963 [4] ignited an impressive development. The chemical synthesis of peptides then became accessible even to researchers without a specific training in organic synthesis.

Currently, a variety of methods is available to accomplish the chemical synthesis of peptides. Among the numerous reference books [5–9] and review articles [10–16], it is worth recalling the comprehensive and detailed account contained in the five-volume Houben–Weyl series [17]; theoretical aspects and experimental procedures, both for protection and activation methods, are described in great detail. Therefore, in view of the ample literature and for the sake of brevity, Section 2.2 of this chapter will briefly describe only a few synthetic methods useful for the assembly of nonstandard peptides. Indeed, peptides designed for use in materials chemistry often contain sterically hindered, noncoded α-amino acids. The high number of protecting groups available will not be reviewed, as there is no need, in general, to use a specific protecting group when dealing with a difficult coupling.

Sections 2.3 to 2.5 will present the basic elements and parameters (amide bond and torsion angles) required to understand peptide conformations and an overview of the known, both common and unusual, peptide 3D structures.

With the aim at facilitating information retrieval, all sections are laid down in a schematic way.

2.2 Synthesis of Difficult Peptide Sequences

Couplings involving noncoded, sterically hindered α-amino acids are often characterized by low yields even after long reaction times. This modest reactivity may lead to racemization of chiral C-activated α-amino acids possessing an H atom on their α-carbon. Racemization (or epimerization) is even easier when the activation involves a peptide segment, because of the high chance of 5(4*H*)-oxazolone formation [7, 9, 18]. Therefore, syntheses via segment condensation have to be planned carefully. In general, racemization-free activation methods should be employed also in the case of a stepwise main-chain elongation.

Among the variety of available coupling reagents, those reported hereafter were selected because of their effectiveness with sterically hindered α-amino acids. The procedures described can be applied both to solution and solid-supported syntheses.

Symmetrical anhydride

- The symmetrical anhydrides of α-amino acids are usually prepared *in situ* by treating an N^{α}-protected amino acid with a half-equivalent of carbodiimide. However, better results in the coupling reactions are obtained when the anhydride is isolated beforehand [16, 17]. A tertiary amine is added to neutralize the N^{α}-protected amino acid liberated during the anhydride aminolysis.

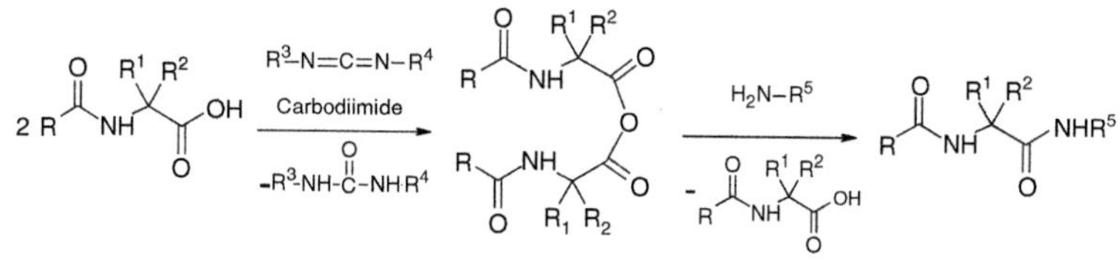

Symmetrical anhydride

- Peptide bond formation proceeds in satisfactory yields and low amounts of side products.
- When expensive amino acids are involved, the need for two equivalents of N^{α}-protected residues to form the symmetrical anhydride represents a serious drawback [14].

EDC/HOAt

- The hydrosoluble carbodiimide EDC greatly facilitates the isolation and purification procedures when operating in solution. It mediates the formation of an active ester between the aminoacyl moiety and the HOAt [19] hydroxyl group.
- HOAt has a greater racemization suppression ability as compared to HOSu [20] and HOBt [21] and it is more effective in catalyzing peptide bond formation. For the latter reason, it is used even when racemization is not a concern. Through its N^2 and N^7 atoms, HOAt assists the (amine) nucleophilic attack on both faces of the planar active ester (this is the conformation that was observed in the crystal state [22, 23]) [19, 24, 25].

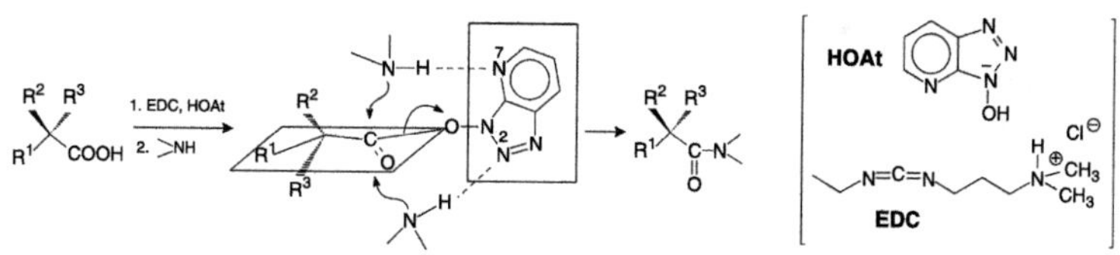

- An active ester with HOAt (or HOBt) forms also when the CIP/HOAt [26], BOP [27], PyBOP [28], HBTU [29], TBTU [30], and HATU [31, 32] activating reagents are used.

HATU

- Among the currently extensively used uronium salts, HATU is the reagent of choice in terms of coupling efficiency [33, 34] and amino acid derivative or peptide optical stability [35, 36]. It is often used in combination with one equivalent of HOAt.

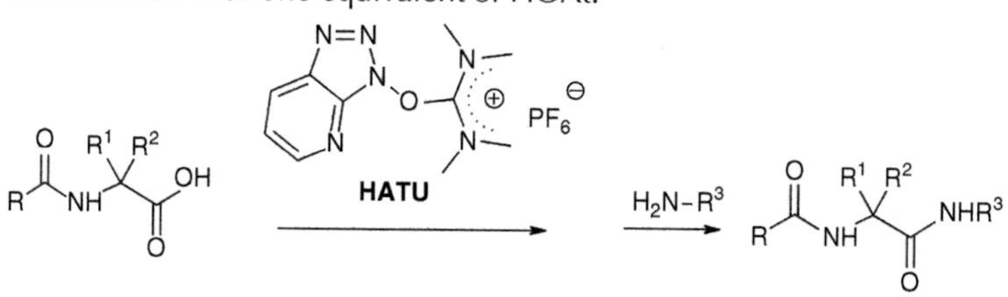

- A major acylating species is believed be the active ester intermediate, as in the case of the EDC/HOAt activation procedure.
- In view of its simpler procedure, the use of HATU represents a good choice for the SPPS incorporation of sterically hindered residues.

α-Aminoacyl fluoride

- The α-aminoacyl fluoride method [37, 38] is more effective than the EDC/HOAt and symmetrical anhydride procedures because of: (i) the small size of the leaving group (the fluoride ion), (ii) the activation of the carboxyl group induced by the high electronegativity of fluorine, and (iii) the slow rate of formation of the poorly reactive 5(4*H*)-oxazolone [38–40].
- N^{α}-protected α-aminoacyl fluorides are easily prepared by treating the corresponding carboxylic acids with cyanuric fluoride [37]. *In situ* preparation is also feasible [41]. However, better results are obtained with previously isolated α-aminoacyl fluorides.

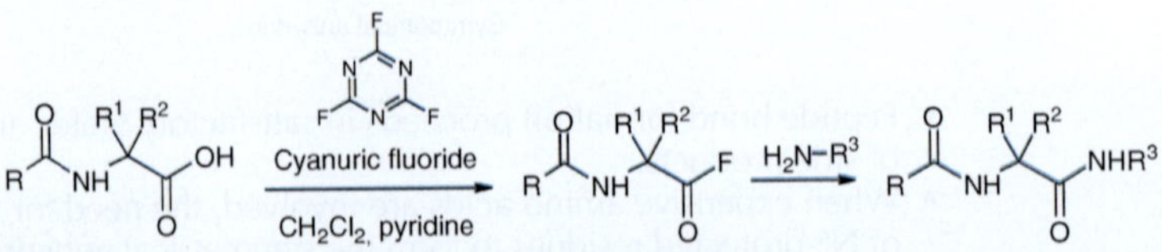

α-aminoacyl fluoride

- Remarkable achievements using this procedure are represented by the syntheses of the octapeptide –[(αMe)Val]$_8$– [42] and the heptapeptide –Iva-(αMe)Val–Iva–(αMe)Val–(αMe)Phe–(αMe)Val–Iva– [43].
- The α-aminoacyl fluoride activation method is also excellent for SPPS protocols: four Aib residues in a row were successfully linked [44], peptaibols were synthesized in high purity and yields [45], and unnatural, sterically hindered amino acids were easily incorporated [46]. Its coupling efficiency was shown to be superior to that of the easy-to-use HATU [39, 47]. In addition, couplings can be performed even in the absence of any base, thus avoiding premature deblocking of the Fmoc N-protecting group or the undesired 5(4*H*)-oxazolone formation [10].

α-Azidoacyl chlorides

- In recent years, the use of α-azidoacyl chlorides in the synthesis of difficult peptide sequences significantly increased [48–51]. These reagents are very efficient in peptide bond formation mainly because they cannot form the slowly reacting 5(4*H*)-oxazolone (because the azide group lacks a carbonyl moiety).
- The azide behaves as an N^{α}-protecting group: it can be easily converted into the free amine by catalytic hydrogenation.

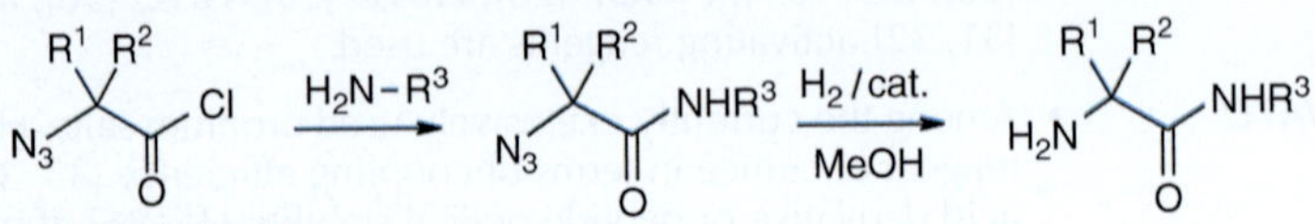

α-azidoacyl chloride

- This coupling procedure requires preparation of appropriate α-azido acid precursors and their subsequent conversion into α-azidoacyl chlorides. For this reason and for the care required in handling the potentially exploding azides, this activating method has not yet found a wider application. However, being probably the most effective among those illustrated here, it appears to be the method of choice for the formation of very difficult amide bonds.

2.3 Peptide (Amide) Bond

Where

- Proteins and enzymes (and a variety of hormones and antibiotics as well) are naturally occurring polypeptide molecules.
- A dipeptide unit, formed by two α-amino acid building blocks, can be depicted as follows:

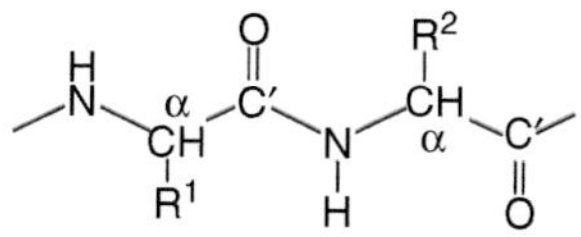

- N, $C^{\alpha}_{sp^3}$, C'_{sp^2}: three atoms per amino acid residue in the main chain (backbone).
- R^1, R^2 : side chains that differentiate the amino acids.
- According to a long-used convention [52, 53], the N-terminus of the main chain (residue 1) should be written on the far left side.

Electronic structure

Resonance of the *secondary* (–CO–NH–) peptide (amide) bond

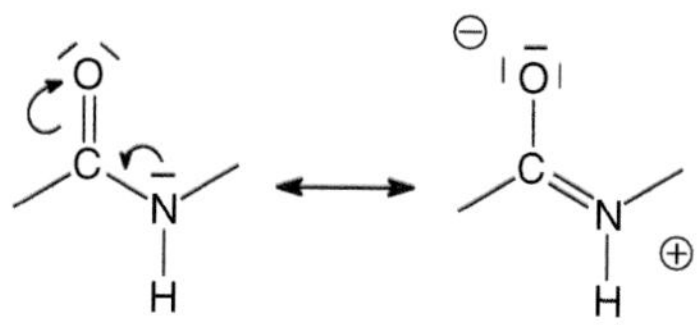

Mean length of C'_{sp^2}–N bond in peptides: 1.34Å (from X-ray diffraction analyses), intermediate between the lengths of C_{sp^3}–N (1.49Å) and C_{sp^2}=N (1.27Å) bonds. Therefore, the C'_{sp^2}–N bond in peptides has a partial double bond character.

Cis/trans isomerism

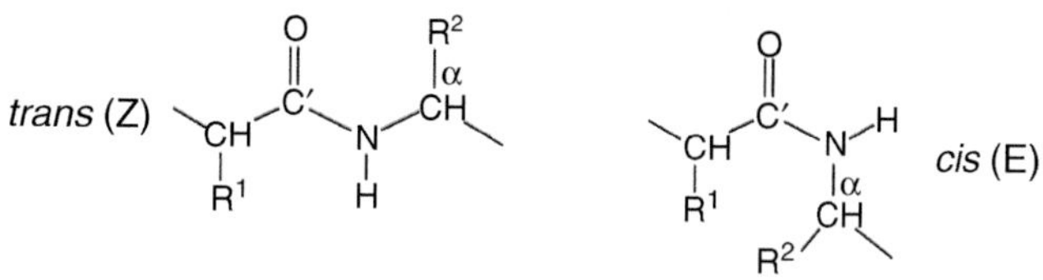

- In the *trans* isomer the $R^2 \cdots O$ repulsion plays a significant role.
- However, in the *cis* isomer the $R^1 \cdots R^2$ repulsion is even more severe.
- For secondary amides, the *trans* isomer is more stable than the *cis* isomer (by ~2 kcal/mole); the energy barrier between the two isomers (rotation about the C′–N bond) is ~18 kcal/mole.

2.4 Peptide Torsion Angles

Stereo-chemical definitions

Peptide torsion angles (dihedral angles; angles between planes).

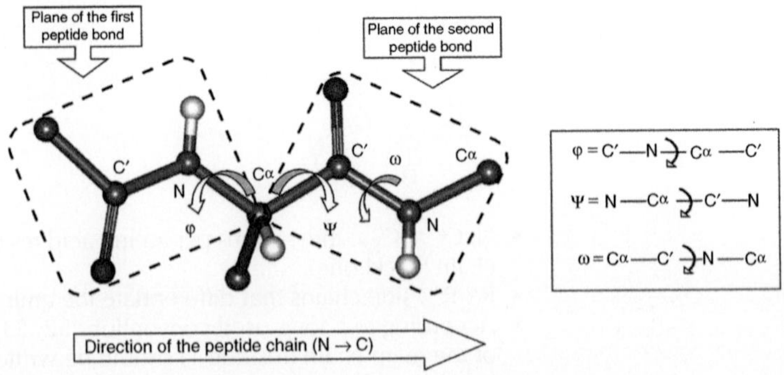

- In the fully extended (zig-zag) peptide conformation, depicted above, the torsion angles φ, ψ, ω are ±180° (all-*trans* conformation), according to the 1970 IUPAC-IUB Commission rules [54]. It is worth pointing out that all structural biochemistry textbooks are using this specific conformation for the illustration of a peptide chain, which, however, has been observed only extremely rarely in naturally occurring peptides and proteins (see below). Before 1970, the notation for φ, ψ, ω torsion angles was from 0° to 360°, as opposed to the current notation from –180° to 180°.
- The central α-amino acid is in the L-configuration (for almost all protein amino acids L ≅ *S*, according to the 1956 Cahn–Ingold–Prelog [55] rule and notation).
- As ≥ 95% of the (mostly secondary) peptide bonds are *trans* (ω = 180°), the conformational problem can be satisfactorily simplified by describing exclusively the sets of φ, ψ torsion angles.
- If a set of φ, ψ torsion angles repeats itself almost identically along a peptide chain, then various types of helical structures may generate, the characteristic parameters of which are:
  - number of amino acid residues per helix turn (*n*);
  - axial translation (*d*), i.e. shift (in Å) along the helix axis per residue;
  - pitch (*p*), where $p = n \times d$, i.e. shift (in Å) along the helix axis per turn.

Ramachandran map

- The 1963 Ramachandran map [56, 57] was based on hard-sphere models.

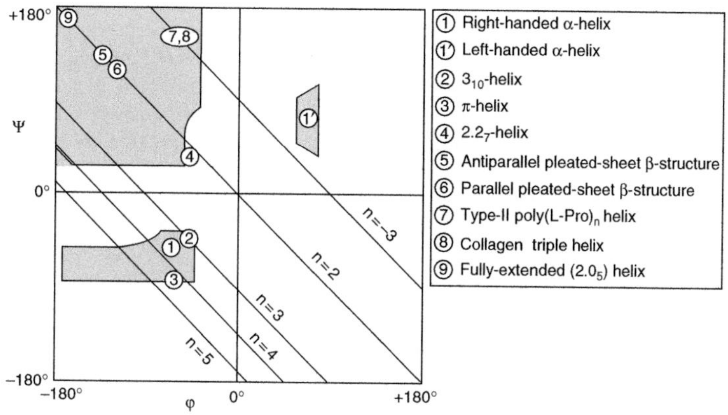

- Subsequently, a variety of conformational energy computations largely confirmed this map.
- This (asymmetric) map is that typical of an L(*S*)-α-amino acid.
- The grey areas (15–20% of the total area) correspond to the *allowed* conformations, i.e. free from severe intramolecular steric interactions (based on van der Waals radii of atoms).
- The area of the allowed conformations for the only achiral protein amino acid (Gly, no R side chain) is much wider (~40% of the total area) and shows a symmetrical pattern with respect to φ, ψ = 0°.
- There are two φ, ψ maps for the only N-alkylated protein amino acid (Pro) as the Xxx–Pro *tertiary* peptide(amide) bond can rather easily accommodate in the unusual *cis* conformation ($\omega = 0°$) besides the common *trans* conformation ($\omega = 180°$).
- The largest population of allowed conformations is seen for φ values of about −60°, as this torsion angle permits the largest separation between the O atom (of the C=O group of the *i* residue) and the R side chain (in particular, the C^β atom) of the $i+1$ residue.
- Most of these helical structures are characterized by a *right*-handed screw sense, as they are positioned below and on the left side of the diagonal with $n = 2$ (where the 'flat', zig-zag helices are found). The only *left*-handed helices are the ($n = -3$) type-II poly-(L-Pro)$_n$ helix and the related collagen triple helix (see below), and the diastereomeric left-handed α-helix (7, 8 and 1′, respectively, in the figure).

2.5 Peptide Secondary Structures

2.5.1 α-Helix

α-Helix: history

- The α-helix (3.6_{13}-helix in the 1950 Bragg–Kendrew–Perutz notation [58]) is the most abundant and stable ordered secondary structure in proteins.
- First proposed in 1950 by Pauling [59, 60], who used the 'bottom-up' approach.
- This helix is termed α because the Pauling proposal was in part based on the X-ray diffraction data of the fibrous protein α-keratin (from wool and hair) published by Atsbury in the 1930s [61], who, however, using the 'top-down' approach, failed to propose correct parameters for the α-helix.
- First experimentally authenticated using X-ray diffraction by Perutz in 1951 [62] and Kendrew in 1960 [63], who solved the crystal structures of the heme-containing, oxygen transporter, proteins hemoglobin and myoglobin, respectively.

α-Helix: structure

- α-Helices are stabilized by intramolecular, backbone···backbone C=O···H–N H-bonds involving 13 atoms (C_{13} form or α-turn). The helical-type α-turn is one of the various pseudo-cyclic forms, first studied in detail by Pavone [64]. It encompasses entirely three amino acid residues (those with the R^2, R^3, and R^4 side chains). The H bond is of the 1←5 type. All –CO–NH– bonds are in the *trans* conformation.

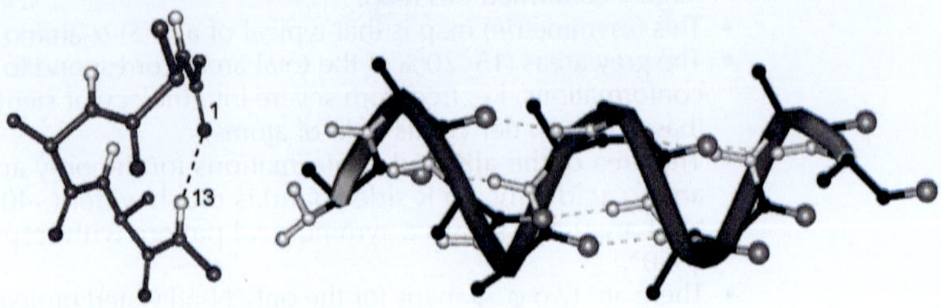

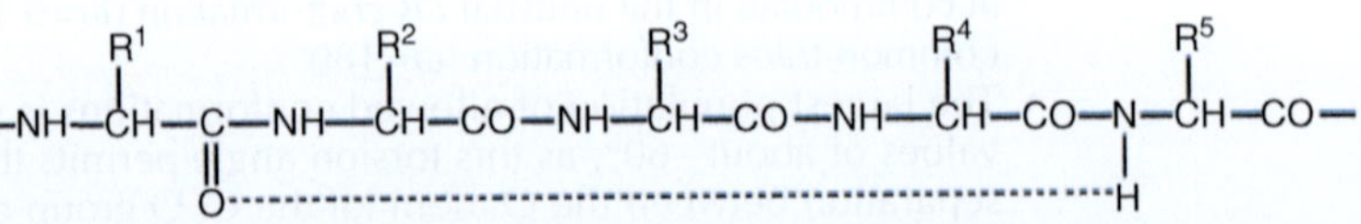

- The (right-handed) α-helical parameters (from the most recent statistical analysis of X-ray diffraction structures at atomic resolution of oligopeptides, published by Toniolo and Benedetti in 1991 [65]) are as follows:

$$\varphi = -63°,\ \psi = -42°,\ n = 3.63,\ d = 1.56\,Å,\ p = 5.67\,Å$$

- It is worth pointing out that the α-helix is *not* characterized by an integer number of amino acids per turn (3.6). This is why Pauling had to fight against the general view held by structural biochemists in the early 1950s to make his proposal accepted by the scientific community (at that time, only polypeptide helices with an integer number of amino acids per turn were considered stable enough).
- A (right-handed) α-helical model of a decapeptide is viewed along the helix axis (the side chains are not eclipsed, but rather they are significantly staggered).

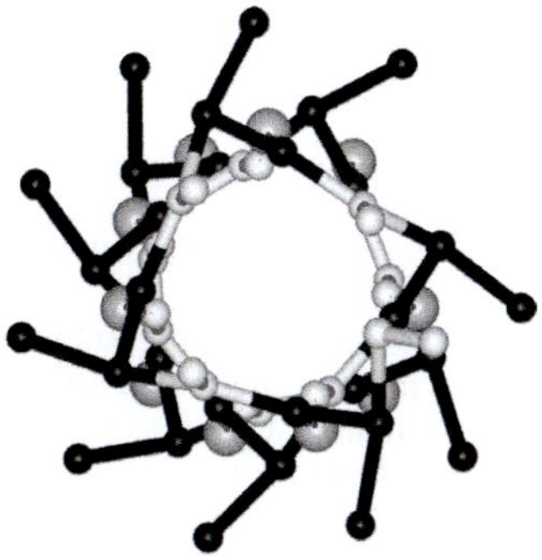

α-Helix: promoting residues

- Several α-amino acids are considered particularly effective α-helix promoters (Blout's classification, 1962 [66]): Ala, Leu, Glu, Lys, Met, Phe, Tyr (their side chains are either linear or γ-branched).
- L-Asp and L-Asn are also helicogenic, although moderately. However, due to their (intramolecular dipole···dipole and H-bonding) side-chain to main-chain interactions, they are frequently found in (*diastereomeric*) *left*-handed α-helical segments (helix 1′ in the Ramachandran map).

α-Helix: aggregates

- Since in the α-helix there are ~*3.5 amino acids per turn*, the smallest integer number characterizing this helix is *7*, which requires two complete α-helical turns. This is the reason why the biologically relevant, amphiphilic (or amphipathic) helices (with one face hydrophobic and the other face hydrophilic) are characterized by *heptad* (*a, b, c, d, e, f, g*) repeats of amino acids, with analogous physicochemical properties at specific positions in the heptad (e.g. in aqueous solutions positions *a* and *d* require hydrophobic residues for antiparallel dimer formation; the hydrophilic positions e and *g*, immediately on the back, reinforce dimer stability via ionic interactions) [67].
- Membrane-active, antibacterial peptides typically fold into amphiphilic α-helices [68].

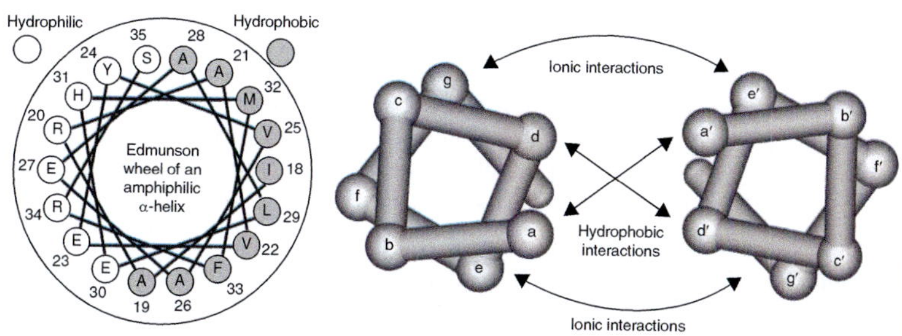

- In the case of α-keratin, in 1952 Crick [69] first suggested a self-association of two α-helices (termed a 'coiled coil' dimer) with an angle between their axes of about 20° and a 'knob-into-holes' packing mode of their side chains (for Leu-rich helices, in 1988 this motif was termed 'Leu zipper' by McNight [70]).

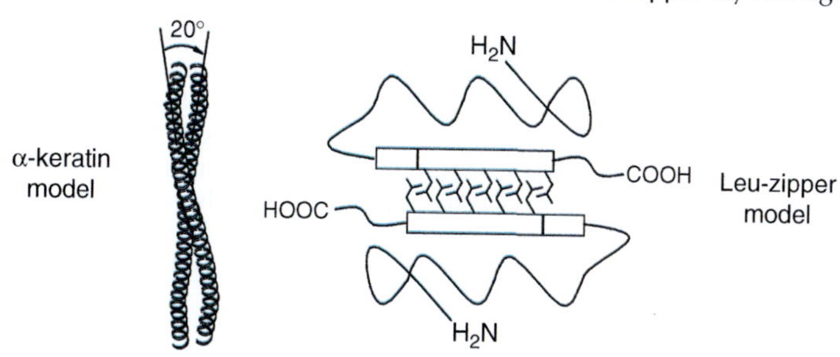

2.5.2 3_{10}-Helix

3_{10}-Helix: history

- First proposed by Taylor in 1941 [71], well before the classical α-helix (this is because the 3_{10}-helix is characterized by an *integer* number (3) of amino acids per turn).
- The (right-handed) type-III β-turn (C_{10} form), according to the 1968 Venkatachalam classification [72], is the building block for the right-handed 3_{10}-helix. Type-I and type-II β-turns, initially called β as they characterize the cross-β structure (see below), were first proposed by Geddes *et al.* in 1968 [73] who termed them type-A and type-B, respectively. A few months later, Venkatachalam [72] studied in detail type-I to type-III β-turns. Other types of β-turns were discussed later on [74].

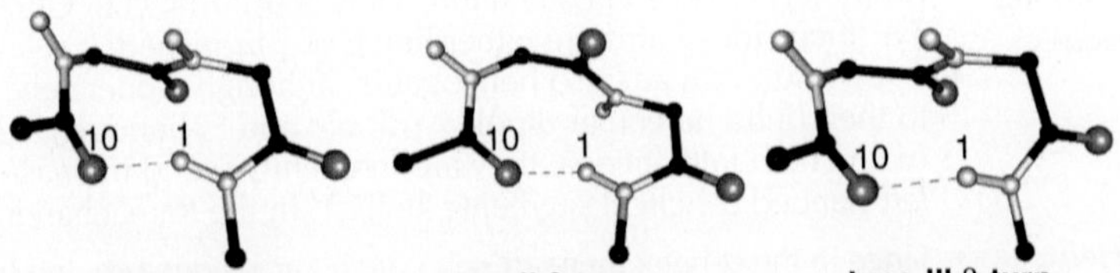

	type-I β-turn	type-II β-turn	type-III β-turn
φ_2, ψ_2	−60°, −30°	−60°, 120°	−60°, −30°
φ_3, ψ_3	−90°, −0°	80°, 0°	−60°, −30°

- The 3_{10}-helix was first experimentally authenticated by Balaram in 1978 [75] by X-ray diffraction analysis of a model, terminally protected, homo-pentapeptide from Aib (α-aminoisobutyric acid), namely Tos–$(Aib)_5$–OMe, where the acceptor of the N-terminal intramolecular H-bond is one of the two oxygen atoms of the *para*-toluenesulfonamide (Tos–NH–) group. The critical main-chain length for 3_{10}-helix formation for a terminally protected $(Aib)_n$ homo-peptide series was found to be $n = 3$ (Benedetti and Toniolo in 1982 [76]).
- 3_{10}-Helical residues represent about 10% of all helical residues in globular proteins [77]. The majority of the 3_{10}-helices are short (3–4 residues) and are mostly located either at the N-terminus or at the C-terminus ('extensions') of α-helices, but some of them have been identified with a length of 7–12 residues. 3_{10}-Helices have been proposed as intermediates in the process of folding of α-helices in globular proteins [78].

3_{10}-Helix: structure

- The (right-handed) 3_{10}-helical parameters (from the Toniolo and Benedetti statistical analysis [65]) are as follows:

$$\varphi = -57°,\ \psi = -30°,\ n = 3.24,\ d = 1.94\ \text{Å},\ p = 6.29\ \text{Å}$$

- This helix is more elongated and less wide as compared to the α-helix. It is also less stable, since the lengths/angles of the intramolecular C=O···H–N H-bonds and the nonbonded steric interactions are less favorable. Remarkably, the experimentally found *n* value (3.24) is not an integer number. It is evident that this helix gains stability from a slight staggering of its side chains (almost one on top of the other after a complete helix turn).
- As mentioned above, 3_{10}-helices are stabilized by intramolecular, backbone···backbone, C=O···H–N H-bonds including 10 atoms (C_{10} form or β-turn). The helical-type (type-III) β-turn is one of the various pseudo-cyclic forms that encompasses entirely two amino acid residues (those with the R^2 and R^3 side chains). The H-bond is of the $1 \leftarrow 4$ type. All –CO–NH– amide bonds are in the *trans* conformation.

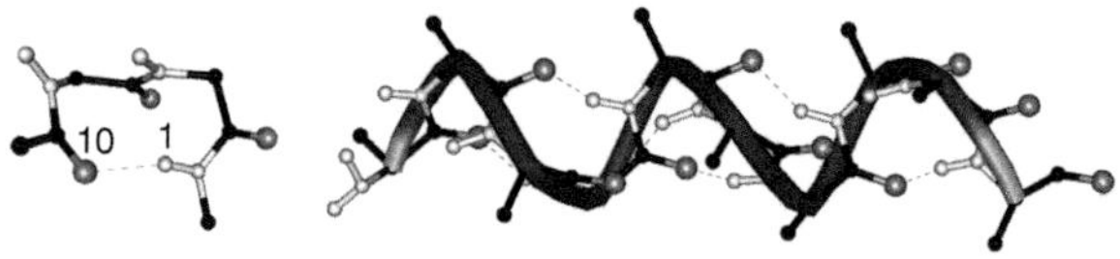

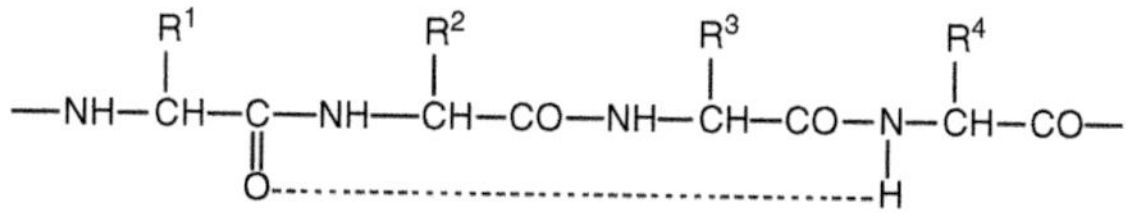

3_{10}-Helix promoting residues

- Because of the Thorpe–Ingold (*gem*-dimethyl) effect [79], Aib is strongly helicogenic and imparts an extremely high crystallinity to its peptides. It also characterizes a family of naturally occurring, membrane-active, peptide antibiotics called 'peptaibols' (Benedetti and Toniolo [80]) or 'peptaibiotics' (Toniolo and Brückner [81]). Upon self-association, some of them form ion-conducting channels in the membranes.
- The right-/left-handed 3_{10}-helical structures of the achiral $(Aib)_n$ ($n = 10, 11$) homo-oligomers have been reported (Benedetti and Toniolo [82], Gessman *et al.* [83]). These are the longest 3_{10}-helices and the longest homopeptides from any amino acid, the 3D structures of which have been solved by X-ray diffraction (at atomic resolution). All of the C$^{\alpha}$-methylated α-amino acids investigated tend to support the 3_{10}-helical structure [79].

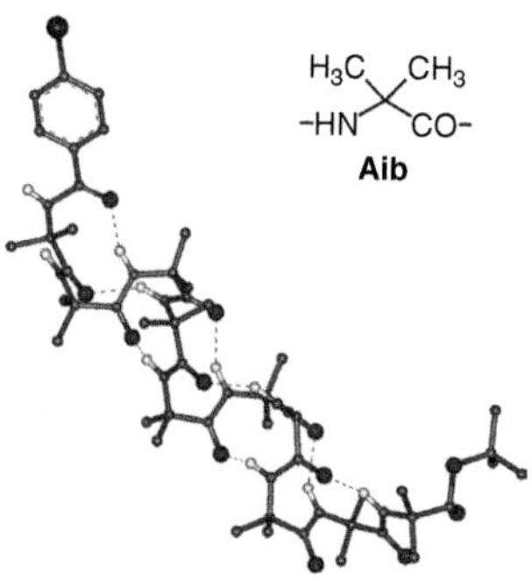

β-Bend ribbon spiral

- A variant of the 3_{10}-helix is the β-bend ribbon spiral, generated by an alternation of a Pro residue (lacking the H-bonding donor NH group) and a strongly helicogenic residue (Aib).
- First observed by Karle and Balaram in 1987 [84] in their X-ray diffraction analysis of the peptaibol zervamycin, it was characterized in detail by X-ray diffraction in several $(Aib–Pro)_n$ model peptides in 1992 (Benedetti and Toniolo [85]).

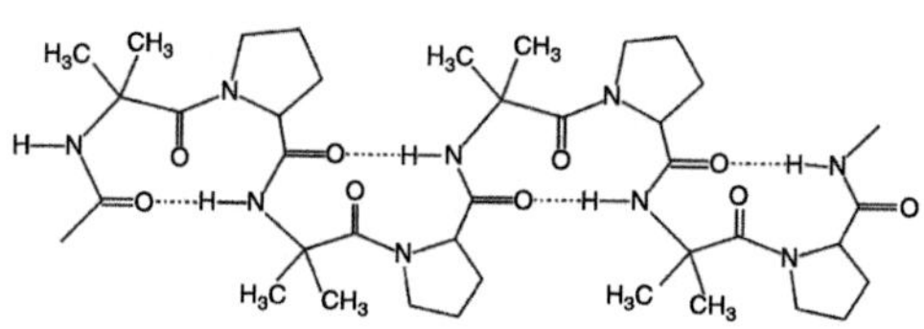

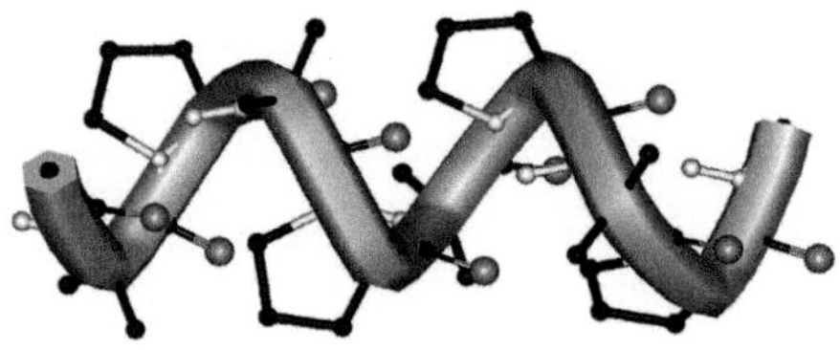

- All Aib–Pro bonds are *trans*. The Pro–Aib bonds deviate markedly ($|\Delta\omega| > 10°$) from the planar *trans* value (180°).

3$_{10}$-Helix: spacer and template

- Amphiphilic 3_{10}-helices can be easily prepared by taking advantage of appropriate amino acid side chains. In view of their almost threefold repeating units, they are even more suitable than α-helices as spacers (bridges) [86] or templates [87] for studies in various areas of chemistry.

2.5.3 2.2$_7$-Helix

γ-Turns

- There are two γ-turn (C_7) conformations for an L-amino acid residue. Both pseudo-cyclic forms include 7 atoms and encompass entirely one amino acid residue (that with the R^2 side chain). The intramolecular H-bond is of the 1 ← 3 type. The internal –CO–NH– amide bond is in the *trans* conformation. The intramolecular H-bond is strongly bent. The ω torsion angles deviate somewhat from the planar *trans* (180°) value. The first proposal and conformational energy computations were published by Némethy and Printz in 1972 [88].
- These types of turns are rare in linear peptides. However, they are quite common in the conformationally forced, small ring, *cyclo*-4- and *cyclo*-5-peptides. In globular proteins the ratio γ-turns/β-turns is ~ 1:7 (the first γ-turn in a globular protein, thermolysin, where the central (R^2) residue is Thr, was reported by Matthews in 1972 [89]).

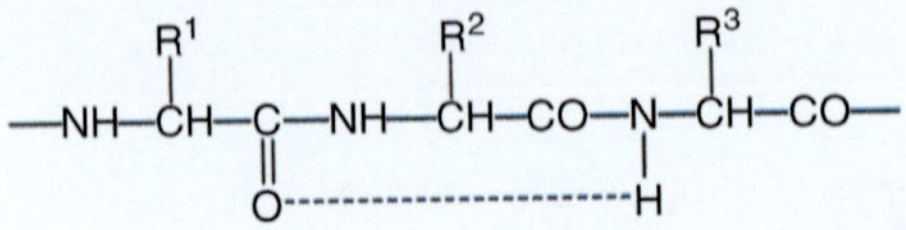

- For an L-residue, the two types of turns are called:
  (I) γ-turn (φ, ψ=70°, –70°) less stable; side-chain R: *axial*,
  (II) inverse γ-turn (φ, ψ=–70°, 70°) more stable; side-chain R: *equatorial*.

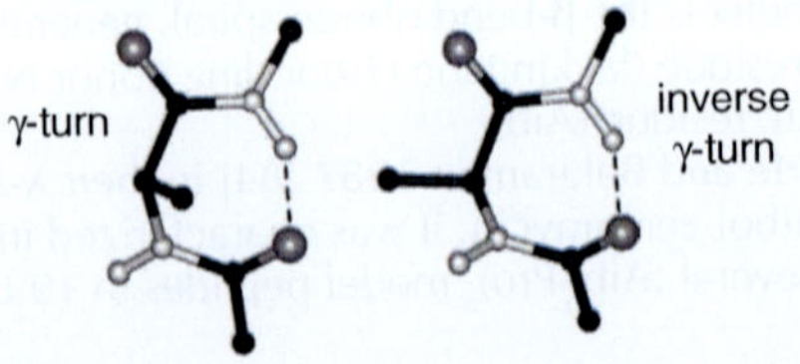

2.2$_7$-Helices (γ-helices)

- Two or more consecutive γ-turns generate a 2.2$_7$(γ)-helix. This helix is tighter and more elongated than the 3_{10}-helix. Its rise per residue (the *d* parameter) is ~ 2.80 Å.

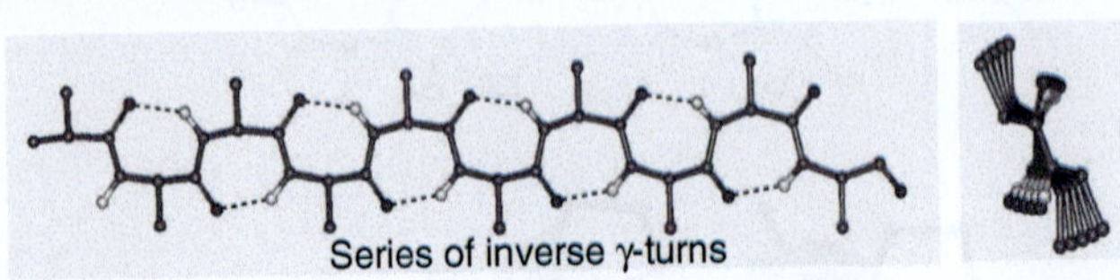

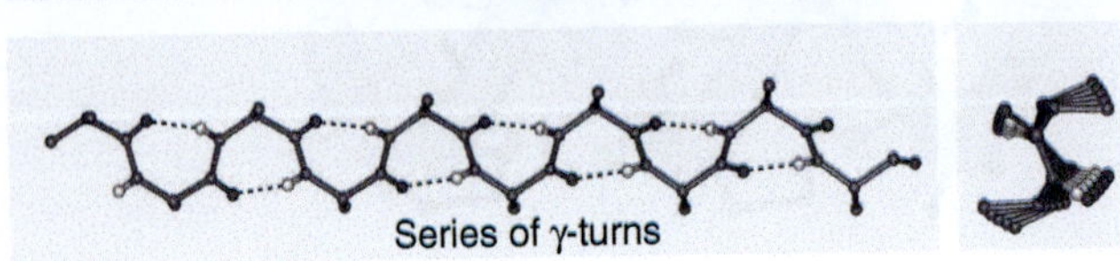

2.2_7-Helix: examples

- Two consecutive inverse γ-turns (the incipient 2.2_7-helix) have been reported in model peptides for the first time by Cativiela in 2005 [90] for the dipeptide heterochiral sequence –L-Pro–D-c_3Dip–. This sequence will only form a γ-bend ribbon spiral (because Pro lacks the H-bonding donor NH group), not an ideal 2.2_7-helix. In any case, this heavily side-chain substituted C^{α}-cyclopropyl amino acid is a promising tool for the construction of the 2.2_7-helix.

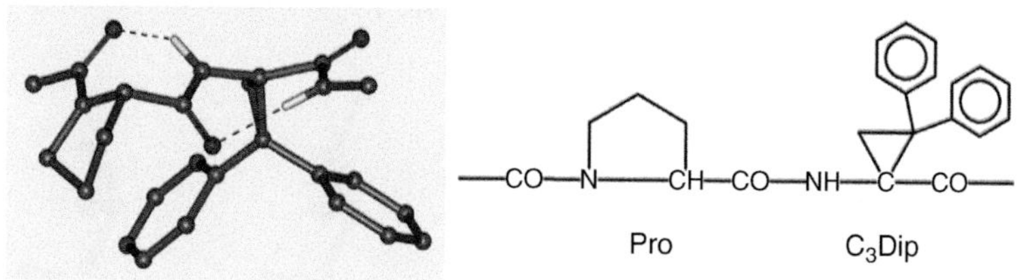

- In globular proteins two examples are known: (i) a repetive γ-turn segment formed by the two residues –Gly–Ile– (a distorted γ-turn followed by a regular inverse γ-turn) [91] and (ii) the serpentine shape of three consecutive, inverse γ-turns formed by the –Thr–Lys–Gln– stretch [92]. In any case, it is quite evident that much work remains to be performed in this specific area of ordered peptide secondary structures.

2.5.4 Pleated-Sheet β-Structures

β-Sheets: history

- The pleated-sheet β-structure is the second most common type of ordered secondary structure in *globular* proteins.
- Both types [*parallel* (β||)- or *antiparallel* (β⊥)-chains] of pleated-sheet β-structures have been proposed by Pauling and Corey in 1951 [93].
- The *pleated*-sheet β-structures (where each residue is extended) are more stable than the *flat*-sheet β-structures (where each residue is *fully* extended, $\varphi = \psi = 180°$) (see below), because in the former less unfavorable intra- and interresidue(s) nonbonded interactions are operative.

β-Sheets: structures

- The *antiparallel*-chain β-sheet structure is more stable (and common) than its parallel-chain counterpart since the directionality of its interchain H-bonds is optimal.

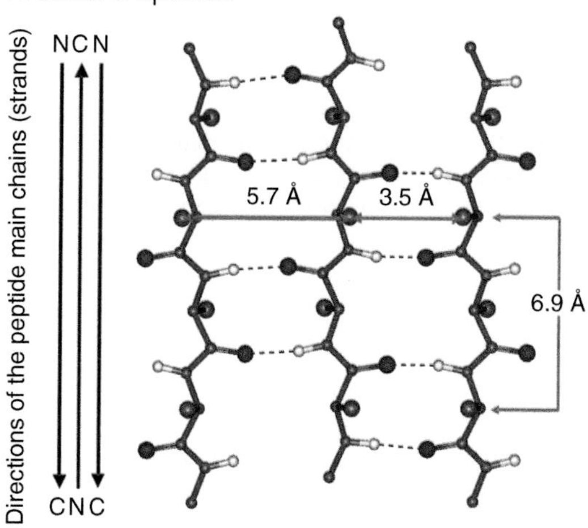

- In the *parallel*-chain β-sheet structure the distances between the C^{β} atoms of the side chains (R) of the residues in register and in adjacent strands repeat themselves identically (~ 4.5 Å). In contrast, in the antiparallel-chain β-sheet structure these same distances strictly alternate between a longer distance (~ 5.7 Å) and a shorter distance (~ 3.5 Å). The short (~ 3.5 Å) distance in the antiparallel-chain β-sheet structure well explains the observation that severely sterically hindered amino acids (e.g. the β-branched Val and Ile) strongly prefer the parallel-chain β-sheet structure (Toniolo in 1978 [94]).

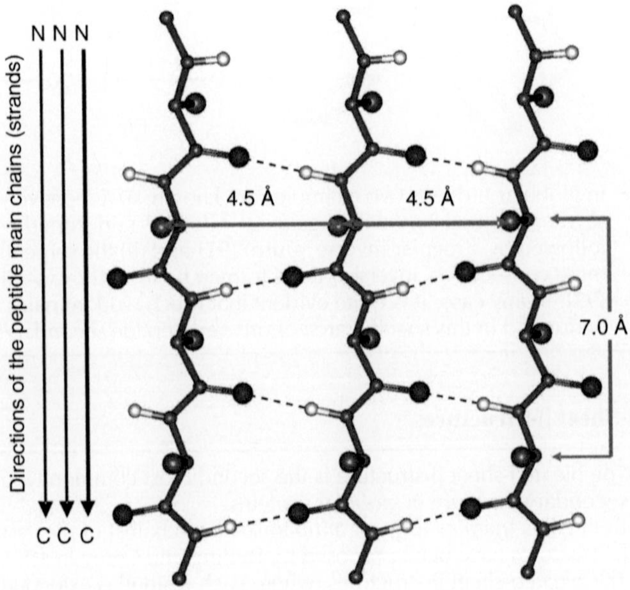

β-Sheets: parameters

- In both types of β-sheet structures the side chains of successive residues alternate (up/down) with respect to the average plane of the main chain. Thus, if the polar/apolar characters of the side chains alternate as well, then *amphiphilic* β-sheet structures may originate [95].
- The helical parameters for the two types of pleated-sheet β-structures are close [95]:

  Antiparallel: φ=–139°, ψ=135°, n=2.00, d=3.47 Å, p=6.94 Å
  Parallel: φ=–119°, ψ=113°, n=2.00, d=3.50 Å, p=7.00 Å

- Due to their integer number of amino acids per turn (n=2), both types of β-sheet structures lie on the corresponding diagonal of the Ramachandran map. The differences of their φ, ψ torsion angles from those of the fully extended structure (see below) explain their slightly wavy appearance.

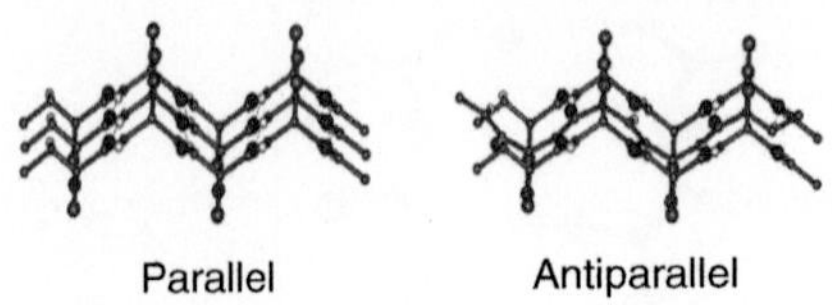

- The pleated-sheet β-structures can be either of the *intra*- or *inter*-molecular type. For example, they can be of the antiparallel-chain type.

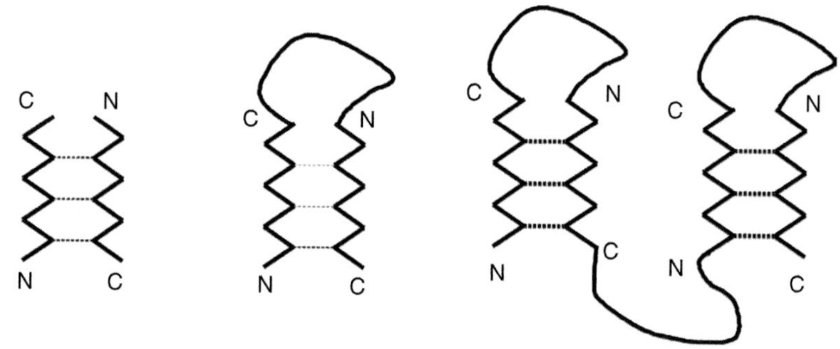

Two strands β-hairpin "Cross-β" or "β-meander"

β-Sheets: promoting residues

- According to the Blout classification (1962) [66], non α-helix forming and, as a consequence, effective β-sheet structure-forming amino acid residues are those with sterically demanding, β-branched, side chains (i.e. Val, Ile, and Thr) and those that can form side-chain to main-chain H-bonds (i.e. Ser, Thr, Cys). While Pro is a β-sheet structure breaker, one of the preferred structures for Gly is the antiparallel-chain β-sheet structure.
- The β-sheet structure does also occur in *fibrous* proteins. A well-known case is that of fibroin [96], the protein characterizing the *Bombix mori* silk, which is rich not only in Gly and Ala but in Ser as well, and largely adopts an amphiphilic antiparallel-chain β-sheet structure (the word *serine* comes from the Greek term 'seros', which means *silk*).

β-Sheets: effects

- Due to their extremely poor solubility, β-sheet structures exhibit low reactivity in peptide bond formation ('difficult sequences'; Mutter and Toniolo [97]).
- More importantly, the β-sheet structures are responsible for the onset of a variety of neurodegenerative 'conformational diseases', such as those characterized by fibril formation and amyloid deposits (Alzheimer, prion, Parkinson, Huntington, Machado ataxia, dementia with Lewy bodies, kuru, Creutzfeld–Jakob, bovine spongiform encephalopathy or 'mad-cow' diseases).

2.5.5 2.0_5-Helix

2.0_5-Helix: structure

- The fully extended peptide conformation, or 2.0_5-helix, with $\varphi = \psi = 180°$, was proposed at an early stage in structural studies of proteins [98]. The repeating motif of the fully extended peptide conformation is the intramolecularly H-bonded form depicted in the figure. The relative disposition of the two dipoles, N–H and C=O, is such that there is obviously some interaction between them. Since these four atoms, together with the central C^{α}-atom, are involved in a pentagonal pseudo-cyclic structure, this conformation is also called the C_5 structure [99].

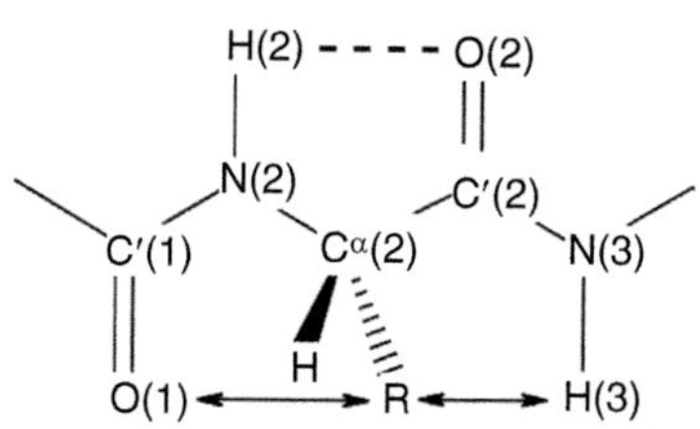

- This type ($i \rightarrow i$) of intramolecular H-bond is the only one among those mentioned in this chapter where the N–H donor group *precedes* the C=O acceptor group in the sequence. All other intramolecular H-bonds are of the $i \leftarrow i+n$ type. In other words, in this helix the usual C=O···H–N H-bond direction has become N–H···O=C.
- The influence of the bulkiness of the lateral substituent can easily be explained by considering the intramolecular nonbonded interactions between the side-chain group R and the preceding C=O and following N–H groups, which induce a warping of this nonsymmetrical structure.
- The bond angles internal to the pentagonal ring are smaller, while those including atoms of the main chain (external to the ring system) are larger than the corresponding average bond angles observed in peptides.
- The intramolecular $N_i \cdots O_i$ distance (2.54 Å) in the H-bonded, fully extended, peptides is much shorter than the corresponding distance usually observed in helical peptides (2.8–3.0 Å) [100, 101].

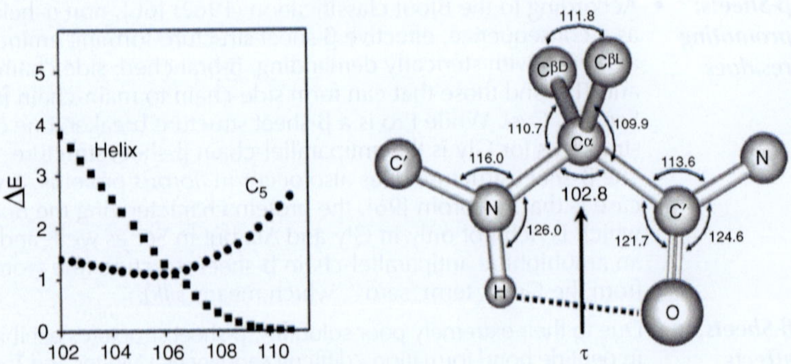

- Interestingly, the critical sp^3 N–C$^\alpha$–C′ bond angle (τ) is dramatically narrowed (from 109.5° to less than 103°).
- From calculations it turns out that the energy ΔE (kcal/mol) of the C_5 conformation of the Ac–Deg–NHMe (Ac, acetyl; NHMe, methylamino) derivative becomes lower than that of the helical conformation when the bond angle τ is < 107° [102].

2.0$_5$-Helix: promoting residues

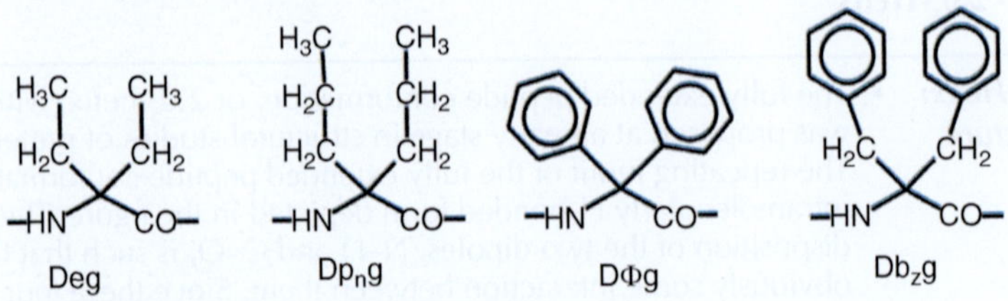

- The highly crystalline nature of peptides rich in the *achiral* C$^\alpha$-tetrasubstituted residues shown above (with two side chains identical and longer than methyls) was exploited for extensive X-ray diffraction analyses [100, 101]. Multiple C_5 conformations are a common observation for these achiral peptides in the crystal state.

  Interestingly:

  (i) The Dp$_n$g and Deg homo-peptides represent the first examples (Toniolo in 1984 [103] and 1988 [102], respectively) in which consecutive C_5 forms (2.0$_5$-helices) have been experimentally observed.

(ii) The N–H and C=O groups characterizing this intramolecularly H-bonded structure are not involved in the intermolecular H-bonding scheme.
(iii) The amino acid side chains of Deg and Dp_ng tend to be fully extended to relieve the unfavorable intramolecular side chain-to-main chain and side chain-to-side chain interactions.
(iv) The axial translation per residue in this helix is ≈ 3.70 Å, the longest possible for a single amino acid, which makes this conformation extremely attractive for its use as a spacer [86, 104].

2.0_5-Helix: natural occurrence

- For coded amino acids, unequivocal verification of the occurrence of the C_5 form has been obtained in the crystal state by X-ray diffraction analyses of a few, favorable compounds, i.e. Gly- and Ala-rich peptides with short side chains [98, 105].
- In globular proteins a repeating C_5 motif has been so far authenticated only in the X-ray diffraction analysis of the $-(Gly)_4-$ sequence of His–*t*RNA–synthetase [106].

2.0_5-Helix: examples

- The X-ray diffraction structure of the 2.0_5-helical, achiral, homo-pentapeptide Tfa–$(Deg)_5$–O*t*Bu (Tfa, trifluoroacetyl) shows five consecutive N–H···O=C intramolecular H-bonds, each giving rise to a C_5 form [102]. This is the longest 2.0_5-helix published to date.

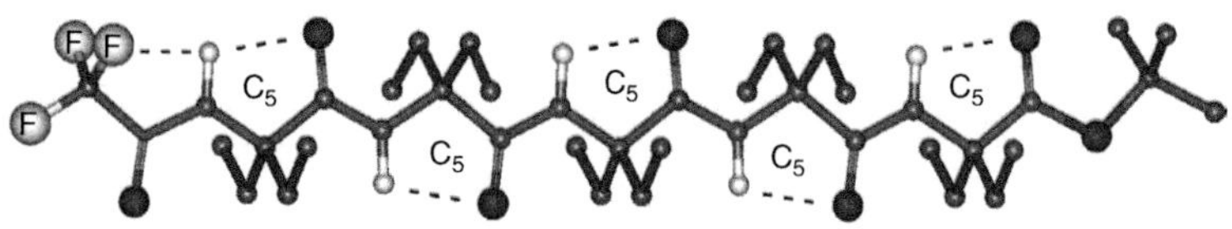

- In recent years, Imawaka, Tanaka, and Suemune (2000) [107] and Crisma *et al.* (2011) [108] clearly demonstrated that even for homo-peptides based on C^α-tetrasubstituted *chiral* α-amino acids the 2.0_5-helix is a common observation, provided that both amino acid side chains are longer than a methyl group (e.g. any C^α-ethylated protein amino acid, those from except Gly and Ala). As an example, the X-ray diffraction structure of the 2.0_5-helical, Tfa/O*t*Bu protected, chiral homo-tripeptide based on C^α-ethyl, C^α-*n*-pentylglycine (Epg) has been reported [108]. Three, consecutive N–H···O=C intramolecular H-bonds, each generating a C_5 form, are observed.

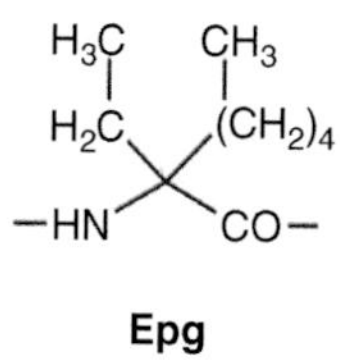

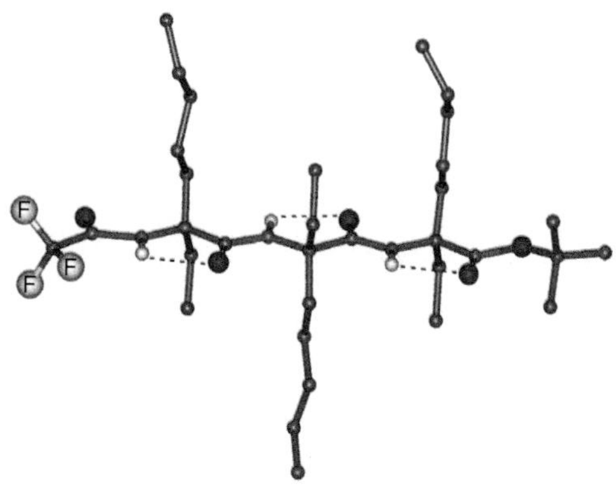

- Moreover, homo-peptides from $C^{\alpha,\beta}$-didehydro-alanine (ΔAla), characterized by sp^2 α- and β-carbon atoms, adopt a 2.0_5-helical structure in solution and in the crystalline state [109]. They are stabilized by two types of intramolecular H-bonds:
(i) N_i–H···O_i=C_i (forming a five-membered ring, typical of the 2.0_5-helix);
(ii) $C\beta_{i+1}$–H···O_i=C_i (forming a six-membered ring, typical of ΔAla peptides).

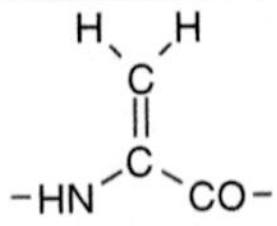

ΔAla

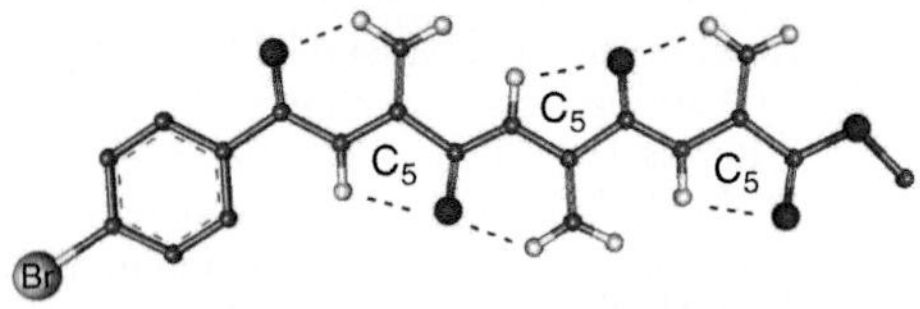

X-ray diffraction structure of *p*BrBz-$(\Delta Ala)_3$-OMe

- These completely flat molecules form planar sheets. They exist in isolation and pack in layers without any significant contribution from intermolecular C=O···H–N H-bonds. These graphene-like molecules may represent potentially active bridges in electron transfer reactions.

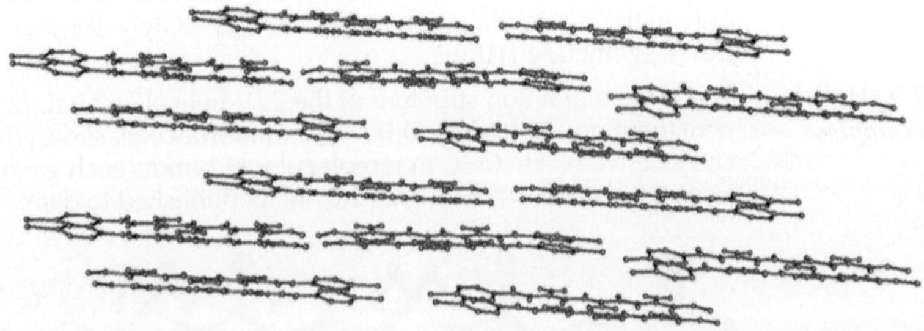

2.5.6 Poly-(L-Pro)$_n$ Helices and Collagen Triple Helix

Poly(L-Pro)$_n$ helices

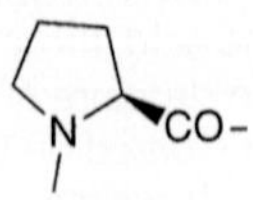

Pro

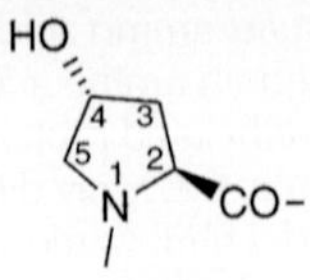

(4R)-Hyp

- The φ torsion angle of Pro and Hyp (4-hydroxyproline) is fixed by the pyrrolidine cyclic structure (to about –70° for the L-enantiomer).

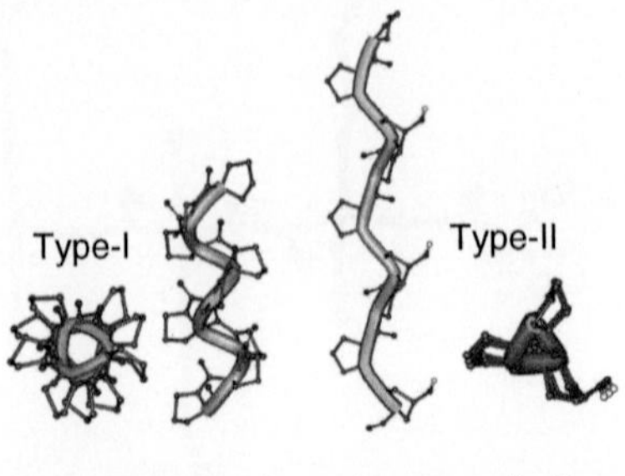

- Type-I and type-II poly(L-Pro)$_n$ helix models [110–112]. Appropriate side-chain replacements may induce an amphiphilic character to type-II poly-(L-Pro)$_n$, a strictly threefold helix. This is the case of the [L-Pro–L-Pro–4(R)–L–Hyp]$_3$ type-II helix represented above (right). The view down the helix axis emphasizes its amphiphilic character.
- Type-I and type-II poly (L-Pro)$_n$ helices are *right*- and *left*-handed, respectively.

- Their sets of φ, ψ torsion angles are close, but they differ substantially because form I has all tertiary peptide bonds in the *cis* conformation ($\omega = 0°$), while in the form II these bonds are all in the *trans* conformation ($\omega = 180°$). Therefore, the less stable form I is remarkably more compact than the more stable form II. The kinetics of the folding and unfolding processes in globular proteins are often governed by the *cis/trans* Xxx–Pro conformational transition [113].
- Because all peptide bonds are tertiary (no NH groups), these two helices are not stabilized by any intramolecular H-bond. Their interconversions (called 'mutarotations' because they were originally followed by looking at variations of the polarimetric values with time) are solvent driven. Type-II poly-(L-Pro)$_n$ is the dimorph largely prevailing in polar solvents.

Poly (L-Pro)$_n$ helices: parameters

The parameters characterizing these two helices are as follows:

	Form I	Form II	
φ (°)	–83	–80	The φ, ψ values are very close.
ψ (°)	158	150	
ω (°)	0	180	Major conformational difference.
n	3.3	–3.0	The sign '-' indicates a *left*-handed helix for the L-enantiomer.
d (Å)	1.9	3.1	Form II is more elongated than form I.
p (Å)	6.3	9.3	

- In L-Pro-containing peptides the value of the L-Pro ψ torsion angles may be either ~ 150° (the preferred *trans'* conformation, *semi*-extended Pro) or ≅ –30° (the less common *cis'* conformation, helical Pro). In globular and transmembrane proteins [113, 114], Pro has indeed been found in 3_{10}-/α-helices, but only at the first two/three positions, because it lacks the H-bonding donor peptide NH group (the NH groups at these positions are not involved in the intramolecular H-bonding schemes of the 3_{10}-/α-helices). If inserted in an internal position of a helix, Pro has been shown to induce a kink of ~ 20°.
- Homo-(L-Pro)$_n$ stretches should be used with caution as spacers or bridges (often termed 'rigid rods'), because their rigidity is questionable in view of the variety of the possible conformations discussed above.

Collagen triple helix

- The most widely distributed fibrous protein in human body is *collagen*. It is characterized by a repeating triplet of amino acids, (L-Pro–Xxx–Gly)$_n$, where Xxx is almost any α-amino acid ('consensus' sequence). Its 3D structure is a 'triple helix coiled coil'. It was first proposed by Ramachandran and Kartha in 1955 [115]. The model was improved by Rich and Crick in the same year [116] (see also Ramachandran in 1956 [117]).
- Each helix is closely related to the left-handed type-II poly(L-Pro)$_n$ conformation. This is why collagen is very rich in Pro. Gly is always present at position 2 of the triplet because it is the least sterically demanding amino acid (in collagen, Gly occurs at each extremely hindered intersection of the three strands of the triple helix). Gly 'point mutations' induce severe 'molecular diseases' in bones [118]. The three helices are parallel to each other and in register.

- The most abundant type of collagen has two α_1 chains (formed by 338 consecutive triplets, about 1050Å long) and one α_2 chain [119]. During the biosynthesis, the triple helix folds from the C-terminus (the rate determining step is the *cis→trans* Xxx–Pro conversion) [120]. There are two interchain H-bonds per triplet (one strong, (Gly)N–H···O=C(Pro), and one water-mediated and weak). About 10% of the Pro residues in collagen are replaced by posttranslationally generated (4R)Hyp residues [121, 122]. This enzyme (Pro hydroxylase)-mediated process is regio- and stereospecific. Their secondary alcohol side-chain groups form additional H-bonds (with peptide C=O groups and water molecules, and between each other), further stabilizing the triple helix structure.

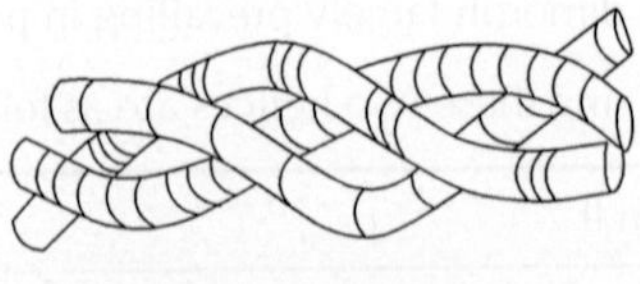

Collagen triple helix

References

[1] Fischer, E. (1907) Synthese von polypeptiden. XVII, Ber. Deutsch. Chem. Gesell. 40: 1754–1767.

[2] Goodman, M., Cai, W., Smith, N.D. (2003) The bold legacy of Emil Fischer. J. Pept. Sci. 9: 594–603.

[3] Sheehan, J.C., Hess, G.P. (1955) A new method of forming peptide bonds. J. Am. Chem. Soc. 77: 1067–1068.

[4] Merrifield, R.B. (1963) Solid phase peptide synthesis. I. The synthesis of a tetrapeptide. J. Am. Chem. Soc. 85: 2149–2154.

[5] Bodanszky, M. (1984) Principles of Peptide Synthesis. Springer, Berlin, Germany.

[6] Bayley, P. D. (1990) An Introduction to Peptide Chemistry. John Wiley & Sons, Ltd, Chichester, UK.

[7] Lloyd-Williams, P., Albericio, F., Giralt, E. (1997) Chemical Approaches to the Synthesis of Peptides and Proteins. CRC Press, Boca Raton, FL, USA.

[8] Jones, J. (2002) Amino Acid and Peptide Synthesis. Oxford University Press, Oxford, UK.

[9] Benoiton, N.L. (2006) Chemistry of Peptide Synthesis. CRC Press, Boca Raton, FL, USA.

[10] Carpino, L.A., Beyermann, M., Wenschuh, H., Bienert, M. (1996) Peptide synthesis via amino acid halides. Acc. Chem. Res. 29: 268–274.

[11] Humphrey, J.M., Chamberlin, A.R. (1997) Chemical synthesis of natural product peptides: coupling methods for the incorporation of noncoded amino acids into peptides. Chem. Rev. 97: 2243–2266.

[12] Elmore, D.T. (2002) Peptide synthesis. In: Amino Acids, Peptides, Proteins. Barrett, G.C., Davies, J.S. (Eds). Specialist Periodical Reports, Vol. 33. The Royal Society of Chemistry, Cambridge, UK, pp. 83–134.

[13] Han, S., Kim, Y. (2004) Recent development of peptide coupling reagents in organic synthesis. Tetrahedron 60: 2447–2467.

[14] Montalbetti, C.A.G.N., Falque, V. (2005) Amide bond formation and peptide coupling, Tetrahedron 61: 10827–10852.

[15] Valeur, E., Bradley, M. (2009) Amide bond formation: beyond the myth of coupling reagents. Chem. Soc. Rev. 38: 606–631.

[16] El-Faham, A., Albericio, F. (2011) Peptide coupling reagents, more than a letter soup. Chem. Rev. 111: 6557–6602.

[17] Goodman, M., Felix, A., Moroder, L., Toniolo, C. (Eds) (2003) Synthesis of peptides and peptidomimetics. In: Houben–Weyl: Methods of Organic Chemistry, Vol. E 22. Thieme, Stuttgart, Germany.
[18] Wipf, P., Heimgartner, H. (1986) Kupplung von Peptiden mit C-terminalen α,α-disubstituierten α-Aminosaeuren via Oxazol-5(4*H*)-one. Helv. Chim. Acta 69: 1153–1162.
[19] Carpino, L.A. (1993) 1-Hydroxy-7-azabenzotriazole. An efficient peptide coupling additive. J. Am. Chem. Soc. 115: 4397–4398.
[20] Wünsch, E., Drees, F. (1966) Zur Synthese des Glucagons, X. Darstellung der Sequenz 22–29. Chem. Ber. 99: 110–120.
[21] König, W., Geiger, R. (1970) Eine neue Methode zur Synthese von Peptiden: Aktivierung der Carboxylgruppe mit Dicyclohexylcarbodiimid unter Zusatz von 1-Hydroxy-benzotriazolen. Chem. Ber. 103: 788–798.
[22] Crisma, M., Valle, G., Moretto, V., Formaggio, F., Toniolo, C., Albericio, F. (1998) Reactive intermediates in peptide synthesis. Molecular and crystal structures of HOAt and HOOBt, and some ester and amide derivatives of HOBt, HOAt and HOOBt. Lett. Pept. Sci. 5: 247–258.
[23] Toniolo, C., Crisma, M., Formaggio, F. (1996) Understanding α-amino acid chemistry from X-ray diffraction structures. Biopolymers, 40: 627–651.
[24] Carpino, L.A., El-Faham, A. Albericio, F. (1995) Efficiency in peptide coupling: 1-hydroxy-7-azabenzotriazole vs. 3,4-dihydro-3-hydroxy-4-oxo-1,2,3-benzotriazine. J. Org. Chem. 60: 3561–3564.
[25] Carpino, L.A., Ferrer, F.J. (2001) The 5,6- and 4,5-benzo derivatives of 1-hydroxy-7-azabenzotriazole. Org. Lett. 3: 2793–2795.
[26] Akaji, K., Tamai, Y., Kiso, Y. (1997) Efficient synthesis of peptaibol using a chloroimidazolidium coupling reagent, CIP. Tetrahedron 53: 567–584.
[27] Castro, B., Dormoy, J.R., Evin, G., Selve, C. (1975) Reactifs de couplage peptidique I (1) – l'hexafluorophosphate de benzotriazolyl *N*-oxy trisdimethylamino phosphonium (B.O.P.). Tetrahedron Lett. 16: 1219–1222.
[28] Frérot, E., Coste, J., Pantaloni, A., Dufour, M., Jouin, P. (1991) PyBOP and PyBroP: two reagents for the difficult coupling of the α,α-dialkyl amino acid Aib. Tetrahedron 47: 259–270.
[29] Dourtoglou, V., Ziegler, J.-C., Gross, B. (1978) L'hexafluorophosphate de *O*-benzotriazolyl-*N*,*N*-tetramethyluronium. Un reactif de couplage peptidique nouveau et efficace. Tetrahedron Lett. 1269–1272.
[30] Knorr, R., Trzeciak, A., Bannwarth, W., Gillessen, D. (1989) New coupling reagents in peptide chemistry. Tetrahedron Lett. 1927–1930.
[31] Carpino, L.A., Imazumi, H., El-Faham, A., Ferrer, F.J., Zhang, C., Lee, Y., Foxman, B.M., Henklein, P., Hanay, C., Mügge, C., Wenschuh, H., Klose, J., Beyermann, M., Bienert, M. (2002) The uronium/guanidinium peptide coupling reagents: finally the true uronium salts. Angew. Chem. Int. Ed. 41: 441–445.
[32] Carpino, L.A., Henklein, P., Foxman, B.M., Abdelmoty, I., Costisella, B., Wray, V., Domke, T., El-Faham, A., Mügge, C. (2001) The solid state and solution structure of HAPyU. J. Org. Chem. 66: 5245–5247.
[33] Carpino, L. A., El-Faham, A., Minor, C.A., Albericio, F. (1994) Advantageous applications of azabenzotriazole (triazolopyridine)-based coupling reagents to solid-phase peptide synthesis. J. Chem. Soc., Chem. Commun. 201–203.
[34] De Zotti, M., Biondi, B., Peggion, C., Park, Y., Hahm, K.-S., Formaggio, F., Toniolo, C. (2011) Synthesis, preferred conformation, protease stability, and membrane activity of heptaibin, a medium-length peptaibiotic. J. Pept. Sci. 17: 585–594.
[35] Carpino, L.A., El-Faham, A., Albericio, F. (1994) Racemization studies during solid-phase peptide synthesis using azabenzotriazole-based coupling reagents. Tetrahedron Lett. 35: 2279–2282.
[36] Spetzler, J.C., Meldal, M., Felding, J., Vedsø, P., Begtrup, M. (1998) Novel acylation catalysts in peptide synthesis: derivatives of *N*-hydroxytriazoles and *N*-hydroxytetrazoles. J. Chem. Soc., Perkin Trans. 1: 1727–1732.
[37] Carpino, L.A., Sadat-Aalaee, D., Chao, H.G., DeSelms, R.H. (1990) [(9-Fluorenylmethyl)oxy] carbonyl (FMOC) amino acid fluorides. Convenient new peptide coupling reagents applicable

to the FMOC/*tert*-butyl strategy for solution and solid-phase syntheses. J. Am. Chem. Soc. 112: 9651–9652.
[38] Carpino, L.A., Mansour, E.M.E., Sadat-Aalaee, D. (1991) *tert*-Butyloxycarbonyl and benzyloxycarbonyl amino acid fluorides. New, stable rapid-acting acylating agents for peptide synthesis. J. Org. Chem. 56: 2611–2614.
[39] Sapia, A.C., Slomczynska, U., Marshall, G.R. (1994) Evaluation of two new coupling agents for incorporation of α,α-dialkylamino acids, such as α-methylalanine, in solid-phase peptide synthesis. Lett. Pept. Sci. 1: 283–290.
[40] Fiammengo, R., Licini, G., Nicotra, A., Modena, G., Pasquato, L., Scrimin, P., Broxterman, Q.B., Kaptein, B. (2001) Duality of mechanism in the tetramethylfluoroformamidinium hexafluorophosphate-mediated synthesis of *N*-benzyloxycarbonylamino acid fluorides. J. Org. Chem. 66: 5905–5910.
[41] Carpino, L.A., El-Faham, A. (1995) Tetramethylfluoroformamidinium hexafluorophosphate: a rapid-acting peptide coupling reagent for solution and solid phase peptide synthesis. J. Am. Chem. Soc. 117: 5401–5402.
[42] Polese, A., Formaggio, F., Crisma, M., Valle, G., Toniolo, C., Bonora, G.M., Broxterman, Q.B., Kamphuis, J. (1996) Peptide helices as rigid molecular rulers: a conformational study of isotactic homo-peptides from α-methyl-α-isopropylglycine, [L-(αMe)Val]$_n$. Chem. Eur. J. 2: 1104–1111.
[43] Gratias, R., Konat, R., Kessler, H., Crisma, M., Valle, G., Polese, A., Formaggio, F., Toniolo, C., Broxterman, Q.B., Kamphuis, J. (1998) First step toward the quantitative identification of peptide 3_{10}-helix conformation with NMR spectroscopy: NMR and X-ray diffraction structural analysis of a fully-developed 3_{10}-helical peptide standard. J. Am. Chem. Soc. 120: 4763–4770.
[44] Wenschuh, H., Beyermann, M., Krause, E., Brudel, M., Winter, R., Schümann, M., Carpino, L.A., Bienert, M. (1994) Fmoc amino acid fluorides: convenient reagents for the solid-phase assembly of peptides incorporating sterically hindered residues. J. Org. Chem. 59: 3275–3280.
[45] Wenschuh, H., Beyermann, M., Haber, H., Seydel, J.K., Krause, E., Bienert, M., Carpino, L. A., El-Faham, A., Albericio, F. (1995) Stepwise automated solid phase synthesis of naturally occurring peptaibols using Fmoc amino acid fluorides. J. Org. Chem. 60: 405–410.
[46] Yokum, T.S., Bursavich, M.G., Gauthier, T., Hammer, R.P., McLaughlin, M.L. (1998) 3_{10}-Helix stabilization via side-chain salt bridges. J. Chem. Soc., Chem. Commun. 1801–1802.
[47] Caciagli, V., Cardinali, F., Bonelli, F., Lombardi, P. (1998) Large-scale production of peptides using the solid phase continuous flow method. Part 2: Preparative synthesis of a 26-mer peptide thrombin inhibitor. J. Pept. Sci. 4: 327–334.
[48] Meldal, M., Juliano, M.A., Jansson, A.M. (1997) Azido acids in a novel method of solid-phase peptide synthesis. Tetrahedron Lett. 2531–2534.
[49] Jost, M., Greie, J.-C., Stemmer, N., Wilking, S.D., Altendorf, K., Sewald, N. (2002) The first total synthesis of efrapeptin C. Angew. Chem. Int. Ed. 41: 4267–4269.
[50] Clayden, J., Castellanos, A., Solà, J., Morris, G.A. (2009) Quantifying end-to-end conformational communication of chirality through an achiral peptide chain. Angew. Chem. Int. Ed. 48: 5962–5965.
[51] Weigelt, S., Huber, T., Hofmann, F., Jost, M., Ritzefeld, M., Luy, B., Freudenberger, C., Majer, Z., Vass, E., Greie, J.-C., Panella, L., Kaptein, B., Broxterman, Q.B., Kessler, H., Altendorf, K., Hollósi, M., Sewald, N. (2012) Synthesis and conformational analysis of efrapeptins. Chem. Eur. J. 18: 478–487.
[52] IUPAC-IUB Commission on Biochemical Nomenclature (1972) Symbols for amino-acid derivatives and peptides. Recommendations (1971). Biochemistry 11: 1726–1732.
[53] Barrett, G.C., Elmore, D.T. (1988) Amino Acids and Peptides. Cambridge University Press, Cambridge, UK.
[54] IUPAC-IUB Commission on Biochemical Nomenclature (1970) Abbreviations and symbols for the description of the conformation of polypeptide chains. Tentative rules (1969). Biochemistry 9: 3471–3479.
[55] Cahn, R.S., Ingold, C., Prelog, V. (1956) The specification of asymmetric configuration in organic chemistry. Experientia 12: 81–94.

[56] Ramachandran, G.N., Ramakrishnan, C., Sasisekharan, V. (1963) Stereochemistry of polypeptide chain configurations. J. Mol. Biol. 7: 95–99.
[57] Ramachandran, G.N., Sasisekharan, V. (1968) Conformation of polypeptides and proteins. Adv. Protein Chem. 23: 283–438.
[58] Bragg, L., Kendrew, J.C., Perutz, M.F. (1950) Polypeptide chain configurations in crystalline proteins. Proc. Roy. Soc. (Lond.) A 203: 321–357.
[59] Pauling, L., Corey, R.B. (1950) Two hydrogen-bonded spiral configurations of the polypeptide chain. J. Am. Chem. Soc. 72: 5349.
[60] Pauling, L., Corey, R.B., Branson, H.R. (1951) The structure of proteins. Two hydrogen-bonded helical configurations of the polypeptide chain. Proc. Natl Acad. Sci. USA 37: 205–211.
[61] Astbury, W.T. (1936) X-ray studies of protein structure. Nature (Lond.) 137: 803–805.
[62] Perutz, M.F. (1951) New X-ray evidence on the configuration of polypeptide chains. Nature (Lond.) 167: 1053–1054.
[63] Kendrew, J.C., Dickerson, R.E., Strandberg, B.E., Hart, R.G., Davies, D.R., Phillips, D.C., Shore, V.C. (1960) Structure of myoglobin. A three-dimensional Fourier synthesis at 2Å resolution. Nature (Lond.) 185: 422–427.
[64] Pavone, V., Gaeta, G., Lombardi, A., Nastri, F., Meglio, O., Isernia, C., Saviano, M. (1996) Discovering protein secondary structures: classification and description of isolated α-turns. Biopolymers 38: 705–721.
[65] Toniolo, C., Benedetti, E. (1991) The polypeptide 3_{10}-helix. Trends Biochem. Sci. 16: 350–353.
[66] Blout, E.R. (1962) The dependence of the conformation of polypeptides and proteins upon amino acid composition. In: Polyamino Acids, Polypeptides, and Proteins. Stahmann, M.A. (Ed.). The University of Wisconsin Press, Madison, WI, USA, pp. 275–279.
[67] Gruber, M., Lupas, A.N. (2003) Historical review. Another 50th anniversary. New periodicities in coiled coils. Trends Biochem. Sci. 28: 679–685.
[68] Peggion, C., Formaggio, F., Crisma, M., Epand, R.F., Epand, R.M., Toniolo, C. (2003) Trichogin: a paradigm for lipopeptaibols. J. Pept. Sci. 9: 679–689.
[69] Crick, F.H.C. (1952) Is α-keratin a coiled coil? Nature (Lond.), 170: 882–883.
[70] Landschulz, W.H., Johnson, P.F., McNight, S.L. (1988) The leucine zipper. A hypothetical structure common to a new class of DNA-binding proteins. Science 240: 1759–1764.
[71] Taylor, H.S. (1941) Large molecules through atomic spectacles. Proc. Am. Phil. Soc. 85: 1–7.
[72] Venkatachalam, C.M. (1968) Stereochemical criteria for polypeptides and proteins. V. Conformation of a system of three linked peptide units. Biopolymers 6: 1425–1436.
[73] Geddes, A.J., Parker, K.D., Atkins, E.D.T., Beighton, E. (1968) Cross-β conformation in proteins. J. Mol. Biol. 32: 343–358.
[74] Lewis, P.N., Momany, F.A., Scheraga, H.A. (1973) Chain reversal in proteins. Biochim. Biophys. Acta 303: 211–229.
[75] Shamala, N., Nagaraj, R., Balaram, P. (1978) The 3_{10}-helical conformation of a pentapeptide containing α-aminoisobutyric acid (Aib): X-ray crystal structure of Tos–$(Aib)_5$–OMe. J. Chem. Soc., Chem. Commun. 996–997.
[76] Benedetti, E., Bavoso, A., Di Blasio, B., Pavone, V., Pedone, C., Toniolo, C., Bonora, G.M. (1982) Peptaibol antibiotics. A study on the helical structure of the 2–9 sequence of emerimicins III and IV. Proc. Natl Acad. Sci. USA 79: 7951–7954.
[77] Barlow, D.J., Thornton, J.M. (1988) Helix geometry in proteins. J. Mol. Biol. 201: 601–619.
[78] Millhauser, G.L. (1995) Views of helical peptides. A proposal for the position of the 3_{10}-helix along the thermodynamic folding pathway. Biochemistry 34: 3873–3877.
[79] Toniolo, C., Crisma, M., Formaggio, F., Peggion, C. (2001) Control of peptide conformation by the Thorpe–Ingold effect (C^{α}-tetrasubstitution). Biopolymers (Pept. Sci.) 60: 396–419.
[80] Benedetti, E., Bavoso, A., Di Blasio, B., Pavone, V., Pedone, C., Crisma, M., Bonora, G. M., Toniolo, C. (1982) Solid-state and solution conformation of homo-oligo (α-aminoisobutyric acids) from tripeptide to pentapeptide: evidence for a 3_{10}-helix. J. Am. Chem. Soc. 104: 2437–2444.
[81] Toniolo, C., Brückner, H. (2009) Peptaibiotics. Fungal Peptides Containing α-Dialkyl α-Amino Acids. Wiley-VCH, Weinheim, Germany.

[82] Pavone, V., Di Blasio, B., Santini, A., Benedetti, E., Pedone, C., Toniolo, C., Crisma, M. (1990) The longest, regular polypeptide 3_{10}-helix at atomic resolution. J. Mol. Biol. 214: 633–635.

[83] Gessman, R., Brückner, H., Petratos, K. (2003) Three complete turns of a 3_{10}-helix at atomic resolution. The crystal structure of Z–(Aib)$_{11}$–OtBu. J. Pept. Sci. 9: 753–762.

[84] Karle, I.L., Flippen-Anderson, J., Sukumar, M., Balaram, P. (1987) Conformation of a 16-residue zervamicin IIA analog peptide containing three different structural features: 3_{10}-helix, α-helix and β-bend ribbon. Proc. Natl Acad. Sci. USA 84: 5087–5091.

[85] Di Blasio, B., Pavone, V., Saviano, M., Lombardi, A., Nastri, F., Pedone, C., Benedetti, E., Crisma, M., Anzolin, M., Toniolo, C. (1992) Structural characterization of the β-bend ribbon spiral: crystallographic analysis of two long (L-Pro–Aib)$_n$ sequential peptides. J. Am. Chem. Soc. 114: 6273–6278.

[86] Toniolo, C., Crisma, M., Formaggio, F., Peggion, C., Broxterman, Q.B., Kaptein, B. (2004) Molecular spacers for physicochemical investigations based on novel helical and extended peptide structures. Biopolymers (Pept. Sci.) 76: 162–176.

[87] Toniolo, C., Crisma, M., Formaggio, F., Peggion, C., Broxterman, Q.B., Kaptein, B. (2005) Peptide β-bend and 3_{10}-helix: from 3D-structural studies to applications as templates. J. Incl. Phenom. Macrocyclic Chem. 51: 121–136.

[88] Némethy, G., Printz, M.P. (1972) The γ-turn, a possible folded conformation of the polypeptide chain. Comparison with the β-turn. Macromolecules 5: 755–758.

[89] Matthews, B.W. (1972) The γ-turn. Evidence for a new folded conformation in proteins. Macromolecules 5: 818–819.

[90] Jiménez, A.I., Ballano, G., Cativiela, C. (2005) First observation of two consecutive γ-turns in a crystalline linear dipeptide. Angew. Chem. Int. Ed. 44: 396–399.

[91] Tsunemi, M., Matsuura, Y., Sakakibara, S., Katsube, Y. (1996) Crystal structure of an elastase-specific inhibitor elafin complexed with porcine pancreatic elastase determined at 1.9 Å resolution. Biochemistry 35: 11570–11576.

[92] Yang, M., Culhane, J.C., Szewczuk, L.M., Gocke, C.B., Brautigam, C.A., Tomchick, D.R., Machins, M., Cole, P.A., Yu, H. (2007) Structural basis of histone demethylation by LSD1 revealed by suicide inactivation. Nat. Struct. Biol. 14: 535–539.

[93] Pauling, L., Corey, R.B. (1951) Configurations of polypeptide chains with favored orientations around single bonds: two new pleated sheets. Proc. Natl Acad. Sci. USA 37: 729–740.

[94] Toniolo, C. (1978) Structural role of valine and isoleucine residues in proteins. A proposal. Macromolecules 11: 437–438.

[95] Walton, A.G. (1981) Polypeptides and Protein Structure. Elsevier, New York, USA.

[96] Heim, M., Römer, L., Scheibel, T. (2010) Hierarchical structures made of proteins. The complex architecture of spider webs and their constituent silk proteins. Chem. Soc. Rev. 39: 156–164.

[97] Mutter, M., Pillai, V.N.R., Anzinger, H., Bayer, E., Toniolo, C. (1981) Conformational aspects in peptide synthesis. In: Peptides 1980. Brunfeldt, K. (Ed.). Scriptor, Copenhagen, Denmark, pp. 660–665.

[98] Toniolo, C. (1980) Intramolecularly H-bonded peptide conformations. CRC Crit. Rev. Biochem. 9: 1–44.

[99] Cung, M.T., Marraud, M., Néel, J. (1972) Étude expérimentale de la conformation de molécules dipeptidiques. Comparaison avec les prévisions théoriques. Ann. Chim. (Paris) 7: 183–209.

[100] Toniolo, C., Benedetti, E. (1991) The fully-extended polypeptide conformation. In: Molecular Conformations and Biological Interactions. Balaram, P., Ramaseshan, S. (Eds). Indian Academy of Sciences, Bangalore, India, pp. 511–521.

[101] Crisma, M., Valle, G., Bonora, G.M., Toniolo, C., Lelj, F., Barone, V., Fraternali, F., Hardy, P.M., Maia, H.L.S. (1991) Preferred conformation of peptides from $C^{\alpha,\alpha}$-symmetrically disubstituted glycines: aromatic residues. Biopolymers 31: 637–641.

[102] Benedetti, E., Barone, V., Bavoso, A., Di Blasio, B., Lelj, F., Pavone, V., Pedone, C., Bonora, G.M., Toniolo, C., Leplawy, M.T., Kaczmarek, K., Redlinski, A. (1988) Structural versatility of peptides from $C^{\alpha,\alpha}$-dialkylated glycines. I. A conformational energy computation

and X-ray diffraction study of homo-peptides from $C^{\alpha,\alpha}$-diethylglycine. Biopolymers 27: 357–371.
[103] Benedetti, E., Toniolo, C., Hardy, P., Barone, V., Bavoso, A., Di Blasio, B., Grimaldi, P., Lelj, F., Pavone, V., Pedone, C., Bonora, G.M., Lingham, I. (1984) Folded and extended structures of homo-oligopeptides from α,α-dialkylated glycines. A conformational energy computation and X-ray diffraction study. J. Am. Chem. Soc. 106: 8146–8152.
[104] Crisma, M., Formaggio, F., Moretto, A., Toniolo, C. (2006) Peptide helices based on α-amino acids. Biopolymers (Pept. Sci.) 84: 3–12.
[105] Birkedal, H., Schwarzenbach, D., Pattison, P. (2002) A fully extended tetrapeptide consisting of natural amino acids. J. Chem. Soc., Chem. Commun. 2812–2813.
[106] Åberg, A., Yaremchuk, A., Tukalo, M., Rasmussen, B., Cusack, S. (1997) Crystal structure analysis of the activation of histidine in *Thermus thermophilus* histidyl-tRNA synthetase. Biochemistry 36: 3084–3094.
[107] Imawaka, N., Tanaka, M., Suemune, H. (2000) The first fully planar C_5 conformation of homo-oligopeptides prepared from a chiral α-ethylated α,α-disubstituted amino acid: (*S*)-butylethylglycine (= (2*S*)-2-amino-2-ethylhexanoic acid). Helv. Chim. Acta 83: 2823–2835.
[108] Crisma, M., Moretto, A., Peggion, C., Panella, L., Kaptein, B., Broxterman, Q.B., Toniolo, C. (2011) Chiral, fully extended helical peptides. Amino Acids 41: 629–641.
[109] Crisma, M., Formaggio, F., Toniolo, C., Yoshikawa, T., Wakamiya, T. (1999) Flat peptides. J. Am. Chem. Soc. 121: 3272–3278.
[110] Mandelkern, L. (1967) Poly-L-proline. In: Poly-α-Amino Acids. Protein Models for Conformational Studies. Fasman, G.D. (Ed.). Dekker, New York, USA, pp. 675–724.
[111] MacArthur, M.W., Thornton, J.M. (1991) Influence of proline residues on protein conformation. J. Mol. Biol. 218: 397–412.
[112] Yaron, A., Naider, F. (1993) Proline-dependent structural and biological properties of peptides and proteins. Crit. Rev. Biochem. Mol. Biol. 28: 31–81.
[113] Eyles, S.J., Gierasch, L.M. (2000) Multiple roles of prolyl residues in structure and folding. J. Mol. Biol. 301: 737–747.
[114] Williams, K.A., Deber, C.M. (1991) Proline residues in transmembrane helices: structural or dynamic role? Biochemistry 30: 8919–8923.
[115] Ramachandran, G.N., Kartha, G. (1955) Structure of collagen. Nature (Lond.) 176: 593–595.
[116] Rich, A., Crick, F.H.C. (1955) Structure of collagen. Nature (Lond.) 176: 915–916.
[117] Ramachandran, G.N. (1956) Structure of collagen. Nature (Lond.), 177: 710–711.
[118] Bryan, M.A., Cheng, H., Brodsky, B. (2011) Sequence environment of mutation affects stability and folding in collagen model peptides of osteogenesis imperfecta. Biopolymers (Pept. Sci.) 96: 4–13.
[119] Bolboaca, S.D., Jäntschi, L. (2008) A structural informatics study on collagen. Chem. Biol. Drug Design 71: 173–179.
[120] Ramachandran, G.N. (Ed.) (1967) Conformation of Biopolymers, Vol. 2. Academic Press, New York, USA, pp. 429–554.
[121] Ananthanarayanan, V.S. (1983) Conformational criteria for, and consequences of proline hydroxylation in collagen. In: Conformation in Biology. Srinivasan, R., Sarma, R.H. (Eds). Adenine Press, Guilderland, NY, USA, pp. 99–111.
[122] Shoulders, M.D., Kotch, F.W., Choudhary, A., Guzci, I.A., Raines, R.T. (2010) The aberrance of the 4S diastereomer of 4-hydroxyproline. J. Am. Chem. Soc. 132: 10857–10865.

3

Conformational Aspects and Molecular Dynamics Simulations of Peptide Hybrid Materials: From Methods and Concepts to Applications

Carlos Alemán[1,2], *Oscar Bertran*[3], *Jordi Casanovas*[4], *Juan Torras*[5], *Guillermo Revilla-López*[1] *and David Zanuy*[1]

[1]*Departament d'Enginyeria Química, E.T.S d'Enginyers Industrials de Barcelona, Universitat Politècnica de Catalunya, Spain*
[2]*Center for Research in Nano-Engineering, Universitat Politècnica de Catalunya, Spain*
[3]*Departament de Física Aplicada, EEI, Universitat Politècnica de Catalunya, Spain*
[4]*Departament de Química, Escola Politècnica Superior, Universitat de Lleida, Spain*
[5]*Departament d'Enginyeria Química, EEI, Universitat Politècnica de Catalunya, Spain*

Acronyms

Aβ	Amyloid beta peptide
Ac_3c	1-Aminocyclopropanecarboxylic acid
Ac_4c	1-Aminocyclobutanecarboxylic acid
Ac_5c	1-Aminocyclopentanecarboxylic acid
Ac_6c	1-Aminocyclohexanecarboxylic acid
Aib	α-Aminoisobutyric acid

Peptide Materials: From Nanostructures to Applications, First Edition. Edited by Carlos Alemán, Alberto Bianco and Mariano Venanzi.

AM1	Austin model 1
CC	Coupled cluster
CI	Configuration interaction
Db_zg	α,α-Dibenzylgycline
Deg	$C^{\alpha,\alpha}$-diethylglycine
DFT	Density functional theory
Dϕg	α,α-Diphenylgycline
GAFF	Generalized AMBER force field
GGA	Generalized gradient approximation
HF	Hartree–Fock
LCAO	Linear combination of atomic orbitals
MD	Molecular dynamics
MIC	Minimum image convention
MM	Molecular mechanics
MP	Møller–Plesset
Nc-aa	Noncoded amino acids
NCAD	Noncoded amino acid data base
NDDO	Neglect of diatomic differential overlap
O*t*Bu	*tert*-Butoxy
PCM	Polarizable continuum model
PEG	Poly(ethylene glycol)
PGDGA	Poly(γ-D-glutamic acid)
PM3	Parametric model number 3
QM	Quantum mechanics
(2R,3R)$c_3$diPhe	(2R,3R)-1-amino-2,3-diphenylcyclopropanecarboxylic acid
RE	Replica exchange
RESP	Restrained electrostatic potential
RMSD	Root-mean-square deviation
SA	Simulated annealing
SCF	Self-consistent field
SE	Semiempirical
SPIO	Superparamagnetic iron oxide
Tfa	Trifluoroacetyl

3.1 Computational Chemistry

Given the current computer power, computational chemistry can nowadays be considered as an important branch of Chemistry. It incorporates our knowledge on theoretical chemistry into efficient computer programs, many commercially available, which can be used to solve a large variety of chemical and biological problems. The range of chemical systems that can be considered is very broad, going from isolated atoms or small molecules to polymers, biological macromolecules or molecular associations in gas, liquid or solid phases, all of them in both static and dynamic situations [1, 2]. In addition, specific graphic software is usually used to interactively visualize, and even manipulate, in an attractive way the results obtained.

Computational chemistry studies can be carried out to complement and interpret experimental data by performing calculations on the structures, properties and reactions of real substances. For example, it is possible to determine molecular geometries in ground and excited states, dipole moments and charge distributions, potential energy surfaces, rate constants and reaction barriers for elementary reactions, or infrared (IR) and nuclear magnetic resonance (NMR) spectra. Even more interestingly, computational chemistry is particularly useful to investigate materials that are too expensive or too difficult to manipulate, or simply to determine properties that are experimentally inaccessible. For example, it may be used to predict the stability, properties, and possibilities of new chemical compounds, to find adequate starting points for laboratory synthetic routes, to explore new reaction mechanisms, to design molecules that interact specifically with others, or to determine what modifications can be done in a given substance in order to improve or modulate some of its properties. Therefore, computer predictions using the appropriated software before experiments may be a great help for laboratory chemicals, saving time and specialized personal and laboratory resources.

Many computational studies are oriented to determine properties of systems with a fixed disposition of the constituent atoms and/or molecules. Other studies, as conformational searches, are primarily aimed to establish which nuclear coordinates correspond to the stable structures of a given system. In any case, the first stage in computational studies is the selection of the theoretical methodology to be employed. Thus, chemical problems can be faced from quantum mechanics (QM) or from classical mechanics, also called molecular mechanics (MM), formalisms. In the former ones chemical systems are described as a set of nuclei and electrons that follow quantum physics principles. In the second one, systems are considered as a set of particles located at the nuclei that behave as expected from classical physics theory. As a consequence of explicitly considering the electrons, QM methods enable the evaluation of properties that depend on the electron distribution (e.g. ionization potentials, electron affinities, and optical spectra) and, in particular, the study of chemical reactions where bonds between atoms are broken and formed. At present time, QM calculations are also frequently used to provide a precise description of the conformational preferences of biomolecules, like small and medium peptides. Although QM studies offer the most accurate description of matter and allow evaluate properties that cannot be analyzed by other methods, their computational requirements are very high, increasing rapidly with the size of the system under study. In consequence, QM calculations are typically restricted to systems involving a few tenths of atoms, while less accurate and less computationally demanding MM methods are used for bigger systems and, in particular, for studies where the electronic structure is not important.

3.2 Quantum Mechanical Calculations: Concepts

QM calculations usually involve the problem of solving the time-independent Schrödinger equation associated to the system under study:

$$\hat{H}\Psi = E\Psi \tag{3.1}$$

where $\hat{H}$ is the Hamiltonian operator containing the kinetic and potential energy of the nuclei and electrons, E is the energy, and Ψ is the wave function that implicitly contains all the measurable information about the system. The second-order differential Schrödinger equation is an eigenvalue equation, whose resolution provides pairs of $\{E,\Psi\}$ values. Between them, the pair with lower E describes the energy and the behavior of the system in its ground state.

Solving the Schrödinger equation is in practice only affordable by using some approximations. Among them, the Born–Oppenheimer approximation, which considers that electrons can adjust almost instantaneously to any change in the positions of the nuclei, is particularly outstanding. Therefore, movement of nuclei and electrons can be considered separately, $\Psi = \psi(\text{nuclei})\cdot\psi(\text{electrons})$, and the energy is evaluated as the sum of the electronic energy at fixed nuclei positions and the repulsion energy of the nuclei. Another usual simplification is the description of ψ(electrons) by molecular orbitals (monoelectronic mathematical functions delocalized over the entire system), which are expressed as lineal combinations of a set of selected mathematical functions centered at the nuclei. This approach is denoted the linear combination of atomic orbitals (LCAO) and the set of functions used to describe the orbitals is named the *basis set*. Both the reliability of the results and the computational resources required to obtain them grow with size and sophistication of the basis set.

In practice, there are three methodologies, which essentially differ in their philosophies, to solve a chemical problem using QM calculations: *ab initio*, semiempirical and density functional theory (DFT) methodologies.

3.2.1 *Ab Initio* Methods

Ab initio methods are grounded on solving the Schrödinger equation without using empirical or semiempirical parameters (i.e. from the first principles of quantum theory [3] with no inclusion of experimental data). As stated before, some approximations are necessary to solve the mathematical problem, different *ab initio* methodologies being proposed depending on their thoroughness. The simplest *ab initio* method is the Hartree–Fock (HF) method, which usually constitutes the first step towards more accurate approximations. Conceptually the method is based on the molecular orbital theory, so that the ground state electronic wave function of an N-electron system is expressed by a single determinant of N spin orbitals (Slater determinant), $|\Psi_0\rangle$. The Slater determinant is an antisymmetrical linear combination of spin molecular orbitals (χ_i) products, which in turn are linear combinations of atomic orbitals (the m functions that constitute the basis set):

$$\chi_i = \sum_{s=1}^{m} c_{is}\phi_s \tag{3.2}$$

According to the variational principle, the energy obtained from an approximated wave function will always be higher than the real energy. Therefore, the best approximated wave function is obtained through the set of coefficients c_{is} that provide the minimum energy for the system:

$$E_0 = \left\langle \Psi_0 \middle| \hat{H} \middle| \Psi_0 \right\rangle \tag{3.3}$$

This set of coefficients is obtained iteratively following the self-consistent field (SCF) method. The resulting wavefunction is the best variational approximation to the ground state of the system and, according to the postulates of QM, can be used to compute measurable physicochemical properties of the system. The major weakness of the HF method is the fact that it does not describe electron–electron repulsion effects properly since the effects of instantaneous electron–electron interactions are only averaged. On that score, *electron correlation*, which is the term used to refer to the instantaneous repulsion between electrons, are not considered in the HF theory. In order to overcome this deficiency, post-HF methods have being designed by introducing corrections to the HF calculations. These methods give more accurate results than HF, even though they are computationally more demanding.

Two of the most used post-HF methods are the configuration interaction (CI) method and the Møller–Plesset (MP) procedure. The CI method [3, 4] is a variational approach based on the fact that the exact wave function for the ground and excited states of an *N*-electron problem can be written as a linear combination of all possible *N*-electron Slater determinants formed from a complete set of spin orbitals. In this way, the electronic state of a system is described as a mixture of all possible electronic states. In practice, as it is not possible to handle infinite (complete) basis sets, full CI results are exact only within the subspace spanned by the orbital basis set. Moreover, even using reduced bases sets, the number of determinants required in a full CI expansion grows factorially with the number of electrons and orbitals. Required computer resources are reduced by truncating the CI to the excitations that represent the more important contributions to electron correlation.

An alternative way to tackle the problem of electron correlation is that proposed into the MP procedure [5], in which the 'true' Hamiltonian is expressed as the sum of the zero order Hamiltonian $\hat{H}_0$ and a perturbation operator $\hat{V}$ that is supposed to be small compared to $\hat{H}_0$:

$$\hat{H} = \hat{H}_0 + \lambda \hat{V} \tag{3.4}$$

As the perturbation parameter λ can vary from 0 to 1, the eigenfunctions and corresponding energies of $\hat{H}$ can be written as a Taylor expansion. In particular, for the ground state:

$$|\Psi\rangle = |\Psi_0\rangle + \lambda |\Psi^{(1)}\rangle + \lambda^2 |\Psi^{(2)}\rangle + \cdots \tag{3.5}$$

$$E = E_0 + \lambda E^{(1)} + \lambda^2 E^{(2)} + \cdots \tag{3.6}$$

where $E^{(1)}$ is the first order correction to the energy, $E^{(2)}$ the second order correction, and so on. The sum of E_0 and $E^{(1)}$ corresponds to the HF energy. Therefore, to improve the HF energy it is necessary to use the MP perturbation theory to at least the second order. This level of theory is referred as MP2 and involves the calculation of

$$E^{(2)} = \langle \Psi_0 | \hat{V} | \Psi^{(1)} \rangle$$

Nowadays, coupled cluster (CC) theory [6, 7] has been developed as one of the most reliable computational methods for electronic structure calculations. The fundamental idea behind this method is the use of an exponential operator that, when applied to a reference function, produces a new wave function that correlates the motions of a specified number of electrons. However, in general CC calculations on peptides are difficult at the present time because of the huge amount of computer resources required by this methodology.

3.2.2 Semiempirical Methods

Semiempirical (SE) methods were firstly developed to study large systems using moderate computational resources. Although these methods are also based on the resolution of the Schrödinger equation, they involve some drastic approximations directed to reduce the time spent in the evaluation and manipulation of the integrals found in *ab initio* HF calculations. In order to compensate for such simplifications, some empirical parameters typically derived from experimental (or *ab initio*) data are introduced, replacing some of the integrals that are completely excluded from the calculations.

SE methods reduce the computational effort considering only the valence electrons of the system explicitly, which move in the field created by the core nuclei and internal electrons. This approach is based on the fact that electrons involved in chemical bonds and other phenomena are mainly those of the valence shell. Furthermore, SE calculations use minimum basis sets comprising *s*, *p* and sometimes *d* Slater type orbitals. Modern SE models are also based on the *neglect of diatomic differential overlap* (NDDO) approximation, in which the overlap between two atomic orbitals of different atoms is set to zero.

Current SE models differ by the further approximations used when evaluating the rest of one- and two-electron integrals, and by the parametrization philosophy. Two of the most used SE methods used to study organic and biorganic systems are the Austin model 1 (AM1) [8] and the parametric model number 3 (PM3) [9], which differ in the treatment of the core–core repulsion and, especially, in the parametrization strategy. These SE models, which are faster and computationally cheaper than the *ab initio* ones, can be applied to molecules of thousands of atoms giving good results when applied to molecules similar to those used in the parametrization process.

3.2.3 Density Functional Theory

Density functional theory (DFT) methods, which represent an alternative to the resolution of the Schrödinger equation, are variational procedures currently used to study the electronic structure (essentially the ground state) of a large number of molecules. Within this theory, the parameter to optimize is the functional that relates the electronic density of the system with its energy (i.e. DFT methods work with the density functional instead of complex many-electron wavefunctions). Thus, computational requirements are low in comparison with HF and post-HF methods, as the wavefunction of an N-electron system depends on $3N$ spatial coordinates, while the density has only three variables. The problem is that, although it is known that it exists, the exact form of the functional connecting the electron density and the energy of the system is unknown. The goal of the different DFT methods is to design functionals linking these two properties.

The DFT is based on the theorems postulated by Hohenberg and Kohn [10], which state that the ground state energy of a nondegenerated electronic system, as well as its electronic properties, are solely defined by its electron density $\rho(r)$. Besides, the variational principle allows the conclusion to be made that the energy of any calculated density is always higher than the energy of the actual density. Accordingly, the electronic energy can be calculated as a simple sum of different contributions that depend on $\rho(r)$:

$$E[\rho(r)] = E^T[\rho(r)] + E^V[\rho(r)] + E^J[\rho(r)] + E^{XC}[\rho(r)] \tag{3.7}$$

where E^T, E^V, E^J, and E^{XC} represent the kinetic energy, the electron–nucleus and nucleus–nucleus potential energy, the Coulombic repulsion between electrons, and the interchange-correlation energy, respectively. On the other hand, Kohn and Sham wrote $\rho(r)$ as the sum of the square modulus of a set of one-electron orthonormal orbitals (Kohn–Sham orbitals) [11]:

$$\rho(r) = \sum_{i=1}^{N} |\phi_i(r)|^2 \tag{3.8}$$

Combining the last two expressions and solving the variational problem should permit the minimum value of the energy to be obtained, which corresponds to the exact ground state electronic density. However, the mathematical formulation of the $E^{XC}[\rho(r)]$ term is unknown.

The difference between various DFT methods is the choice of the functional used to represent the exchange-correlation energy. Some functionals, such as the VWN [12] and the local correlation functional of Perdew [13], assume that the density can be treated as a uniform electron gas. Others, such as Becke 95 [14], Perdew 86 [15], Perdew–Burke–Ernzerhof [16], Perdew–Wang 91 (PW91) [17], and the popular Lee–Yang–Par (LYP) [18], use gradient corrected functionals (*generalized gradient approximation*, GGA), in which the exchange-correlation energy depends not only on the electron density but also on derivatives of the density. $E^{XC}[\rho(r)]$ can also be evaluated using hybrid density functionals, which combine a conventional GGA method with a percentage of HF exchange. Some of these functionals are the B3LYP [18, 19], which is the most popular, B3PW91 [17, 19], MPW1K [20], O3LYP [18, 21], X3LYP [17, 18, 22], and BMK [23]. In summary, DFT methods provide better results than the HF and some post-HF *ab initio* methods at a lower computational cost.

3.2.4 Solvent Effects in Quantum Mechanical Calculations

Since many phenomena take place in solution, an important goal of QM methods is to evaluate how the solvent affects the behavior of the system (e.g. polarization and dipole orientation effects and specific solute–solvent interactions). There are two approaches to treat these effects: those considering explicit solvent molecules and those that represent the solvent as a continuous medium. Explicit solvent models are the most accurate but very demanding from a computational point of view, since a realistic description of the solvated system requires a large number of explicit molecules [24, 25]. Consequently, explicit solvent treatment is more frequent in MM calculations than in QM ones.

Solvent effects in QM calculations are usually evaluated through continuum models (i.e. the solute is treated at the QM level while the solvent is described as an infinite dielectric medium). In particular, the *polarizable continuum model* (PCM) [26] has been very successful. The PCM considers the solvent as a uniform dielectric medium and the solute placed in a solvent cavity defined through interlocking van der Waals spheres centered at each nucleus of the solute. Solvent polarization effects are accounted through virtual point charges located on the cavity surface, which are proportional to the derivative of the solute electrostatic potential. In PCM calculations the interaction potential between the solvent and solute is considered as a perturbation $\hat{V}_R$ of the Hamiltonian $\hat{H}_0$ for the isolated solute system:

$$\left(\hat{H}_0 + \hat{V}_R\right)\Psi = E\Psi \tag{3.9}$$

From an energetic point of view, the solvation free energy (ΔG_{solv}), which is the energy necessary to transfer the solute from the gas phase to the solvent, is estimated as the sum of three terms:

$$\Delta G_{solv} = \Delta G_{cav} + \Delta G_{vdW} + \Delta G_{ele} \tag{3.10}$$

where ΔG_{cav} is the energy involved in the creation of the solute cavity within the solvent, ΔG_{vdW} is the contribution due to the solute–solvent van der Waals interactions, and ΔG_{ele} is the electrostatic component due to the polarization of the solvent.

3.3 Quantum Mechanical Calculations on Hybrid Peptide Materials: Some Examples

QM calculations have been largely used to determine the intrinsic conformational properties of small peptides involving synthetic amino acids, also denoted noncoded amino acids (nc-aa). Thus, many of these amino acids can be viewed as hybrid molecules formed by proteinogenic α-amino acids and organic fragments, which are introduced to restrict the conformational properties. These restrictions can be exploited in many fields such as biomedicine, biotechnology, biomaterials, and nanobiology [27].

Among the nc-aa with severe conformational restrictions are the α,α-dialkylated (also named *quaternary*) amino acids. Tetrasubstitution at C^{α} introduces constraints in the backbone dihedral angles, stabilizing particular elements of the peptide secondary structure. The simplest α,α-dialkylated amino acid is α-aminoisobutyric acid (Aib, in Figure 3.1), that is, α,α-dimethylgycline or α-methylalanine. Replacement of each α-hydrogen in glycine (Gly) by a methyl group results in a drastic reduction of the available conformational space [28]. However, much complex conformational restrictions are expected when the two α-hydrogens of Gly are replaced by bulky phenyl or benzyl groups, the resulting α-amino acids being denoted α,α-diphenylgycline (Dϕg in Figure 3.1) and α,α-dibenzylgycline (Db_zg in Figure 3.1), respectively.

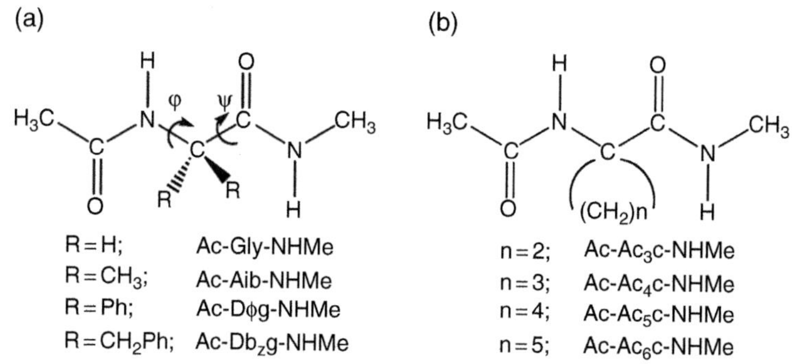

Figure 3.1 *Chemical structure of small peptides studied using DFT calculations and containing: (a) glycine (Gly), α-aminoisobutyric (Aib), α,α-diphenylglycine (Dϕg), and α,α-dibenzylglycine (Db$_z$g) and (b) 1-aminocycloalkane-1-carboxylic acids*

DFT calculations on small model peptides (Figure 3.1a) formed by Aib [28], Db$_z$g [29], and Dϕg [30] residues blocked at the ends by acetyl and *N*-methylamine groups (Ac and NHMe, respectively) indicated that the benzyl side groups restrict significantly the backbone conformation of Ac–Db$_z$g–NHMe. Thus, the only accessible backbone conformations are those associated with fully extended, γ-turn and semifolded motifs, while both Ac–Aib–NHMe and Ac–Dϕg–NHMe show a higher conformational flexibility. The destabilizing energy contribution found when the methyl groups of Ac–Aib–NHMe replace the α-hydrogen atoms of Ac–Gly–NHMe was significantly more pronounced when the phenyl and, especially, the benzyl groups of Ac–Dϕg–NHMe and Ac–Db$_z$g–NHMe, respectively, substitute the methyl groups of Ac–Aib–NHMe. These features are clearly reflected in Figure 3.2a, which compares the accessible conformations of Ac–Aib–NHMe, Ac–Db$_z$g–NHMe and Ac–Dϕg–NHMe with those of Ac–Gly–NHMe in the ϕ,ψ Ramachandran map.

On the other hand, nc-aa 1-aminocycloalkane-1-carboxylic acids (Ac$_n$c, where *n* indicates the size of the ring) are the result of $C^\alpha \leftrightarrow C^\alpha$ cyclization whereby dialkylated glycine residues with cyclic side chains are formed. The conformational properties and electronic characteristics of *N*-acetyl-*N*′-methylamide derivatives of the cyclopropane [31] (Ac$_3$c), cyclobutane [32] (Ac$_4$c), cyclopentane [33] (Ac$_5$c), and cyclohexane [34] (Ac$_6$c) analogues (Figure 3.1b) were fully characterized using *ab initio* and DFT QM calculations. Comparison of the results obtained for these four hybrid peptides precisely reflect how the conformational flexibility of the backbone increases with the size of the ring (Figure 3.2b). Thus, maps obtained for Ac–Ac$_3$c–NHMe and Ac–Ac$_4$c–NHMe, which are relatively similar, are significantly constrained with respect to that of the peptide containing the simplest α,α-dialkylated amino acid (Ac–Aib–NHMe, Figure 3.2a). In contrast, although Ac–Ac$_6$c–NHMe tends to adopt folded conformations, as is usual in Ac$_n$c dipeptides, its conformational freedom is higher than that of the corresponding Ac$_3$c, Ac$_4$c, and Ac$_5$c analogues. This partial flexibility is mainly the consequence of the lack of strain in the cyclohexane ring. Finally, Ac–Ac$_5$c–NHMe presents an intermediate situation, which is characterized by the fact that each accessible backbone conformation is compatible with different arrangements of the cyclopentane ring.

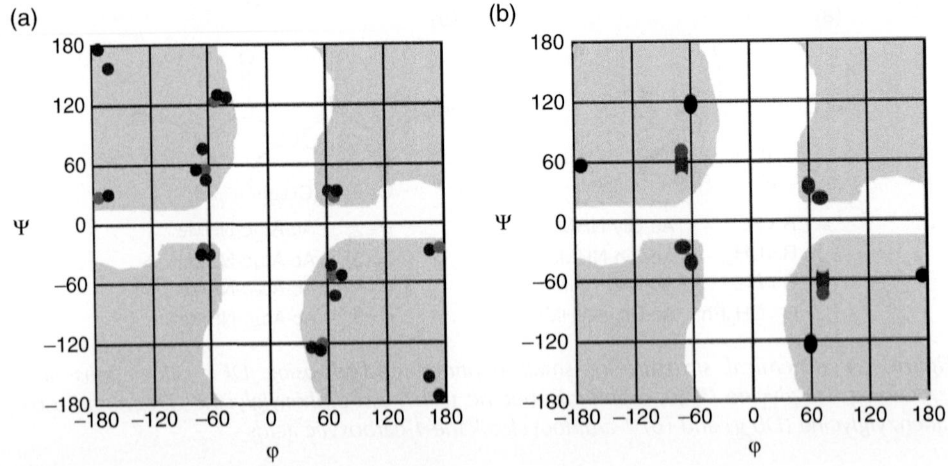

Figure 3.2 *Ramachandran maps indicating the position in the ϕ,ψ-map of the minimum energy conformations predicted by DFT calculations for peptides containing nc-aa. The regions with flesh color correspond to the sterically accessible regions of the ϕ,ψ-map for the Ac–Gly–NHMe dipeptide. (a) Positions of the minimum energy conformations found for Ac–Aib–NHMe (red), Ac–Dϕg–NHMe (green), and Ac–Db$_z$g–NHMe (dark blue). The energy gap between the minima of the lowest and highest energy is 3.4, 9.8, and 4.6 kcal/mol, respectively (taken from references [28, 29], and [30], respectively). (b) Minimum energy conformations found for Ac–Ac$_n$c–NHMe dipeptides (taken from references [31–33], and [34]): yellow indicates the positions shared by Ac$_n$c-containing dipeptides with n = 3, 4, 5, and 6; pink refers to the positions shared by Ac$_n$c-containing dipeptides with n = 4, 5, and 6; red corresponds to the positions shared by Ac–Ac$_5$c–NHMe and Ac–Ac$_6$c–NHMe; and dark red are positions exclusively found for Ac–Ac$_6$c–NHMe (see color plate figure)*

3.4 NCAD: An Information Management System of Quantum Mechanical Calculations on Noncoded Amino Acids for Peptide Design

Nc-aa are frequently conceived not only to emulate natural diversity but to enhance and create novel properties for the building blocks of peptides and proteins. Thus, physicochemical properties of nc-aa are open to rational design, as can be seen in the number of publications reporting their synthesis and characterization [35–41]. Properties attaining conformational tendencies and protease-mediated degradation resistance are especially remarkable in protein and peptide engineering. Systematic experimental investigations on these topics are frequently avoided since the synthesis of nc-aa is highly resource-consuming at both human and economic levels. That is why theoretical chemistry, and especially accurate QM calculations, surges as a major tool to slash costs down by refining both methods and goals. The combination of experimental and theoretical research has boosted the available knowledge in the field.

The scattering of information related to nc-aa, which involves a high number of papers and publication types, has been on the rise since the initial studies in this field. Moreover, QM studies are frequently based not only on different formalisms but also on different computational levels for the same formalism. This multiplicity of information represents a

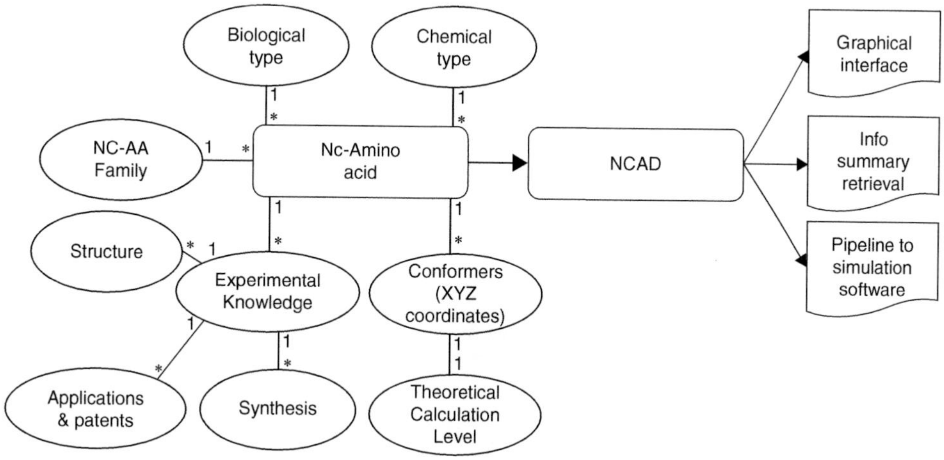

Figure 3.3 *Scheme depicting the nc-aa information organization in the NCAD and the information retrieval that users obtain from the data base; 1 means a unique correspondence between two fields and * means multiple correspondence*

serious drawback for most of the researchers working in fields different from theoretical research (e.g. synthetic chemistry, structure determination, and physicochemical properties). This scenario highlights the need for an information management system that not only compiles the relevant knowledge on nc-aa but also structures, and offers it in a user-friendly way.

The NCAD (*noncoded amino acid data base*) is, to the best of our knowledge, the only freely available information management resource to compile, organize, and graphically present information about nc-aa [42, 43]. The NCAD was designed to store structural data of nc-aa derived from QM calculations, altogether with their experimental knowledge in the fields of synthesis, structural characterization (NMR, IR, X-ray crystallography, etc.), biomedical and material engineering applications (drug and novel materials development), and industrial patents. As can be seen in Figure 3.3, the NCAD focuses on peer-reviewed reported structures of nc-aa obtained by using QM methods and offers the dihedral angle values and the Cartesian coordinates. Any other knowledge referred to every nc-aa and reported in indexed publications is also linked to theoretical information. Besides, the NCAD contains other characteristic parameters that have been specially derived or found for the incorporation of the nc-aa in the data base (e.g. α-carbon chirality, accessible structural motifs, molecular weight, empirical formula, IUPAC rules-based name, and stereo representation of the molecule). An NCAD installation pack can be freely downloaded from the following web site: http://recerca.upc.edu/imem/index.htm.

The NCAD is implemented in an installation pack together with data base consult software and a molecular visualization tool based on the RasWin [44] program (Figure 3.4a). This visualization tool also allows different structural data to be measured in the *xyz* files resulting from theoretical calculations and, if available, in the *xyz* file of the crystallographic structures. The data base is not only user-friendly with regard to information retrieval but also with the search criteria. As can be seen in Figure 3.4b, the NCAD enables a multifactor search according to different nc-aa-related parameters: molecular formula, molecular

(a)

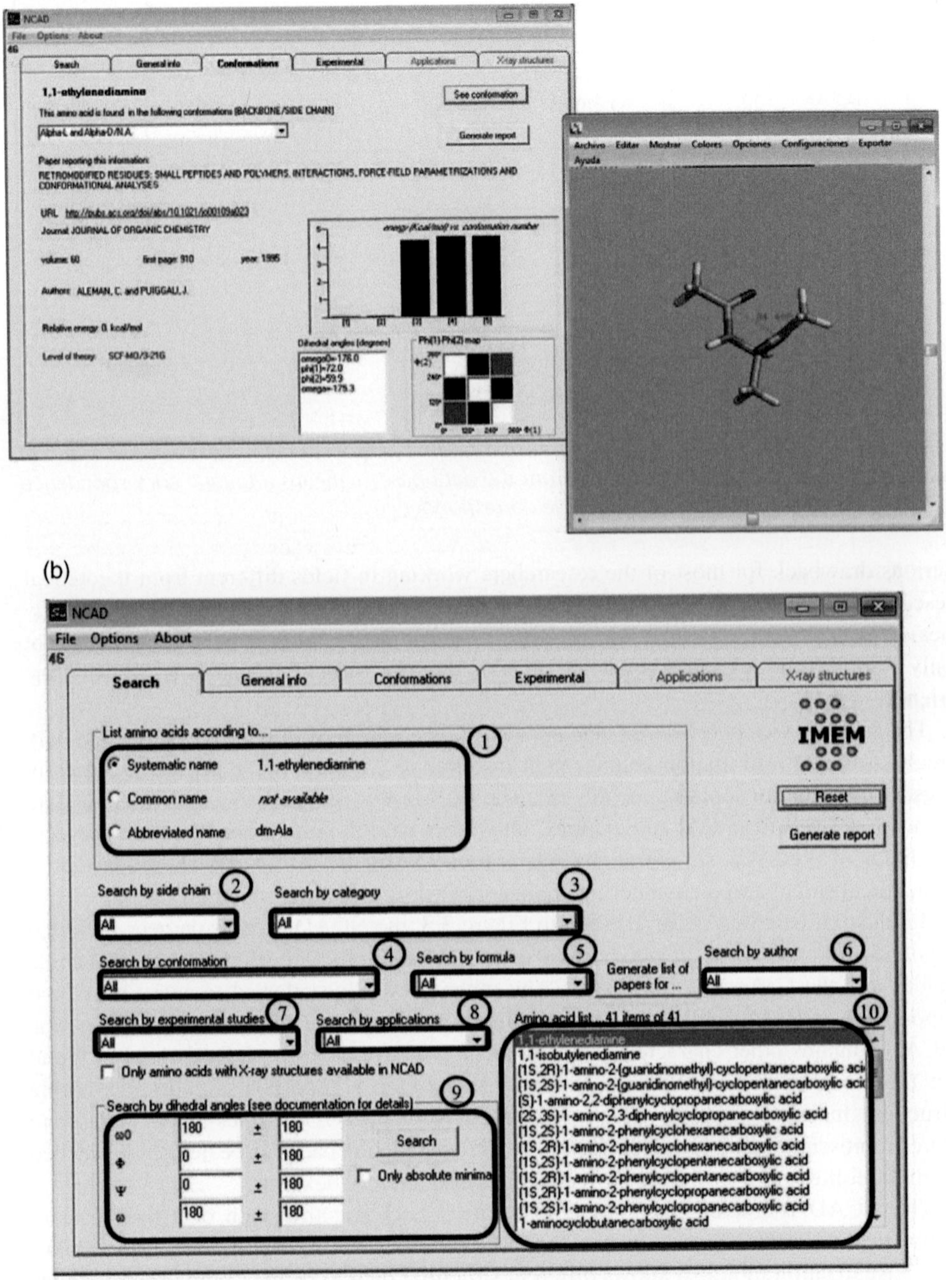

Figure 3.4 (a) Visualization tool embedded in the NCAD. (b) Search screen of NCAD. The searching criteria are: (1) systematic, common, or abbreviated name; (2) side chain chemical nature; (3) nc-aa type; (4) secondary structure; (5) chemical formula; (6) author of the published paper; (7) availability and type of experimental studies; (8) availability and type of applications; (9) dihedral angle value; and (10) IUPAC systematic name

weight, nc-aa common name, molecule alias, side chain nature, paper author, secondary structure, dihedral angle value, availability of experimental studies, and availability of material science and biomedical applications. Multifactor searches enable refined queries that render the suitable candidates for substitution of naturally occurring amino acids in many different systems. Figure 3.5 shows some representative nc-aa included in the data base, which have been categorized according to their chemical structure.

The usefulness of the NCAD as a tool in computer aided drug design was recently demonstrated by proposing different targeted mutations able to reduce the conformational flexibility of methione-enkephalin (Tyr–Gly–Gly–Phe–Met) [42]. This peptide, which is known for its activity as an opioid neurotransmitter, has a high conformational variability due to its short length. Although such flexibility allows the peptide to adopt conformations that mediate the binding to a number of receptors, the affinity towards some of them could be enhanced by limiting the degrees of freedom through the selective substitution of natural amino acids by constrained nc-aa. The criteria used for the selection of nc-aa to replace the Gly and Tyr residues without altering the biologically active conformation of the peptide were: (i) nc-aa having a chemical nature similar to that of the natural amino acids (i.e. aromatic, aliphatic, charged, etc.) and (ii) nc-aa having a similar conformational behavior, that is, adopting similar secondary structures. The nc-aa selected to replace Gly and Tyr were α-aminoisobutyric acid (Aib) and (2R,3R)-1-amino-2,3-diphenylcyclopropanecarboxylic acid [(2R,3R)$c_3$diPhe], respectively, a detailed description of such achievements being reported in reference [41]. More applications related to Materials Engineering, in which nc-aa are used to design helical structures with hydrogen bonding patterns different from that of the α-helix, are detailed in reference [43].

The NCAD is under continuous upgrading since new nc-aa categories are being included continuously with the aim of embracing all the hundreds of nc-aa molecules that have been calculated using QM methods. Besides, new software applications are under development in order to achieve a pipeline of files with the major simulation software packages. Moreover, software to embed nc-aa residues in already folded macromolecules and to refine side-chain conformation is being designed and implemented.

3.5 Molecular Mechanics Calculations: Concepts

MM methods are based on mathematical models that assume a molecule as a collection of balls and springs representing, respectively, particles and bonds. Thus, this method ignores the electronic motions and each particle is described by a radius, hardness and net charge. In MM calculations, systems can be described according to two different models (Figure 3.6a): (i) the all-atom model, in which each particle corresponds to an atom, and (ii) the united atom model, which describes a very small group of atoms as a single particle (pseudoatom) without producing any significant damage in the chemical information of the system (e.g. the CH_3 and CH_2 groups of aliphatic side chains and the CH of aromatic rings are represented using a single particle approach).

In coarse-graining models a relative large number of atoms (e.g. a side group or even a complete amino acid) are described as a single soft particle. This reduction in the number of degrees of freedom of the system, which is in detriment of its chemical description,

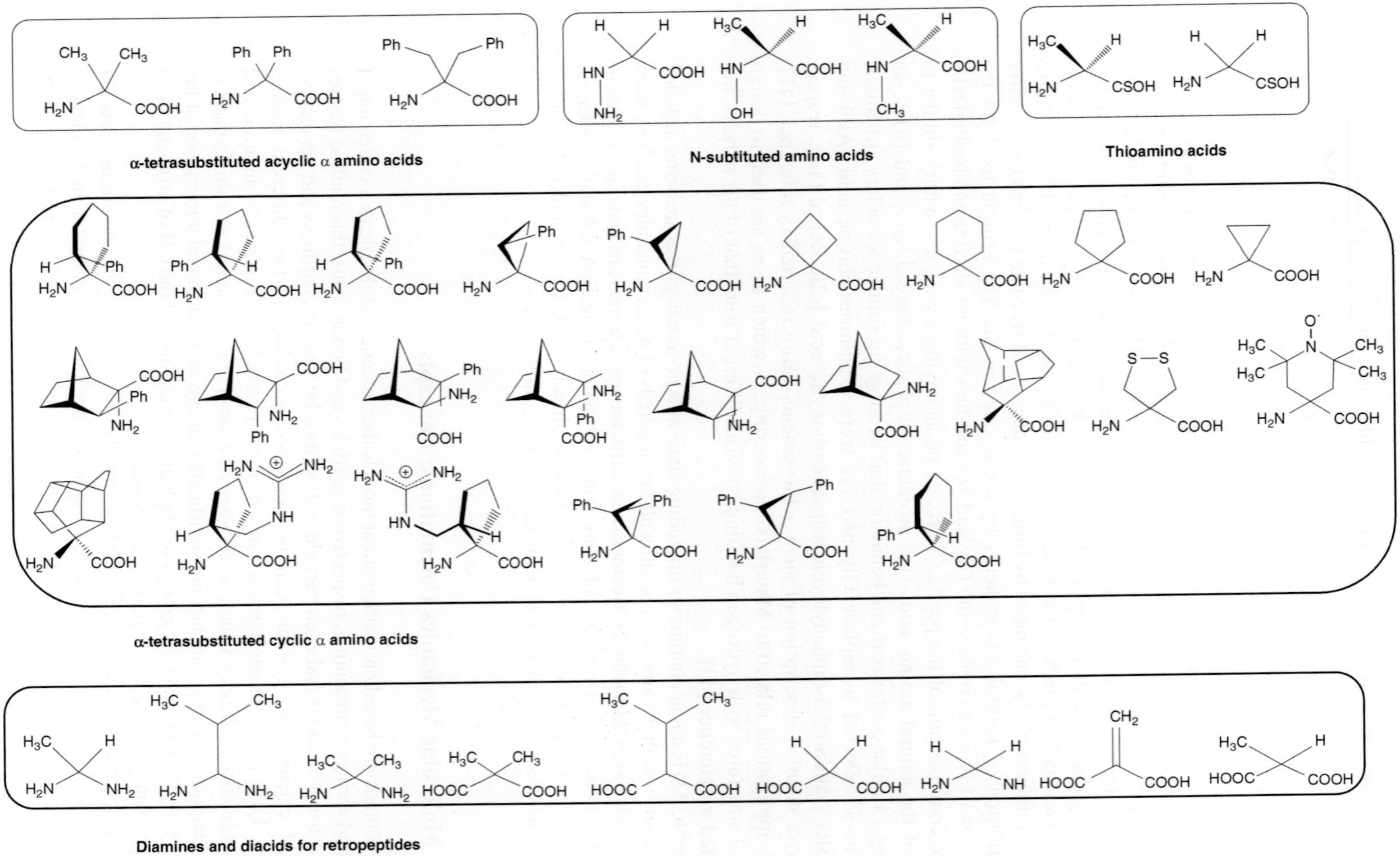

Figure 3.5 Representative nc-aa included in the NCAD

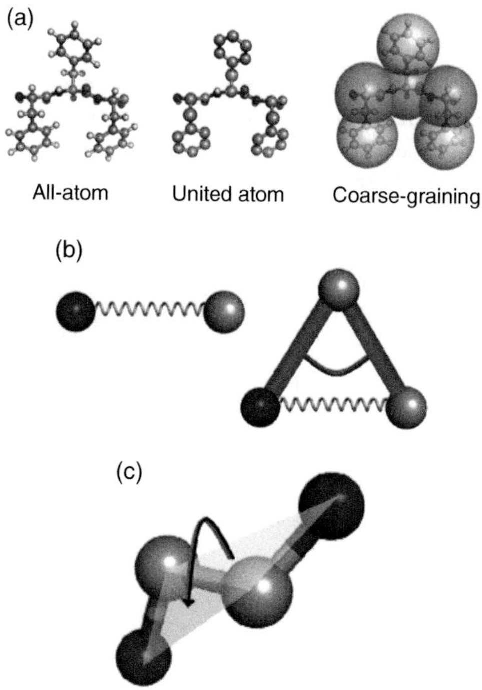

Figure 3.6 (a) Schematic representation of triphenylalanine using all-atom, united atom, and coarse-graining models. (b) Description of bond stretching and bond angle bending using a spring. (c) Description of a dihedral angle able to make 360° turns

enables larger simulations than for all-atom and united models. Specifically, the latter approaches are practically limited to simulation times and system sizes less than 500 ns and 20 nm, respectively, whereas the simulated time and length scales can increase several orders of magnitude (mesoscopic scales) with simulations based on coarse-graining models. Thus, elimination of both degrees of freedom and fine interaction details provokes a reduction of the required computational resources, allowing the time- and length-scales gap to be diminished between computational and experimental methods of studying peptide-based materials.

Although calculations of systems described using all-atom, united atom, and coarse-graining models can be carried out using the same algorithms, the laws that describe the interactions between particles are completely different. Thus, the potentials used to describe interactions in all-atom and united atom models are transferable and based on classical mechanics, whereas those needed by coarse-graining models have to be derived for each specific case and are not transferable [45, 46]. On the other hand, MM can be used for different procedures typically associated with molecular modelling and computational chemistry, like computation of the energy of a molecular system, energy minimization, Monte Carlo, and molecular dynamics (MD). In this chapter we focus on energy minimization and MD techniques combined with all-atom and united atom models, which are the most frequent in peptide studies.

3.5.1 Force Fields

The basis of all MM simulation methods based on all-atom and united atom models is a potential energy function $\Phi(\boldsymbol{R})$. This function gives the potential energy of a system containing N particles and depends on the position, $\vec{r}_i$, of each particle i, which is denoted by the $3N$-dimensional vector $\boldsymbol{R}=(\boldsymbol{r}_1, \boldsymbol{r}_2, \ldots, \boldsymbol{r}_N)$. Ideally one would like to obtain this energy by QM but in practice this is not possible for systems larger than some hundreds of atoms. In these cases $\Phi(\boldsymbol{R})$ is expressed as a series of analytical terms: a so-called force field. The analytical form of a force field typically relates the chemical common sense with classical mechanics, and for a system of N particles has the form:

$$\Phi(\boldsymbol{R})=\sum_{bonds}\frac{1}{2}K_b(b-b_0)^2+\sum_{angles}\frac{1}{2}K_\theta(\theta-\theta_0)^2+\sum_{dihedrals}K\left[1+\cos(n\varphi-\delta)\right]$$
$$+\sum_{impropers}\frac{1}{2}K_\xi(\xi-\xi_0)^2+\sum_{pairs(i,j)}\left(\frac{C_{12}(i,j)}{r_{ij}^{12}}-\frac{C_6(i,j)}{r_{ij}^{6}}+\frac{q_iq_j}{4\pi\varepsilon_0 r_{ij}}\right) \quad (3.11)$$

The first two terms represent the covalent bond stretching and bond angle energies, respectively, which are described as the elastic deformation of a spring (Figure 3.6b). The next two terms describe the torsional terms: the first term, known as the torsion potential, relates to the energy barriers associated with the rotation around the single covalent bonds (Figure 3.6c) and those contributions are adjusted to a Fourier expansion of cosines. The second term takes into account the geometry distortions that rigid bond geometries can undergo, such as movements out of the plane in molecules that present conjugated π electrons. This term can be expressed in some force fields as a Fourier expansion. The last term, the nonbonded part, models the interaction between atoms that are three or more bonds away plus all intermolecular interactions. It is composed of the van der Waals interaction, described through a Lennard–Jones potential, and the Coulomb interaction. The Lennard–Jones term for intramolecular third-neighbor interactions is usually reduced by a half [47].

The parameters K_b, b_0, K_θ, θ_0, K_ξ, ξ_0, K_φ, n, δ, C_{12}, C_6, and q_i are the force field parameters. QM can be used to obtain some of these (e.g. molecular geometries, torsional potentials, and atomic charges) whereas others are usually obtained empirically [48–50].

On the other hand, at this point the existence of QM/MM methods should be mentioned. In this hybrid technique the chemically most relevant region is treated at the QM level, while the rest of the system is described by MM potentials. QM/MM formalisms differ, besides the MM potential and the QM level of theory employed, in both how the two parts are connected and the technique applied to model the system as a whole. Regarding the first, there are essentially two ways to connect the QM and MM parts: (i) the link atom scheme, which uses a monovalent atom (i.e. usually hydrogen) to cap the unsaturated QM atoms, and (ii) methods based on the use of localized orbitals in the boundaries separating the QM and MM parts. On the other hand, the QM/MM formalisms can be combined with all the techniques described in the next sections and typically used in MM simulations (e.g. energy minimization and molecular dynamics). A more detailed description of the hybrid QM/MM formalism is not provided in this chapter since its use in the field of peptides is relatively scarce compared to those of individual QM and MM methods.

3.5.2 Energy Minimization

A minimization procedure, which does not take into account the temperature, is frequently used as a previous step of the MD simulation (i.e. it allows a suitable initial state of the system to be obtained). Energy minimization methods can be divided into different classes depending on the order of the derivative used for locating a minimum on the energy surface. Among them, different first derivative techniques like the steepest descent or the conjugate gradient, which make use of a gradient function, are the most employed when the energy is described by a potential energy function like that of Equation (3.11).

3.5.3 Molecular Dynamics

The aim of MD is to reproduce the time-dependent motional behavior of a molecule. It is assumed that the atoms or pseudoatoms in the molecular system interact with each other according to the rules of the used force field. Thus, the system is described as a system of point masses moving in an effective field. In the MD method the motion of the point charges is governed by Newton's equation of motion:

$$\frac{dv_i(t)}{dt} = m_i^{-1}\boldsymbol{F}_i\left(\{\boldsymbol{r}_i(t)\}\right) \tag{3.12}$$

which are integrated numerically for all particles interacting through a known potential like that displayed in Equation (3.11). The particle positions and velocities are represented by $\boldsymbol{r}_i(t)$ and $\boldsymbol{v}_i(t)$, respectively. The forces, $\boldsymbol{F}_i$, are obtained from the potential functions through

$$\boldsymbol{F}_i = -\frac{\partial\Phi(\boldsymbol{R})}{\partial r_i}$$

A simple finite Taylor expansion of $\boldsymbol{r}_i(t)$ at the time point $t=\Delta t$ yields

$$\boldsymbol{r}_i(t+\Delta t) = \boldsymbol{r}_i(t_n) + \left.\frac{d\boldsymbol{r}_i}{dt}\right|_{t_n}(\Delta t) + \left.\frac{d^2\boldsymbol{r}_i}{dt^2}\right|_{t_n}\frac{(\Delta t)^2}{2!} + \left.\frac{d^3\boldsymbol{r}_i}{dt^3}\right|_{t_n}\frac{(\Delta t)^3}{3!} + O(\Delta t^4) \tag{3.13}$$

A linear combination of the expressions for $\boldsymbol{r}_i(t+\Delta t)$ and $\boldsymbol{r}_i(t-\Delta t)$ leads to a simpler expression for the position after the time lapse Δt:

$$\boldsymbol{r}_i(t+\Delta t) = 2\boldsymbol{r}_i(t) - \boldsymbol{r}_i(t-\Delta t) - \left[\left(\frac{1}{m}\right)\nabla\Phi(\boldsymbol{r}_i(t))\right]\Delta t^2 + O(\Delta t^4) \tag{3.14}$$

This is the main frame of the so-called Verlet algorithm [51]. This method does not explicitly give the instantaneous velocities, which often are necessary for the calculation of certain physical quantities like the kinetic energy. This can create technical challenges in MD simulations, because kinetic energy and instantaneous temperatures at time *t* cannot be calculated for a system until the positions are known at time $t+\Delta t$. This deficiency can either be dealt with using the velocity Verlet algorithm or the leapfrog variant (see below) [52], or estimating the velocity using the position terms and the mean value theorem:

$$v(t)=\frac{\boldsymbol{r}(t+\Delta t)-\boldsymbol{r}(t-\Delta t)}{2\Delta t}+error\ (\Delta t^2) \tag{3.15}$$

The Verlet velocity method and the leapfrog algorithm compute the instantaneous velocities using the corresponding Taylor expansions of the velocity terms (analogous to Equation (3.13)). The only difference between both algorithms is that the former computes velocities at the same time step as the new positions while the later computes the new velocities at a half-step time (i.e. $t=\Delta t/2$).

The time step Δt should be at least of the same order of magnitude as the fastest movement present in the system. Typically, in atomistic representations such movement corresponds to the vibration of the molecular bonds, with $\Delta t=1$ fs for all-atom models that consider the motion of hydrogen atoms, while $\Delta t=2$ fs for united atom models as well as all-atom models in which the hydrogen atoms are considered frozen. The sequential collection of coordinates and velocities obtained by this procedure is called *trajectory*.

3.5.4 Boundary Conditions, Pair-List and Long-Range Interactions

If a solid, liquid, or solution is modelled, it is a common practice to minimize the wall effects. The most used way to do this is by applying periodic boundary conditions (PBC). This means that the atoms of the system are put into a box, which is treated as if it is surrounded by a number of images of itself (e.g. 26 if the simulation box is orthogonal – cubic or orthorhombic), as is displayed in Figure 3.7.

This approach is based on the minimum image convention (MIC). For any given particle inside the box, its position becomes the centre of a new box with dimensions identical to the simulation box. It is assumed that this particle only interacts with those particles that are within this region. In practice, for an orthogonal box of side length L, the interacting particles would lie within the range $-½L$ and $½L$. In practice the number of interacting particles is lower than what MIC would impose, since the effect nonbonding potentials can be neglected for long distances. This means that a cut-off distance, shorter than half the box size, can be applied when calculating the nonbonding terms. Such a strategy allows obtaining a list of interacting pairs to be obtained within the cut-off radius at any time of the simulation (*pair-list*) [53]. In order to save time, the pair-list is updated every 5 or 10 time steps, instead of after each step. However, longer time steps for updating may produce sudden jumps in the force exerted on an atom when another atom enters in its cut-off radius.

The truncation of electrostatic terms (i.e. cut-off distance smaller than 15 Å), which is a long-range potential, can produce drastic divergences on the computed physical properties, such as the molecular structure or the dipole–dipole correlation. In order to overcome this limitation, a more sophisticated treatment of long-range electrostatic interactions is required. The most frequently used procedures are the reaction field method and the Ewald summation technique. In the reaction field method the solvent outside the cut-off range is approximated by replacing it with an infinite continuum of uniform permittivity (i.e. the interaction beyond the cut-off distance is assumed to be equivalent to that of the bulk

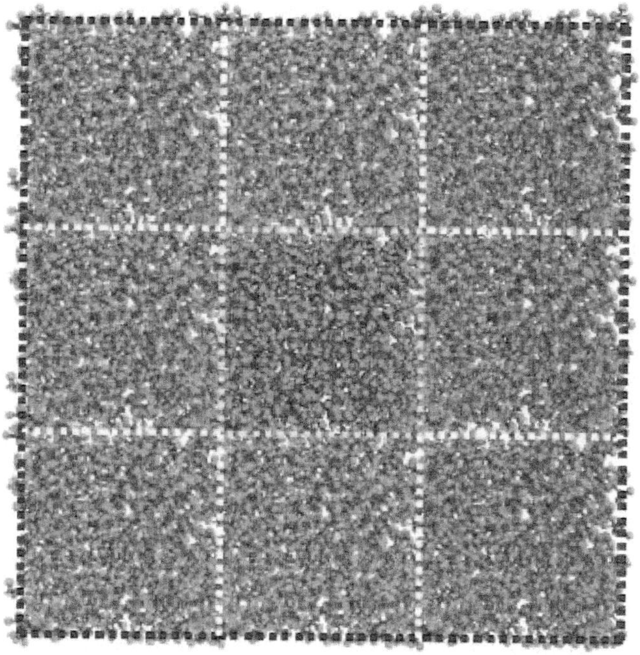

Figure 3.7 *Schematic definition of periodic boundary conditions: the simulation box (lighted) is surrounded by a number of images of itself (see color plate figure)*

solvent) [54]. In the Ewald summation technique, which is a periodic lattice method, the simulation box containing the molecular system is surrounded by a number of infinite copies of itself [55].

This strategy efficiently computes the electrostatic interactions between a particle and all the neighbors among its periodic images. The space is divided in two subdomains, one that contains the particles within the cut-off distance (where coulomb energy is directly computed) and a reciprocal space beyond it, in which the electrostatic interaction is computed by means of the electrostatic potential generated by the particles present in the reciprocal space. Using a reciprocal lattice facilitates a fast convergence of electrostatic potential, since very long distances in the real space rapidly tend to small values in the reciprocal lattice.

3.5.5 Temperature and Pressure

When the system has been placed in the simulation box, random velocities $\boldsymbol{v}_i(t-\Delta t/2)$ are assigned according to a Maxwellian distribution for the desired temperature of the simulation. When Newton's equations of motion are integrated the total energy is conserved and, if the volume is held constant, the simulation will generate a microcanonical ensemble (N, V, E). For comparison with experiments, other ensembles are more appropriate: (N, V, T), in which the temperature is kept constant (canonical ensemble), or (N, P, T), with constant pressure and temperature (isobaric–isothermal ensemble). The temperature of a system consisting of N particles is calculated using the average kinetic energy per degree of freedom:

$$\frac{1}{2}kN_fT = \sum_{i=1}^{N}\frac{1}{2}m_iv_i^2 \tag{3.16}$$

where m_i and v_i are the mass and velocity of particle i. N_f is equal to $3N - N_c$ where N_c is the total number of constraints in the system. These comprise local constraints (e.g. bond lengths), but also global ones (e.g. due to the possible removal of translational and rotational motion of the centre of masses).

The pressure is calculated with the virial theorem, applied to the centers of mass of all particles. Intramolecular contributions cancel out. The internal pressure P is taken equal to the macroscopic pressure:

$$P = \frac{2}{3V}\left(E_{\text{kin}} - \frac{1}{2}\sum_{\text{pairs(i,j)}}\vec{r}_{ij}\vec{F}_{ij}\right) \tag{3.17}$$

where $\boldsymbol{r}_{ij}$ and $\boldsymbol{F}_{ij}$ denote the distance and force between the center of mass of particles i and j.

The Berendsen algorithm [56], also named the weak-coupling algorithm, is among the most usual methods to control the temperature and pressure. The Berendsen temperature coupling mimics weak coupling with first-order kinetics to an external heat bath with given temperature T_0. Thus, the coupling is achieved by multiplying the velocities after each dynamics step by

$$\lambda = \left[1 + \frac{\Delta t}{\tau_T}\left(\frac{T_0}{T(t - \Delta t/2)} - 1\right)\right]^{1/2} \tag{3.18}$$

where τ_T is the temperature coupling constant. The effect of this algorithm is that a deviation of the system temperature from T_0 is slowly corrected according to

$$\frac{dT}{dt} = \frac{T_0 - T}{\tau} \tag{3.19}$$

which means that a temperature deviation decays exponentially with a time constant τ_T. This method has the advantage that the strength of the coupling can be varied and adapted to the user requirement: for equilibration purposes the coupling time can be taken quite short (e.g. 0.01 ps), but for reliable equilibrium runs it can be taken much longer (e.g. 0.5 ps), in which case it hardly influences the conservative dynamics.

The Berendsen pressure coupling rescales the coordinates and box vectors every step, or every n_p steps, multiplying by

$$\mu = \left[1 + \frac{\beta\Delta t}{\tau_P}\left(P(t) - P_0\right)\right]^{1/2} \tag{3.20}$$

Since the pressure relaxation time τ_P is an adjustable parameter, an accurate value for the compressibility of the system is not required. This has the effect of a first order relaxation of the pressure towards a given reference pressure P_0:

$$\frac{dP}{dt} = \frac{P_0 - P}{\tau_P} \tag{3.21}$$

3.6 Molecular Dynamics Simulations on Peptides

In spite of the imposed molecular geometry restrictions (i.e. those associated with the molecular connectivity and the relative rigidity of the amide bond), a large number of conformational possibilities needs to be considered for the identification of the preferences of peptides. Thus, the conformational preferences of a peptide are determined by exploring the energy landscape through a reliable search procedure. Application of systematic conformational search procedures becomes impractical if the number of flexible dihedral angles, including both backbone (φ and ψ, Figure 3.1a) and side groups (χ_i), is very large. Moreover, it should be noted that, frequently, the conformational preferences of peptides are not defined exclusively by the global minimum because of the contribution of the vibrational energy levels to the statistical weight of each conformation. This limitation can only be overcome by performing an exhaustive exploration of the energy landscape of each peptide to identify all the highly populated conformations.

In this section we present some practical aspects (i.e. construction of the molecular system, description of useful exploration techniques based on MD, data mining, and analysis) and explicit examples related to the computational simulation of peptides, either alone or as a part of hybrid materials.

3.6.1 Construction of the Molecular Model

Peptide chains are built from their primary structure definitions, coordinates being taken from predefined data bases with all amino acid structures that are usually contained in modern software packages. Because of the variety and complexity of systems that can be currently modeled using MD techniques, force field parameters (see Section 5.1) are specialized. Within the context of proteins and peptides simulation, AMBER [57], CHARMM [58], GROMOS [59], and OPLS-AA [60] are the most widely used force fields.

Sometimes, a new residue beyond the twenty usual amino acids should be incorporated into the simulation model (e.g. a docking molecule or an nc-aa). In that case, a specific residue derivation has to be performed in order to fit the new fragment in the force field libraries. For example, incorporation of the new fragment into the AMBER force field requires explicit definition of its molecular architecture in the residue library, extrapolation of some force field parameters (i.e. stretching and bending force constants, torsional potentials, and van der Waals parameters can be directly transferred in many cases) from the generalized AMBER force field (GAFF) [61], and parameterization of the rest of the force field parameters. Explicit parametrization of the electrostatic parameters is always needed, in the case of the AMBER force field this parameters being obtained using the restrained

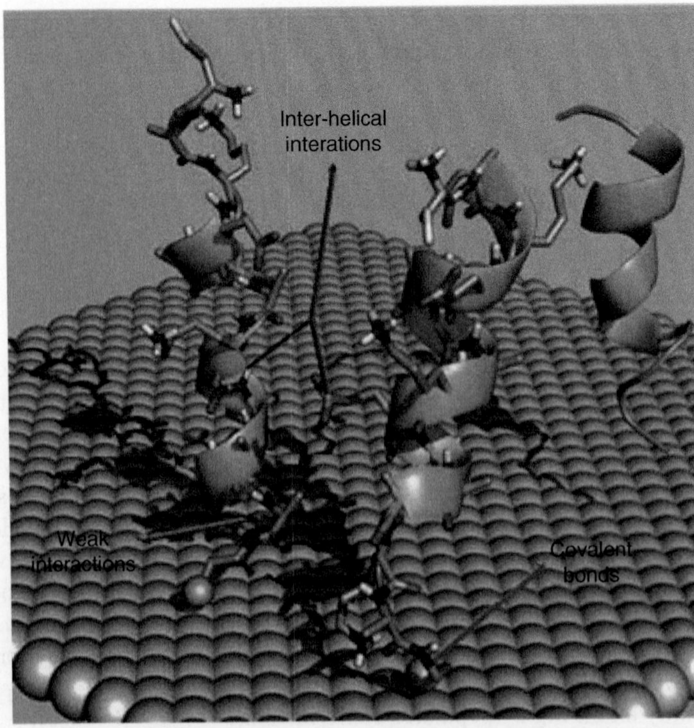

Figure 3.8 *Molecular scheme depicting the main interactions typically found in hybrid systems formed by helical peptides attached to inorganic surfaces*

electrostatic potential (RESP) strategy [62, 63]. This methodology derives the atom-centered point charges from a fit to the QM-computed molecular electrostatic potential at the molecular surface.

The modeling of new hybrid materials made of inorganic/organic–peptide conjugates is an emerging field in the frontier between Material Science and Biological Chemistry. Successful development of this field requires integrative approaches that combine concepts of very different disciplines. For example, modeling of peptides tethered to surfaces should be carried out using molecular models able to capture effects like the surface charge dependencies, the peptide dipole moment, the peptide size, the solution ionic strength, etc. [64]. For this purpose, the surface model should be defined as a single crystal, a simple nanoparticle, or a complex nanostructure, with specific definition and parameterization of surface–peptide interactions being required in all cases (Figure 3.8) [64–68].

3.6.2 Practical Strategies for the Application of Molecular Dynamics Techniques

MD simulations, in spite of being restricted to predefined starting points, should be able to overcome energy barriers and explore different regions of the conformational space. A common strategy used in MD to reach different regions of the conformational space is to perform the simulation at high unrealistic temperatures. This procedure avoids getting stuck in a local minimum of the conformational space and several structures are extracted from the trajectory to proceed with the energy minimization at regular intervals of simulation time. However, identification of the global energy minimum as well as of the more representative local minima can be carried out using smarter procedures based on MD. Among them, simulated annealing (SA) and replica exchange (RE) are two of the most widely used.

SA-MD [69] is a computational strategy that mimics the annealing technique in metallurgy, which involves heating and controlled cooling of materials to increase the size of its crystals and reduce their defects. In the conformational search, the internal energy is the cost function used to evaluate different structures while the temperature plays the role of control parameter. Thus, starting from a very high temperature, the system is allowed to reach the thermal equilibrium. Initially, the high temperatures facilitate the visit of very different conformational states, even those of high energy, since the barriers that separate them are easily crossed. This ensures a very reasonable exploration range. When the temperature is forced to decrease, the stable conformational states become more probable. Ideally, at the absolute zero the system should reach the global minimum of energy, even though, in practice, the SA-MD technique does not always converge into the most stable region at the end of the simulation.

In order to overcome the limitations of the SA-MD, a modified procedure based on minimization of the structures generated during the SA-MD cycles has been proposed [70]. Accordingly, the modified SA-MD technique was found to be robust enough to obtain conformations close in energy to the global minimum but located at different regions of the potential energy hypersurface [65, 71]. In this modified strategy, the practical process can be summarized as follows: (a) starting from an initial structure, this is quickly heated until a high temperature is reached and, subsequently, slowly cooled to an intermediate temperature; (b) after reaching the thermal equilibrium, a set of structures are extracted and minimized; (c) the minima resulting from this process are stored in a rank-ordered library of unique minimum energy conformations; (d) the lowest energy structure is used as the initial conformation of a new SA-MD cycle, which starts at (a). The process is completed when the last SA-MD cycle does not contribute with any new structure to the library of a unique minimum energy conformation.

RE-MD, also called parallel tempering MD, is another technique focused on the identification of the global minimum energy conformation, overcoming the multiple-minima problem [72]. This procedure was initially proposed by Swendsen and Wang [73], which involved a set of Monte Carlo simulations at different temperatures, being later extended to MD algorithms. The RE-MD method, which was recently reviewed by Earl and Deem [74], consists of the simulation of a given number of replicas of the system under study, trajectories being run at different temperatures for the same time length. Every given time, a replica swap is attempted between two different temperatures, using a Metropolis criterion for the exchange acceptance. Each replica at a high temperature samples a large number of conformations in the phase space while each replica at a low temperature searches in a local region of the phase space. The exchange of complete configurations between replicas at consecutive temperatures is the key to obtain a good conformational sampling. In fact, the lower temperature replicas have access to a wide region of phase space because of the inclusion of the higher temperature replicas. Parallel execution of different temperature replicas requires a huge amount of computational resources with respect to other conformational search methods, like SA-MD. Moreover, the efficiency of RE-MD grows with the number of replicas and, therefore, with the computational cost. In spite of this, if enough computer resources are available, this technique may allow a sampling of the phase space more complete than procedures based on a single temperature simulation, like, for example, the conventional SA-MD.

3.6.3 Analysis of the Simulation Data

Once the simulation has been carried out, extraction of the desired information in the most useful way requires the suitable processing of the recorded trajectory. Some of the most common analyses are discussed in this section.

3.6.3.1 Root-Mean-Square Deviation (RMSD)

The RMSD is typically used to measure the similarity between different structures from a statics (average value) or dynamics (temporal evolution) point of view. Since the secondary structure of peptides is defined by the position of the backbone atoms, the RMSD determines the average distance between the backbone atoms of the two compared structures.

3.6.3.2 End-to-End Distance

The end-to-end distance in a peptide is defined as the distance separating the N- and C-termini. This value is equivalent to the sum of all vectors connecting consecutive atoms along the peptide backbone. The temporal evolution of the end-to-end distance allows the identification of important conformational reorganizations when large changes are detected.

3.6.3.3 Accumulated Ramachandran Plots

Graphical representation of the cumulative plot of the dihedral angles labeled φ and ψ is used to ascertain the secondary structures visited by each residue during the trajectory.

3.6.3.4 Analysis of Hydrogen Bonds

IUPAC defines a hydrogen bond as '*an attractive interaction between a hydrogen atom from a molecule or a molecular fragment X–H in which X is more electronegative than H, and an atom Y in the same or a different molecule, in which there is evidence of bond formation*' [75]. This type of intermolecular interaction, which is stronger than the van der Waals one, is frequently responsible for secondary and tertiary structures of peptides and proteins. Thus, detailed analyses of hydrogen bonds involved on the MD trajectory (e.g. time series, occupancy, and lifetimes of defined hydrogen bonds) are used to illustrate the source of stabilization of a given conformation. The geometric parameters usually employed to define the existence of hydrogen bonds are: X–H$\cdots$Y lower than 2.50Å and $\angle$X–H$\cdots$Y higher than 120º, even though the exact values depend on the chemical nature of X and Y.

3.6.3.5 Cluster Analysis

Clustering is a useful methodology commonly employed to put some light among the huge amount of structural information provided not only by long MD trajectories but also by search procedures like SA-MD and RE-MD. Clustering is the task of categorizing a set of structures into specific groups, named clusters. All the structures that belong to a specific cluster have similar metrics values (e.g. RMSD, distance matrix error, existence of salt bridges and/or hydrogen bonds, secondary structure, and/or dihedral angles), making them easier to be identified from other cluster structures. It is worth noting that clustering does not follow a universal set of criteria but depends on the particular features that characterize each studied conformational population, which conditions the metrics used to define a cluster and the efficiency to find them [76, 77].

3.6.4 Peptide Dynamics

MD has been widely used to study the conformational preferences of peptides, some examples being described in this section. More specifically, the examples selected for this section correspond to the helical conformation of poly(γ-D-glutamic acid), abbreviated PGDGA, its stoichiometric complexes formed with dodecyltrimethylammonium cations, and some synthetic peptides that have been proposed as potential candidates for applications related to the construction of molecular machines.

PGDGA (Figure 3.9a) is an exopolypeptide naturally produced by a variety of species of genus *Bacillus*, like *Bacillus lichenformis* and *Bacillus subtilis* [78]. This poly(γ-amino acid) has been extensively investigated in terms of its synthesis, biochemistry, and production under different environmental conditions [79–81]. The helical conformation of PGDGA was early investigated by atomic-resolution computer simulations [82, 83]. It was found that PGDGA forms a left-handed helix with 19-membered ring hydrogen bonds set between the CO of the amide group i and the NH of the amide group $i+3$, this conformation being fully compatible with the optical rotatory dispersion experiments [82]. More recently, MD simulations were carried out on relatively small homopeptides to study the influence of both the molecular length and the temperature on the stability of such helical secondary structures [84]. Simulations were performed considering two peptides made of 10 and 20

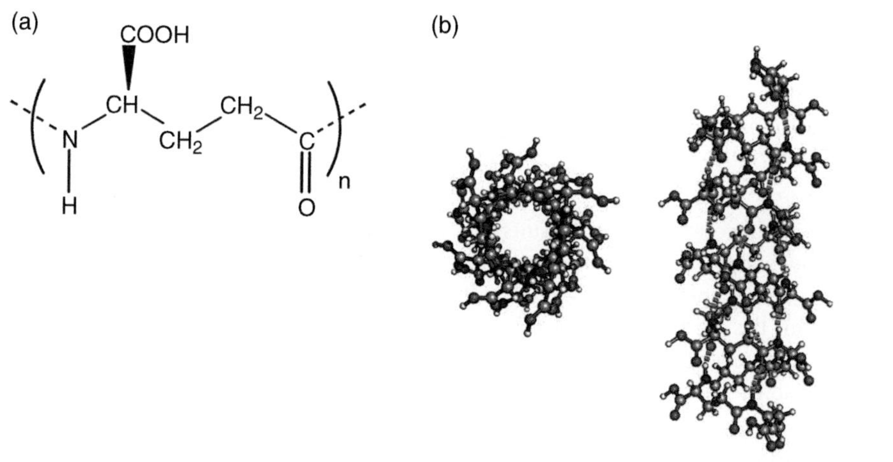

Figure 3.9 *(a) Chemical structure of poly(γ-D-glutamic acid). (b) Equatorial (left) and axial (right) helical conformations derived for PGDGA using atomic-resolution computer simulations*

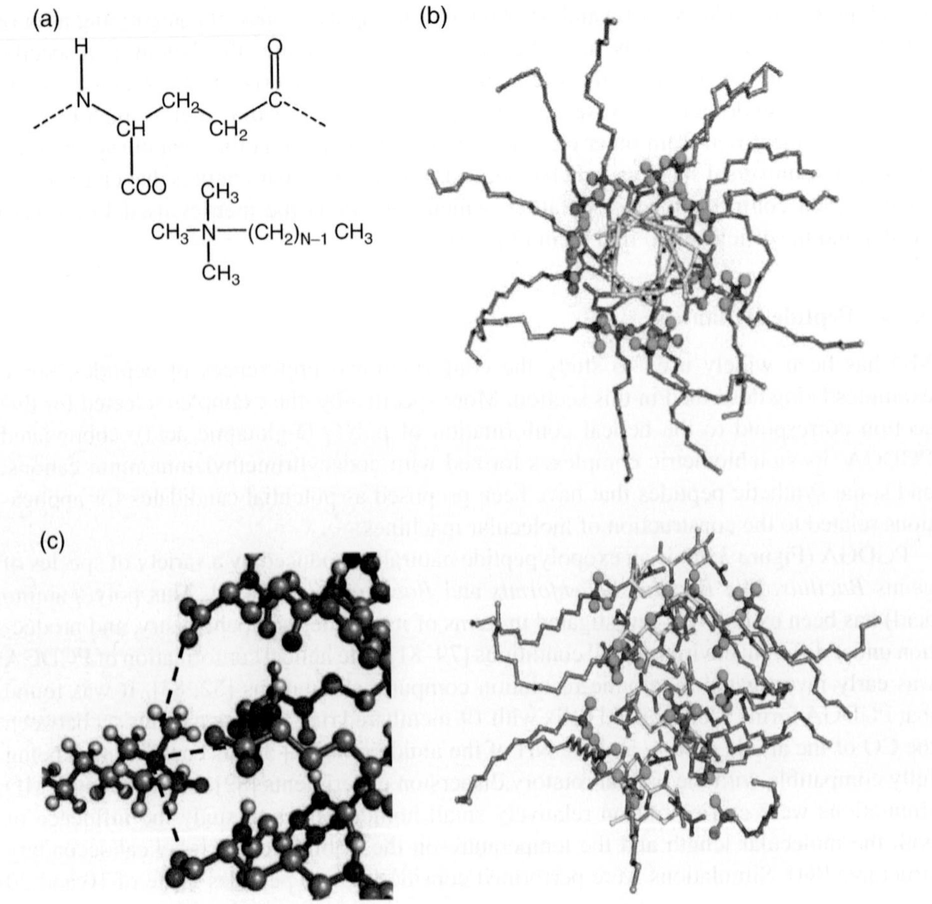

Figure 3.10 *(a) Chemical structure of self-assembled stoichiometric complexes formed by dodecyltrimethylammonium cations and ionized poly(γ-D-glutamic acid). (b) Axial (top) and equatorial (bottom) views of the helical arrangement adopted by the poly(γ-D-glutamic acid) in the complexes. (c) Detailed view displaying the multiple surfactant··· carboxylate interaction (i.e. the polar head group of the each surfactant molecule interacts simultaneously with several carboxylate groups)*

residues in aqueous solution. Results reveal that, independently of the temperature, the helix is not stable for the smallest homopeptide, precluding the use of small oligopeptides in technological applications where the helical conformation plays a crucial role. On the other hand, the helix was found to be stable for the homopeptide made of 20 residues (Figure 3.9b), even though the temperature induced a conformational transition from the helical state to the random coil. In spite of this analysis, the energy of the whole system indicated that the helix is more stable than the random coil state by about 5 kcal/mol.

Within this context, MD simulations were also used to investigate the supramolecular organization of self-assembled stoichiometric complexes formed by dodecyltrimethylammonium cations and ionized PGDGA in chloroform solution (Figure 3.10a) [85]. Results indicated that the polypeptide chain tends to adopt an α-like helix conformation (Figure 3.10b) similar to that obtained for PGDGA in aqueous solution. This secondary structure was found to be responsible for the homogeneous distribution of the surfactant molecules around the peptide chain, enabling the attractive interactions to be maximized between oppositely charged groups while minimizing the repulsions between ions of equal charge. A multiple interaction pattern involving surfactant···carboxylate interaction, and in some cases surfactant···amide interactions, was identified for the polar head group of the surfactant molecules (Figure 3.10c). However, this multiplicity, which was previously observed in surfactant···polypeptide complexes made of poly(α-glutamic acid) [86, 87], does not alter the stability of the helical conformation.

On the other hand, among potentially useful candidates to build organic frames for molecular machines, peptides and polypeptides stand out over many other flexible macromolecules. To design peptide-based molecular devices with selective conformational features, a major goal is to obtain detailed information on the relationships between their properties and external factors. The activity of many research groups focused on this field allowed the assessment of the conformational tendencies of a large number of peptides depending on their chemical composition. This effort involved both their 3D structural diversity and the chemical features of the amino acid building blocks [88]: in particular, such studies included not only standard residues but synthetic nc-aa, as well as multiple variations of their basic structure. Within this context, the conformational tendencies of $C^{\alpha,\alpha}$-diethylglycine (Deg)-based peptides (Figure 3.11a) were recently studied in chloroform solution using MD simulations [89]. Specifically, the conformational effects of breaking the symmetry of the host Tfa–$(Deg)_5$–OtBu (Tfa, trifluoroacetyl; OtBu, tertbutoxy) pentapeptide with punctual replacements at different sequence positions of one Deg residue by its corresponding guest chiral analog, L-α-aminobutyric acid (L-Abu; Figure 3.11a), were examined by simulating the following peptides: Tfa–$(Deg)_5$–OtBu, Tfa–$(Deg)_2$–L-Abu–$(Deg)_2$–OtBu, Tfa–$(Deg)_3$–L-Abu–Deg–OtBu, and Tfa–$(Deg)_4$–L-Abu–OtBu. Simulations showed that only the Deg homopeptide is able to stabilize a 2.0_5 helix (Figure 3.11b), even though a kinked arrangement with all the Deg residues adopting a fully extended conformation was found to be stable when the L-Abu residue is introduced in the middle of the sequence (Figure 3.11c). On the other hand, when the L-Abu residue is closer to the C-end of the sequence, the peptide chain prefers a partially folded 3_{10}-helix (Figure 3.11d). Additional simulations on Tfa–$(Deg)_3$–L-Abu–$(Deg)_3$–OtBu highlighted that, when the size of the Deg segments increases, their tendency to adopt a 2.0_5 helix predominates over the preferred folded conformation of L-Abu. The overall picture, which was extracted after more than 300 ns of MD simulation, was that breaking the α-carbon symmetry of achiral Cα-tetrasubstituted amino acids is a promising strategy to build up polypeptides with modulated conformational tendencies.

3.6.5 Hybrid Peptide Dynamics

Hybrid biomaterials are considered as a promising new way to attain goals that include: environmental friendship, energy-use efficiency, inexpensiveness, reliability, endurance, self-organization, and biocompatibility. A hybrid material combines either two different compounds or parts of them to generate a chemical 'chimera' that has new emerging properties. Hybrid materials are also found in nature (i.e. bone, cartilage, teeth) and are open to a

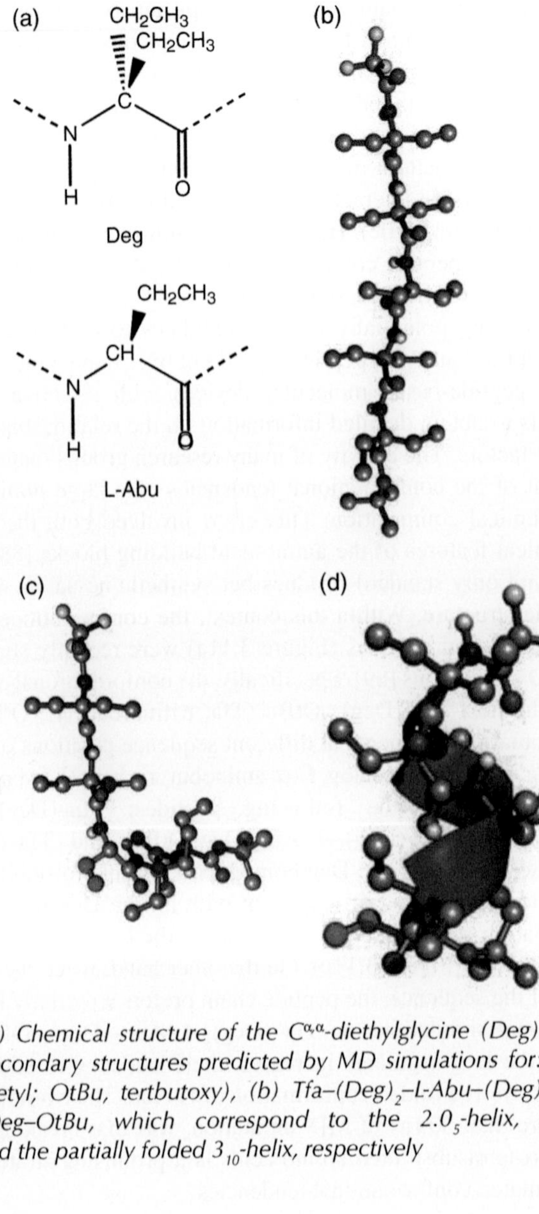

Figure 3.11 *(a) Chemical structure of the $C^{\alpha,\alpha}$-diethylglycine (Deg) and L-α-aminobutyric acid (L-Abu). Secondary structures predicted by MD simulations for: (b) Tfa–(Deg)$_5$–OtBu (Tfa, trifluoroacetyl; OtBu, tertbutoxy), (b) Tfa–(Deg)$_2$–L-Abu–(Deg)$_2$–OtBu, and (c) Tfa–(Deg)$_3$–L-Abu–Deg–OtBu, which correspond to the 2.0$_5$-helix, the kinked extended arrangement, and the partially folded 3$_{10}$-helix, respectively*

wide range of possible combinations according to their chemical nature and properties. The size scale for these materials is highly variable, going from nanometer to millimeter, thus aiming at an endless number of possible applications and material types to be considered.

According to the chemical nature of the interaction, hybrid biomaterials are categorized into two main groups:

- *Class I.* These are hybrid materials whose parts are held together by 'weak' chemical interactions such as a hydrogen bond and van der Waals interactions.

- *Class II.* These are materials whose parts are bonded by 'strong' covalent, ionic, and/or metallic linkages.

Class I hybrids are typically used as sensing tools because the formation of weak interactions favors their application in detection (i.e. the sensing molecules and the detected substance form the hybrid conjugate). According to the chemical nature of their parts, class II hybrids can be sorted into organic–organic, organic–inorganic, and inorganic–inorganic. These classes embed many materials as potentially hybrid, even though this section is specifically focused on peptides attached to inorganic surfaces (i.e. class II, organic–inorganic) and peptide–polymer conjugates (i.e. class II, organic–organic).

3.6.5.1 Dynamics of Peptides Tethered to Inorganic Surfaces

Hybrid systems made of peptides tethered to a surface or nanoparticle can be very attractive because of their potential applications, especially in biomedicine. An interesting example can be found in the small pentapeptide named CREKA (Cys–Arg–Glu–Lys–Ala), which specifically homes to tumors by binding to fibrin and fibrin-associated clotted plasma proteins (e.g. fibronectin) in tumor vessels [90]. Thus, this peptide, which was identified by screening phage-displayed libraries to discover specific targets in tumor vessels [91], does not recognize tumors grown in mutant mice null for fibrogen or in mice lacking plasma fibronectin that becomes covalently bound to fibrin during blood clotting. In addition, it was established that CREKA linked to amino dextran-coated superparamagnetic iron oxide (SPIO) nanoparticles not only binds to blood and plasma but also effectively induces further localized tumor clotting [90]. This amplification system enhanced homing of the nanoparticles in a mouse tumor model without causing clotting or other obvious side effects elsewhere in the body, producing enhancement of tumor imaging. It was further found that the chemical nature of the nanoparticle is not important for this activity because both CREKA-SPIO nanoparticles and CREKA-coated liposomes were found to cause clotting in the tumor vessels [90].

Recently the energy landscape and bioactive conformation of CREKA and the same peptide labeled with fluorescein, the fluorescent dye attached to the Cys through a flexible linker, were determined using the modified SA-MD procedure described in the previous section [66, 92]. In addition, a system that mimics the peptide attached to the nanoparticle by tethering it to a surface was also considered [66]. More specifically, the sulfhydryl group of the Cys residue was attached to one spherical particle at the center of a surface made of 10×10 particles. The artificial surface represented a hard and reactive wall. Results indicated that the conformational profile of the REKA sequence is similar in all cases, the Cys residue providing distinctive features in each case. The bioactive conformation proposed for CREKA, which is displayed in Figure 3.12a, presents both a β-turn motif and strong interactions involving the side groups of Arg, Glu, and Lys. This conformation, which was found to be very stable by performing additional MD simulations, is appropriate to form intermolecular interactions with a receptor. Thus, its shape looks like a pocket with the charged groups pointing outwards. Moreover, this bioactive conformation was found to be among the most stable conformations of CREKA independently of the density of peptides attached to the surface [68] and the chemical nature of the surface [67].

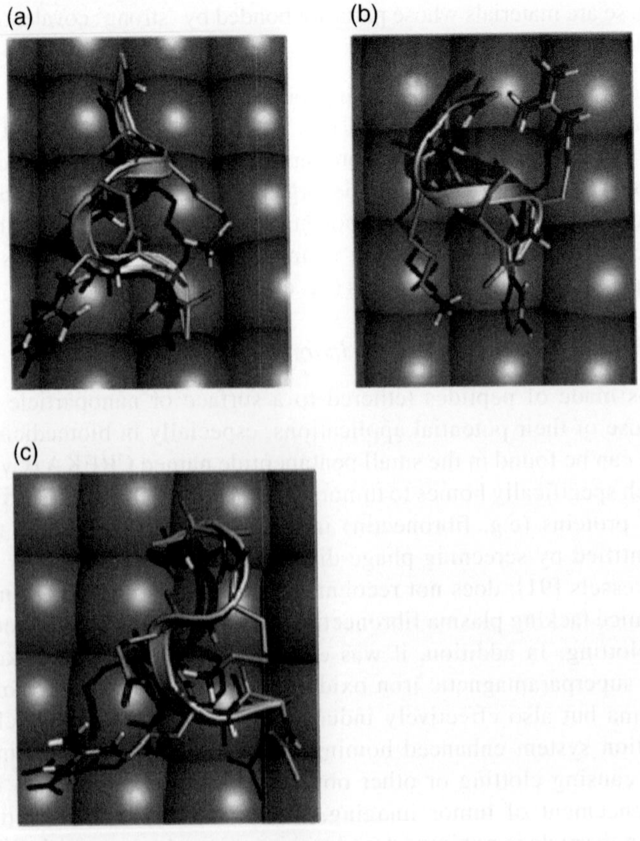

Figure 3.12 *(a) Bioactive conformation proposed for CREKA and its analogs, (b) CR-αMeE-KA, and (c) CR-NMeE-KA (αMeE and NMeE refer to α-methyl- and N-methyl-glutamic acid, respectively) attached to a nanoparticle. These conformations were determined by applying the modified SA-MD procedure (see color plate figure)*

Very recently the bioactive conformation of CREKA was used to design some analogs using computational tools, which were generated by replacing an amino acid in the sequence by a nonproteinogenic counterpart [71, 93, 94]. Thus, the goodness of the designed CREKA analogs was defined by determining their energy landscape and bioactive conformations through intensive application of the modified SA-MD and RE-MD procedures (some examples are displayed in Figure 3.12b). Although the initial aims were both to enhance the tumor imaging activity of CREKA by stabilizing the bioactive conformation and to impart protection against proteolytic resistance by incorporating nc-aa, a much more interesting application from a medical point of view was found for some of the designed analogs [93]. Specifically, combination of the self-amplifying homing system formed by CREKA-coated SPIO with nanoparticles coated with another tumor-homing peptide and nanoparticles with an elongated shape allowed not only effective imaging of the tumors but also produced extensive inhibition of tumor growth by obstructing tumor circulation through blood clotting. The efficacy of this theranostic nanosystem was increased by replacing some residues in CREKA by nonproteinogenic counterparts (i.e. *N*-methyl and Cα-methyl analogs), which increased the stability of the peptide in the circulation. Thus, such CREKA analogs substantially improved the tumor-homing efficiency and increased the blockage of tumor vessels. Treatment of mice with prostate cancer with multiple doses of theranostic nanoparticles with CREKA analogs induced tumor necrosis and a highly significant reduction in tumor growth [93].

3.6.5.2 Dynamics of Peptide–Polymer Conjugates

Among class II organic–organic hybrid materials, bioinspired block copolymers are of fundamental importance due to their huge variety. These materials have been extensively reviewed as '*macromolecular chimeras*' [95–97] due to the frequent combination of biological fragments (e.g. protein segments, peptides, and nucleic acids) and synthetic polymers (e.g. polyesters, polyamides, and conducting polymers). Moreover, they might combine either two biological fragments or two synthetic polymers as long as they resemble biological macromolecules. Many promising applications for these hybrids have been reported in many different fields such as biomedicine [98, 99], material science [100, 101], and nanomedicine [102, 103].

Within this context, peptide-containing amphiphiles have recently attracted special attention since they combine the specificity of peptide interactions (e.g. targeted binding to proteins) with properties inherent to amphiphiles, such as the capacity to self-assemble into complex ordered aggregates, which can range from micelles to nanofibrils [104–106]. Thus, systems based on fragments of amyloid beta peptide (hereafter named Aβ) present a high propensity to form fibrils (e.g. the KLVFF segment, which is part of the hydrophobic core from residues 16 to 20, has been identified as one of the key sequences for the fibrillization of the full Aβ peptide) [107]. Recently a new hybrid material was obtained by conjugating the FFKLVFF peptide, which is Aβ(16–20), KLVFF, extended by two phenylalanine (Phe) residues, with the highly hydrophilic poly(ethylene glycol), hereafter abbreviated to PEG [108]. While FFKLVFF is highly hydrophobic and does not dissolve in water, its conjugation with PEG leads to an amphiphilic molecule that forms self-assembled fibrillar structures in dilute aqueous solution and nematic and hexagonal columnar liquid crystal phases at higher concentration [109, 110].

The presence of a charged residue in the amyloidogenic core of the peptide segment precluded the complete understanding of which residues held the key to the association process. In an effort to understand the role of different intermolecular interactions in the self-assembly, a new hybrid system was prepared by combining PEG with a peptide comprising four consecutive Phe residues (i.e. FFFF-PEG) [111]. The lack of charged or polar residues in the peptide sequence renders FFFF-PEG an ideal model system comprising a short but very hydrophobic peptide moiety. Moreover, conjugation of PEG conferred amphiphilic properties to the new material.

A combination of QM calculations and MD simulations was recently used to examine the intrinsic conformational properties of the FFFF segment and how its conjugation with the hydrophilic PEG affects the conformational preferences of both the hydrophobic peptide and the inner hydrodynamic properties of the polymer [112]. Specifically, the inner conformational properties of the hydrophobic peptide were investigated through an extensive and systematic conformational search using DFT calculations. A building-up strategy, in which the conformational exploration can be framed to those regions of the

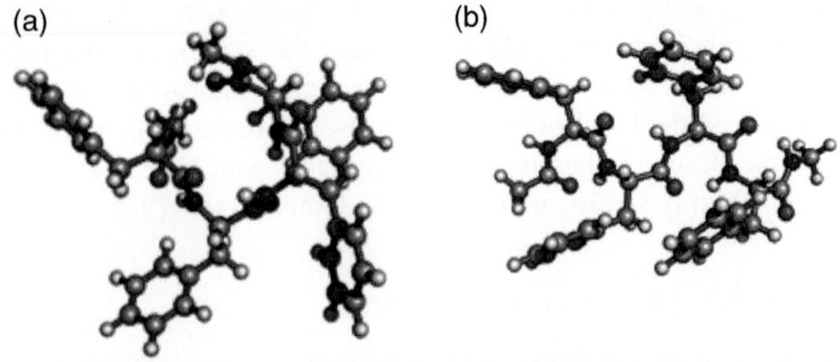

Figure 3.13 Lowest energy minimum conformations predicted by QM calculations for a tetrapeptide comprising four consecutive Phe residues (FFFF): (a) a 2.2_7-ribbon structure, in which the four residues adopt a γ-turn motif, and (b) a helix formed from alternate conventional and reverse γ-turn motifs

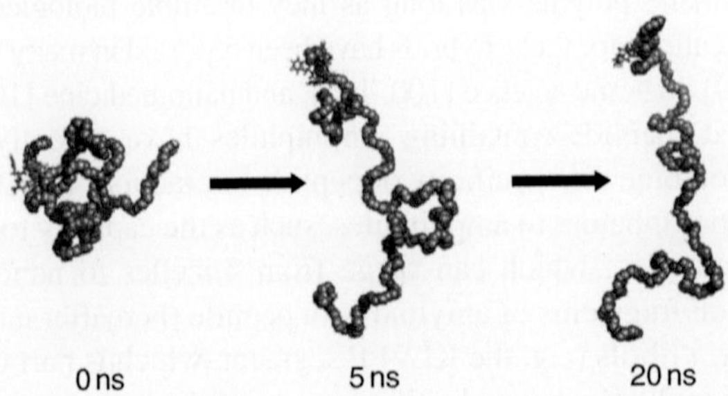

Figure 3.14 Structures of the FFFF-PEG amphiphile taken at the beginning of the simulation and after 5.0 and 20.0 ns of MD

energy hypersurface that enclose the lower energy structures, depicted a scenario that showed a very marked propensity towards arrangements that are the outcome of expanding turn-like conformations throughout the peptide chain. Thus, the most stable arrangements found for the FFFF peptide featured regular organizations that either propagated a γ-turn motif from residue one to four, forming a 2.2_7-ribbon structure (Figure 3.13a), or alternate conventional and reverse γ-turns that made up a helical-like arrangement (Figure 3.13b).

The structural organization of the complete amphiphile, FFFF-PEG, was characterized in dilute aqueous solution by means of MD simulations [112]. Analysis of the structural dynamics of the PEG segment revealed a radical conformational change, going from a sphere-like organization to a partially unfolded state with an elongated shape. Thus, in order to maximize the contacts with the polar solvent, PEG rapidly unfolds in aqueous solution, this behaviour being analogous to that described for this polymer when it is not part of an amphiphile [113, 114]. These features were monitored by analyzing the temporal evolution of the three spatial components of the radius of gyration ($R_{G,xx}$, $R_{G,yy}$, and $R_{G,zz}$) through the MD trajectory. It was noted that $R_{G,xx} \approx R_{G,yy} \approx R_{G,zz}$ at the first stages of the simulation conformation, this situation changing to $R_{G,xx} \approx R_{G,yy} < R_{G,zz}$ at the end of the simulation. The conformational changes undergone by the PEG chain are illustrated in Figure 3.14, which shows selected snapshots recorded during the trajectory.

Comparison of the results provided by the MD simulation of the FFFF-PEG system with those obtained for the two individual fragments indicated that the conformational preferences of peptide inserted in the amphiphile do not seem to be influenced by the hydrophilic tail [112]. Starting from a completely extended conformation, the peptide rapidly evolved towards a structure very similar to the more stable conformations predicted by QM for the isolated peptide. The hydrodynamic parameters (i.e. radius of gyration and hydrodynamic radius) shown by the hydrophilic counterpart were slightly influenced by the hydrophobic moiety, which affects the number of waters of hydration. In spite of this, it was considered that both counterparts organize themselves as independent modules, which is in agreement with the reported experimental observations when the amphiphilic material aggregates [111]. This behavior showed that this amphiphilic material is capable of independently preserving the inner physical properties of both opposite counterparts and anticipates interesting applications as a self-assembled matrix for encapsulating hydrophobic molecules in polar environments.

3.7 Summary

This chapter presents the principles and concepts of the main computational methodologies currently used to explore the conformational preferences of peptides and peptide-containing hybrid systems. Some of these methods are based on rigorous QM calculations that are very demanding from a computational point of view, being consequently limited to systems of small size. In spite of this limitation, they are currently used to examine systematically the conformational preferences of peptides containing less than four residues. Within this context, QM studies on nc-aa are particularly important because the conformational properties of these systems, which can be considered as hybrid molecules containing conventional α-amino acids and organic fragments, are frequently very different from those of proteinogenic α-amino acids. Moreover, QM calculations on nc-aa recently resulted in a data base denoted NCAD, which can be considered a very useful tool for the targeted design of synthetic mutations on peptides and proteins. The main characteristics and concepts related to this data base have also been described in this chapter. On the other hand, the principles of conventional MD as well as of the sampling techniques based on this methodology (i.e. SA-MD and RE-MD) have been discussed. Their applicability to simply study the structural dynamics as well as to explore the conformational preferences of different types of peptide and peptide-containing hybrid materials (e.g. peptides tethered to inorganic surfaces, self-assembled stoichiometric complexes formed by charged polypeptides and surfactants, and peptide–polymer conjugates) has been illustrated with numerous examples. Both concepts and examples have been presented considering a pedagogic level suitable for advanced graduation students and nonexpert postgraduation researchers interested in the application of computational tools to investigate peptide materials.

Acknowledgements

This work has been supported by MICINN and FEDER (MAT2009-09138) and Generalitat de Catalunya (Research Group 2009 SGR 925 and XRQTC). Support for the research of C.A. was received through the prize 'ICREA Academia' for excellence in research funded by the Generalitat de Catalunya.

References

[1] Lewars, E. (2004) Computational Chemistry: Introduction to the Theory and Applications of Molecular and Quantum Mechanics. Kluwer Academic Publishers, Dordrecht.

[2] Jensen, F. (2007) Introduction to Computational Chemistry, 2nd edition. John Wiley & Sons, Ltd, Chichester.

[3] Szabo, A., Ostlund, N.S. (1982) Modern Quantum Chemistry: Introduction to Advanced Electronic Structure Theory. Macmillan Publishing Co., Inc., New York.

[4] Sherrill, C.D., Schaefer, H.F. (1999) The configuration interaction method: advances in highly correlated approaches. Adv. Quant. Chem. 34: 143–269.

[5] Møller, C., Plesset, M.S. (1934) Note on an approximation treatment for many-electron systems. Phys. Rev. 46: 618–622.

[6] Coester, F., Kümmel, H. (1960) Short-range correlations in nuclear wave functions. Nucl. Phys. 17: 477–485.

[7] Paldus, J., Čížek, J., Shavitt, I. (1972) Correlation problems in atomic and molecular systems. IV. Extended coupled-pair many-electron theory and its applications to the BH_3 molecule. Phys. Rev. A 5: 50–67.

[8] Dewar, M.J.S., Zoebisch, E.G., Healy, E.F., Stewart, J.J.P. (1985) AM1: a new general purpose quantum mechanical molecular model. J. Am. Chem. Soc. 107: 3902–3909.

[9] Stewart, J.J.P. (1989) Optimization of parameters for semi-empirical methods I-method. J. Comp. Chem. 10: 209–220.

[10] Hohenberg, P., Kohn, W. (1964) Inhomogeneous electron gas. Phys. Rev B 136: B864–B871.

[11] Kohn, W., Sham, L.J. (1965) Self-consisting equations including exchange and correlation effects. Phys. Rev. 140: A1133–A1138.

[12] Vosko, S.H., Wilk, L., Nusair, M. (1980) Accurate spin-dependent electron liquid correlation energies for local spin density calculations: a critical analysis. Can. J. Phys. 58: 1200–1211.

[13] Perdew, J.P., Zunger, A. (1981) Self-interaction correction to density-functional approximations for many-electron systems. Phys. Rev. B 23: 5048–5079.

[14] Becke, A.D. (1996) Density-functional thermochemistry. IV. A new dynamical correlation functional and implications for exact-exchange mixing. J. Chem. Phys. 104: 1040–1046.

[15] Perdew, J.P. (1986) Density-functional approximation for the correlation energy of the inhomogeneous electron gas. Phys. Rev. B 33: 8822–8824.

[16] Perdew, J.P., Burke, K., Ernzerhof, M. (1986) Generalized gradient approximation made simple. Phys. Rev. Lett. 77: 3865–3868.

[17] Perdew, J.P., Wang, Y. (1992) Accurate and simple analytic representation of the electron gas correlation energy. Phys. Rev. B 45: 13244–13249.

[18] Lee, C., Yang, W., Parr, R.G. (1988) Development of the Colle–Salvetti correlation-energy formula into a functional of the electron density. Phys. Rev. B 37: 785–789.

[19] Becke, A.D. (1993) A new mixing of Hartree–Fock and local density-functional theories. J. Chem. Phys. 98: 1372–1377.

[20] Lynch, B.J., Fast, P.L., Harris, M., Truhlar, D.G. (2000) Adiabatic connection for kinetics. J. Phys. Chem. A 104: 4811–4815.

[21] Schultz, N.E., Zhao, Y., Truhlar, D.G. (2005) Density functionals for inorganometallic and organometallic chemistry. J. Phys. Chem. A 109: 11127–11143.

[22] Xu, X., Goddard, W.A. (2004) The X3LYP extended density functional for accurate descriptions of nonbond interactions, spin states, and thermochemical properties. Proc. Natl Acad. Sci. USA 101: 2673–2677.

[23] Boese, A.D., Martin J.M.L. (2004) Development of density functionals for thermochemical kinetics. J. Chem. Phys. 121: 3405–3416.

[24] Alemán, C. (1999) Hydration of cytosine using combined discrete/SCRF models: influence of the number of discrete solvent molecules. Chem. Phys. 244: 151–162.

[25] Siboulet, B., Marsden, C.J., Vitorge, P. (2006) A theoretical study of uranyl solvation: explicit modelling of the second hydration sphere by quantum mechanical methods. Chem. Phys. 326: 289–296.

[26] Miertuš, S., Scrocco, E., Tomasi, J. (1981) Electrostatic interaction of a solute with a continuum. A direct utilization of ab initio molecular potentials for the prevision of solvent effects. Chem. Phys. 55: 117–129.

[27] Alemán, C., Zanuy, D., Jiménez, A.I., Cativiela, C., Haspel, N., Zheng, J., Casanovas, J., Wolfson, H., Nussinov, R. (2006) Concepts and schemes for the re-engineering of physical protein modules: generating nanodevices via targeted replacements with constrained amino acids. Phys. Biol. 3: S54–S62.

[28] Alemán, C. (1997) Conformational properties of α-amino acids disubstituted at α-carbon. J. Phys. Chem. B 101: 5046–5050.

[29] Casanovas, J., Zanuy, D., Nussinov, R., Alemán, C. (2007) Intrinsic conformational characteristics of α,α-diphenylglycine. J. Org. Chem. 72: 2174–2181.

[30] Casanovas, J., Nussinov, R., Alemán, C. (2008) Intrinsic conformational preferences of Cα·α-dibenzylglycine. J. Org. Chem. 72: 4205–4211.

[31] Alemán, C., Jiménez, A.I., Cativiela, C., Pérez, J.J., Casanovas, J. (2002) Influence of the phenyl side chain on the conformation of cyclopropane analogues of phenylalanine. J. Phys. Chem. B 106: 11849–11858.

[32] Casanovas, J., Zanuy, D., Nussinov, R., Alemán, C. (2006) Identification of the intrinsic conformational properties of 1-aminocyclobutane-1-carboxylic acid. Chem. Phys. Lett. 429: 558–562.

[33] Alemán, C., Zanuy, D., Casanovas, J., Cativiela, C., Nussinov, R. (2006) Backbone conformational preferences and pseudorotational ring puckering of 1-aminocyclopentane-1-carboxylic acid. J. Phys. Chem. B 110: 21264–21271.

[34] Rodríguez-Ropero, F., Zanuy, D., Casanovas, J., Zanuy, D., Nussinov, R., Alemán, C. (2008) Application of 1-aminocyclohexane carboxylic acid to protein nanostructure computer design. J. Chem. Inf. Model. 48: 333–343.

[35] Hughes, A.B. (2009) Amino Acids, Peptides and Proteins in Organic Chemistry, Volume 1: Origins and Synthesis of Amino Acids. Wiley-VCH: Weinheim.

[36] Jiménez, A.I., Vaquero, V., Cabezas, C., López, J.C., Cativiela, C., Alonso, J.L. (2011) The singular gas-phase structure of 1-aminocyclopropanecarboxylic acid (Ac(3)c). J. Am. Chem. Soc. 133: 10621–10628.

[37] Sayago, F.J., Laborda, P., Calaza, M.I., Jiménez, A.I., Cativiela, C. (2011) Access to the cis-fused stereoisomers of proline analogues containing an octahydroindole core. Eur. J. Org. Chem. 11: 2011–2028.

[38] Cativiela, C., Ordóñez, M. (2009) Recent progress on the stereoselective synthesis of cyclic quaternary alpha-amino acids. Tetrahedron Asymm. 20: 1–63.

[39] Calaza, M.I., Cativiela, C. (2008) Stereoselective synthesis of quaternary proline analogues. Eur. J. Org. Chem. 20: 3427–3448.

[40] Cativiela, C., Díaz-De-Villegas, M.D. (2007) Recent progress on the stereoselective synthesis of acyclic quaternary alpha-amino acids. Tetrahedron Asymm. 18: 569–623.

[41] Ordóñez, M., Cativiela, C. (2007) Stereoselective synthesis of gamma-amino acids. Tetrahedron Asymm. 18: 3–99.

[42] Revilla-López, G., Torras, J., Curcó, D., Casanovas, J., Calaza, M.I., Zanuy, D., Jiménez, A.I., Cativiela, C., Nussinov, R., Grodzinski, P., Alemán, C. (2010) NCAD, a database integrating the intrinsic conformational preferences of non-coded amino acids. J. Phys. Chem. B 114: 7413–7422.

[43] Revilla-López, G., Rodríguez-Ropero, F., Curcó, D., Torras, J., Calaza, M.I., Zanuy, D., Jiménez, A.I., Cativiela, C., Nussinov, R., Alemán, C. (2011) Integrating the intrinsic conformational preferences of noncoded alpha-amino acids modified at the peptide bond into the noncoded amino acids database. Proteins 79: 1841–1852.

[44] Sayle, R. A., Milner-Whilte, E.J. (1995) Rasmol – biomolecular graphics for all. Trends Biochem. Sci. 20: 374–382.

[45] Curcó, D., Nussinov, R., Alemán, C. (2007) Coarse-grained representation of β-helical protein building blocks. J. Phys. Chem. B 111: 10538–10549.

[46] Curcó, D., Nussinov, R., Alemán, C. (2007) Coarse-graining the self-assembly of β-helical protein building blocks. J. Phys. Chem. B 111: 14006–14011.

[47] Billeter, M., Howard, A.E., Kuntz, I.D., Kollman, P.A. (1988) A new technique to calculate low-energy conformations of cyclic molecules utilizing the ellipsoid algorithm and molecular dynamics: application to 18-crown-6. J. Am. Chem. Soc. 110: 8385–8391.

[48] Alemán, C., Orozco, M. (1994) PAPQMD/AM1 parametrization of the bonded term of aromatic biomolecules. Biopolymers 34: 941–955.

[49] Alemán, C., Luque, F.J., Orozco, M. (1993) A new scaling procedure to correct semiempirical MEP and MEP-derived properties. J. Comput. Aided Mol. Design 7: 721–742.

[50] Alemán, C., Canela, E.I., Franco, R., Orozco, M. (1991) A new strategy for the evaluation of force parameters from quantum mechanical computations. J. Comput. Chem. 12: 664–674.

[51] Hockney, R.W., Eastwood, J.W. (1981) Computer Simulations Using Particles. McGraw Hill, New York.

[52] Verlet, L. (1967) Computer 'experiments' on classical fluids: I. Thermodynamical properties of Lennard–Jones molecules. Phys. Rev. 159: 98–103.

[53] van Gunsteren, W.F., Berendsen, H.J.C, Colonna, F., Perahia, D., Hollenberg, J.P., Lellough, D. (1984) On searching neighbours in computer simulations of macromolecular systems. J. Comput. Chem. 5: 272–279.

[54] Mazur, M.H., McCammon, J.A., Lybrand, T.P. (1990) New methodology for conjugate additions of allylic ligands to .alpha.,.beta.-unsaturated ketones: synthetic and spectroscopic studies. J. Am. Chem. Soc. 112: 4411–4419.

[55] Darden, T., York, D., Pedersen, L. (1993) Particle mesh Ewald: an N·long(N) method for Ewald sums in large systems. J. Chem. Phys. 98: 10089–10092.

[56] Berendsen, H.J.C., Postma, J.P.M., van Gunsteren, W.F., DiNola, A., Haak, J.R. (1984) Molecular-dynamics with coupling to an external bath. J. Chem. Phys. 81: 3684–3690.

[57] Cornell, W.D., Cieplak, P., Bayly, C.I., Gould, I.R., Merz, K.M., Ferguson, D.M., Spellmeyer, D.C., Fox, T., Caldwell, J.W., Kollman, P.A. (1995) A second generation force field for the simulation of proteins, nucleic acids, and organic molecules. J. Am. Chem. Soc. 117: 5179–5197.

[58] MacKerell, A.D., Bashford, D., Bellott, M., Dunbrack, R.L., Evanseck, J.D., Field, M.J., Fischer, S., Gao, J., Guo, H., Ha, S., Joseph-McCarthy, D., Kuchnir, L., Kuczera, K., Lau, F.T.K., Mattos, C., Michnick, S., Ngo, T., Nguyen, D.T., Prodhom, B., Reiher, W.E., Roux, B., Schlenkrich, M., Smith, J.C., Stote, R., Straub, J., Watanabe, M., Wiórkiewicz-Kuczera, J., Yin, D., Karplus, M. (1998) All-atom empirical potential for molecular modeling and dynamics studies of proteins. J. Phys. Chem. B 102: 3586–3616.

[59] Oostenbrink, C., Villa, A., Mark, A.E., van Gunsteren, W.F. (2004) A biomolecular force field based on the free enthalpy of hydration and solvation: the GROMOS force-field parameter sets 53A5 and 53A6. J. Comput. Chem. 25: 1656–1676.

[60] Jorgensen, W.L., Maxwell, D.S., Tirado-Rives, J. (1996) Development and testing of the OPLS all-atom force field on conformational energetics and properties of organic liquids. J. Am. Chem. Soc. 118: 11225–11236.

[61] Wang, J., Wolf, R.M., Caldwell, J.W., Kollman, P.A., Case, D.A. (2004) Development and testing of a general Amber force field. J. Comput. Chem. 25: 1157–1174.

[62] Bayly, C.I., Cieplak, P., Cornell, W., Kollman, P.A. (1993) A well-behaved electrostatic potential based method using charge restraints for deriving atomic charges: the RESP model. J. Phys. Chem. 97: 10269–10280.

[63] Cieplak, P., Cornell, W.D., Bayly, C., Kollman, P.A. (1995) Application of the multimolecule and multiconformational RESP methodology to biopolymers: charge derivation for DNA, RNA, and proteins. J. Comput. Chem. 16: 1357–1377.

[64] Gray, J.J. (2004) The interaction of proteins with solid surfaces. Curr. Opin. Struct. Biol. 14: 110–115.

[65] Sarikaya, M., Tamerler, C., Jen, A.K.Y., Schulten, K., Baneyx, F. (2003) Molecular biomimetics: nanotechnology through biology. Nat. Mater. 2: 577–585.

[66] Zanuy, D., Flores-Ortega, A., Casanovas, J., Curcó, D., Nussinov, R., Alemán, C. (2008) The energy landscape of a selective tumor-homing pentapeptide. J. Phys. Chem. B 112: 8692–8700.

[67] Curcó, D., Revilla-López, G., Alemán, C., Zanuy, D. (2011) Atomistic modeling of peptides bound to a chemically active surface: conformational implications. J. Pepti. Sci. 17: 132–138.

[68] Curcó, D., Zanuy, D., Nussinov, R., Alemán, C. (2011) A simulation strategy for the atomistic modeling of flexible molecules covalently tethered to rigid surfaces: application to peptides. J. Comput. Chem. 32: 607–619.

[69] Kirkpatrick, S., Gelatt, C.D., Vecchi, M.P. (1983) Optimization by simulated annealing. Science 220: 671–680.

[70] Baysal, C., Meirovitch, H. (1999) Efficiency of simulated annealing for peptides with increasing geometrical restrictions. J. Comput. Chem. 20: 1659–1670.

[71] Zanuy, D., Flores-Ortega, A., Jiménez, A.I., Calaza, M.I., Cativiela, C., Nussinov, R., Ruoslahti, E., Alemán, C. (2009) *In silico* molecular engineering for a targeted replacement in a tumor-homing peptide. J. Phys. Chem. B 113: 7879–7889.

[72] Sugita, Y., Okamoto, Y. (1999) Replica-exchange molecular dynamics method for protein holding. Chem. Phys. Lett. 314: 141–151.

[73] Swendsen, R.H., Wang, J.-S. (1986) Replica Monte Carlo simulation of spin glasses. Phys. Rev. Lett., 57: 2607–2609.

[74] Earl, D.J., Deem, M.W. 2005. Parallel tempering: theory, applications, and new perspectives. Phys. Chem. Chem. Phys. 7: 3910–3916.

[75] Arunan, E., Desiraju, G.R., Klein, R.A., Sadlej, J., Scheiner, S., Alkorta, I., Clary, D.C., Crabtree, R.H., Dannenberg, J.J., Hobza, P., Kjaergaard, H.G., Legon, A.C., Mennucci, B., Nesbitt, D.J. (2011) Definition of the hydrogen bond. Pure Appl. Chem. 83: 1637–1641.

[76] Sowdhamini, R., Blundell, T.L. (1995) Structure comparison of native and mutant human recombinant FKBP12 complexes with the immunosuppressant drug FK506 (tacrolimus). Protein Sci. 4: 506–520.

[77] Rui, X., Wunsch II, D. (2005) Survey of clustering algorithms. Neural Networks, IEEE Transactions on 16: 645–678.

[78] Thorne, C.B., Gómez, C.G., Noyes, H.E., Housewright, R.D. (1954) Production of glutamyl polypeptide by *Bacillus subtilis*. J. Bacteriol. 68: 307–315.

[79] Ko, Y., Gross, R.A. (1998) Effects of glucose and glycerol on γ-poly(glutamic acid) formation by *Bacillus licheniformis* ATCC 9945a. Biotechnol. Bioeng. 57: 430–437.

[80] Cromwick, A.-M., Gross, R.A. (1995) Effects of manganese (II) on *Bacillus licheniformis* ATCC 9945A physiology and gamma-poly(glutamic acid) formation. Int. J. Biol. Macromol. 17: 259–267.

[81] Pérez-Camero, G., Congregado, F., Bou, J.J., Muñoz-Guerra, S. (1999) Biosynthesis and ultrasonic degradation of bacterial poly(gamma-glutamic acid). Biotechnol. Bioeng. 63: 110–115.

[82] Zanuy, D., Alemán, C., Muñoz-Guerra, S. (1998) On the helical conformation of un-ionized poly(γ-D-glutamic acid). Int. J. Biol. Macromol. 23: 175–184.

[83] Zanuy, D., Alemán, C., Muñoz-Guerra, S. (2000) A microscopic view of a helical poly(γ-peptide): molecular dynamics simulations of a 20-residue un-ionized poly(γ-D-glutamic acid) in water. Macromol. Theory Simul. 9: 543–549.

[84] Zanuy, D., Alemán, C. (2001) Poly(γ-glutamic acid) in aqueous solution: molecular dynamics simulations of 10- and 20-residue chains at different temperatures. Biomacromolecules 2: 651–657.

[85] Zanuy, D., Alemán, C. (2007) Molecular dynamics study of complexes of poly(glutamate) and dodecyltrimethylammonium. Biomacromolecules 8: 663–671.

[86] Zanuy, D., Casanovas, J., Alemán, C. (2004) Thermal stability of the secondary structure of poly(α,L-glutamate) in self-assembled complexes as studied by molecular dynamics in chloroform solution. J. Am. Chem. Soc. 126: 704–705.

[87] Zanuy, D., Alemán, C. (2003) Molecular dynamics simulation of surfactant·poly(α,L-glutamate) complexes in chloroform solution: influence of the chemical constitution of the surfactant in the molecular organization. Langmuir 19: 3987–3995.

[88] Crisma, M., Formaggio, F., Moretto, A., Toniolo, C. (2006) Peptide helices based on alpha-amino acids. Biopolymers 84: 3–12.

[89] Torras, J., Zanuy, D., Crisma, M., Toniolo, C., Bertran, O., Alemán, C. (2008) Correlation between symmetry breaker position and the preferences of conformationally constrained homopeptides: a molecular dynamics investigation. Biopolymers (Pept. Sci.) 90: 695–706.

[90] Simberg, D., Duza, T., Park, J.H., Essier, M., Pilch, J., Zhang, L., Derfus, AM., Yang, M., Hoffman, R.M., Bathia, S., Sailor, M.J., Ruoslahti, E. (2007) Biomimetic amplification of a nanoparticle homing to tumors. Proc. Natl Acad. Sci. 104: 932–936.

[91] Hoffman, J.A., Laakkonen, P., Porkka, K., Bernasconi, M., Ruoslahti, E. (2004) *In vivo* and *ex vivo* selections using phage-displayed libraries. In: Phage Display: A Practical Approach. Oxford University Press, Chapter 10, pp. 171–191.

[92] Zanuy, D., Curcó, D., Nussinov, R., Alemán, C. (2009) Influence of the dye presence on the conformational preferences of CREKA, a tumor homing linear pentapeptide. Biopolymers (Pept. Sci.) 92: 83–93.

[93] Agemy, L., Sugahara, K.N., Kotamraju, V.R., Gujraty, K., Girard, O.M., Kono, Y., Mattrey, R.F., Park, J.-H., Sailor, M.J., Jimenez, A.I., Cativiela, C., Zanuy, D., Sayago, F.J., Alemán, C., Nussinov, R., Ruoslahti, E. (2010) Nanoparticle-induced molecular blockade in human prostate cancer. Blood 116: 2847–2856.

[94] Revilla-López, G., Torras, J., Nussinov, R., Alemán, C., Zanuy, D. (2011) Exploring the energy landscape of a molecular engineered analog of a tumor-homing peptide. Phys. Chem. Chem. Phys. 13: 9986–9994.

[95] Börner, H.G., Schladd, H. (2007) Bioinspired functional block copolymers. Soft Matter 3: 394–408.

[96] Klok, H.A. (2009) Peptide/protein-synthetic polymer conjugates: Quo Vadis. Macromolecules 42: 7990–8000.

[97] Kumar, R.J., MacDonald, M.J., Singh, T.B., Waddington, L.J., Holmes, A.B. (2011). Hierarchical self-assembly of semiconductor functionalized peptide alpha-helices and optoelectronic properties. J. Am. Chem. Soc. 133: 8564–8573.

[98] Chen, C., Constantinou, A., Deonarain, M. (2011) Modulating antibody pharmacokinetics using hydrophilic polymers. Expert Opin. Drug. Del. 8: 1221–1236.

[99] Yoncheva, K., Momekov, G. (2011) Antiangiogenic anticancer strategy based on nanoparticulate systems. Expert Opin. Drug. Del. 8: 1041–1056.

[100] Börner, H.G. (2009) Strategies exploiting functions and self-assembly properties of bioconjugates for polymer and materials sciences. Prog. Polym. Sci. 34: 811–851.

[101] Hamley, I.W. (2010) Liquid crystal phase formation by biopolymers. Soft Matter 6: 1863–1871.

[102] Kwak, M., Hermann, A. (2010) Nucleic acid/organic polymer hybrid materials: Synthesis, superstructures, and applications. Angew. Chem., Int. Ed. Engl. 49: 8574–8587.

[103] Liu, T.Y., Hu, S.H., Liu, D.M., Chen, S.Y., Chen, I.W. (2009) Biomedical nanoparticle carriers with combined thermal and magnetic responses. Nano Today 4: 52–65.

[104] Klok, H.A. (2005) Biological–synthetic hybrid block copolymers: combining the best from two worlds. J. Polym. Sci.: Part A: Polym. Chem. 43: 1–17.

[105] Borner, H.G., Schlaad, H. (2007). Bioinspired functional block copolymers. Soft Matter 3: 394–408.

[106] Löwik, D.W.P.M., Ayres, L., Smeenk, J.M., van Hest, J.C.M. (2006) Synthesis of bio-inspired hybrid polymers using peptide synthesis and protein engineering. Adv. Polym. Sci. 202: 19–52.

[107] Hamley, I.W. (2007) Peptide fibrillization. Angew. Chem., Int. Ed. 46: 8128–8147.

[108] Krysmann, M.J., Castelletto, V., Kelarakis, A., Hamley, I.W., Hule, R.A., Pochan, D.J. (2008). Self-Assembly and hydrogelation of an amyloid peptide fragment. Biochemistry 47: 4597–4605.

[109] Krysmann, M.J., Castelletto, V., Hamley, I.W. (2007) Fibrillisation of hydrophobically modified amyloid peptide fragments in an organic solvent. Soft Matter 3: 1401–1406.

[110] Hamley, I.W., Krysmann, M.J., Castelletto, V., Kelarakis, A., Noirez, L., Hule, R.A., Pochan, D. (2008). Nematic and columnar ordering of a PEG–peptide conjugate in aqueous solution. Chem. Eur. J. 14: 11369–11375.

[111] Castelletto, V., Hamley, I.W. (2009) Self assembly of a model amphiphilic phenylalanine peptide/polyethylene glycol block copolymer in aqueous solution. Biophys. Chem. 141: 169–174.

[112] Zanuy, D., Hamley, I.W., Alemán, C. (2011) Modeling the tetraphenylalanine-PEG hybrid amphiphile: from DFT calculations on the peptide to molecular dynamics simulations on the conjugate. J. Phys. Chem. B 115: 8937–8946.

[113] Yang, X., Su, Z., Wu, D., Hsu, S.L., Stidham, H.D. (1997) Raman analysis of a conformational distribution of poly(ethylene oxide) and its model compound in the liquid state. Macromolecules 30: 3796–3802.

[114] Smith, G.D., Bedrov, D., Borodin, O. (2000) Conformations and chain dimensions of poly(ethylene oxide) in aqueous solution: a molecular dynamics simulation study. J. Am. Chem. Soc. 122: 9548–9549.

4

Peptronics: Peptide Materials for Electron Transfer

Emanuela Gatto and Mariano Venanzi

Department of Chemical Sciences and Technologies, University of Roma Tor Vergata, Italy

Acronyms

ACN	Acetonitrile
AFM	Atomic force microscopy
ANQ	Anthraquinone
BuCN	Buthylacetonitrile
CA	Chronoamperom etry
CS	Charge separated
CV	Cyclic voltammetry
DCE	Dichloroethane
DMA	Dimethylacetamide
DOS	Density of states
ECZ	*N*-ethylcarbazolyl
EIS	Electrochemical impedance spectroscopy
ESR	Electron spin resonance
ET	Electron transfer
FCWD	Franck–Condon weighted density
HB	Hydrogen bond
HOMO	Highest occupied molecular orbital
IR-RAS	Infrared reflectance anisotropy spectroscopy

Peptide Materials: From Nanostructures to Applications, First Edition. Edited by Carlos Alemán, Alberto Bianco and Mariano Venanzi.

LUMO	Lowest unoccupied molecular orbital
MLCT	Metal-to-ligand charge transfer
MV^{2+}	Methyl viologen
PET	Photoinduced electron transfer
PG	Photocurrent generation
PTZ	Phenothiazine
SAM	Self-assembled monolayer
SE	Superexchange
STM	Scanning tunneling microscopy
STS	Scanning tunneling spectroscopy
TEOA	Triethanolamine
TOAC	2,2,6,6-Tetramethylpiperidine-1-oxyl-4-amino-4-carboxylic acid
UHV	Ultra high vacuum
ZCP	Zero current potential

4.1 Introduction

Peptide materials are nowadays of great technological interest for several reasons. First of all modern peptide chemistry allows relatively long peptides to be obtained easily at low cost, high yields and high purity. What is mostly important for advanced applications, peptides can be designed to populate specific secondary structures (α-helix, β-sheet, β-hairpin) and to feature unique self-assembly properties for building up nano- and mesoscopic structures (nanotubes, nanowires, fibrils, dendrimers). Furthermore, peptides can easily be derivatized with smart groups, endowing the peptide scaffold with specific functionalities (cargo capacities, electro- and photoactive units). Last but not least, peptides are chiral molecules, capable of controlling enantioselective molecular recognition and generating chiral supramolecular architectures.

Let us only cite the example of peptide foldamers, i.e. peptide molecules with well-defined and predictable folding properties [1]. One of the most successful strategies to achieve this goal has been the synthesis of nonproteinogenic amino acids characterized by tetra-substituted C^{α}-atoms. This has the effect of blocking the φ and ψ torsional angles of a given residue to specific values that strictly depend on the steric hindrance properties of the C^{α}-substituents, thus determining the peptide secondary structure. Therefore 3_{10}-helices, β-hairpins, 2.0_5 and other unusual conformations have been obtained by proper selection of these synthetic building blocks [2].

Among the ordered peptide structures, α-helix is considered to play an important role in electron transfer (ET) mediation, and its efficiency in long range ET has been reported to be significantly larger than that of alkyl chains [3]. The better ET efficiency across helical peptides is explained by the ordered coupling of hydrogen-bonded amide sites. In particular, the amide n and π orbitals are thought to represent the stepping stones on the pathway that electrons (or holes) pursue while flowing across the peptide chain [4–6]. Although less efficient than phenylen-ethynylene or phenylen-vinylene oligomers, helical peptides present several advantages as components for molecular electronics, because of well-specified and controlled secondary structure, facile molecular design and arrangement of functional groups, unique self-assembly properties (*bottom-up* approach).

The donor(D)–bridge(B)–acceptor(A) scheme, firstly introduced by Aviram and Ratner [7], provides a unifying framework for discussion of a broad range of charge-transfer processes in solution and in the solid state, and it will be adopted in this chapter.

However, while ET studies have partially clarified the mechanisms operating in D–B–A systems in solution, the understanding of such processes in hybrid systems, i.e. electroactive peptides immobilized on conductive or semiconductive surfaces, is still inadequate. For this reason, ET in solution will be discussed at first, introducing the basic theory of ET through peptide spacers, while the second section of the chapter will be dedicated to ET in peptide systems linked to a solid (metal or semiconductor) substrate. Seminal works presenting the most convincing achievements in the field will be discussed in both cases.

4.2 Electron Transfer through Peptide Scaffolds in Solution

4.2.1 Theoretical Background

For introducing the basic theoretical models of ET across peptide matrices, we refer to a D–B–A system, where D and A are an electron donating and electron acceptor groups, respectively, and B is representing a linear peptide chain.

ET between D–A sites connected by a molecular bridge is now fairly well understood [8–10]. There are several factors affecting the efficiency of the ET process, first of all the distance and orientation of the D–A pair. However, these geometrical parameters are determined by the secondary structure attained by the peptide bridge and by the dynamics of the backbone and of the side chains on the time scale of the ET process. Both structural and dynamical features are ultimately determined by the amino acid composition of the peptide bridge.

A well established mechanism for peptide ET is *bridge assisted superexchange* [4–6]. In this case, the bridge serves only as a medium, and electrons or holes never reside on the bridge. Superexchange (SE), a coherent tunneling process, is mediated by virtual states associated to electron (D^+–B^-–A) or hole (D–B^+–A^-) transfer. In the absence of stepping sites, biologically relevant ET is limited to a maximum distance of about 20Å.

Following the Marcus–Hush theory of ET reactions [8–10], the ET rate constant is given by

$$k_{ET} = \left(\frac{\pi}{\hbar^2 \lambda kT}\right)^{\frac{1}{2}} H_{DA}^2 \exp\left[-\frac{(\lambda + \Delta G^\circ)^2}{4\lambda RT}\right] \tag{4.1}$$

where H_{DA}^2 is the electronic coupling matrix, associated to the probability that an electron tunnels through the D–A potential barrier, ΔG° is the thermodynamic driving force of the ET reaction, i.e. the difference in the oxidation potentials of the D–A pair and λ, the so-called reorganization energy, i.e. the energy needed to bring the nuclei from the equilibrium position of the reactants to the equilibrium position of the products.

These parameters can be evaluated through spectroscopic studies of charge-transfer band intensities and energies (H_{DA}, λ), structural and vibrational frequency differences (λ)

and electrochemical or other thermodynamic measurements ($\Delta G°$). The reorganization energy λ is usually factorized in an outer contribution, associated to the reorganization of solvent modes around a charge separated (CS) state, and an inner contribution, associated to nuclear (vibrational) reorganization of the D–A pair, i.e. $\lambda = \lambda_{in} + \lambda_{out}$.

In a quantum-based description of ET, Equation (4.1) can be rewritten as [11]

$$k_{ET} = \frac{1}{\hbar^2} H_{DA}^2 (FCWD) \tag{4.2}$$

where FCWD is the Franck–Condon weighted density of vibronic states, i.e.

$$FCWD = \left(\frac{1}{4\pi\lambda kT} \right)^{1/2} \exp\left[-\frac{(\lambda + \Delta G°)^2}{4\lambda RT} \right] \tag{4.3}$$

In the case of a molecular bridge comprising *n* bridging units, i.e. the *n* amino acids of a polypeptide chain, the D–A electronic coupling matrix can be factorized in terms of the coupling between the donor and the first bridging unit (H_{D1}), the intersite coupling matrix G_{1n} and the coupling matrix between the terminal bridging unit and the electron acceptor group (H_{nA}):

$$H_{DA} = H_{D1} G_{1n} H_{nA} \tag{4.4}$$

The intersite coupling matrix is usually expressed as a Green function product evaluated at the donor energy state. From a quantal viewpoint, SE is, therefore, a coherent tunneling process through molecular orbitals mediated by D^+–B^-–A (electron transfer) or D–B^+–A^- (hole transfer) virtual states (Figure 4.1a) [12].

In the case of weak coupling between the bridging sites, a perturbation model can be applied to the computation of the electronic coupling matrix, obtaining [13]

$$H_{DA} = \frac{H_{D1} H_{nA}}{E_D - E_{n1}} \prod_{i=1}^{n-1} \frac{H_{i,i+1}}{E_D - E_{i+1}} \tag{4.5}$$

For *n* identical bridging units, Equation (4.5) simplifies to

$$H_{DA} = \frac{H_{D1} H_{BB}^{n-1} H_{nA}}{(E_D - E_B)^n} \tag{4.6}$$

Within this scheme, the electronic coupling matrix is found to decrease exponentially with the D–A distance r_{DA} [14, 15]:

$$H_{DA} = H_{DA}^0 \exp\left[-\beta (r_{DA} - r_0) \right] \tag{4.7}$$

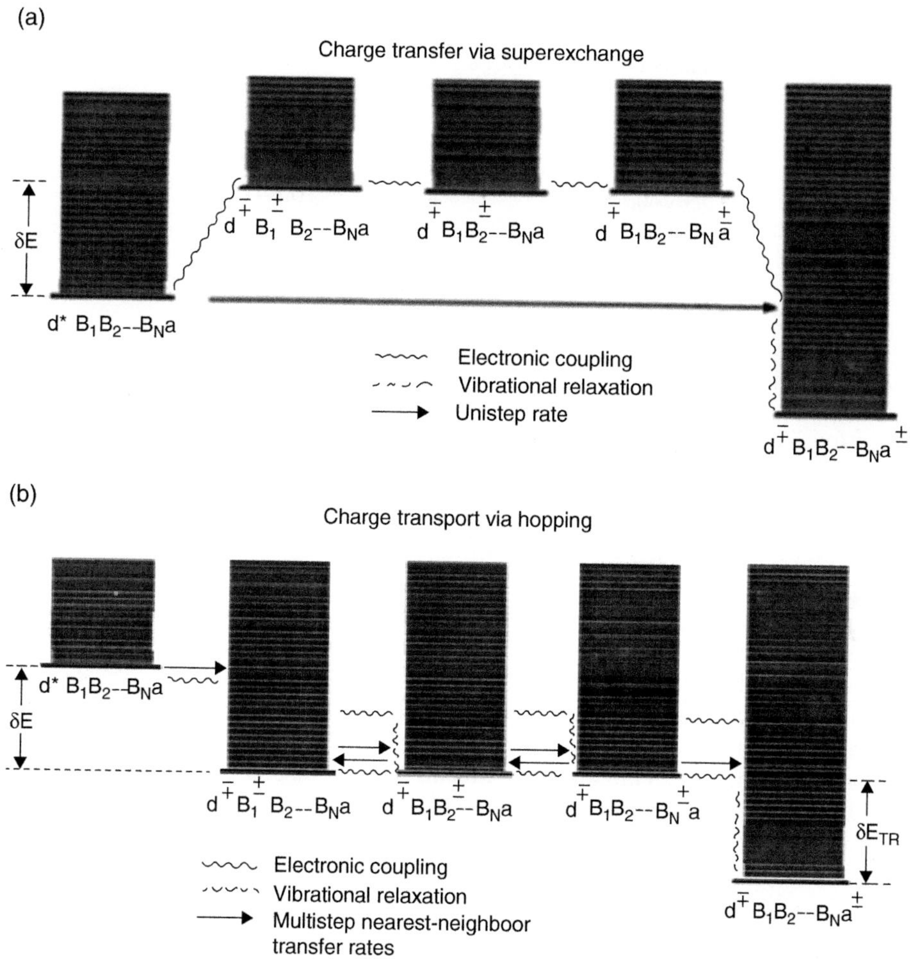

Figure 4.1 *(a) Schematic drawing of the superexchange mechanism in a D–(B)$_n$–A system. (b) Schematic drawing of the hopping mechanism in a D(B)$_n$A system. Reproduced with permission from reference [12]. Copyright (1998) National Academy of Sciences, USA*

where H_{DA}^{0} is the value of the electronic coupling matrix at the D–A contact distance r_0 and β the coupling strength parameters, i.e. the inverse of the distance at which the electronic coupling matrix decrease by a factor e. From perturbation theory calculations [13]:

$$\beta = \frac{2}{l}\ln\left(\frac{E_D - E_B}{H_{BB}}\right) \tag{4.8}$$

where l is the length of the bridging unit. For long range ET, β should be small, which is favoured by strong coupling between the bridging units (large H_{BB}) and small energy difference between the donor and virtual states (small E_D–E_B). In case of resonant tunneling

Table 4.1 *Electronic coupling parameter for different molecular spacers*

Spacer	β (Å⁻¹)
Saturated hydrocarbon	1
DNA duplexes	0.8–0.6
Unsaturated hydrocarbon	0.2
Oligopeptides	1.1–0.2

(null energy gap), a region of distance independence or little fall-off with distance of rates is expected, giving rise to the concept of molecular wires.

Therefore, the prevision of the SE model is that the ET rate constant would decrease exponentially with the D–A distance:

$$k_{ET} = k_{tun} = k_{ET}^{0} e^{-\beta r_{DA}} \tag{4.9}$$

This exponential decrease has been verified in a series of D–A systems using appropriate β values that can be considered as typical of the different molecular spacers (Table 4.1).

Conjugated molecular bridges like oligothiophenes or carotenoids have been shown to have slower distance-dependent rate constants with respect to saturate molecular spacers (alkanes, peptides). However, seminal works by Gray and Winkler [3, 16, 17] demonstrated that, for a series of Ru-modified Azurin proteins, long-distance ET can occur over more than 20Å, showing an exponential decrease with the D–A distance.

Although the exponential decrease of the ET efficiency foreseen by SE could be judged unfavourable for using peptides as molecular wires, definitely smaller decay constants at a long distance have been reported: 0.18Å^{-1} for helical oligoprolines [18], 0.02–0.04Å^{-1} for α-helical peptides [19] and 0.05Å^{-1} for collagen-like triple helices [20].

A possible alternative mechanism, which could enable ET over such a long distance, is *hopping ET*, where electrons are relayed with a moderate ET distance dependence by hopping through regularly arranged sites, which are electronically coupled to each other [12]. The transferred electron resides on the bridge for a finite time and the ET process may be viewed as a series of discrete steps. In the case of peptide spacers, amide bonds and side-chain aromatic units (tryptophan, tyrosine) have been proposed as possible hopping sites [21]. In agreement with the hopping mechanism, electrons (or holes) reside on the bridge and diffuse by hopping between bridge sites, the D^{+}–B^{-}–A or D–B^{+}–A^{-} being real, populated states (Figure 4.1b).

At odds with the SE model, based on tunneling through the bridge molecular states, hopping ET is a thermally activated process, involving nuclear motion and electron hopping over the barrier. Within this frame

$$k_{ET} = k_{ET}^{0} \exp\left(-\frac{E_a}{kT}\right) \tag{4.10}$$

It can be shown that, in the diffusive hopping limit, the electronic conduction is typical of ohmic ET and the ET rate constant is inversely proportional to the D–A separation [7]:

$$k_{ET} = k_{hop} \propto \frac{1}{N}\exp\left(-\frac{\Delta E_{DB}}{RT}\right) \tag{4.11}$$

where N is the number of hopping sites and ΔE_{DB} is the energy difference between the donor and the bridging units. It is now generally accepted that SE and hopping mechanisms can operate in a parallel and competitive way [22], i.e.

$$k_{ET} = k_{SE} + k_{hop} \tag{4.12}$$

Interestingly, from the data reported in Table 4.1, it can be noted that the β values of different peptide chains span the full range from saturated to unsaturated hydrocarbon chains. This finding clearly reflects what is already found in ET proteins, where only a *multiple pathway approach* could explain the efficiency and the long-distance character of the ET process [23].

By this approach, several mechanisms should be taken into account besides through-bond ET, comprising through-space tunneling, hydrogen bond (HB) assisted ET and hopping between aromatic side chains. Therefore, the matrix element for the electronic coupling, H_{DA}, should be evaluated by the product of the coupling terms for jumps along a σ bond and jumps across a space, eventually mediated by aromatic groups or HB shortcuts. The pathway that gives the largest product is taken as the optimum one. Refinement of this model has been recently proposed by taking protein dynamics into account, improving the agreement with experimental data [24, 25].

An alternative model based on a molecular motion-assisted ET mechanism in which a peptide bridge thermally fluctuates to accelerate ET has been also proposed [26]. However, the combination of the two, i.e. a molecular-motion-assisted hopping mechanism is also thought to be plausible [27].

A unified treatment of SE and hopping (sequential) ET has been attempted by Petrov *et al.* [28] based on a common set of parameters (energy gaps, driving forces, intersite couplings). In the case of oligoproline spacers, it was demonstrated that the ET efficiency and distance dependence can be accounted for by considering concomitant SE and bridge-assisted hopping mechanisms. The energy gap for the thermal activation of hopping ET was found to be around 0.21 eV, which is close to that reported for DNA molecules [12]. However, for a small energy gap of the order of 0.2–0.3 eV, the influence of the SE mechanism remains large as well. Petrov *et al.* restrict their approach to the case where the intrasite relaxation is faster than the time characteristic of the ET process, i.e. $\tau_{rel} \ll \tau_{ET}$, allowing for a formulation of the problem in terms of a set of coupled kinetic equations [28]. Since the characteristic time τ_{rel} is typically of the order of 0.1–10 ps, the proposed theory is valid for those ET processes where the rates do not exceed values of 10^{11}–$10^{13}\,s^{-1}$, i.e. it is appropriate for nonadiabatic bridge-mediated ET. In this case, the ET reaction proceeds along the D–B–A electronic active coordinates weakly coupled to quasi-equilibrium distribution of vibrations. The general solution of the coarse-grained kinetic equations is characterized by a double exponential kinetics that can be used to discuss the different regimes of bridge-assisted ET [28].

4.2.2 Seminal Experimental Results

4.2.2.1 Hopping versus Superexchange Mechanism

The competition between SE and hopping mechanisms with the D–A pair separation has been thoroughly investigated, especially in view of the great ET efficiency at long distances of some biological systems. To assess the influence of aromatic amino acids on long-distance ET Cordes *et al.* [29] synthesized model nonapeptides functionalized with an ET D–A pair. The aromatic electron donor at the C-terminus was separated from the electron acceptor tyrosine residue at the N-terminus by two proline triplets about 20Å away. A further amino acidic spacer, the side chain of which was alternatively constituted by an aliphatic or an aromatic group, was introduced in the fifth position of the peptide chain, somewhat halfway between the D–A pair. Interestingly, with alanine or leucine as central amino acids only 1% of the tyrosine was oxidized by intramolecular ET, but the tyrosine radical yield grows to 27 and 36% when 2,4-dimethoxy- or 2,4,6-trimethoxyphenylalanine, respectively, was used as the central amino acidic spacer. The simultaneous existence of all the oxidized aromatic intermediates, i.e. the tyrosil radical (A), the dialkoxyphenyl alanine radical cation (D) and the 2,4-dimethoxy- or 2,4,6-trimethoxyphenylalanine radical cation (B), was proven spectroscopically by nanosecond transient absorption measurements (Figure 4.2).

This finding clearly indicates that the aromatic radical cation of the central amino acid is an efficient and stable intermediate during the ET process, suggesting that a two-step hopping mechanism predominates. The introduction of a 'stepping stone' along the ET pathway leads to a 20- to 30-fold increase of the ET rate, emphasizing the contribution of the hopping mechanism to long-range ET with respect to SE. The latter drives ET in the case of all-aliphatic amino acid spacers.

Facing the same issue, Malak *et al.* [18] introduced a series of D–B–A systems based on peptide bridging units, which showed a clear transition from an SE to a hopping mechanism with increasing D–A separation. The D–A pair was represented by a 4-carboxy-4′-methyl-2,2′-bipyridine complex of Ruthenium(II) [$(bpy)_2Ru^{II}L$] and a pentammine-Ruthenium(III) [$apyRu^{III}(NH_3)_5$], respectively, while the spacer was an oligoproline chain, with the number of proline residues varying from 0 to 9. Different spectroscopic techniques were used to study the ET process from the picosecond to the millisecond time scales, experiencing D–A distances in the range of 8.7–32Å. Electron pulse radiolysis was used to determine the ET rate constants for the reaction $Ru^{II}(L^{-})–Pro_n–apyRu^{III} \rightarrow Ru^{II}(L)–Pro_n–apyRu^{II}$, where $n=3–9$, while for shorter bridges ($n=0–4$) the rate constants for the photoinduced ET $Ru^{II}(L^{*})–Pro_n–apyRu^{III} \rightarrow Ru^{III}(L)–Pro_n–apyRu^{II}$ were determined. Time resolved fluorescence and femtosecond transient absorption were used to measure the time decay of the excited and intermediate species. The dependence of the ET rate constant on the D–A separation distance is shown in Figure 4.3.

For the shorter peptides, $n=0–4$, the distance dependence is typical of an ET process mediated by SE ($\beta=1.4$Å^{-1}), while for the longer peptides, $n=5–9$, a much weaker distance

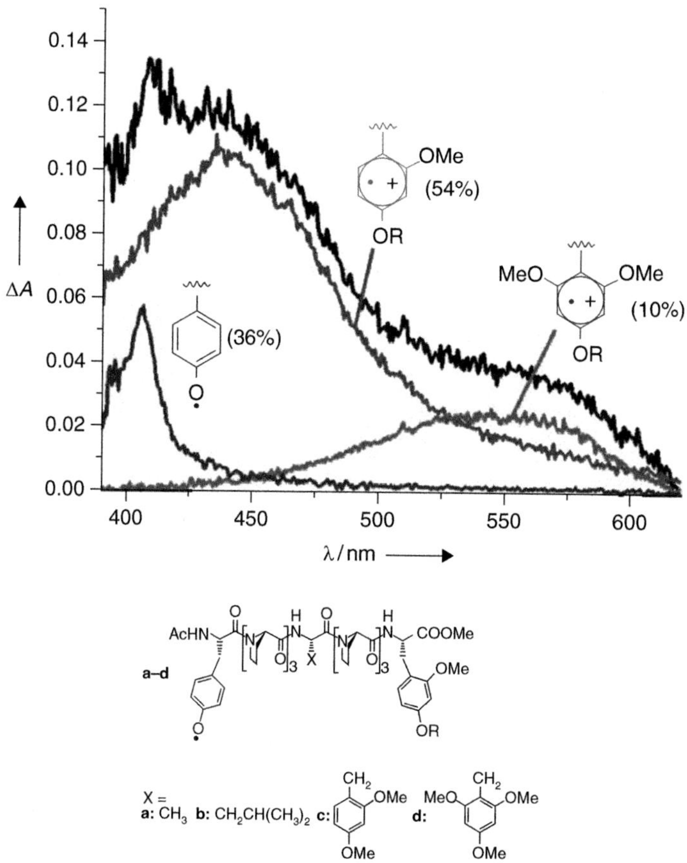

Figure 4.2 *Transient absorption spectrum of the oligoproline nonapeptide with X = 2,4,6-trimethoxyphenylalanine 40 ns after the laser flash (black line). The spectra of the tyrosil radical (acceptor), the dialkoxyphenyl alanine radical cation (donor) and the 2,4,6-trimethoxyphenylalanine radical cation (spacer) were also reported for comparison. Reproduced with permission from reference [29]. Copyright (2008) John Wiley & Sons, Ltd*

dependence (β=0.18 Å^{-1}) was observed, the latter value being more consistent with an electron hopping mechanism. It therefore appears that SE predominates for D–A separation distances lower than 20 Å, while the hopping mechanism is the only efficient contributor to very long range ET.

4.2.2.2 Hydrogen-Bond-Assisted Electron Transfer

Besides the specific amino acidic composition of the bridge, hydrogen bonds have been also recognized to increase the ET rate across the peptide chain, a feature lacking in the oligoproline systems discussed so far. Antonello *et al.* [30] investigated electrochemically

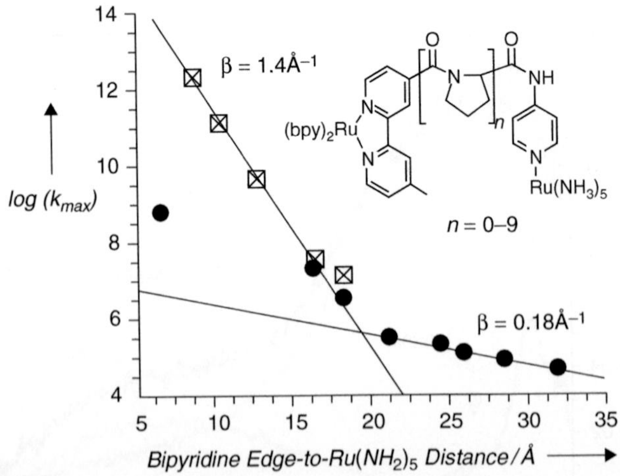

Figure 4.3 *Plot of log(k_{ET}) from radiolysis (circles) and photoinduced ET (squares) experiments versus the $Ru^{II} \rightarrow Ru^{III}$ edge distance for n=0–9 prolines. Reproduced with permission from reference [5]. Copyright (2005) John Wiley & Sons, Ltd*

induced ET in D–B–A systems, where D was a phthalimide radical anion ($E°$=–1.35V), A a peroxide group ($E°$=–1.21V) and B was an $(Aib)_n$ homopeptide with *n* varying from 0 to 6. Structural studies demonstrated the incipient formation of a 3_{10}-helix even for *n*=3. The increase of the peptide length not only resulted in a greater D–A separation distance, but also increased the number of H-bonded residues. CV measurements showed a positive $E°$ shift with increasing peptide length (0.1V for the formation of the adjacent H-bond), suggesting that H-bonding interaction significantly lowered the LUMO energy of the phthalimide group. It was found that the ET rates for *n*=2 and *n*=3 were essentially the same, or even slightly larger than the ET rate measured for *n*=1. Only for *n*=4 could a marked decrease of the ET rate constant be observed. Interestingly, kinetic data indicated some periodicity of the obtained rate constants with the lengthening of the peptide chain, reflecting the modulation of the D–A distance imposed by the peptide helical structure. The conclusion was that for the shorter compounds of the series (*n*=1–3) the increasing D–A distance was balanced by the positive contribution of H-bond mediated pathways to the ET process. In agreement with this idea, the maximum rate observed for *n*=3 should be ascribed to the fact that the same residue is H-bonded to both D and A.

When a *p*-cyanobenzamide group ($E°$=–1.70V) was substituted to the phthalimide donor, a $\Delta G°$ more negative by 0.35V was obtained [31]. This should have increased the ET rate by SE (in the positive Marcus region) and decreased the donor–bridge energy gap, favouring the establishment of hopping pathways, too. The results confirmed the trend observed in the case of the phthalimide donor, showing rates slightly dependent upon distance for the shorter homologues ($1 \leq n \leq 3$) and absolute values greater than the phthalimide–peroxide pair. Temperature-dependent experiments revealed that the two D–A systems investigated have comparable activation energies (E_a=0.34eV and 0.37eV for the phthalimide–peroxide and *p*-cyanobenzamide–peroxide pair, respectively). This finding led the authors to conclude that an H-bond-assisted SE was predominating in both cases.

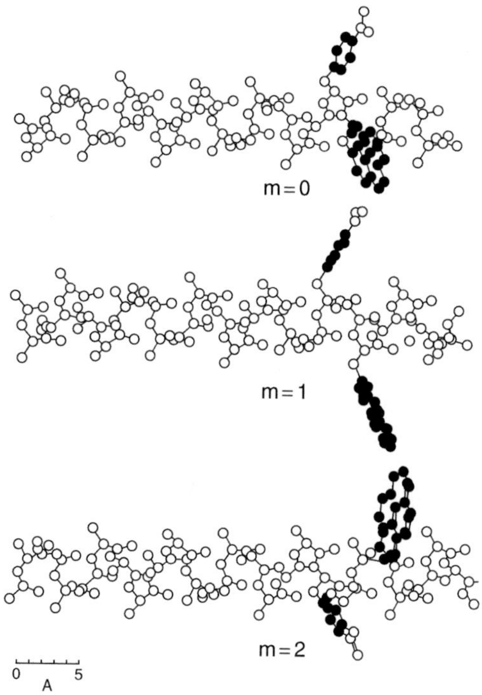

Figure 4.4 Most probable side-chain orientation of the three polyglutamates investigated from conformational calculations. Reproduced with permission from reference [33]. Copyright (1991) American Chemical Society

4.2.2.3 Photoinduced Electron Transfer

Photoinduced ET (PET) in peptide models has been one of the methods of election for the comprehension of the molecular details of ET processes, the *holy grail* being the explanation of the primary mechanisms governing photosynthetic processes. The field was pioneered by Sisido and coworkers [32, 33], who synthesized α-helical poly(γ-benzyl L-glutamates) containing a photoactive core formed by an L-*p*(dimethyl1amino)phenylalanine (D) and an L-1-pyrenylalanine (A) as an electron D–A pair, separated by 0, 1 or 2 L-alanine (B) bridging units. The helical conformation of the polypeptide modulates the ET efficiency constraining the D–A pair to attain a definite distance and orientation. Due to the peptide secondary structure, the D–A pair in the $m=2$ polypeptide, although separated by a larger number of bonds, is located closer than the D–A pair in the $m=1$ polypeptide, the centre-to-centre (edge-to-edge) distances predicted from conformational calculations being 9.0Å (5.5Å) and 13.2Å (9.4Å), respectively. Besides separation distance, the relative orientation of the D–A pair is also affected by the helical structure of the peptide backbone, so that in the $m=0$ polypeptide the D–A pair attains a head-to-tail orientation, but shows a face-to-face arrangement in the $m=2$ polypeptide (Figure 4.4).

The rates of PET in different compounds were evaluated from time-resolved fluorescence experiments. At 20 °C in tetrahydrofuran, the experimental PET rate constants were $3.8\times10^{7}\,s^{-1}$, $0.4\times10^{6}\,s^{-1}$ and $10.2\times10^{7}\,s^{-1}$ for $m=0$, 1 and 2, respectively. These results clearly indicate that the ET rates do not simply depend on the number of bonds separating the D–A pair, but that the 3D spatial distance between the two groups should be properly taken into account. This finding strongly suggests that *through-space* ET is the predominant mechanism, at least at these short distances. Assuming an exponential dependence of the ET rate constant on the D–A edge-to-edge distance (Equation (4.9)), $\beta = 1.3\,\text{Å}^{-1}$ was obtained, typical of SE. Polypeptides with $m=0$ and $m=2$, showing similar D–A distances but very different orientations, feature comparable ET rate constants, suggesting that ET was not very sensitive to the orientation of the D–A pair. Temperature-dependent experiments provided activation enthalpies of a few kcal/mol, showing that large conformational changes, i.e. correct positioning of the D–A pair, are needed to trigger the ET reaction.

The same poly(γ-benzyl L-glutamate) α-helical scaffold was functionalized by Sisido *et al.* [34] with an L-1-pyrenylalanine (D) and L-*p*-nitrophenylalanine (A) pair separated by $n=0$ to 8 γ-ciclohexyl L-glutamate units. The rates of PET from the pyrenyl excited state to the nitrophenyl group were evaluated from the decay curves of pyrenyl fluorescence recorded at different temperatures from −58 to +30 °C. Also in this case, the rate constants show a complex dependence on the number of amino acids separating the D–A pair, reflecting the helical geometry of the peptide backbone. In particular, an increase in the ET rates with an increase in the number of spacing amino acids from 1 to 2 and from 5 to 6 was observed. When the most stable conformers, obtained by conformational calculations, were taken into account, the ET rate constants exhibited a simple exponential dependence on the edge-to-edge distance between the two active probes with a distance decay factor $\beta = 0.66 \pm 0.1\,\text{Å}^{-1}$, suggesting that ET predominantly occurs via a through-space mechanism made possible by SE-assisted tunneling. Temperature-dependent experiments proved that ET rate constants are only weakly *T*-dependent for most compounds, the activation enthalpies being less than 1 kcal/mol. The very low activation enthalpy indicates that the ET process is determined by pure electronic effects, with only a weak contribution by reorganization energy terms originating from activated conformational changes or solvent coordination. The optimum ET pathways from pyrenyl to nitrophenyl were searched by applying a tunneling pathway model, and the relative values of the ET matrix elements were evaluated for each polypeptide. Figure 4.5 shows the optimum ET pathways evaluated from the computed structures obtained by conformational energy calculations.

For most of the structures reported in Figure 4.5, it appears that HB pathways provided a significant contribution to the ET process, with several single and two through-HB jumps identified.

4.2.2.4 *Secondary Structure and Electric Dipole Effects*

The effect of the peptide secondary structure on the ET efficiency was theoretically investigated by Shin *et al.* [35] in a D–B–A system, where D is a C-terminal aminopyridine group and A an N-terminal 4-carbonyl-2,2′-bipyridine. The peptide bridge was a homopolyglycine chain, with the number of bridging units varying between 0 and 6. The peptide chain was constrained to attain different secondary structures: α-helix, 3_{10}-helix, β-sheets and polyproline I and II helical structures. In helical peptides, the ground state molecular dipole is a linear function of the number of amino acid residues and strongly depends, in magnitude and direction, on the secondary structure adopted by the peptide. The molecular

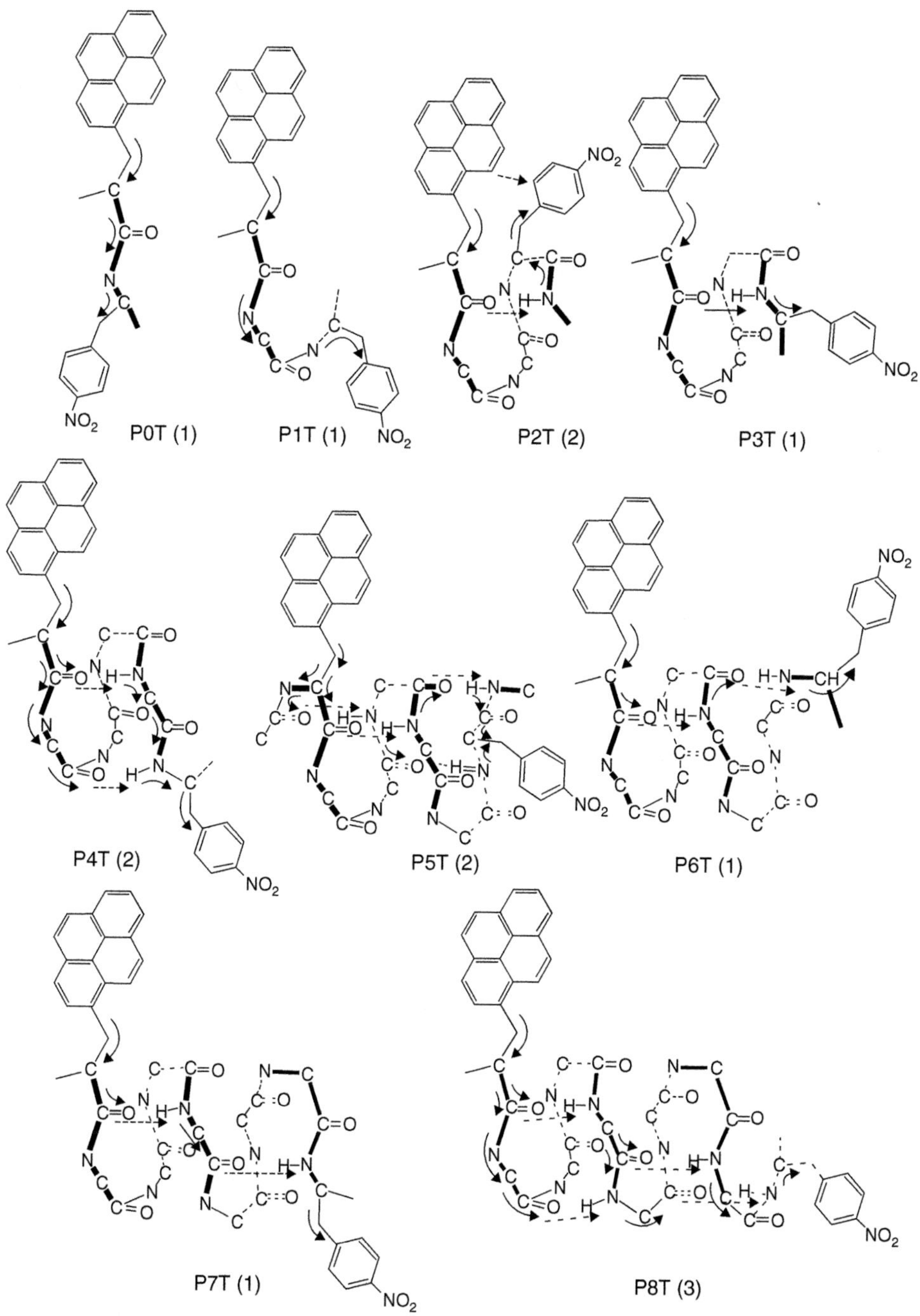

Figure 4.5 *Optimum ET pathways for the investigated PmT polypeptides. In parentheses the number of equivalent or nearly equivalent pathways is also indicated. Reproduced with permission from reference [34]. Copyright (2001) American Chemical Society*

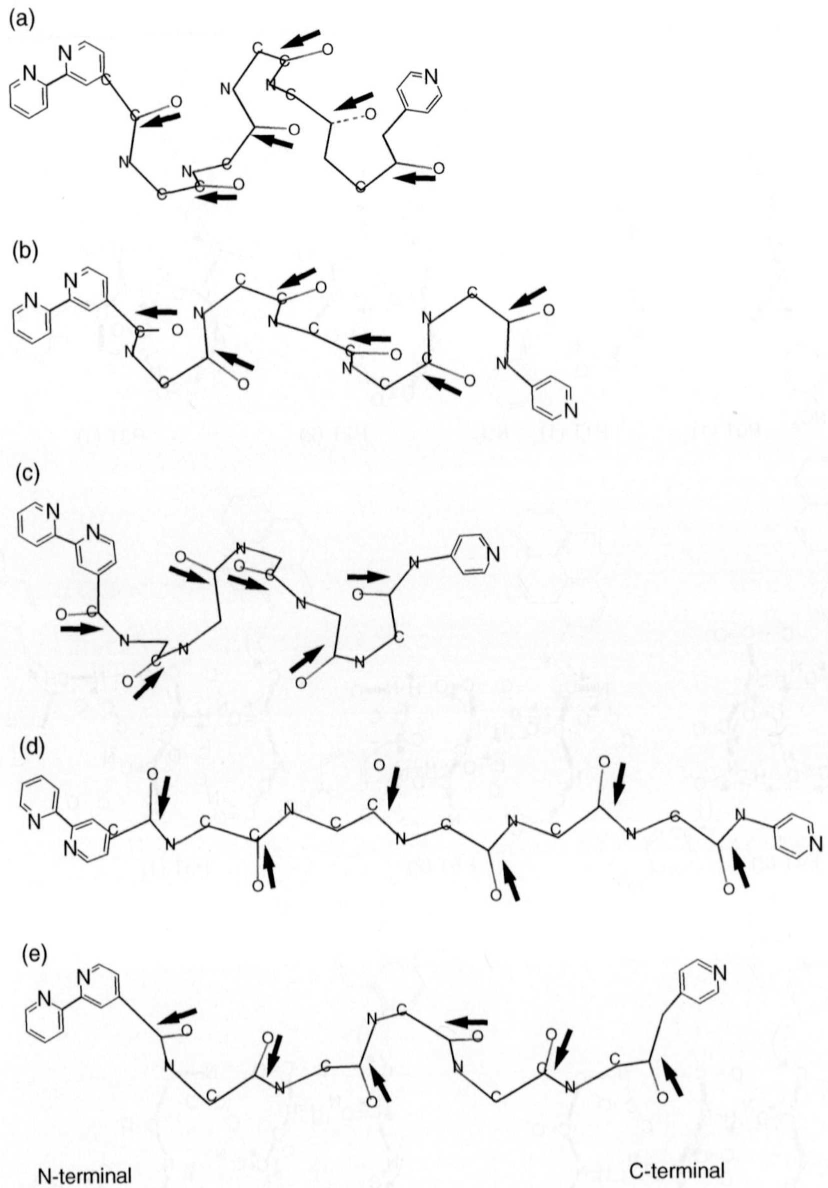

Figure 4.6 Structures and the local dipole moments for the (gly)$_5$ residues in the (a) α-helix, (b) 3_{10}-helix, (c) polyproline I-helix, (d) β-strand and (e) polyproline II-helix conformations. Reproduced with permission from reference [35]. Copyright (2003) American Chemical Society

dipole increment per residue correlates very well with the directional cosine measuring the orientation of the carbonyl groups with respect to the helical axis of the different secondary structures and can be well approximated by the sum of the local dipoles of the backbone amide groups (Figure 4.6).

The dipole moment of an isolated peptide bond is estimated to be 3.5 D. However, in the polyproline I- and II-helix and in the β-strand structures, the molecular dipole increases by 3.9, 1.5 and 0.25 D per residue, respectively. The molecular dipoles of α- and 3_{10}-helices were estimated to increase by 5.0 and 4.5 D per additional residue, respectively, but in the opposite direction of the helical axis with respect to the former structures (Figure 4.6). The charge-transfer transition energies and the electronic coupling matrix H_{DA} were found to be markedly affected by the magnitude and direction of the dipole generated by the peptide secondary structure. A plot of $\ln(H_{DA})$ versus the charge-transfer distance, assumed to be proportional to the dipole moment change between the ground and the charge-transfer states, showed that the ET efficiency decreases along the series polyproline II > α-helix > 3_{10}-helix > β-strand > polyproline I, the electronic coupling parameter β being 0.69, 1.17, 1.26, 1.39 and 1.49 Å^{-1}, respectively. It should be noted, however, that the different topologies of these secondary structures result in very different through-space D–A distances for an equal number of bridging units. The translational distance per residue changes from 1.50 Å for an α-helix to 2.0 Å for a 3_{10}-helix, 3.12 Å for polyproline-II and 3.1 Å for polyglycine-II. In the systems discussed above, the five residues through-space distance is 9.6 Å for an α-helix compared with 16.4 Å in the polyproline II-helix and 19.0 Å for the β-strand structure.

The effect of the ground state electric dipole associated to the peptide secondary structure on the ET rate was firstly verified by Galoppini and Fox through PET studies on D–B–A systems [36–38], formed by an *N,N*-dimethyl-*p*-aniline (D) and a 2-pyrenil-alanine (A) group functionalizing a helical oligopeptide formed by Ala–Aib–Ala triplets. Two of these triplets separated the D–A pair, and one additional Ala–Aib–Ala triplet was added at both the N- and C-termini to avoid unfolding terminal effects that could affect the ET efficiency. Two peptides with the same amino acidic sequence, but differing on the reverse position of the D–A pair along the dipolar helix, were therefore synthesized.

As a consequence of the reverse positioning of the D–A pair along the peptide helix, photoinduced ET generates a CS state (D^{+}–A^{-}), which is oriented along the dipole electric field in one of the two peptides and against this field in the other. The α-helix macrodipole was estimated to generate an electrostatic field along the helix axis of 10^9 V/m^2, producing an effective positive charge at the N-terminus and an effective negative charge at the C-terminus of about 0.8×10^{-19} C. For ET driving forces in the normal Marcus region, the ET rate will be fastened when the (D^{+}–A^{-}) CS state is oriented against this field. Actually, it was found that ET rates producing (D^{+}–A^{-}) antiparallely oriented with respect to the electric field generated by the helix dipole were 5–27 times larger than those measured in the case of 'along-field' CS states, depending on the solvent and temperature. Significantly, the ratio between the two ET rate constants decreases with increases in the solvent dielectric constant (from 27 in THF to 5 in methanol) and the temperature (from 70 at 228 K to 27 at 295 K). Although changes in these variables affect the ET rates in many aspects, both medium and temperature are expected to reduce the influence of the electrostatic potential generated by the helix macrodipole. When guanidinium carbonate was added to THF solutions of the two peptides, the ratio of the ET rates decreased from 27 to 1, while in 2,2,2-trifluoroethanol, the solvent in which the peptides exist as random coils, the two rate constants are almost comparable. This finding clearly indicates that, upon unfolding the peptide helix, the misalignment of the carbonyl dipoles disrupts the helix dipole effect, levelling the two ET rates.

The effect of the dipole electric field on the rate of the intramolecular PET reaction generated by a peptide 3_{10}-helix functionalized with a D–A pair was also investigated by Gatto and coworkers [39]. As in the peptide model studied by Fox and Galoppini, the D–A position along the helix axis was reversed, but the octapeptide spacer was exclusively formed by $C^{\alpha,\alpha}$-tetrasubstituted residues, i.e. Aib and (α-methyl)*N*-valine. The ratio between the ET rate constants associated to 'against field' versus 'along field' ET in acetonitrile was only 2.1 as compared to 8 found in the case of α-helical peptides. Besides unfolding terminal effects that probably lower the ET rate constants ratio for the former compounds, these results confirm the minor influence of the 3_{10}-helix dipole on the directionality of the ET process, most likely ascribable to the distortion of the H-bond network stabilizing this secondary structure.

4.2.2.5 Solvent Dependence

The ET solvent dependence has been investigated in details by Meyer and coworkers [40] for a 13-residue oligopeptide functionalized with a triad of photoactive probes, mimicking an artificial photosynthetic unit (Figure 4.7).

In the oligoproline assembly, the ruthenium polypyridyl complex acts as a molecular antenna, capturing electromagnetic excitation in the visible region and producing a metal-to-ligand charge transfer (MLCT) excited state. The phenotiazine (PTZ) group features electron donor properties and the anthraquinone (ANQ) moiety acts as an electron acceptor group. Two excited state processes originate from the MLCT excited state: ET to the anthraquinone electron acceptor or ET from the phenotiazine group. The energetically most stable state is the ($PTZ^{+\bullet}$–$ANQ^{-\bullet}$) CS state, where the oxidized phenotiazine and the reduced anthraquinone are separated by five proline spacers. In the final step, the CS state returns to the ground state by back ET. From electrochemistry and photophysical experiments it was shown that reductive ET ($PTZ \rightarrow Ru^{III}b_2m^-$) is thermodynamically favoured with respect to oxidative ET by $\Delta G°$ values of –0.44, –0.36, –0.34 and –0.24 eV in dichloroethane (DCE), butylacetonitrile (BuCN), acetonitrile (ACN) and

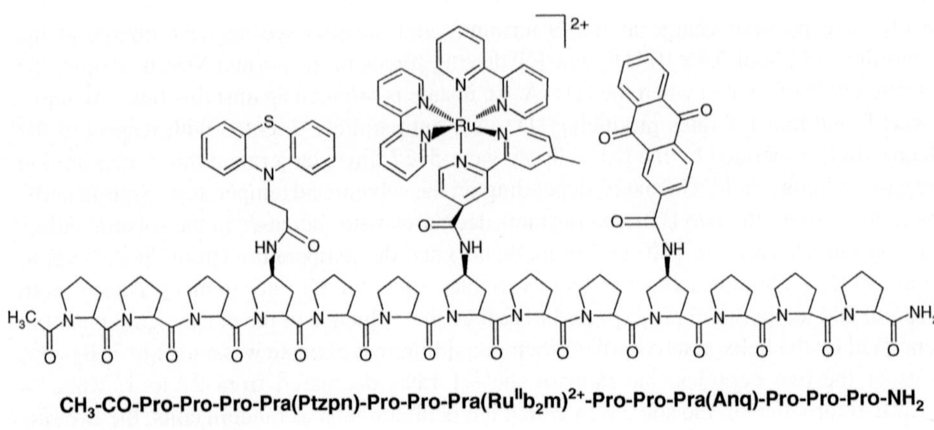

CH$_3$-CO-Pro-Pro-Pro-Pra(Ptzpn)-Pro-Pro-Pra(Ru^{II}b$_2$m)$^{2+}$-Pro-Pro-Pra(Anq)-Pro-Pro-Pro-NH$_2$

Figure 4.7 *Oligoproline assembly. Ru^{II}b$_2$m: ruthenium polypyridyl complex; Ptzpn: phenotiazine; Anq: anthraquinone. Reproduced with permission from reference [40]. Copyright (2004) American Chemical Society*

dimethylacetamide (DMA), respectively. This step is followed by fast $Ru^{II}b_2m^{-\cdot} \rightarrow ANQ$ ET, promoted by a favourable free energy of –0.11 to –0.48 eV depending on the solvent. The quantum yield for formation of the CS state was shown to be strongly solvent dependent, varying from 33% (DCE), 54% (BuCN), 86% (ACN) and 43% (DMA). The decay of the CS state is determined by direct $ANQ^{-\cdot} \rightarrow PTZ^{+\cdot}$ back ET, a highly exoergonic step that occurs in the Marcus inverted region ($|\Delta G^{\circ}| > \lambda$) with an efficiency also strongly solvent dependent.

4.3 Electron Transfer through Supported Peptide Matrices

In this section ET processes across peptide matrices supported on conductive (metal or semiconductor) substrates will be discussed, adopting the same frame used for introducing ET in solution, i.e. D–B–A systems with B representing a peptide spacer. However, a different 'electron source' (D) or 'electron sink' (A) will be introduced. Hybrid systems like peptide/electrode, electrode/peptide/electrode, nanoparticle/peptide, nanoparticle/peptide/electrode have been investigated, with the peptide bridge eventually functionalized with electro- and/or photoactive groups [41].

The mechanisms describing ET in these systems rely on the same superexchange (coherent tunneling) and hopping mechanisms already described for ET in solution (Figure 4.8).

Knowledge on ET at the molecule/metal interface has been greatly advanced by introducing electro- or photoactive molecules into self-assembled monolayers (SAMs). The SAM organization imposes a three-dimensional order on the charge-transfer bridge, so that even flexible chains can be studied at a fixed donor–acceptor distance.

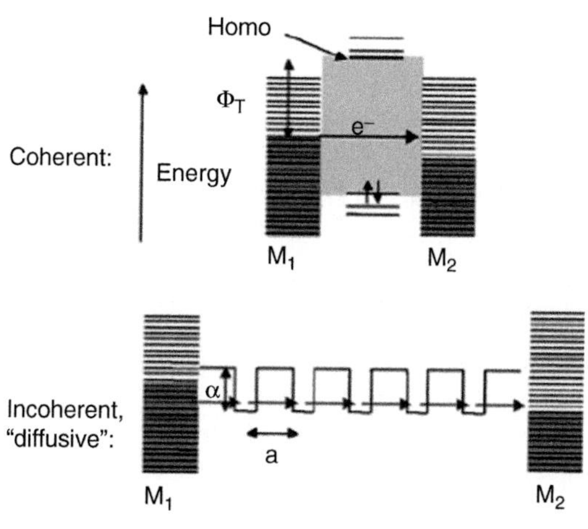

Figure 4.8 Superexchange (coherent tunneling) and hopping mechanism for ET in a metal–molecular bridge–metal assembly. Reproduced with permission from reference [41]. Copyright (2004) American Chemical Society

4.3.1 Theoretical Background

4.3.1.1 Adsorbate–Metal Electrode

The key issues involved in interfacial ET concern the *structure*, i.e. which atoms are present at the interface and where they are located, the *electronic configuration*, i.e. the electronic states of the molecules at the interface and how they are coupled to the electronic states of the metal substrate, and the ET dynamics, i.e. the time scale for electron transport and for response of the system to ET [22].

As regards the electronic configuration at the electrode/molecule interface, several aspects should be taken into account: (a) the extent of charge distribution due to adsorption or chemical linkage to the surface, (b) the alignment of molecular orbital energies to the metal Fermi energy and (c) the coupling between the discrete molecular states and the manifold of the continuous metal bands. Charge redistribution at the molecule/metal interface produces an electrostatic field that heavily affects ET, mimicking the rectifying effect of a Schottky barrier. Surface work function measurements with nanometric resolution quantitatively probed this effect [42, 43].

The rate constant for nonadiabatic interfacial ET from a metal electrode to a bound redox group can be expressed as [44, 45]:

$$k_{ET} = \frac{\rho_M(E_F)}{\hbar}\left(\frac{\pi}{\lambda kT}\right)^{1/2}\int_{-\infty}^{\infty}\left|H_{DA}(\varepsilon)\right|^2 \exp\left[-\frac{\left(\lambda+\left(E_{app}-E^{\circ}\right)e-\varepsilon\right)^2}{4\lambda k_B T}\right] f(\varepsilon)\,d\varepsilon \quad (4.13)$$

where $\rho_M(E_F)$ is the effective density of electronic states near the Fermi energy (E_F) of the electrode, E_{app} is the applied potential, $E°$ is the reduction potential of the redox species, ε is the electrode energy relative to the Fermi level and $f(\varepsilon)$ is the Fermi–Dirac distribution of occupied states in the metal. Equation (4.13) is the analogue of Equation (1.1) for an adsorbate–metal system. Both equations express nonadiabatic ET rate constants in terms of reorganization energy, free energy driving force and electronic coupling matrix.

In SAMs composed of organic thiols bound to gold, an exponential distance dependence of k_{ET} as a function of length and nature of the bridge has been demonstrated, with results similar to those inferred from ET studies in solution. For alkanes, β lies in the range 0.9–1.3 Å^{-1}, while for oligophenylene β is 0.5–0.7 Å^{-1} [46]. Peptide-based SAMs have been shown to have β values as low as 0.12 Å^{-1} from electrochemical experiments across oligoproline chains [47].

4.3.1.2 Metal–Bridge–Metal Systems

While in homogeneous and interfacial ET, rate constants are the experimental observables quantifying the process, resistance (or conductance) is the measured parameter for ET in metal–bridge–metal (M–B–M) systems. The electronic conductance is given by the Landauer equation [48, 49], which relates the conductance G to the electron transmission coefficient T:

$$G = \frac{2e^2}{h}T(E_F) \quad (4.14)$$

where E_F is the Fermi level of the contacts at zero applied potential. For $T=1$, $G=G_0=2e^2/h=12.9\ \text{k}\Omega^{-1}$, the maximum conductance per molecule. Like rate constants, the distance dependence of conductance can be shown to follow an exponential decrease:

$$G = G_0 \exp(-\beta r_{DA}) \tag{4.15}$$

or an 'ohmic' behaviour:

$$G \propto \frac{1}{N}\exp\left(\frac{-\Delta E_{DB}}{RT}\right) \tag{4.16}$$

where, as in the case of homogeneous ET (Equation (4.11)), N is the number of intervening bridging units.

There are several experimental arrangements to test the conductance of the bridge in M–B–M systems: (i) *break junctions*, in which single molecules are trapped between metal electrodes [48, 49], (ii) conducting AFM tip on SAMs immobilized on gold substrates [50], (iii) metal–bridge–nanoparticle-conducting tip [51]. It was found that, at low bias (<50 mV), the absolute magnitude of molecular conductance in M–B–M systems is close to what can be inferred from measurements of interfacial ET rate constants.

The most important parameter in M–B–M systems is the alignment of the metal Fermi energy (E_F) to that of the molecular HOMO and LUMO levels of the bridge as a function of the applied potential. A suitable voltage bias can bring a molecular level into resonance with a metallic conduction state producing a large turn-on of the current. In the off-resonance condition the alignment of E_F affects the tunneling decay rate β and this exponentially affects the current intensity.

However, the exact resonance conditions are difficult to assess, due to involvement of an additional barrier at the electrode/molecule interface. This is reminiscent of the Schottky barrier φ at the metal/semiconductor contact, i.e. the difference between the metal Fermi level E_F and the semiconductor conduction band E_c, $\varphi=E_c-E_F$. Paralleling the molecular HOMO–LUMO energy difference to the conduction–valence gap of a semiconductor, the idea is that metal states having energy within that of the gap may tunnel to the upper or to the lower energy state by electron or hole transfer.

Density functional studies on phenyldithiolate connected to two Au contacts demonstrated that dipole effects, caused by the electrostatic character of the Au–S linkage, affect the line-up of the M–B–M system [52]. These considerations are also related to the applied voltage profile, i.e. the I/V response of M–B–M systems under bias, as the largest potential drop has been found to occur for atoms near the electrode surface [53].

Let us consider an M–B–M system with equal metal electrodes (same E_F), displaced from each other by the energy eV, where V is the external voltage bias (Figure 4.9). Assuming that the molecular levels are not perturbed by the applied potential V, electrons that can transfer through the molecular bridge are those within the energy window between the left (E_F^L) and right (E_F^R) electrode energy Fermi levels. In the case of symmetric bridge and bridge/electrode contacts, the left electrode energy will move down by $eV/2$ ($E_F^L = E_F - eV/2$), while the right electrode will move up by the same quantity ($E_F^R = E_F + eV/2$). This produces a symmetric I/V curve, i.e. $I(+V)=-I(-V)$ [41].

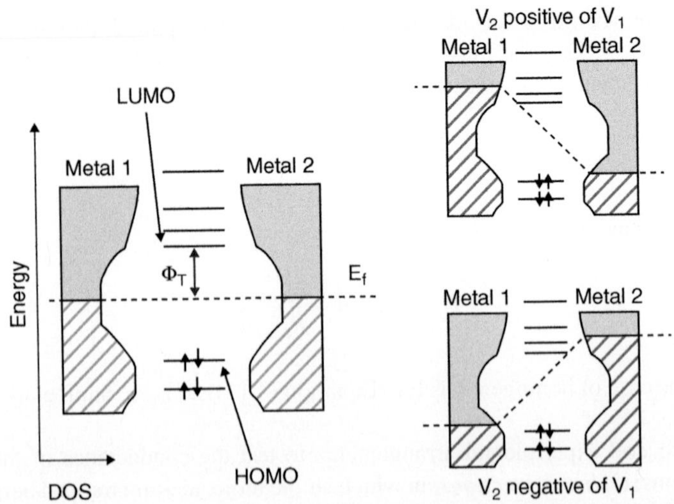

Figure 4.9 Alignment of energy levels in a metal–molecule–metal device. Reproduced with permission from reference [41]. Copyright (2004) American Chemical Society

On the other hand, if the bridge is asymmetric or different left/right electrode/molecule interfaces occur, an asymmetry parameter η can be introduced, so that for the left electrode energy one obtains $E_F^L = E_F - \eta eV$ and for the right electrode $E_F^R = E_F + (1-\eta)eV$. The difference between the left and right electrode is eV, as required, but the I/V curves are asymmetric as η differs from 1. For the extreme case of $\eta=0$, the left electrode resonates with the molecular states, while the right electrode is off-resonance by ± 1 eV (this might model an M–B–M system with a molecular bridge covalently linked to the left electrode and weakly interacting with the right electrode). The magnitude of the field in I/V experiments is in the range 10^7–10^8 V/m, comparable to that commonly used in ET experiments in solution ($10^9/D_s$ V/m, where D_s is the solvent dielectric constant).

A relation correlating conductance and ET rates was recently proposed [54–56], i.e.

$$G = e^2 k_{ET}\left(E_F\right)\rho_M\left(E_F\right) \tag{4.17}$$

where ρ_M is the electrode density of states. In the case of coherent tunneling (high energy gap between the D–A states and the intermediate bridge CT states), the ET rate constant can be expressed as

$$k_{ET} = \frac{2\pi}{\hbar}\left|H_{DA}\right|^2 DOS \tag{4.18}$$

where DOS is the effective density of states, corresponding to FCWD of vibronic states (Equation (4.3)) for homogeneous ET. When ET from a molecular adsorbate to a metal electrode takes place, an additional factor γ_M related to the electronic manifold of the metal must be introduced, i.e.

$$DOS = \gamma_M (FCWD) \tag{4.19}$$

For a gold electrode $\gamma_{Au}=0.024$ and $\rho_{Au}=0.3\,eV^{-1}$.

This approach clearly shows that the contacts are as important as the bridge molecular structure in governing ET.

4.3.2 Seminal Experimental Results

4.3.2.1 Electrochemical Studies

Electrochemical studies of peptides immobilized on electrodes led to fundamental achievements in the comprehension of charge-transfer across peptides. In particular, SAMs functionalized with redox active species have contributed powerfully to characterize ET processes at nanoscale.

The two-dimensional quasi-crystalline arrangement of peptide SAMs minimizes the conformational freedom of the single peptide chains. Electrochemical techniques, such as cyclic voltammetry (CV), chronoamperometry (CA) and electrochemical impedance spectroscopy (EIS) have been used extensively to study the ET kinetics of peptide SAMs immobilized on conductive surfaces.

The influence of HB to the ET efficiency was investigated by Kraatz *et al.* [57] in H-bonded peptide assemblies mimicking the interstrand collagene structure. In particular, tri-, hexa- and nonapeptides comprising 1, 2 and 3 Pro–Pro–Gly triads, respectively, were functionalized at the N-terminus by a cysteamine group for immobilization on gold and a ferrocenoyl group at the C-terminus as a redox active probe. Each single peptide strand adopts a left-handed polyproline-II structure and the three peptide strands were found to assemble into a supramolecular right-handed triple helical structure by forming interpeptide H-bonds between the amide NH of each glycine residue and the carbonyls of proline residues on adjacent peptide strands, as shown in Figure 4.10.

As a result, a hydrophilic corridor on the interior and a hydrophobic outer surface with all the aliphatic residues pointing towards the exterior of the triple helix is formed. CD spectra showed that while the hexa- and nonapeptides adopted a collagen-like structure, the tripeptide existed predominantly as a monomer in solution. ET rates were obtained by CA and EIS measurements. A small kinetic effect, estimated by the ratio of the rate constants measured in water and in D_2O, was observed, ranging from $\frac{k_{ET}^{H_2O}}{k_{ET}^{D_2O}} = 1.5$ for the tripeptide to 1.1 for the hexapeptide to 1.2 for the nonapeptide. This effect does not depend on structural differences among the three peptide films and was ascribed to the influence of H-bonding in the ET process. Furthermore, the k_{ET} exhibit an inverse linear dependence with increasing peptide length, suggesting that through-space tunneling is not the major ET pathway. This conclusion was strengthened by the observation that the experimental rates are 4 orders of magnitude larger than those expected for a pure SE mechanism.

Long-range ET over 4 nm was investigated by Morita and Kimura [58] in two helical hexadecapeptides, both functionalized by a redox active ferrocenyl group and a lipoic acid

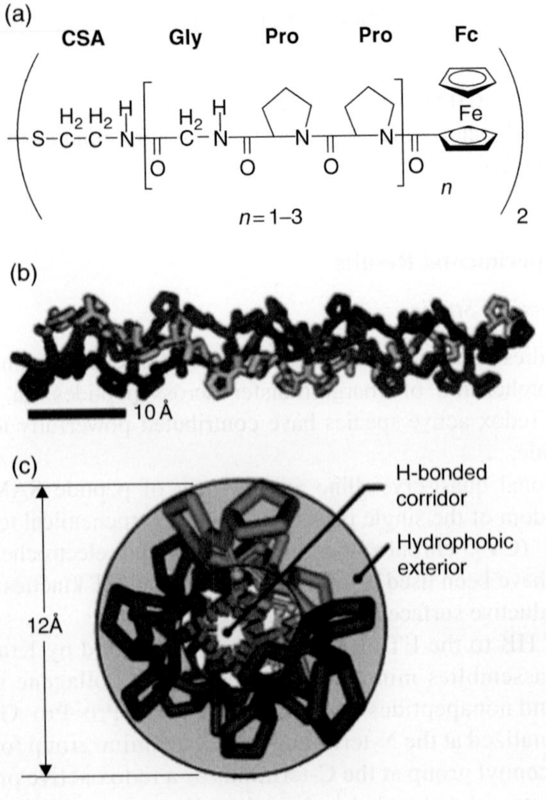

Figure 4.10 *(a) Molecular structure of the peptide investigated; (b) supramolecular coiled-coil structure adopted by a Pro–Pro–Gly repeat unit; (c) along helical axis view of the coiled-coil structure, showing the H-bonded inner corridor and the hydrophobic outer surface. Reproduced with permission from reference [57]. Copyright (2004) American Chemical Society (see color plate figure)*

for chemical linking to a gold electrode. The two peptides differ only in the position of the two functional groups that were alternatively exchanged at the N- and at the C-termini, so that the helix macrodipole pointed in opposite directions to gold. CA experiments showed that the ET rate constant for the peptide SAM in which the helix dipole points with its positive end to the gold electrode is threefold larger than the other (2.0 s^{-1} versus 0.68 s^{-1}). The values of the experimental rate constants are larger by 2–3 orders of magnitude with respect to those calculated in the case of a pure SE mechanism (k_{ET} ~ 0.003 s^{-1}), assuming an exponential dependence on the distance and an electronic coupling parameter equal to 1.1 Å^{-1} per CH_2 group and 0.66 Å^{-1} per peptide residue [58]. This finding suggests that the hopping mechanism predominates at such large distances, the amide bonds being suggested as the most probable hopping sites. Notably, the formal potential of the ferrocenyl group in the two peptide SAMs (0.643 and 0.633 V, respectively), measured by CV experiments, is not affected by the electrostatic field of the helix dipole. This suggests that the main effect of the helix dipole is lowering, by the positive charge at the N-terminus, the activation barrier at the interface between the gold surface and the peptide SAM, i.e. at the Au–S contact. Lowering the barrier height will lead to an acceleration of the rate-determining step of ET between gold and the nearest amide group.

In agreement with previous findings, a very weak dependence of the ET rates on the applied overpotential was found for two octadecapeptides functionalized with a ferrocenyl unit at more than 40Å distance from a gold electrode [59]. This result strongly suggests that the hopping mechanism also predominates in the present systems. Accordingly, ET rate constants calculated by assuming a pure SE mechanism are orders of magnitude lower than those obtained by the experimental data (i.e. 42 s^{-1} versus 0.0005 s^{-1} and 28 s^{-1} versus 0.06 s^{-1} for the two octadecapeptides investigated). Also in this case, amide groups were suggested to be the plausible hopping sites, because they are regularly arranged and close to each other, i.e. they are strongly electronically coupled. The carrier is considered to be a hole because the amide LUMO (–1.2 V) is so high that electron injection from the HOMO (–5.1 eV) of ferrocene to the amide group is very unlikely.

Two ET steps should be considered at the electrode/peptide SAM interface: (i) ET from the amide HOMO level (–6.5 eV) to the S-atom (–5.5 eV) and (ii) ET from the S-atom to gold (–5.1 eV). The electrostatic field of the helix dipole was found to affect the ET process (k_{ET} = 42 s^{-1} for the peptide functionalized with ferrocenyl at the C-terminus compared with k_{ET} = 28 s^{-1} for the N-terminus derivative). More effective was the substitution of a phenylene spacer to a methylene group, fastening the ET rate constant to 257 s^{-1}. From the reported values, it appears that electron hopping among the amide groups is the rate determining step. Nevertheless, ET at the peptide/gold contact, where the applied potential drops dramatically with respect to the smooth decrease along the peptide chain, should also be taken into account.

The rectification properties of helical hexadecapeptides, formed by eight repeats of the L-Leu–Aib dyad, and engineered in an M–B–M junction, were also investigated by Kitagawa *et al.* [60] by ultra-high vacuum scanning tunneling microscopy (UHV-STM) experiments. Molecular rectifiers are the most important components in molecular electronics for potential application as diode elements and in memory storage devices. The bare hexadecapeptide and the same hexadecapeptide, functionalized at the C-terminus by a redox-active ferrocenyl group, were alternatively embedded in a dodecanethiol SAM forming isolated peptide bundles, as clearly imaged by STM measurements. A metal/SAM/STM tip junction was therefore assembled, the *I*/*V* profile of which can be measured by scanning tunneling spectroscopy (STS) experiments. Typical *I*/*V* profiles of the dodecanethiol and of the bicomponent peptide/dodecanethiol SAMs are shown in Figure 4.11.

It should be noted that the *I*/*V* response of the dodecanethiol SAM exhibits a linear current response at low bias and an exponential response at higher biases, this slight asymmetry being ascribed to the asymmetry of the molecule/electrode contact. More asymmetric responses were obtained in the case of the peptide/dodecanethiol SAM, with currents at negative bias nearly fivefold larger than those measured for the alkanethiol SAM. The presence of the ferrocenyl group also enhances the asymmetry of the *I*/*V* profile by a factor 2 with respect to the bare hexadecapeptide/dodecanethiol SAM. Notably, currents measured at positive biases are comparable for all the systems investigated.

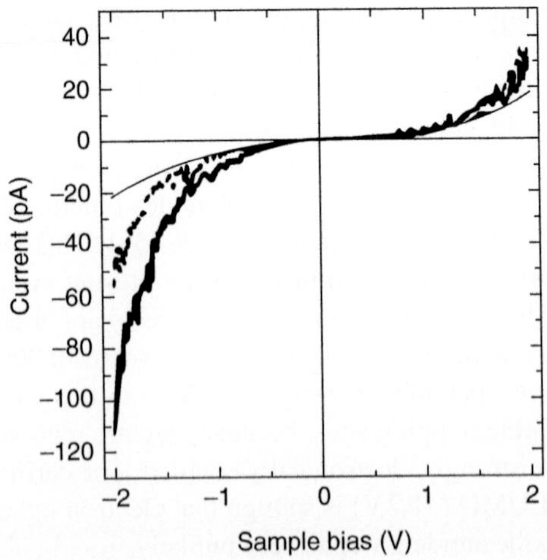

Figure 4.11 *I/V curves of dodecanethiol (thin solid line), hexadecapeptide/dodecanethiol (dashed line), and ferrocenyl-hexadecapeptide/dodecanethiol (thick solid line). Reproduced with permission from reference [60]. Copyright (2005) American Chemical Society*

The Simmons equation models ET through a metal–insulator–metal junction at an applied bias potential V as an electron tunneling process through a rectangular barrier of length d and height φ [61]:

$$I=\frac{Ce}{4\pi^2\hbar d^2}\left\{\left(\varphi-\frac{eV}{2}\right)\exp\left[-\frac{2\sqrt{2m}}{\hbar}\alpha\sqrt{\left(\varphi-\frac{eV}{2}\right)}d\right]\right.$$
$$\left.-\left(\varphi+\frac{eV}{2}\right)\exp\left[-\frac{2\sqrt{2m}}{\hbar}\alpha\sqrt{\left(\varphi+\frac{eV}{2}\right)}d\right]\right\} \tag{4.20}$$

C, α and ϕ can be used as adjustable parameters in the fitting of the experimental data. At low bias Equation (4.20) can be approximated by

$$I=\frac{Ce^2\sqrt{2m\phi}}{h^2 d}\alpha V\exp\left[-\frac{2\sqrt{2m}}{\hbar}\alpha\sqrt{\phi}d\right] \tag{4.21}$$

the attenuation factor β being as follows:

$$\beta=\frac{2\sqrt{2m}}{\hbar}\alpha\sqrt{\phi} \tag{4.22}$$

Values of β of $1.1\,\text{Å}^{-1}$ and $0.75\,\text{Å}^{-1}$ for the dodecanethiol and the bare hexadecapeptide/dodecanethiol SAMs were found, respectively, in agreement with literature data.

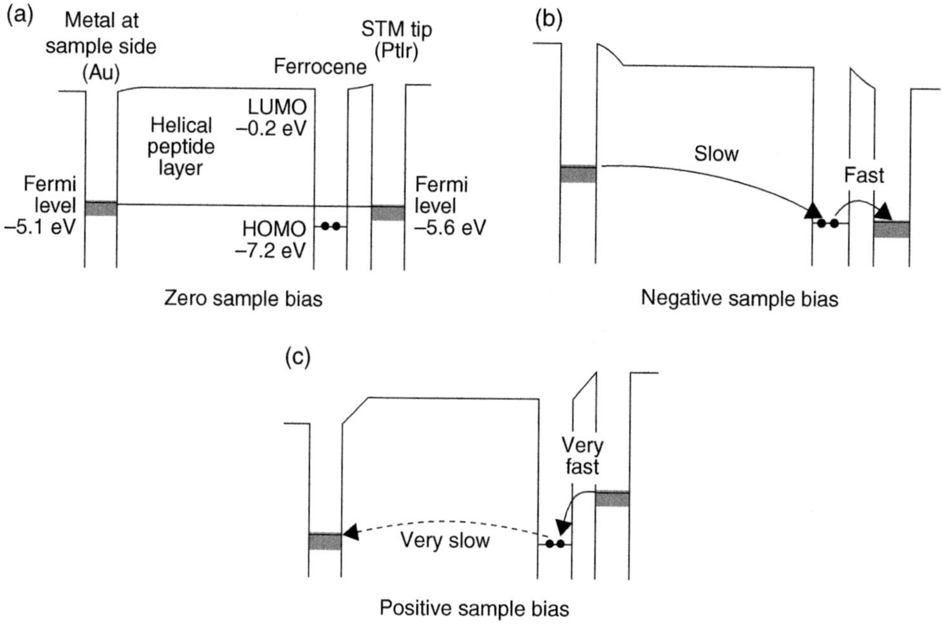

Figure 4.12 *Energy diagrams for electron transmission in the metal–SAM–STM tip junction (a) at zero sample bias, (b) at negative sample bias and (c) at positive sample bias. Reproduced with permission from reference [60]. Copyright (2005) American Chemical Society*

Let us consider in detail this example. The ferrocenyl group, separated from the metal by the peptide SAM and, hence, too far to be electronically coupled with it, adds to the system HOMO and LUMO energy levels positioned at –0.7 and –0.2 eV with respect to the vacuum, respectively. The Fermi levels of the metal underneath the monolayer (Au) and the tip (PtIr) are –5.1 eV and –5.6 eV, respectively, while the HOMO and LUMO of the amide groups are –9.0 eV and –0.4 eV far away from the metal energy levels. The energy levels of the system at zero bias are schematized in Figure 4.12a, showing that while the ferrocenyl LUMO energy is well above the Fermi level of the two metal contacts, the HOMO energy is positioned just below them.

When a suitable negative bias is applied to the gold electrode, the STM tip potential is lowered with respect to the ferrocene HOMO level (Figure 4.12b), giving rise to a fast ferrocene→(adjacent) tip ET, followed by slow ET from (distant) gold to ferrocene. In the case of a suitable positive bias, the gold electrode can be driven to resonate with the ferrocene HOMO state, reversing the ET direction (Figure 4.12c). However, the driving force for the ferrocene→tip ET is greater than for the reverse ET, leading to faster ET at negative bias with respect to that at positive bias, explaining the rectification properties of the investigated M–B–M system.

The conduction properties of two dodecapeptides, denoted as F4 and F2, comprising repeated Lys–Aib–Ala–Aib quadruplets, were investigated by UHV-STM experiments in a metal–SAM–STM tip arrangement by Kimura and coworkers [62]. F4 and F2 were functionalized by four and two ferrocene units, respectively, two of which are positioned at the peptide chain termini, with both derivatized at the N-terminus by a lipoic group for immobilization on gold. F0, i.e. the same peptide chain devoid of ferrocene units, was also investigated as a reference compound. IR-RAS and ellipsometry measurements showed that while F4 and F2 take a vertical orientation on gold, F0 was almost randomly oriented in the SAM. Accordingly, STM measurements revealed the formation of a densely packed SAM from F4 and F2 and a disordered monolayer from F0. STM imaging of bicomponent SAMs formed by deposition of a 1:10 peptide/dodecanethiol solution showed peptide bundles protruding from the alkanethiol monolayer. *I*/*V* measurements on F2/alkanethiol or F4/alkanethiol bicomponent SAMs found that the incorporation of ferrocene units results in fivefold larger current flows with respect to that measured for F0, suggesting electron hopping through the ferrocene units. However, contrasting results with earlier findings have been found: (i) the *I*/*V* response of the F2 SAM is nearly symmetric for negative and positive bias. In this case the difference in the ET driving force is probably balanced by the helix dipole effect on the ET process; (ii) the ET rates of F2 and F4 are comparable, despite the presence in the latter compound of two additional ferrocene units, which are very likely to act as hopping sites. The problem in this case could arise from the orientation of the intervening ferrocene units probably not being correctly relayed along the α-helical scaffold.

The effects of different constituent amino acids, molecular orientation and molecular packing on long-range ET through a helical peptide monolayer were investigated by Takeda and coworkers [27]. In particular, helical peptides having a thiophenyl group at the N-terminus and a redox-active ferrocene moiety at the C-terminus were immobilized on gold electrodes. The ET rate dependence on the peptide sequence was studied by investigating ET from ferrocene to gold in two hexadecamer peptides composed of eight L-Leu–Aib and L-Ala–Aib repeats. The bulky isobutyl side chain of Leu makes the peptide backbone separation in the Leu–Aib SAM larger than the separation realized in the Ala–Aib SAM. Moreover, the Leu–Aib peptide forms a tightly packed monolayer through interdigitation among the Leu side chains. CA and EIS measurements showed that ET is definitely slower in the Leu–Aib peptide SAM than in the Ala–Aib peptide SAM, i.e. the standard ET rate constants from EIS being $46 \pm 5\,s^{-1}$ versus $603 \pm 289\,s^{-1}$ and $50 \pm 6\,s^{-1}$ versus $889 \pm 300\,s^{-1}$ from CA, respectively. The slower ET rates can be interpreted as the result of the larger separation among the peptide chains in Leu–Aib peptides giving rise to less probable electron tunneling or suppression of interchain hole hopping among the amide groups. A tight molecular packing could also be responsible for dynamical effects, i.e. restricted molecular motions lowering the ET process in a tight monolayer. Electrochemical experiments, i.e. CA and EIS, on bicomponent Ala–Aib SAMs suggest that ET is accelerated as the monolayer becomes looser to a certain extent. The proposed mechanism is that, following ET from the nearest amide group to gold via the sulfur atom to form a cation radical of the amide group (hole), the hole hops across the amide groups until it reaches the ferrocene moiety. This mechanism should be favoured by a certain flexibility of the peptide chain.

Interestingly, Schlag and coworkers [63–65] demonstrated by theoretical calculations that the dihedral angles of the peptide backbone significantly change to form an active conformation when a hole is transferred from one amide group to the other. In the active conformation, a strong electronic coupling is attained between the neutral and cationic radical of amide groups, the activation barrier becoming almost negligible.

Local vibrational motions at the interface and along the peptide backbone may be hindered by tight molecular packing of the monolayer hampering hole hopping and lowering the overall ET rate.

Gatto *et al.* [66] have recently reported on an hexapeptide, comprising three Aib, two Ala and a 2,2,6,6-tetramethylpiperidine-1-oxyl-4-amino-4-carboxylic acid (TOAC) residues. TOAC is characterized by a nitroxide group, a stable radical species widely used as a probe in ESR [67] and fluorescence studies [68, 69], which endows the molecule with peculiar redox activity. CV measurements carried out at different sweep rates showed that the oxidative peak of TOAC linearly increased with the scan rate, indicating that the observed CV peaks arose from surface-bound TOAC groups. The measured standard potential (0.60 V versus SCE) was found to be very similar to that measured in solution, suggesting a relatively small interfacial potential drop through the peptide layer.

The coverage of the electrode surface by the peptide SAM can be obtained by analysing the dependence of the CV peak intensity (I_p) on the voltage scan rate (v) through the equation

$$I_p(v) = \frac{Nn^2F^2}{4RT}v \tag{4.23}$$

where N is the number of redox-active sites on the surface, n the number of electrons transferred and F the Faraday constant. From the slope of I_p versus v, a surface coverage of 14.4×10^{-11} mol/cm^2 was obtained, in good agreement with that estimated by assuming closely packed 3_{10}-helix peptide chains (16.6×10^{-11} mol/cm^2). CA experiments were also carried out for characterizing the ET process from TOAC to the gold electrode. The through-space distance was estimated to be around 17 Å, obtained by considering the lipoic acid in all-*trans* conformation and the peptide as a rigid 3_{10}-helix (2.0 Å rise per residue) vertically oriented with respect to the normal surface by 40°, as suggested by IR-RAS measurements [70]. Experimental *I*–*t*, i.e. current intensity versus time, curves in the 2–200 μs time region were collected at positive (TOAC oxidation) and negative (TOAC$^+$ reduction) overpotentials, as shown in the Tafel plot reported in Figure 4.13.

Interestingly, the Tafel plot shows a marked asymmetry of the k_{ET} values obtained in oxidative (anodic, k_a) and reductive (cathodic, k_c) conditions, the ET standard rate constant extrapolated at zero overpotentials being $k_{ET}^0 = 9.2\pm0.1\ \mathrm{s}^{-1}$. This effect can be ascribed to the electrostatic field associated to the helix dipole, which favours anodic ET to the gold surface. However, the ratio between the anodic and cathodic rate constants ($k_a/k_c = 1.19\pm0.05$) was found to be definitely smaller than that observed in α-helical peptides, probably because of the distorted H-bond pattern of 3_{10}-helix with respect to α-helix.

Asymmetric ET through a helical polyalanine SAM chemisorbed on a gold surface was also studied by Sek and coworkers [71]. The polyalanine scaffold was functionalized by a ferrocenecarbonyl electroactive group at the N-terminus and a cysteamine linker at the C-terminus for binding to gold substrates. The ET efficiency through peptide molecules was characterized by electrochemical techniques, i.e. CV and impedance spectroscopy, and STS on a bicomponent SAM formed by ferrocene-poly(Ala) and *n*-octadecanethiol. The standard rate constant for ET through the polypeptide, determined from the Tafel plot by extrapolating to zero overpotential, was found to be $k_{ET}^0 = 0.30\pm0.08\ \mathrm{s}^{-1}$. Assuming an

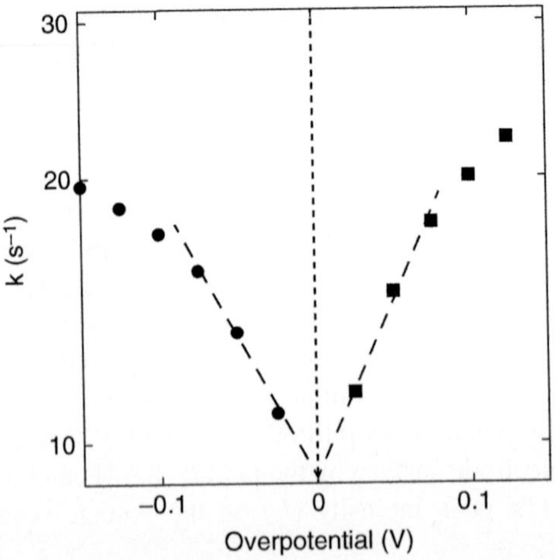

Figure 4.13 Dependence of the ET rate constant on the applied overpotential for TOAC oxidation (filled squares) and TOAC⁺ reduction (filled circles) at the gold electrode modified by the peptide SAM. Reproduced with permission from reference [66]. Copyright (2008) John Wiley & Sons, Ltd

exponential dependence of the ET rate constant from the ferrocene–gold distance and a through-space distance of 28.5 Å, a tunneling coefficient $\beta = 0.73 \pm 0.01$ Å^{-1} was obtained. Significantly, a marked asymmetry of the Tafel plot was observed, the reductive process, i.e. ET from gold to ferrocene (cathodic current, k_c), being favoured with respect to the oxidative process, i.e. ET from ferrocene to gold (anodic current, k_a). It was found that $k_c/k_a = 1.43 \pm 0.07$. Significantly, this is opposite to what was found in references [60] and [66], because the peptides were linked to the gold electrodes with the helix dipole pointing to the opposite direction, i.e. at the N-terminus in references [60] and [66] and at the C-terminus in reference [71]. This finding therefore confirms the role of the electrostatic field associated to the macrodipole of helical peptides on the ET rate. STS experiments carried out on the single-component SAM formed by H–(Ala)$_{14}$NH–(CH$_2$)–SH, provided a comparable value for the tunneling coefficient, i.e. $\beta = 0.69 \pm 0.12$ Å^{-1}. Current/bias voltage experiments also found that currents at positive biases were 2–3 times smaller than those measured at negative biases, in agreement with electrochemical results.

The effect of the linker molecule on monolayer formation and long-range ET in a helical peptide SAM was recently investigated by Kimura and coworkers [72]. The linker structure was found to influence the monolayer packing, orientation, dynamics and peptide surface density. Furthermore, the electronic structure of the linker modified the electronic coupling between the redox unit and the electrode, affecting the ET rate through the SAM/metal junction. In this study, helical peptides functionalized by a ferrocenyl redox unit at the C-terminus and three different linkers at the N-terminus for immobilization on gold were synthesized. The linkers were 4-thiobenzoic acid (NL), 3-fluoro-4-thiobenzoic acid (FL) and 2-methoxy-4-thiobenzoic acid (ML) in the acetyl-protected form, while the peptide chain was a hexadecamer formed by alternating sequences of L-Leu and Aib. The ET rate constants were determined by CA and EIS. In the latter technique ET rates were determined from the measurement of the electron transfer resistance (R_{ET}) and capacitance (C_{ET}) of the peptide layer by the equation

$$k_{ET}^{0} = \frac{1}{2R_{ET}C_{ET}} \tag{4.24}$$

Comparable values were obtained for NL, FL and ML SAMs, respectively, i.e. k_{ET}^{0} = 11.3±1.9 s^{-1}, 12.0±2.0 s^{-1} and 14.4±0.8 s^{-1}. Since the HOMOs of the linkers and the amide groups are closer to the gold Fermi level than their LUMOs, the ET process is thought to proceed via hole charge carriers for both tunneling and hopping mechanisms.

Ultra-long-range ET across a peptide SAM composed of 80-mer α-helices was investigated by Kimura *et al.* [73]. The peptide building block comprised 40 consecutive Ala–Aib dyads, a lipoic group at the N-terminus for immobilization on gold and a redox-active ferrocene unit at the C-terminus. From the molecular length, estimated to be 134 Å, and the tilt angle of the peptide chain in the SAM, as determined by IR-RAS measurements (37±8°), the thickness of the peptide SAM was evaluated to be 95±10 Å, to be compared with the 75±12 Å thickness obtained by ellipsometry measurements. A reversible peak associated to the ferrocenium/ferrocene redox pair with a formal potential of 0.45 V clearly demonstrated that ET across the peptide SAM over such a long distance is possible. Standard ET rate constants were determined by EIS measurements, showing values spread out over 2 orders of magnitude (0.1–10 s^{-1}). This finding is taken as strong evidence of a predominant hopping mechanism, because according to a pure SE tunneling mechanism, the ET rate constants should attain totally negligible values (10^{-21}–10^{-47} s^{-1}). Following this view, an electron is transferred from the N-terminal amide group to gold, generating an amide cation radical (hole) that hops through the amide groups to reach the C-terminus, where it is finally reduced from the ferrocene unit. The intramolecular pathway, characterized by hopping across the amide sites of a single peptide chain, was thought to predominate due to the strong electronic coupling between neighbouring amide groups. However, HB-mediated or through-space intermolecular pathways should also be considered. Temperature-dependent EIS measurements allowed determination of the ET activation energy (E_a = 0.57±0.11 eV), which compares very well with the theoretical value of E_a = 0.63 eV. Interestingly, EIS experiments indicate that ET is accelerated as the capacitance increases, with a characteristic saturation effect at high capacitances. This finding suggests that the dynamics of the peptide chains could positively affect the ET rate (*activated hopping* mechanism), facilitated by loose packing of the monolayer. This contribution emphasizes the role of helical peptides as electron mediators over very long distances (up to 120 Å), paving the way for their use as molecular wires.

4.3.2.2 Conductance of Single Peptide Molecules

The conductance of individual peptide molecules has been recently investigated by Tao and coworkers [74, 75] in an M–B–M system, where M stands for the two gold electrodes and B for different oligopeptides of formula Cys–Cysteamine (1), Cys–Gly–Cysteamine (2), Cys–Gly–Cys (3) and Cys–(Gly)$_2$–Cysteamine (4), alternatively bridging the two metal contacts. The thiol groups of Cys and Cysteamine provided two S-terminals for binding to Au electrodes. Interestingly, the amine group of cysteamine in 1, 2 and 4 positions and the

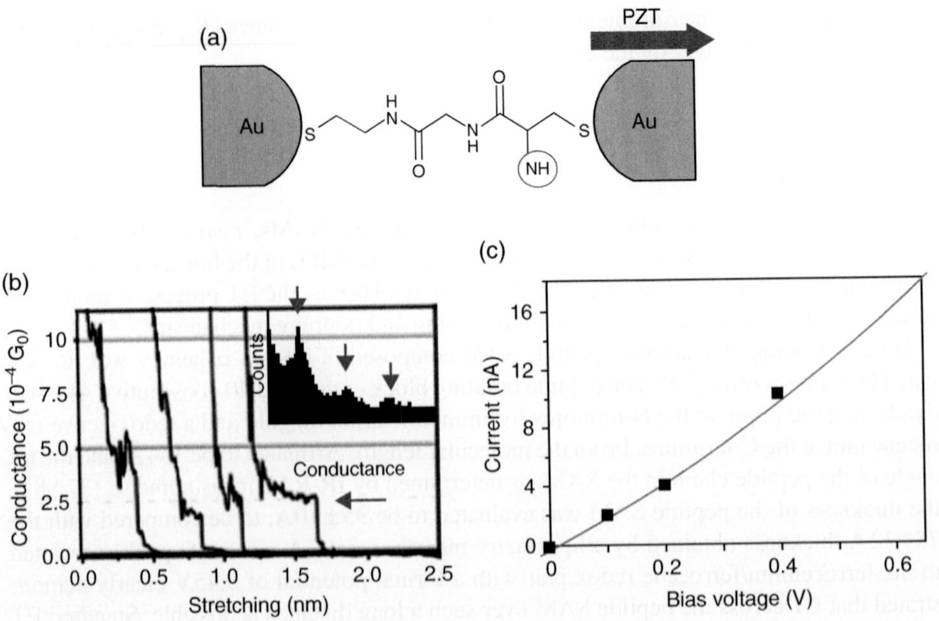

Figure 4.14 *(a) Schematic illustration of a molecular junction formed by separating two electrodes. (b) Several typical conductance curves of peptide 1 during the formation of molecular junctions. The discrete changes in the conductance are due to the breakdown of individual molecules. The inset in (b) is a conductance histogram constructed from over 500 individual conductance curves, showing peaks near integer multiples of* $2.5 \times 10^{-4}\,G_0$*. (c) I–V characteristic of peptide 1 determined from conductance histograms at different bias voltages. Reproduced with permission from reference [75]. Copyright (2005) American Chemical Society*

Cys carboxyl group in 3 position can be protonated by changing the pH of the solution, making it possible to study the pH dependence of the charge-transfer process. Individual molecular junctions can be created by repeatedly moving the two gold electrodes in and out of contact in a 1 mM solution of the sample molecules. A piezoelectric transducer was able to pull the second electrode (an STM conductive tip) in on/off contact with a peptide SAM covalently linked to a gold electrode, controlling the formation of single molecular junctions, which were counted by the number of steps in the conductance between the two electrodes (Figure 4.14).

Due to variations in the microscopic detail of the molecule–electrode contact, a statistical analysis over many measurements is generally needed to obtain significant results. A typical conductance histogram (see the inset of Figure 4.14b) shows well-defined peaks near 1 times, 2 times and 3 times the fundamental conductance value, ~$2.5 \times 10^{-4}\,G_0$. Furthermore, conductance measurements as a function of pH revealed that the maximum conductance decreased from 2.5×10^{-4} to $1.2 \times 10^{-4}\,G_0$ when the solution pH was increased from pH 2 to pH 13. Notably, a sigmoid-like titration curve for peptides 1 and 2, and a complex behaviour for peptide 3, where both the carboxyl and amine groups determine the protonation equilibria, could be observed. The presence of charges along the peptide bridge

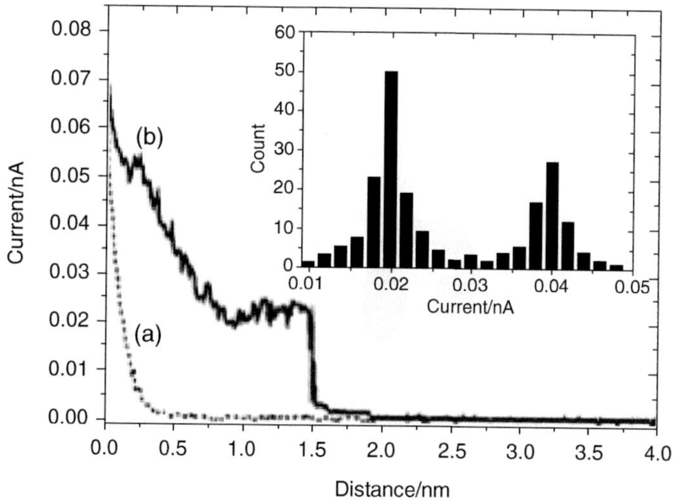

Figure 4.15 *Current–distance curve for bare (a) and peptide-modified (b) gold substrate (bias voltage of −0.3 V). Inset: histograms of the current values obtained for Au–peptide–Au junctions at the bias of −0.3 V. Reproduced with permission from reference [76]. Copyright (2005) American Chemical Society*

clearly affects the tunneling barrier and thus the conductance, lowering the barrier for ET at pH 2, where the amine and carboxyl groups are protonated.

In agreement with a coherent tunneling mechanism, the peptide conductance (G) showed an exponential dependence on the peptide length (L), i.e.

$$G = A\exp\left[-bL\right] \tag{4.25}$$

with $A = 0.15\,G_0$ and $b = 0.87$ (±0.07) Å^{-1} per residue.

A gold/peptide SAM/STM tip arrangement was also used by Sek *et al.* [76, 77] for studying the electrical behaviour of molecular junctions incorporating an α-helical peptide. A tetradecamer peptide was functionalized at both terminal ends with two thiol groups, which can be chemically linked to the gold substrate and the gold-coated STM tip. The S-atom of the N-terminal Cysteine was protected with an acetamidomethyl group, while a Cysteamine linker was inserted at the C-terminus. A dodecanedithiol SAM was also prepared and used as a reference compound. The junctions were formed by placing the gold STM tip in contact with the monolayer linked to a gold electrode and currents in the order of 50–500 pA were obtained. The initial resistance of the tip–sample gap was observed to vary between 5×10^8–5×10^9 Ω. When the STM tip was lifted while keeping a constant *xy* position, a current as a function of distance was recorded. Reliable count statistics were obtained by repeating the procedure for at least 200 times for each bias voltage. Typical current–distance curves were reported for the bare and peptide-modified substrates in Figure 4.15.

While the bare gold substrate shows a fast exponential decay of the current with the distance from the tip, a current plateau was observed for gold substrates modified by a peptide SAM. The current was found to fall abruptly when the gold/SAM/tip contact broke away. Significantly, the maxima of the histogram obtained by reporting the number of counts as a function of the measured current (inset of Figure 4.15) are located at multiples of the current fundamental value. A reasonable explanation of this result is that consecutive peaks in the histogram correspond to conductivity through one or two molecules trapped between the tip and the substrate. Furthermore, the dependence of the current on the applied bias was measured for the peptide and the dodecanethiol SAM between –0.4 and +0.4 V. Interestingly, asymmetric *I*/*V* curves were obtained only for the peptide SAM, confirming the rectification properties of helical peptide spacers. In particular, lower currents were measured in the case of positive applied voltage, inducing negative tunneling currents, i.e. electron flow from the tip to the substrate. The experiments showed that ET from the substrate to the tip is more efficient, the currents recorded for negative bias being up to four times larger than those recorded for the positive bias. This finding was ascribed to the electrostatic field of the helix dipole, producing an effective positive charge at the N-terminus linking the STM tip.

4.3.2.3 Photocurrent Generation across Peptide Self-Assembled Monolayers

Photocurrent generation (PG) experiments on peptide self-assembled monolayers covalently linked to a gold electrode via Au–S linkages have been thoroughly carried out by Kimura and coworkers [78]. To this aim, a tridecapeptide, composed of Ala–Aib dyad repeats, functionalized by an *N*-ethylcarbazolyl (ECz) group at the N-terminus and a disulfide group at the C-terminus, was synthesized. The Ala–Aib dyad was chosen for the high propensity to attain helical conformations, ECz as a photosensitizer group and the disulfide, introduced via a lipoic acid, for covalent linking to the Au electrode. The same peptide scaffold was functionalized with ECz at the C-terminus and the disulfide at the N-terminus for investigating the effect of the electrostatic field associated to the helix macrodipole on PG efficiency. A first effect of the helix macrodipole is on the coverage of the gold electrode, which is larger for the peptide SAM linked to gold through the N-terminus. This is because the electrostatic field of the helix macrodipole, directed from the C-terminus (negative) to the N-terminus (positive), stabilizes the Au^+–S^- linkage.

PG measurements were carried out using an electrochemical cell in a standard three-electrode configuration with the gold-coated substrate acting as the working electrode, Ag/AgCl as the reference electrode and a Pt wire as the auxiliary electrode. The mechanism of photoactivated generation of a cathodic current in an electrochemical cell with an electron acceptor in solution (methyl viologen, MV^{2+}) proceeds through the following steps. Upon photoexcitation of ECz, ET from ECz* to MV^{2+} readily occurs (step I), followed by ET from gold to the ECz radical cation (step II). The reduced viologen is then oxidized at the auxiliary electrode closing the circuit and giving rise to a net electronic current (step III). Importantly, the PG efficiency is larger for a peptide SAM, the helix dipole of which points to the aqueous phase so as to accelerate the gold→ECz ET. The action spectrum, i.e. the excitation wavelength dependence of PG intensity, overlapped the absorption spectrum of ECz, confirming the photosensitizing role of carbazole. In the presence of an electron acceptor in solution (triethanolamine, TEOA, 50 mM) an anodic current was measured, as a result of ECz*→Au ET, followed by TEOA→$ECz^{+\cdot}$ ET. The first step in the anodic condition and the second step in the cathodic condition is most likely to represent the rate-limiting steps of PG because of the long distance between the photoactive probe and the gold surface. Significantly, in the case of an anodic photocurrent the peptide SAM showing the larger efficiency is the one with the helix dipole pointing with its positive end to the gold surface. A decrease in the anodic photocurrent was observed when increasing the negative bias to the gold electrode, reaching an apparent zero current status at a certain potential (zcp, zero current potential). In agreement with Equation (4.13), a negative bias reduces the energy gap between the oxidation potential of ECz* and the Fermi level of gold, resulting in a lower anodic photocurrent. Interestingly, at applied potentials larger than zcp, a cathodic current was measured, indicating that a zero current status can be obtained when cathodic and anodic currents balance each other.

Measurements of photocurrent efficiency as a function of the applied potential showed that the ECz–Au electronic coupling is greater for the peptide SAMs with respect to an alkanethiol SAM of comparable thickness, the experimental β values being 0.58–0.60 $Å^{-1}$ for the two peptide SAMs as compared to 0.9–1.1 $Å^{-1}$ reported for alkanethiol SAMs. A third effect can be ascribed to the helix dipole, i.e. the negative shift of the zcp for the N-terminally linked SAM (zcp = –0.30 V) in contrast to the positive shift measured for the C-terminally linked SAM (zcp = –0.11 V), as compared to the zcp of an apolar alkanethiol SAM (zcp = –0.16 V). A negatively shifted zcp results in a larger driving force for PG in anodic conditions. The expected potential shift induced by the dipole moment of the helical peptide ($\delta\varphi$) can be calculated by [79]

$$\delta\varphi = \frac{\mu n \cos\gamma}{\varepsilon_0 \varepsilon_m \left(1 + 9\alpha n^{3/2}\right)} \tag{4.26}$$

where μ is the helix dipole moment, n the number of peptide molecules per unit area, γ the tilt angle of the helix axis from the surface normal, ε_0 the vacuum permittivity, and ε and α the dielectric constant and the polarizability of the peptide SAM, respectively.

On the basis of Equation (4.26), the shifts in the zcp were evaluated to be –0.95 V and +0.40 V for the two peptide SAMs investigated, as compared to the experimental values of –0.14 V and +0.05 V. The calculated values, however, should be corrected for polarization effects ascribable to the Au^+–S^- linkage and the charge transfer process [42].

Kimura and coworkers [80] also carried out PG experiments on peptide SAMs composed of 3_{10}-helical nonapeptides functionalized with photoactive naphthyl units. Four different nonapeptides were synthesized, all derivatized at the N-terminus with a lipoic acid for chemical linkage to a gold electrode. The nonapeptide chain comprised three Ala–Aib–Aib triads and the peptides differ in the number of Ala residues functionalized with a naphthyl unit: none in the reference compound (SSA3B), one at the N-terminus in SSNA2B, one at the C-terminus in SSA2NB and three in SSN3B. In the energy-minimized 3_{10}-helical conformation of SSN3, the three naphthyl groups are spaced in a linear array along the

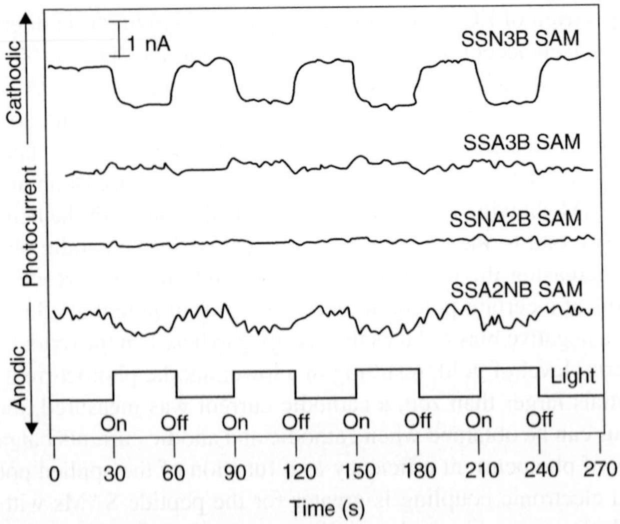

Figure 4.16 *Photocurrent generation upon light irradiation at 280 nm at 0 V applied potential in the anodic condition (50 mM TEOA solution). Reproduced with permission from reference [80]. Copyright (2004) American Chemical Society*

helical axis with face-to-face orientation, the centre-to-centre distances between adjacent naphthyl groups being in the range 6.1–6.4 Å. The photocurrents measured for these nonapeptides in anodic conditions (TEOA as the electron donor in solution) are reported in Figure 4.16.

A significant photocurrent was observed only in the case of the SSN3B SAM (2.1% PG efficiency), while no anodic photocurrent was generated by SSA3B or SSNA2B and only a weak current (roughly one-third of SSN3B) was detected for SSA2NB. This finding clearly emphasizes the role of the naphthyl chromophore in sensitizing the ET process. Interestingly, SSNA2B does not contribute to the photocurrent, indicating that the nonapeptide forms a densely packed film, so that the TEOA electron donor cannot diffuse through the SAM to reach the naphthyl group at the N-terminus. The high photocurrent efficiency of SSN3B strongly supports the idea that electron hopping among the linearly relayed naphthyl groups effectively speeds up the ET process. Furthermore, electron hopping should be promoted by effective mediation of the helical peptide backbone through additional HB pathways and electrostatic dipole effects.

For real application of molecular electronics, three important issues should be faced: (i) attain the full comprehension of the electronic properties at a molecular level; (ii) develop individual molecular modules such as molecular wires, diodes, transistors and other devices that control electronic response via external stimuli; (iii) integrate the molecular modules into a highly ordered structure. A molecular photodiode system, composed of two helical peptides forming a mixed SAM on a gold substrate, was designed by Yasutomi *et al.* [81]. The two peptide elements carry two different photoactive chromophores that can be excited selectively, giving rise to a photocurrent, the direction of which can be switched from anodic to cathodic by choosing the wavelength of the irradiating light. Because of the electrostatic field generated by the helix dipole, helical peptides carrying a sensitizer can act as a molecular photodiode, the direction of the photocurrent being determined, at a certain range of the applied potentials, by the direction of the dipole moment. Specifically, two hexadecapeptides, denoted as SSL16ECz and Rul16SS, were formed by alternating sequences of L- or D-leucines and α-aminoisobutyric acids. SSL16ECz carried a disulfide group at the N-terminus and an ethylcarbazolyl (ECz) group at the C-terminus, while Rul16SS carries a tris(2,2′-bipyridine)ruthenium(II) complex (Ru) at the N-terminus and a disulfide group at the C-terminus. L-Leucine and D-leucine were alternatively chosen to form the respective peptide backbones to favour phase separation in their mixed SAMs and, at the same time, to minimize unwanted photoinduced ET between the ECz groups and the Ru groups. When the two peptides were co-assembled on a gold substrate via an Au–S linkage, the dipole moment of SSL16ECz was directed from the monolayer surface to gold, whereas that of Rul16SS pointed in the opposite direction. As we have already discussed, ET along the dipole moment in a helical peptide is much faster than that against the dipole moment. Therefore, anodic PG, i.e. ET from the SAM outer surface to gold, will be promoted by excitation of the ECz group of SSL16ECz, whereas cathodic PG by ET from gold to the monolayer surface will be promoted by excitation of the Ru group of Rul16SS.

In both cases, the rate determining step is ET between the chromophore and gold, which was evaluated to be of the order of $10^6\,s^{-1}$, which is much lower (about 5 orders of magnitude lower) than that for the diffusion-controlled reaction between the chromophore and the redox species in solution. Both zcp (–0.13 V for SSL16ECz and +0.03 V for Rul16SS) and surface potential (–0.15 V for SSL16ECz and +0.35 V for Rul16SS) of the two single-component SAMs were found to be significantly different because of the dipole moment effect. The optimum potential for switching the photocurrent was therefore selected to be –0.05 V, where the anodic photocurrent by the SSL16ECz SAM and the cathodic photocurrent by the Rul16SS SAM are about the same. Photocurrent switching was therefore performed on the bicomponent SSL16ECz/Rul16SS SAM at the applied potential of –0.05 V in the mixed TEOA/MV^{2+} solution by selective, alternating photoirradiation at 351 nm (ECz excitation, anodic current) and 459 nm (Ru excitation, cathodic current). The associated action spectrum is simply given by the overlap of the absorption spectra of ECz (from 310 to 400 nm) and Ru (from 400 to 520 nm).

Yasutomi *et al.* [82] also showed that the photocurrent direction can be reversibly shifted between anodic and cathodic by changing the pH of the solution, by using a 16mer helical peptide composed of an alternating sequence of L-Leu–Aib dyads and functionalized at the N-terminus by a disulfide group for immobilization on gold and L-3-(3-*N*-ethylcarbazolyl) alanine at the C-terminus as a photoactive group (ECzCOOH). IR-RAS experiments revealed that the peptide SAM attained a vertical orientation on the surface and CV blocking experiments in an aqueous $K_4[Fe(CN)_6]$ solution confirmed the tight packing of the SAM. At low pH, upon photoexcitation of ECz in an aqueous solution containing an electron donor (TEOA), anodic PG was promoted due to photoinduced ET from ECz* to gold (Figure 4.17).

This process was accelerated by the electrostatic field associated to the helix dipole (positive end at the N-terminus). At high pH this process was enhanced by the negative charge attained by the terminal carboxylate group, the PG quantum yields at pH = 10 being 2.0% and 0.07% for the carboxylate and esterified (–COOEt) peptide SAMs, respectively.

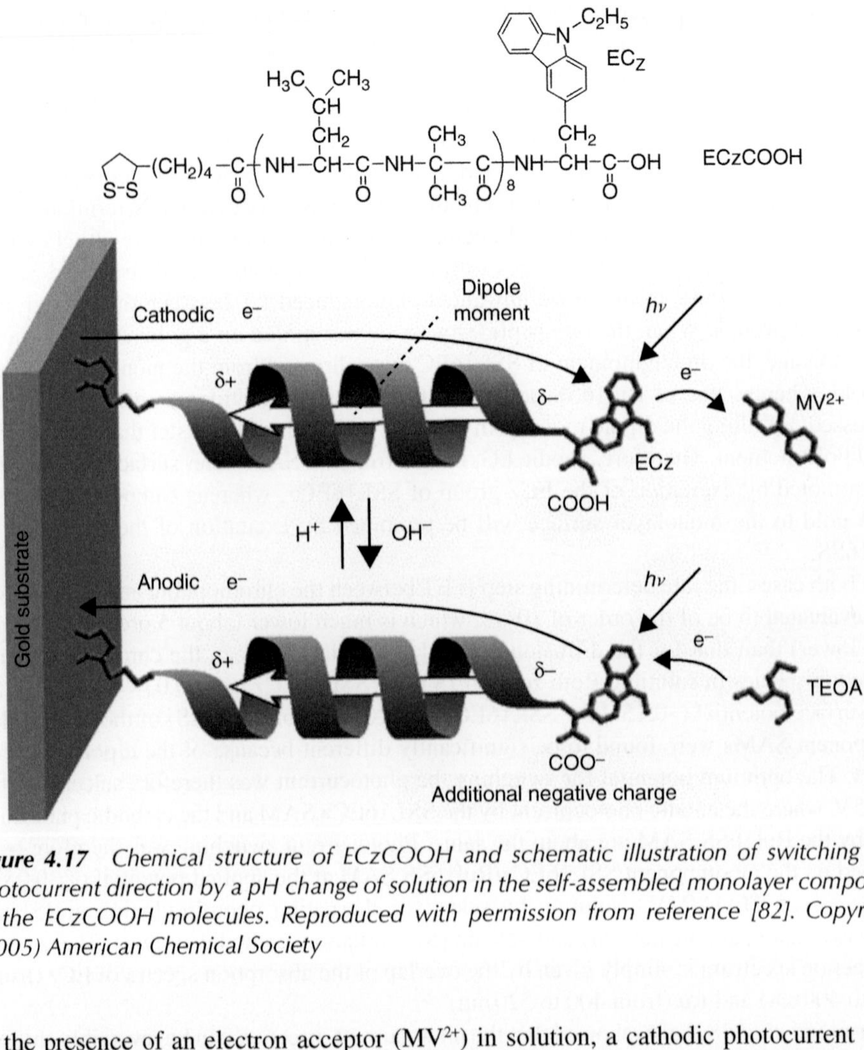

Figure 4.17 Chemical structure of ECzCOOH and schematic illustration of switching the photocurrent direction by a pH change of solution in the self-assembled monolayer composed of the ECzCOOH molecules. Reproduced with permission from reference [82]. Copyright (2005) American Chemical Society

In the presence of an electron acceptor (MV^{2+}) in solution, a cathodic photocurrent was obtained at low pH and at negative applied potential. It is worth noting that the pH-induced switch of PG was fully reversible.

The introduction of conformationally constrained residues in the peptide chain allowed control of the peptide secondary structure, modulating the separation distance between the photoactive group and the gold surface. A conformationally constrained hexapeptide, predominantly composed of Aib residues, was shown to form a densely packed SAM when immobilized on a gold substrate by Au–S chemisorption through the disulfide group of an N-terminal lipoic acid [83, 84]. The high content of Aib residues is responsible for the rigid 3_{10}-helix structure attained by the hexapeptide, denoted in the following as SSA4WA. A Trp (W) amino acid was introduced in the peptide chain as a photoactive molecule for fluorescence and PG studies [66]. When a gold interdigitated microelectrode coated by the chemisorbed peptide layer was immersed in an electrolytic solution containing an electron donor species (TEOA 50 mM), an anodic current was readily

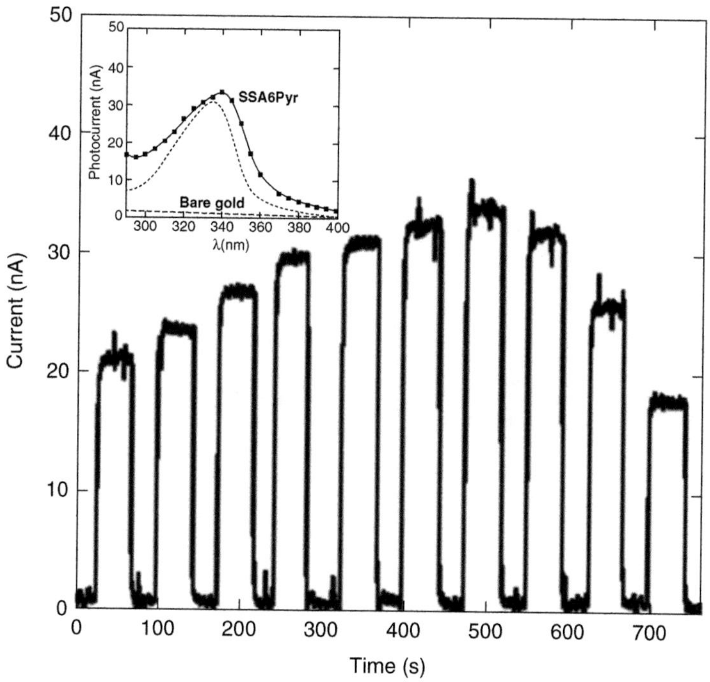

Figure 4.18 *On/off cycles of electronic current upon photoexcitation of the Pyr unit in SSA6Pyr at different excitation wavelengths (λ_{max}=340 nm). The electronic current intensities depend on the Pyrene absorption spectrum, as shown by the excitation spectrum of SSA6Pyr reported in the inset. Inset: photocurrent action spectrum and excitation spectrum of SSA6Pyr in the same experimental conditions (slit width = 15 nm). For comparison, the photocurrent action spectrum of the bare gold electrode is also reported. Reproduced with permission from reference [86]. Copyright (2011) John Wiley & Sons, Ltd*

obtained upon photoexcitation of the Trp sensitizer (*antenna effect*). In agreement with literature data, a slightly negative zcp of –0.15 V was measured, typical of helical peptides functionalized by an S-containing group at the N-terminus. The same peptide matrix was very recently used to investigate the PG efficiency of mono- and bicomponent peptide SAMs on gold substrates [85].

The PG efficiency of two peptide foldamers, one composed of six Aib residues, a lipoic group at the N-terminus for binding to gold and a Pyrene antenna chromophore at the C-terminus (SSA6Pyr), and the other by eight C^{α}-tetrasubstituted residues and a Pyrene unit at the N-terminus (A8Pyr), was investigated in our group [85–87]. A bare hexapeptide, lacking the Pyrene antenna chromophore, was also synthesized as a reference compound (SSA6). The anodic current generated upon photoexcitation of the SSA6Pyr at different excitation wavelengths in the presence of an electron donor (TEOA) in solution is shown in Figure 4.18. The photocurrent efficiency at λ_{ex} = 340 nm was found to be equal to 0.05%.

In the inset of Figure 4.18, the action spectrum of SSA6Pyr, closely overlapping the Pyrene absorption spectrum recorded under the same conditions (slit width = 15 nm), was also reported. The antenna effect of the Pyrene sensitizer was verified by measuring the PG

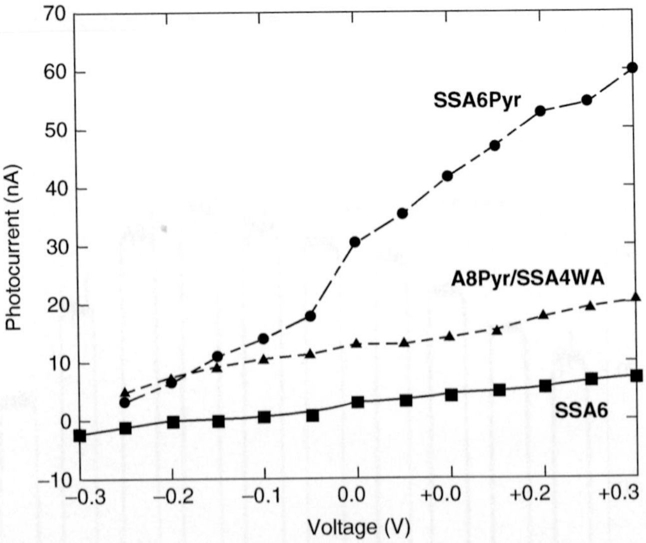

Figure 4.19 *Photocurrent generated by SSA6Pyr (full circles), A8Pyr/SSA4WA (full triangles) or SSA6 (full squares) peptide SAMs by excitation at 340 nm as a function of the applied bias potential. Reproduced with permission from reference [86]. Copyright (2011) John Wiley & Sons, Ltd*

for a series of bicomponent peptide SAMs (1:10 SSA6Pyr/SSA6, 1:1 A8Pyr/SSA4WA, 1:1 A8Pyr/SSA6). CV blocking experiments showed that the bicomponent SAM formed by SSA4WA, chemically linked to the gold substrate, and A8Pyr, intercalated into the SSA4WA SAM, was also densely packed [85]. The antiparallel arrangement of the two peptides, stabilized by intermolecular dipole–dipole interactions, leads both the Trp and Pyrene groups in contact with the electrolyte solution. Interestingly, the photocurrent measured upon excitation of the Pyrene group is definitely minor for the SAMs containing the adsorbed A8Pyr peptide with respect to the chemisorbed SSA6Pyr SAM. To investigate in detail the role of the peptide/gold contact in determining the efficiency of the ET process, the dependence of the anodic photocurrent on the applied potential was analysed at the Pyrene maximum excitation wavelength ($\lambda = 340$ nm) for the SSA6Pyr and the A8Pyr/SSA4WA peptide SAMs (Figure 4.19) [86]. The same analysis was also carried out for the SAM formed by the Pyrene-lacking peptide SSA6 as a control experiment.

In all cases, a decrease of the anodic photocurrent intensity was observed upon reducing the positive bias voltage to the working electrode. This result is due to the fact that the less positive is the applied bias and the smaller is the energy gap between the oxidation potentials of the excited Pyr and the gold electrode Fermi level, i.e. the ET driving force in the normal Marcus region. The slope of the photocurrent/voltage (P/V) curve depends on the coupling between the HOMO/LUMO orbitals of the antenna chromophore and the electronic DOS of the metal mediated by the peptide bridge. Unfortunately, the observed linear dependence of the photocurrent on the applied potential does not allow per se to distinguish between a coherent tunneling SE and a diffusive hopping mechanism, because both ET models were shown to be linearly dependent on the applied potential at low voltages.

From the data reported in Figure 4.19, it appears that the slope of the P/V curve measured for the SSA6Pyr SAM is definitely steeper than that obtained for the A8Pyr/SSA4WA SAM, a finding ascribed to the different types of contact established by the two SAMs at the peptide/gold interface (*junction effect*). The $Au^{\delta+}$–$S^{\delta-}$ junction allows a through-bond ET pathway, with a relatively low ET activation barrier at the gold–peptide interface. This pathway is not operative for A8Pyr, which lacks the disulfide group and is firmly embedded in the bicomponent SAM by favourable interchain interactions with SSA4WA or SSA6. In the latter case, a direct Pyr*→Au ET across the A8Pyr peptide backbone would necessarily require a through-space step from the peptide N-terminus to the gold surface, an event characterized by a relatively high activation energy. Furthermore, the helix dipole of A8Pyr in the bicomponent SAM is directed towards the negative end, pointing to the gold electrode, so as to frustrate ET in that direction. A possible alternative pathway may proceed through initial intermolecular ET from the excited Pyr to Trp (Pyr*+Trp→$Pyr^{+\bullet}$+$Trp^{-\bullet}$) or to the amide site of a nearby peptide chain, followed by ET across the Au–S linked peptide chain from $Trp^{-\bullet}$ or the amide radical anion to gold. This second step should also be favoured by the electrostatic field generated by the peptide helix. In agreement with this idea, the photocurrent efficiency of the A8Pyr/SSA4WA SAM (IPCE = 0.02%) was found to be definitely greater than the one measured for the A8Pyr/SSA6 SAM (ICPE = 0.006%), suggesting a predominant contribution of the Trp group to the intermolecular ET process. In any case, the photocurrents generated by both A8Pyr/SSA4WA and A8Pyr/SSA6 SAMs are definitely greater than that measured for the SSA6 SAM (IPCE = 0.002%), emphasizing the role of Pyrene as a photosensitizer even in the case of a peptide not covalently linked to the gold surface.

4.4 Conclusions and Perspectives

Electron transfer across peptide scaffolds is an extremely intriguing and complex subject, ranging from natural photosynthesis to the design of new hybrid materials for molecular electronics. If important achievements have been reached in the comprehension of the mechanisms governing ET in peptide models in solution, ET in peptide/solid substrates (peptide SAMs coating metal electrodes, peptide–nanoparticles conjugates, peptide wires on chips) still requires a dedicated scientific effort. The process at the peptide/metal or semiconductor junction, the modification of the electronic properties of the surface induced by physi- or chemisorbed peptide films, the coupling of photo- or electroactive probes to the conductive terminal through the peptide layer and the identification of the charge carrier (electron/hole) across the peptide chain should be considered as still open questions worthy of investigation. As often happens, this contribution probably raises more questions than answers.

Acknowledgements

Stimulating discussions and critical reading of the manuscript by Prof. B. Pispisa are gratefully acknowledged. This contribution was funded by the Italian Ministry for Education, University and Research (MIUR), PRIN2010 2010FM738P.

References

[1] Gellman, S.H. (1998) Foldamers: a manifesto. Acc. Chem. Res. 31: 173–180.

[2] Toniolo, C., Crisma, M., Formaggio, F., Peggion, C. (2001) Control of peptide conformation by the Thorpe–Ingold effect (C^{α}-tetrasubstitution). Biopolymers (Pept. Sci.) 60: 396–419.

[3] Beratan, D.N., Onuchic, J.N., Winkler, J.R., Gray, H.B. (1992) Electron-tunneling pathways in proteins. Science 258: 1740–1741.

[4] Isied, S.S., Ogawa, M.Y., Wishart, J.F. (1992) Peptide-mediated intramolecular electron transfer: long-range distance dependence. Chem. Rev. 92: 381–394.

[5] Long, Y.-T., Abu-Irhayem, E., Kraatz, H.B. (2005) Peptide electron transfer: more questions than answers. Chem. Eur. J. 11: 5186–5194.

[6] Cordes, M., Giese, B. (2009) Electron transfer in peptides and proteins. Chem. Soc. Rev. 38: 892–901.

[7] Aviram, A., Ratner, M.A. (1974) Molecular rectifiers. Chem. Phys. Lett. 29: 277–283.

[8] Hush, N.S. (1985) Distance dependence of electron transfer rates. Coord. Chem. Rev. 64: 135–157.

[9] Newton, M.D., Sutin, N. (1984) Electron transfer reactions in condensed phases. Annu. Rev. Phys. Chem. 35: 437–480.

[10] Marcus, R.A., Sutin, N. (1985) Electron transfer in chemistry and biology. Biochim. Biophys. Acta 811: 265–322.

[11] Bolton, J.R., Archer, M.D. (1991) Basic electron transfer theory. In: Electron Transfer in Inorganic, Organic and Biological Systems, Adv. Chem. Ser. 228. American Chemical Society, Washington, DC, pp. 7–23.

[12] Jortner, J., Bixon, M., Langenbacher, T., Michel-Beyerle, M.E. (1998) Charge transfer and transport in DNA. Proc. Natl. Acad. Sci. USA 95: 12759–12765.

[13] Newton, M.D. (1991) Quantum chemical probes of electron transfer kinetics: the nature of donor–acceptor interactions. Chem. Rev. 91: 767–792.

[14] McConnell, H.M. (1961) Intramolecular charge-transfer in aromatic free radicals. J. Chem. Phys. 35: 508–515.

[15] Hopfield, J.J. (1974) Electron transfer between biological molecules by thermally activated tunneling. Proc. Natl Acad. Sci. USA 71: 3640–3644.

[16] Gray, H.B., Winkler, J.R. (2003) Electron transfer through proteins. Q. Rev. Biophys. 36: 341–372.

[17] Gray, H.B., Winkler, J.R. (2005) Long-range electron transfer. Proc. Natl Acad. Sci. USA, 102: 3534–3539.

[18] Malak, R.A., Gao, Z., Wishart, J.F., Isied, S.S. (2004) Long-range electron transfer across peptide bridges: the transition from electron superexchange to Hopping. J. Am. Chem. Soc. 126: 13888–13889.

[19] Kai, M., Takeda, K., Morita, T., Kimura, S. (2008) Distance dependence of long-range electron transfer through helical peptides. J. Pept.Sci. 14: 192–202.

[20] Dey, S.K., Long, Y.T., Chowdhury, S., Sutherland, T.C., Mandal, H.S., Kraatz, H.B. (2007) Study of electron transfer in ferrocene-labeled collagen-like peptides. Langmuir 23: 6475–6477.

[21] Stubbe, J., Nocera, D.G., Yee, C.S., Chang, C.Y. (2003) Radical initiation in the class I ribonucleotide reductase: long-range proton-coupled electron transfer? Chem. Rev. 103: 2167–2202.

[22] Adams, D.M., Brus, L., Chidsey, C.E., Creager, S., Creutz, C., Kagan, C.R., Kamat, P.V., Lieberman, M., Lindsay, S., Marcus, R.A., Metzger, R.M., Michel-Beyerle, M.E., Miller, J.R., Newton, M.D., Rolison, D.R., Sankey, O., Schanze, K.S., Yardley, J., Zhu, X. (2003) Charge transfer on the nanoscale: current status. J. Phys. Chem. B 107: 6668–6697.

[23] Onuchic, J.N., Beratan, D.N., Winkler, J.R., Gray, H.B. (1992) Pathway analysis of protein Electron Transfer reactions. Annu. Rev. Biophys. Biomol. Struct. 21: 349-377.

[24] Kourtis, S.S., Balabin, I.A., Kawatsu, T., Beratan, D.N. (2005) Protein dynamics and electron transfer. Electronic decoherence and non-Condon effects. Proc. Natl Acad. Sci. USA 102: 3552–3557.

[25] Prytkova, T.R., Kurnokov, I.V., Beratan, D.N. (2007) Coupling coherence distinguishes structure sensitivity in protein electron transfer. Science 315: 622–625.
[26] Orlowski, G.A., Chowdhuri, S., Kraatz, H.B. (2007) Reorganization energies of ferrocene–peptide monolayers. Langmuir 23: 12765–12770.
[27] Takeda, K., Morita, T., Kimura, S. (2008) Effects of monolayer structures on long-range electron transfer in helical peptide monolayer. J. Phys. Chem. B 112: 12840–12850.
[28] Petrov, E.G., Shevchenko, Y.V., Teslenko, V.I. (2001) Nonadiabatic donor-acceptor electron transfer mediated by a molecular bridge: a unified theoretical description of the superexchange and hopping mechanism. J. Chem. Phys. 115: 7107–7122.
[29] Cordes, M., Köttgen, A., Jasper, C., Jacques, O., Boudebous, H., Giese, B. (2008) Influence of amino acid side-chains on long-distance electron transfer in peptides: electron hopping via 'stepping stones'. Angew. Chem. 47: 3461–3463.
[30] Antonello, S., Formaggio, F., Moretto, A., Toniolo, C., Maran, F. (2003) Anomalous distance dependence of electron transfer across peptide spacers. J. Am. Chem. Soc. 125: 2874–2875.
[31] Polo, F., Antonello, S., Formaggio, F., Toniolo, C., Maran, F. (2005) Evidence against the hopping mechanism as an important electron transfer pathway for conformationally constrained oligopeptides. J. Am. Chem. Soc. 127: 492–493.
[32] Sisido, M., Tanaka, R., Inai, Y., Imanishi, Y. (1989) Photoinduced electron transfer on a single α-helical polypeptide chain. J. Am. Chem. Soc. 111: 6790–6796.
[33] Inai, T., Sisido, M., Imanishi, Y. (1991) Photoinduced electron transfer on a single α-helical polypeptide chain. Evidence of through-space mechanism. J. Phys. Chem. 95: 3847–3851.
[34] Sisido, M., Hoshino, S., Kusano, H., Kuragaki, M., Makino, M., Sasaki, H., Smith, T.A., Ghiggino, K.P. (2001) Distance dependence of photoinduced electron transfer along α-helical polypeptides. J. Phys. Chem. B 105: 10407–10415.
[35] Shin, Y.K., Newton, M.D., Isied, S.S. (2003) Distance dependence of electron transfer across peptides with different secondary structures: the role of peptide energetic and electronic coupling. J. Am. Chem. Soc. 125: 3722–3732.
[36] Galoppini, E., Fox, M.A. (1996) Effect of the electric field generated by the helix dipole on photoinduced intramolecular electron transfer in dichromophoric α-helical peptides. J. Am. Chem. Soc. 118: 2299–2300.
[37] Batchelder, T.L., Fox III, R.J., Meier, M.S., Fox, M.A. (1996) Intramolecular excited state electronic coupling along an α-helical peptide. J. Org. Chem. 61: 4206–4209.
[38] Fox, M.A., Galoppini, E. (1997) Electric field effect on electron transfer rates in dichromophoric peptides: the effect of helix unfolding. J. Am. Chem. Soc. 119: 5277–5285.
[39] Gatto, E., Porchetta, A., Stella, L., Guryanov, I., Formaggio, F., Toniolo, C., Kaptein, B., Broxterman, Q.B., Venanzi, M. (2008) Conformational effects on the electron transfer efficiency in peptide foldamers based on α,α-disubstituted glycyl residues. Chem. Biodiv. 5: 1263–1278.
[40] Striplin, D.R., Reece, S.Y., McCafferty, D.G., Wall, C.G., Friesen, D.A., Erickson, B.W., Meyer, T.J. (2004) Solvent dependence of intramolecular electron transfer in a helical oligoproline assembly. J. Am. Chem. Soc. 126: 5282–5291.
[41] McCreery, R.L. (2004) Molecular electronic junctions. Chem. Mat. 16: 4477–4496.
[42] Miura, Y., Kimura, S., Kobayashi, S., Iwamoto, M., Imanishi, Y., Umemura, J. (1999) Negative surface potential produced by self-assembled monolayers of helix peptides oriented vertically to a surface. Chem. Phys. Lett. 315: 1–6.
[43] Palermo, V., Palma, M., Samorì, P. (2006) Electronic characterization of organic thin films by Kelvin probe force microscopy. Adv. Mater. 18: 145–164.
[44] Gosavi, S., Marcus, R.A. (2000) Non adiabatic electron transfer at metal surfaces. J. Phys. Chem. B 104: 2067–2072.
[45] Smalley, J.F., Feldberg, S.W., Chidsey, C.E.D., Linford, M.R., Newton, M.D., Liu, Y.-P. (1995) The kinetics of electron transfer through ferrocene-terminated alkanethiol monolayers on gold. J. Phys. Chem. 99: 13141–13149.
[46] Holmlin, R.E., Haag, R., Chabinye, M.L., Ismagilov, R.F., Cohen, A.E., Terfort, A., Rampi, M.A., Whitesides, G.M. (2001) Electron transport through thin organic films in metal–insulator–metal junctions based on self-assembled monolayers. J. Am. Chem. Soc. 123: 5075–5085.

[47] Galka, M.M., Kraatz, H.B. (2002) Electron transfer studies in self-assembled monolayers of helical Ferrocene–oligoproline–Cystamine bound to gold. Chem. Phys. Chem. 3: 356–359.
[48] Reed, M.A., Zhou, C., Muller, C.J., Burgin, T.P., Tour, J.M. (1997) Conductance of a molecular junction. Science 278: 252–254.
[49] Kerqueris, C., Bourgoin, J.-P., Palacin, S., Esteve, D., Urbina, C., Magoga, M., Joachain, C. (1999) Electron transport through a metal–molecule–metal junction. Phys. Rev. B 59: 12505–12513.
[50] Wold, J., Frisbie, C.D. (2000) Formation of metal–molecule–metal tunnel junctions: microcontacts to alkanethiol monolayers with a covalently AFM tip. J. Am. Chem. Soc. 122: 2970–2971.
[51] Cui, X.D., Primak, A., Zarate, X., Tomfohr, J., Sankey, O.F., Moore, A.L., Moore, T.A., Gust, D., Harris, G., Lindsay, S.M. (2001) Reproducible measurement of single-molecule conductivity. Science 294: 571–574.
[52] Xue, Y., Datta, S., Ratner, M.A. (2001) Charge transfer and 'band lineup' in molecular electronic devices: a chemical and numerical interpretation. J. Chem. Phys. 115: 4292–4299.
[53] Li, J., Tomfohr, J.K., Sankey, O.F. (2003) Theoretical study of carotene as a molecular wire. Physica E: Low-dimensional Systems and Nanostructures 19: 133–138.
[54] Tian, W., Datta, S., Hong, S., Reifenberger, R., Henderson, J.I., Kubiak, C.P. (1998) Conductance spectra of molecular wires. J. Chem. Phys. 109: 2874–2882.
[55] Segal, D., Nitzan, A., Ratner, M.A., Davis, W.B. (2000) Activated conduction in microscopic molecular junctions. J. Phys. Chem. B 104: 2790–2793.
[56] Segal, D., Nitzan, A., Davis, W.B., Wasiliewski, M.R., Ratner, M.A. (2000) Electron transfer in bridged molecular systems. A steady-state analysis of coherent tunneling and thermal transitions. J. Phys. Chem. B 104: 3817–3829.
[57] Kraatz, H.B., Bediako-Amoa, I., Gyepi-Garbrah, S.H., Sutherland, T.C. (2004) Electron transfer through H-bonded peptide assemblies. J. Phys. Chem. B 108: 20164–20172.
[58] Morita, T., Kimura, S. (2003) Long-range electron transfer over 4 nm governed by an inelastic hopping mechanism in self-assembled monolayers of helical peptides. J. Am. Chem. Soc. 125: 8732–8733.
[59] Watanabe, J., Morita, T., Kimura, S. (2005) Effects of dipole moment, linkers and chromophores at side chains on long-range electron transfer through helical peptides. J. Phys. Chem. B 109: 14416–14425.
[60] Kitagawa, K., Morita, T., Kimura, S. (2005) Molecular rectification of a helical peptide with a redox group in the metal–molecule–metal junction. J. Phys. Chem. B 109: 13906–13911.
[61] Simmons, J.G. (1963) Generalized formula for the electron tunnel effect between similar electrodes separated by a thin insulating film. J. Appl. Phys. 34: 1793–1803.
[62] Kitagawa, K., Morita, T., Kimura, S. (2005) Electron transfer in metal–molecule–metal junction composed of self-assembled monolayers of helical peptides carrying redox-active Ferrocene units. Langmuir 21: 10624–10631.
[63] Baranov, L.Y, Schlag, E.W. (1999) New mechanism for facile charge transport in polypeptides. Z. Naturf. 54: 387–396.
[64] Schlag, E.W., Scheu, S.Y., Yang, D.Y., Selzle, H.L., Lin, S.H. (2000) Charge conductivity in peptides: dynamic simulations of a bifunctional model supporting experimental data. Proc. Natl Acad. Sci. USA 97: 1068–1072.
[65] Schlag, E.W., Scheu, S.Y., Yang, D.Y., Selzle, H.L., Lin, S.H. (2007) Distal charge transport in peptides. Angew. Chem. Int. Ed. 46: 3196–3210.
[66] Gatto, E., Stella, L., Formaggio, F., Toniolo, C., Lorenzelli, L., Venanzi, M. (2008) Electroconductive and photocurrent generation properties of self-assembled monolayers formed by functionalized, conformationally constrained peptides on gold electrodes. J. Pept. Sci. 14: 184–191.
[67] Milov, A.D., Tsvetkov, Y.D., Formaggio, F., Crisma, M., Toniolo, C., Raap, J. (2001) The secondary structure of a membrane-modifying peptide in a supramolecular assembly studied by PELDOR and CW-ESR spectroscopies. J. Am. Chem. Soc. 123: 3784–3789.
[68] Pispisa, B., Palleschi, A., Stella, L., Venanzi, M., Toniolo, C. (1998) A nitroxide derivative as a probe for conformational studies of short linear peptides in solution. J. Phys. Chem. B 102: 7890–7898.

[69] Venanzi, M., Gatto, E., Bocchinfuso, G., Palleschi, A., Stella, L., Baldini, C., Formaggio, F., Toniolo, C. (2006) Spectroscopic and molecular mechanics investigation of peptide folding dynamics. A time-resolved study from the nanosecond to the microsecond time regime. J. Phys. Chem. B 110: 22843–22841.

[70] Wen, X., Linton, R.W., Formaggio, F., Toniolo, C., Samulski, E.T. (2004) Self-assembled monolayers of hexapeptides on gold surface: characterization and orientation distribution analysis. J. Phys. Chem. A 108: 9673–9681.

[71] Sek, S., Tolak, A., Misicka, A., Palys, B., Bilewics, R. (2005) Asymmetry of electron transmission through monolayers of helical polyalanine adsorbed on gold surfaces. J. Phys. Chem. B 109: 18433–18438.

[72] Arikuma, Y., Takeda, K., Morita, T., Ohmae, M., Kimura, S. (2009) Linker effects on monolayer formation and long-range electron transfer in helical peptide monolayers. J. Phys. Chem. B 113: 6256–6266.

[73] Arikuma, Y., Nakayama, H., Morita, T., Kimura, S. (2011) Ultra-long-range electron transfer through a self-assembled monolayer on gold composed of 120Å-long α-helices. Langmuir 27: 1530–1535.

[74] Xu, B., Xiao, X., Tao, N. (2004) Measurements of single-molecule electromechanical properties. J. Am. Chem. Soc. 125: 16164–16165.

[75] Xiao, X., Xu, B., Tao, N. (2005) Conductance titration of single peptide molecules. J. Am. Chem. Soc. 126: 5370–5371.

[76] Sek, S., Swiatek, K., Misicka, A. (2005) Electrical behavior of molecular junctions incorporating α-helical peptides. J. Phys. Chem. B 109: 23121–23124.

[77] Sek, S. (2007) Two metal–molecule binding modes for peptide molecules junctions. J. Phys. Chem. C 111: 12860–12865.

[78] Morita, T., Kimura, S., Kobayashi, S. (2000) Photocurrent generation under a large dipole moment formed by self-assembled monolayers of helical peptides having an *N*-ethylcarbazolyl group. J. Am. Chem. Soc. 122: 2850–2859.

[79] Ito, E., Iwamoto, M. (1997) Electronic density of states in metal/polyimide Langmuir–Blodgett film interface and its temperature dependence. J. Appl. Phys. 81: 1790–1797.

[80] Yanagisawa, K., Morita, T., Kimura, S. (2004) Efficient photocurrent generation by self-assembled monolayers composed of 3_{10}-helical peptides carrying linearly spaced naphthyl groups at the side chains. J. Am. Chem. Soc. 126: 12780–12781.

[81] Yasutomi, S., Morita, T., Imanishi, Y., Kimura, S. (2004) A molecular photodiode system that can switch photocurrent direction. Science 304: 1944–1947.

[82] Yasutomi, S., Morita, T., Kimura, S. (2005) pH-controlled switching of photocurrent detection by self-assembled monolayer of helical peptides. J. Am. Chem. Soc. 127: 14564–14565.

[83] Venanzi, M., Pace, G., Palleschi, A., Stella, L., Castrucci, P., Scarselli, M., De Crescenzi, M., Formaggio, F., Toniolo, C., Marletta, G. (2006) Densely-packed self-assembled monolayers on gold surfaces from a conformationally constrained helical hexapeptide. Surf. Sci. 600: 409–416.

[84] Pace, G., Venanzi, M., Castrucci, P., Scarselli, M., De Crescenzi, M., Palleschi, A., Stella, L., Formaggio, F., Toniolo, C., Marletta, G. (2006) Static and dynamic features of a helical hexapeptide chemisorbed on a gold surface. Mat. Sci. Eng. C 26: 918–923.

[85] Gatto, E., Porchetta, A., Scarselli, M., De Crescenzi, M., Formaggio, F., Toniolo, C., Venanzi, M. (2012) Playing with peptides: how to build a supramolecular peptide nanostructure by exploiting helix···helix macrodipole interactions. Langmuir 28: 2817–2826.

[86] Gatto, E., Caruso, M., Porchetta, A., Toniolo, C., Formaggio, F., Crisma, M., Venanzi, M. (2011) Photocurrent generation through peptide-based self-assembled monolayers on a gold surface: antenna and junction effects. J. Pept. Sci. 17: 124–131.

[87] Gatto, E., Stella, L., Baldini, C., Toniolo, C., Formaggio, F., Venanzi, M. (2009) Photocurrent generation in peptide-based self-assembled monolayers on gold electrodes. Superlattices Microstruct. 46: 34–39.

Part II

Peptide Nanostructures

5

Molecular Architecture with Peptide Assembling for Nanomaterials

Shunsaku Kimura and Motoki Ueda

Department of Material Chemistry, Graduate School of Engineering, Kyoto University, Japan

Acronyms

ABC	Accelerated blood clearance
DGBE	Diethylene glycol bis(3-aminopropyl) ether
EPR	Enhanced permeability and retention
FDA	Food and Drug Administration
NIRF	Near-infrared fluorescence
PEG	Poly(ethylene glycol)
PTD	Protein translation domain
RGD	Arg–Gly–Asp
TEM	Transmission electron microscopy

5.1 Introduction

Self-assembling of amphiphilic peptides in aqueous medium provides molecular assemblies with various morphologies such as micelles, worm-like micelles, fibers, tubes, and vesicles. Peptide vesicles in aqueous medium are discussed intensively here because the structures can be generally well specified on a molecular basis, while the molecular packing in hydrophobic cores of micelles and worm-like micelles are generally obscure. Fibers and tubes are attractive as functional materials, but most of them are out of control about length

Peptide Materials: From Nanostructures to Applications, First Edition. Edited by Carlos Alemán, Alberto Bianco and Mariano Venanzi.

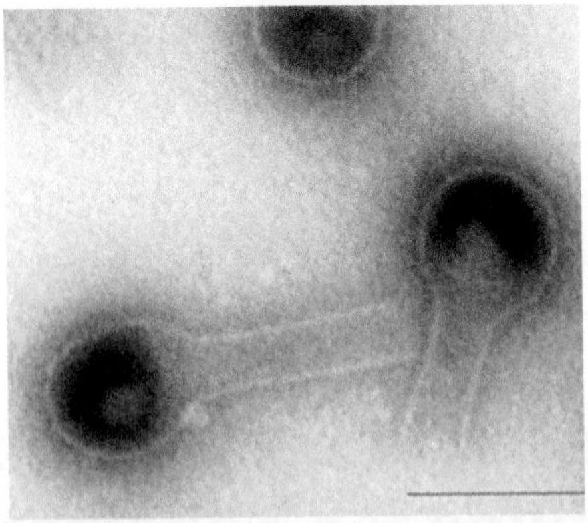

Figure 5.1 TEM images of a nano round-bottom flask shaped morphology prepared from a mixture of right- and left-handed helices. Reproduced with permission from reference [1]. Copyright (2004) The Royal Society of Chemistry

and diameter, frequently accompanied by a wide size distribution. One of the characteristic points of using peptides as assembling units is well-characterized molecular dimensions of the peptides leading to a well-organized structure at macro levels. The vesicular structures thus sensitively reflect peptide designs based on their intrinsic properties, which point is the main theme of this chapter.

Another reason for focus on vesicles is a new aspect in the molecular assembling. Recently, an advanced morphology, where tubulation on peptide vesicles generated a nano round-bottom flask shape morphology, was reported using phase separation in one membrane (Figure 5.1) [1]. Two different morphologies of a tube and a vesicle are linkable to create a joined morphology, which will widen the scope of molecular assembling science. The key points for obtaining the advanced morphology are tight and regular molecular packing of helices and stereo-complex formation of right- and left-handed helices in membrane, which are easily attained by helical peptides. Vesicular assemblies prepared from peptides are described as follows, principally according to a time series.

5.2 Peptide Vesicles

The first report on vesicle formation using peptides or polypeptides with specified primary structures (coacervates prepared from polypeptides, which were obtained by thermal polymerization of amino acids, are excluded from the criterion) is a capsule with diameters ranging from 0.2 μm to 10 μm and a shell thickness of 170–250 nm, which was prepared from diketopiperazine-based tetrapeptides [2]. However, it is a microcapsule rather than a vesicle because the membrane structure was not clarified on a molecular basis. Mixed assemblies of bilayer-type structures and vesicles with diameters ranging from tens to hundreds of nanometers were obtained with poly(styrene)$_{40}$-*b*-poly(isocyano-L-Ala–L-Ala)$_{10}$ [3]. The size distribution of the molecular assemblies was wide, probably due to irregular molecular packing (entanglement) of a poly(styrene) block in the hydrophobic core. Carboxyl groups of the dipeptides were used for the hydrophilic block.

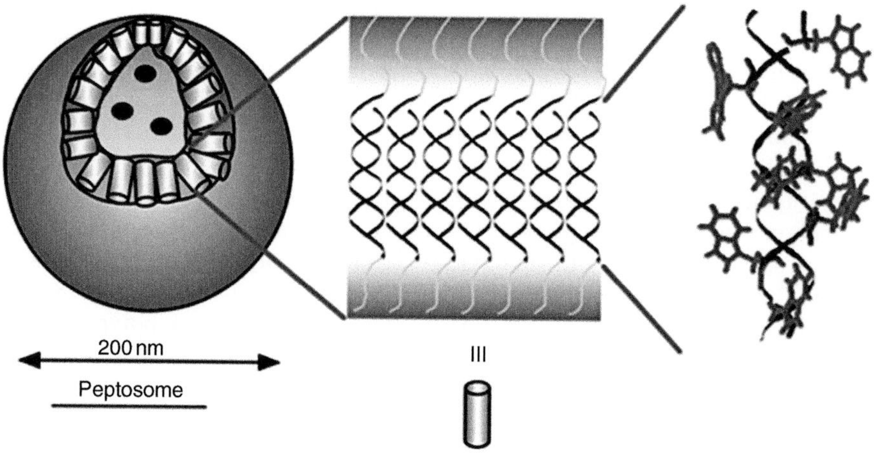

Figure 5.2 Schematic illustration of peptosome prepared from a conjugate of gramicidin A and PEG. The gramicidin A blocks took a double helical structure to constitute the hydrophobic core region of the vesicular membrane. Reproduced with permission from reference [6]. Copyright (2004) The Royal Society of Chemistry (see color plate figure)

5.2.1 Peptosome

Our group focused on a good self-assembling property of helices, which self-assembled on gold with vertical orientation under suitable preparative conditions for molecular assembling [4]. Hydrophobic helical peptides can be a driving force for regular self-assembling in water. For example, H–(Ala–Abi)$_8$–OBzl yielded a spherical assembly of 30–40 nm with a narrow distribution as a first example of peptide vesicles with a well-defined size [5]. Spherical assemblies were also prepared from a PEG–gramicidin A conjugate (Figure 5.2) [6] and a lactobionic acid–(Ala–Aib)$_4$ conjugate [7], which were identified as vesicles with a single wall composed of a double helix and interdigitated helices, respectively, by frozen-hydrated/cryo-transmission electron microscopy (cryo-TEM). Since the vesicular structure and the size of about 100 nm diameter are similar to those of liposome, vesicles prepared from peptides are named 'peptosome'. Polymer vesicles, polymersomes, are prepared from synthetic amphiphilic block polymers such as poly(styrene)$_{200}$-*b*-poly(acrylic acid)$_8$ [8]. A distinct point of peptosomes from general polymersomes is the well-defined structure of the hydrophobic domain of vesicular membranes. On the other hand, hydrophobic polymer chains in polymersomes are generally entangled in the hydrophobic core region. For example, helices aligned side by side are packed tightly in the case of peptosomes prepared from a PEG–gramicidin A conjugate and a lactobionic acid–(Ala–Aib)$_4$ conjugate.

5.2.2 Polypeptide as a Hydrophilic Block (AB Type and ABA Type)

A stimuli-responsive vesicle was prepared from poly(butadiene)-*b*-poly(L-Glu) copolymers, which formed a vesicle of about 120 nm diameter. Poly(L-Glu) took a α-helical conformation at pH 4.5, but a coil conformation at pH 11.5 [9]. With variation of the block lengths of the copolymers, the sizes of the self-assemblies were adjustable, ranging from 30 nm to 110 nm while keeping robustness against change of ionic strength [10]. Poly(butadiene)-*b*-poly(L-Glu) copolymers were also analyzed by other groups [11]. When replacing the synthetic polymer, poly(butadiene), with polyester, synthetic block poly(ester-amide)s can become biodegradable biomaterials, which may be applicable to medical fields. Amphiphilic poly(Lys)-*b*-poly(caprolactone) block copolymers were synthesized using FDA-approved tin(II) octanoate as catalyst. The copolymer with a poly(Lys) to polycaprolactone ratio of 1:1 formed vesicles having a size of around 120 nm [12]. Triblock polypepitdes, poly$(Lys)_m$-*b*-poly$(Glu(OBzl))_n$-*b*-poly$(Lys)_m$ with $m=64$, 131, or 134 and $n=64$, 137, or 181, self-assembled into vesicles of about 260–300 nm diameter. The triblock polypeptides are considered as two A(B/2) diblock copolypeptides with an antiparallel orientation, where the chain ends of the B/2 blocks are covalently connected. The ABA terpolypeptides easily formed a monolayer membrane similar to the bilayer formed by the diblock copolypepitdes. It was found that mixtures of plasmid DNA and the triblock copolypeptides also formed vesicular structures containing both encapsulated and condensed DNA [13]. Poly$(butadiene)_{107}$-*b*-poly$(Lys)_{27}$ also showed a pH responsive property which originates partially from secondary structure changes within the polypeptide chain. Poly(butadiene)-*b*-poly(Lys) formed vesicles with a diameter of 184 nm under acidic conditions ($2<pH<5.5$), which shrank to 144 nm at higher pH ($5.5<pH<10$) as a result partially of the α-helix to coil secondary structure transition in the molecular assembly. Further, when the temperature was raised over 40 °C, the hydrodynamic size of vesicles at pH 10.9 increased by a factor of two due to the helix-sheet transition, a process that was kinetically controlled [14].

5.2.3 Block Polypeptides Having a Hydrophobic Polypeptide

Deming *et al.* developed a synthetic methodology that allows syntheses of block copolypeptides of high precision using 2,2′-bipyridyl-*Ni*-1,5-cyclooctadiene as a catalyst for polymerization of α-aminoacid-*N*-carboxyanhydride [15]. Amphiphilic diblock copolypeptides composed of the ethylene glycol-modified Lys for the hydrophilic block and Leu for the hydrophobic block were prepared to show fully helical conformation. By mixing Lys residues in the hydrophobic block, the hydrophobic helical block responded to pH change. The induced helix-to-coil transition below pH 9 led to porous membranes.

5.2.4 Other ABA Triblock Copolymers

Using Ac–Ser–Ala–Gly–Ala–Gly–Glu–Gly–Ala–Gly–Ala–Gly–Ser–Gly–OH as an initiator for the atom transfer radical polymerization of methyl methacrylate yields an ABA triblock copolymer [16]. The peptide sequence is hydrophilic to form a β hairpin structure. Vesicles with diameters ranging from 100 nm to 300 nm were obtained from the triblock copolymer. ABA triblock copolymers containing a short hydrophilic middle block also formed vesicles. Poly(Lys(Z))-*b*-poly[diethylene glycol bis(3-aminopropyl) ether](DGBE)-*b*-poly(Lys(Z)) having DGBE to Lys(Z) ratios ranging from 1:6 to 1:60 changed diameters from 80 nm to 320 nm when the Lys(Z) block length was increased [17].

5.2.5 Hyper-branched Polymers and Dendrimers

A hyper-branched Lys is known to constitute a dendron, which was chemically modified with alkyl chains to yield a lipidic cationic poly(Lys) dendron. The dendron was found to self-assemble into a vesicular structure of about 300 nm diameter [18]. To a third-generation poly(Lys) dendron, several chains of tetradecanoic acid were attached. Vesicular structures of 240 nm diameter (polydispersity index of 0.36) were found to be stable against Triton X-100, which transformed vesicular structures into surfactant–dendron mixed micelles at a concentration as high as 8 mM [19]. A lipidic cationic Lys-based dendron also formed a vesicular structure, and the solute leakage was influenced by addition of cholesterol [20]. A dendritic block was also used as a hydrophilic head group of amphiphiles with a well-defined structure and a molecular shape. Poly(γ-*n*-dodecyl-L-Glu) was connected with a hydroxyl-terminated polyester dendritic block, which was self-assembled into vesicular structures of 600 nm diameter with a bilayer structure of hydroxyl dendritic blocks at the membrane surface and the flexible alkyl side chains in the membrane [21]. Amide dendron–dipeptide conjugates, where Phe–Phe or Gly–Gly was used at the focal point, were also found to self-assemble into vesicles [22].

5.2.6 Triskelion Structure

Anther unique molecular shape is that of a triskelion structure, which can also be used for peptide amphiphiles. A three-arm helix peptide, which possesses three 16mer helical peptides with using tris(3-aminopropyl)amine as a core, was synthesized (Figure 5.3). Notably, the three-arm helix peptide, which has molecular weight of 5228, was dispersed easily in water even though this compound has no explicit hydrophilic moiety in the molecule. Dynamic light scattering measurement and transmission electron microscopy images of water dispersions revealed that spherical assemblies with a diameter of 70 nm were formed. The central tertiary amine moiety in the three-arm helix peptide should act as a hydrophilic part due to the convergent point of the three negatively polarized dipoles of three helices (Figure 5.4) [23]. A similar but much smaller peptide, where three Trp–Trp were connected to tris(2-aminoethyl)amine, was also found to form vesicular structures in a mixed solution of methanol and water [24].

5.2.7 Cyclic Peptide as Template for Amphiphilicity

Cyclic peptides have two opposite faces above and below the cyclic skeleton, which can be designed to be hydrophilic and hydrophobic. A typical example of such cyclic peptide facial amphiphile is gramicidin S, which is a naturally occurring cyclic peptide antibiotic. A cyclic hexapeptide having a sequence of alternating Arg and Tyr modified with perfluroinated alkyl chain was synthesized. The cyclic peptide self-assembled into vesicles with diameters of about 120 nm. The vesicular membrane constitutes a bilayer of the hydrophobic interior of fluoroalkyl chains and the hydrophilic exterior of guanidinium groups. The vesicles were shown to be taken up by HeLa cells due to the cell-penetrating property of guanidinium clusters on the surface [25].

5.2.8 Lipid-like Structure

The lipid-shaped molecules composed of poly(Glu)$_{16}$ with two hydrophobic tails of octadecane, cholesterol, or silsesquioxane at the C terminal also self-assembled into vesicles. The aggregation number did not change with a pH change [26].

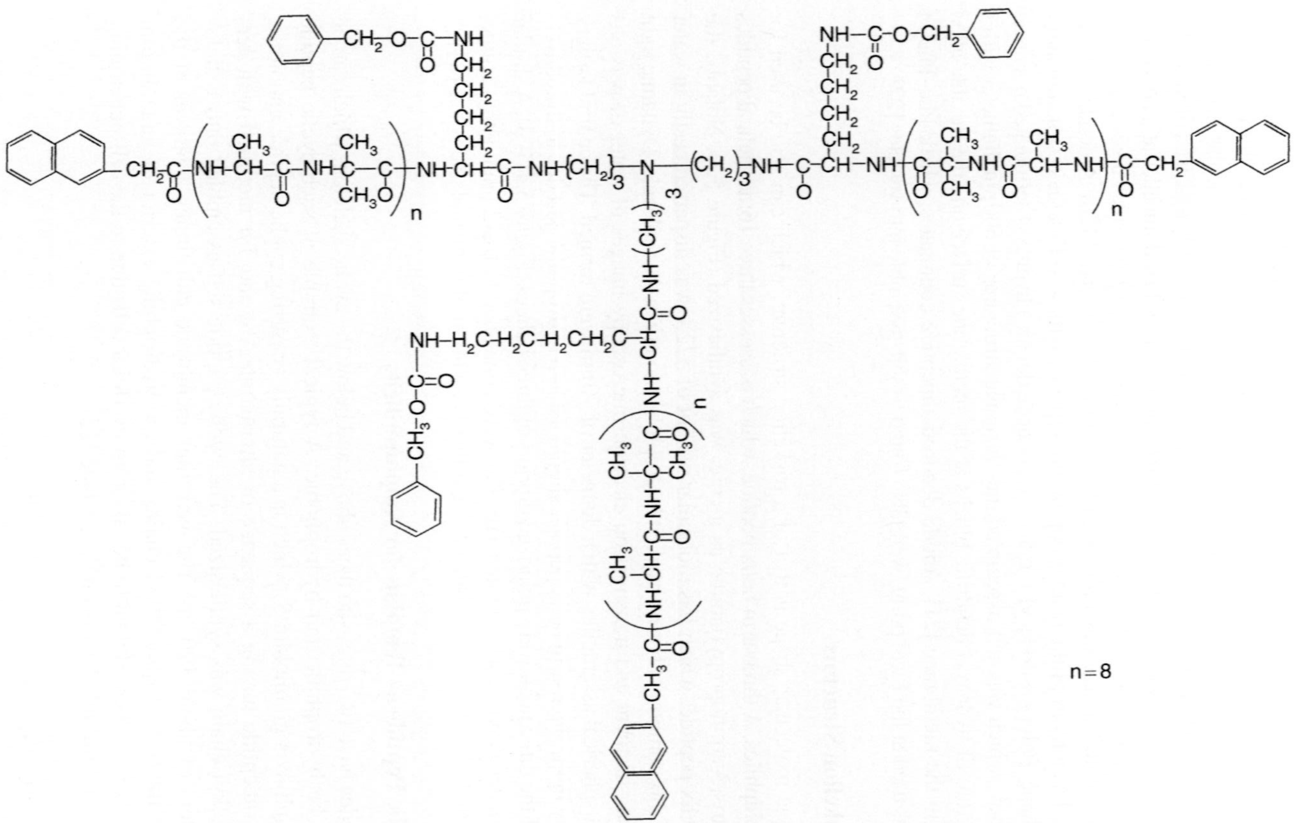

n=8

Figure 5.3 *A triskelion-type peptide with a molecular weight of 5228, which self-assembled into vesicles with using the tertiary amino group as a hydrophilic group. Reproduced with permission from reference [23]. Copyright (2004) Nature Group Publishing*

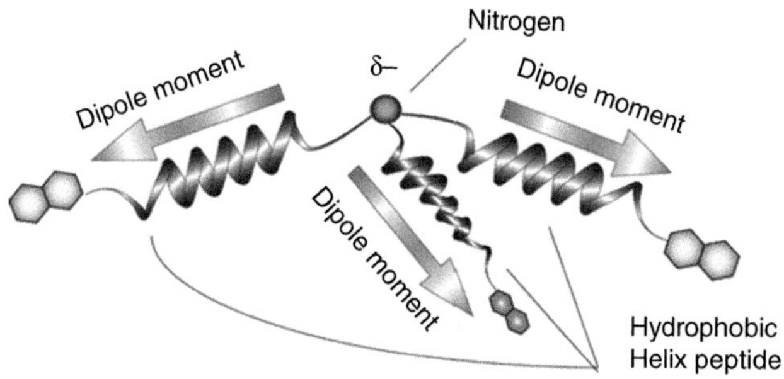

Figure 5.4 The C terminals of three helices converge on the tertiary amino group to enhance the hydrophilic property of the N moiety in the triskelion-type peptide. Reproduced with permission from reference [23]. Copyright (2004) Nature Group Publishing

5.3 Peptide Building Blocks

It is surprising to know that it is possible to adopt a wide range of peptide chain lengths for self-assembling. However, those peptides showing good self-assembling properties seem equipped with conformation restraint and rigidity.

5.3.1 Oligopeptides

A pH-responsive morphology change was reported with $(His)_2$–$(Val)_6$, which possesses a pI of 6.92. At a low pH region, $(His)_2$–$(Val)_6$ took vesicles and tubes, but reassembled into large membranous sheets at pH above the pI [27]. It is suggested that molecular assemblies composed of positively charged peptides may be able to deliver negatively charged nucleic acids. A class of short lipid-like amphiphilic peptides was reported. Ac–$(Ala)_6$–Asp and others formed tubes and vesicles. These peptide detergents have the ability to solubilize and stabilize various membrane proteins [28]. Ac–Ala–Ala–Val–Val–Leu–Leu–Leu–Trp–Glu_n–COOH with $n=2$ or 7 was found to form vesicles with diameters of about 120 nm, suggesting that the mass increase of the hydrophilic moiety in the amphiphiles had no influence on the size and shape of the vesicles [29]. The decapeptide took predominantly the poly(Pro) type II conformation, which enables intermolecular hydrogen bonding in the assembly to stabilize the membrane with an interdigitated structure [30]. Introduction of Cys residues, Ac–Ala–Cys–Val–Cys–Leu–Cys–Leu–Trp–Glu–Glu–COOH, was effective in improving the physical stability of peptide vesicles by covalently crosslinking the oligopeptides via disulfide bonds. The Cys-containing peptide self-assembled into vesicles with a diameter of about 150 nm. Upon covalent crosslinking, the vesicles showed resistance to DMF. On the other hand, the release of calcein from the crosslinked vesicles was comparable to that of the noncrosslinked vesicles, indicating that the covalent bonds did not increase the barrier function of the vesicular membranes towards small hydrophilic molecules [31]. The peptide vesicles were internalized by cells predominantly via adsorptive macropinocytosis. Water-insoluble phthalocyanines were quantitatively entrapped within the membranes, showing an active nanocarrier for the intracellular delivery of photosensitizers for photodynamic therapy [32].

5.3.2 Dipeptides

Phe–Phe, which is a recognition motif of the Alzheimer's β-amyloid polypeptide, self-assembled into ordered and discrete nanotubes. The introduction of a thiol group into the Phe–Phe peptide altered its assembly from tubular to spherical particles similar to those formed by Gly–Gly [33]. The significant stability of the peptide nanostructures is consistent with the structural stability of amyloid fibrils. It was shown that dilution of a peptide–nanotube dispersion resulted in the formation of a vesicle, where the morphology transition occurred reversibly upon a change in the peptide concentration [34]. Concentration-dependent transformation from vesicles to nanotubes at neutral pH was found with 6-aminohexanoic acid–Tyr–Glu. The vesicles could encapsulate a potent anticancer drug, which was released in the presence of Ca^{2+} ions [35]. The Phe–Phe peptide nanotubes also changed the morphology to vesicles upon electrostatical association with ssDNA, which easily entered into cells [36]. Using constrained areas of microporous membranes for self-assembling, Phe–Phe self-assembled into microvesicles [37]. Dipeptides containing a conformationally constrained α,β-dehydrophenylalanine (ΔPhe) residue were also reported. Glu–ΔPhe and Lys–ΔPhe formed vesicles with diameters of 400 nm and 560 nm, respectively, with moderate sized polydispersity. The vesicles showed resistance against proteinase K digestion, and uptake into HeLa cells was demonstrated [38]. A small zwitterion molecule, which was synthesized by condensation of Ala and guanidinocarbonyl pyrrole carboxylic acid, also assembled into vesicles with a diameter of about 60 nm in DMSO. The zwitterion molecule formed ion-paired dimers with a kink of ca. 38°, which aggregated into membranes [39]. Twitter ions, where the two charges interact simultaneously with each other, can generate strong molecular assemblies [40].

5.3.3 β-Peptides

β-Peptides are known to form regular helical structures, which can be used as a hydrophobic block of amphiphiles. Tetramer, pentamer, and hexamer of *trans*-2-amioncyclohexanecarboxylic acid were found to self-assemble into vesicles with diameters ranging from 20 nm to 180 nm. β-Peptides took helical conformation to associate in the vesicular membranes [41]. For example, a hexamer of *cis*-2-aminocyclohexanecarboxylic acid having a free amine at the N terminal and amide at the C terminal self-assembled into vesicles with an average diameter of 100 nm. On the other hand, other β-peptides composed of *cis*-2-aminocyclopentanecarboxylic acids and diexo-3-aminobicyclo[2.2.1]hept-5-ene-2-carboxylic acids did not show any self-association, suggesting that small changes in the side-chain chemistry have a significant influence on the self-assembling behavior through stable compact helical conformation and adequately hydrophobic side-chains [42].

5.3.4 Naturally Occurring Peptides

Naturally occurring peptides were also found to produce vesicles. A microbially produced siderophore, marinobactin E, facilitates Fe(III) acquisition through coordination of Fe(III) by the marinobactin headgroup. Marinobactin E self-assembled into micelles due to the amphiphilic property of conjugation of a six-amino-acid peptide with palmitic acid. Upon coordination of Fe(III), micelles were transformed to vesicles [43]. Bridging the headgroups of multiple marinobactin E by Fe(III) should change the volume balance between the hydrophilic moiety and the hydrophobic moiety to induce the morphology transition. Another naturally occurring sequence from tomato bushy stunt virus was utilized to prepare vesicles. A 24mer β-annulus peptide, Ile–Asn–His–Val–Gly–Gly–Thr–Gly–Gly–Ala–Ile–Met–Ala–Pro–Val–Ala–Val–Thr–Arg–Gln–Leu–Val–Gly–Ser, was synthesized. The peptide has a sticky end of Val–Thr–Arg–Gln–Leu–Val–Gly–Ser for self-assembling. Three molecules associated together to make a triangular structure with three sticking segments from the three corners. Then, eight units of the triangular structure, where 24 molecules of the peptide in total were involved, self-assembled into virus-like nanocapsules with diameters of ca. 50 nm [44].

5.4 Peptide Architecture

The association of peptide molecules in water can be driven by several factors. A typical driving force is hydrophobic interaction between hydrophobic segments. The molecular exteriors of the associating segments should be designed to fit each other about the concave–convex relationship reminiscent of the leucine zipper motif. Other factors of complementary sequence, ion complex, stereocomplex, dipole interaction, and hydrogen bonds are still effective in molecular assembling.

5.4.1 Protein Cages

Functionalization of vesicles can be achieved easily by starting from protein cage nanoparticles. The dodecameric DNA binding protein from starved cells (Dps) protein cage was chemically modified with phenanthrolene via the free-sulfhydryl groups at the cage surface. With the addition of Fe(II), the protein cages assembled into higher order structures of coordination polymers [45]. Ferritin incorporates cations inside, a property that was used for encapsulation of positively charged gadolinium chelator. The exterior of apoferritin was chemically modified with dextran to suppress delivery to the liver. As a result, the dextran-modified apoferritin with encapsulation of the magnetic resonance imaging contrast agent showed a proton relaxivity 10 times higher than that of standard clinically used Gd contrast agents (Figure 5.5) [46].

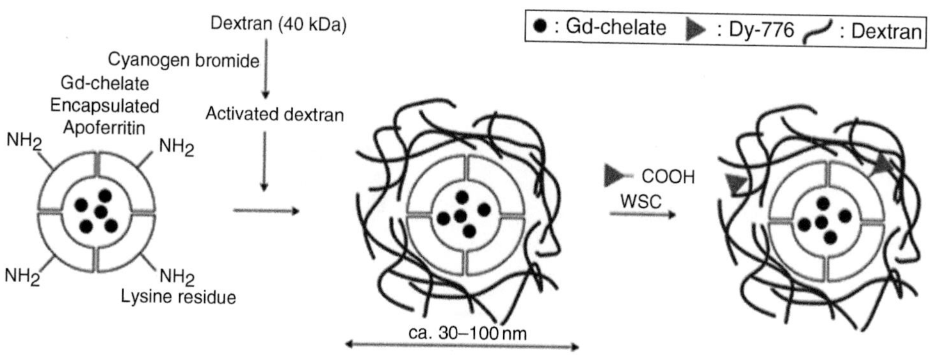

Figure 5.5 *A magnetic resonance imaging contrast agent using apoferritin. The surface was coated with dextran to avoid entrapment in the liver. Apoferritin is effective for condensation of the cationic Gd-chelates inside. Reproduced with permission from reference [46]. Copyright (2011) Wiley-Blackwell*

5.4.2 Ion-Complex for Self-Assembling

Generally vesicle membranes bear a hydrophobic property because they are formed by hydrophobic interaction among the hydrophobic block of amphiphiles. On the other hand, ion-complex formation is also available for membrane preparation. Mixing of a pair of oppositely charged block copolymers, PEG-*b*-poly(Asp) and PEG-*b*-poly(Lys), produced vesicles with diameters up to 10 μm, which are named 'PICsome'. The PICsome membrane is semipermeable, which has been found to be favorable for carriers of therapeutic compounds. Further, the use of organic solvents is unnecessary for the preparation [47].

5.4.3 Stereo-Complex for Self-Assembling

Helical peptides can take either a right-handed or a left-handed sense by choosing L-residues or D-residues. When two types of helices are mixed, a stereo-complex among them may be formed due to tight molecular packing between the right-handed and left-handed helices, as proved by the stereo-complex of poly(L-lactic acid) and poly(D-lactic acid). A mixture of poly(Sar)$_{27}$-*b*-(L-Leu–Aib)$_6$ (right-handed helix) and poly(Sar)$_{27}$-*b*-(D-Leu–Aib)$_6$ (left-handed helix) self-assembled in saline to planar sheets, which transformed the morphology into vesicles upon heating the solution at 90 °C for 10 min. The diameter of vesicles was about 200 nm with a narrow size distribution (Figure 5.6) [48].

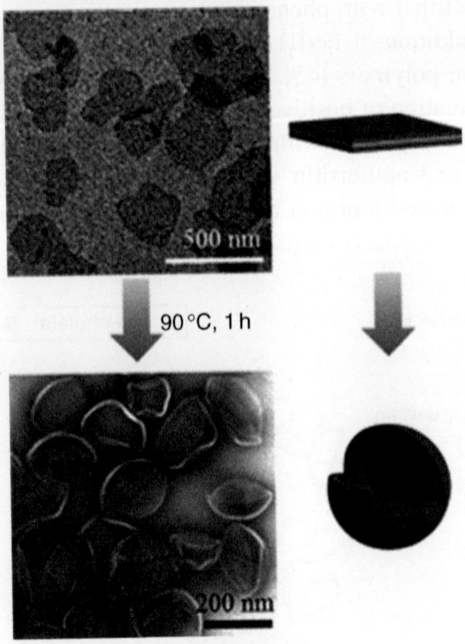

Figure 5.6 A mixture of poly(Sar)$_{27}$-b-(L-Leu–Aib)$_6$ (right-handed helix) and poly(Sar)$_{27}$-b-(D-Leu–Aib)$_6$ (left-handed helix) (1/1 mol/mol) self-assembled into planar sheets, which were converted to vesicles upon heating. The size distribution was very narrow. Reproduced with permission from reference [48]. Copyright (2011) The Royal Society of Chemistry

5.4.4 Inside-out Morphology Transformation

An inside-out morphology change of vesicles was also reported. A zwitter ionic diblock copolypeptide, poly(Glu)$_{15}$-*b*-poly(Lys)$_{15}$, was synthesized. The copolypeptide self-assembled into vesicles with diameters of 110 nm and 175 nm at pH 3–4 and pH 11–12, respectively. The vesicles can be reversibly produced as a function of pH in pure water, which are named 'schizophrenic vesicles' [49].

5.5 Function of Peptide Assemblies

Usage of vesicles as a microreactor was shown by preparation of gold nanoparticles. Poly(ethylene glycol)$_{77}$-*b*-(Leu–Aib)$_{12}$–naphthylalanine–OBzl self-assembled into vesicles with diameters of ca. 135 nm. In the presence of 0.1 wt% $HAuCl_4$, gold nanoparticles with diameters of ca. 5 nm were formed. Notably, a reference compound without the naphthylalanine residue could not yield gold nanoparticles. It is suggested that the cluster of the negative partial charge at the C terminal of the 25mer helix dipole was directed to the inner water phase to promote the reductive reaction in the vesicles (Figure 5.7) [50]. Block copolypeptide vesicles were used as a template into an inorganic material. In this case, an individual supramolecular structure was translated into an individual inorganic nanoparticle. For example, using vesicles prepared from poly(Lys)$_{23}$-*b*-poly(Phe)$_{23}$ as a template, silica hollow spheres were synthesized by just the addition of orthosilicic acid solution within 5 minutes [51].

A condensation reaction without any condensation reagents was allowed in a vesicular system. Vesicles were prepared from an amphiphilic *N*-phosphorylamino acid. While keeping vesicles in solution, the formation of dipeptide was confirmed. *N*-phosphorylamino acid should activate itself through an intramolecular phosphoric-carboxylic mixed anhydride intermediate, which can be subjected to a nucleophilic attack of the carboxylic hydroxyl of another phosphyrlamino acid in the vesicle to form dipeptide via a carboxylic anhydride intermediate. The condensation reaction is realized with the help of a suitable orientation and packing of amphiphiles at the vesicle/water interface together with certain conformational freedom in the vesicular bilayer system [52].

Functionalization of polypeptide vesicles was demonstrated with poly(Arg)$_{60}$-*b*-poly(Leu)$_{20}$, which formed vesicles with diameters ranging from 50 nm to a few μm by extruding through different polycarbonate filters. A nonamer of Arg is known to have an ability to transport

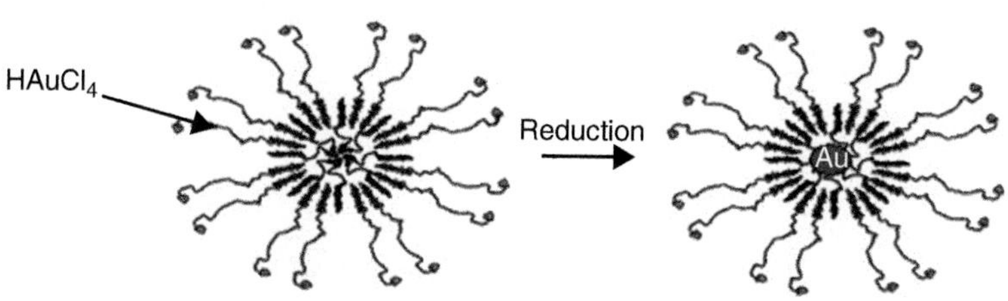

Figure 5.7 The inner surface of vesicles prepared from poly(ethylene glycol)$_{77}$-b-(Leu–Aib)$_{12}$–naphthylalanine–OBzl is considered to be negatively charged due to the C teminals of helices (partially negatively charged dipoles), which may promote a reduction reaction of Au^{3+} ions to gold nanoparticles. Reproduced with permission from reference [50]. Copyright (2004) Elsevier

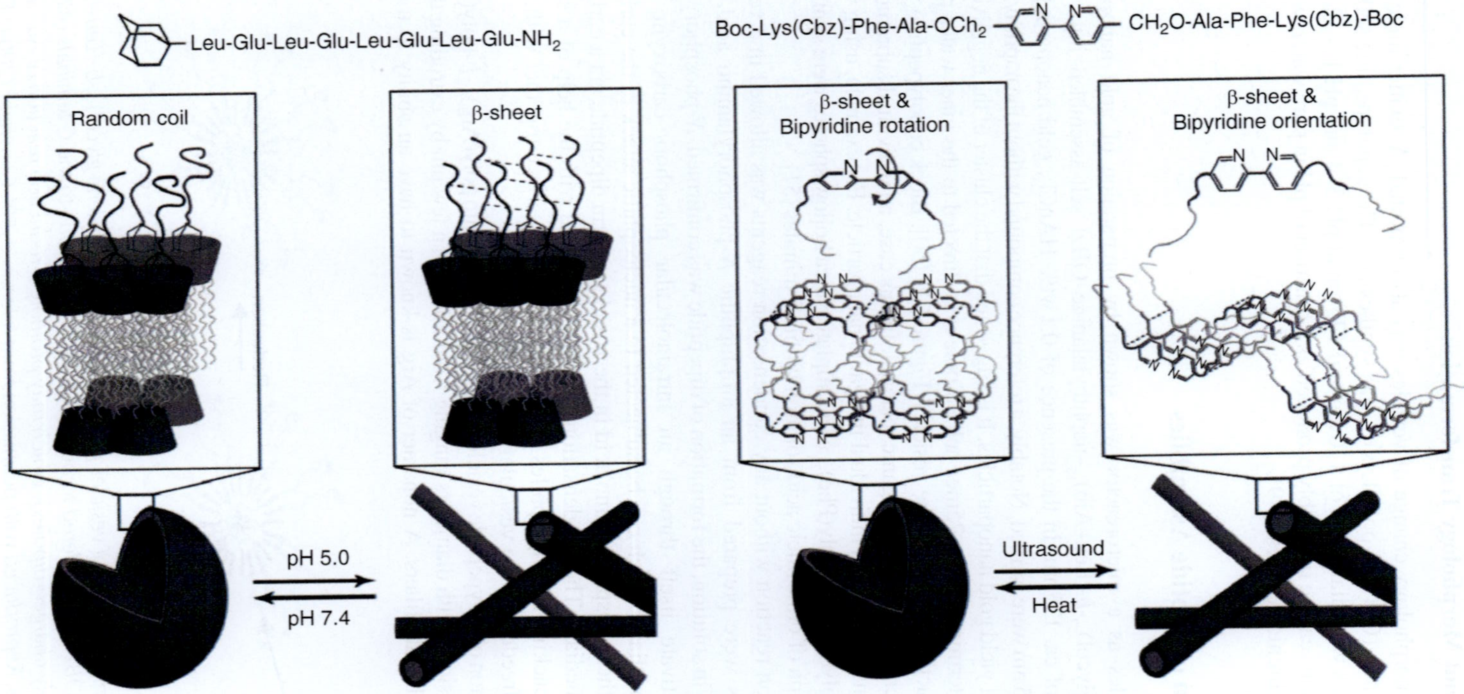

Figure 5.8 *The schematic illustrations of the transformations between vesicles and fibers by stimulations of pH change and heating. Redrawn from references [57] and [58] (see color plate figure)*

cargoes into cells due to the continuous guanidium residues, which constitute an essential sequence of protein-transduction domains (PTD) and the Tat sequence. The vesicles prepared from poly(Arg)$_{60}$-*b*-poly(Leu)$_{20}$ thus showed efficient intracellular delivery. Strikingly, the vesicles were minimally cytotoxic despite the fact that homopolyarginines over 20 residues generally show cytotoxicity. The condensed chain assembly at the vesicular surface is most likely to prevent free diffusion of the polycations to cell surfaces [53].

Surface functionalization with the RGD-peptide sequence, which is a cell adhesion motif, of poly(ethylene oxide)-*b*-poly(butadiene) vesicles induced a change from unilamellar vesicles to multilamellar vesicles. The attractive interactions between RGD moieties are considered to be operative in the assemblies [54]. Incorporation of photolabile 2-nitrophenylalanine or fluorescein-conjugated Lys at the junction moiety of a diblock copolymer of poly(ethylene glycol)-*b*-poly(caprolactone) endowed the vesicles with properties of UV-induced release of encapsulated biocytin or a fluorescence labeling agent [55].

Surface functionality can also be provided to vesicles with a noncovalent method. Poly(γ-benzyl Glu)-*b*-Gly(Glu–Ile–Ala–Ala–Leu–Glu–Lys)$_3$–NH_2 formed vesicles with the bilayer membrane. The vesicles can incorporate Ac–(Lys–Ile–Ala–Ala–Leu–Lys–Glu)$_3$–Gly-*b*-poly(ethylene glycol) with a coiled-coil formation between the acidic 22mer peptide and the basic 22mer peptide. With an appropriate design of the peptides forming the coiled-coil region, therefore, the surface can be functionalized [56].

Self-assembly prepared from multiple components is promising for obtaining dynamic and adaptive supramolecular materials. A β-cyclodextrin derivative with dodecyl chains self-assembled into cyclodextrin vesicles, which acted as a host vesicle for adamantine modified octapeptides. The octapeptide was designed to constitute an amphiphilic β-sheet domain at pH 5.0, but a random coil at pH 7.4. Using a reversible morphological change of the multicomponent supramolecular complex from vesicles to fibers, a pH-triggered release of the encapsulated contents was observed [57]. Morphological transformation between nanofibers and vesicles was also attained by using bipyridine–tripeptide self-assembly. The bipyridine moiety, which was flanked by β-sheet forming tripeptides, changed the conformation by ultrasound treatment to promote formation of dense bilayer structures into nanofibers. On the other hand, heating recovered the mobile property of bipyridine groups to lead to the formation of vesicles [58]. The schematic illustrations of these transformations between vesicles and fibers are shown in Figure 5.8.

5.6 Tumor Imaging with Peptide Nanocarrier

As mentioned in the introduction, peptides can be designed to self-assemble for a specific function on the basis of the molecular structure-assembling relationship. To demonstrate this point in detail, a series of peptide designs for development of peptide nanocarriers for tumor imaging is described as follows.

Antibody drugs have been launched and are widely used because of their reduced adverse effects based on their targeting ability to specific diseased sites. Typical examples are Rituxan® for non-Hodgkin lymphoma and Herceptin® for metastatic breast cancer. Difficulties of antibody drugs in general clinical treatments, however, are (i) their high cost for production, (ii) selection of target sites preferably to specific cell markers and differentiation antigens without affecting important cell functions, and (iii) preparation of human antibody. While antibody specifically binds to a specific target molecule or site, nanocarriers with sizes in the range from 10 nm to a few hundreds of nm accumulate generally in solid tumors due to the enhanced permeability and retention (EPR) effect [59]. The EPR effect explains the accumulation mechanism as a result of leaky angiogenesis blood tubes and immature lymphatic system in tumor regions. Since the EPR effect appears in all types of solid tumors, nanocarriers have an advantage for targeting solid tumors without using any specific guiding molecules to deliver correctly to tumor regions. On the other hand, there are several points requested for nanocarriers to be applied for clinical usages.: (i) high accumulation rate in solid tumors, (ii) biocompatibility, (iii) unchanged disposition upon multiple dosages, and (iv) penetration into solid tumors against high interstitial pressure. The strategies responding to these issues for our case are explained as follows.

We postulate the following four designing principles for nanocarriers to accumulate in solid tumors.

1. Size selection and hydrophilic surface: 10 nm is a lower limit of the nanocarrier size to avoid fast excretion from the kidney and 100 nm is an upper limit to suppress recognition of the reticuloendothelial system such as the liver and spleen and opsonization as well. Further, surroundings of nanocarriers are totally hydrophilic to prevent interaction with plasma proteins, an association that influences the disposition of nanocarrirs.
2. Biodegradable polymers: Only a limited number of polymers are allowed for administration to humans. For clinical application, the (polymer) components of nanocarriers should be biodegradable and non-toxic, including their metabolized species as well. PEG, which is admitted by the FDA for clinical usage, is a very useful hydrophilic polymer to decorate proteins and nanocarriers endowed with a long sustaining property in the bloodstream and organs. On the other hand, it is pointed out that PEG is nonbiodegradable and may cause a high rate of infusion reaction, even though PEG is considered to be nontoxic due to fast excretion from the kidney.
3. Stealth property: PEGylated liposome is called as stealth liposome because *in vivo* the retention time in blood of liposome is significantly increased by surface modification with PEG chains. Hydrophilicity and dynamic chain mobility of PEG explain the long circulation in the body due to prevention from triggering the compliment system and trap by the endothelial system. However, Dams *et al.* pointed out that disposition of the stealth liposome at second dosage differed from that at first dosage [60]. The stealth liposome is recognized by the dendritic cell in the spleen, which triggers the immune system. Indeed, the production of IgM for PEG chains was detected and upregulated at the first dosage of the stealth liposome. The change of disposition upon multiple dosages is called the accelerated blood clearance (ABC) phenomenon (Figure 5.9). We showed two ways for nanocarriers to evade from the immune system: size reduction down to 10–20 nm and coating with a hydrophilic polymer brush structure instead of a mushroom structure. In the case of the stealth liposome, the PEG chains are expected to take a mushroom-shaped conformation to make the surface hydrophilic and dynamic. On the other hand, when the hydrophilic polymer chains take extended conformation and are packed densely at the surface, the hydrophilic shell of nanocarriers is composed of a polymer brush structure, which does not allow interaction with proteins in plasma (Figure 5.10).
4. Size reduction and PGE1: Generally anticancer drugs have a difficulty to infiltrate solid tumors due to the high interstitial pressure. Nanocarriers of 30 nm diameter were found

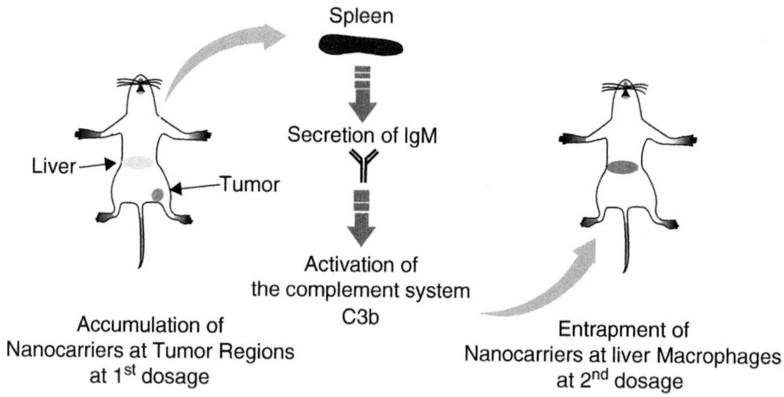

Figure 5.9 *Accelerated blood clearance phenomenon appears in a tumor-bearing mouse with multiple dosages of nanocarriers*

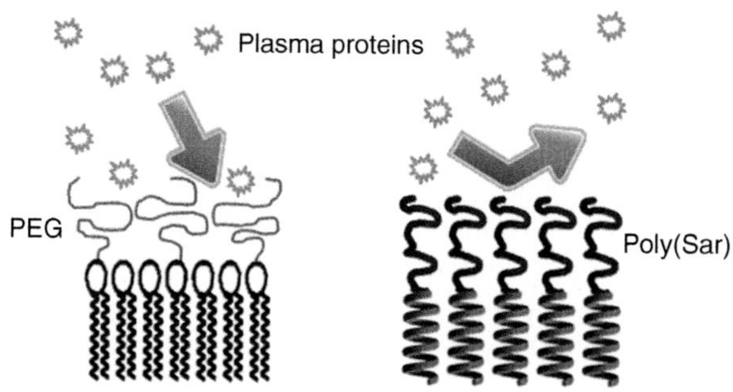

Figure 5.10 *Schematic illustration of a mushroom structure and a polymer brush structure of hydrophilic polymer chains around nanocarriers. Stealth liposomes use PEG chains of a mushroom structure, which are hydrophobic and dynamic. Poly(Sar) chains in peptosomes are forced to take a polymer brush structure of high density and less free volume owing to the regular and tight packing of helices at the hydrophobic region*

to accumulate even in pancreatic cancer, suggesting that this small size may be effective against the interstitial pressure. Further, simultaneous administration of PGE1 and others have been proposed to overcome the interstitial pressure.

We started to search for nanocarriers that satisfy these points for imaging solid tumors. As described in the initial part of this chapter that polypeptides have advantages for preparation of nanocarriers through molecular assembling of properly designed amphiphilic polypeptides. The first distinct point of our strategy to prepare polypeptide molecular assemblies is usage of hydrophobic helical peptide block in amphiphilic polypeptides. Tight packing of α-helices is a general structural motif found in proteins. Helices have a tendency to associate together in an antiparallel way for favorable dipole–dipole interaction to form a bundle structure. Parallel packing of helices is also available as found in the Leu-zipper motif. Because formation of molecular assemblies is driven by hydrophobic interaction among the hydrophobic block of amphiphiles, usage of helical peptides should increase physical stability of molecular assemblies. Further, the association of helices takes place regularly, which leads to molecular assemblies having well-defined structures. Using this regular molecular packing of helices as a scaffold, we try to construct a hydrophilic polymer-brush structure at surroundings of the molecular assemblies.

The second point is to use poly(Sar) as a hydrophilic polypeptide block. Poly(Sar) is as hydrophilic as PEG. Amide bonds in poly(Sar) can take both *trans* and *cis* configurations in order to expand the available conformation. The smallest two hydrogens at Cα also endow the molecular chain with a dynamic property. The differences between poly(Sar) and PEG are biodegradable or synthetic and thick or thin about the main chains. Biodegradable polypeptides have an advantage for clinical usage and the thick chain of poly(Sar) is preferable for preparation of a densely packed polymer-brush structure.

A mixture of poly(Sar)$_{43}$-*b*-poly(Glu(OMe))$_{18}$ and poly(Sar)$_{65}$-*b*-poly(Glu(OMe))$_{18}$ self-assembled into vesicles of about 100 nm diameter with a moderate size distribution. The vesicles were labeled with a near-infrared fluorescence (NIRF) probe and were subjected to *in vivo* tumor imaging by NIRF. *In vivo* retention in the blood experiment showed long circulation of the vesicles, which accumulated in tumor regions and kept the vesicular structure [61]. However, for clinical usage, the size distribution of nanocarriers should be narrow.

We directed our research to use more specified peptide molecules where molecular lengths of helical hydrophobic blocks are uniform, which would result in the preparation of molecular assemblies with a narrow size distribution. Poly(Sar)$_{27}$-*b*-(L-Leu–Aib)$_6$ was synthesized and dispersed in saline. The formation of curved sheets was observed. Upon heating the solution at 90 °C for 10 min, the curved sheets changed morphology to a tube structure of 60 nm diameter and 200 nm length. The size distribution was very narrow, because the tube structure was formed simply by sticking the hydrophobic edges of the curved sheets, resulting in a narrow size distribution. On the other hand, a mixture of poly(Sar)$_{27}$-*b*-(L-Leu–Aib)$_6$ (right-handed helix) and poly(Sar)$_{27}$-*b*-(D-Leu–Aib)$_6$ (left-handed helix) self-assembled in saline to planar sheets. Upon heating the solution at 90 °C for 10 min, the planar sheets were torn off to generate a vesicular structure with a homogeneous size distribution of 200 nm diameter. Electron diffraction from the membrane composed of a mixture of poly(Sar)$_{27}$-*b*-(L-Leu–Aib)$_6$ and poly(Sar)$_{27}$-*b*-(D-Leu–Aib)$_6$ revealed regular and tight molecular packing of the helices as a checkerboard design [48]. The morphology difference of the curved or the planar sheets also supports the interpretation of regular packing of one type of helix or a mixture of right-handed and left-handed helices. Further, tight packing of right-handed and left-handed helices has been reported as a stereo-complex, such as a mixture of poly(L-lactic acid) and poly(D-lactic acid), indicating that the concept of a stereo-complex can be expanded widely to helices composed of various chemical species. Vesicles prepared from a mixture of poly(Sar)$_{27}$-*b*-(L-Leu–Aib)$_6$ and poly(Sar)$_{27}$-*b*-(D-Leu–Aib)$_6$ of 200 nm diameter may be too large for tumor imaging. To reduce the size, a mixture of poly(Sar)$_{10}$-*b*-(L-Leu–Aib)$_6$ and poly(Sar)$_{27}$-*b*-(D-Leu–Aib)$_6$ was examined. The mixture produced vesicles of around 120 nm diameter with a narrow distribution, but the vesicles were trapped in the liver and failed in tumor imaging. Probably, the short poly(Sar) chains did not make the surroundings of vesicles hydrophilic enough for imaging. Since vesicles prepared from a mixture of poly(Sar)$_{43}$-*b*-poly(Glu(OMe))$_{18}$ and poly(Sar)$_{65}$-*b*-poly(Glu(OMe))$_{18}$ accumulated in tumor sites, the minimum length of poly(Sar) chains may be 43mer for nanocarriers to accumulate in tumor regions.

The size of the sheets can be adjusted by changing the hydrophilic block length. For example, poly(Sar)$_{27}$-*b*-(L-Leu-Aib)$_6$ afforded curved sheets with sides of 200 nm length, which were downsized to sheets with sides of a few tens of nm length by using poly(Sar)$_{60}$-*b*-(L-Leu–Aib)$_6$. The hydrodynamic diameter of the sheets was 50 nm. The sheets were labeled with a near-infrared fluorescence chromophore and were applied for tumor imaging of a tumor bearing mouse. The tumor region was clearly imaged by fluorescence from the planar sheets, which accumulated in the tumor region due to the EPR effect. An important finding when using these sheets for tumor imaging is that the ABC phenomenon was significantly reduced. The tumor region was imaged by fluorescence even at the second injection of the planar sheets. Koide *et al.* reported that polymer micelles prepared from PEG-*b*-poly(Asp(β-benzyl)) of 50 nm promoted rapid clearance of a following injection of PEGylated liposome due to the ABC phenomenon, but polymer micelles of 10 nm (PEG-*b*-poly(Asp(β-nonyl)) and 30 nm (PEG-*b*-poly(Asp(β-pentyl)) did not influence the disposition of PEGylated liposome that was injected after three days from the polymer micelle injection [62].

The size of nanocarrier is therefore considered to be crucial for evading the ABC phenomenon. However, there should be another factor to suppress the ABC phenomenon, when we consider that two different results were obtained from the two types of nanocarrier of 50 nm. The sheets of 50 nm prepared from poly(Sar)$_{60}$-*b*-(L-Leu–Aib)$_6$ reduced the ABC phenomenon, but PEG-*b*-poly(Asp(β-benzyl)) of 50 nm induced it. The difference may be due to the different structures of the hydrophilic layer surrounding the nanocarriers. In the case of the sheets composed of poly(Sar)$_{60}$-*b*-(L-Leu–Aib)$_6$, the hydrophilic poly(Sar) layer is expected to take a polymer-brush structure due to the scaffold effect of hydrophobic helical peptides, which were regularly packed in the hydrophobic region of the sheet structure. On the other hand, PEG chains surrounding the polymer micelle should take rather a mushroom-shaped conformation, because the main chain of PEG is thinner than poly(Sar), yielding more void space with PEG at the hydrophilic shell of the polymer micelle than with poly(Sar), allowing PEG chains to take a compact configuration (a mushroom structure). Further, the hydrophobic core of the polymer micelle should not be well regulated and lose the scaffold effect, which was observed with the helix association in the sheet structure.

5.7 Perspectives

Peptide molecules are designed to take defined conformation, which is an essential point in providing scaffolds for functional groups to be aligned in the most appropriate way to elicit effects from the molecule and the molecular assemblies. The structural motifs are abundantly found in proteins. Further, we can enrich the content of available structures by using unnatural amino acids and D-amino acids as components, which may lead to the scope to produce nanomaterials beyond what mother Nature creates. For example, the stereo-complex formation between right- and left-handed helices to generate a novel morphology of a nano round-bottom flask [1] is one clue for further advancements. Another important point of peptide nanomaterials is its environmentally benign property. Biodegradability of peptides is also advantageous for applications in medical fields. Taken together, peptides will provide many nanomaterials in real use.

References

[1] Ueda, M., Makino, A., Imai, T., Sugiyama, J., Kimura, S. (2011) Tubulation on peptide vesicles by phase-separation of a binary mixture of amphiphilic right-handed and left-handed helical peptides. Soft Matter 7: 4143–4146.
[2] Bergeron, R.J., Phanstiel, O., Yao, G.W., Milstein, S., Weimar, W.R. (1994) Macromolecular self-assembly of diketopiperazine tetrapeptides. J. Am. Chem. Soc. 116: 8479–8484.
[3] Cornelissen, J.J.L.M., Fischer, M., Sommerdijk, N.A.J.M., Nolte, R.J.M. (1998) Helical superstructures from charged poly(styrene)-poly(isocyanodipeptide) block copolymers. Science 280: 1427–1430.
[4] Kimura, S. (2008) Molecular dipole engineering: new aspects of molecular dipoles in molecular architecture and their functions. Org. Biomol. Chem. 6: 1143–1148.
[5] Fujita, K., Kimura, S., Imanishi, Y. (1999) Spherical self-assembly of a synthetic alpha-helical peptide in water. Langmuir 15: 4377–4379.
[6] Kimura, S., Kim, D.H., Sugiyama, J., Imanishi, Y. (1999) Vesicular self-assembly of a helical peptide in water. Langmuir 15: 4461–4463.
[7] Kimura, S., Muraji, Y., Sugiyama, J., Fujita, K., Imanishi, Y. (2000) Spontaneous vesicle formation by helical glycopeptides in water. J. Colloid Interf. Sci. 222: 265–267.
[8] Zhang, L.F., Eisenberg, A. (1995) Multiple morphologies of crew-cut aggregates of polystyrene-*b*-poly(acrylic acid) block-copolymers. Science 268: 1728–1731.
[9] Checot, F., Lecommandoux, S., Gnanou, Y., Klok, H.A. (2002) Water-soluble stimuli-responsive vesicles from peptide-based diblock copolymers. Angew. Chem. Int. Edit. 41: 1339–1343.
[10] Checot, F., Brulet, A., Oberdisse, J., Gnanou, Y., Mondain-Monval, O., Lecommandoux, S. (2005) Structure of polypeptide-based diblock copolymers in solution: stimuli-responsive vesicles and micelles. Langmuir 21: 4308–4315.
[11] Kukula, H., Schlaad, H., Antonietti, M., Forster, S. (2002) The formation of polymer vesicles or 'peptosomes' by polybutadiene-block-poly(L-glutamate)s in dilute aqueous solution. J. Am. Chem. Soc. 124: 1658–1663.
[12] Motala-Timol, S., Jhurry, D., Zhou, J.W., Bhaw-Luximon, A., Mohun, G., Ritter, H. (2008) Amphiphilic poly(L-lysine-*b*-caprolactone) block copolymers: synthesis, characterization, and solution properties. Macromolecules 41: 5571–5576.
[13] Iatrou, H., Frielinghaus, H., Hanski, S., Ferderigos, N., Ruokolainen, J., Ikkala, O., Richter, D., Mays, J., Hadjichristidis, N. (2007) Architecturally induced multiresponsive vesicles from well-defined polypeptides. Formation of gene vehicles. Biomacromolecules 8: 2173–2181.
[14] Gebhardt, K.E., Ahn, S., Venkatachalam, G., Savin, D.A. (2008) Role of secondary structure changes on the morphology of polypeptide-based block copolymer vesicles. J. Colloid Interf. Sci. 317: 70–76.
[15] Bellomo, E.G., Wyrsta, M.D., Pakstis, L., Pochan, D.J., Deming, T.J. (2004) Stimuli-responsive polypeptide vesicles by conformation-specific assembly. Nat. Mater. 3: 244–248.
[16] Ayres, L., Hans, P., Adams, J., Lowik, D.W.P.M., van Hest, J.C.M. (2005) Peptide-polymer vesicles prepared by atom transfer radical polymerization. J. Polym. Sci. Pol. Chem. 43: 6355–6366.
[17] Sun, J., Shi, Q., Chen, X.S., Guo, J.S., Jing, X.B. (2008) Self-assembly of a hydrophobic polypeptide containing a short hydrophilic middle segment: vesicles to large compound micelles. Macromol. Chem. Phys. 209: 1129–1136.
[18] Al-Jamal, K.T., Sakthivel, T., Florence, A.T. (2003) Dendrisomes: cationic lipidic dendron vesicular assemblies. Int. J. Pharm. 254: 33–36.
[19] Al-Jamal, K.T., Sakthivel, T., Florence, A.T. (2005) Solubilisation and transformation of amphipathic lipidic dendron vesicles (dendrisomes) into mixed micelles. Colloid Surface A 268: 52–59.
[20] Al-Jamal, K.T., Sakthivel, T., Florence, A.T. (2005) Dendrisomes: vesicular structures derived from a cationic lipidic dendron. J. Pharm. Sci. – US 94: 102–113.
[21] Tian, L., Nguyen, P., Hammond, P.T. (2006) Vesicular self-assembly of comb-dendritic block copolymers. Chem. Commun. 3489–3491.
[22] Lee, J., Kim, J.M., Yun, M., Park, C., Park, J., Lee, K.H., Kim, C. (2011) Self-organization of amide dendrons with focal dipeptide units. Soft Matter 7: 9021–9026.

[23] Fukuda, M., Sugiyama, J., Morita, T., Kimura, S. (2006) Fully hydrophobic artificial protein but water dispersible due to large dipole. Polym. J. 38: 381–386.
[24] Ghosh, S., Reches, M., Gazit, E., Verma, S. (2007) Bioinspired design of nanocages by self-assembling triskelion peptide elements. Angew. Chem. Int. Edit. 46: 2002–2004.
[25] Chung, E.K., Lee, E., Lim, Y.B., Lee, M. (2010) Cyclic peptide facial amphiphile preprogrammed to self-assemble into bioactive peptide capsules. Chem.-Eur. J. 16: 5305–5309.
[26] Ray, J.G., Ly, J.T., Savin, D.A. (2011) Peptide-based lipid mimetics with tunable core properties via thiol-alkyne chemistry. Polym. Chem. – UK 2: 1536–1541.
[27] von Maltzahn, G., Vauthey, S., Santoso, S., Zhang, S.U. (2003) Positively charged surfactant-like peptides self-assemble into nanostructures. Langmuir 19: 4332–4337.
[28] Yang, S.J., Zhang, S.G. (2006) Self-assembling behavior of designer lipid-like peptides. Supramol. Chem. 18: 389–396.
[29] van Hell, A.J., Costa, C.I.C.A., Flesch, F.M., Sutter, M., Jiskoot, W., Crommelin, D.J.A., Hennink, W.E., Mastrobattista, E. (2007) Self-assembly of recombinant amphiphilic oligopeptides into vesicles. Biomacromolecules 8: 2753–2761.
[30] van Hell, A.J., Klymchenko, A., Burgers, P.P., Moret, E.E., Jiskoot, W., Hennink, W.E., Crommelin, D.J.A., Mastrobattista, E. (2010) Conformation and intermolecular interactions of SA2 peptides self-assembled into vesicles. J. Phys. Chem. B 114: 11046–11052.
[31] van Hell, A.J., Crommelin, D.J.A., Hennink, W.E., Mastrobattista, E. (2009) Stabilization of peptide vesicles by introducing inter-peptide disulfide bonds. Pharm. Res. – Dord. 26: 2186–2193.
[32] van Hell, A.J., Fretz, M.M., Crommelin, D.J.A., Hennink, W.E., Mastrobattista, E. (2010) Peptide nanocarriers for intracellular delivery of photosensitizers. J. Control Release 141: 347–353.
[33] Reches, M., Gazit, E. (2004) Formation of closed-cage nanostructures by self-assembly of aromatic dipeptides. Nano Lett. 4: 581–585.
[34] Yan, X.H., Cui, Y., He, Q., Wang, K.W., Li, J.B., Mu, W.H., Wang, B.L., Ou-yang, Z.C. (2008) Reversible transitions between peptide nanotubes and vesicle-like structures including theoretical modeling studies. Chem. – Eur. J. 14: 5974–5980.
[35] Naskar, J., Banerjee, A. (2009) Concentration dependent transformation of oligopeptide based nanovesicles to nanotubes and an application of nanovesicles. Chem. – Asian J. 4: 1817–1823.
[36] Yan, X.H., He, Q., Wang, K.W., Duan, L., Cui, Y., Li, J.B. (2007) Transition of cationic dipeptide nanotubes into vesicles and oligonucleotide delivery. Angew. Chem. Int. Edit. 46: 2431–2434.
[37] Huang, R.L., Su, R.X., Qi, W., Zhao, J., He, Z.M. (2011) Hierarchical, interface-induced self-assembly of diphenylalanine: formation of peptide nanofibers and microvesicles. Nanotechnology 22: 245609.
[38] Mishra, A., Panda, J.J., Basu, A., Chauhan, V.S. (2008) Nanovesicles based on self-assembly of conformationally constrained aromatic residue containing amphiphilic dipeptides. Langmuir 24: 4571–4576.
[39] Rehm, T., Stepanenko, V., Zhang, X., Wurthner, F., Grohn, F., Klein, K., Schmuck, C. (2008) A new type of soft vesicle-forming molecule: an amino acid derived guanidiniocarbonyl pyrrole carboxylate zwitterion. Org. Lett. 10: 1469–1472.
[40] Rehm, T.H., Schmuck, C. (2010) Ion-pair induced self-assembly in aqueous solvents. Chem. Soc. Rev. 39: 3597–3611.
[41] Martinek, T.A., Hetenyi, A., Fulop, L., Mandity, I.M., Toth, G.K., Dekany, I., Fulop, F. (2006) Secondary structure dependent self-assembly of beta-peptides into nanosized fibrils and membranes. Angew. Chem. Int. Edit. 45: 2396–2400.
[42] Mandity, I.M., Fulop, L., Vass, E., Toth, G.K., Martinek, T.A., Fulop, F. (2010) Building beta-peptide H10/12 foldamer helices with six-membered cyclic side-chains: fine-tuning of folding and self-assembly. Org. Lett. 12: 5584–5587.
[43] Owen, T., Pynn, R., Martinez, J.S. and Butler, A. (2005) Micelle-to-vesicle transition of an iron-chelating microbial surfactant, marinobactin E. Langmuir 21: 12109–12114.
[44] Matsuura, K., Watanabe, K., Matsuzaki, T., Sakurai, K., Kimizuka, N. (2010) Self-assembled synthetic viral capsids from a 24-mer viral peptide fragment. Angew. Chem. Int. Edit. 49: 9662–9665.

[45] Broomell, C.C., Birkedal, H., Oliveira, C.L.P., Pedersen, J.S., Gertenbach, J.A., Young, M. and Douglas, T. (2010) Protein cage nanoparticles as secondary building units for the synthesis of 3-dimensional coordination polymers. Soft Matter 6: 3167–3171.
[46] Makino, A., Harada, H., Okada, T., Kimura, H., Amano, H., Saji, H., Hiraoka, M., Kimura, S. (2011) Effective encapsulation of a new cationic gadolinium chelate into apoferritin and its evaluation as an MRI contrast agent. Nanomed.–Nanotechnol. 7: 638–646.
[47] Koide, A., Kishimura, A., Osada, K., Jang, W.D., Yamasaki, Y., Kataoka, K. (2006) Semipermeable polymer vesicle (PICsome) self-assembled in aqueous medium from a pair of oppositely charged block copolymers: physiologically stable micro-/nanocontainers of water-soluble macromolecules. J. Am. Chem. Soc. 128: 5988–5989.
[48] Ueda, M., Makino, A., Imai, T., Sugiyama, J., Kimura, S. (2011) Transformation of peptide nanotubes into a vesicle via fusion driven by stereo-complex formation. Chem. Commun. 47: 3204–3206.
[49] Rodriguez-Hernandez, J., Lecommandoux, S. (2005) Reversible inside-out micellization of pH-responsive and water-soluble vesicles based on polypeptide diblock copolymers. J. Am. Chem. Soc. 127: 2026–2027.
[50] Nishikawa, H., Morita, T., Sugiyama, J., Kimura, S. (2004) Formation of gold nanoparticles in microreactor composed of helical peptide assembly in water. J. Colloid Interf. Sci. 280: 506–510.
[51] Jan, J.S., Lee, S.J., Carr, C.S., Shantz, D.F. (2005) Biomimetic synthesis of inorganic nanospheres. Chem. Mater. 17: 4310–4317.
[52] Wang, H.Y., Li, Y.M., Xiao, Y., Zhao, Y.F. (2005) Condensation properties of vesicles formed from an amphiphilic *N*-phosphorylamino acid. J. Colloid Interf. Sci. 287: 307–311.
[53] Holowka, E.P., Sun, V.Z., Kamei, D.T., Deming, T.J. (2007) Polyarginine segments in block copolypeptides drive both vesicular assembly and intracellular delivery. Nat. Mater. 6: 52–57.
[54] Zupancich, J.A., Bates, F.S., Hillmyer, M.A. (2009) Synthesis and self-assembly of RGD-functionalized PEO-PB amphiphiles. Biomacromolecules 10: 1554–1563.
[55] Katz, J.S., Zhong, S., Ricart, B.G., Pochan, D.J., Hammer, D.A., Burdick, J.A. (2010) Modular synthesis of biodegradable diblock copolymers for designing functional polymersomes. J. Am. Chem. Soc. 132: 3654–3655.
[56] Marsden, H.R., Handgraaf, J.W., Nudelman, F., Sommerdijk, N.A.J.M., Kros, A. (2010) Uniting polypeptides with sequence-designed peptides: synthesis and assembly of poly(gamma-benzyl L-glutamate)-*b*-coiled-coil peptide copolymers. J. Am. Chem. Soc. 132: 2370–2377.
[57] Versluis, F., Tomatsu, I., Kehr, S., Fregonese, C., Tepper, A.W.J.W., Stuart, M.C. A., Ravoo, B.J., Koning, R.I., Kros, A. (2009) Shape and release control of a peptide decorated vesicle through pH sensitive orthogonal supramolecular interactions. J. Am. Chem. Soc. 131: 13186–13187.
[58] Ke, D.M., Zhan, C.L., Li, A.D.Q., Yao, J.N. (2011) Morphological transformation between nanofibers and vesicles in a controllable bipyridine-tripeptide self-assembly. Angew. Chem. Int. Edit. 50: 3715–3719.
[59] Maeda, H., Greish, K., Fang, J. (2006) The EPR effect and polymeric drugs: a paradigm shift for cancer chemotherapy in the 21st century. Polymer Therapeutics Ii: Polymers as Drugs, Conjugates and Gene Delivery Systems 193: 103–121.
[60] Dams, E.T.M., Laverman, P., Oyen, W.J.G., Storm, G., Scherphof, G.L., Van der Meer, J.W.M., Corstens, F.H.M., Boerman, O.C. (2000) Accelerated blood clearance and altered biodistribution of repeated injections of sterically stabilized liposomes. J. Pharmacol. Exp. Ther. 292: 1071–1079.
[61] Tanisaka, H., Kizaka-Kondoh, S., Makino, A., Tanaka, S., Hiraoka, M., Kimura, S. (2008) Near-infrared fluorescent labeled peptosome for application to cancer imaging. Bioconjugate Chem. 19: 109–117.
[62] Koide, H., Asai, T., Hatanaka, K., Urakami, T., Ishii, T., Kenjo, E., Nishihara, M., Yokoyama, M., Ishida, T., Kiwada, H., Oku, N. (2008) Particle size-dependent triggering of accelerated blood clearance phenomenon. Int. J. Pharm. 362: 197–200.

6

Principles of Shape-Driven Nanostructure Design via Self-Assembly of Protein Building Blocks

Idit Buch[1], *Chung-Jung Tsai*[2], *Carlos Alemán*[3,4] *and Ruth Nussinov*[1,2]

[1]*Department of Human Molecular Genetics and Biochemistry, Sackler Institute of Molecular Medicine, Sackler Faculty of Medicine, Tel Aviv University, Israel*

[2]*Center for Cancer Research Nanobiology Program, SAIC-Frederick, Inc., National Cancer Institute, NCI-Frederick, USA*

[3]*Departament d'Enginyeria Química, E.T.S. d'Enginyeria Industrial de Barcelona, Universitat Politècnica de Catalunya, Spain*

[4]*Center for Research in Nano-Engineering, Universitat Politècnica de Catalunya, Spain*

Acronyms

CA	Capsid
EM	Electron microscopy
FF	Diphenylalanine
LRR	Leucine-rich repeats
MD	Molecular dynamics

Peptide Materials: From Nanostructures to Applications, First Edition. Edited by Carlos Alemán, Alberto Bianco and Mariano Venanzi.

6.1 Introduction

Self-assembly implies a process where the individual units contain the information needed to build a template for an organization composed of these units. It is one of the most important strategies in nature for obtaining complex functional structures. There are numerous biological examples of self-assembly: self-assembly of phospholipids into multilamellar vesicles; the assembly of peptides or proteins in amyloid fibrils in disease; the assembly of proteins into viral coats; assemblies of RNA and protein molecules into the ribosomes; multimolecular complexes controlling DNA replication and RNA transcription; the use of DNA origami to form DNA polyhedra; the tubulins association into microtubule protofilaments; and more. The richness of examples in biology is inspiring, suggesting that we can draw on biological structures and their parts as components in designed functional systems [1–5]. In particular, the naturally occurring regular self-assembled systems are usually remarkably stable. Had the systems not been in their thermodynamically most stable state, they would not have self-assembled in a regular, ordered fashion. Instead, they would have formed amorphous aggregates. Ordered self-assembly requires good shape complementarity and high physicochemical stability. Nonetheless, while self-assembly is advantageous as a strategy, in practice using self-assembly in design introduces an immense challenge. Key questions like what is the mechanism of the self-assembly and how to control the self-assembly toward the desired shape as the single most stable structure and avoid amorphous states are still open [6–9]; further, there are questions like which favorable interactions should be introduced and which unit shapes to select for the engineering of the target shape. It is the weak nonbonded interactions, such as hydrogen bonds, van der Waals forces, and π–π interactions that govern the ordered hierarchical molecular assembly. Atomic-level modifications via advanced synthetic organic chemistry techniques may further target specific regions to enhance the favorable interactions between the units [10–13]. Self-assembly is currently a practical approach for building a wide variety of nanostructures and several researchers devised a 'bottom-up' strategy [14–16]. This requires an understanding of the individual molecular building blocks, their structures, assembling properties, and dynamic behavior.

Here, we focus on a few specific experimentally self-assembled multicopy systems with each repeat being a unit of operation. Since the fundamental rules that govern these assemblies are not yet understood, theory and computations are not able to predict the single most stable self-assembled structure, only some range of potential candidates. The problem is compounded by the inaccuracy of the energy function and the computational time scales. From the computational standpoint, the weak interacting forces occasionally cause the self-assembling molecules to get trapped in local minima. Nevertheless, a strategy that preserves the native interfaces [17–20] is expected to have high chances of occurring in experiment, since such association is inherently more stable, with an energy gap separating it from all other potential 3D arrangements.

6.2 Self-Assembly into Preferred Shapes

6.2.1 Why Does a Given Building Block Prefer to Self-Assemble into a Particular Shape?

What is the driving force toward a particular organization and when does *ordered* self-assembly occur? In principle, if the building blocks are identical and their concentration is

(a)

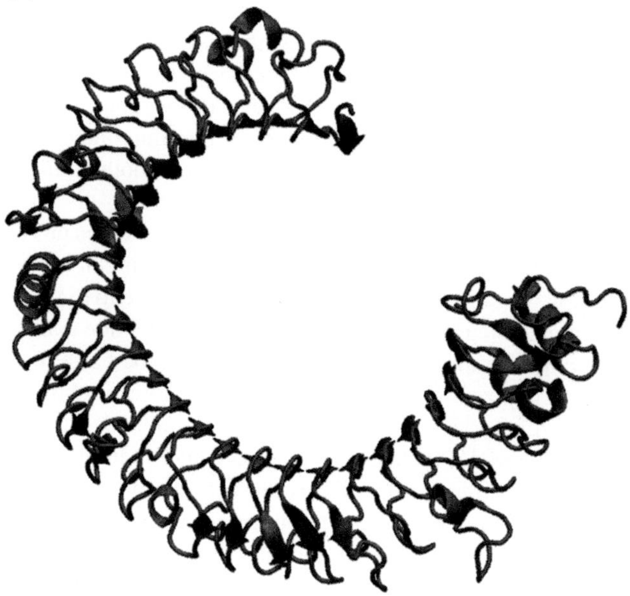

(b)

Figure 6.1 *Examples of two distinct geometrical closures: (a) the Toll-like receptor 3 (pdb code: 2A0Z), where the disc-like shapes do not close into a ring due to aberrant termini; (b) the surfactant-like peptide AcV$_6$D, which self-assembles into a nanotube. The acetylated N-termini point into the membrane-like double layer, whereas the charged aspartic acid at the C-termini point toward the interior and exterior of the tube. A third of the nanotube is seen. The coordinates were produced using CHARMM Image Facility [23]*

high, structural propagation can be indefinite. However, in biological systems edge units often have defects, and these prevent such an undesirable perpetuation. To use the units in design, these edge irregularities should be removed. One such example is the Toll-like receptor 3 ligand-binding domain, which consists of leucine-rich repeats (LRRs). In this protein 23 LRRs form a flat horseshoe-shaped structure [21, 22] (Figure 6.1a). The 2.4 Å resolution crystal structure illustrates irregular N- and in particular C-terminal capping, which prevent full circularization. In addition, the irregular 'top' and 'bottom' surfaces prevent stacking of these disk-like structures. While this strategy is essential *in vivo*, the design towards a closed structure must modify or even remove these irregularities. The surfactant-like peptide AcV$_6$D [24] presents an example where such irregularities are not present. This short peptide self-assembles into a tube and forms a closed membrane-like shape (Figure 6.1b). Another example of a neat closure, with no irregularities, is the microtubule [25, 26]. *In vivo*, microtubule assemblies always form a tube. At the same time, the interdigitation of the microtubule building blocks, the tubulins, prevents closure into vesicles. This can be attributed mainly to the 3D shape of the tubulins. A similar situation exists in the flagellar motor protein [27, 28]. As opposed to microtubules and the flagellar motor protein, the nonfunctional diphenylalanine (FF) forms both a tube and a vesicle in the test tube [29–31].

Thus, some building blocks favor an association into *either* tubes or vesicles, whereas others can form *both* tubes and vesicles. For the latter case, it can be expected that the preferred organization is concentration-dependent. Consistently, in the case of the diphenylalanine peptide, experiments revealed that at high peptide concentration, a tube was formed; at low concentration, a vesicle [32]. Since a tube has a lower curvature, it is the preferred FF organization: the less curved the more nonbonded interactions. A tube can be analyzed using a single wedge that has nonbonded interactions with its neighboring wedges. This wedge dictates the inner and outer radii of the tube. Per specific wedge, decreasing the curvature of the tube is the natural consequence of the increase in its radii. As a result, new interwedge interactions at the outer surface may be formed and existing ones strengthened. Increasing the curvature of the tube achieves the opposite effect [33].

A tube has a one-dimensional curvature, whereas a vesicle has two. When the concentration is low, the assembly process cannot proceed in one dimension to bury a sufficiently large hydrophobic surface. Consequently, the limited extent of a nonpolar buried surface area in a short tube drives the system to closure since a vesicle has a larger buried surface area. On the other hand, when the concentration is high, there is no such pressure on the system. A long tube buries an extensive nonpolar surface. Thus, when the concentration is high, there is a high chance of vesicles fusing to form a tube. The capsid (CA) protein of the HIV-1 provides an appropriate example. *In vitro*, it was observed to assemble into both cylinders and cones and a mutant of the CA monomer also formed spheres [34]. For the tube, hexameric arrangements of the CA monomers were observed; on the other hand, the less favorable pentameric organization leads to formation of closed cones and sphere shapes, as was also observed in the capsid protein of other retroviruses [35, 36]. The 2D curvatures of the cones or spheres have fewer interactions than the tubes with a one-dimensional curvature. *In vivo*, mature functional viral capsids form the closed conical/spherical geometries through the introduction of the pentameric defects, leading to capsid closure. Mechanistically, self-assembly of repeating units can also lead to a planar sheet. Depending on the interactions, the sheet may have a tendency to curve in one dimension; under such circumstances it will form a tube. On the other hand, if it has a tendency to curve in two dimensions, it will form a vesicle. The prevailing conditions dictate the preferred associations whether the repeating units will curve in one or in two dimensions [6, 17]. Figure 6.2 illustrates the concepts of tube and vesicle formation by self-assembly of repeating units. The final structure is mainly influenced by the curvature of a short linker that interconnects these units covalently.

There are many ways in which the building blocks can interact, yet the limited number of preferred interactions is the main reason for a regular self-assembly. For chain-linked units, as in repeat proteins [38, 39], the covalent linkage dictates the preferred organization.

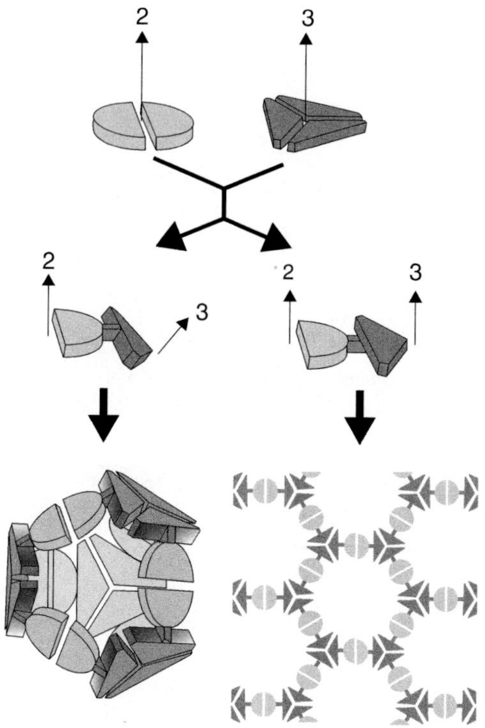

Figure 6.2 An illustration of the self-assembly of homo-oligomers into various geometries. Two homo-oligomers (in this example: a homo-dimer (light-grey) and a homo-trimer (black)) are fused covalently by a short linker (grey) to create monomers that have two self-assembling domains (second row). The shape and orientation of the linker affects the monomer, such that a curved orientation may lead to the formation of a nanocage (third row), whereas a flat orientation may lead to the formation of a sheet, which may close into a helical filament or a tube. Reproduced with permission from reference [37]. Copyright (2001) National Academy of Sciences, USA

On the other hand, for units that are not chain-connected, the specificity of the interactions between the repeating units determines the final geometry. The heat shock cages architecture and the viruses provide good examples [40, 41]. For simplicity, Figure 6.3 presents an illustration of the mechanism of tube formation with a single-walled carbon nanotube as an example. Starting with a graphene (Figure 6.3a), a carbon nanotube can be formed by rolling up the carbon sheet to merge the two edges, as shown in Figure 6.3b. A tube is identified by two integer numbers (N1, N2). N1 refers to the number of cells used to wrap up the tube and N2 refers to the number of cells sliding along the edge vector after tube closure [42]. The lattice-based tube can also be viewed as two different helical tubes (Figures 6.3c and d). Starting from the helix, the growth process toward tube formation can be described in terms of mathematical operations as rotational duplications by the tube axis and as translational duplications along the tube axis.

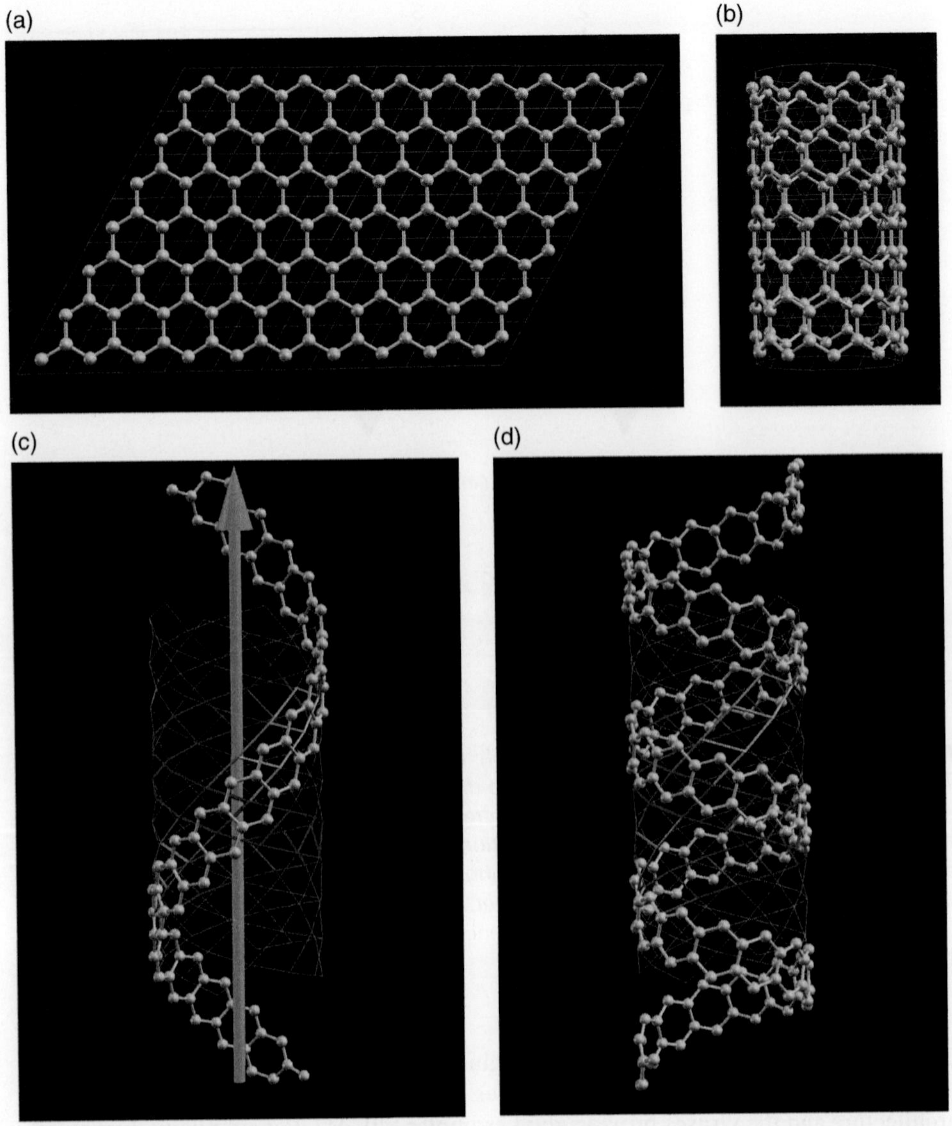

Figure 6.3 *An illustration of the mechanisms of tube formation using a single-walled carbon nanotube as an example. (a) A planar sheet of carbon atoms is represented by 10 × 7 cells (in thin cyan lines) with two carbon atoms in every 2D lattice cell. (b) 10 cells are used to wrap up the tube with no helical twist or rise; therefore this is a (10,0) nanotube [18]. (c) and (d) The lattice-based tube can also be viewed as two different helical tubes when viewed along the two edges of the lattice. Both tubes shown are identified as a (10, 4) nanotube with N1 cells highlighted in magenta and N2 cells highlighted in pink. The helices of the hexagonal carbon around the tube axis are clearly displayed. In (c), the helix growth process can be described in terms of rotational duplications (four times) by the tube axis and in (d) as translational duplications (one time) along the tube axis. Note that the carbon atoms in (c) and (d) are the rotational/translational units respectively (see color plate figure)*

6.2.2 The Self-Assembly Formation Mechanism – A Lesson from Lipid Tubules

While very few theories have been proposed for protein nanotube formation, this is not the case for lipids [43, 44]. The formation of lipid tubules and their intermediates have been the focus of many studies, with the chiral diacetylenic phospholipids (containing phospholipid triple bonds) serving as a model system. Differential phase-contrast microscopy under slow cooling conditions showed that diacetylenic phospholipids self-assemble into spherical multilamellar vesicles. These rapidly transform into helical ribbons. The rapid transformation was interpreted to imply that the cylindrical phases provide the principal material supporting the helix formation rather than the saturated solution. The outer layer of the multilamellar hollow coiled helical structures presents a handedness that is related to the phospholipid chirality, suggesting that these are intermediate structures in lipid tubule formation. This growth of helical ribbons from spherical vesicles at high temperatures was observed to be a dominant tubule formation mechanism. Surprisingly, even with a pure enantiomer of the diacetylenic phospholipid [DC(8,9)PC] the probabilities of left- or right-handed nucleation and ribbon formation in the initial process are roughly equal, although there is a slight tendency toward right-handedness, with a metastable left-handed state. However, they are uniform without helical kinks. Rapid elongation of the helix occurs simultaneously with their broadening and closure. Finally, outer coaxial layers are added in a much slower process, leading to the formation of the final lipid tubule. The slower formation of these layers, the outcome of accretion of lipid molecules from solution, was proposed to reflect the molecular chirality. Hence, overall, the rates of longitudinal and vertical helix growth are very different in the lipid tubule formation; the helices can follow left- or right-handedness; and the outer layers have uniform handedness. With current lipid tubules, experimental data permit only the derivation of a simplified model; the model already allows some predictions to be made of the self-assembly process. Understanding the mechanism of formation of protein nanotubes may benefit from these propositions as similar principles apply.

6.2.3 Experimental Results

Table 6.1 presents a summary of examples from the literature where protein nanotubes have been observed and Figure 6.4 presents the corresponding monomer structures of these where available. The HIV-1 CA protein (Figure 6.4a) has only a single layered tube. The outer diameter is in the range of 41.2 nm to 50 nm. The crystal structure of the wild-type monomer is available, as well as a reconstructed image from cryo-EM at 9 Å resolution, which provides its hexameric and dimeric association into a tube [46]. In the case of α-lactalbumin [54, 55], the structure of the entire monomer is available (Figure 6.4b). However, only a fragment of the monomer forms the tube following proteolysis, and it is unclear which. Like the HIV-1 CA protein, it forms a single-layered tube. However, no tubular structure is available. For FF the monomer crystal structure is available [48] (Figure 6.4c), but no atomic resolution structure is available for the tube. FF has a wide range of inner/outer tube diameters (Table 6.1). A model of the FF tube and a possible peptide nanotube design scheme are discussed in Section 6.3.2. The bacterial flagellar protein filament (Figure 6.4d) has a complete atomic model reconstructed from a cryo-EM image. Data are also available for the

Table 6.1 *Examples of protein monomers that self-assemble into a nanotube*

Protein Building block	Number of layers (wall thickness)	Inner diameter (nm)	Outer diameter (nm)	Monomer experimental details	Tubular experimental details
HIV-1 CA protein	Single	13.0–21.8	41.2–50.0	X-ray diffraction [45]	Cryo-EM, 9.0 Å resolution [46]
α-Lactalbumin	Single	8.7	21	X-ray diffraction [47]	No
Diphenylalanine (FF)	Multiple	300–1800	400–2000	X-ray diffraction of the hexagon [48]	No
Flagellar	Single	2	12–25	Cryo-EM, 4.0 Å resolution [49]	Cryo-EM, 4.0 Å resolution [50]
Surfactant-like peptides (AcV_6D)	Double	22–42	30–50	No	TEM, low resolution [51]
Microtubules	Single	15	25	Electron diffraction, 3.7 Å resolution [52]	Cryo-EM, 28.0 Å resolution [53]

surfactant-like peptide AcV_6D (Figure 6.4e), including TEM and dynamic light scattering, which provide dimensions and shapes. Figure 6.1b presents a top view of approximately a third of the tube. Considerable data are available for the microtubule, including the crystal structure of the αβ tubulin heterodimer (Figure 6.4f), two-dimensional crystalline structures, and three-dimensional cryo-EM images.

The monomers depicted in Figure 6.4 illustrate examples of different sizes, architectures, shapes, and physicochemical properties that can assemble into tubes or vesicles. The tube/vesicle formation mechanism is different for each of these. The HIV CA protein self-assembly may initially grow in a planar fashion, with fluctuations allowing closure via a collision–diffusion mechanism. This may explain the different radii of the tubes and cones. Like the HIV CA, the α-lactalbumin, the flagellar protein, and the tubulins assemble into a single layer due to their large size and bulky fold. However, in tubulins the preferred interactions are already curved: the association between α-tubulin and β-tubulin defines the curvature. Consequently, tubulins form tubes of limited radius. The formation mechanism of the surfactant-like peptides is different. Since they possess very few amino acids, their assembly highly depends on the physicochemical properties of their surface (where all amino acids are exposed). Hence, the AcV6D peptide forms a double-layered tube where the neutralized N-termini of the inner layer reside against the neutralized N-termini of the outer layer. Finally, as opposed to the above examples, the formation mechanism of FF is dictated by its small size; its two amino acids allow multilayer lamination into a thick tube wall, with variable radii. Hence, if the observed radius is always fixed, as in microtubules, the self-assembly curves as it proceeds; on the other hand, if the radius is variable, as in CA or FF, different tube formation mechanisms are possible, depending on the monomer size and flexibility.

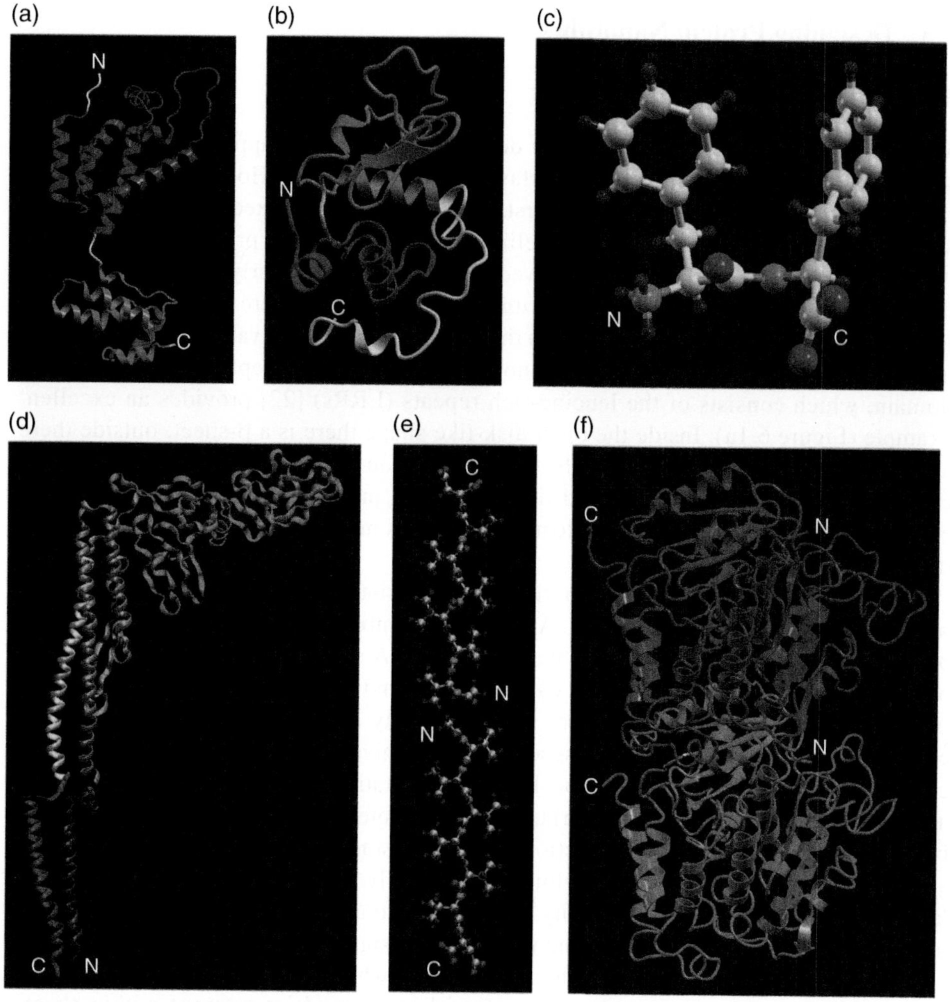

Figure 6.4 *Experimental examples of natural protein monomers that self-assemble to create nanotubes. (a) The HIV-1 CA monomer (pdb code: 1e6j) [45]. The N-terminal domain (in red) self-assembles to create a hexamer, and the C-terminal domain (in green) self-assembles into a dimer. A short linker of six residues (in white) connects the domains. (b) The full atomic structure of α-lactalbumin (pdb code: 1a4v) [47]. The colors (red, green, yellow) represent different interacting domains. (c) A model of the detailed atomic structure of diphenylalanine. The phenylalanine rings are perpendicular to each other. (d) The bacterial flagellar protein monomer (pdb code: 1ucu) [49]. Each color represents a different domain. (e) The double-layered surfactant-like peptide AcV_6D. The hydrophilic heads point outwards whereas the long acetylated hydrophobic tails point inward, towards each other. (f) The structure of the αβ-tubulin heterodimer (pdb code: 1tub) [52]. Each color represents a monomer. Both monomers form the asymmetric unit of the microtubule (see color plate figure)*

6.3 Designing Protein Nanotubes

6.3.1 Shape-Driven Design

While any protein block can be used for design, repeat proteins appear good candidates since they already inherently possess favorable associations. However, using repeat proteins may present two problems: first, the radius will be fixed; second, it will be small as in microtubules or in the Flagellin. In principle, repeat proteins have a superhelical structure resembling that observed in lipid tubules. Clearly, if the curvature is high, the helicity increases; however, unlike lipid tubules, there is no tube closure, necessitating interface modification. On the other hand, if the curvature is low, flat rings are created. Ring stacking creates a nanotube. The toll-like receptor 3 ligand binding domain, which consists of the leucine-rich repeats (LRRs) [22] provides an excellent example (Figure 6.1a). Inside the open disk-like shape there is a β-sheet; outside there are α-helices. The curvature is small, the radius is fairly fixed and the native interface can be used to complete the circular design. Hence, in this superhelical repeat protein case, an open tube is formed. For closure, mutations need to be introduced to redesign the interface.

Interface design follows the hallmarks of the association of protein building blocks in protein folding, design, and binding. Among all potential associations, the most favorable geometrically and energetically will take place. A stable structure will be obtained only if the association has an energy gap between it and all alternative associations. This is the design goal: an interface more stable than any alternative association. The key chemical elements stabilizing the associations are universal [56]: hydrophobic patches, good geometric complementarity, packing, electrostatic interactions, hydrogen bonds, and no unsatisfied charges. The interbuilding block interactions should possess stability beyond that observed in native functional proteins, since the designs frequently necessitate rigidity. As illustrated by amyloid self-assembly, a strong favorable intermolecular association is the key to durability under conditions such as high temperatures and nonphysiological pH. A successful nanostructure design should dictate an ordered self-assembly as compared to an amorphous aggregate, which implies a preferred mode of interaction. In amyloid assembly, the preferred interstrand interactions within sheets derive from the backbone hydrogen bonds and side-chain interactions. Between sheets tight shape complementarity and hydrophobic interactions are key factors. The principles of folding and design are similar: both aim at preferred conformational states and their interactions.

Using protein building blocks in nanodesign presents advantages as well as drawbacks [7]. Protein parts provide an infinite number of possibilities for shapes and chemistry. The nanostructure can result from self-assembly along one axis only, as in circular short peptides [57], or self-assembly along two axes, as in the surfactant-like peptide [24, 51]. Modifications can be introduced into the repeat protein such as covalent binding of two or more parts. It is challenging to predict the structure following such modifications. In the case of combining two repeat parts through a short linker, the resulting nanostructure will highly depend on the linker – its preferred orientation and rigidity [17]. Simulations can be used to reduce the large number of potential constructs and accelerate the nanodesign [20].

6.3.2 Structural Properties of Protein Nanotubes and a Design Scheme

A nanotube design strategy should (i) lead to a well-ordered structure; (ii) arrange the repeating units along the helical axes; (iii) form a hollow tube with dimensions specified by inner and outer diameters; (iv) be able to lead successfully to either a single- or a multilayered tube; and (v) lead to unique structures. The tube construction scheme below uses carbon nanotubes for demonstration. Figure 6.5 presents the carbon nanotube construction scheme, using a 2D wrapping system. Figure 6.6 illustrates various aspects of the nanotube construction. In Figure 6.6a, several examples of single-walled carbon nanotubes are illustrated. In all nanotubes, eight cells are used to complete one full round, but each nanotube has a different helicity extent. As the helicity extent increases, the inner and outer diameters decrease. As Figure 6.6b illustrates, when wrapping a planar 2D lattice on to the surface of a cylinder the geometrical transformation leads to an inherent geometrical distortion, which is most significant in protein nanotubes. The magnitude of this distortion at the tube surface is proportional to the curvature of the constructed tube. Even if a molecular distortion correction is implemented, a tube will always display a higher density toward its interior and a lower density toward its exterior. The distance between the highest and the lowest density regions determines the tube width, which directly affects the upper and lower limit of the nanotube diameter. Figure 6.6c illustrates an example of a multiwalled carbon nanotube where each layer has a different number of cells in one full round, depending on the diameter. As an example of

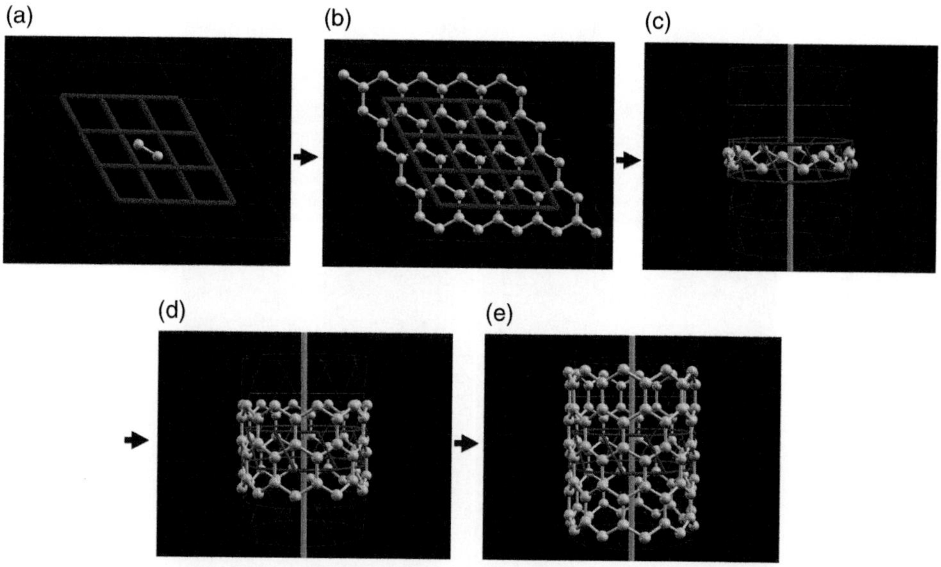

Figure 6.5 *The computational construction protocol of a single-walled carbon nanotube. (a) Two carbons are placed on a planar sheet. The middle frame highlights (in grey) the unit cell. The cell parameters are a = 2.459 Å, b = 2.459 Å, γ= 120°. (b) The unit cell is duplicated in the 2D plane using P6 symmetry. Two covalent bonds link each carbon atom to two neighboring cells. (c) A single row of ten cells is wrapped around a cylinder to create one full round of a nanotube. (d) Two more rounds of cells are added along the tube axis direction. (e) Another two rounds are added to make the tube height of five cells*

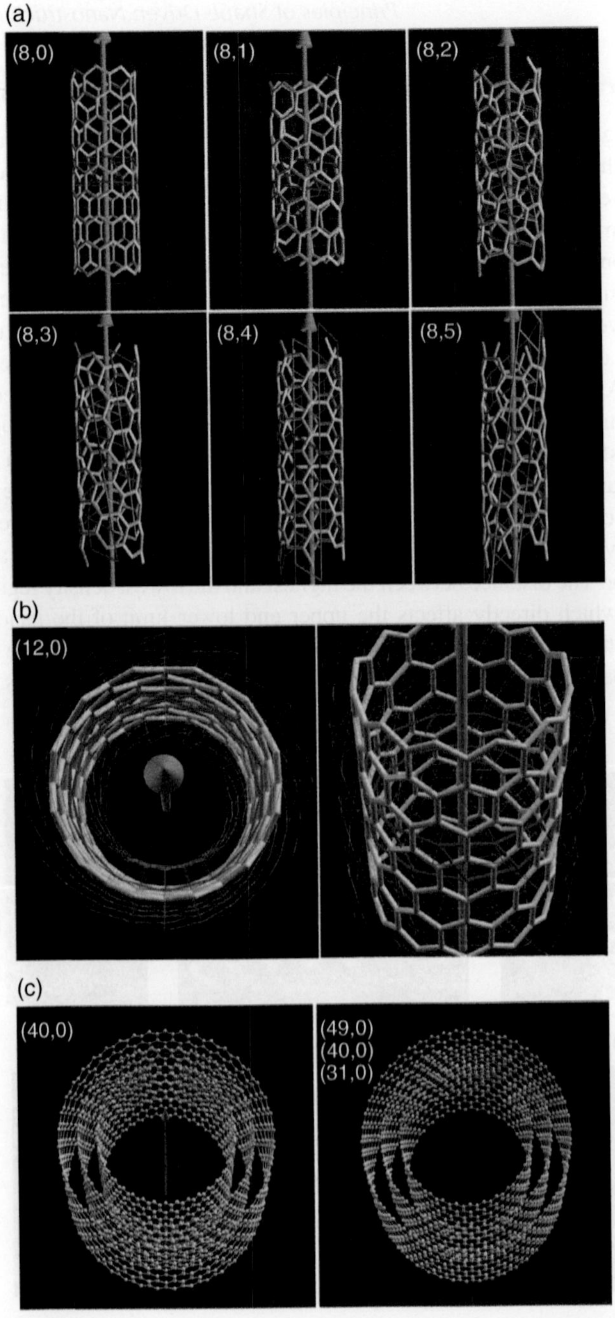

Figure 6.6 *Computational aspects of design and construction demonstrated on carbon nanotubes. (a) Six examples of single-walled carbon nanotubes having eight cells that wrap one full round and a different helicity extent [0–5]. As the helicity extent increases, the diameter decreases, and the nanotube becomes more fibrillar in shape. (b) The distortion formed as a result of the tube curvature. The inner area of the tube wall is more condensed than the outside area. (c) As a result of the distortion, in a multiwalled carbon nanotube, the number of cells needed to complete one full round decreases with the inner diameter and increases with the outer one (see color plate figure)*

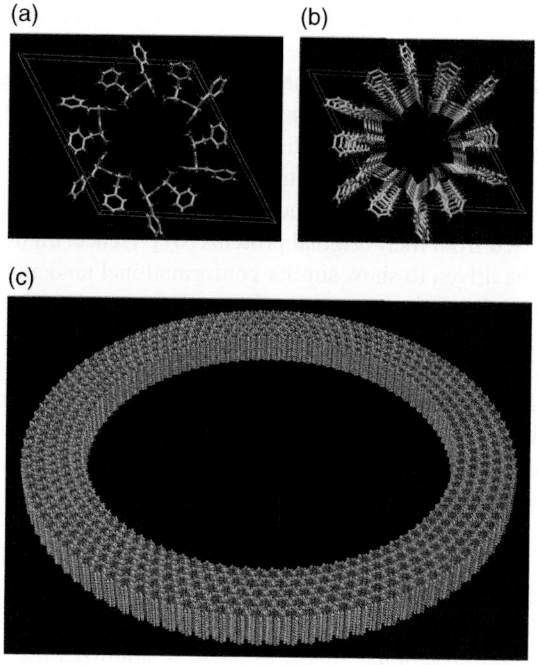

Figure 6.7 An atomic model of the diphenylalanine nanotube. (a) The unit cell consists of six dipeptides in a hexagonal arrangement such that the backbones face the tube interior while the hydrophobic side chains face its exterior. (b) The lamination of the unit cell along the hexagonal axis to create a tiny pore. (c) The FF nanotube with the tiny pores parallel to the tube axis. The width of the illustrated nanotube is 50 nm (see color plate figure)

a multiwalled peptide nanotube that was observed in the lab, we illustrate the atomic model of the diphenylalanine nanotube in Figure 6.7. A unit cell that consists of six diphenylalanine peptides arranged in a hexagonal shape (Figure 6.7a) self-assembles with identical cells along the tube axis direction to create a pore [48] (Figure 6.7b) and a nanotube with an approximate width of 50 nm. The nanotube is comprised of tiny pores parallel to the tube axis [33] (Figure 6.7c). Hence, the FF nanotube is in fact a unique example for a multiwalled perforated peptide nanotube.

6.3.3 Incorporation of Nonproteinogenic Amino Acids

Nanostructure design involves the ability to predictably manipulate the properties of the self-assembly of autonomous units with preferred conformational states. Above, we focused on design strategy using peptides and protein parts that can be assembled in a bottom-up approach into architectures that may differ from those found in nature. Proteins are particularly suitable candidates for nanostructure design since they provide diverse molecular shapes that can be assembled into a broad range of scaffolds [58]. A similar strategy was followed using nucleic acids [59, 60].

We have followed this strategy to design self-assembled β-helical proteins that contain a tubular or fibrillar motif in their folds. We selected two-turn repeat building blocks from native left-handed β-helices that presented relatively high stability compared to other candidates. Copies of the structural unit were stacked with no covalent linkage between them. The stability of these self-assembled organizations was assessed by all-atom molecular dynamics simulations. We observed that a structural model based on their self-assembly was a stable nanotube construct. While no direct experimental evidence for the stability of such a construct is available, it was demonstrated that, under appropriate conditions, short segments possessing a high density of secondary structure motifs can be conformationally stable even when excised from their original proteins [61]. Hence, on their own, secondary structure repeats can be driven to show similar conformational tendencies.

We have also used an alternative strategy to obtain a construct based on stacked β-helical motifs [62]. Rather than pick the most stable building blocks as described above, we strove to identify the motifs that would contribute most dominantly to the physicochemical stability of the *unfolded* states. These are likely to be the most populated when excised from the parent protein. Such a two-turn β-helical building block was stabilized by the substitutions of naturally occurring amino acid with nonproteinogenic amino acids in key positions. Our goal was to overpopulate the folded conformation by designing synthetic amino acids whose preferred conformational states would coincide with the preferred native conformation of the peptide at that position, and at the same time these residues would be conformationally restricted. Such a strategy would stabilize the flexible regions of the peptide, particularly the β-helix turns. A conformationally restricted amino acid limits the accessible conformational space; as such, it stabilizes the motif and the self-assembled construct. Constrained amino acids further alter the potential energy surface of the neighboring residues, enhancing the folded state and therefore the design [10, 63]. In our case, the native turn sequence consisted of Pro–Gly–Val; Gly was replaced by the cyclic nonproteinogenic residues Ac_3c (1-aminocyclopropane-1-carboxylic acid) and Ac_5c (1-aminocyclopentane-1-carboxylic acid). Simulations of the native sequence, the peptides with the substituted synthetic residues, and the constructs were performed, and the crystalline structures of five peptides of Pro–Xaa–Val or Xaa–Val sequences have been solved by X-ray diffraction analysis [62]. Later, they were compared with theoretical predictions to ensure a proper synthetic residue choice. Both theory and experiment indicate that the Pro–Ac_nc–Val sequences exhibit a high propensity to adopt turn-like conformations; interestingly, this propensity is not affected to any significant extent by the chemical environment. Hence, such a strategy leads to shape-driven nanodesign, when the shape is stabilized by conformationally constrained residues.

6.3.4 MD Simulations as a Testing Tool for Novel Designs

Computer simulations play an increasingly important role in the biomolecular sciences. Although continuous progress is being made in the development of experimental analysis methods, computational techniques are currently the only possibility for atomic-resolution analyses of many complex biological mechanisms.

MD simulation is a well-established method for studying protein dynamics [64], and it was proven to be useful in addressing the self-assembly of protein systems [65, 66]. In all-atom techniques, all atoms in the system, including the solvent, are represented explicitly. In principle this should permit a highly accurate characterization of complex physicochemical processes. In practice, the simulation of self-assembly processes, starting from random initial conditions and attempting to reach time scales similar to the experimental ones, is still beyond all-atom MD simulations capabilities, due to the intrinsic difficulties of these techniques to sample the conformational space efficiently. However, important information can be gained from all-atom simulations of much shorter time scales, as these often filter out constructions that are unstable. In particular, we have devised an all-atom MD simulation approach to validate self-assembled protein nanotubes [23]. The approach utilizes the fact that in such systems an asymmetric protein unit of relatively small size self-assembles with identical neighboring units into a nanotube, through cylindrical symmetry operations. Hence, only the asymmetric unit is simulated, taking into account its nonbonded interactions with the neighboring cells every preconfigured number of simulation timesteps. Recently, this approach was extended to the simulation of icosahedral structures, this time using full rotational symmetry [67]. Nevertheless, any construction shown to be stable by MD simulations must still be assessed by wet-lab experiments. For a comprehensive description of the theory and applications of MD simulations, see Chapter 3 [68].

6.4 Summary and Outlook

The principles of folding, binding, and self-assembly in nanostructure design are similar. For a successful shape-driven building block-based nanodesign the conformation of the building block in the self-assembly should be similar to its preferred conformation when isolated in solution; that is, the building block should be in its native conformational state exhibiting durability to environmental conditions. Nanotubes essentially involve two dimensional lattices, resembling a flat sheet rolled into a cylindrical shape. Thus, when using building blocks to create a sheet new interfaces need to be engineered. The designed new interface between the building blocks should be highly favored, with few preferred alternates, leading to an energy gap and driving the self-assembly into the desired state. Alternatively, native interfaces could be retained by covalently linking two different monomers taken from homo-oligomeric assemblies. The challenge is in the choice of the monomers and in the design of the peptide linker such that the designed shape would be obtained. Our compilation illustrates the richness of naturally occurring protein building block shapes that can be used to form nanotubes with various properties.

Here we have largely focused on nanotubes and the mechanism of nanotubule formation. However, in addition to constructing protein and peptide nanotubes [69], cages have also been designed following naturally occurring cage architectures [70–72]. Raman and coworkers [19] have designed polyhedra consisting of linked natural and engineered peptides based on principles similar to those of Yeates and coworkers [17, 37] In their construct, one domain is naturally occurring and the other contained modification. The authors attempted to fix the geometry by introducing an intramolecular disulfide bridge between the helical domains. However, under certain conditions, the disulfide bridges formed intermolecularly, leading to nonhomogeneous cage architectures.

The intricacy and complexity of biological systems and the repertoire of shapes and properties that they provide create vast design opportunities for nanobiology. Here we presented several potential shape-based self-assembly design strategies. To best illustrate some of our designs we used carbon nanotubes. By inserting synthetic nonnatural residues at judiciously selected positions, or synthetic peptide linkers, the stability of the constructs may be further enhanced. Such broadening of the chemical space may lead to an almost unlimited range of nanosystems and architectures and consequently applications.

Acknowledgements

This project has been funded in whole or in part with Federal funds from the National Cancer Institute, National Institutes of Health, under contract number HHSN261200800001E. The content of this publication does not necessarily reflect the views or policies of the Department of Health and Human Services, and nor does mention of trade names, commercial products, or organizations imply endorsement by the US Government. This research was supported (in part) by the Intramural Research Program of the NIH, National Cancer Institute, and Center for Cancer Research. Support for the research of CA was through the DIUE (contract number 2009SGR925) and the prize 'ICREA Academia' for excellence in research funded by the Generalitat de Catalunya.

References

[1] Kol, N., Adler-Abramovich, L., Barlam, D., Shneck, R.Z., Gazit, E., Rousso, I. (2005) Self-assembled peptide nanotubes are uniquely rigid bioinspired supramolecular structures. Nano Lett. 5: 1343–1346.

[2] Tysseling-Mattiace, V.M., Sahni, V., Niece, K.L., Birch, D., Czeisler, C., Fehlings, M.G., Stupp, S.I., Kessler, J.A. (2008) Self-assembling nanofibers inhibit glial scar formation and promote axon elongation after spinal cord injury. J. Neurosci. 28: 3814–3823.

[3] Zhang, S. (2008) Designer self-assembling peptide nanofiber scaffolds for study of 3-D cell biology and beyond. Adv. Cancer Res. 99: 335–362.

[4] Webber, M.J., Tongers, J., Newcomb, C.J., Marquardt, K.T., Bauersachs, J., Losordo, D.W., Stupp, S.I. (2011) Supramolecular nanostructures that mimic VEGF as a strategy for ischemic tissue repair. Proc. Natl Acad. Sci. USA 108: 13438–13443.

[5] Zhang, D.Y., Seelig, G. (2011) Dynamic DNA nanotechnology using strand-displacement reactions. Nat. Chem. 3: 103–113.

[6] Reches, M., Gazit, E. (2006) Molecular self-assembly of peptide nanostructures: mechanism of association and potential uses. Current Nanoscience 2: 105–111.

[7] Tsai, C.J., Zheng, J., Zanuy, D., Haspel, N., Wolfson, H., Aleman, C., Nussinov, R. (2007) Principles of nanostructure design with protein building blocks. Proteins 68: 1–12.

[8] Besenius, P., Portale, G., Bomans, P.H., Janssen, H.M., Palmans, A.R., Meijer, E.W. (2010) Controlling the growth and shape of chiral supramolecular polymers in water. Proc. Natl Acad. Sci. USA 107: 17888–17893.

[9] Morris, K., Serpell, L. (2010) From natural to designer self-assembling biopolymers, the structural characterisation of fibrous proteins and peptides using fibre diffraction. Chem. Soc. Rev. 39: 3445–3453.

[10] Zheng, J., Zanuy, D., Haspel, N., Tsai, C.J., Aleman, C., Nussinov, R. (2007) Nanostructure design using protein building blocks enhanced by conformationally constrained synthetic residues. Biochemistry 46: 1205–1218.

[11] Zanuy, D., Jimenez, A.I., Cativiela, C., Nussinov, R., Aleman, C. (2007) Use of constrained synthetic amino acids in beta-helix proteins for conformational control. J. Phys. Chem. B 111: 3236–3242.

[12] Pouget, E., Fay, N., Dujardin, E., Jamin, N., Berthault, P., Perrin, L., Pandit, A., Rose, T., Valery, C., Thomas, D., Paternostre, M., Artzner, F. (2010) Elucidation of the self-assembly pathway of lanreotide octapeptide into beta-sheet nanotubes: role of two stable intermediates. J. Am. Chem. Soc. 132: 4230–4241.

[13] Altunbas, A., Pochan, D.J. (2011) Peptide-based and polypeptide-based hydrogels for drug delivery and tissue engineering. Top. Curr. Chem. 135: 135–167.

[14] Capito, R.M., Azevedo, H.S., Velichko, Y.S., Mata, A., Stupp, S.I. (2008) Self-assembly of large and small molecules into hierarchically ordered sacs and membranes. Science 319: 1812–1816.

[15] Douglas, S.M., Dietz, H., Liedl, T., Hogberg, B., Graf, F., Shih, W.M. (2009) Self-assembly of DNA into nanoscale three-dimensional shapes. Nature 459: 414–418.
[16] Tarabout, C., Roux, S., Gobeaux, F., Fay, N., Pouget, E., Meriadec, C., Ligeti, M., Thomas, D.M.I.J., Besselievre, F., Buisson, D.A., Verbavatz, J.M., Petitjean, M., Valery, C., Perrin, L., Rousseau, B., Artzner, F., Paternostre, M., Cintrat, J.C. (2011) Control of peptide nanotube diameter by chemical modifications of an aromatic residue involved in a single close contact. Proc. Natl Acad. Sci. USA 108: 7679–7684.
[17] Yeates, T.O., Padilla, J.E. (2002) Designing supramolecular protein assemblies. Curr. Opin. Struct. Biol. 12: 464–470.
[18] Tsai, C.J., Zheng, J., Nussinov, R. (2006) Designing a nanotube using naturally occurring protein building blocks. PLoS Comput. Biol. 2: e42.
[19] Raman, S., Machaidze, G., Lustig, A., Aebi, U., Burkhard, P. (2006) Structure-based design of peptides that self-assemble into regular polyhedral nanoparticles. Nanomedicine 2: 95–102.
[20] Buch, I., Tsai, C.J., Wolfson, H.J., Nussinov, R. (2011) Symmetry-based self-assembled nanotubes constructed using native protein structures: the key role of flexible linkers. Protein Pept. Lett. 18: 362–372.
[21] Choe, J., Kelker, M.S., Wilson, I.A. (2005) Crystal structure of human Toll-like receptor 3 (TLR3) ectodomain. Science 309: 581–585.
[22] Bell, J.K., Botos, I., Hall, P.R., Askins, J., Shiloach, J., Segal, D.M., Davies, D.R. (2005) The molecular structure of the toll-like receptor 3 ligand-binding domain. Proc. Natl Acad. Sci. USA 102: 10976–10980.
[23] Buch, I., Brooks, B.R., Wolfson, H.J., Nussinov, R. (2009) Computational validation of protein nanotubes. Nano Lett. 9: 1096–1102.
[24] Adams, D.J., Holtzmann, K., Schneider, C., Butler, M.F. (2007) Self-assembly of surfactant-like peptides. Langmuir 23: 12729–12736.
[25] Gardner, M.K., Hunt, A.J., Goodson, H.V., Odde, D.J. (2008) Microtubule assembly dynamics: new insights at the nanoscale. Curr. Opin. Cell. Biol. 20: 64–70.
[26] Barbier, P., Dorleans, A., Devred, F., Sanz, L., Allegro, D., Alfonso, C., Knossow, M., Peyrot, V., Andreu, J.M. (2010) Stathmin and interfacial microtubule inhibitors recognize a naturally curved conformation of tubulin dimers. J. Biol. Chem., 285: 31672–31681.
[27] Minamino, T., Imada, K., Namba, K. (2008) Molecular motors of the bacterial flagella. Curr. Opin. Struct. Biol. 18: 693–701.
[28] Chen, S., Beeby, M., Murphy, G.E., Leadbetter, J.R., Hendrixson, D.R., Briegel, A., Li, Z., Shi, J., Tocheva, E.I., Muller, A., Dobro, M.J., Jensen, G.J. (2011) Structural diversity of bacterial flagellar motors. Embo J. 30: 2972–2981.
[29] Reches, M., Gazit, E. (2003) Casting metal nanowires within discrete self-assembled peptide nanotubes. Science 300: 625–627.
[30] Huang, R., Su, R., Qi, W., Zhao, J., He, Z. (2011) Hierarchical, interface-induced self-assembly of diphenylalanine: formation of peptide nanofibers and microvesicles. Nanotechnology 22: 245609.
[31] Wang, M., Du, L., Wu, X., Xiong, S., Chu, P.K. (2011) Charged diphenylalanine nanotubes and controlled hierarchical self-assembly. ACS Nano 5: 4448–4454.
[32] Song, Y., Challa, S.R., Medforth, C.J., Qiu, Y., Watt, R.K., Pena, D., Miller, J.E., van Swol, F., Shelnutt, J.A. (2004) Synthesis of peptide-nanotube platinum-nanoparticle composites. Chem. Commun. (Camb.): 1044–1045.
[33] Gorbitz, C.H. (2006) The structure of nanotubes formed by diphenylalanine, the core recognition motif of Alzheimer's beta-amyloid polypeptide. Chem. Commun. (Camb.): 2332–2334.
[34] Ganser-Pornillos, B.K., Yeager, M., Sundquist, W.I. (2008) The structural biology of HIV assembly. Curr. Opin. Struct. Biol. 18: 203–217.
[35] Cardone, G., Purdy, J.G., Cheng, N., Craven, R.C., Steven, A.C. (2009) Visualization of a missing link in retrovirus capsid assembly. Nature 457: 694–698.
[36] Yeager, M. (2011) Design of *in vitro* symmetric complexes and analysis by hybrid methods reveal mechanisms of HIV capsid assembly. J. Mol. Biol. 410: 534–552.
[37] Padilla, J.E., Colovos, C., Yeates, T.O. (2001) Nanohedra: using symmetry to design self assembling protein cages, layers, crystals, and filaments. Proc. Natl Acad. Sci. USA 98: 2217–2221.

[38] Main, E.R., Lowe, A.R., Mochrie, S.G., Jackson, S.E., Regan, L. (2005) A recurring theme in protein engineering: the design, stability and folding of repeat proteins. Curr. Opin. Struct. Biol. 15: 464–471.

[39] Barrick, D., Ferreiro, D.U., Komives, E.A. (2008) Folding landscapes of ankyrin repeat proteins: experiments meet theory. Curr. Opin. Struct. Biol. 18: 27–34.

[40] Janner, A. (2008) Comparative architecture of octahedral protein cages. I. Indexed enclosing forms. Acta Crystallogr. A 64: 494–502.

[41] Johnson, J.E. (2008) Multi-disciplinary studies of viruses: the role of structure in shaping the questions and answers. J. Struct. Biol. 163: 246–253.

[42] Tsai, C.J., Nussinov, R. (2011) A unified convention for biological assemblies with helical symmetry. Acta Crystallogr. D. Biol. Crystallogr. 67: 716–728.

[43] Stachowiak, J.C., Hayden, C.C., Sasaki, D.Y. (2010) Steric confinement of proteins on lipid membranes can drive curvature and tubulation. Proc. Natl Acad. Sci. USA 107: 7781–7786.

[44] Hughson, F.M., Reinisch, K.M. (2010) Structure and mechanism in membrane trafficking. Curr. Opin. Cell. Biol. 22: 454–460.

[45] Monaco-Malbet, S., Berthet-Colominas, C., Novelli, A., Battai, N., Piga, N., Cheynet, V., Mallet, F., Cusack, S. (2000) Mutual conformational adaptations in antigen and antibody upon complex formation between an Fab and HIV-1 capsid protein p24. Structure 8: 1069–1077.

[46] Ganser-Pornillos, B.K., Cheng, A., Yeager, M. (2007) Structure of full-length HIV-1 CA: a model for the mature capsid lattice. Cell 131: 70–79.

[47] Chandra, N., Brew, K., Acharya, K.R. (1998) Structural evidence for the presence of a secondary calcium binding site in human alpha-lactalbumin. Biochemistry 37: 4767–4772.

[48] Gorbitz, C.H. (2001) Nanotube formation by hydrophobic dipeptides. Chemistry 7: 5153–5159.

[49] Yonekura, K., Maki-Yonekura, S., Namba, K. (2003) Complete atomic model of the bacterial flagellar filament by electron cryomicroscopy. Nature 424: 643–650.

[50] Yonekura, K., Maki-Yonekura, S., Namba, K. (2005) Building the atomic model for the bacterial flagellar filament by electron cryomicroscopy and image analysis. Structure 13: 407–412.

[51] Vauthey, S., Santoso, S., Gong, H., Watson, N., Zhang, S. (2002) Molecular self-assembly of surfactant-like peptides to form nanotubes and nanovesicles. Proc. Natl Acad. Sci. USA 99: 5355–5360.

[52] Nogales, E., Wolf, S.G., Downing, K.H. (1998) Structure of the alpha beta tubulin dimer by electron crystallography. Nature 391: 199–203.

[53] Tan, D., Rice, W.J., Sosa, H. (2008) Structure of the kinesin13–microtubule ring complex. Structure 16: 1732–1739.

[54] Graveland-Bikker, J.F., Schaap, I.A., Schmidt, C.F., de Kruif, C.G. (2006) Structural and mechanical study of a self-assembling protein nanotube. Nano Lett. 6: 616–621.

[55] Ipsen, R., Otte, J. (2007) Self-assembly of partially hydrolysed alpha-lactalbumin. Biotechnol. Adv. 25: 602–605.

[56] Dong, H., Paramonov, S.E., Aulisa, L., Bakota, E.L., Hartgerink, J.D. (2007) Self-assembly of multidomain peptides: balancing molecular frustration controls conformation and nanostructure. J. Am. Chem. Soc. 129: 12468–12472.

[57] Dhathathreyan, A., Nair, B.U. (2010) Influence of sequence on the self-assembly of peptide nanoribbons on silicon substrates. J. Phys. Chem. B 114: 16650–16654.

[58] Kumar, P., Pillay, V., Modi, G., Choonara, Y.E., du Toit, L.C., Naidoo, D. (2011) Self-assembling peptides: implications for patenting in drug delivery and tissue engineering. Recent Pat. Drug. Deliv. Formul. 5: 24–51.

[59] Krishnan, Y., Simmel, F.C. (2011) Nucleic acid based molecular devices. Angew. Chem. Int. Ed. Engl. 50: 3124–3156.

[60] Afonin, K.A., Bindewald, E., Yaghoubian, A.J., Voss, N., Jacovetty, E., Shapiro, B. A., Jaeger, L. (2010) *In vitro* assembly of cubic RNA-based scaffolds designed in silico. Nat. Nanotechnol. 5: 676–682.

[61] Cortajarena, A.L., Mochrie, S.G., Regan, L. (2011) Modulating repeat protein stability: the effect of individual helix stability on the collective behavior of the ensemble. Protein Sci. 20: 1042–1047.

[62] Ballano, G., Zanuy, D., Jimenez, A.I., Cativiela, C., Nussinov, R., Aleman, C. (2008) Structural analysis of a beta-helical protein motif stabilized by targeted replacements with conformationally constrained amino acids. J. Phys. Chem. B 112: 13101–13115.

[63] Zanuy, D., Rodriguez-Ropero, F., Haspel, N., Zheng, J., Nussinov, R., Aleman, C. (2007) Stability of tubular structures based on beta-helical proteins: self-assembled versus polymerized nanoconstructs and wild-type versus mutated sequences. Biomacromolecules 8: 3135–3146.

[64] Brooks, B.R., Brooks 3rd, C.L., Mackerell Jr, A.D., Nilsson, L., Petrella, R.J., Roux, B., Won, Y., Archontis, G., Bartels, C., Boresch, S., Caflisch, A., Caves, L., Cui, Q., Dinner, A.R., Feig, M., Fischer, S., Gao, J., Hodoscek, M., Im, W., Kuczera, K., Lazaridis, T., Ma, J., Ovchinnikov, V., Paci, E., Pastor, R.W., Post, C.B., Pu, J.Z., Schaefer, M., Tidor, B., Venable, R.M., Woodcock, H.L., Wu, X., Yang, W., York, D. M., Karplus, M. (2009) CHARMM: the biomolecular simulation program. J. Comput. Chem. 30: 1545–1614.

[65] Vijayaraj, R., Sundar Raman, S., Mahesh Kumar, R., Subramanian, V. (2010) Studies on the structure and stability of cyclic peptide based nanotubes using oligomeric approach: a computational chemistry investigation. J. Phys. Chem. B 114: 16574–16583.

[66] Hamley, I.W., Nutt, D.R., Brown, G.D., Miravet, J.F., Escuder, B., Rodriguez-Llansola, F. (2010) Influence of the solvent on the self-assembly of a modified amyloid beta peptide fragment. II. NMR and computer simulation investigation. J. Phys. Chem. B 114: 940–951.

[67] Roy, A., Post, C.B. (2011) Microscopic symmetry imposed by rotational symmetry boundary conditions in molecular dynamics simulation. J. Chem. Theory Comput. 7: 3346–3353.

[68] Alemán, C., Bertran, O., Casanovas, J., Torras, J., Revilla-López, G., Zanuy, D. (2012) Conformational aspects and molecular dynamics simulations of peptide hybrid materials: from methods and concepts to applications. In: Peptide Materials: From Nanostructures to Applications. Venanzi, M., Alemán, C. Bianco, A. (Eds). John Wiley & Sons, Ltd, Chichester.

[69] Matsumura, S., Uemura, S., Mihara, H. (2005) Construction of biotinylated peptide nanotubes for arranging proteins. Mol. Biosyst. 1: 146–148.

[70] Ochoa, W.F., Chatterji, A., Lin, T., Johnson, J.E. (2006) Generation and structural analysis of reactive empty particles derived from an icosahedral virus. Chem. Biol. 13: 771–778.

[71] Suci, P.A., Berglund, D.L., Liepold, L., Brumfield, S., Pitts, B., Davison, W., Oltrogge, L., Hoyt, K.O., Codd, S., Stewart, P.S., Young, M., Douglas, T. (2007) High-density targeting of a viral multifunctional nanoplatform to a pathogenic, biofilm-forming bacterium. Chem. Biol. 14: 387–398.

[72] Kramer, R.M., Li, C., Carter, D.C., Stone, M.O., Naik, R.R. (2004) Engineered protein cages for nanomaterial synthesis. J. Am. Chem. Soc. 126: 13282–13286.

7

Peptide-Based Soft Spherical Structures

K. Vijaya Krishna[1], *Nidhi Gour*[1] *and Sandeep Verma*[1,2]

[1]*Department of Chemistry and* [2]*DST Unit of Excellence on Soft Nanofabrication, Indian Institute of Technology Kanpur, India*

Acronyms

Aib	2-Aminoisobutyric acid
CP	Cationic polymer
ΔPhe	α,β-Dehydrophenylalanine
FF	Diphenylalanine
FITC	Fluorescein isothiocyanate
Phg	Phenylglycine
PHGGG	Pro–His–Gly–Gly–Gly
WW	Ditryptophan, Trp–Trp

7.1 Introduction

Peptides are versatile precursors for controlled synthesis of nanostructures having a distinct advantage of ease of synthesis, scale up and biocompatibility [1–4]. The chemistry of amino acid building is well established, making chemical manipulation of precursor peptides a facile process, with a possibility of fine-tuning gross morphology and generating

Peptide Materials: From Nanostructures to Applications, First Edition. Edited by Carlos Alemán, Alberto Bianco and Mariano Venanzi.

soluble nanostructures. It is important to realize that a broad three-stage control of precursor properties could be achieved in peptide-based materials. The stages are as follows:

1. *Side-chain and backbone modifications*: Functionalization of amino acid side-chains and chemical modifications of C- or N-terminus are versatile methods to modulate properties of peptide building blocks [5, 6]. These modifications at times are applied to gain control over solubility properties or to make the conjugates respond to externally applied stimuli of pH and salt concentration. Such chemical modifications are also valuable in post-synthetic labeling of conjugates for specific applications such as tracking the movement of a peptide nanostructure within the confines of cellular environment.
2. *Role of secondary structures*: This parameter is extremely important to control gross morphology of peptide-based self-assembled structures. Introduction of amino acid residues supporting α-helical or β-sheet features allow fine-tuning of the structural outcome of peptide conjugate, which could be later translated to construct various superstructures [7, 8].
3. *Tuning of noncovalent interactions*: It is possible to devise strategies involving conjugation of peptides building blocks to functionalized organic linkers in order to maximize noncovalent interactions such as hydrogen bonding, π–π and van der Waals interactions, to arrive at kinetically and thermodynamically stable structures. Richness of peptide chemistry is able to support synthetic explorations around this theme to generate novel peptide conjugates [9, 10].

In recent years, efficient and elegant methodologies have been developed for molecular confinement of biomolecules and other materials inside peptide nanostructures. The surge in peptide-based soft matter research is due to specific advantages that this system offers while delivering guest molecules or the payload. Due to nanometric dimensions, it is possible to envisage direct introduction of such systems into systemic circulation, without the apparent risk of blocking blood vessels. Suitably protected peptide-based soft structures can remain intact for a longer time and deliver guest molecules at desired sites more effectively.

This chapter deals with some recent examples in the design and application of peptide-based nanostructures.

7.2 Short Peptide Sequences

Ease of chemical synthesis and bulk preparation are possibly two main considerations when using short peptide sequences for the construction of peptide nanostructures. Multiple low-energy, noncovalent interactions, such as hydrogen bonding, van der Waals and π–π interactions, could be invoked for the stabilization of hierarchically complex one-, two- and three-dimensional self-assembled structures, from relatively simple peptide building blocks.

Diphenylalanine (Phe–Phe, FF) dipeptide, the core recognition motif of Alzheimer's β-amyloid polypeptide, has attracted considerable attention for the construction of peptide-base soft structures. In a pioneering study by Reches and Gazit, it was found that FF self-assembles to form nanotubes, which also serve as excellent sacrificial templates for fabrication of silver nanowires through a reductive process [11]. An interesting morphological transition was reported by this group when they used phenylglycine (Phg) as the dipeptide motif [12]. Phg offers constrained rotational freedom around the C–C bond and increased steric hindrance compared to the Phe residue. These attributes were ascribed to the formation of nanoscale vesicles from Phg–Phg dipeptides, which were able to withstand extreme chemical conditions. Curiously, a similar morphological switch was also observed when the Phe–Phe dipeptide was conjugated to a cysteine residue at the N-terminus [12]. This tripeptide self-assembled into soft nanospheres suggesting a role of disulfide bridge formation in the closure of nanotubes, leading to the emergence of spherical structures. Simple N-protected FF acid was also shown to self-assemble into nanospheres with an associated photoluminescence behavior [13]. The absorption experiments confirmed quantum dot-like properties of these nanospheres upon self-assembly [14] (see Figure 7.1).

Huang *et al.* reported the formation of soft vesicles with FF dipeptide by merely changing the surface used to study the assembly process [15]. The role of interface-induced self-assembly was ascribed to morphological transition of FF fibers to vesicles. A host of factors such as surface properties, for example surface tension, relative hydrophilicity/hydrophobicity of the medium, peptide–substrate and solvent–substrate interactions, may have a direct impact on the morphological consequences of the diphenylalanine sequence, including the reversibility of fiber-to-vesicle nanostructural transformation [16, 17] (see Figure 7.2).

Li *et al.* reported a morphological transition of chemically modified cationic FF peptide into vesicles for cellular transport of oligonucleotides [18] (see Figure 7.3). These positively charged modified peptide nanotubes, when complexed with oligonucleotides, exhibit a transition in morphology to vesicles upon dilution, at physiological pH.

A fluorescent dye tagged ssDNA was complexed with cationic dipeptide nanotubes and imaged by confocal laser scanning microscopy to reveal nanotubular structures. Further dilution of this solution revealed green fluorescent vesicles, indicating that the ssDNA–peptide complex transformed into a vesicular structure. Further experiments confirmed the

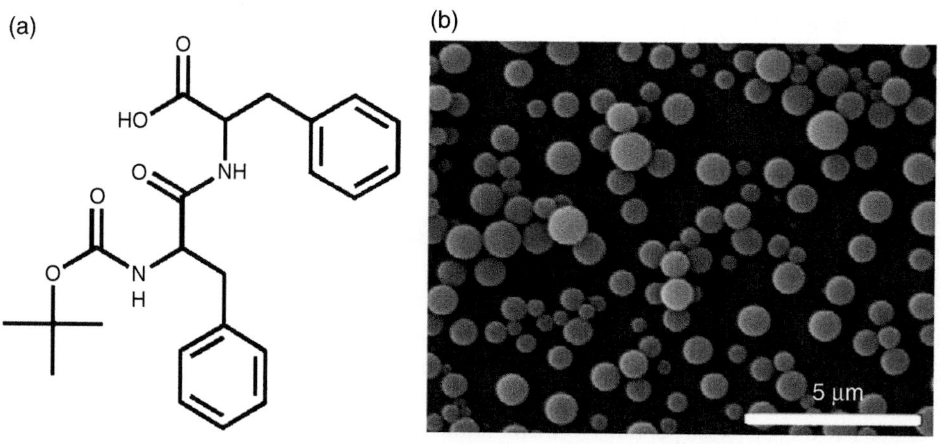

Figure 7.1 (a) Structure of N-protected FF acid; (b) SEM images of peptide nanospheres. Reproduced with permission from reference [14]. Copyright (2009) American Institute of Physics

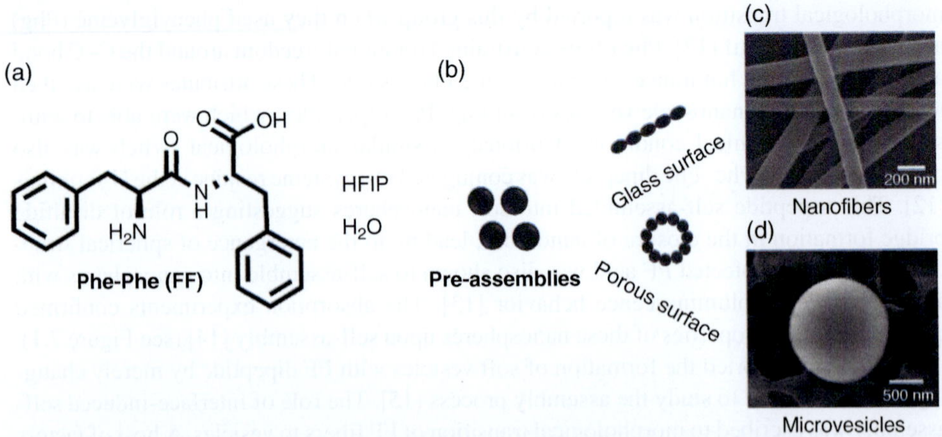

Figure 7.2 Proposed mechanism for peptide self-assemblies: (a) diphenylalanine (FF) peptide; (b) pre-assemblies of FF dipeptide; (c) SEM images of nanofibers formed on glass surface; and (d) microvesicles formed on microporous cellulose ester surface. Reproduced with permission from reference [16]. Copyright (2011) IOP Publishing

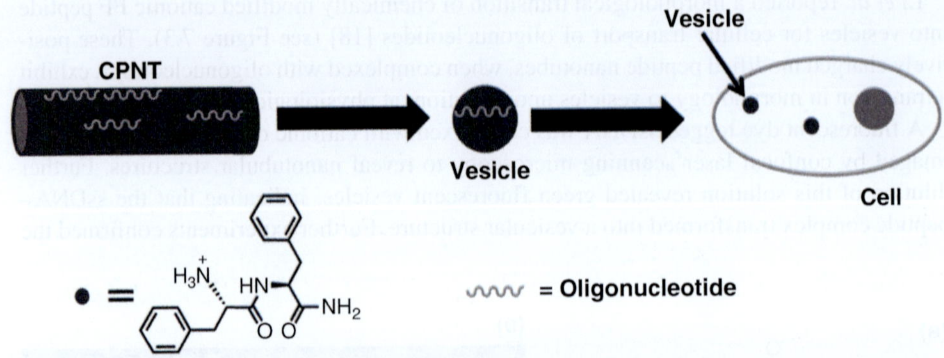

Figure 7.3 Proposed transition of the cationic dipeptide nanotubes into vesicles for oligonucleotide delivery. Reproduced with permission from reference [18]. Copyright (2007) John Wiley & Sons, Ltd

accumulation of the peptide–ssDNA complex inside cytoplasm, while maintaining cell viability during the delivery process (Figure 7.4).

Conversely, examples of vesicle-to-rod (fibers) transformations are also described – for instance, a morphological transformation from sphere-to-rod by fusion of tripeptide conjugates, containing a synthetic rigid linker of abiotic origin [19]. A synthetic aromatic linker was employed to increase crystallinity of tripeptide conjugates and to engineer an extended peptide backbone. These peptides formed hollow vesicles, with rough morphology, at high concentrations, but dilution of the stock solutions resulted in coalescence of vesicular structures to afford sphere-to-rod transition (see Figure 7.5).

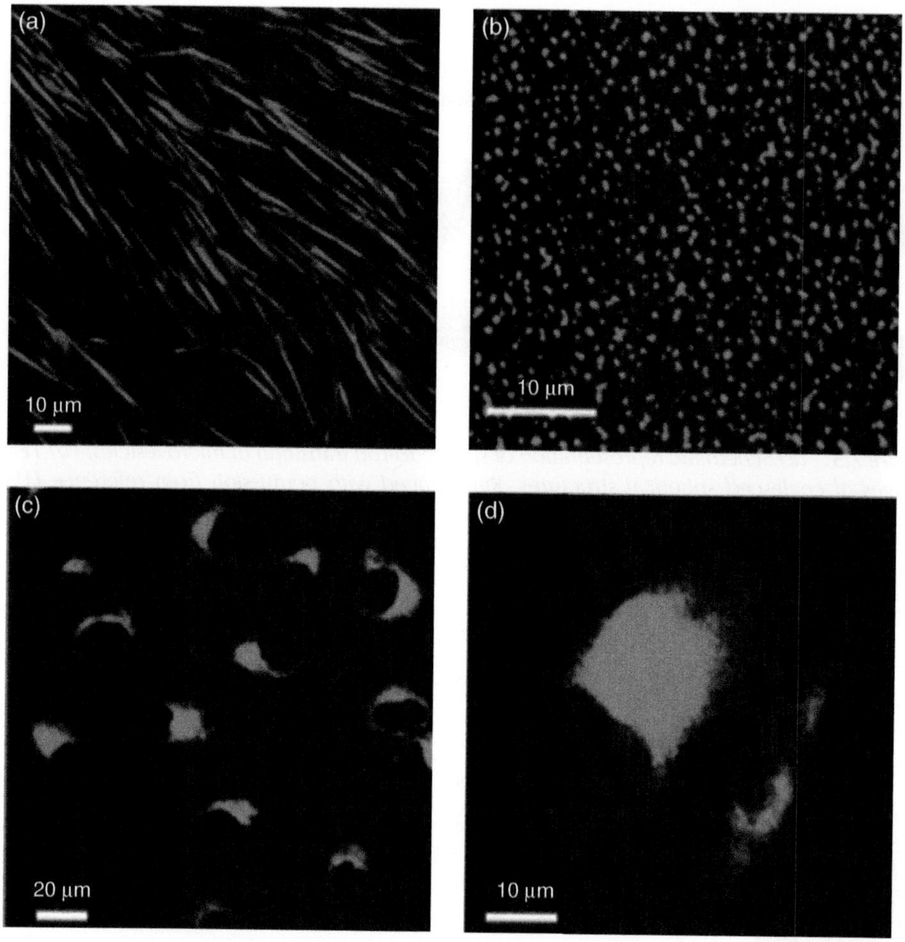

Figure 7.4 *Confocal laser scanning microsopy images of: (a) fluorescently labeled peptide nanotubes–ssDNA complex; (b) nanospheres upon dilution; (c) and (d) cellular uptake of peptide nanospheres–ssDNA complex by HeLa cells. Reproduced with permission from reference [18]. Copyright (2007) John Wiley & Sons, Ltd (see color plate figure)*

In another study, Yao *et al.* reported morphological transitions in β-sheet forming peptides linked with a bipyridine spacer [20] (see Scheme 7.1). The presence of a bipyridine unit resulted in a 'U'-shaped structure and a vesicular morphology upon self-assembly. However, ultrasonication of the vesicle solution for a brief period resulted in the formation of an opaque gel, which showed the presence of fibers when analyzed with microscopy techniques. Interestingly, this transformation was thermally reversible. It was reasoned that reversibility between vesicles and fibers directly results from conformational changes induced by bipyridine groups forcing β-sheet-like bilayers to adopt either loose- or tightly-packed structures. The ultrasound-induced rotation in the bipyridine groups was implicated as a driving force for self-assembly.

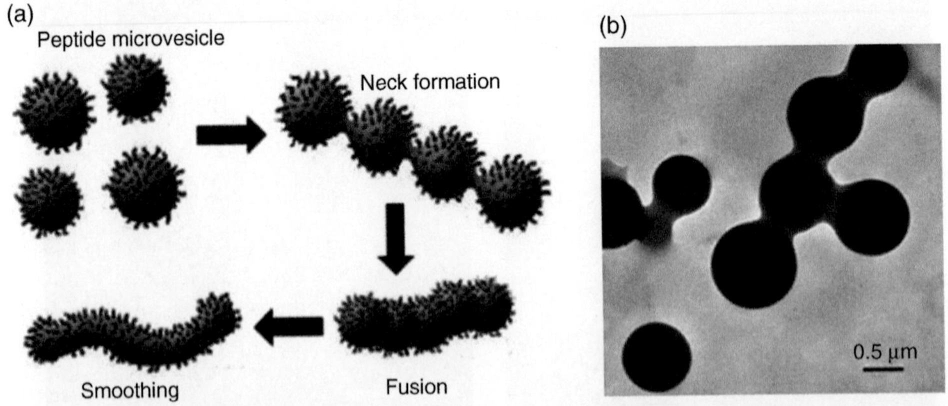

Figure 7.5 (a) Schematic representation of sphere-to-rod transition of microvesicles; (b) TEM images of coalesced spherical structures. Reproduced with permission from reference [19]. Copyright (2011) American Chemical Society

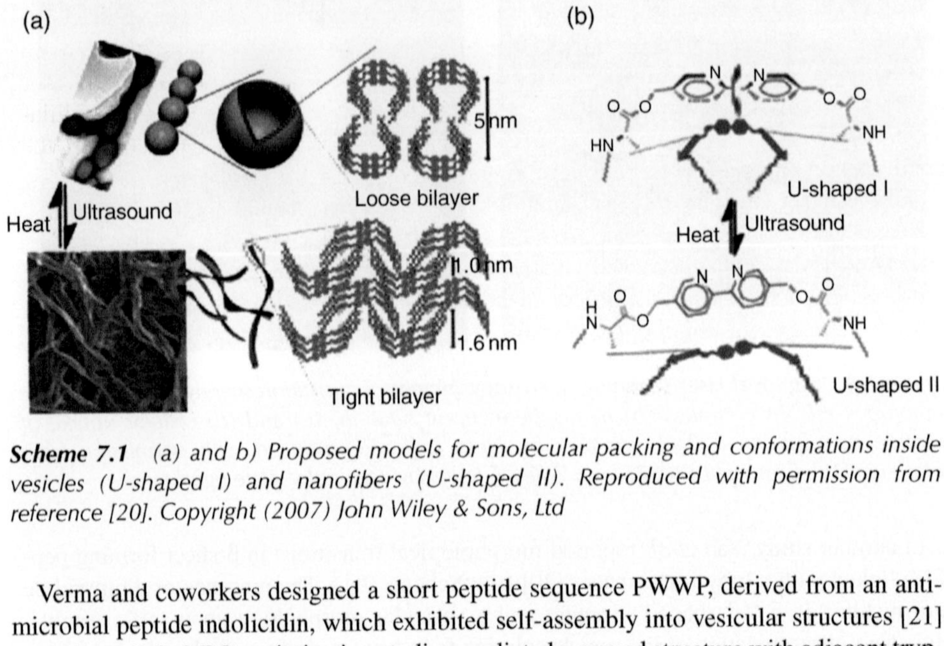

Scheme 7.1 (a) and b) Proposed models for molecular packing and conformations inside vesicles (U-shaped I) and nanofibers (U-shaped II). Reproduced with permission from reference [20]. Copyright (2007) John Wiley & Sons, Ltd

Verma and coworkers designed a short peptide sequence PWWP, derived from an antimicrobial peptide indolicidin, which exhibited self-assembly into vesicular structures [21] (see Figure 7.6). MM^+ optimization studies predicted a curved structure with adjacent tryptophan residues being involved in π–π stacking interactions. Structural characterization revealed the formation of self-assembled spherical aggregates, as confirmed by microscopy studies. Light scattering analysis and stability studies confirmed homogeneity of size and robust nature of these vesicles. However, interaction with monovalent cations, such as Na^+ and K^+ ions, resulted in disruption of vesicle morphology and release of encapsulated Rhodamine B dye, suggesting a role of π–π and cation-π interactions in vesicle stability and disruption, respectively.

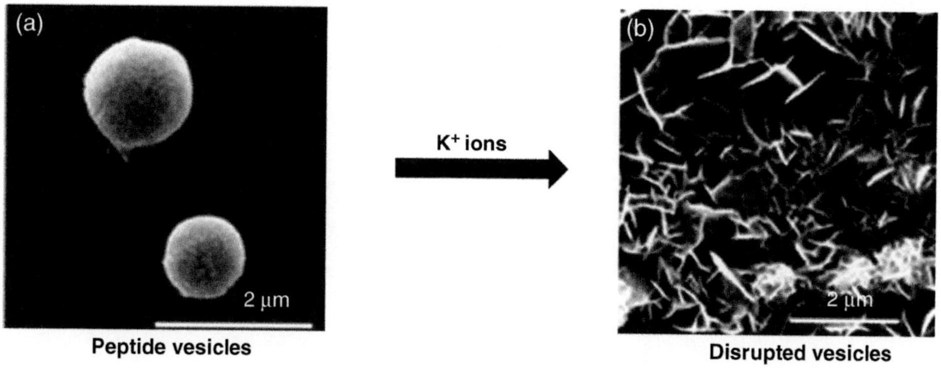

Figure 7.6 *(a) SEM image of PWWP solution after 5 min incubation; (b) KCl-mediated disruption of vesicles. Reproduced with permission from reference [21]. Copyright (2007) The Royal Society of Chemistry*

The role of the Trp–Trp dipeptide was exploited again by attaching a ditryptophan unit around a *tren* scaffold [22]. The design of this conjugate was inspired from the three-legged structure of a clathrin monomer unit, which forms well ordered cage-like nanostructures on self-assembly [23, 24]. The MM^+ optimized structure revealed an overlap of indole aromatic rings, suggesting a crucial role of π–π stacking interactions. Homogeneous multilamellar spherical structures were observed and the mechanism was attributed to the interdigitation of ditryptophan arms (see Figure 7.7).

Encapsulation and release of Rhodamine B dye was facilitated at mildly acidic pH, which upon further acidification led to deformation and vesicle coalescence. This stimuli-responsive observation was accounted to nitrogen protonation causing electrostatic repulsions that would render the soft structures leaky in the acidic environment.

As an extension to ditryptophan-based spherical structures, a synthetic C_3-symmetric structure was further modified to introduce a thiol functional group. This modification was engineered to explore stable interaction with gold nanoparticles [25] (see Figure 7.8), followed by exploitation of plasmonic heating of gold nanoparticles on exposure to electromagnetic irradiation [26–29]. Gold nanoparticle-decorated soft spherical structures were briefly exposed to sunlight, which led to deformation of their morphology. This approach released encapsulated Rhodamine B dye, thus offering an entry into sunlight-responsive soft spherical structures.

Ditryptophan-based soft structures were also subjected to an interesting interfacial self-assembly experiment where peptides affording tubular structures were added to the spherical structures [30]. Co-incubation of C_3-symmetric conjugate $(WW)_3$Tren with FF and PHGGG-Cap-PHGGG peptides afforded hybrid assemblies, where individual fragments retained their gross self-assembled morphologies that were formed with the help of hydrogen bonding and interdigitation of aromatic residues (see Figure 7.9).

Such a design could be applied for the generation of complex self-assembled aggregates, which could not be obtained by a singular molecule.

The Trp–Trp dipeptide was conjugated to the biotin carboxylic acid group to create a vitamin–peptide hybrid structure (see Figure 7.10). The choice of tryptophan residue was

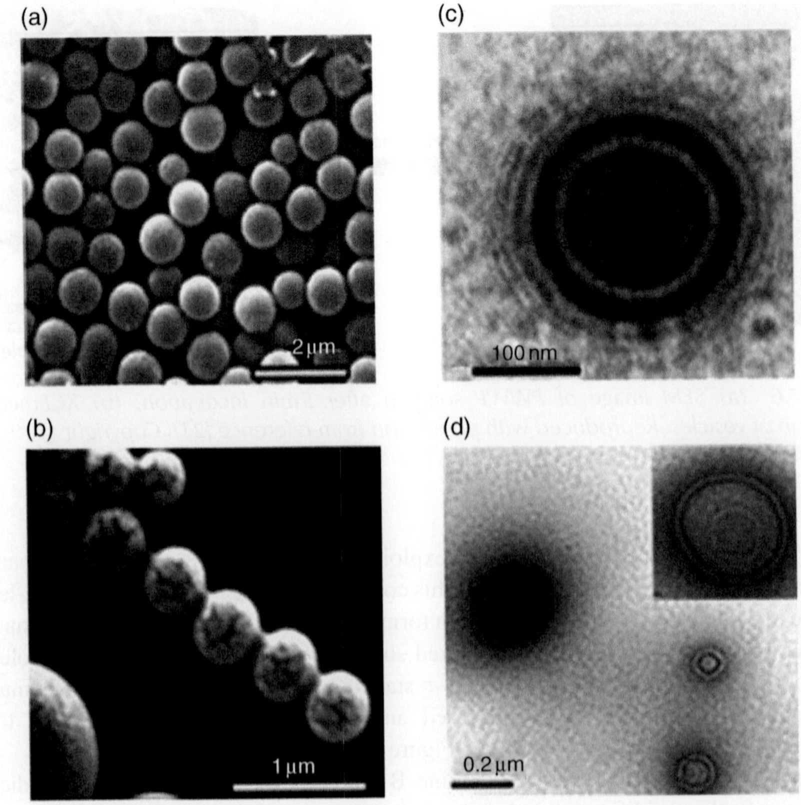

Figure 7.7 Self-assembled structures formed by C_3-symmetric ditryptophan conjugate: (a) SEM micrograph of spherical structures; (b) beads-on-a-string arrangement; (c) and (d) HR-TEM micrographs. Reproduced with permission from reference [22]. Copyright (2007) John Wiley & Sons, Ltd

dictated by the fact that molecular recognition and high affinity biotin–avidin interaction is supported by four crucial tryptophan residues at the biotin binding site in avidin protein. Thus it was of interest to see how self-assembly of biotin will be influenced by the presence of tryptophan. Biotin afforded fibrous structures in solution, but interestingly biotin–TrpTrp conjugation resulted in the formation of soft vesicles, with appreciable size homogeneity [31].

The stability of biotin-containing vesicles was confirmed by their interaction with a focused Ga^+ ion beam, thereby suggesting reasonable structural stability and robust nature of these structures. A 500×500 nm section was milled off the surface to reach the interior core of the spherical structure (Figure 7.11a). These microspheres were also amenable to ion beam assisted bisection suggesting structural integrity, which allows them to withstand ion currents for short time periods (Figure 7.11b). It was also possible to mill these microspheres into different shapes, such as a through-and-through square hole, thus again revealing a highly stable core organization (Figure 7.11c).

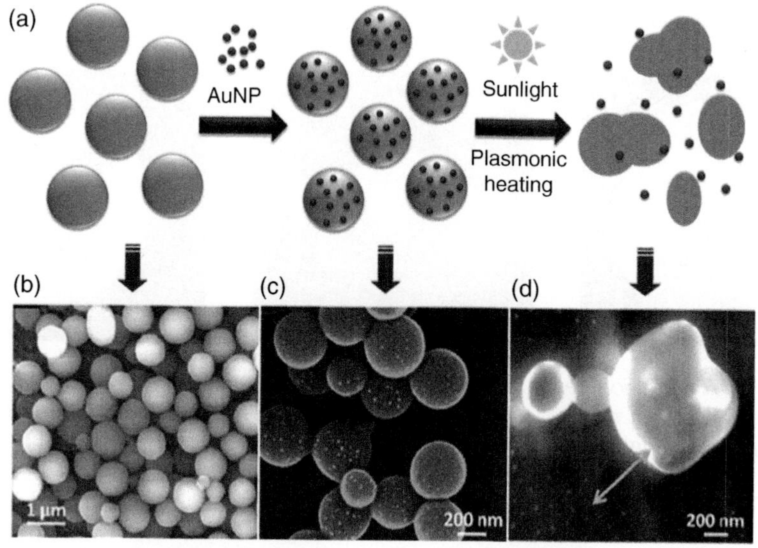

Figure 7.8 (a) Schematic of sunlight-responsive soft spherical nanostructures. SEM micrographs of (b) unmodified thiolated vesicles; (c) gold nanoparticles on the vesicle surface; and (d) disruption of vesicles on exposure to sunlight. Adapted and reproduced with permission from reference [21]. Copyright (2010) The Royal Society of Chemistry

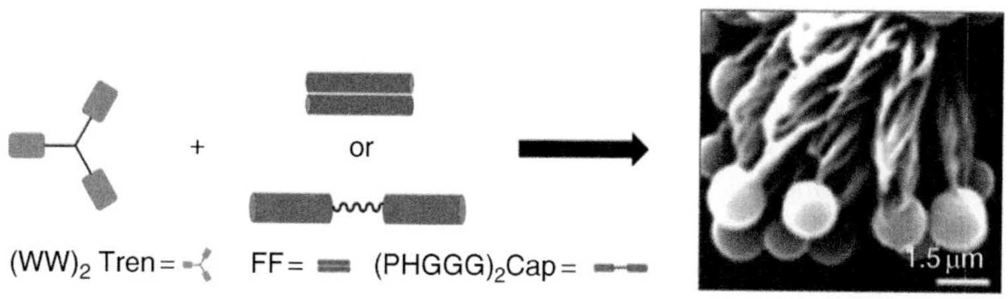

Figure 7.9 Schematic of peptide fiber growth from the vesicle surface. Adapted and reproduced with permission from reference [30]. Copyright (2008) John Wiley & Sons, Ltd

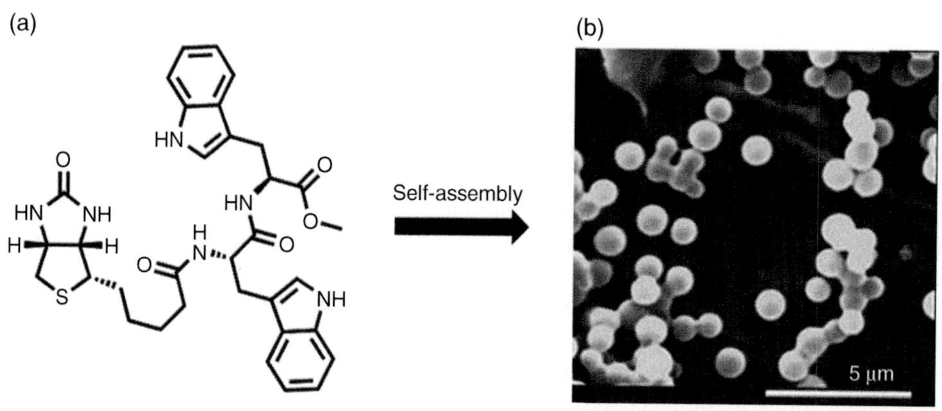

Figure 7.10 (a) Molecular structure of biotinylated ditryptophan methyl ester; (b) SEM image of vesicular self-assembly. Adapted and reproduced with permission from reference [31]. Copyright (2008) John Wiley & Sons, Ltd

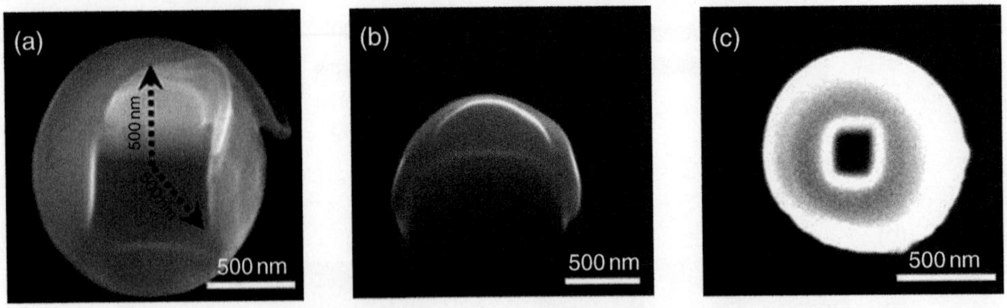

Figure 7.11 Focused ion beam milling of spherical structures: (a) milling of a 500 × 500 nm cross-section; (b) bisection; (c) creating a square hole. Reproduced with permission from reference [31]. Copyright (2008) John Wiley & Sons, Ltd

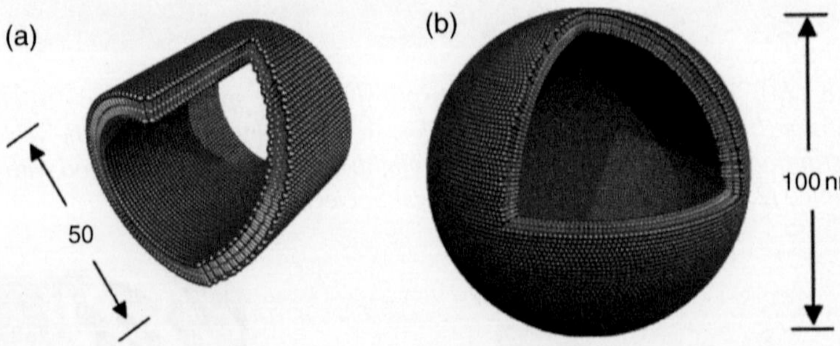

Figure 7.12 Molecular model of structures formed from amphiphilic peptides: (a) peptide nanotube; (b) peptide nanovesicle. Reproduced with permission from reference [33]. Copyright (2002) American Chemical Society

7.3 Amphiphilic Peptides

Zhang *et al.* [32, 33] and Stupp *et al.* [34] provided early examples of self-assembly of 'surfactant-like peptides' with non-polar tails and polar head groups (see Figure 7.12). A number of variations were explored with hydrophobic tails consisting of non-polar amino acids like glycine, alanine, valine, leucine, or isoleucine, and hydrophilic head groups consisting of cationic or anionic amino acids [35]. The shape and extent of electrostatic repulsions between the head groups governs the extent of curvature in the supramolecular structure, while the occurrence of a bilayer-like arrangement was proposed in the aqueous environment. Thus, the overall morphology of soft structures in this particular design depends on factors such as the area of charged head groups, length of the hydrophobic tail and volume occupied [36].

These results paved the way for extensive work on peptide amphiphiles for fabrication of soft structures and various biological applications. In a recent example, Mastrobattista *et al.* reported self-assembly of amphiphilic oligopeptides into vesicles and they described their applications as delivery vehicles in photodynamic therapy [37] (Figure 7.13). Two self-assembling peptides SA2 and SA7, negatively charged at neutral pH, were developed containing different numbers of glutamic acid residues [38]. The formation

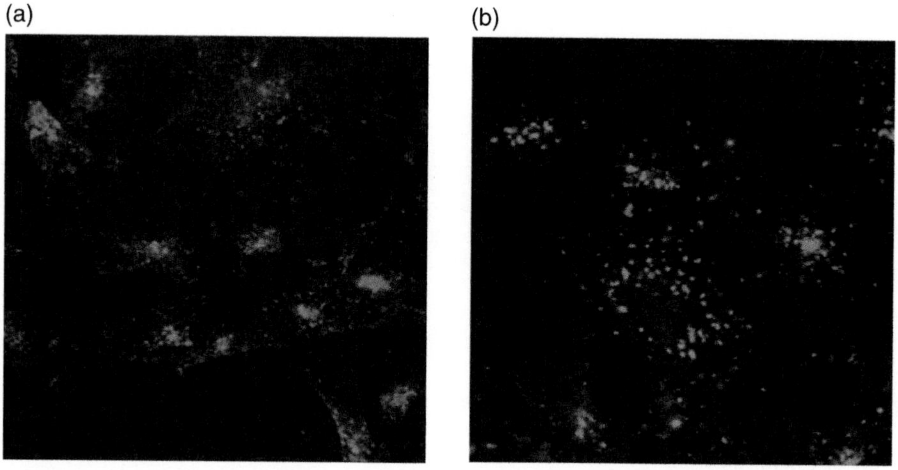

Figure 7.13 Cell uptake of vesicle-forming SA2C3 peptides: (a) co-localization of peptides (blue) with transferrin (green) was absent; (b) co-localization of peptides (blue) with dextran 10000 (green) is visualized as multiple spots of dotted patterned co-localization (cyan). Reproduced with permission from reference [37]. Copyright (2010) Elsevier (see color plate figure)

of vesicles was pH-dependent and reversible, and it was determined from experimental data and theoretical calculations that hydrophobic interactions played a critical role in vesicle stability, in addition to the hydrogen bonding network [39]. An improved design was engineered by introducing a thiolated amphiphilic peptide. As a specific example of delivery, a poorly soluble phthalocyanine photosensitizer was encapsulated within the cavity of amphiphilic peptide vesicles. Incorporation of photosensitizers in cross-linked or non-cross-linked vesicles was applied to photodynamic therapy, where reactive oxygen species were generated upon irradiation of light. Cytotoxicity and cell internalization experiments confirmed the localization of phthalocyanine-loaded vesicles inside endocytic compartments [37].

Conjugation of glutamic acid and lysine residues to α,β-dehydrophenylalanine (ΔPhe) resulted in anionic and cationic amphiphilic peptide vesicles [40]. These peptides self-assembled into vesicles with an average diameter of 200 nm and 280 nm for Glu–ΔPhe and Lys–ΔPhe, respectively. The introduction of unnatural amino acid analogue provides structural stability [41] and increased proteolytic stability [42], which was verified by co-incubation with a highly nonspecific proteolytic enzyme, proteinase K, for over 24 h. The vesicles were extremely stable to the degradative action of proteinase K.

In another example, cationic peptide amphiphiles having responsive stimuli towards metal ions were described [43]. In this report, a rare phenomenon of multivesicular architectures, i.e. vesicle inside vesicle, was observed (see Figure 7.14). Water-soluble synthetic anionic dipeptides of glutamic acid and of hydrophobic amino acids (valine, norvaline, and leucine) retain their cationic nature over a range of pH (2–12) and reveal multivesicular architecture, as observed in TEM [44]. However, incubation with Ca^{2+} ions resulted in vesicle rupturing. Encapsulation studies and cytotoxicity experiments demonstrated the utility of these vesicles as efficient carriers for cell internalization experiments [45–48].

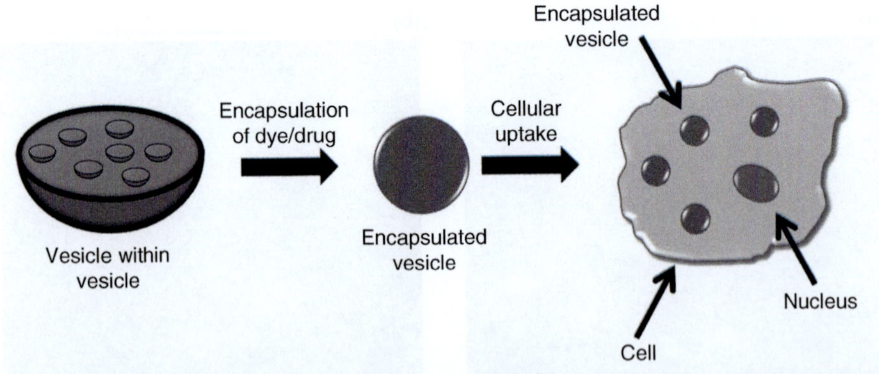

Figure 7.14 Schematic description of vesicle-in-vesicle system of multivesicular architecture of peptides. Reproduced with permission from reference [43]. Copyright (2011) The Royal Society of Chemistry

In a design inspired by peptide-tRNA interaction, Strazewski *et al.* reported vesicle formation from an amphiphilic peptide–RNA conjugate [49]. A lipophilic peptide fragment was chosen as it could serve as an anchor for transient immobilization of tRNA on lipidic bilayer vesicles [50, 51]. The peptide was mainly constituted of alanine and glutamic acid, which was conjugated to tRNA at its 3′-end. The resulting amphiphilic peptidyl–tRNA conjugate demonstrated the potential to self-assemble spontaneously into highly polydispersed vesicles, as analyzed by light scattering and atomic force microscopy.

Luis *et al.* have described the design of C_2-symmetric gemini amphiphilic pseudopeptides, which self-assemble to give different morphologies [52, 53] (see Figure 7.15). These pseudopeptides contain two amino acid moieties connected through their C-terminus by a flexible spacer, while the amino termini are appended with hydrophobic tails. Such a design was introduced to harness beneficial effects of both hydrogen bonding and van der Waals interactions. Interestingly, a strong correlation between the chemical structure and chemical environment was evident. Amorphous peptide-based material was obtained in chloroform, fibrous nanostructures in methanol or aqueous methanol, while vesicular morphology was obtained in acidic or aqueous environment. Minor modification in the structure to fine-tune hydrophobicity versus hydrophilicity leads not only to vesicle formation but also to biocompatible hydrogels, which could have potential applications in cellular delivery and other nanobiotechnological applications [52].

Stupp *et al.* have designed buckled vesicle systems with diacetylene functionalized peptide amphiphiles, inspired by the similarity of shape and dimensions to viral capsids [54, 55]. Diacetylene groups could be polymerized to afford a polydiacetylene backbone through favorable topotactic reactions, which fixes various interacting domains in amphiphilic peptides within the supramolecular ordered state [56]. Many other groups have also successfully employed a diacetylene strategy to fix the position of peptide amphiphiles, within the confines of hierarchically ordered ensembles, in order to create novel functional materials with colorimetric transitions [57–59].

Kim *et al.* reported morphological transitions of diacetylene functionalized D-Ala–D-Ala dipeptide on vancomycin binding [60]. This design is based on the fact that vancomycin, a glycopeptide antibiotic, inhibits bacterial cell wall synthesis as it specifically binds

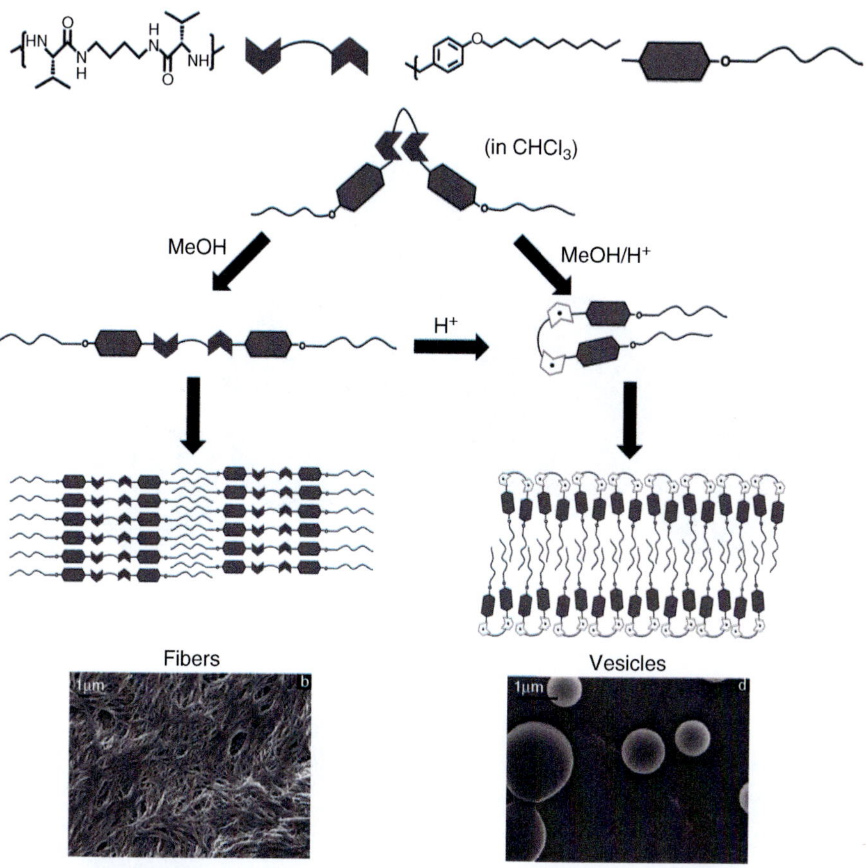

Figure 7.15 Schematic representation of self-assembling behavior of the gemini amphiphilic pseudopeptides in different conditions. Reproduced with permission from reference [52]. Copyright (2011) Royal Society of Chemistry

to D-Ala-D-Ala dipeptide present in peptidoglycans [61]. Application of a heating and cooling cycle resulted in ribbon-like morphology from amphiphilic peptide dispersions. In a dramatic morphological transformation, a vesicle-like morphology was obtained when D-Ala–D-Ala dipeptide containing amphiphile interacted with vancomycin. As a further confirmation, the diacetylene moiety was photopolymerized upon UV irradiation to reveal spherical fluorescent dots, due to red-phase fluorescence of polydiacetylenes.

In a recent study, Tovar *et al.* showed photopolymerization of oligopeptides with diacetylenes units. Alignment of randomly self-assembled nanofibers into noodle-like hydrogels was demonstrated, without affecting their supramolecular structure. Further experiments with electrostatic force microscopy proved nanofiber–AFM tip interaction at an applied potential due to induced charge due to free electron transfer between peptide molecules upon polymerization induced bridging [62].

Meier *et al.* reported observation of self-assembled vesicles of an amphiphilic peptide comprising a lysine containing tripeptide constituting its polar end and gramicidin A

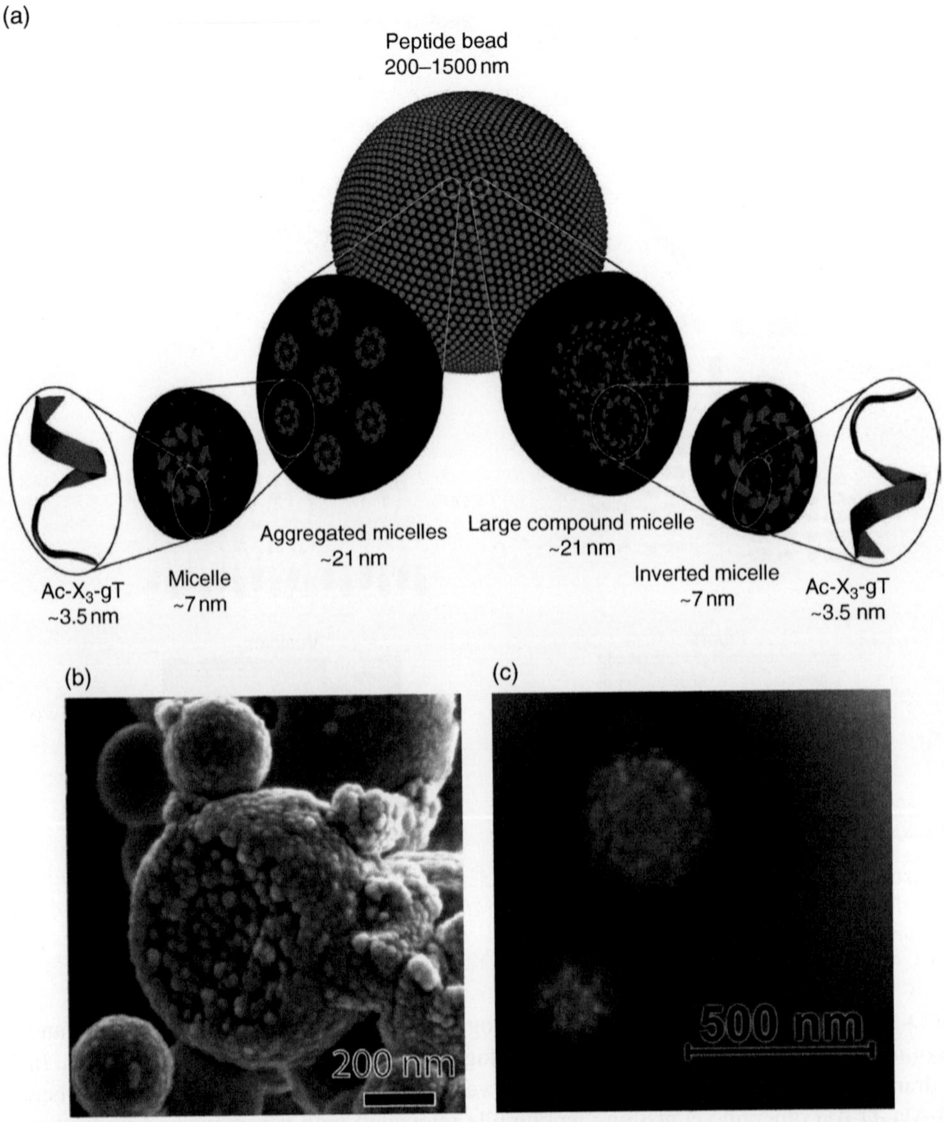

Figure 7.16 *(a) Schematic drawing shows self-assembly of smaller spherical subunits into peptide beads; (b) SEM image of a peptide bead revealing smaller spherical subunits in its core; (c) TEM image of peptide beads. Reproduced with permission from reference [63]. Copyright (2011) American Chemical Society (see color plate figure)*

constituting its nonpolar end [63]. Resulting peptide vesicles showed nanosize peptide substructures within their interiors when probed by transmission electron microscopy and high-resolution scanning electron microscopy (Figure 7.16). Such nanocompartments, within a microscopic vesicle, could be applied as next generation delivery vehicles.

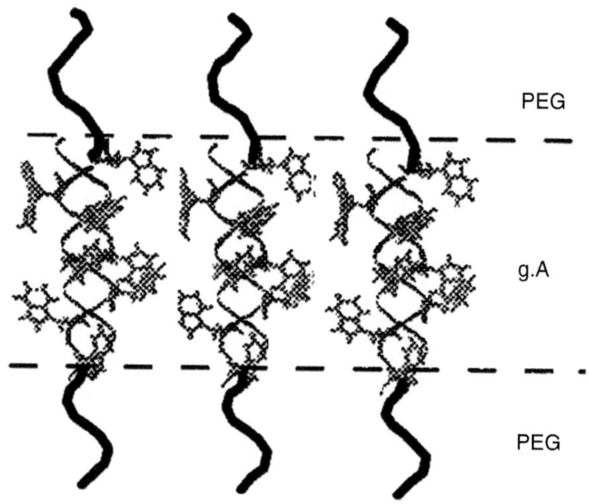

Figure 7.17 Illustration of a helix-peptide membrane of gramicidin A–PEG vesicles. Reproduced with permission from reference [64]. Copyright (1999) American Chemical Society

7.4 Peptide–Polymer Hybrids

Peptosomes are specialized hybrids, which contain amphiphilic block copolymers of two polymeric units, where one or both the polymeric units are polypeptides, with completely different properties. The prime motivation behind development of such systems was the possibility of designing molecular chimeras from biologically relevant materials and to provide superior stability to existing liposomal systems. Peptosomes offer a greater variety in design as they are able to incorporate information required to direct the assembly of functional nanomaterials, without the need for post-synthetic functionalization. Structurally, these soft structures allow translation of certain structural features of proteins in their morphology and they tend to respond favorably to external stimuli such as solvent, pH, and temperature, to name a few. They could be considered as stable, bioinspired counterpart of liposomes, but with greater clearance time and interesting properties such as cell adhesion, biosensing, and controlled nanoscale ordering.

Kimura *et al.* described the conjugation of gramicidin A [64] (see Figure 7.17), a naturally occurring peptide antibiotic, with a hydrophilic polyethylene glycol fragment resulting in an amphiphilic block copolymer [65]. Dynamic light scattering analysis and cryo-electron microscopy of this amphiphilic block copolymer revealed nanosized peptide vesicles, with a membrane thickness of 5 nm. The α-helical features engendered in peptide conjugates were invoked for the formation of vesicles [66]. Interestingly, it was possible to transform multilamellar vesicles obtained through this process to unilamellar vesicles upon sonication. These vesicles were able to encapsulate dye and release it on adition of a non-ionic detergent Triton X-100.

Further reports on self-assembly of glycopeptide conjugates with well-defined helical conformation was described for a hydrophobic α-helix forming peptide with $(Ala–Aib)_n$ as a repeating unit [67]. The resulting peptide was functionalized with lactobionic acid, an

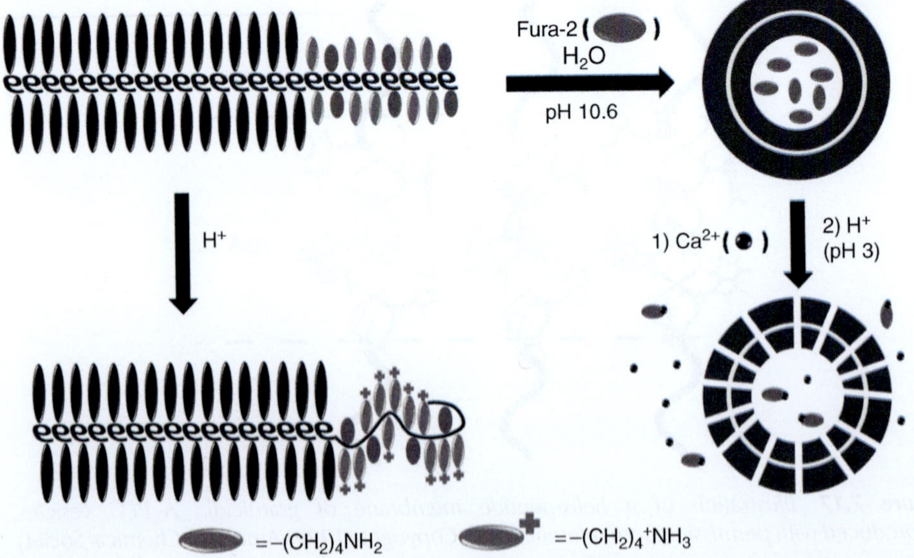

Figure 7.18 *Schematic drawing of pH-responsive lysine–leucine peptosomes and their change in conformation with pH followed by release of entrapped Fura-2 dye. Adapted and reproduced with permission from reference [75]. Copyright (2004) Nature Publishing Group*

uncharged disaccharide, to eliminate undesired electrostatic interactions at its hydrophilic end. It was found that resulting peptosome critically depended on the ratio of block copolymeric units [68].

Lecommandoux *et al.* [69, 70] and Förster *et al.* [71] provided early reports about the stimuli-responsive behavior of peptosomes. They described synthesis of polybutadiene-*b*-poly(L-glutamate) copolymers made by anionic polymerization and subsequent ring-opening polymerization. The size of peptosomes was manipulated by reversible conformation change as a function of both pH [72], and ionic strength and synthesis of 'shape-persistent nano-capsules' were achieved by polybutadiene cross-linking [70, 73, 74].

Deming *et al.* reported vesicle formation of ordered peptosomes from ethylene glycol modified amino acid residues [75] (see Figure 7.18). These peptosomes adopted a stable rod-like conformation in solution due to a strong helix forming tendency of the polypeptide blocks [76]. Introduction of lysine residue, in addition to leucine residues, rendered the peptosomes pH-sensitive and leaky. This was proved by the release of Fura-2 dye in the presence of Ca^{2+} ions at different pH values. Release of fluorescent dye was attributed to the membrane porosity or complete disruption of the vesicle surface, through lysine protonation.

Functionalization of liposomes, polymerosomes, and peptosomes with polyethylene glycol leads to increased retention in the reticulo-endothelial system [77, 78]. However, the use of polyethylene glycol also results in side effects, nonbiodegradability, and toxicity problems [79]. Thus, the need for a natural and biocompatible alternative led to the use of polysacrosine, a polymer of naturally occurring amino acid [80]. Kimura *et al.* designed peptosomes of polysacrosine and poly(γ-methyl-L-glutamate) as hydrophilic and hydrophobic blocks,

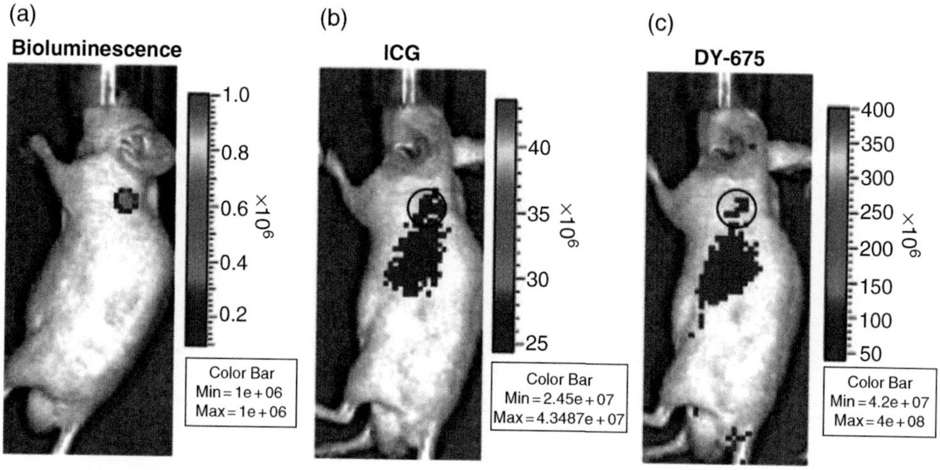

Figure 7.19 *In vivo cancer imaging using labeled peptosome: (a) bioluminescence image of SUIT2/EF-luc xenografts; (b) fluorescence image with Indocyanin green; (c) fluorescence image of DY-675. Reproduced with permission from reference [81]. Copyright (2005) American Chemical Society (see color plate figure)*

respectively [81]. As seen in previous examples, the formation and size of vesicles were shown to depend on the ratio of the hydrophobic and hydrophilic blocks.

An interesting application of soft peptide structures was devised considering vascular physiology. It is known that submicrometer-sized defects occur on vascular walls due to rapid angiogenesis and, moreover, slow clearance of foreign particles by the lymph system causes an enhanced permeability and retention effect [82]. In order to exploit this phenomenon, Kimura *et al.* carefully tuned peptosomes in the 100 nm range and encapsulated FITC-dextran in these peptosomes (Figure 7.19). Near-infrared fluorescence imaging of these peptosomes allowed absorption in near-infrared and far-red light, with no interfering absorption from biological sources [83–85]. Efficient retention of peptosomes at the tumor sites in a mouse model was ascertained even after 72 h of injection time.

In another study, Deming *et al.* used unmodified poly(L-lysine)-*b*-poly(L-leucine) peptosomes to demonstrate encapsulation of dextran and their selective permeability in sucrose and salt solutions [86]. It was shown that vesicle morphology remained unaltered at low salt concentrations, perhaps due to permeability of both water and salt across the membrane, but rupturing was observed at high salt concentrations due to reduced electrostatic interactions between lysine chains of polymer, causing flattening of vesicles and eventual precipitation. Cell internalization experiments demonstrated the possibility of co-localization of peptosomes with dye labeled dextran in epithelial and endothelial cell lines. A related study further reported peptosomes of poly(L-arginine)-*b*-poly(L-leucine), which showed properties similar to lysine–leucine peptosomes (Figure 7.20) [87–93].

Lecommandoux *et al.* reported formation of schizophrenic vesicles formed by water soluble block copolymers of oppositely charged peptides [92] (see Figure 7.21). A zwitterionic diblock copolymer poly(L-glutamic acid)$_{15}$-*b*- poly(L-lysine)$_{15}$ was synthesized where poly(L-glutamic acid) was fully protonated at acidic pH, resulting in a change from

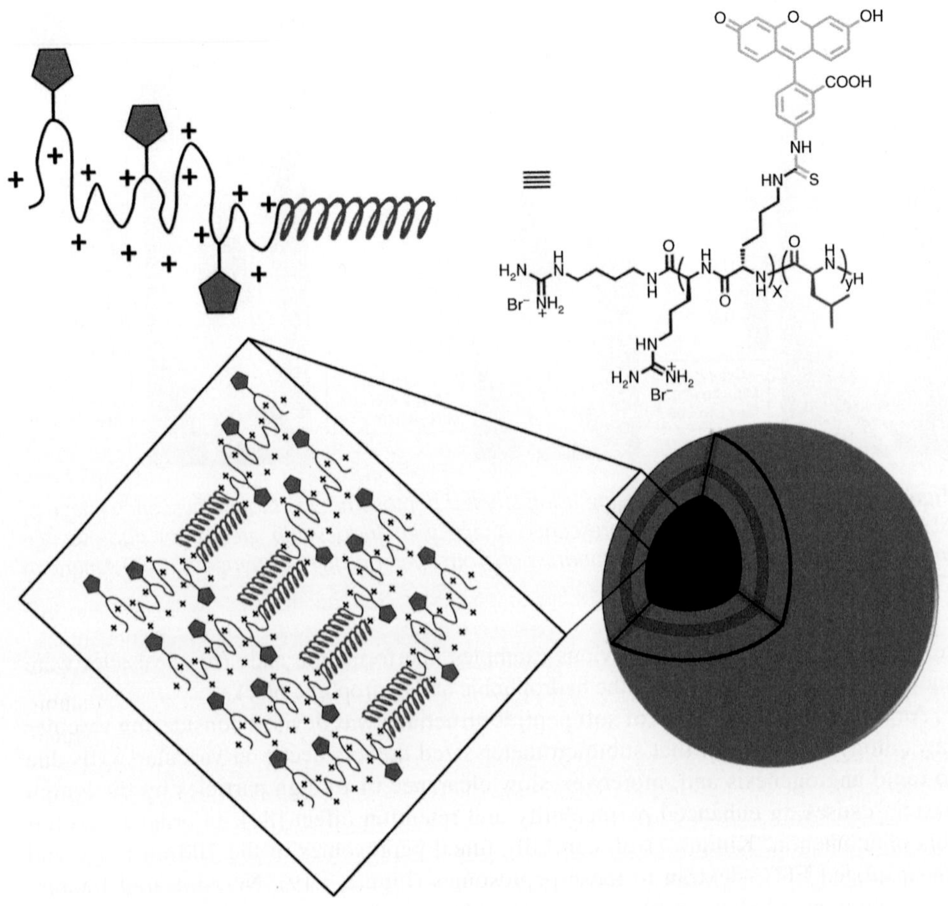

Figure 7.20 Design strategy for arginine–leucine vesicles. Adapted and reproduced with permission from reference [87]. Copyright (2004) Nature Publishing Group

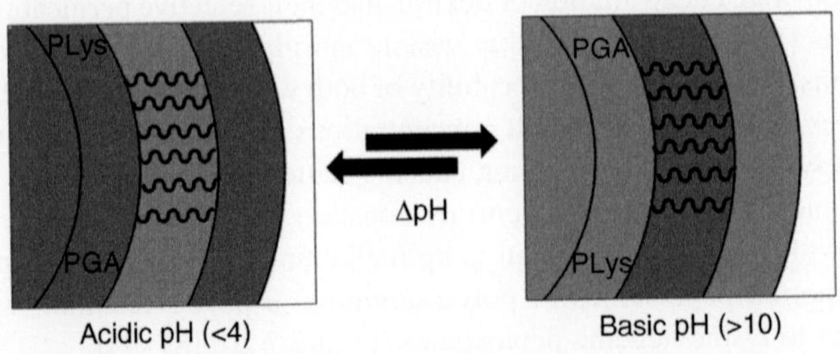

Figure 7.21 Schematic representation of the self-assembly of diblock copolymer poly(L-glutamic acid)$_{15}$-b- poly(L-lysine)$_{15}$ into vesicles. Reproduced with permission from reference [92]. Copyright (2005) American Chemical Society

a charged coil to a more compact α-helical structure, while ε-amino groups of lysine are protonated, giving it a charged character. Similarly, under basic conditions, the poly(L-lysine) block is neutralized, while the poly(L-glutamic acid) block is negatively charged. Such a pH-induced transformation was also monitored by ^1H NMR spectra.

Kataoka *et al.* designed a new class of PICsomes by mixing oppositely charge block co-polymers of the same amino acid [94]. The major difference in the design was that they used the same polymeric unit polyethyleneglycol-poly(*β*-benzyl-L-aspartate) and functionalized it to get two polymers of opposite charges. The gross morphology of PICsomes was hollow, with a core-shell architecture [95, 96]. Yan *et al.* reported a unique concept of complex peptosomes [97]. They described the synthesis of polymeric peptide vesicles by one-step self-assembly of hyperbranched poly(L-lysine) polymers. These peptosome vesicles were obtained by simply mixing poly(L-lysine), a cationic polymer, and H_x–COOH, a hyperbranched polymer prepared from polyester boltron H_x (Scheme 7.2). This simple strategy not only allowed control over size by varying the concentration of two polymers but also resulted in vesicles with unusual size and stability, over a range of pH and long shelf-life.

Jing *et al.* reported an exciting development by demonstrating hemoglobin encapsulation in vesicles, comprising poly(L-lysine)-*b*-poly(L-phenylalanine), in aqueous solutions [98]. This encapsulation offered a protective environment in plasma to improve the stability of these oxygen carrier molecules, which are crucial components of blood substituents [99]. This polymer self-assembles into vesicles in aqueous medium due to a α-helix forming tendency of a poly(L-phenylalanine) segment, with the overall structure being stabilized by hydrophobic π–π stacking interactions. The encapsulation of carbonylated hemoglobin was carried out at pH 5.8, which was followed by irradiation of vesicles to convert it into corresponding oxyhemoglobin. This conversion also suggests vesicle permeability to dissolved oxygen in aqueous solution. Such applications will be useful in designing artificial blood substituents with improved stability.

Recent advances in design concerns biological mimicry with a greater degree of tunability [100–102]. Lecommandoux *et al.* reported solution-phase behavior of a new amphiphilic dextran-block–poly(γ-benzyl L-glutamate) copolymer (see Figure 7.22), which revealed vesicle formation, with a membrane thickness of 20 nm, in aqueous medium [103]. Vesicle organization was primarily governed by a α-helical conformation, further stabilized by intramolecular hydrogen bonding favoring strong side-by-side interactions between peptide bond dipoles. These self-assembled nanostructures were projected as new generation delivery systems having a high affinity for glycoproteins [104].

7.5 Future Outlook

Peptide building blocks are clearly superior in design and are amenable to facile chemical modifications leading to improvement in stability. Such systems offer enormous advantage in delivery applications owing to tunability of design and a responsive nature towards externally applied stimuli and their biocompatible constituents. Site-specific targeting of these soft structures will have to be addressed in order to broaden further the scope of such delivery vehicles for real-life medicinal applications.

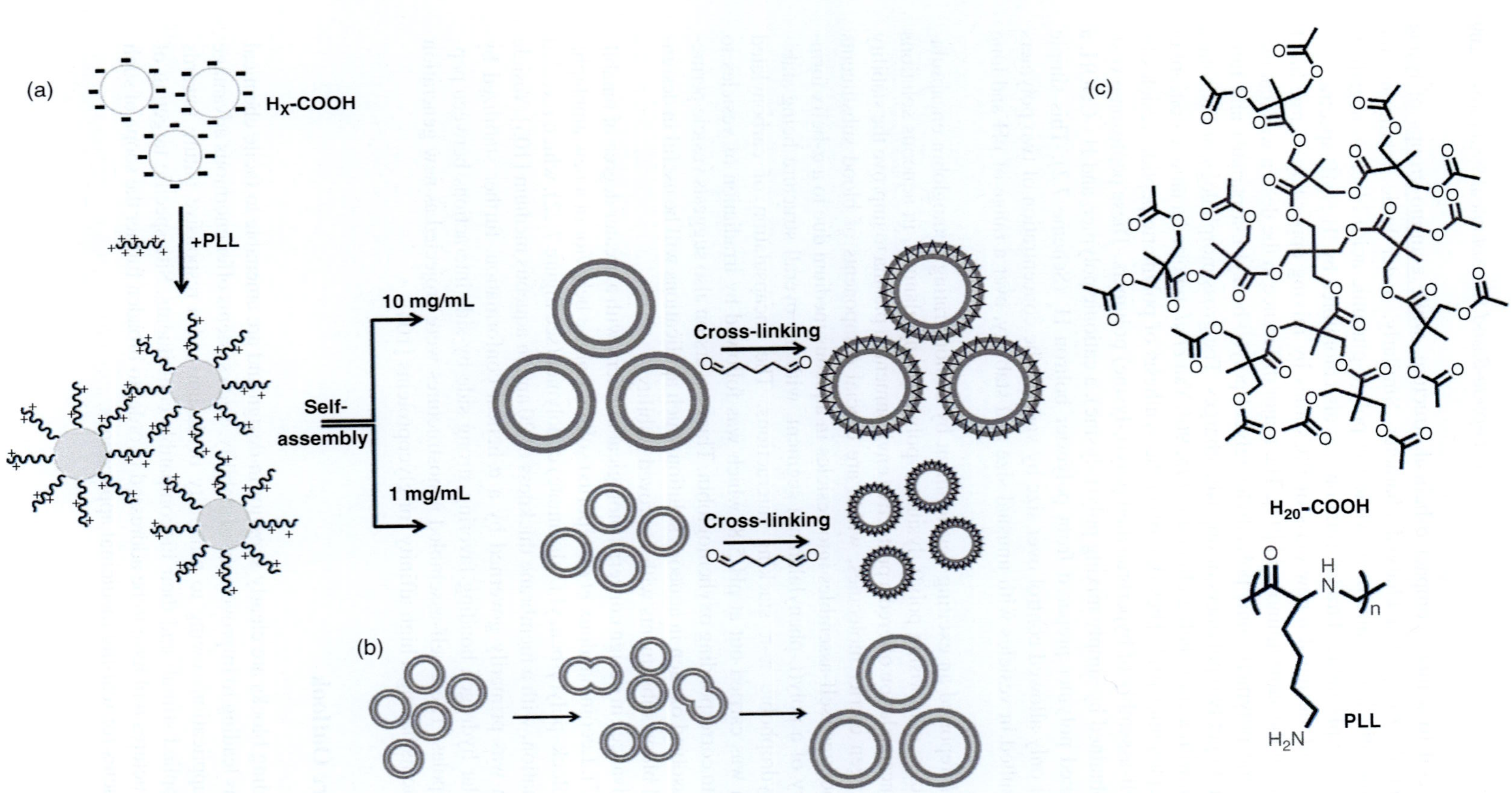

Scheme 7.2 Design strategy for complex peptosomes: (a) complex self-assembly process of H_x–COOH and PLL at different polymer concentrations; (b) formation of giant CPs by the fusion of small CPs; and (c) molecular structures of PLL and H_{20}–COOH. Adapted with permission from reference [97]. Copyright (2011) American Chemical Society

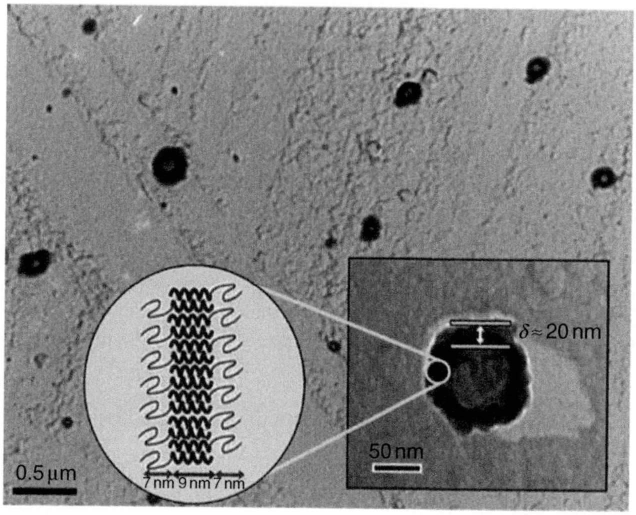

Figure 7.22 Electron micrograph of a dried dispersion of dextran-block-PBLG copolymer and schematic representation of the membrane of a vesicle-like structure. Reproduced with permission from reference [103]. Copyright (2009) John Wiley & Sons, Ltd

References

[1] Matson, J.B., Zha, R.H., Stupp, S.I. (2011) Peptide self-assembly for crafting functional biological materials. Curr. Opin. Solid State Mater. Sci. 15: 225–235.
[2] Cavalli, S., Albericio, F., Kros, A. (2010) Amphiphilic peptides and their cross-disciplinary role as building blocks for nanoscience. Chem. Soc. Rev. 39: 241–263.
[3] Hamley, I.W. (2011) Self-assembly of amphiphilic peptides. Soft Matter 7: 4122–4138.
[4] Kushner, A.M., Guan, Z. (2011) Modular design in natural and biomimetic soft materials. Angew. Chem., Int. Ed. 50: 9026–9057.
[5] Deska, J., Kazmaier, U. (2008) Peptide backbone modifications. Curr. Org. Chem. 12: 355–385.
[6] Li, X. (2011) Click to join peptides/proteins together. Chem. Asian J. 6: 2606–2616.
[7] Ackbarow, T., Buehler, M.J. (2009) Alpha-helical protein domains unify strength and robustness through hierarchical nanostructures. Nanotechnology 20: 075103.
[8] Smeenk, J.M., Otten, M.B.J., Thies, J., Tirell, D.A., Stunnenberg, H.G., Van Hest, J.C.M. (2005) Controlled assembly of macromolecular β-sheet fibrils. Angew. Chem., Int. Ed. 44: 1968–1971.
[9] Sheu, S.-Y., Yang, D.-Y., Selzle, H.L., Schlag, E.W. (2003) Energetics of hydrogen bonds in peptides. Proc. Natl Acad. Sci. USA 100: 12683–12687.
[10] Gazit, E. (2002) Mechanistic studies of the process of amyloid fibrils formation by the use of peptide fragments and analogues: Implications for the design of fibrillization inhibitors. Curr. Med. Chem. 9: 1725–1735.
[11] Reches, M., Gazit, E. (2003) Casting metal nanowires within discrete self-assembled peptide nanotubes. Science 300: 625–627.
[12] Reches, M., Gazit, E. (2004) Formation of closed-cage nanostructures by self-assembly of aromatic dipeptides. Nano Lett. 4: 581–585.
[13] Abramovich, L.A., Gazit, E. (2008) Controlled patterning of peptide nanotubes and nanospheres using inkjet printing technology. J. Pept. Sci. 14: 217–223.
[14] Amdursky, N., Molotskii, M., Gazit, E., Rosenman, G. (2009) Self-assembled bioinspired quantum dots: Optical properties. Appl. Phys. Lett. 94: 261907 (1–3).

[15] Huang, R., Su, R., Qi, W., Zhao, J., He, Z., (2011) Solvent and surface controlled self-assembly of diphenylalanine peptide: From microtubes to nanofibers. Soft Matter 7: 6418–6421.
[16] Huang, R., Su, R., Qi, W., Zhao, J., He, Z., (2011) Hierarchical, interface-induced self-assembly of diphenylalanine: Formation of peptide nanofibers and microvesicles. Nanotechnology 22: 245609 (1–7).
[17] Sun, B.Q., Sirringhaus, H. (2006) Surface tension and fluid flow driven self-assembly of ordered ZnO nanorod films for high-performance field effect transistors. J. Am. Chem. Soc. 128: 16231–16237.
[18] Yan, X., He, Q., Wang, K., Duan, L., Cui, Y., Li, J. (2007) Transition of cationic dipeptide nanotubes into vesicles and oligonucleotide delivery. Angew. Chem., Int. Ed. 46: 2431–2434.
[19] Maity, S., Jana, P., Maity, S.K., Haldar, D. (2011) Fabrication of hollow self-assembled peptide microvesicles and transition from sphere-to-rod structure. Langmuir 27: 3835–3841.
[20] Ke, D., Zhan, C., Li, A.D.Q., Yao, J. (2011) Morphological transformation between nanofibers and vesicles in a controllable bipyridine-tripeptide self-assembly. Angew. Chem., Int. Ed. 50: 3715–3719.
[21] Ghosh, S., Singh, S.K., Verma, S. (2007) Self-assembly and potassium ion triggered disruption of peptide-based soft structures. Chem. Commun. 2296–2298.
[22] Ghosh, S., Reches, M., Gazit, E., Verma, S. (2007) Bioinspired design of nanocages by self-assembling triskelion peptide elements. Angew. Chem., Int. Ed. 46: 2002–2004.
[23] Fotin, A., Cheng, Y., Sliz, P., Grigorieff, N., Harrison, S.C., Kirchhausen, T., Walz, T. (2004) Molecular model for a complete clathrin lattice from electron cryomicroscopy. Nature 432: 573–579.
[24] Ferguson, M.L., Prasad, K., Sackett, D.L., Boukari, H., Lafer, E.M., Nossal, R. (2006) Conformation of a clathrin triskelion in solution. Biochemistry 45: 5916–5922.
[25] Barman, A.K., Verma, S. (2010) Sunlight mediated disruption of peptide-based soft structures decorated with gold nanoparticles. Chem. Commun. 46: 6992–6994.
[26] Kang, B., Mackey, M.A., El-Sayed, M.A. (2010) Nuclear targeting of gold nanoparticles in cancer cells induces DNA damage, causing cytokinesis arrest and apoptosis. J. Am. Chem. Soc. 132: 1517–1519.
[27] Park, J.-H., Von Maltzahn, G., Xu, M.J., Fogal, V., Kotamraju, V.R., Ruoslahti, E., Bhatia, S.N., Sailor, M.J. (2010) Cooperative nanomaterial system to sensitize, target, and treat tumors. Proc. Natl Acad. Sci. USA 107: 981–986.
[28] Cole, J.R., Mirin, N.A., Knight, M.W., Goodrich, G.P., Halas, N.J. (2009) Photothermal efficiencies of nanoshells and nanorods for clinical therapeutic applications. J. Phys. Chem. C 113: 12090–12094.
[29] Sperling, R.A., Rivera Gil, P., Zhang, F., Zanella, M., Parak, W.J. (2008) Biological applications of gold nanoparticles. Chem. Soc. Rev. 37: 1896–1908.
[30] Ghosh, S., Verma, S. (2008) Templated growth of hybrid structures at the peptide–peptide interface. Chem.-Eur. J. 14: 1415–1419.
[31] Joshi, K.B., Verma, S. (2008) Ditryptophan conjugation triggers conversion of biotin fibers into soft spherical structures. Angew. Chem., Int. Ed. 47: 2860–2863.
[32] Vauthey, S., Santoso, S., Gong, H., Watson, N., Zhang, S. (2002) Molecular self-assembly of surfactant-like peptides to form nanotubes and nanovesicles. Proc. Natl Acad. Sci. USA 99: 5355–5360.
[33] Santoso, S., Hwang, W., Hartman, H., Zhang, S. (2002) Self-assembly of surfactant-like peptides with variable glycine tails to form nanotubes and nanovesicles. Nano Lett. 2: 687–691.
[34] Hartgerink, J.D., Beniash, E., Stupp, S.I. (2001) Self-assembly and mineralization of peptide–amphiphile nanofibers. Science 294: 1684–1688.
[35] Von Maltzahn, G., Vauthey, S., Santoso, S., Zhang, S. (2003) Positively charged surfactant-like peptides self-assemble into nanostructures. Langmuir 19: 4332–4337.
[36] White, L.R., Israelachvili, J.N., Ninham, B.W. (1976) Dispersion interaction of crossed mica cylinders: a reanalysis of the Israelachvili–Tabor experiments. J. Chem. Soc. Faraday Trans. 72: 1525–1568.
[37] Van Hell, A.J., Fretz, M.M., Crommelin, D.J.A., Hennink, W.E., Mastrobattista, E. (2010) Peptide nanocarriers for intracellular delivery of photosensitizers. J. Cont. Rel. 141: 347–353.

[38] Van Hell, A.J., Costa, C.I.C.A., Flesch, F.M., Sutter, M., Jiskoot, W., Crommelin, D.J.A., Hennink, W.E., Mastrobattista, E. (2007) Self-assembly of recombinant amphiphilic oligopeptides into vesicles. Biomacromolecules 8: 2753–2761.
[39] Van Hell, A.J., Klymchenko, A., Burgers, P.P., Moret, E.E., Jiskoot, W., Hennink, W.E., Crommelin, D.J.A., Mastrobattista, E. (2010) Conformation and intermolecular interactions of SA2 peptides self-assembled into vesicles. J. Phys. Chem. B 114: 11046–11052.
[40] Mishra, A., Panda, J.J., Basu, A., Chauhan, V.S. (2008) Nanovesicles based on self-assembly of conformationally constrained aromatic residue containing amphiphilic dipeptides. Langmuir 24: 4571–4576.
[41] Pathak, S., Singh Chauhan, V. (2011) Rationale-based, *de novo* design of dehydrophenylalanine-containing antibiotic peptides and systematic modification in sequence for enhanced potency. Antimicrob. Agents Chemother. 55: 2178–2188.
[42] Koley, P., Pramanik, A. (2011) Nanostructures from single amino acid-based molecules: Stability, fibrillation, encapsulation, and fabrication of silver nanoparticles. Adv. Funct. Mat. 21: 4126–4136.
[43] Naskar, J., Roy, S., Joardar, A., Das, S., Banerjee, A. (2011) Self-assembling dipeptide-based nontoxic vesicles as carriers for drugs and other biologically important molecules. Org. Biomol. Chem. 9: 6610–6615.
[44] Chiu, H.-C., Lin, Y.-W., Huang, Y.-F., Chuang, C.-K., Chern, C.-S. (2008) Polymer vesicles containing small vesicles within interior aqueous compartments and pH-responsive transmembrane channels. Angew. Chem., Int. Ed. 47: 1875–1878.
[45] Morimoto, B.H., Koshland Jr, D.E. (1994) Conditional activation of cAMP signal transduction by protein kinase C. The effect of phorbol esters on adenylyl cyclase in permeabilized and intact cells. J. Biol. Chem. 269: 4065–4069.
[46] Hu, W., Onuma, T., Birukawa, N., Abe, M., Ito, E., Chen, Z., Urano, A. (2008) Change of morphology and cytoskeletal protein gene expression during dibutyryl cAMP-induced differentiation in C6 glioma cells. Cell. Mol. Neurobiol. 28: 519–528.
[47] Kurino, M., Fukunaga, K., Ushio, Y., Miyamoto, E. (1996) Cyclic AMP inhibits activation of mitogen-activated protein kinase and cell proliferation in response to growth factors in cultured rat cortical astrocytes. J. Neurochem. 67: 2246–2255.
[48] Haghighat, N., McCandless, D.W., Geraminegad, P. (2000) Responses in primary astrocytes and C6-glioma cells to ammonium chloride and dibutyryl cyclic-AMP. Neurochem. Res. 25: 277–284.
[49] Terenzi, S., Biała, E., Nguyen-Trung, N.Q., Strazewski, P. (2003) Amphiphilic 3'-peptidyl–RNA conjugates. Angew. Chem., Int. Ed. 42: 2909–2912.
[50] Percot, A., Zhu, X.X., Lafleur, M. (1999) Design and characterization of anchoring amphiphilic peptides and their interactions with lipid vesicles. Biopolymers, 50: 647–655.
[51] Smith, T.C. (2001) Obcells as proto-organisms: membrane heredity, lithophosphorylation, and the origins of the genetic code, the first cells, and photosynthesis. J. Mol. Evol. 53: 555–595.
[52] Rubio, J., Alfonso, I., Burguete, M.I., Luis, S.V. (2011) Stimulus responsive self-assembly of gemini amphiphilic pseudopeptides. Soft Matter 7: 10737–10748.
[53] Rubio, J., Alfonso, I., Burguete, M.I., Luis, S.V. (2012) Interplay between hydrophilic and hydrophobic interactions in the self-assembly of a gemini amphiphilic pseudopeptide: From nano-spheres to hydrogels. Chem. Commun. 48: 2210–2212.
[54] Greenfield, M.A., Palmer, L.C., Vernizzi, G., De La Cruz, M.O., Stupp, S.I. (2009) Buckled membranes in mixed-valence ionic amphiphile vesicles. J. Am. Chem. Soc. 131: 12030–12031.
[55] Hsu, L., Cvetanovich, G.L., Stupp, S.I. (2008) Peptide amphiphile nanofibers with conjugated polydiacetylene backbones in their core. J. Am. Chem. Soc. 130: 3892–3899.
[56] Cui, H., Webber, M.J., Stupp, S.I. (2010) Self-assembly of peptide amphiphiles: From molecules to nanostructures to biomaterials. Biopolymers 94: 1–18.
[57] Löwik, D.W.P.M., Shklyarevskiy, I.O., Ruizendaal, L., Christianen, P.C.M., Maan, J.C., Van Hest, J.C.M. (2007) A highly ordered material from magnetically aligned peptide amphiphile nanofiber assemblies. Adv. Mater. 19: 1191–1195.
[58] Biesalski, M., Tu, R., Tirrell, M.V. (2005) Polymerized vesicles containing molecular recognition sites. Langmuir 21: 5663–5666.

[59] Van Den Heuvel, M., Löwik, D.W.P.M., Van Hest, J.C.M. (2010) Effect of the diacetylene position on the chromatic properties of polydiacetylenes from self-assembled peptide amphiphiles. Biomacromolecules 11: 1676–1683.

[60] Park, K.H., Lee, J.-S., Park, H., Oh, E.-H., Kim, J.-M. (2007) Vancomycin-induced morphological transformation of self-assembled amphiphilic diacetylene supramolecules. Chem. Commun. 410–412.

[61] Kahne, D., Leimkuhler, C., Lu, W., Walsh, C. (2005) Glycopeptide and lipoglycopeptide antibiotics. Chem. Rev. 105: 425–448.

[62] Diegelmann, S.R., Hartman, N., Markovic, N., Tovar, J.D. (2012) Synthesis and alignment of discrete polydiacetylene–peptide nanostructures J. Am. Chem. Soc. 134: 2028–2031.

[63] De Bruyn Ouboter, D., Schuster, T.B., Mantion, A., Meier, W. (2011) Hierarchical organization of purely peptidic amphiphiles into peptide beads. J. Phys. Chem. C 115: 14583–14590.

[64] Kimura, S., Kim, D.-H., Sugiyama, J., Imanishi, Y. (1999) Vesicular self-assembly of a helical peptide in water. Langmuir 15: 4461–4463.

[65] Duclohier, H. (2010) Antimicrobial peptides and peptaibols, substitutes for conventional antibiotics. Curr. Pharm. Des. 16: 3212–3223.

[66] Demizu, Y., Doi, M., Sato, Y., Tanaka, M., Okuda, H., Kurihara, M. (2011) Screw-sense control of helical oligopeptides containing equal amounts of L- and D-amino acids. Chem.-Eur. J. 17: 11107–11109.

[67] Shemesh, D., Domcke, W. (2011) Effect of the chirality of residues and γ-turns on the electronic excitation spectra, excited-state reaction paths and conical intersections of capped phenylalanine-alanine dipeptides. Chem. Phys. Chem. 12: 1833–1840.

[68] Fuks, G., Mayap Talom, R., Gauffre, F. (2011) Biohybrid block copolymers: Towards functional micelles and vesicles. Chem. Soc. Rev. 40: 2475–2493.

[69] Chécot, F., Lecommandoux, S., Gnanou, Y., Klok, H.-A. (2002) Water-soluble stimuli-responsive vesicles from peptide-based diblock copolymers. Angew. Chem., Int. Ed. 41: 1339–1343.

[70] Chécot, F., Lecommandoux, S., Klok, H.-A., Gnanou, Y. (2003) From supramolecular polymersomes to stimuli-responsive nano-capsules based on poly(diene-*b*-peptide) diblock copolymers. Eur. Phys. J. E 10: 25–35.

[71] Kukula, H., Schlaad, H., Antonietti, M., Förster, S. (2002) The formation of polymer vesicles or 'peptosomes' by polybutadiene-block-poly(L-glutamate)s in dilute aqueous solution. J. Am. Chem. Soc. 124: 1658–1663.

[72] Bertrand, N., Fleischer, J.G., Wasan, K.M., Leroux, J.-C. (2009) Pharmacokinetics and biodistribution of N-isopropylacrylamide copolymers for the design of pH-sensitive liposomes Biomaterials 30: 2598–2605.

[73] Erhardt, R., Böker, A., Zettl, H., Kaya, H., Pyckhout-Hintzen, W., Krausch, G., Abetz, V., Müller, A.H.E. (2001) Janus micelles. Macromolecules 34: 1069–1075.

[74] Rheingans, O., Hugenberg, N., Harris, J.R., Fischer, K., Maskos, M. (2000) Nanoparticles built of cross-linked heterotelechelic, amphiphilic poly(dimethylsiloxane)-*b*-poly(ethylene oxide) diblock copolymers. Macromolecules 33: 4780–4790.

[75] Bellomo, E.G., Wyrsta, M.D., Pakstis, L., Pochan, D.J., Deming, T.J. (2004) Stimuli-responsive polypeptide vesicles by conformation-specific assembly. Nature Mater. 3: 244–248.

[76] Sun, V.Z., Li, Z., Deming, T.J., Kamei, D.T. (2011) Intracellular fates of cell-penetrating block copolypeptide vesicles. Biomacromolecules 12: 10–13.

[77] Immordino, M.L., Dosio, F., Cattel, L. (2006) Stealth liposomes: Review of the basic science, rationale, and clinical applications, existing and potential. Int. J. Nanomed. 1: 297–315.

[78] Guo S., Huang L. (2011) Nanoparticles escaping RES and endosome: Challenges for siRNA. Delivery for cancer therapy. J. Nanomater. 742895: 1–12.

[79] Dancey, G., Violet, J., Malaroda, A., Green, A.J., Sharma, S.K., Francis, R., Othman, S., Parker, S., Buscombe, J., Griffin, N., Chan, P.-S., Malhotra, A., Woodward, N., Ramsay, A., Ross, P., Lister, T.A., Amlot, P., Begent, R., McNamara, C. (2009) A phase I clinical trial of CHT-25: A ^{131}I-labeled chimeric anti-CD25 antibody showing efficacy in patients with refractory lymphoma. Clin. Cancer Res. 15: 7701–7710.

[80] Akira, M., Shinae, K., Ryo, Y. (2009) Near-infrared fluorescence tumor imaging using nanocarrier composed of poly(L-lactic acid)-block–poly(sarcosine) amphiphilic polydepsipeptide. Biomaterials 30: 5156–5160.

[81] Tanisaka, H., Kizaka-Kondoh, S., Makino, A., Tanaka, S., Hiraoka, M., Kimura, S. (2008) Near-infrared fluorescent labeled peptosome for application to cancer imaging. Bioconj. Chem. 19: 109–117.

[82] Vladimir, T. (2011) Tumor delivery of macromolecular drugs based on the EPR effect. Adv. Drug Del. Rev. 63: 131–135.

[83] Ntziachristos, V., Bremer, C., Weissleder, R. (2003) Fluorescence imaging with near-infrared light: New technological advances that enable *in vivo* molecular imaging. Eur. Radiol. 13: 195–208.

[84] Hsu, A.R., Hou, L.C., Veeravagu, A., Greve, J.M., Vogel, H., Tse, V., Chen, X. (2006) *In vivo* near-infrared fluorescence imaging of integrin αvβ3 in an orthotopic glioblastoma model. Mol. Imaging Biol. 8: 315–323.

[85] Christian, N.A., Milone, M.C., Ranka, S.S., Li, G., Frail, P.R., Davis, K.P., Bates, F.S., Therien, M.J., Ghoroghchian, P.P., June, C.H., Hammer, D.A. (2007) Tat-functionalized near-infrared emissive polymersomes for dendritic cell labeling. Bioconjug. Chem. 18: 31–40.

[86] Holowka, E.P., Pochan, D.J., Deming, T.J. (2005) Charged polypeptide vesicles with controllable diameter. J. Am. Chem. Soc. 127: 12423–12428.

[87] Holowka, E.P., Sun, V.Z., Kamei, D.T., Deming, T.J. (2007) Polyarginine segments in block copolypeptides drive both vesicular assembly and intracellular delivery. Nature Mater. 6: 52–57.

[88] Rothbard, J.B., Jessop, T.C., Wender, P.A. (2005) Adaptive translocation: The role of hydrogen bonding and membrane potential in the uptake of guanidinium-rich transporters into cells. Adv. Drug Deliv. Rev. 57: 495–504.

[89] Futaki, S. (2005) Membrane-permeable arginine-rich peptides and the translocation mechanisms. Adv. Drug Deliv. Rev. 57: 547–558.

[90] Brooks, H., Lebleu, B., Vivès, E. (2005) Tat peptide-mediated cellular delivery: Back to basics. Adv. Drug Deliv. Rev. 57: 559–577.

[91] Wadia, J.S., Dowdy, S.F. (2005) Transmembrane delivery of protein and peptide drugs by TAT-mediated transduction in the treatment of cancer. Adv. Drug Deliv. Rev. 57: 579–596.

[92] Rodríguez-Hernández, J., Lecommandoux, S. (2005) Reversible inside-out micellization of pH-responsive and water-soluble vesicles based on polypeptide diblock copolymers. J. Am. Chem. Soc. 127: 2026–2027.

[93] Ranquin, A., Versées, W., Meier, W., Steyaert, J., Van Gelder, P. (2005) Therapeutic nanoreactors: Combining chemistry and biology in a novel triblock copolymer drug delivery system. Nano Lett. 5: 2220–2224.

[94] Koide, A., Kishimura, A., Osada, K., Jang, W.-D., Yamasaki, Y., Kataoka, K. (2006) Semipermeable polymer vesicle (PICsome) self-assembled in aqueous medium from a pair of oppositely charged block copolymers: Physiologically stable micro-/nanocontainers of water-soluble macromolecules. J. Am. Chem. Soc. 128: 5988–5989.

[95] Daniel, W., Stiriba, S.-E., Holger, F. (2010) Hyperbranched polyglycerols: From the controlled synthesis of biocompatible polyether polyols to multipurpose applications. Acc. Chem. Res. 43: 129–141.

[96] Kataoka, K., Harada, A., Nagasaki, Y. (2001) Block copolymer micelles for drug delivery: Design, characterization and biological significance. Adv. Drug Deliv. Rev. 47: 113–131.

[97] Guo, B., Shi, Z., Yao, Y., Zhou, Y., Yan, D. (2009) Facile preparation of novel peptosomes through complex self-assembly of hyperbranched polyester and polypeptide. Langmuir 25: 6622–6626.

[98] Sun, J., Huang, Y., Shi, Q., Chen, X., Jing, X. (2009) Oxygen carrier based on hemoglobin poly(L-lysine)-block–poly(L- phenylalanine) vesicles. Langmuir 25: 13726–13729.

[99] Chang, T.M.S. (2006) Blood substitutes based on nanobiotechnology. Trends Biotechnol. 24: 372.

[100] Bromley, E.H.C., Channon, K., Moutevelis, E., Woolfson, D.N. (2008) Peptide and protein building blocks for synthetic biology: From programming biomolecules to self-organized biomolecular systems. ACS Chem. Biol. 3: 38–50.

[101] Zhang, S. (2003) Fabrication of novel biomaterials through molecular self-assembly. Nat. Biotechnol. 21: 1171–1178.

[102] Hamley, I.W. (2003) Nanotechnology with soft materials. Angew. Chem., Int. Ed. 42: 1692–1712.

[103] Schatz, C., Louguet, S., Meins, J.-F.L., Lecommandoux, S. (2009) Polysaccharide-block–polypeptide copolymer vesicles: Towards synthetic viral capsids. Angew. Chem., Int. Ed. 48: 2572–2575.

[104] Ahmed, F., Photos, P.J., Discher, D.E. (2006) Polymersomes as viral capsid mimics. Drug Dev. Res. 67: 4–14.

Part III

Peptide Conjugates and Hybrid Materials

8

Peptide-Based Carbon Nanotube Dispersal Agents

Anton S. Klimenko and Gregg R. Dieckmann

Department of Chemistry and Alan G. MacDiarmid NanoTech Institute, The University of Texas at Dallas, USA

Acronyms

AA	Amino acid
Ac	Acetylated N-terminus of a peptide
CNTs	Carbon nanotubes
$CONH_2$	Amidated C-terminus of a peptide
MWNTs	Multiwalled CNTs
NTCs	Nontubular carbonaceous species
RCPs	Reversible cyclic peptides
SPF	SPX peptide with a Phe residue
SPW	SPX peptide with a Trp residue
SPX	Extended peptide with the AA sequence Ac–XVVVVVKK–$CONH_2$ (where X represents an aromatic AA)
SPY	SPX peptide with a Tyr residue
SWNTs	Single-walled CNTs

Peptide Materials: From Nanostructures to Applications, First Edition. Edited by Carlos Alemán, Alberto Bianco and Mariano Venanzi.

8.1 Introduction

This chapter highlights recent work in the use of peptide-based dispersal agents developed for the functionalization of carbon nanotubes (CNTs) for a broad variety of applications. Peptide dispersal agents of CNTs are generally categorized by the secondary structure they adopt; therefore, the chapter will be divided into micellar, helical, β-strand, cyclic, and amorphous peptides (Figure 8.1). Although there are numerous examples in the literature where naturally occurring 'globular' proteins have also been successfully utilized to functionalize CNTs (e.g., see references [1–15]), those studies will not be highlighted here since the focus of this chapter is on 'designed' peptide-based systems.

Over the last decade, CNTs have become a research focus for a variety of disciplines, ranging from chemistry and engineering to biology and medicine (see reviews [16–26] and references therein). Their unique properties stem from the carbon sp^2 hybridization and the bond strain exerted by the chirality and curvature of the nanotube (Figure 8.2). Their properties, though advantageous for their potential applications, make CNTs difficult to work with in aqueous and other environments. Following their synthesis, CNTs are heavily bundled. In order to access singly dispersed nanotubes, they must be dispersed and debundled, most commonly through sonication [27–34]. Covalent functionalization methods, such as carboxylation, can make CNTs more soluble and/or impart other useful functionalities [35–39]. The effects are normally modulated through the type and extent of functionalization as a function of time. These methods may result in an improved manipulation of CNTs, but they disrupt the conjugation within the system, thereby impacting the innate properties of the CNTs. However, noncovalent functionalization methods have been successfully used to coat the surface of CNTs while preserving the conjugation (Figure 8.3). These methods employ the use of dispersal agents, including surfactants [40–50], polymers [51–59], DNA molecules [60–69], proteins [1–15], and polypeptides [27–34, 70–73]. Noncovalent methods are not only designed to retain the natural properties of CNTs, but also provide a greater degree of control over the system.

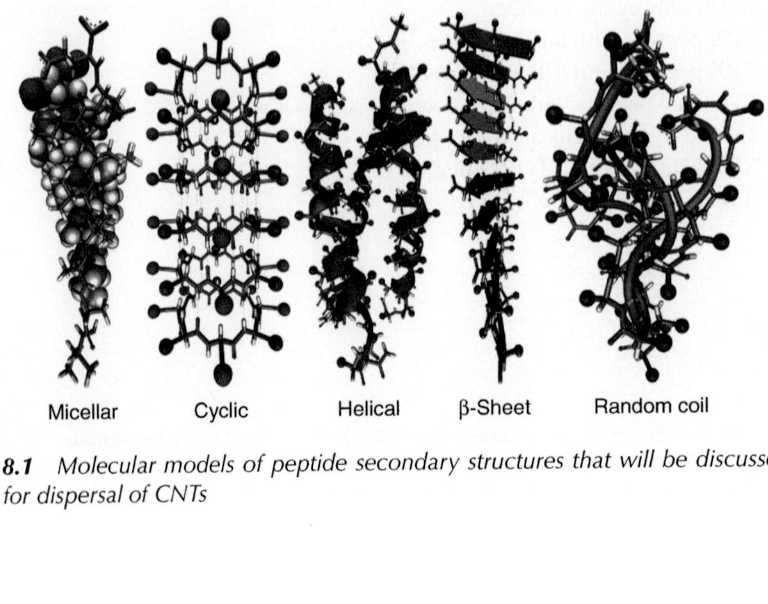

Figure 8.1 Molecular models of peptide secondary structures that will be discussed in this review for dispersal of CNTs

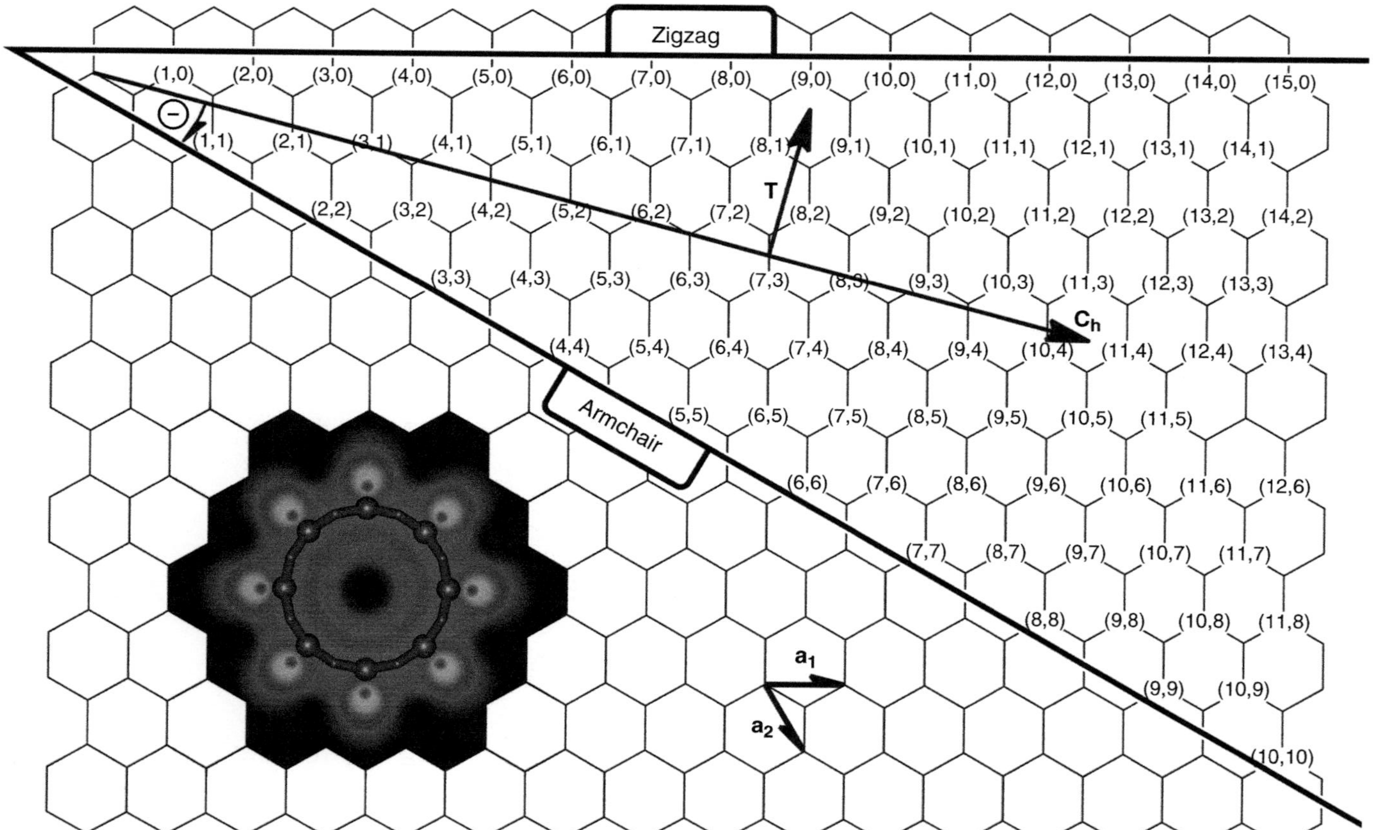

Figure 8.2 *A schematic diagram of a graphene sheet derived from an 'unrolled' CNT with the basis vectors labeled for identification of CNT chiral indices (n, m). The vector **T** represents the axis parallel to the length of the CNT. The vector* $\boldsymbol{C_h}(\overrightarrow{\boldsymbol{C_h}} = \boldsymbol{n}\overrightarrow{\boldsymbol{a_1}} + \boldsymbol{m}\overrightarrow{\boldsymbol{a_2}})$ *is perpendicular to **T** and dictates the CNT chirality, ranging from (n, 0) for zigzag CNTs to (n, m=n) for armchair CNTs. The CNT diameter* $(d = {}^{l}\!/_{\pi}\sqrt{n^2+nm+m^2}$, where $l = 2.48\,\text{Å})$ *is a function of the CNT chirality, while the effective diameter* (d_e) *is slightly greater, as it takes into account the electron density. The heat map (insert) represents an electron density diagram of an (8, 0) CNT at a single cross-section taken along the* $\boldsymbol{C_h}$ *vector (dark grey=highest electron density; light grey=lowest electron density)*

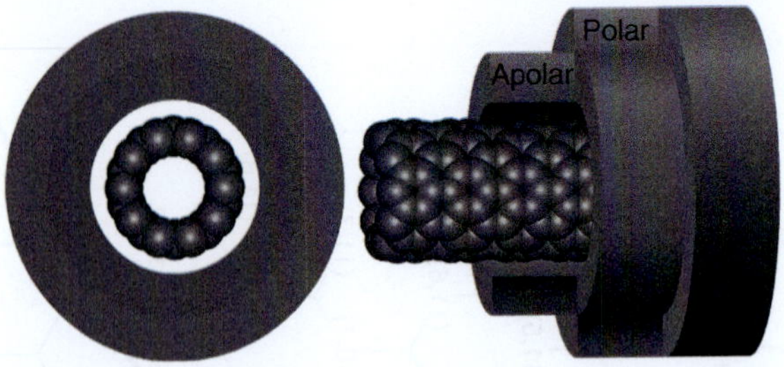

Figure 8.3 Schematic views of general noncovalent dispersal agent interactions with a CNT. In an aqueous environment, the dispersal agent rearranges to present its least polar segments to the CNT and its most polar segments to water

Polypeptides (referred to here as peptides) are polymers characterized by a sequence of amino acid (AA) monomers coupled together via amide bonds. Peptides are synthesized with full control over peptide sequence and length [74–76]; however, peptides can also be synthesized using step-growth polymerization, which has limited control over the resulting sequence, length, or potential branching [77, 78]. Classical peptides have a polydispersity index of 1 and range in length from 2 to approximately 100 residues. The definition of peptides has expanded considerably from a monodispersed, linear chain of amino acids to include a countless number of AA chains, branched and linear, with varying polydispersity indices.

The variability of peptides stems from the AAs from which they are composed, as well as the sequence of those AAs. An 'amino acid' incorporated into a synthetic peptide can be any molecule with a primary or secondary amine and a carboxylic acid functional group, which makes it possible to incorporate a wide variety of molecules into the peptide chain. The secondary structure of a peptide can take on many forms, the most common being random coil, helix, β-strand, extended (micellar), and cyclic (Figure 8.1). Examples can be found where all of these peptide secondary structures have been used in peptides to disperse CNTs (both single-walled CNTs (SWNTs) and multiwalled CNTs (MWNTs) [28, 79]) with varying degrees of success. All peptide dispersal agents function as surfactants in the sense that they function to lower the surface tension of water around the nanotubes. They have effective hydrophobic and hydrophilic surfaces that are used to interact with both the CNTs and water [28, 31–34, 70, 72, 73, 80–82]. The classification of these molecules is, therefore, based on the secondary structure of the respective peptides. This chapter will discuss each category in detail, with a focus placed on the root of their interactions with themselves and the CNTs, as well as the benefits of using each type.

8.2 α-Helical Surfactant Peptides

Helices are one of the first and best-studied protein secondary structures. Helical peptides draw their roots from biological proteins, where many forms of helices are common. Current designed helical peptides usually base their primary structure on the library of known peptides [83–87]. The secondary and quaternary structures, as well as the associated thermodynamic properties, have been well established. Small alterations to the sequence of these known peptides tailor them to suit intended functions [28, 31, 33, 34, 71, 87–94].

A typical α-helix, the most common helical form, consists of a seven-residue repeating sequence which completes approximately two turns of the helix, best demonstrated by a helical wheel diagram (Figure 8.4) [71, 87–93]. An amphiphilic helix presents two different faces, a more hydrophobic face (*a* and *d* positions) and a more hydrophilic face (*b*, *c*, *e*, *f* positions). This duplicity, in conjunction with AAs that increase helical propensity, drives

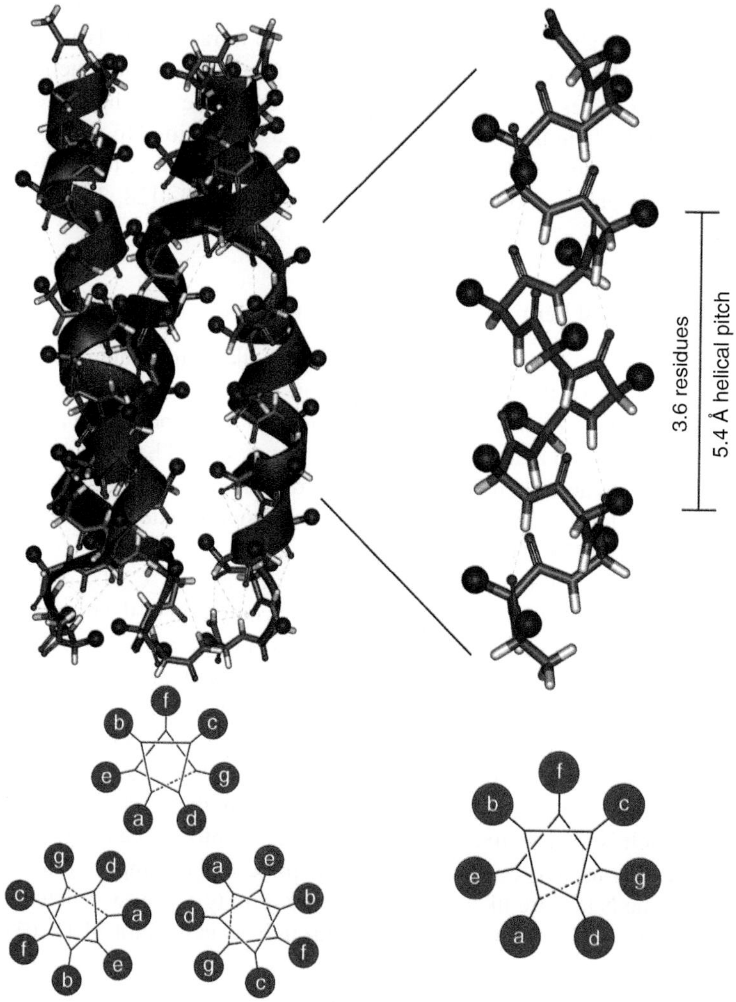

Figure 8.4 *(Top left) Molecular model of a trimeric α-helical supercoil. (Top right) Enhanced structural view of a segment in one of the supercoil's substituent α-helices. Below the models are helical wheel diagrams of the two structures summarizing apolar (red) and polar (blue) AA regions; the 'a' through 'g' labels indicate the AA positions in the heptad. The supercoil helical wheel diagram accents the positioning of the helices with apolar AAs facing the core of the supercoil (see color plate figure)*

the folding of these peptides. Helical peptides do not typically fold as a single helix in solution, but rather through association with another hydrophobic surface, most often another α-helix, decreasing the entropy of both [71, 87–89, 91–93, 95–102]. The structure is further stabilized by hydrogen bonding (H-bonding) between the AAs in the peptide backbone (residues i and $i+4$). Folding can also be induced by altering the solvent conditions in a way that decreases the H-bonding capacity of the solvent [87, 89, 91–93].

The tertiary structure of helices is commonly in the form of supercoiled n-mers, n ranging from 2 to 4 (higher order aggregates are possible, though not common), which can have parallel or antiparallel directionality. In the process of supercoiling, the hydrophobic area is accommodated in a way that disrupts the intra-peptide H-bonding on the other, solvent-exposed side of the helix. The interaction between these helical peptides and the surface of a CNT is similar to supercoiling in the sense that the helix deforms to accommodate the CNT (Figure 8.5d), causing similar disruptions to the H-bonding of the hydrophilic face. However, CNTs cannot coil in return, making it necessary for the helices to distort to a greater extent with smaller diameter CNTs [28, 31, 33, 34, 71, 90, 97, 98, 102].

8.2.1 Model for Helical Peptide Dispersion of Nanotubes

In the absence of CNTs, a helical peptide solution is likely to contain an equilibrium between three isoforms of the peptide: random coil (monomer), helical monomer, and helical n-mer (Figure 8.5). However, the helical monomer is not present at a high concentration. The random coil and helical monomer populations should have the highest affinity for the CNT surface because their hydrophobic residues are more accessible. The helical n-mer will have the lowest affinity because the hydrophilic exterior screens potential hydrophobic interactions [28, 31, 33, 34, 71].

The initial conditions of the dispersion consist of bundled CNTs and nontubular carbonaceous species (NTCs) surrounded by a high concentration of the dispersal agent. In our current model for the process of CNT dispersal by peptides, as the shearing forces created by sonication creates exposed singular CNT regions (likely at the bundle ends where 'fraying' can occur), the dispersal agent moves in to coat the exposed CNT surfaces, inhibiting re-aggregation. As this process continues, eventually the CNT is excised from the bundle and becomes extensively coated by peptide and monodispersed. During this process, the peptide that binds to the CNT surface rearranges to form the most stable conformation it can achieve, which is the α-helix. The overall peptide affinity to the CNT, as well as the inter-peptide interactions, serves to stabilize the peptide–CNT complex [28, 31, 33, 34, 70–72, 90, 103]. The NTCs are likely to have a lower affinity for most structured peptides and are subsequently removed from the solution through centrifugation [104, 105].

8.2.2 Peptide–Nanotube Interactions

For helical peptides, there are three types of interactions important in dispersing CNTs: intra-peptide interactions, peptide–peptide interactions, and peptide–nanotube hydrophobic interactions. Intra-peptide interactions are the primary affecters of helical structure. Inter-peptide interactions also impact monomer folding and are important in the assembly of the peptides and in their spatial arrangement. Finally, peptide–nanotube interactions

Figure 8.5 Molecular model of a random coil (a), α-helical monomer (b), supercoil (c) and α-helical peptide-coated CNT (d). All of these species may be found in equilibrium with one another in a CNT dispersion

determine the peptide's ability to move and conform around the nanotube surface, as well as impact monomer folding.

Intra-peptide interactions are important in determining the helical propensity of the peptide. By designing peptides to optimize these interactions, a higher degree of control over the helical propensity of the dispersal agent may be attained [71, 88, 90, 95–102]. This in turn expands the concentration range under which the dispersions may be successful. Highly helical peptides are not always great dispersal agents, since a folded monomer is likely to self-associate, yet the monomer is more effective at CNT dispersal than the aggregate. Peptides with a lower helical propensity, while still able to disperse nanotubes, may be less likely to attain an organized folded structure when bound to the nanotube surface [28, 31, 33, 34, 71, 90, 97, 98, 102]. The challenge is to fine-tune intra-peptide interactions to suit the desired function.

The interaction of peptides with one another in the presence of CNTs plays an important role in CNT dispersal. Inter-peptide interactions control the assembly of helical peptides and

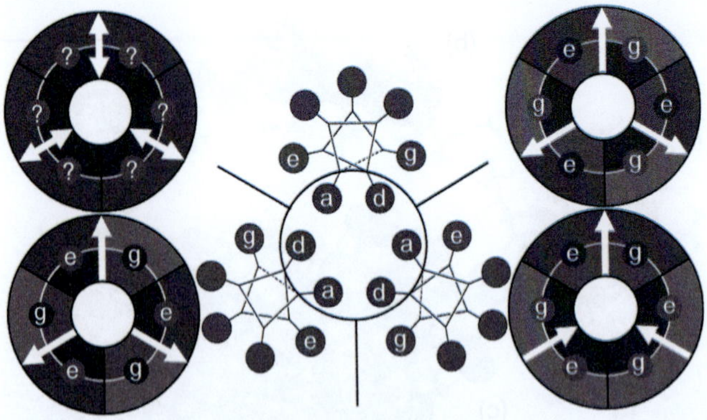

Figure 8.6 *An illustration of how AA charge in the adjacent e and g positions impacts the orientation of α-helices in a supercoil or while coiled about a CNT. In the case of uniform charges in the e/g positions (top left), the helices are oriented in a random fashion. In the case of opposite charges in the e/g positions (bottom left), the helices are driven to a parallel arrangement. Uneven distributions of anions and cations designed to form complimentary pairs can be driven to form longitudinally aligned tertiary structures in either parallel or antiparallel conformations (top and bottom right, respectively). This discussion does not address the important role that the a and d hydrophobic positions can play in peptide–peptide orientations and the peptide association state*

their arrangement both in the supercoil and on the CNT surface. The primary forces involved in inter-peptide helical interactions are hydrophobic, H-bonding, and charge–charge interactions. The *e* and *g* positions of the helix are most involved in these interactions, with the *b* and *c* positions involved to a lesser degree (Figure 8.6) [88]. Charged residues in these locations help dictate whether the helix forms a parallel or antiparallel arrangement, as well as influencing the association state of the peptide. Peptides having the same charge in both *e* and *g* positions experience significant repulsion, thereby decreasing the association state. The consequence for nanotube dispersal is a decrease in the observed peptide–CNT binding affinity, thereby accenting the effects of π–π and/or hydrophobic interactions with the nanotube. Conversely, oppositely charged residues at *e* and *g* positions lead to a higher propensity for self-association, which diminishes the specificity of peptide–CNT hydrophobic and π–π interactions. The observed peptide aggregation number depends on the size of these charged groups and the bulkiness of the hydrophobic core [88, 90, 97, 98, 102].

Of the three potential interactions, without which the peptide would not disperse CNTs, those between the peptide and the nanotube are the most important. These interactions can be broken down into three components: aliphatic interactions, aromatic interactions, and secondary interactions from the *e* and *g* side chains (the convention used here is that the *a*/*d* hydrophobic face of the helical peptide, the one that makes up the bundle core, is also used to interact with the CNTs). Though not all of them are essential to achieving a stable dispersion, each has an impact on the dispersion quality, as well as the structure of the peptide–CNT complex [70–72, 90, 97, 98, 102].

Aliphatic chains that compose the hydrophobic surface of the helical peptide include both alkyl AAs in the *a* and *d* positions, as well as alkyl linkers of the hydrophilic AAs in the *e* and *g* positions. They contribute an oily layer between the peptide and the nanotube

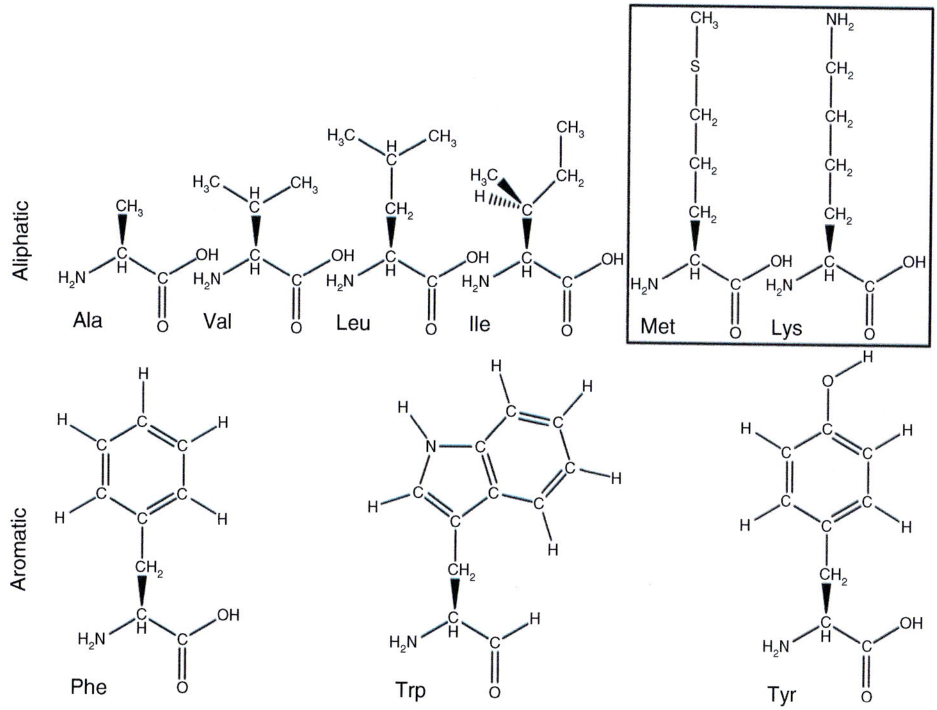

Figure 8.7 *Examples of side chains of naturally occurring AAs that interact in either an aliphatic or aromatic fashion. The side chains in the box can interact in an aliphatic manner despite having polar or charged polar components*

surface (Figure 8.7) [27, 28, 31, 33, 34, 71, 90, 97, 98, 102, 103, 106]. This is important in allowing the peptide mobility on the nanotube surface. They also serve to decrease the void volume, the volume in the hydrophobic region not taken up by atoms in the complex, and is consequently filled with water molecules, thereby decreasing the overall energy of the system. Aliphatic groups allow for efficient packing about the CNT [70–72, 90, 97, 98, 102].

Aromatic residues in the helical core increase the peptide–nanotube affinity and the efficiency of the peptide as a dispersal agent (Figure 8.7). The interaction of the aromatic AA and CNT, which is proportional to the number of substituent aromatic rings, is much stronger than the interactions involving aliphatic groups [28, 31, 33, 34, 71, 73, 80–82, 90, 97, 98, 102, 103]. It is necessary to balance the overall affinity of the component AAs with the peptide's helical propensity. By overloading the hydrophobic core with high affinity AAs like Trp, it is possible to perturb helix and supercoil formation (Figure 8.7), in which case the peptide might bind to the CNT surface in a more random fashion. Conformational rearrangement to a helix can then occur on the CNT surface due to favorable inter-peptide interactions.

8.2.3 Peptide–Nanotube Complex Structure

The structure the peptide forms with the CNTs is likely to be consistent with the types of structures it forms in supercoils (Figure 8.4) [33, 34, 82, 90]. A peptide consisting of L-AAs forms a right-handed α-helix and a left-handed supercoil. The supercoil itself can be either antiparallel or parallel. The association state for these supercoils tends to be small, with few going beyond tetramers [82, 90]. When a CNT is incorporated into the hydrophobic core of a helical supercoil, the aggregate stability is enhanced through creation of an enlarged hydrophobic surface area, similar to increasing the *a*/*d* side-chain length which normally limits the aggregation state of α-helices. For CNTs that have a diameter comparable to the size of the α-helix, the CNT can accommodate *n* perfectly helical peptides around it. As the CNT diameter increases, each helix must distort to account for the additional surface area, until the CNT diameter is large enough to accommodate $n+1$ peptides (Figure 8.8a). These distortions can include: lateral expansion of the side chains, distortion of the helical backbone to a larger diameter and/or increase of the contact angle with the CNT. The resulting peptide–CNT complex should be most stable for CNTs with diameters that accommodate an integral number of peptides in such a way that the helices do not have to distort either via expansion to cover more surface area or through contraction to fit in more peptides [28, 31, 33, 34, 90]. As the helices become more distorted, the number of stabilizing

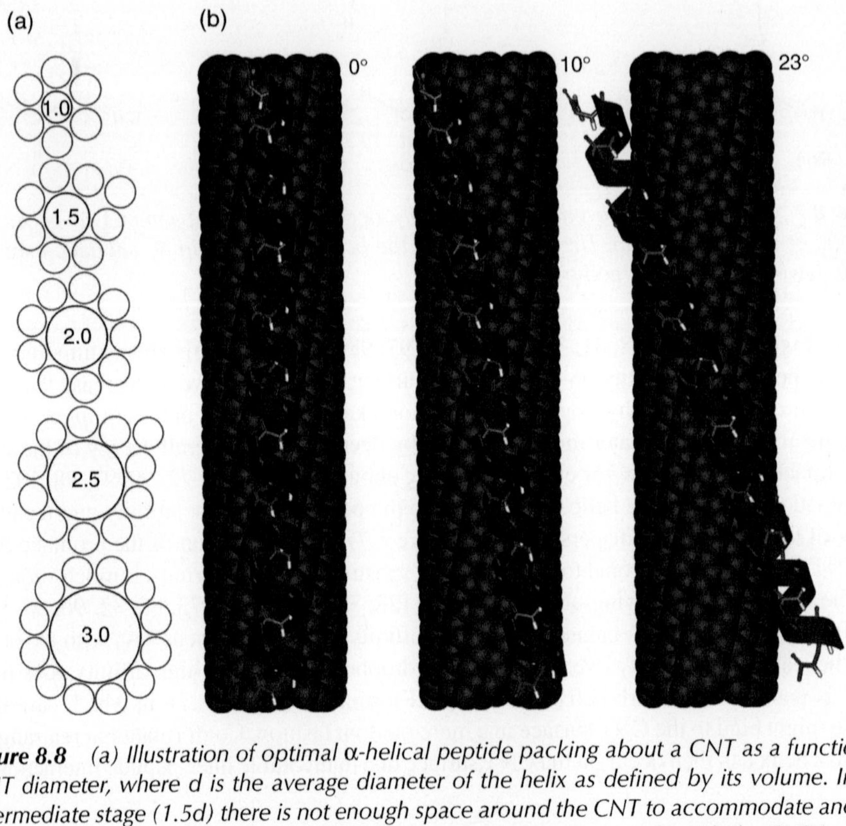

Figure 8.8 *(a) Illustration of optimal α-helical peptide packing about a CNT as a function of CNT diameter, where d is the average diameter of the helix as defined by its volume. In the intermediate stage (1.5d) there is not enough space around the CNT to accommodate another α-helix. The peptide is forced to deviate from its optimum contact angle with the CNT in order to cover a larger circumference; the contact angle is the deviation of the peptide from the CNT axis (b). The distance each set of helices covers along the CNT length is decreased and is compensated by additional helices (see color plate figure)*

intra-peptide backbone H-bonds decreases to accommodate the curvature of the CNT as well as the hydrophobic residues (Figure 8.8b). As a consequence the complex loses stability, which can be reflected by its binding affinity or the on/off rate of exchange. The stability of this complex also depends on the helical propensity of the peptide and the length of the hydrophobic appendages in the *a*, *d* and *e*, *g* positions. The bulkiness of these hydrophobic appendages also dictates how close the peptides can get to the CNTs. This in turn determines the number of helices that comfortably fit around the CNT and, therefore, the CNT diameters that result in the most stable complexes.

Factors that alter the helical propensity of the peptide have several effects on the peptide–CNT complex. Alterations that enhance the propensity in the *a* and *d* positions have positive effects on the peptide–CNT affinity and the overall stability of the dispersion. These changes include inter-peptide bridges, shorter side-chain length and any other alterations that minimize water access to the H-bonding in the backbone [82, 95, 96, 99–101]. Conversely, measures that decrease helical propensity in the *a* and *d* positions disrupt peptide–CNT interactions and creates a less stable dispersion. These measures include hydrophilic groups in the *a* and *d* positions and backbone positions that are *n*-methylated [82, 95, 96, 99–101]. Increasing the propensity in the *b*, *c*, and *f* positions rigidify the helix, decreasing its ability to distort to accommodate tubular structures. This results in the peptide favoring more planar carbon species. The denaturing of those regions usually has a minimal effect on the overall stability of the dispersion. Changes to the *e* and *g* positions have limited effects on the dispersion stability, although they can increase and decrease the helical propensity of the peptide [90].

In addition to the first layer of α-helices that are on the CNT, a secondary assembly of α-helices can potentially form around the first layer [90]. These helices are not driven by hydrophobic interactions, but rather charge–charge interactions involving primarily the *f* position but also the *b* and *c* positions of the primary layer helices. Depending on the charge distribution, they can form assemblies either parallel or antiparallel to the primary helices, or they can coil in a left or right handed fashion about the peptide–CNT complex. This superstructure serves to stabilize the underlying complexes and can be used to further functionalize the system, as has been shown elegantly in a recent paper by DeGrado *et al.* [90].

8.3 β-Strand Surfactant-Like Peptides

Of the four primary peptide secondary structures used for CNT dispersal, the β-strand is in some ways the most widely used and the least explicitly utilized class of peptide [70, 72, 107], and are rather used in conjunction with other peptide forms. This is largely due to the high self-association constant of amphiphilic β-strand peptides, resulting in a low population available for CNT binding and therefore low dispersal efficiency. Their underlying structural influences are similar to those of both extended and cyclic peptides, and therefore examples of them will be included in those upcoming sections.

In one set of studies, peptides comprised of a β-strand between two hydrophilic capping termini were designed that successfully dispersed CNTs (Figure 8.9) [70, 72]. The dispersibility of these peptides is proportional to the size of the β-strand region, as well as the length of the hydrophilic termini. With increasing termini length, however, the peptides

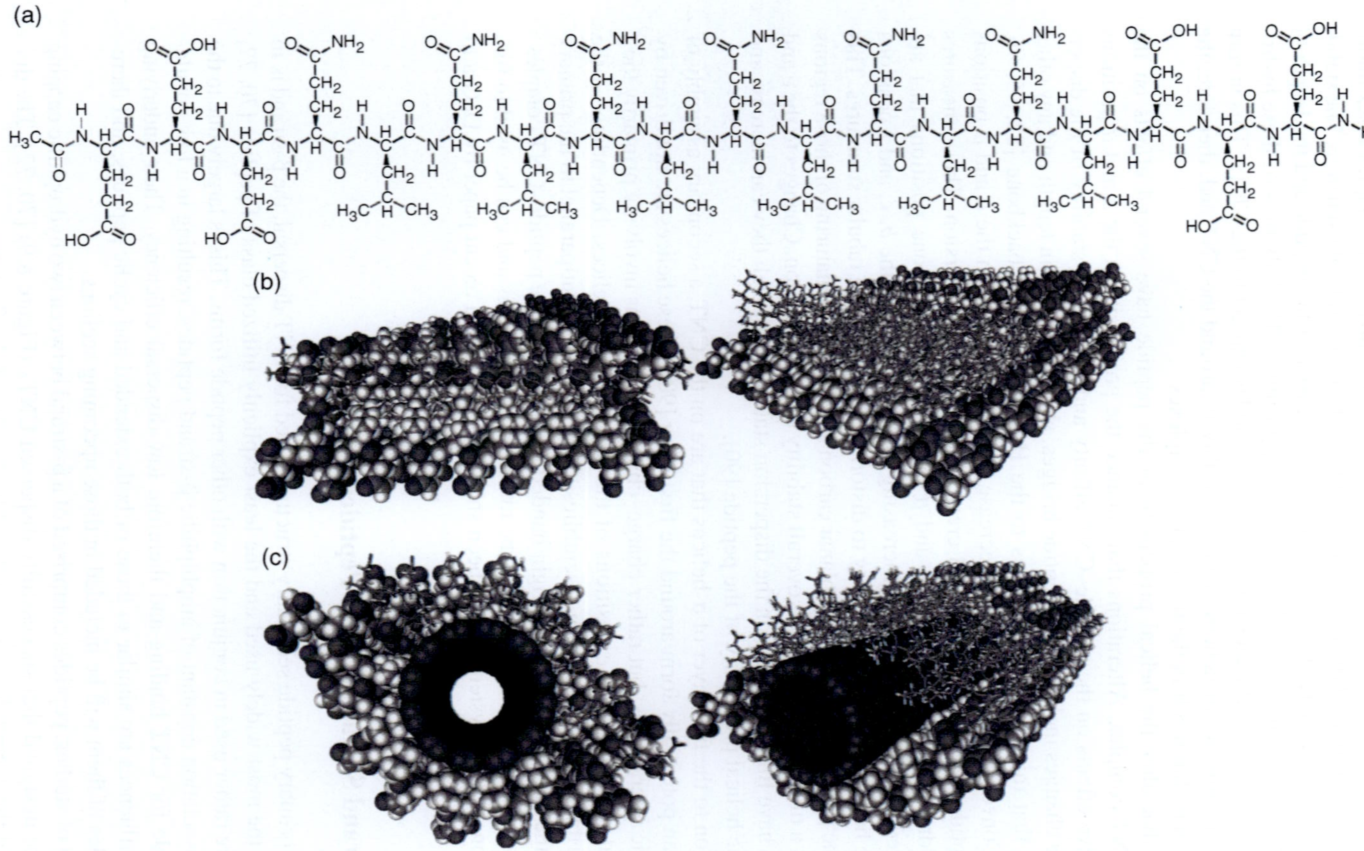

Figure 8.9 *Structure and interactions of a β-sheet surfactant peptide ($Ac–E_3(LQ)_6E_3–NH_2$). (a) Skeleton schematic of a single peptide strand, with blue polar side chains, red apolar side chains, and green peptide bonds that participate in β-sheet hydrogen bonding. (b) Space-filling and stick depiction of a β-sheet structure forming though self-aggregation in the absence of SWNTs. (c) Space-filling and stick depiction of a potential β-sheet structure formed by the peptide around a CNT (see color plate figure)*

lose the ability to form β-structures and successfully disperse CNTs. The hydrophobic region of the β-sheet can be composed of any aromatic or aliphatic AAs, with interactions consistent with those described in Section 8.2.1. The overall structure of this β-sheet is further stabilized by extensive H-bonding commonly found in β-peptides. These peptides are proposed to form a sandwich-like assembly with the CNT, resulting in H-bonds parallel to the tube axis. However, similar to α-helical peptides, these peptides self-associate in solution, meaning that the peptide–CNT complex is always in equilibrium with peptide aggregates, which could decrease the efficiency of the CNT dispersion. Overall, these peptides have been shown to form stable dispersions and result in monodisperse CNTs.

8.4 Extended Peptides

Extended or linear peptides are the simplest class of peptide dispersal agents, as they do not take on a well-defined secondary structure, even in the presence of CNTs. Unlike their counterparts, these peptides are devoid of secondary or tertiary structural features, leaving only their sequence to drive the formation of their unique quaternary structures [27]. The governing principle that drives the formation of these structures is amphiphilicity. Each peptide has a hydrophobic region that surrounds the CNTs, while the hydrophilic region serves to break the surface tension of the water around the CNT [27]. Unlike many of the other peptide dispersal agents, the synthesis of these peptides is not constrained by the necessity for a well-defined sequence and therefore can employ cheaper, bulk synthetic methods.

The primary structure can include a number of natural and unnatural AAs, as well as N- and/or C-terminal functionalization by a multitude of structures. They can be conceptualized as block-copolymers with well-defined subunits [27]. Each block can serve specific functions, and the number and nature of the subunits drive the formation of a wide variety of quaternary structures ranging from simple spherical micelles to extended micelles and fibers, annular and other complex micelles, and even liposomes and derivatives thereof [106]. Unlike their classical detergent counterparts, these micelles are highly structured, as is the case with most peptide-based assemblies. The composition of these quaternary structures can be both homo- and heterogeneous, depending on their function. Due to the linear nature of these peptides, the size and nature of the side chains plays a primary role in the quaternary structure of the resulting aggregates, more so in the hydrophobic core region.

The overarching principle that makes these structures suitable for CNT interaction is the presence of a hydrophobic region. The interactions can be through either aliphatic or aromatic groups as described in section 2.1. The difference between these peptides and the other peptide dispersal agents is the interactions both within each block and between peptides [27]. The interaction common to all peptidic segments of these molecules is the backbone H-bonding. This results in extended organization throughout the surfactant structure. This bonding can result in the patterning of the surfactant about the surface of the CNTs that can resemble β-sheet structures (Figure 8.10). This pattern can be either parallel to the tube length and perpendicular to its surface, in a ring perpendicular to the tube length, or spiraling about the CNT surface [32, 81].

The general structure of these peptides includes one end that interacts with the CNTs, a middle segment that interacts with other peptides, and a head group that interacts with the

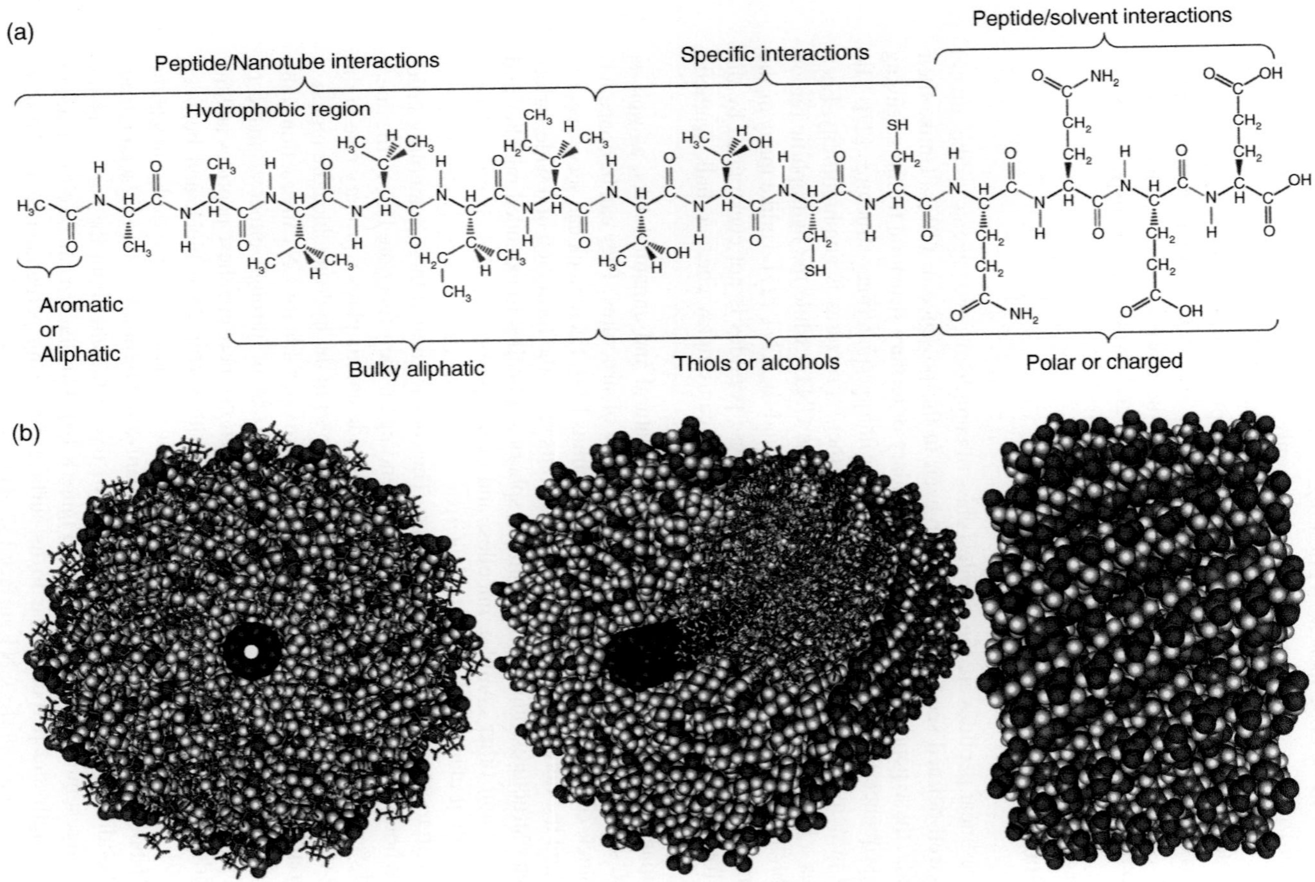

Figure 8.10 *Structure and interactions of an extended surfactant peptide. (a) Skeleton schematic of a single peptide strand, with blue polar AAs, red apolar AAs, and purple AAs integral in controlling peptide distribution and interpeptide interactions. (b) Space-filling and stick depiction of a surfactant peptide ($Ac–A_2V_2L_2E_2–NH_2$) in a β-sheet-like structure around a CNT. Extensive β-like H-bonding exists along the length of the CNT (see color plate figure)*

solvent [32, 81]. The hydrophobic core tends to be less important than the other two segments because the portion that actually interacts with the CNT has a very small cross-section. The central region tends to be the most important because the steric interaction of the peptide with itself tends to significantly affect the stability of the resulting structure. Attempts to increase the stability of such dispersions include the incorporation of cysteine residues into that core, which form disulfide bonds, thereby locking the peptide on to the CNT surface [27].

There have been many attempts to alter the central segments of these peptides in order to enhance inter-peptide interactions. The modifications include changing the length of this region, altering the size of the aliphatic side chains of the AAs in this region, and adding cysteine residues in order to encourage inter-peptide crosslinking via disulfide bonding [27]. However, these alterations only impact the equilibrium between free and bound peptide, and do not significantly enhance or hinder the actual dispersal ability of the peptide.

Extended peptides are an efficient vector for testing the effects of varying aromatic systems on the electronic properties of SWNTs. By systematically modifying the N-terminus of a simple base peptide SPX (AA sequence: Ac–XVVVVVKK–$CONH_2$, where Ac and $CONH_2$ represent the acetylated N-terminus and amidated C-terminus, respectively) at the X position with aromatic carboxylic acids, it is possible to observe the effects of aromatic rings interacting directly with the π-network of the CNTs [32, 81]. It was found that the natural AA side chains of Phe (SPF), Tyr (SPY), and Trp (SPW) resulted in absorbance peak shifts in the van Hoff singularities associated with semiconducting SWNTs, as well as shifts in the relative abundances of SWNTs dispersed. Through circular dichroism, the peptides were found to be ordered, to an extent of 60%, but lacking a single major secondary structure. Raman spectroscopy suggests that the peptides alone in solution have a 35:65 ratio of α-helix:β-sheet content. The dispersions formed using these peptides showed a significant susceptibility to salting effects. In the absence of buffers, SEM characterization displayed a random distribution of small CNT fibers, whereas buffered dispersions resulted in highly ordered and uniform bundles [32].

8.5 Amorphous Peptides

The simplest peptides to bind to SWNTs are amorphous peptides with sequences not designed to form either specific secondary structures or specific peptide aggregates. Such peptides rely on aromatic residues to drive CNT interactions, while the remaining sequence serves to stabilize the peptide–CNT complex, repel other CNTs, and prevent reaggregation [73, 80]. The aromatic side chains of Phe, Tyr, and Trp display high affinities for conjugated surfaces, like CNTs, due to their hydrophobic character and capacity for π–π electronic interactions [73, 80]. Sequences for such peptides are often found through high throughput assays as opposed to *de novo* design. One such technique is phage display, a selection technique that depends on the screening of peptide libraries with the goal of finding those with a high affinity for CNT surfaces. The sequences HWKHPWGAWDTL, HWSAWWIRSNQS, HYKHPYGAYDTL, and HYSAYYIRSNQS displayed on phage have been shown to successfully bind MWNTs [79]. The resultant structures are plaques composed of multiple phage microspheres bound together by MWNT aggregates. It has been found that, despite their capacity for CNT binding, there is little evidence that these peptides show specificity for tubular materials over planar or other allotropes of carbon that can be found in raw CNT powders.

Quantum-mechanical studies of the interaction of such sequences demonstrate the amorphic nature of these peptides, even on a CNT surface [80]. The aromatic residues (especially Trp) appear to drive the majority of the peptide–CNT interaction, while polar residues are oriented into the solution. These peptides have a high degree of freedom to rearrange on the surface and obtain the most stable conformation [73, 80]. The study included a focus on defects in the CNT conjugation in the form of carboxyl and hydroxyl groups, which were shown to have a limited effect on the peptide affinity. Peptide–CNT binding was found to be highest at the aromatic residues, which were found to avoid the defect areas to maximize the contact area with the conjugated substrate. In contrast, the polar side chains were observed to form H-bonds with the local defects, driving some of the peptide towards those regions. Overall, despite their ability to disperse CNTs with potential for biocompatibility, amorphous peptides have limited utility as dispersal agents due to their lack of control over arrangement on CNTs. The availability of other peptides or DNA capable of doing the same makes them an inefficient alternative as dispersal agents.

8.6 Cyclic Peptides

A more recent approach to the dispersal and stabilization of CNTs in aqueous conditions with peptides is through the use of cyclic peptides. The peptide design is originally based on peptides with alternating L- and D-AAs, which form small peptide rings with an even number of residues (Figure 8.11) [108–114]. The result is that the side chains of these peptides reside on one face of the peptide backbone, which encourages an annular shape under appropriate conditions. Cyclic peptides have been found to form stacked tube-like structures in various solvents through formation of inter-peptide backbone H-bonds (Figure 8.11) [109–117]. This extensive H-bonding network is brought about by a decrease in both the side-chain solubility, as well as the peptide backbone H-bonding capacity with the solvent.

The internal cavity of these peptide nanotubes presents a conjugated and hydrophobic surface, an ideal environment for enveloping CNTs. The process by which a cyclic peptide could surround a CNT would involve sliding peptides on to the CNT from the ends. Thermodynamic simulations have shown that in aqueous conditions there should be a favorable driving force (ΔG) for cyclic peptides to encircle a CNT versus free in solution [118]. In order to be an effective dispersal agent, the cyclic peptide must completely coat the surface of the CNT. A relatively short (100 nm) CNT would require approximately 200 rings to cover its surface. Using sonication as a dispersal method, the process of encircling CNTs as described above would not happen fast enough to outcompete nonspecific binding of peptides to the CNT surface. Therefore, despite the thermodynamic suitability of cyclic peptides for CNT dispersal, there exists a kinetic barrier that results from the one-dimensional nature of CNTs which prevents a complete coating. Attempts made to disperse SWNTs with cyclic peptides have shown that the resulting dispersions contain a considerable amount of nontubular carbonaceous material and are not stable to dialysis, designed to remove excess peptide not encircling the SWNTs [119].

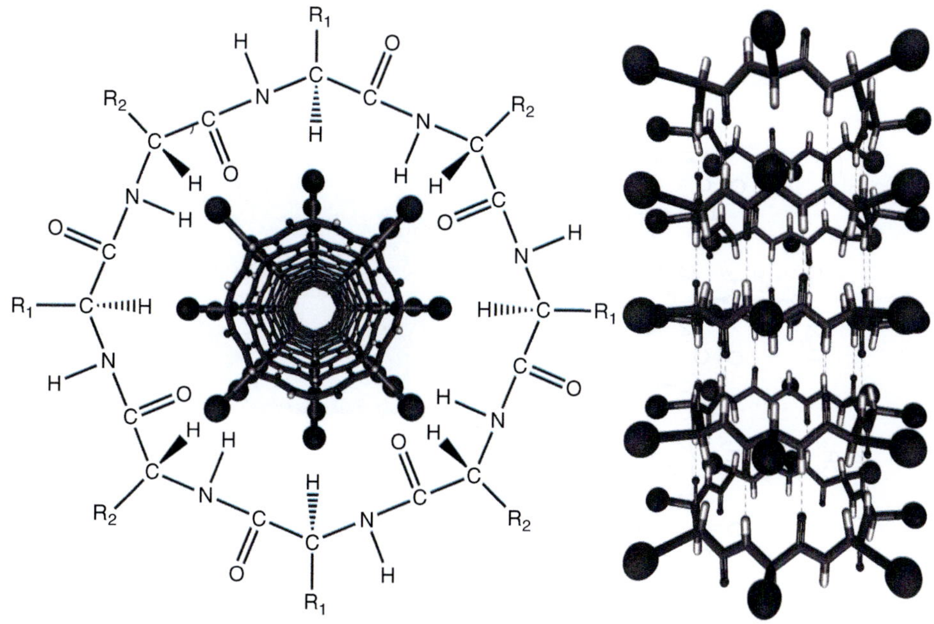

Figure 8.11 *(Left) Atomic structure and molecular model (inside) representations of a molecular model representation of cyclic peptides assembled into a nanotube. (Right) Orthogonal view of cyclic peptide nanotube showing extensive H-bonding along its length. AA side chains R_1 and R_2 correspond to* L *and* D *chirality positions, respectively*

8.6.1 Reversible Cyclic Peptides

To overcome the kinetic barrier, a closable linkage was introduced into the cyclic peptide backbone to allow the peptide to switch between a linear form and a head-to-tail cyclized form. This results in a linear peptide that can become cyclized around a CNT once the appropriately functionalized termini are brought together and form the covalent linkage [29, 30, 119]. The linkage can be reversible or irreversible, as long as the bond can be formed easily and preferably under mild conditions.

The initial peptide designs by Dieckmann *et al.*, called reversible cyclic peptides (RCPs), consist of an alternating sequence of L-/D-AAs with Lys as the charged group, a Tyr chromophore, and thiol functionalized N- and C-termini, which result in disulfide bond formation upon oxidation (Figure 8.12) [29, 30, 119]. The oxidizing agent is dissolved oxygen and the oxygen on the surface of the CNTs upon their immersion in the aqueous dispersal medium. RCPs have been shown to form stable SWNT dispersions in aqueous conditions and are stable to prolonged dialysis. The resulting dialyzed dispersions show a lack of free peptide in solution, as well as very little other small particle contaminants (Figure 8.13) [30, 119].

The spectroscopic properties of these RCP/SWNT complexes suggest that there is a considerable amount of interaction between the π–network of the SWNTs and the conjugated peptide bonds, resulting in a doping effect [30, 119]. This interaction is strong enough to shift the van Hoff singularities attributed to individual SWNT chiralities, although each

(a)

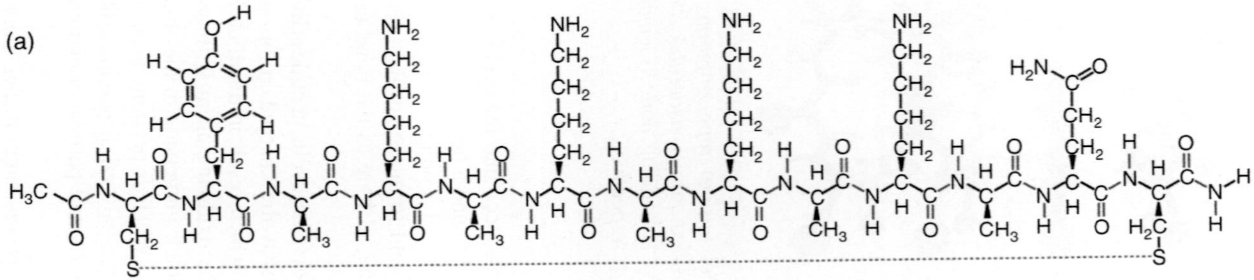

(b)

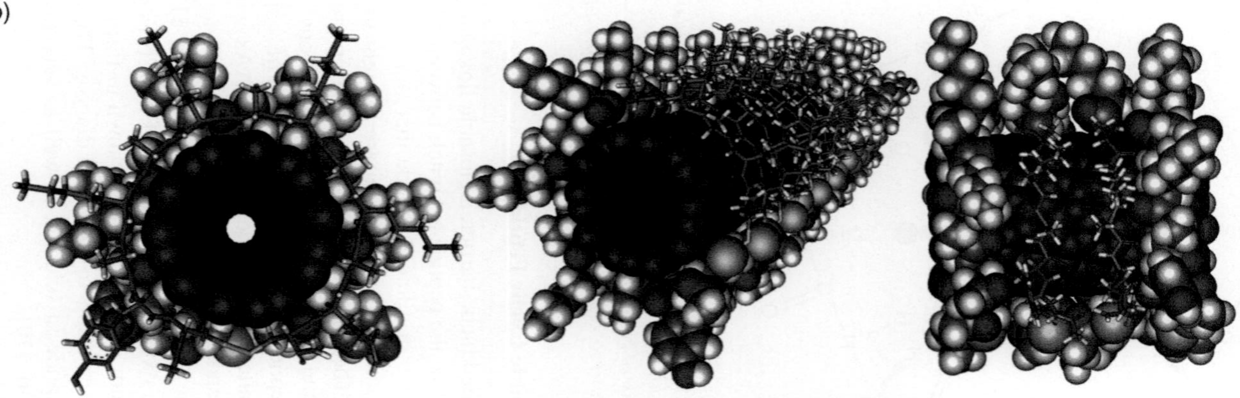

Figure 8.12 *Structure and interactions of a reversible cyclic peptide RC5–Cys. (a) Skeleton schematic of a single peptide strand, with blue polar side chains, red apolar side chains, and green peptide bonds that participate in β-sheet-like H-bonding. (b) Space-filling and stick depiction of oxidized RCP rings covering the surface of a CNT (see color plate figure)*

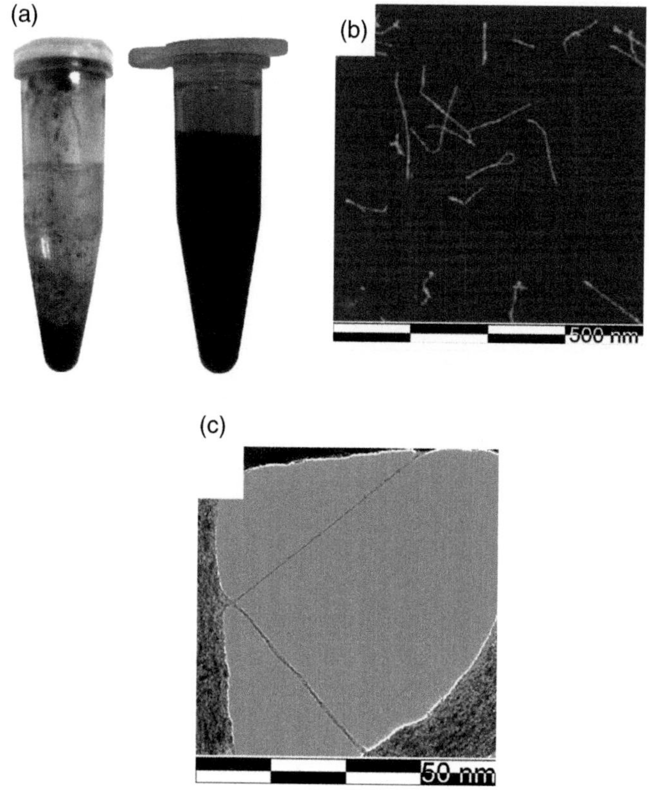

Figure 8.13 SWNTs dispersed with RC7–Cys–LL viewed (a) optically following sonication of $SWNT_{CP}$ with and without peptide, respectively; (b) by AFM and diluted 1:50 with DI water (scale bar 500 nm); and (c) by TEM and diluted 1:10 with DI water (scale bar 50 nm)

type of CNT is affected to a different degree. The resultant UV–Vis–NIR peaks appear red-shifted from those observed in SDS–SWNT dispersions (Figure 8.14). There are also observed small changes in the relative populations of SWNTs. Atomic force microscopy of these CNT dispersions (with saturating peptide concentrations) shows continuous coverage of the nanotubes by the RCPs. The AFM shows that the CNTs appear to be straighter as compared to CNTs dispersed with other agents. This effect is attributed to the H-bonding between adjacent peptide rings [30, 119].

8.7 Summary and Outlook

There is an increasing need to find effective methods of dispersing and organizing CNTs for a multitude of applications. Peptides, a class of molecules with a demonstrated ability to form highly ordered structures, are proving to be an effective scaffold for building complex CNT-based systems. Each example of a peptide-based CNT dispersal agent in this

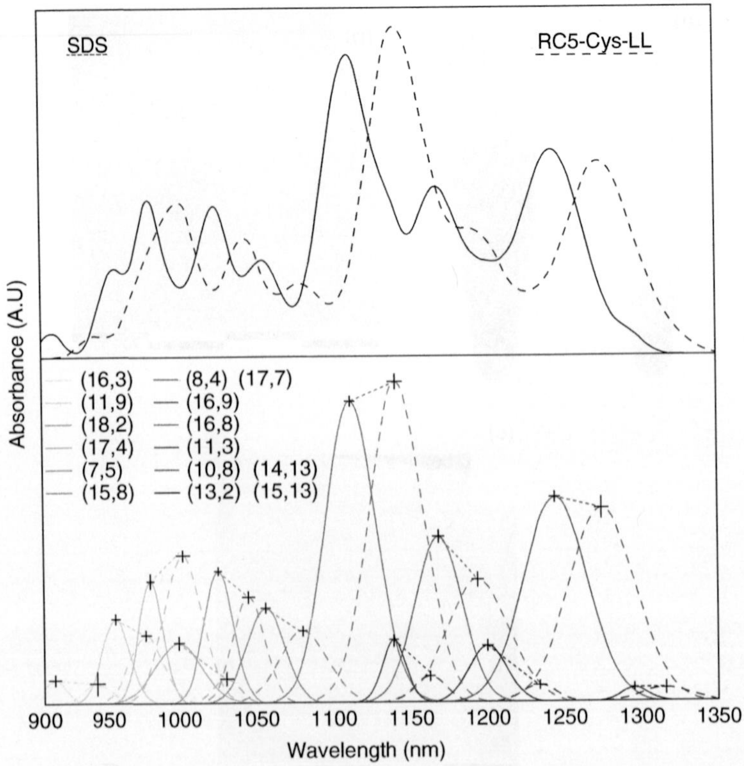

Figure 8.14 *Compilation and deconvolution of UV/Vis/NIR spectra for RC5–Cys–LL dispersed SWNTs after 100 000 g centrifugation and dialysis against 3 L of deionized water. The deconvoluted peaks attributed to van Hove singularities of various SWNT chiralities are designated by color (bottom). The summation of each set of van Hove singularities for the corresponding peptides (top). All spectra were background subtracted to remove contributions from π-plasmon, NTC, and bundled CNT absorption features (see color plate figure)*

chapter is built on similar principals. In each case, backbone H-bonding plays a pivotal role in the resultant structures, both in the presence and absence of CNTs. These H-bonds impart a degree of order that is not present in most other classes of dispersal agents. Amorphous peptides are capable of dispersing all types of apolar or aromatic materials, independent of size. Helical peptides create local segments of order around the circumference of a CNT. β-strand peptides can create a bidirectional disparity between the *x* and *y* axes of the CNT. Extended peptides can create axial order along the CNT length and provide an effective platform for testing doping effects on CNTs. Cyclic peptides impart a rigidity to CNT constructs and have high stability and high surface charge density.

The next stage in peptide–CNT research is likely to focus on the use of combinations of peptides with other molecules to achieve higher order structures. The addition of other materials can enhance the final composite by allowing each subunit to perform the function for which it is best suited. With each passing year, the field of CNT research is becoming more accessible, allowing for better methodologies for constructing and analyzing new

CNT-based materials. Innovative thinking and hard work will realize the implementation of CNT-based materials in the foreseeable future.

Acknowledgements

The authors would like to thank Dinushi R. Samarajeewa and Dr Inga H. Musselman (both from The University of Texas at Dallas, or UTD) for the collection and assistance with interpretation of AFM data (Figure 8.13), and Pooja Bajaj (UTD), Jane Nguyen (UTD), Dr Musselman (UTD) and Dr Christopher Gilpin (The University of Texas Southwestern Medical Center, Dallas, TX, USA) for the collection and assistance with interpretation of TEM data (Figure 8.13).

References

[1] Balavoine, F., Schultz, P., Richard, C., Mallouh, V., Ebbesen, T.W., Mioskowski, C. (1999) Helical crystallization of proteins on carbon nanotubes: a first step towards the development of new biosensors. Angew. Chem. Int. Ed. 38: 1912–1915.

[2] Tsang, S.C., Davis, J.J., Green, M.L.H., Hill, H.A.O., Leung, Y.C., Sadler, P.J. (1995) Immobilization of small proteins in carbon nanotubes: high-resolution transmission electron-microscopy study and catalytic activity. Chem. Comm.: 1803–1804.

[3] Erlanger, B.F., Chen, B.-X., Zhu, M., Brus, L. (2001) Binding of an anti-fullerene IgG monoclonal antibody to single wall carbon nanotubes. Nano Lett. 1: 465–467.

[4] Bradley, K., Briman, M., Star, A., Grüner, G. (2004) Charge transfer from adsorbed proteins. Nano Lett. 4: 253–256.

[5] Lin, Y., Allard, L.F., Sun, Y.-P. (2004) Protein-affinity of single-walled carbon nanotubes in water. J. Phys. Chem. B 108: 3760–3764.

[6] Elkin, T., Jiang, X., Taylor, S., Lin, Y., Gu, L., Yang, H., Brown, J., Collins, S., Sun, Y.-P. (2005) Immuno-carbon nanotubes and recognition of pathogens. Chem. Biochem. 6: 640–643.

[7] MacDonald, R.A., Laurenzi, B.F., Viswanathan, G., Ajayan, P.M., Stegemann, J.P. (2005) Collagen–carbon nanotube composite materials as scaffolds in tissue engineering. J. Biomed. Mater. Res. 74A: 489–496.

[8] Bottini, M., Cerignoli, F., Tautz, L., Rosato, N., Bergamaschi, A., Mustelin, T. (2006) Adsorption of Streptavidin onto single-walled carbon nanotubes: application in fluorescent supramolecular nanoassemblies. J. Nanosci. Nanotech. 6: 3693–3698.

[9] Karajanagi, S.S., Yang, H., Asuri, P., Sellitto, E., Dordick, J.S., Kane, R.S. (2006) Protein-assisted solubilization of single-walled carbon nanotubes. Langmuir 22: 1392–1395.

[10] Matsuura, K., Saito, T., Okazaki, T., Ohshima, S., Yumura, M., Iijima, S. (2006) Selectivity of water-soluble proteins in single-walled carbon nanotube dispersions. Chem. Phys. Lett. 429: 497–502.

[11] Nepal, D., Geckeler, K.E. (2006) pH-Sensitive dispersion and debundling of single-walled carbon nanotubes : lysozyme as a tool. Small 2: 406–412.

[12] Song, L., Meng, J., Zhong, J., Liu, L., Dou, X., Liu, D., Zhao, X., Luo, S., Zhang, Z., Xiang, Y., Xu, H., Zhou, W., Wu, Z., Xie, S. (2006) Human fibrinogen adsorption onto single-walled carbon nanotube films. Colloids and Surfaces B: Biointerfaces 49: 66–70.

[13] Guiseppi-Elie, A., Lei, C.H., Baughman, R.H. (2002) Direct electron transfer of glucose oxidase on carbon nanotubes. Nanotechnology 13: 559–564.

[14] Davis, J.J., Green, M.L.H., Hill, H.A.O., Leung, Y.C., Sadler, P.J., Sloan, J., Xavier, A.V., Tsang, S.C. (1998) The immobilisation of proteins in carbon nanotubes. Inorganica Chimica Acta 272: 261–266.

[15] Karajanagi, S.S., Vertegel, A.A., Kane, R.S., Dordick, J.S. (2004) Structure and function of enzymes adsorbed onto single-walled carbon nanotubes. Langmuir 20: 11594–11599.

[16] Cole, M.W., Crespi, V.H., Dresselhaus, M.S., Dresselhaus, G., Fischer, J.E., Gutierrez, H.R., Kojima, K., Mahan, G.D., Rao, A.M., Sofo, J.O., Tachibana, M., Wako, K., Xiong, Q. (2010) Structural, electronic, optical and vibrational properties of nanoscale carbons and nanowires: a colloquial review. Journal of Physics. Condensed Matter: An Institute of Physics Journal 22: 334201.

[17] Lacerda, L., Raffa, S., Prato, M., Bianco, A., Kostarelos, K. (2007) Cell-penetrating CNTs for delivery of therapeutics. Nano Today 2: 38–43.

[18] Kostarelos, K., Bianco, A., Prato, M. (2009) Promises, facts and challenges for carbon nanotubes in imaging and therapeutics. Nature Nanotechnology 4: 627–633.

[19] Wang, J. (2005) Carbon-nanotube based electrochemical biosensors: a review. Electroanalysis 17: 7–14.

[20] Lin, Y., Taylor, S., Li, H.P., Fernando, K.A.S., Qu, L.W., Wang, W., Gu, L.R., Zhou, B., Sun, Y.P. (2004) Advances toward bioapplications of carbon nanotubes. Journal of Materials Chemistry 14: 527–541.

[21] Yang, W., Thordarson, P., Gooding, J.J., Ringer, S.P., Braet, F. (2007) Carbon nanotubes for biological and biomedical applications. Nanotechnology 18: 412001.

[22] Lu, W., Lieber, C.M. (2007) Nanoelectronics from the bottom up. Nature Materials 6: 841–850.

[23] Feldman, A.K., Steigerwald, M.L., Guo, X., Nuckolls, C. (2008) Molecular electronic devices based on single-walled carbon nanotube electrodes. Accounts of Chemical Research 41: 1731–1741.

[24] Liu, Z., Tabakman, S., Welsher, K., Dai, H. (2009) Carbon nanotubes in biology and medicine: *in vitro* and *in vivo* detection, imaging and drug delivery. Nano Research 2: 85–120.

[25] Li, C., Thostenson, E.T., Chou, T.-W. (2008) Sensors and actuators based on carbon nanotubes and their composites: a review. Composites Science and Technology 68: 1227–1249.

[26] Vairavapandian, D., Vichchulada, P., Lay, M.D. (2008) Preparation and modification of carbon nanotubes: review of recent advances and applications in catalysis and sensing. Analytica Chimica Acta 626: 119–129.

[27] Arnold, M.S., Guler, M.O., Hersam, M.C., Stupp, S.I. (2005) Encapsulation of carbon nanotubes by self-assembling peptide amphiphiles. Langmuir 21: 4705–4709.

[28] Dieckmann, G.R., Dalton, A.B., Johnson, P.A., Razal, J., Chen, J., Giordano, G.M., Munoz, E., Musselman, I.H., Baughman, R.H., Draper, R.K. (2003) Controlled assembly of carbon nanotubes by designed amphiphilic peptide helices. J. Am. Chem. Soc. 125: 1770–1777.

[29] Ortiz-Acevedo, A., Dieckmann, G.R. (2004) Synthesis of reversible cyclic peptides. Tetrahed. Lett. 45: 6795–6798.

[30] Ortiz-Acevedo, A., Xie, H., Zorbas, V., Sampson, W.M., Dalton, A.B., Baughman, R.H., Draper, R.K., Musselman, I.H., Dieckmann, G.R. (2005) Diameter-selective solubilization of single-walled carbon nanotubes by reversible cyclic peptides. J. Am. Chem. Soc. 127: 9512–9517.

[31] Poenitzsch, V.Z., Musselman, I.H. (2006) Atomic force microscopy measurements of peptide-wrapped single-walled carbon nanotube diameters. Microsc. Microanal. 12: 221–227.

[32] Samarajeewa, D.R., Dieckmann, G.R., Musselman, I.H. (2010) Effect of surfactant peptides on electronic properties of single-walled carbon nanotubes. Microscopy and Microanalysis 16: 454–455.

[33] Zorbas, V., Ortiz-Acevedo, A., Dalton, A.B., Yoshida, M.M., Dieckmann, G.R., Draper, R.K., Baughman, R.H., Jose-Yacaman, M., Musselman, I.H. (2004) Preparation and characterization of individual peptide-wrapped single-walled carbon nanotubes. J. Am. Chem. Soc. 126: 7222–7227.

[34] Zorbas, V., Smith, A.L., Ortiz-Acevedo, A., Xie, H., Dalton, A.B., Dieckmann, G.R., Draper, R.K., Baughman, R.H., Musselman, I.H. (2005) Importance of aromatic content for peptides/single-walled carbon nanotube interactions. J. Am. Chem. Soc. 127: 12323–12328.

[35] Hirsch, A. (2002) Functionalization of single-walled carbon nanotubes. Angew. Chem. Int. Ed. 41: 1853–1859.

[36] Fu, K., Sun, Y.-P. (2003) Dispersion and solubilization of carbon nanotubes. J. Nanosci. Nanotech. 3: 351–364.

[37] Balasubramanian, K., Burghard, M. (2005) Chemically functionalized carbon nanotubes. Small 1: 180–192.

[38] Banerjee, S., Hemraj-Benny, T., Wong, S.S. (2005) Covalent surface chemistry of single-walled carbon nanotubes. Adv. Mater. 17: 17–29.

[39] Tasis, D., Tagmatarchis, N., Bianco, A., Prato, M. (2006) Chemistry of carbon nanotubes. Chem. Rev. 106: 1105–1136.

[40] O'Connell, M.J., Bachilo, S.M., Huffman, C.B., Moore, V.C., Strano, M.S., Haroz, E.H., Rialon, K.L., Boul, P.J., Noon, W.H., Kittrell, C., Ma, J., Hauge, R.H., Weisman, R.B., Smalley, R.E. (2002) Band gap fluorescence from individual single-walled carbon nanotubes. Science 297: 593–596.

[41] Vaisman, L., Wagner, H.D., Marom, G. (2006) The role of surfactants in dispersion of carbon nanotubes. Adv. Colloid Interface Sci. 128: 37–46.

[42] Islam, M.F., Rojas, E., Bergey, D.M., Johnson, A.T., Yodh, A.G. (2003) High weight fraction surfactant solubilization of single-wall carbon nanotubes in water. Nano Letters 3: 269–273.

[43] Moore, V.C., Strano, M.S., Haroz, E.H., Hauge, R.H., Smalley, R.E., Schmidt, J., Talmon, Y. (2003) Individually suspended single-walled carbon nanotubes in various surfactants. Nano Letters 3: 1379–1382.

[44] Matarredona, O., Rhoads, H., Li, Z.R., Harwell, J.H., Balzano, L., Resasco, D.E. (2003) Dispersion of single-walled carbon nanotubes in aqueous solutions of the anionic surfactant NaDDBS. J. Phys. Chem. B 107: 13357–13367.

[45] Zhang, M.N., Su, L., Mao, L.Q. (2006) Surfactant functionalization of carbon nanotubes (CNTs) for layer-by-layer assembling of CNT multi-layer films and fabrication of gold nanoparticle/CNT nanohybrid. Carbon 44: 276–283.

[46] Yu, J., Grossiord, N., Koning, C.E., Loos, J. (2007) Controlling the dispersion of multi-wall carbon nanotubes in aqueous surfactant solution. Carbon 45: 618–623.

[47] Rastogi, R., Kaushal, R., Tripathi, S.K., Sharma, A.L., Kaur, I., Bharadwaj, L.M. (2008) Comparative study of carbon nanotube dispersion using surfactants. Journal of Colloid and Interface Science 328: 421–428.

[48] Haggenmueller, R., Rahatekar, S.S., Fagan, J.A., Chun, J., Becker, M.L., Naik, R.R., Krauss, T., Carlson, L., Kadla, J.F., Trulove, P.C., Fox, D.F., DeLong, H.C., Fang, Z., Kelley, S.O., Gilman, J.W. (2008) Comparison of the quality of aqueous dispersions of single wall carbon nanotubes using surfactants and biomolecules. Langmuir 24: 5070–5078.

[49] White, B., Banerjee, S., O'Brien, S., Turro, N.J., Herman, I.P. (2007) Zeta-potential measurements of surfactant-wrapped individual single-walled carbon nanotubes. J. Phys. Chem. C 111: 13684–13690.

[50] Wang, H. (2009) Dispersing carbon nanotubes using surfactants. Current Opinion in Colloid and Interface Science 14: 364–371.

[51] O'Connell, M.J., Boul, P., Ericson, L.M., Huffman, C., Wang, Y., Haroz, E., Kuper, C., Tour, J., Ausman, K.D., Smalley, R.E. (2001) Reversible water-solubilization of single-walled carbon nanotubes by polymer wrapping. Chem. Phys. Lett. 342: 265–271.

[52] Star, A., Stoddart, J.F., Steuerman, D., Diehl, M., Boukai, A., Wong, E.W., Yang, X., Chung, S.W., Choi, H., Heath, J.R. (2001) Preparation and properties of polymer-wrapped single-walled carbon nanotubes. Angewandte Chemie-International Edition 40: 1721–1725.

[53] Breuer, O., Sundararaj, U. (2004) Big returns from small fibers: a review of polymer/carbon nanotube composites. Polymer Composites 25: 630–645.

[54] Coleman, J.N., Khan, U., Blau, W.J., Gun'ko, Y.K. (2006) Small but strong: a review of the mechanical properties of carbon nanotube–polymer composites. Carbon 44: 1624–1652.

[55] Peng, C., Zhang, S., Jewell, D., Chen, G.Z. (2008) Carbon nanotube and conducting polymer composites for supercapacitors. Progress in Natural Science – Materials International 18: 777–788.

[56] Liu, P. (2005) Modifications of carbon nanotubes with polymers. European Polymer Journal 41: 2693–2703.

[57] Moniruzzaman, M., Winey, K.I. (2006) Polymer nanocomposites containing carbon nanotubes. Macromolecules 39: 5194–5205.

[58] Prencipe, G., Tabakman, S.M., Welsher, K., Liu, Z., Goodwin, A.P., Zhang, L., Henry, J., Dai, H. (2009) PEG branched polymer for functionalization of nanomaterials with ultralong blood circulation. J. Am. Chem. Soc. 131: 4783–4787.

[59] Prickett, W.M., Van Rite, B.D., Resasco, D.E., Harrison, R.G. (2011) Vascular targeted single-walled carbon nanotubes for near-infrared light therapy of cancer. Nanotechnology 22: 455101.

[60] Zheng, M., Jagota, A., Semke, E.D., Diner, B.A., Mclean, R.S., Lustig, S.R., Richardson, R.E., Tassi, N.G. (2003) DNA-assisted dispersion and separation of carbon nanotubes. Nature Mat. 2: 338–342.

[61] Zheng, M., Jagota, A., Strano, M.S., Santos, A.P., Barone, P., Chou, S.G., Diner, B.A., Dresselhaus, M.S., Mclean, R.S., Onoa, G.B., Samsonidze, G.G., Semke, E.D., Usrey, M.L., Walls, D.J. (2003) Structure-based carbon nanotube sorting by sequence-dependent DNA assembly. Science 302: 1545–1548.

[62] Chou, S.G., Ribeiro, H.B., Barros, E.B., Santos, A.P., Nezich, D., Samsonidze, G.G., Fantini, C., Pimenta, M.A., Jorio, A., Plentz, F., Dresselhaus, M.S., Dresselhaus, G., Saito, R., Zheng, M., Onoa, G.B., Semke, E.D., Swan, A.K., Unlu, M.S., Goldberg, B.B. (2004) Optical characterization of DNA-wrapped carbon nanotube hybrids. Chem. Phys. Lett. 397: 296–301.

[63] Zheng, M., Rostovtsev, V.V. (2006) Photoinduced charge transfer mediated by DNA-wrapped carbon nanotubes. J. Am. Chem. Soc. 128: 7702–7703.

[64] Zheng, M., Semke, E.D. (2007) Enrichment of single chirality carbon nanotubes. J. Am. Chem. Soc. 129: 6084–6085.

[65] Yang, R., Tang, Z., Yan, J., Kang, H., Kim, Y., Zhu, Z., Tan, W. (2008) Noncovalent assembly of carbon nanotubes and single-stranded DNA: an effective sensing platform for probing biomolecular interactions. Analytical Chemistry 80: 7408–7413.

[66] Tu, X., Zheng, M. (2008) A DNA-based approach to the carbon nanotube sorting problem. Nano Res. 1: 185–194.

[67] Tu, X., Walker, A.R.H., Khripin, C.Y., Zheng, M. (2011) Evolution of DNA sequences toward recognition of metallic armchair carbon nanotubes. J. Am. Chem. Soc. 133: 12998–13001.

[68] Roxbury, D., Tu, X., Zheng, M., Jagota, A. (2011) Recognition ability of DNA for carbon nanotubes correlates with their binding affinity. Langmuir 27: 8282–8293.

[69] Xiao, Z., Wang, X., Xu, X., Zhang, H., Li, Y., Wang, Y. (2011) Base- and structure-dependent DNA dinucleotide–carbon nanotube interactions: molecular dynamics simulations and thermodynamic analysis. J. Phys. Chem. C 115: 21546–21558.

[70] Bakota, E.L., Aulisa, L., Tsyboulski, D.A., Weisman, R.B., Hartgerink, J.D. (2009) Multidomain peptides as single-walled carbon nanotube surfactants in cell culture. Biomacromolecules 10: 2201–2206.

[71] O'Brien, E.P., Stan, G., Thirumalai, D., Brooks, B.R. (2008) Factors governing helix formation in peptides confined to carbon nanotubes. Nano Lett. 8: 3702–3708.

[72] Tsyboulski, D.A., Bakota, E.L., Witus, L.S., Rocha, J.D.R., Hartgerink, J.D., Weisman, R.B. (2008) Self-assembling peptide coatings designed for highly luminescent suspension of single-walled carbon nanotubes. J. Am. Chem. Soc. 130: 17134–17140.

[73] Wang, S., Humphreys, E.S., Chung, S.-Y., Delduco, D.F., Lustig, S.R., Wang, H., Parker, K.N., Rizzo, N.W., Subramoney, S., Chiang, Y.-M., Jagota, A. (2003) Peptides with selective affinity for carbon nanotubes. Nature Mat. 2: 196–200.

[74] Pedersen, S.L., Tofteng, A.P., Malik, L., Jensen, K.J. (2012) Microwave heating in solid-phase peptide synthesis. Chem. Soc. Rev. 41: 1826–1844.

[75] Mitchell, A.R. (2008) Bruce Merrifield and solid-phase peptide synthesis: a historical assessment. Biopolymers 90: 175–184.

[76] Amblard, M., Fehrentz, J.-A., Martinez, J., Subra, G. (2006) Methods and protocols of modern solid phase peptide synthesis. Molecular Biotechnology 33: 239–254.

[77] Binauld, S., Hawker, C.J., Fleury, E., Drockenmuller, E. (2009) A modular approach to functionalized and expanded crown ether based macrocycles using click chemistry. Angwe. Chemie Inter. Ed. 48: 6654–6658.

[78] Grieshaber, S.E., Farran, A.J.E., Lin-Gibson, S., Kiick, K.L., Jia, X. (2009) Synthesis and characterization of elastin-mimetic hybrid polymers with multiblock, alternating molecular architecture and elastomeric properties. Macromolecules 42: 2532–2541.

[79] Parmleya, S.F., Smith, G.P. (1988) Antibody-selectable filamentous fd phage vectors: affinity purification of target genes. Gene 73: 305–318.

[80] Walsh, T.R., Tomasio, S.M. (2010) Investigation of the influence of surface defects on peptide adsorption onto carbon nanotubes. Molecular BioSystems 6: 1707–1718.

[81] Xie, H., Becraft, E.J., Baughman, R.H., Dalton, A.B., Dieckmann, G.R. (2008) Ranking the affinity of aromatic residues for carbon nanotubes by using designed surfactant peptides. J. Pept. Sci. 14: 139–151.

[82] Xie, H., Ortiz-Acevedo, A., Baughman, R.H., Draper, R.K., Musselman, I.H., Dalton, A.B., Dieckmann, G.R. (2005) Peptide cross-linking modulated stability and assembly of peptide-wrapped single-walled carbon nanotubes. J. Mater. Chem. 15: 1734–1741.

[83] DeGrado, W.F., Summa, C.M., Pavone, V., Nastri, F., Lombardi, A. (1999) *De novo* design and structural characterization of proteins and metalloproteins. Annu. Rev. Biochem. 68: 779–819.

[84] Lombardi, A., Bryson, J.W., DeGrado, W.F. (1996) *De novo* design of heterotrimeric coiled coils. Biopolym. 40: 495–504.

[85] Schneider, J.P., Lombardi, A., DeGrado, W.F. (1998) Analysis and design of three-stranded coiled coils and three-helix bundles. Fold. Design. 3: R29–R40.

[86] Oakley, M.G., Hollenbeck, J.J. (2001) The design of antiparallel coiled coils. Current Opinion in Structural Biology 11: 450–457.

[87] Doig, A.J. (2008) Stability and design of alpha-helical peptides. In: Molecular Biology of Protein Folding, Part A. Conn, P.M. (Ed.). Elsevier Academic Press, Inc., San Diego, CA, pp. 1–52.

[88] Apostolovic, B., Danial, M., Klok, H.-A. (2010) Coiled coils: attractive protein folding motifs for the fabrication of self-assembled, responsive and bioactive materials. Chem. Soc. Rev. 39: 3541–3575.

[89] Floudas, G., Spiess, H.W. (2009) Self-assembly and dynamics of polypeptides. Macromolecular Rapid Communications 30: 278–298.

[90] Grigoryan, G., Kim, Y.H., Acharya, R., Axelrod, K., Jain, R.M., Willis, L., Drndic, M., Kikkawa, J.M., DeGrado, W.F. (2011) Computational design of virus-like protein assemblies on carbon nanotube surfaces. Science 332: 1071–1076.

[91] Kopecek, J., Yang, J. (2009) Peptide-directed self-assembly of hydrogels. Acta Biomaterialia 5: 805–816.

[92] Oladepo, S.A., Xiong, K., Hong, Z., Asher, S.A. (2011) Elucidating peptide and protein structure and dynamics: UV resonance raman spectroscopy. J. Phys. Chem. Lett. 2: 334–344.

[93] Reiner, A. (2011) Triplet–triplet energy transfer studies on conformational dynamics in peptides and a protein. J. Pept. Sci. 17: 413–419.

[94] Dieckmann, G.R., McRorie, D.K., Lear, J.D., Sharp, K.A., DeGrado, W.F., Pecoraro, V.L. (1998) The role of protonation and metal chelation preferences in defining the properties of mercury-binding coiled coils. J. Mol. Biol. 280: 897–912.

[95] Chang, C.-F., Zehfus, M.H. (1996) The effect of *N*-methylation on helical peptides. Peptide Science 40: 609–616.

[96] Chang, C.-F., Zehfus, M.H. (1998) Effects of backbone modification on helical peptides: the reduced carbonyl modification. Biopolymers 46: 181–193.

[97] Chiu, C.-C., Dieckmann, G.R., Nielsen, S.O. (2009) Role of peptide–peptide interactions in stabilizing peptide-wrapped single-walled carbon nanotubes: a molecular dynamics study. Biopolym. (Pept. Sci.) 92: 156–163.

[98] Fan, W.J., Zeng, J., Zhang, R.Q. (2009) Quantum mechanical quantification of weakly interacting complexes of peptides with single-walled carbon nanotubes. J. Chem. Theory Comput. 5: 2879–2885.

[99] Henchey, L.K., Jochim, A.L., Arora, P.S. (2008) Contemporary strategies for the stabilization of peptides in the [alpha]-helical conformation. Current Opinion in Chemical Biology 12: 692–697.

[100] Kim, Y.-W., Grossmann, T.N., Verdine, G.L. (2011) Synthesis of all-hydrocarbon stapled [alpha]-helical peptides by ring-closing olefin metathesis. Nat. Protocols 6: 761–771.

[101] Kim, Y.-W., Kutchukian, P.S., Verdine, G.L. (2010) Introduction of all-hydrocarbon i, $i+3$ staples into α-helices via ring-closing olefin metathesis. Organic Letters 12: 3046–3049.

[102] Wallace, E.J., D'Rozario, R.S.G., Sanchez, B.M., Sansom, M.S.P. (2010) A multiscale simulation study of carbon nanotube interactions with designed amphiphilic peptide helices. Nanoscale 2: 967–975.

[103] Hartleb, H., Kroeker, K., Hertel, T. (2012) Density gradient ultracentrifugation and stability of SWNT–peptide conjugates. Chem. Phys. Lett. 535: 131–135.

[104] Jacobsen, N.S., Pantano, P. (2011) Determining the percentages of semi-conducting and metallic single-walled carbon nanotubes in bulk soot. Carbon 49: 1998–2006.

[105] Wang, R., Mikoryak, C., Li, S., Bushdiecker II, D., Musselman, I.H., Pantano, P., Draper, R.K. (2011) Cytotoxicity screening of single-walled carbon nanotubes: detection and removal of cytotoxic contaminants from carboxylated carbon nanotubes. Molecular Pharmaceutics 8: 1351–1361.

[106] Angelikopoulos, P., Schou, K., Bock, H. (2010) Surfactant-induced forces between carbon nanotubes. Langmuir 26: 18874–18883.

[107] Jeong, W.-j., Lim, Y.-b. (2012) Combination self-assembly of beta-sheet peptides and carbon nanotubes: functionalizing carbon nanotubes with bioactive beta-sheet block copolypeptides. Macromolecular Bioscience 12: 49–54.

[108] Ghadiri, M.R., Granja, J.R., Milligan, R.A., Mcree, D.E., Khazanovich, N. (1993) Self-assembling organic nanotubes based on a cyclic peptide architecture. Nature 366: 324–327.

[109] Hartgerink, J.D., Clark, T.D., Ghadiri, M.R. (1998) Peptide nanotubes and beyond. Chem. Eur. J. 4: 1367–1372.

[110] Pavone, V., Benedetti, E., Diblasio, B., Lombardi, A., Pedone, C., Tomasich, L., Lorenzi, G.P. (1989) Regularly alternating Lp peptides. III. Hexacyclic peptides from valine or phenylalanine. Biopolymers 28: 215–223.

[111] Khazanovich, N., Granja, J.R., McRee, D.E., Milligan, R.A., Ghadiri, M.R. (1994) Nanoscale tubular ensembles with specified internal diameters – design of a self-assembled nanotube with a 13-angstrom pore. J. Am. Chem. Soc. 116: 6011–6012.

[112] Buriak, J.M., Ghadiri, M.R. (1997) Self-assembly of peptide based nanotubes. Mat. Sci. Eng. C 4: 207–212.

[113] Bong, D.T., Clark, T.D., Granja, J.R., Ghadiri, M.R. (2001) Self-assembling organic nanotubes. Angew. Chem. Int. Ed. 40: 988–1011.

[114] De Santis, P., Morosetti, S., Rizzo, R. (1974) Conformational analysis of regular enantiomeric sequences. Macromolecules 7: 52–58.

[115] Amorin, M., Castedo, L., Granja, J.R. (2003) New cyclic peptide assemblies with hydrophobic cavities: the structural and thermodynamic basis of a new class of peptide nanotubes. J. Am. Chem. Soc. 125: 2844–2845.

[116] Horne, W.S., Stout, C.D., Ghadiri, M.R. (2003) A heterocyclic peptide nanotube. J. Am. Chem. Soc. 125: 9372–9376.

[117] Amorin, M., Castedo, L., Granja, J.R. (2008) Folding control in cyclic peptides through *N*-methylation pattern selection: formation of antiparallel β-sheet dimers, double reverse turns and supramolecular helices by 3α,γ cyclic peptides. Chem. Eur. J. 14: 2100–2111.

[118] Chiu, C.-c., Maher, M.C., Dieckmann, G.R., Nielsen, S.O. (2010) Molecular dynamics study of a carbon nanotube-binding reversible cyclic peptide. ACS Nano 4: 2539–2546.

[119] Becraft, E.J., Klimenko, A.S., Dieckmann, G.R. (2009) Influence of alternating L-/D-amino acid chiralities and disulfide bond geometry on the capacity of cysteine-containing reversible cyclic peptides to disperse carbon nanotubes. Biopolymers (Peptide Science) 92: 212–221.

9

Nanosized Vectors for Transfection Assembled from Peptides and Nucleic Acids

Burkhard Bechinger

Institut de Chimie, CNRS, Université de Strasbourg, France

Acronyms

AAV	Adeno-associated virus
*ATR FTIR	Attenuated total reflection Fourier transform infrared
CPP	Cell penetrating peptide
DOPE	1,2-Dioleoyl-*sn*-glycero-3-phosphoethanolamine
DOTAP	1,2-Dioleoyl-3-trimethylammonium-propane
DT	Diphtheria toxin
HA	Hemeagglutinin
ITC	Isothermal titration calorimetry
*MAS	Magic angle spinning
MEND	Multifunctional envelope-type nano device
NLS	Nuclear localisations sequence
*NMR	Nuclear magnetic resonance
pDNA	Plasmid DNA
PEG	Polyethyleneglycol
PEI	Poly(ethyleneimine)
*siRNA	Small interfering RNA

*The abbreviations are also explained at first occurrence in the text except the more common ones, which are labelled by a star in the list above.

Peptide Materials: From Nanostructures to Applications, First Edition. Edited by Carlos Alemán, Alberto Bianco and Mariano Venanzi.

9.1 Introduction

The self-assembling properties of biomacromolecules form the basis of many well-defined structures and organelles found in cellular systems such as lipid bilayers [1] and well-organised macromolecular complexes [2]. Taking nature as a template, supramolecular chemistry has been created based on the self-organisation and assembly by noncovalent interactions [3], or polyelectrolytes by the layer-by-layer deposition of polymers through electrostatic interactions [4]. This chapter focuses on the controlled condensation of highly anionic nucleic acid polymers through the addition of cationic counter parts, predominantly polypeptides. Furthermore, the alterations and additions that are necessary to direct them to their cellular targets, into the cells and towards specific organelles are discussed. The resulting nano-devices and the specific modifications of their composition and their physicochemical characteristics enhances the therapeutic potential of the nucleic acids.

DNA and RNA are used in biomedical research and developed for applications such as gene therapy or medical treatments using whole plasmids or short oligonucleotide sequences [5, 6]. Due to the sequence information these molecules carry they have the potential to be used in a highly specific and selective manner and with great efficiency. However, in order to be useful as a medical treatment the nucleic acids have to be delivered through the body, where they need to reach the correct target. Once they arrive at the tissue or organs of destination the siRNA, antisense nucleotides and DNA have to enter the cells. In order to do so they have to cross membrane barriers, thereby reaching the cytoplasm, where they either need to develop their activities or from where their journey continues towards the nucleus or yet other organelles [7].

In order to transfer highly charged molecules such as nucleic acids across the membrane and into the cell interior special devices have been designed. On the one hand evolution has established high barriers to prevent the unwanted exchange of genetic information but on the other hand nature has also developed specialised machineries that are made for transferring nucleic acids into foreign cells. Viruses are such devices and their size ranges from a few to hundreds of nanometres. Despite their relative simplicity they come in large varieties carrying single- or double-stranded DNA or RNA. For example, whereas the adenovirus (AV) is relatively large and carries several kbp of double-stranded DNA its satellite viruses, adeno-associated virus (AAV) is only about 22 nm in size and made of a few different proteins and a single-stranded DNA molecule. To complete a viral life cycle AAV needs the help of other viruses. Therefore, it is considered to be nontoxic and exploited for gene transfection in medical applications. Furthermore, during the infectious cycle the AAV capsid is released and attaches to the nuclear pore to transfer the genomic DNA directly into the nucleus [8, 9]. Other viruses have also been used and these follow different pathways for infection [8, 10, 11]. Although viral delivery remains the most widely used approach in clinical trials it also comes with considerable side effects [12]. Therefore, 'nonviral gene delivery systems' are highly desired and are continuously being developed and tested [13].

Such artificial systems have to fulfil a variety of tasks and the present review will discuss these one by one [7, 12, 14]:

- First the anionic nucleic acid polymer has to be compacted in order to be transported more easily to its target including passage of blood vessels, entry into the cell and access to the target organelle. At the same time such compaction provides protection from the abundant enzymes that degrade polypeptides and polynucleotides.

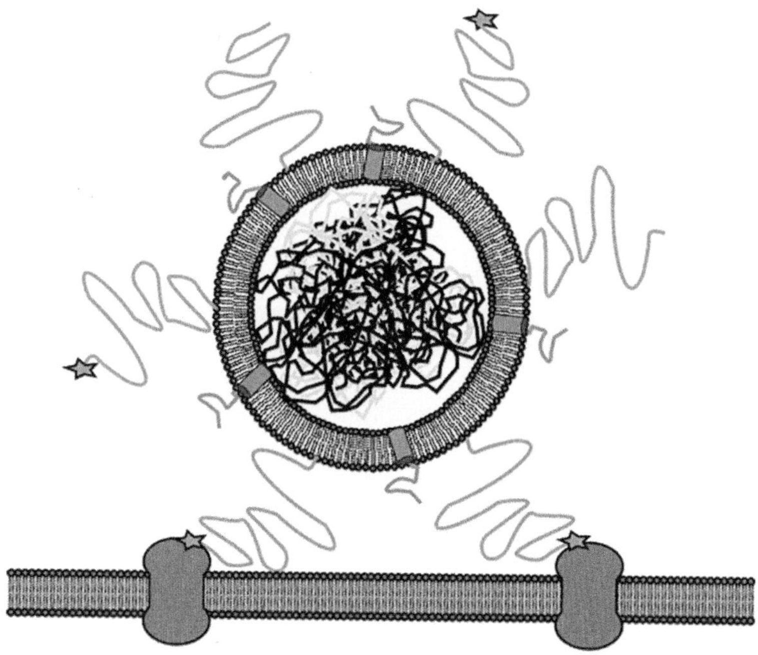

Figure 9.1 *The interaction of a multifunctional envelope-type nano device (MEND) with a specific receptor (blue) is shown. The complex of DNA (black) and a cationic polymer (yellow) is enveloped by a lipid bilayer (pink), which has been functionalised by peptides crossing the bilayer (pink), PEG tentacles (orange) and targeting ligands (green stars). The MEND particles are typically 300 nm in diameter and exhibit a negative surface charge due to the acidic lipids of the membrane [15](see color plate figure)*

- Second, the complexes have to attach to the cells and this should preferentially be a selective process (Figure 9.1).
- Third the complexes have to enter the cells and, in many cases, endosomal pathways are used (Figure 9.2).
- This results in the fourth requirement, namely efficient endosomal escape (Figure 9.2) rather than enzymatic degradation of the nucleic acids within this organelle.
- Fifth, whereas the cytoplasm is the destination of some cargo such as proteins or siRNA (Figure 9.2), other nucleic acids have to be further transported to other organelles such as the nucleus for plasmidic DNA used in gene therapeutic applications.
- Finally, the nucleic acids have to be liberated to be available for interaction with other macromolecules and/or the read-out of information (Figure 9.2).

Whereas many of the above considerations have been defined in the context of DNA and gene delivery, the focus of attention has recently shifted to short RNA polymers as they bear promise for a multitude of potential biomedical applications. Nevertheless, many of the considerations developed for DNA apply equally well to RNA transport although important details may vary. Differences include the size of the resulting complexes and their final destination.

The idea of designing nano-vectors is to assemble individual modules, each providing a selected property required for the delivery of a selected cargo to a specific target

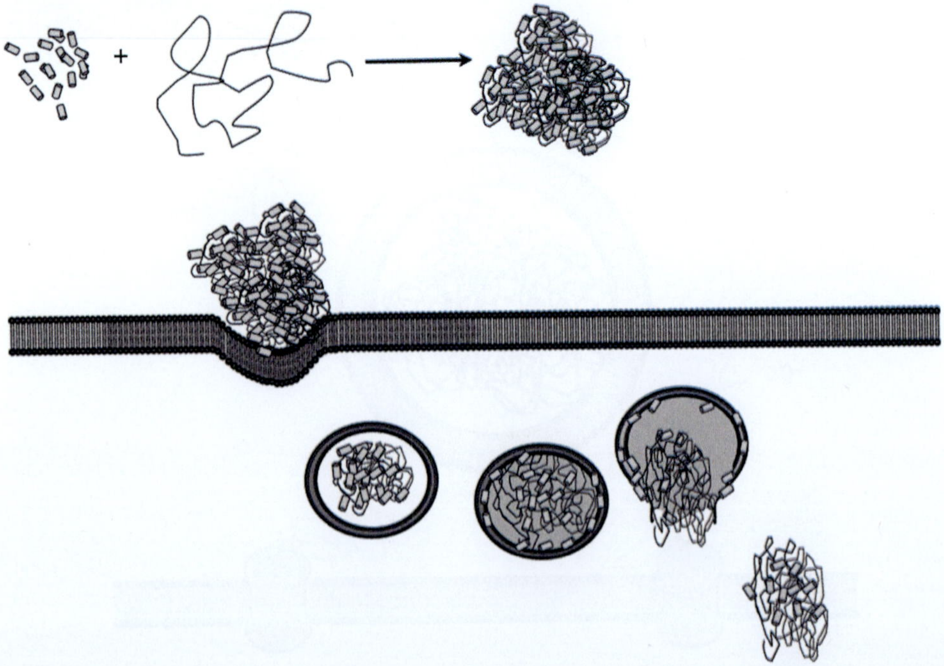

Figure 9.2 The formation (first row), endoscytosis (second row) and endosomal escape (bottom row) of an LAH4–DNA transfection complex is shown. These complexes form spontaneously through electrostatic attraction when the two polymers are mixed. The peptide has been shown active in the transfection of plasmids of 7.6 kbp [16] or of a 21-base siRNA [17]. The resulting structures are typically 100 nm to micrometers in size and carry a slightly positive surface charge [18]. The peptides are shown in yellow and the nucleic acids in black. The resulting complexes are taken up by endocytosis. Acidification of this organelle results in the release of about half of the peptides from the DNA and at least some of them become available for interactions with the membrane. This liberation of peptide from the complex results in membrane lysis and the release of DNA (or siRNA) into the cytoplasm (see colour plate figure)

(e.g. Figure 9.1). These modules can assure highly specific interactions and they have been used to tune the following properties: hydrophilicity and hence solubility, the size and specificity of targeting groups concomitant with the biodistribution, biocompatibility and biodegradation of the nano-particles as well as interactions between the drug and the carrier [19].

9.2 Condensation of Nucleic Acids by Cationic Peptides and Other Macromolecules

Many of the rationally designed transfection agents are based on synthetic cationic molecules. It is thought that these interact with the DNA in a manner that is related to the packing of the nucleic acid by histones. Thereby the cationic compounds induce the compaction of the anionic biopolymer [20–22]. To this end a great variety of molecules is used including synthetic lipids, polymers, dendrimers, nanoparticles or peptides [12, 13]. Indeed, cationic liposomes have been introduced early on and the complexes they form in combination with nucleic acids are called 'lipoplexes' [23]. Furthermore, a number of peptides have been designed that not only carry an excess of cationic charges for interaction with DNA but in addition exhibit a multitude of additional useful characteristics during the transfection process [16]. Such peptides and their applications will be discussed in more detail in the following sections where the additional requirements for efficient transfection will be discussed.

As the review focuses on peptide–nucleic acid complexes, some of the properties of peptides and their preparation shall be discussed shortly. Indeed, peptides bear a number of advantages when future biomedical applications are taken into consideration, such as the ease of product identification and quality control as well as the possibility of well-established production processes on the small and large scale, which are all important properties. Furthermore, their characteristics can easily be manipulated by changing the peptide sequence or by inclusion of non-natural amino acid residues. Indeed, when combining cationic peptides and nucleic acids many of them spontaneously self-associate into complexes by electrostatic interactions, and some of these sequences have been shown to exhibit excellent capacities to deliver nucleic acids into cells [7, 24]. On the other hand, peptides are often considered comparatively expensive for large-scale applications, although it should be borne in mind that production costs per unit can decrease by orders of magnitude when prepared in industrial settings.

9.3 The Size and Shape of Transfection Complexes

When interacting with DNA many of the nonviral vectors form nano- to microsized complexes (Figures 9.1 and 9.2). Whereas some associate in a spontaneous fashion (Figure 9.2) [16] others are designed in such a manner that they assemble in a directed manner to achieve well-defined shapes and size [25] (Figure 9.1). In order to improve the efficiency and the properties of nonviral transfection complexes their resistance to biological environments such as the conditions found in the bloodstream and their ease of passage to the target sites is considered. In this context it should be noted that although transfection has become a major tool, which is by now routinely applied in research, major improvements are still required for *in vivo* and medical applications.

As a consequence much of the present research in the nano-biosciences is directed at systematically designing mechanisms for targeted drug release, in order to maximise the therapeutic efficacy and to avoid nonwanted side effects (reviewed in reference [26]). In this context tuning the size, shape and surface properties are important for the bioavailability and the blood circulation time of nanostructures [19, 27]. Therefore one has to decide if a smaller particle is preferred, which is characterised by lower cargo capacity but also better dissolution rates [28]. This signifies that even within the nanoscale a good choice of the optimal size and shape of such a complex has to be made [29–31]. If, for example, free circulation also through capillaries has to be assured in order to reach the respective target cells, the size of the complexes should not exceed 70 nm [30]. However, an advantage of somewhat larger systems (<200 nm) is the longer half-lives in the bloodstream [29].

Nanosized complexes following this size limitation have been shown to be internalised by endosomes [33], whereas particles exceeding 200 nm are quickly eliminated by phagocytes [34]. Interestingly, by choosing the correct size of liposomal carriers (100–200 nm) their accumulation at tumour tissues has been observed [35]. Two additional examples should be mentioned where the assembly of transfection complexes has been controlled by β-sheet scaffolds [36] or by shaping delivery vehicles in the form of carbon nano-needles made of carbon nano-tubes, which carried in addition functionalised surfaces [37].

Protection from degradation in the bloodstream has been achieved by coating the surface of the complexes with polyethyleneglycol (PEG). The idea stems from liposomal carriers where so-called 'stealth liposomes' have been created to evade the immune system [38]. In the meantime this modification has also been introduced to prolong the half-life of more complex edifices (Figure 9.1) [25]. Protecting nanoparticles with PEG extends their blood circulation time, improves size uniformity, reduces nonspecific interactions with macrophages [14] and confers protection from nucleases [39]. Unfortunately, it also decreases cellular uptake, which needs compensation by, for example, chemical modification that allows for the removal of the PEG chains once the vector reaches its target size. However, the addition of peptides that favour the cellular uptake can re-establish efficient transfection (Figure 9.1) [14]. An alternative coverage of the transfection complexes with human serum albumin has also been shown to be efficient [40]. From comparison with previous studies it has been suggested that albumin plays a potential role in triggering internalisation into the cells as well as functioning as a fusogenic protein that destabilises the endosomes [40].

9.4 Cellular Targeting by Specific Ligands

In order to deliver the nanovectors to specific receptors or cellular structures the surface of the nanostructure has been functionalised with specific molecules including peptides [41, 42]. Such functionalisation can also include the selection of intracellular targets but this will be discussed later in this review [43, 44]. In order to overcome some of the hurdles that hamper efficient nucleic acid delivery, peptides are designed and/or included into complex delivery systems. These include high-affinity ligands for selected receptors of the target cells (Figure 9.1), which can result in much more efficient uptake of the transfection complexes. Ligands that have been used and proven useful include growth factors or transferrin [45, 46], a 12-residue peptide that is specific for cancer cells [47], an HIV receptor protein [48] or antibodies [49]. In addition, other parameters controlling the interactions between the particle and the cell should be considered, such as the particle size, shape and surface charge [33, 50].

9.5 Enhancing the Cellular Uptake of Nanocomplexes

In order to enhance the cellular uptake of transfection complexes and other cargo, cell penetrating peptides (CPPs) are used. These peptides are also called 'protein translocation domains' (PTDs). There is an ongoing debate about the possible pathways of cellular uptake that is efficient when these short cationic peptide sequences are used, but one possible uptake mechanism discussed is through endocytosis (Figure 9.2) [14].

Although cell penetrating peptides (CPPs) by themselves do not achieve efficient transfection of big hydrophilic cargo such as nucleic acids, they have been shown to stimulate cellular uptake of the complexes [14, 50]. Their positive charge, the size of the complexes and specific interactions with the lipids [51] make the transfection complexes more susceptible for cellular uptake even in the presence of anionic molecules [52]. Although the overall charge of the complex is an important parameter, and an optimum exists for the ratio of peptide/transfection reagent [18, 53], polylysines are less efficient in this regard when compared to polyargines. This indicates that the latter amino acid exhibits, beyond charge, additional favourable properties for the transfection process [54]. Also when CPPs are used the uptake can be optimised by tuning the size of the particles [55].

As a consequence, arginine-rich peptides such as the human immunodeficiency virus-derived TAT peptide (residues 48 to 60) or oligoarginines are often included in transfection mixtures [14]. For example, TAT as well as other CPPs have been shown to ameliorate the transfection activities of poly(ethyleneimine) *in vitro* (PEI; cf. below for details) [56, 57]. Furthermore, hybrid peptides comprising sequences of CPPs were created thereby combining a number of different functionalities [14]. It should be noted that when various functionalities are combined with each other it is difficult to predict the resulting topology of the supramolecular aggregate. For example, when TAT is added to a complex of cationic lipids and pDNA the lipoplexes continue to exhibit a negative surface charge even at very high peptide concentrations, an observation that could not be expected [53].

Furthermore, complexes between nucleic acids and cationic liposomes spontaneously form through electrostatic attraction (Figure 9.2). These 'lipoplexes' also exhibit very efficient transfection activities [23] and they are commercially available for the laboratory. The addition of polyglutamates changes the surface charge of cationic polymers, and in some cases makes them less toxic and decreases agglutination phenomena. Furthermore, in real applications other molecules may associate with the complexes, which changes the physicochemical properties of the transfection complexes and therefore their interactions with cellular surfaces. For example, it has been observed that the absorption of plasma proteins confers a positive surface charge to particles made of SiO_2 covered with TAT when incubated in Dulbecco's medium and a negative charge in Dulbecco's modified Eagles medium/fetal calf serum [55]. As such modifications alter the accessibility to the cellular surface or its ligands they may contribute to the often observed decrease of transfection activities in the presence of serum, in addition to unspecific adsorption (neutralisation) and enzymatic degradation that may occur in such complex environments.

9.6 Assuring Endosomal Escape

Many transfection complexes use endosomal pathways to enter the cellular environment (Figure 9.2), but unfortunately many of them get trapped at this stage. Therefore the endosomal escape is a critical step for the design of effective transfection systems and has been subject to several recent reviews [12, 14, 50, 58]. For example, whereas polylysine when mixed with DNA performs poorly during transfection trials adding fusogenic anionic peptides much ameliorates efficiency [59]. As it is difficult to design *ab initio* fusogenic or

Table 9.1 *Sequences of some cationic peptides mentioned in this review*

HA2 (1-23)	GLFGA	IAGFI	ENGWE	GMIDG	WYG				
INF7	GLFEA	IEGFI	ENGWE	GMIDG	WYG				
GALA	WEAAL	AEALA	EALAE	HLAEA	LAEAL	EALAA			
KALA	WEAKL	AKALA	KALAK	HLAKA	LAKAL	KACEA			
Hel 11-7	KLLKL	LLKLW	KKLLK	LLK					
MPG	GALFL	GFLGA	AGSTM	GAWSQ	PKSKR	KV			
ppTG20	GLFRA	LLRLL	RSLWR	LLLRA					
Vpr52-96	DTWTG	VEALI	RILQQ	LLFIH	FRIGC	RHSRI	GIIQQ	RRTRN	GASKS
LAH4	KKALL	ALALH	HLAHL	ALHLA	LALKK	A			
LAH4-L1	KKALL	AHALH	LLALL	ALHLA	HALKK	A			
LAH4-L2	KKALL	ALALH	HLALL	ALHLA	HALKK	A			
Penetratin	RQIKI	WFQNR	RMKWK	K					
TAT 48-60	GRKKR	RQRRR	AHQ						

membrane lytic activities naturally occurring peptides are used as a template. For example, the N-terminal segment of the HA-2 domain of the Influenza virus hemagglutinin protein (Table 9.1) becomes fusogenic in a pH-dependent manner [60]. Thereby, under acidic conditions, as they occur in the later stages of the endosome, they facilitate the passage of the nucleic acids into the cytosol, probably by partial disruption of the membrane. It is thereby avoided that the polylysine/DNA complexes are entrapped in the endolysosomal compartment. By using chloroquine as a biochemical marker it was shown that acidification is a necessary step for the efficient endosomal escape by some peptides [16]. Furthermore, biophysical studies have shown that the HA-2 mediated permeabilisation step is related to the protonation of the side chain carboxyl groups of the peptides, which induces a random coil to α-helix conformational change [60].

The GALA peptide (Table 9.1) is water soluble at neutral conditions and undergoes a random coil-to-helix transition upon a decrease in pH [61]. The helical amphipathic structure shows a high affinity for membranes [61] where pores are formed in a cooperative manner [62, 63]. Both the GALA and the HA-2 fusogenic peptides have been used as additives to various types of DNA transfection complexes to enhance their activity [61, 64–66].

INF7 is an HA-2 derivative known to form pores in liposomes with an increased activity at acidic pH and which performs better than the parent peptide (Table 9.1). It has been successfully used to enhance the activity of cationic liposomes as well as polymer transfection complexes [67–69]. Addition of INF7 to a transfection mixture of PEG and a DNA-binding peptide has resulted in a 50-fold increase in cell transfection activity [70]. INF7 also improved siRNA delivery by nanoparticles [71].

Therefore, anionic fusogenic peptides provide a valuable strategy to promote endosomal escape and it was sought to combine several beneficial activities of the peptide additives in using them as a complexation agent, a helper for endosomal uptake and mediators of pH-dependent membrane permeabilisation. KALA is such a multifunctional peptide for gene delivery [72] and was created by replacing the glutamines of GALA by lysines (Table 9.1). This amphipathic sequence was designed to interact with lipid bilayers at acidic pH.

In a related manner, the anionic amino acids of the fusogenic peptide JTS-1 were replaced by arginines and ppTG20 was created (Table 9.1) [73]. Other amphiphilic peptides used during transfection are Hel 11-7 [74], MPG and Vpr52-96, where the last two are derived from viral protein sequences (Table 9.1).

Membrane fusion and endosomal release can also be promoted by fusogenic lipids such as 1, 2-dioleoyl-*sn*-glycero-3-phosphoethanolamine (DOPE) [75], which have been mixed into lipoplexes [76]. At low pH the phosphatidyethanolamine lipids exhibit a tendency to adopt inverted hexagonal phases, which is thought to be an intermediate of membrane fusion (mechanisms reviewed in reference [77]).

Alternatively, large proteinaceous transporters such as the T-domain of diphtheria toxin have been used for transmembrane transport [78–80]. Upon acidification the protein exhibits membrane insertion and can accomplish the passage of large cargo across the lipid bilayer. The low pH conditions result in large conformational changes that are needed to drive the delivery of cargo into the cells [81].

Yet another strategy takes advantage of the so-called ‘proton sponge effect’ [13, 58, 82]. The underlying idea is based on the capacity of polyethylenimines (PEI) to accept a large number of protons, thereby buffering the endosomes when the associated pumps work to achieve a decrease in pH. The massive proton influx into the endosome is accompanied by the passive influx of counterions such as chloride. The accumulation of ions inside the endosomes is concomitant with osmotic swelling and finally the disruption of the organelle concomitant with the escape of the transfection complexes. The design of other polymers with high transfection capacities is based on the same mechanism and has been pursued by coupling histidines on to a polylysine [83]. The conjugate simulates the effects of PEI as the imidazole group has a pK_a value of about 6 and therefore changes its protonation state during endosomal acidification.

Incorporation of endosome-disrupting polymers, PEI-containing lipids or multiple imidazoles in the delivery system by conjugation or complex formation has become a design feature of nucleic acid delivery systems where the cationic compounds at the same time bind and compact anionic polymers [13, 58]. As this electrostatic association is reversible the nucleic acids can easily be liberated once they reach their destination [84].

Unfortunately toxic effects have also been observed for cationic PEI complexes [85]. Therefore, other compounds were tested such as combinations of TAT with polyhistidines [86]. Although less efficient in transfection assays they are also much less toxic than PEI. Compromises have to be found and the best choice depends on the desired application. Finally, it is thought that the surface decoration with PEI has additional beneficial effects by protecting the particles from serum [87].

9.7 A Family of Multifunctional Peptide Sequences

Based on the ensemble of ideas presented so far a new family of peptides has been developed. These peptides are cationic, complex nucleic acids and also very potent in endosomal release activities [16]. The amino acid sequences consist of alanines and leucines, which are flanked by two lysines at each terminus (LAH4-peptides in Table 9.1). This uniform composition is interrupted by four histidines, which are arranged in such a manner as to line up on one side of an α-helical structure [88, 89]. Structural studies confirm that the peptides indeed adopt α-helical conformations when forming peptide oligomers [90], when interacting with DNA [84], lipid bilayers [91] or detergent micelles [89]. Probably the resulting amphipathic nature of this secondary structure is favourable to homo- and heteromeric interactions with other biomacromolecules and thereby enhances the biological activities of the peptide.

These designed LAH4 sequences are highly efficient in transfecting DNA [16] and siRNA [17] and they are also potent antimicrobial compounds [91–94], thereby reflecting their initial design using magainins as a template [36]. Notably, the peptides are more efficient in the delivery of siRNA when compared to commercial compounds including Lipofectamine, DOTAP (1,2-dioleoyl-3-trimethylammonium-propane) and polyethylenimine [28]. Furthermore, when added to bacteria as complexes with DNA they retain their antimicrobial properties and because of their dual functionality they are of potential interest to treat hereditary diseases such as cystic fibrosis, where *Pseudomonas aeruginosa* infections need to be kept under control [16]. Only recently a novel application has been published since these peptides were used to transport polypeptide-based vaccines adjuvanted with Toll-like receptor 9 agonist CpG oligonucleotides (CpG) [95]. Furthermore, Vectofusin, a derivative of LAH4-L1 (Table 9.1), has been shown to much enhance viral transduction of eukaryotic cell, with high potential for therapeutic applications [118].

Little is known about the molecular structure of transfection complexes. The problem lies in the often amorphous nature of these complexes, which prevents X-ray crystallographic investigations, as well as in the large size of the complexes, which makes multidimensional solution NMR investigations impossible. In order to gain insight into the structural requirements underlying good transfection activities we therefore performed a series of biochemical and biophysical investigations [92, 96–98]. Gel shift experiments and the use of inhibitors of endosomal transport and acidification indicate that a large peptide–DNA complex forms and provide evidence that these enter the cells via an endosomal pathway (Figure 9.2) [16]. When the association equilibria of LAH4 and DNA was investigated in quantitative detail using isothermal titration calorimetry (ITC) it becomes evident that at pH 7.5 up to one peptide is associated with an average 2 base pairs of DNA [84]. Association occurs in the 10 micromolar range and the thermodynamic signature of the ITC experiment suggests that it is driven by electrostatic interactions. At this pH the histidines are uncharged, indicating that the peptide interacts through the lysine side chains with the negatively charged phosphates of the DNA. With multiple lysines in the sequences and their localisation at the termini the peptide is able to bridge distant parts of the DNA and thereby condense this extended polymer. Complex formation is reversible, assuring that the genetic information can be made available to the cell once the nucleic acids reach their final destination.

After endosomal uptake of the transfection complex the pH of this compartment is decreased, which results in the four histidine side chains becoming cationic [88]. Together with the four lysines and the N-terminus the overall charge of the peptides doubles to 8 or 9. When the association of the LAH4 peptides with DNA is investigated at pH 5 fewer peptides are associated with the complex, indicating that within the endosomes about half of the peptides that were initially enclosed in the compartment are released from the DNA [84]. At acidic pH the thermodynamic signature of the ITC experiments also shows that interactions occur through a combination of hydrophobic, van der Waals and electrostatic interactions and these alterations, when compared to neutral conditions, indicate that considerable structural rearrangements occur within the complex. At the same time the peptides that are released from the complex become available to interact with endosomal membranes and if the concentration increases a threshold disrupts their integrity (Figure 9.2).

In this context it is interesting to note that pH-dependent membrane alignments have been detected using solid-state NMR and ATR FTIR spectroscopies [88, 99]. The peptides are transmembrane at pH 7 but orient along the surface at acidic pH. More recent data show that this pH-dependent transition of the peptide from surface-oriented to transmembrane is associated with a shift of the helical regions of the peptide from C-terminal to a more N-terminal localisation [89].

When the size and the surface charge of the complexes was investigated by light scattering and zeta-sizer techniques the complexes were found to carry a positive surface charge density and to exhibit hydrodynamic diameters in the 100 nm to micrometre range [18]. Importantly, the hydrodynamic diameter is dependent on the environmental conditions during the preparation, indicating that there are means to control the size of the complexes. When the ^{13}C chemical shift of the peptide in the complex were analysed using magic angle spinning solid-state NMR spectroscopy a predominatly α–helical conformation was found also in the transfection complex [84]. This technique was also used to identify residues of the peptide that are close to the DNA phosphates. Whereas such contacts could be identified for the lysine side chain amines, where a distance of about 5Å is measured, all the other side chains seemed beyond the detection limit [7].

9.8 Delivery to the Nucleus and Other Intracellular Compartments

Specific ligands have also been used to control trafficking of cargo within the cell interior (similar to the situation illustrated in Figure 9.1). For example, so-called nuclear localisation sequences (NLSs) have been shown to enhance transport into the nucleus [100, 101]. These signals are recognised by proteins of the nuclear import machinery and the NLS, thereby channelled towards the nuclear pore complex [102]. In this manner the cytostatic decapeptide named PM10 could be delivered in an enhanced fashion [44]. An alternative approach consists in coupling a DNA segment recognised by the nuclear factor kappa B (NF kappa B) transcription factor to the cargo [103]. This protein can be made to shuttle between the cytoplasm and the nucleus, thereby increasing the nuclear import of plasmid DNA and the concomitant expression of a reporter gene. It is also interesting to note that the surface charge of the particle has an influence on its intracellular localisation such that the complexes remain trapped in lysosomes or are targeted to the cytoplasm and mitochondria [43]. The possibilities of targeting cargo to the mitochondria has been reviewed in reference [104].

9.9 Combining Different Functionalities into Complex Nanovectors

As cystine bonds are stable in the bloodstream but rapidly degrade when they reach intracellular compartments, they offer a strategy that allows for the controlled delivery of cargo into the cytoplasm [105, 106]. Interpeptide cross-linking using cysteines has been used to refine the transfection process of a peptide sequence containing arginines and histidines for nucleic acid binding and endosomal escape [107]. Cystine bridges have also been used to create reduction-sensitive polymers.

By covalently linking functional domains into hybrid peptides their separation into different compartments is prevented during the transfection process [108]. For example, a variety of functionalities are present in the HA2–p53–R11 sequence, where HA2 is the hemagglutinin fusion peptide (Table 9.1), the p53 protein induces apoptosis and R11 is a cell-penetrating peptide consisting of 11 arginines [109].

Surface recognition (Figure 9.1) and endosomal escape (Figure 9.2) are combined in a proteinacious construct of three functional domains where an antibody confers target cell specifity, the diphtheria toxin T domain (DT-T) facilitates endosomal escape and domain derived from the yeast GAL4 protein assures DNA binding. This protein–DNA complex was rendered cationic with polylysine. The complex built assured delivery of a reporter gene which was 25 times improved when compared to polylysine alone [110]. As polylysine by itself is not very active the resulting efficiency still needs to be increased further.

When the DT-T domain was linked to Streptavidin and a complex formed with PEI polyplexes an order of magnitude improved transfection efficiency was monitored when compared to PEI alone [111]. Notably in these constructs the details of the conjugation have been shown to be important. Although these are relatively modest increases in cargo delivery, the studies are conceptually interesting in view of the ease with which proteins can be designed and recombined using modern molecular biology approaches.

Finally, a protein was designed consisting of four tandem repeats of H2A, a histone-derived DNA condensing domain, a targeting motif for the Erb2 receptor protein, GALA and a proteolytic cleavage site to remove the targeting motif from the substrate [112]. The latter ensures an improvement of endosomal release. A number of other complexes where different functionalities are mixed in the context of liposomal delivery systems have been described in the literature [113–115].

Finally, a layer-by-layer nanovector for nucleic acid transfection has been developed that combined a number of modules discussed in this chapter [25]. The so-called multifunctional envelope-type nano device (MEND) has been designed to allow encapsulation of condensed plasmid DNA, siRNA, protein or else inside a liposomal coat (Figure 9.1). The liposomal coat is made of the fusogenic DOPE phospholipid and negatively charged lipids, such as phosphatidic acid or cholesteryl hemisuccinate, to enhance endosomal escape. To compensate for this negative surface charge, which would otherwise interfere with cellular uptake, cationic peptides such as stearilated R8 are added [14] and thereby the core of a nanoparticle is created. The surface of this structure is thereafter functionalised by different molecules such as the attachment of PEG as a protective layer and/or ligands for specific cellular receptors including, for example, transferrin [25]. Whereas PEG helps in longer blood circulation times the ligand augments the association of the MEND with cells. A further refinement of MENDs is achieved by linking the PEG to the liposomal surface using a connection that can be cleaved enzymatically in particular in association with tumour cells [116]. Furthermore, the fusogenic GALA has been added, which results in synergistic enhancements of transfection activities [117]. The MEND particles are not cytotoxic and have been shown to be highly efficient in transfecting HeLa and A549 cells [25].

In summary, peptides maintain many of their functions also in transfection complexes or in hybrid sequences and they can be used to tune the functionality of transfection complexes by having important properties in condensing nucleic acids, helping cell recognition, initiating endocytosis and assuring endosomal escape and intracellular targeting. Furthermore, peptide scaffolds have been created to shape the nanocomplexes. Although the field of nonviral transfection is in fast development the efficiency still needs to be increased for *in vivo* applications.

Acknowledgements

The financial support by *Vaincre la Mucoviscidose* and the *Agence Nationale de la Recherche* (project TRANSPEP and LabEx Chemisty of Complex Systems) is gratefully acknowledged.

References

[1] Singer, S.J., Nicolson, G.L. (1972) The fluid mosaic model of the structure of cell membranes. Science 175: 720–731.
[2] Dunkle, J.A., Cate, J.H. (2010) Ribosome structure and dynamics during translocation and termination. Annu. Rev. Biophys. 39: 277–244.
[3] Lehn, J.M. (1995) Supramolecular Chemistry – Concepts and Perspectives. Wiley-VCH, Weinheim, Germany.
[4] Leguen, E., Chassepot, A., Decher, G., Schaaf, P., Voegel, J.C., Jessel, N. (2007) Bioactive coatings based on polyelectrolyte multilayer architectures functionalized by embedded proteins, peptides or drugs. Biomol. Eng. 24: 33–41.
[5] Bareford, L.A., Swaan, P.W. (2007) Endocytic mechanisms for targeted drug delivery. Adv. Drug Deliver Rev. 59: 748–758.
[6] Schaffert, D., Wagner, E. (2008) Gene therapy progress and prospects: synthetic polymer-based systems. Gene Ther. 15: 1131–1138.
[7] Bechinger, B., Vidovic, V., Bertani, P., Kichler, A. (2011) A new family of peptide-nucleic acid nanostructures with potent transfection activities. J. Pept. Sci. 17: 88–93.
[8] Smith, A.E., Helenius, A. (2004) How viruses enter animal cells. Science 304: 237–242.
[9] Trotman, L.C., Mosberger, N., Fornerod, M., Stidwill, R.P., Greber, U.F. (2001) Import of adenovirus DNA involves the nuclear pore complex receptor CAN/Nup214 and histone H1. Nat. Cell. Biol. 3: 1092–1100.
[10] Hughson, F.M. (1995) Structural characterization of viral fusion proteins. Curr. Biol. 5: 265–274.
[11] Lakadamyali, M., Rust, M.J., Zhuang, X.W. (2004) Endocytosis of influenza viruses. Microbes Infect. 6: 929–936.
[12] Nakase, I., Kobayashi, S., Futaki, S. (2010) Endosome-disruptive peptides for improving cytosolic delivery of bioactive macromolecules. Biopolymers 94: 763–770.
[13] Pack, D.W., Hoffman, A.S., Pun, S., Stayton, P.S. (2005) Design and development of polymers for gene delivery. Nat. Rev. Drug Discov. 4: 581–593.
[14] El-Sayed, A., Futaki, S., Harashima, H. (2009) Delivery of macromolecules using arginine-rich cell-penetrating peptides: ways to overcome endosomal entrapment. Aaps J. 11: 13–22.
[15] Kogure, K., Akita, H., Harashima, H. (2007) Multifunctional envelope-type nano device for non-viral gene delivery: concept and application of Programmed packaging. J. Control Release 122: 246–251.
[16] Kichler, A., Leborgne, C., März, J., Danos, O., Bechinger, B. (2003) Histidine-rich amphipathic peptide antibiotics promote efficient delivery of DNA into mammalian cells. Proc. Natl Acad. Sci. USA 100: 1564–1568.
[17] Langlet-Bertin, B., Leborgne, C., Scherman, D., Bechinger, B., Mason, A.J., Kichler, A. (2010) Design and evaluation of histidine-rich amphipathic peptides for siRNA delivery. Pharm. Res. 27: 1426–1436.
[18] Kichler, A., Leborgne, C., Danos, O., Bechinger, B. (2007) Characterization of the gene transfer process mediated by histidine-rich peptides. J. Mol. Med. 85: 191–201.
[19] Qiu, L.Y., Bae, Y.H. (2006) Polymer architecture and drug delivery. Pharm. Res. 23: 1–30.
[20] Behr, J. P. (1993) Synthetic gene-transfer vectors. Acc. Chem. Res. 26: 274–278.
[21] Li, W., Szoka Jr, F.C. (2007) Lipid-based nanoparticles for nucleic acid delivery. Pharm. Res. 24: 438–449.
[22] Mintzer, M.A., Simanek, E.E. (2009) Nonviral vectors for gene delivery. Chem. Rev. 109: 259–302.

[23] Felgner, P.L., Gadek, T.R., Holm, M., Roman, R., Chan, H.W., Wenz, M., Northrop, J.P., Ringold, G.M., Danielsen, M. (1987) Lipofection – a highly efficient, lipid-mediated dna-transfection procedure. Proc. Natl Acad. Sci. USA 84: 7413–7417.
[24] Deshayes, S., Morris, M., Heitz, F., Divita, G. (2008) Delivery of proteins and nucleic acids using a non-covalent peptide-based strategy. Adv. Drug Deliv. Rev. 60: 537–547.
[25] Hatakeyama, H., Akita, H., Harashima, H. (2011) A multifunctional envelope type nano device (MEND) for gene delivery to tumours based on the EPR effect: a strategy for overcoming the PEG dilemma. Adv. Drug Deliver. Rev. 63: 152–160.
[26] Goldberg, M., Langer, R., Jia, X.Q. (2007) Nanostructured materials for applications in drug delivery and tissue engineering. J. Biomat. Sci. – Polym. E 18: 241–268.
[27] Svenson, S., Tomalia, D.A. (2005) Commentary – dendrimers in biomedical applications – reflections on the field. Adv. Drug Deliver. Rev. 57: 2106–2129.
[28] Rabinow, B.E. (2004) Nanosuspensions in drug delivery. Nat. Rev. Drug Discov. 3: 785–796.
[29] Ishida, O., Maruyama, K., Sasaki, K., Iwatsuru, M. (1999) Size-dependent extravasation and interstitial localization of polyethyleneglycol liposomes in solid tumor-bearing mice. Int. J. Pharm. 190: 49–56.
[30] Kong, G., Braun, R.D., Dewhirst, M.W. (2000) Hyperthermia enables tumor-specific nanoparticle delivery: effect of particle size. Cancer Res. 60: 4440–4445.
[31] Vinogradov, S.V., Bronich, T.K., Kabanov, A.V. (2002) Nanosized cationic hydrogels for drug delivery: preparation, properties and interactions with cells. Adv. Drug. Deliver Rev. 54: 135–147.
[32] Hawley, A.E., Davis, S.S., Illum, L. (1995) Targeting of colloids to lymph-nodes – influence of lymphatic physiology and colloidal characteristics. Adv. Drug Deliver. Rev. 17: 129–148.
[33] Ogawara, K., Yoshida, M., Furumoto, K., Takakura, Y., Hashida, M., Higaki, K., Kimura, T. (1999) Uptake by hepatocytes and biliary excretion of intravenously administered polystyrene microspheres in rats. J. Drug Target 7: 213–221.
[34] Stolnik, S., Illum, L., Davis, S.S. (1995) Long circulating microparticulate drug carriers. Adv. Drug Deliver. Rev. 16: 195–214.
[35] Hughson, F.M. (1995) Molecular mechanisms of protein-mediated membrane-fusion. Curr. Opin. Struct. Biol. 5: 507–513.
[36] Lim, Y.B., Lee, E., Yoon, Y.R., Lee, M.S., Lee, M. (2008) Filamentous artificial virus from a self-assembled discrete nanoribbon. Angew. Chem. Int. Edit. 47: 4525–4528.
[37] Prato, M., Kostarelos, K., Bianco, A. (2008) Functionalized carbon nanotubes in drug design and discovery. Acc. Chem. Res. 41: 60–68.
[38] Winterhalter, M., Frederik, P.M., Vallner, J.J., Lasic, D.D. (1997) Stealth (R) liposomes: from theory to product. Adv. Drug Deliver. Rev. 24: 165–177.
[39] Lee, S.H., Kim, S.H., Park, T.G. (2007) Intracellular siRNA delivery system using polyelectrolyte complex micelles prepared from VEGF siRNA–PEG conjugate cationic fusogenic peptide. Biochem. Bioph. Res. Co. 357: 511–516.
[40] Simoes, S., Slepushkin, V., Pires, P., Gaspar, R., de Lima, M.C.P., Duzgunes, N. (2000) Human serum albumin enhances DNA transfection by lipoplexes confers resistance to inhibition by serum. BBA – Biomembranes 1463: 459–469.
[41] Katz, E., Willner, I. (2004) Integrated nanoparticle–biomolecule hybrid systems: synthesis, properties, applications. Angew. Chem. Int. Edit. 43: 6042–6108.
[42] Loi, M., Marchio, S., Becherini, P., Di Paolo, D., Soster, M., Curnis, F., Brignole, C., Pagnan, G., Perri, P., Caffa, I., Longhi, R., Nico, B., Bussolino, F., Gambini, C., Ribatti, D., Cilli, M., Arap, W., Pasqualini, R., Allen, T. M., Corti, A., Ponzoni, M., Pastorino, F. (2010) Combined targeting of perivascular endothelial tumor cells enhances anti-tumor efficacy of liposomal chemotherapy in neuroblastoma. J. Controlled Release 145: 66–73.
[43] Panyam, J., Labhasetwar, V. (2003) Biodegradable nanoparticles for drug and gene delivery to cells and tissue. Adv. Drug Deliver. Rev. 55: 329–347.
[44] Yoshikawa, T., Sugita, T., Mukai, Y., Yamanada, N., Nagano, K., Nabeshi, H., Yoshioka, Y., Nakagawa, S., Abe, Y., Kamada, H., Tsunoda, S., Tsutsumi, Y. (2008) Organelle-targeted delivery of biological macromolecules using the protein transduction domain: potential applications for peptide aptamer delivery into the nucleus. J. Mol. Biol. 380: 777–782.

[45] Bareford, L.M., Phelps, M.A., Foraker, A.B., Swaan, P.W. (2008) Intracellular processing of riboflavin in human breast cancer cells. Mol. Pharmaceut. 5: 839–848.
[46] Simoes, S., Slepushkin, V., Pires, P., Gaspar, R., de Lima, M. C. P., Duzgunes, N. (1999) Mechanisms of gene transfer mediated by lipoplexes associated with targeting ligands or pH-sensitive peptides. Gene Therapy 6: 1798–1807.
[47] Negishi, Y., Omata, D., Iijima, H., Takabayashi, Y., Suzuki, K., Endo, Y., Suzuki, R., Maruyama, K., Nomizu, M., Aramaki, Y. (2010) Enhanced laminin-derived peptide AG73-mediated liposomal gene transfer by bubble liposomes and ultrasound. Mol. Pharmaceut. 7: 217–226.
[48] Duzgunes, N., Pretzer, E., Simoes, S., Slepushkin, V., Konopka, K., Flasher, D., de Lima, M.C.P. (1999) Liposome-mediated delivery of antiviral agents to human immunodeficiency virus-infected cells. Mol. Membr. Biol. 16: 111–118.
[49] Mohr, L., Schauer, J.I., Boutin, R.H., Moradpour, D., Wands, J.R. (1997) Targeted gene delivery to hepatocellular carcinoma cells using a monoclonal antibody based nonviral gene delivery system. Hepatology 26: 941.
[50] Juliano, R.L., Nakagawa, O. (2012) Cellular uptake and intracellular trafficking of antisense and siRNA oligonucleotides. Bioconjugate Chemistry 23: 147–157.
[51] Tang, M., Waring, A.J., Lehrer, R.I., Hong, M. (2008) Effects of guanidinium-phosphate hydrogen bonding on the membrane-bound structure and activity of an arginine-rich membrane peptide from solid-state NMR spectroscopy. Angew. Chem. Int. Ed. Engl. 47: 3202–3205.
[52] Khalil, I.A., Kogure, K., Futaki, S., Hama, S., Akita, H., Ueno, M., Kishida, H., Kudoh, M., Mishina, Y., Kataoka, K., Yamada, M., Harashima, H. (2007) Octaarginine-modified multifunctional envelope-type nanoparticles for gene delivery. Gene Ther. 14: 682–689.
[53] Hyndman, L., Lemoine, J.L., Huang, L., Porteous, D.J., Boyd, A.C., Nan, X.S. (2004) HIV-1 Tat protein transduction domain peptide facilitates gene transfer in combination with cationic liposomes. J. Control Release 99: 435–444.
[54] El-Sayed, A., Khalil, I.A., Kogure, K., Futaki, S., Harashima, H. (2008) Octaarginine- and octalysine-modified nanoparticles have different modes of endosomal escape. J. Biol. Chem. 283: 23450–23461.
[55] Mao, Z.W., Wan, L., Hu, L., Ma, L., Gao, C.Y. (2010) Tat peptide mediated cellular uptake of SiO(2) submicron particles. Colloid Surface B 75: 432–440.
[56] Huang, R.Q., Pei, Y.Y., Jiang, C. (2007) Enhanced gene transfer into brain capillary endothelial cells using Antp-modified DNA-loaded nanoparticles. J. Biomed. Sci. 14: 595–605.
[57] Rudolph, C., Plank, C., Lausier, J., Schillinger, U., Muller, R.H., Rosenecker, J. (2003) Oligomers of the arginine-rich motif of the HIV-1 TAT protein are capable of transferring plasmid DNA into cells. J. Biol. Chem. 278: 11411–11418.
[58] Midoux, P., Pichon, C., Yaouanc, J.J., Jaffres, P.A. (2009) Chemical vectors for gene delivery: a current review on polymers, peptides and lipids containing histidine or imidazole as nucleic acids carriers. Brit. J. Pharmacol. 157: 166–178.
[59] Wagner, E., Zenke, M., Cotten, M., Beug, H., Birnstiel, M.L. (1990) Transferrin-polycation conjugates as carriers for DNA uptake into cells. Proc. Natl Acad. Sci. USA 87: 3410–3414.
[60] Plank, C., Oberhauser, B., Mechtler, K., Koch, C., Wagner, E. (1994) The influence of endosome-disruptive peptides on gene transfer using synthetic virus-like gene transfer systems. J. Biol. Chem. 269: 12918–12924.
[61] Li, W.J., Nicol, F., Szoka, F.C. (2004) GALA: a designed synthetic pH-responsive amphipathic peptide with applications in drug and gene delivery. Adv. Drug Deliver. Rev. 56: 967–985.
[62] Choi, H.S., Huh, J., Jo, W.H. (2006) Electrostatic energy calculation on the pH-induced conformational change of influenza virus hemagglutinin. Biophys. J. 91: 55–60.
[63] Nicol, F., Nir, S., Szoka, F.C. (2000) Effect of phospholipid composition on an amphipathic peptide-mediated pore formation in bilayer vesicles. Biophys. J. 78: 818–829.
[64] Moradpour, D., Schauer, J.I., Zurawski, V.R., Wands, J.R., Boutin, R.H. (1996) Efficient gene transfer into mammalian cells with cholesteryl-spermidine. Biochem. Bioph. Res. Co. 221: 82–88.
[65] Simoes, S., Slepushkin, V., Gaspar, R., de Lima, M.C.P., Duzgunes, N. (1999) Successful transfection of lymphocytes by ternary lipoplexes. Bioscience Rep. 19: 601–609.
[66] Vaysse, L., Burgelin, I., Merlio, J.P., Arveiler, B. (2000) Improved transfection using epithelial cell line-selected ligands and fusogenic peptides. BBA – Gen. Subjects 1475: 369–376.

[67] Funhoff, A.M., van Nostrum, C.F., Koning, G.A., Schuurmans-Nieuwenbroek, N.M.E., Crommelin, D.J.A., Hennink, W.E. (2004) Endosomal escape of polymeric gene delivery complexes is not always enhanced by polymers buffering at low pH. Biomacromolecules 5: 32–39.

[68] Jiang, X., Lok, M.C., Hennink, W.E. (2007) Degradable-brushed pHEMA–pDMAEMA synthesized via ATRP and click chemistry for gene delivery. Bioconjugate Chemistry 18: 2077–2084.

[69] van Rossenberg, S.M.W., Sliedregt-Bol, K.M., Meeuwenoord, N.J., van Berkel, T.J.C., van Boom, J.H., van der Marel, G.A., Biessen, E.A.L. (2002) Targeted lysosome disruptive elements for improvement of parenchymal liver cell-specific gene delivery. J. Biol. Chem. 277: 45803–45810.

[70] Moore, N.M., Sheppard, C.L., Barbour, T.R., Sakiyama-Elbert, S.E. (2008) The effect of endosomal escape peptides on *in vitro* gene delivery of polyethylene glycol-based vehicles. J. Gene Med. 10: 1134–1149.

[71] Mykhaylyk, O., Vlaskou, D., Hammerschmid, E., Zelphati, O., Resenecker, J., Plank, C. (2008) siRNA magnetofection *in vitro*. Hum. Gene Ther. 19: 1170.

[72] Wyman, T.B., Nicol, F., Zelphati, O., Scaria, P.V., Plank, C., Szoka Jr, F.C. (1997) Design, synthesis, and characterization of a cationic peptide that binds to nucleic acids and permeabilizes bilayers. Biochemistry 36: 3008–3017.

[73] Gottschalk, S., Sparrow, J.T., Hauer, J., Mims, M.P., Leland, F.E., Woo, S.L., Smith, L.C. (1996) A novel DNA–peptide complex for efficient gene transfer and expression in mammalian cells. Gene Ther. 3: 448–457.

[74] Niidome, T., Takaji, K., Urakawa, M., Ohmori, N., Wada, A., Hirayama, T., Aoyagi, H. (1999) Chain length of cationic alpha-helical peptide sufficient for gene delivery into cells. Bioconjug. Chem. 10: 773–780.

[75] Zhou, X.H., Huang, L. (1994) DNA transfection mediated by cationic liposomes containing lipopolylysine – characterization and mechanism of action. BBA – Biomembranes 1189: 195–203.

[76] Farhood, H., Serbina, N., Huang, L. (1995) The role of dioleoyl phosphatidylethanolamine in cationic liposome-mediated gene-transfer. BBA – Biomembranes 1235: 289–295.

[77] Hafez, I.M., Cullis, P.R. (2001) Roles of lipid polymorphism in intracellular delivery. Adv. Drug Deliver. Rev. 47: 139–148.

[78] Hu, V.W., Holmes, R.K. (1984) Evidence for direct insertion of fragment-A and fragment-B of diphtheria-toxin into model membranes. J. Biol. Chem. 259: 2226–2233.

[79] Kagan, B.L., Finkelstein, A., Colombini, M. (1981) Diphtheria toxin fragment forms large pores in phospholipid bilayer membranes. Proc. Natl Acad. Sci. USA JID – 7505876 78: 4950–4954.

[80] Liger, D., Nizard, P., Gaillard, C., van der Spek, J.C., Murphy, J.R., Pitard, B., Gillet, D. (1998) The diphtheria toxin transmembrane domain as a pH sensitive membrane anchor for human interleukin-2 and murine interleukin-3. Prot. Eng. 11: 1111–1120.

[81] Chenal, A., Prongidi-Fix, L., Perier, A., Vernier, G., Lambotte, S., Fragneto, G., Bechinger, B., Gillet, D., Forge, V., Ferrand, M. (2009) Deciphering membrane insertion of the diphtheria toxin T domain by specular neutron reflectometry and solid-state NMR spectroscopy. J. Mol. Biol. 391: 872–883.

[82] Boussif, O., Lezoualc'h, F., Zanta, M.A., Mergny, M.D., Scherman, Demeneix, B., Behr, J.P. (1995) A versatile vector for gene and oligonucleotide transfer into cells in culture and *in vivo*: polyethylenimine. Proc. Natl Acad. Sci. USA 92: 7297–7301.

[83] Midoux, P., Monsigny, M. (1999) Efficient gene transfer by histidylated polylysine/pDNA complexes. Bioconjug. Chem. 10: 406–411.

[84] Prongidi-Fix, L., Sugewara, M., Bertani, P., Raya, J., Leborgne, C., Kichler, A., Bechinger, B. (2007) Self-promoted uptake of peptide/ DNA transfection complexes. Biochemistry – US 46: 11253–11262.

[85] Fischer, D., Li, Y.X., Ahlemeyer, B., Krieglstein, J., Kissel, T. (2003) *In vitro* cytotoxicity testing of polycations: influence of polymer structure on cell viability and hemolysis. Biomaterials 24: 1121–1131.

[86] Lo, S.L., Wang, S. (2008) An endosomolytic Tat peptide produced by incorporation of histidine and cysteine residues as a nonviral vector for DNA transfection. Biomaterials 29: 2408–2414.

[87] Yang, Y., Lo, S.L., Yang, J.Y., Yang, J., Goh, S.S.L., Wu, C.X., Feng, S.S., Wang, S. (2009) Polyethylenimine coating to produce serum-resistant baculoviral vectors for *in vivo* gene delivery. Biomaterials 30: 5767–5774.

[88] Bechinger, B. (1996) Towards membrane protein design: pH dependent topology of histidine-containing polypeptides. J. Mol. Biol. 263: 768–775.

[89] Georgescu, J., Verly, R.M., Bechinger, B. (2010) NMR structures of the histidine-rich peptide LAH4 in micellar environments: membrane insertion, pH-dependent mode of antimicrobial action and DNA transfection. Biophys. J. 99: 2507–2515.

[90] Marquette, A., Mason, A.J., Bechinger, B. (2008) Aggregation and membrane permeabilizing properties of designed histidine-containing cationic linear peptide antibiotics. J. Pept. Sci. 14: 488–495.

[91] Vogt, T.C.B., Bechinger, B. (1999) The interactions of histidine-containing amphipathic helical peptide antibiotics with lipid bilayers: the effects of charges and pH. J. Biol. Chem. 274: 29115–29121.

[92] Mason, A.J., Bechinger, B., Kichler, A. (2007) Rational design of vector and antibiotic peptides using solid-state NMR. Mini. Rev. Med. Chem. 7: 491–497.

[93] Mason, A.J., Gasnier, C., Kichler, A., Prevost, G., Aunis, D., Metz-Boutigue, M.H., Bechinger, B. (2006) Enhanced membrane disruption and antibiotic action against pathogenic bacteria by designed histidine-rich peptides at acidic pH. Antimicrob. Agents Chemother. 50: 3305–3311.

[94] Mason, A.J., Moussaoui, W., Abdelrhaman, T., Boukhari, A., Bertani, P., Marquette, A., Shooshtarizaheh, P., Moulay, G., Boehm, N., Guerold, B., Sawers, R. J.H., Kichler, A., Metz-Boutigue, M.H., Candolfi, E., Prevost, G., Bechinger, B. (2009) Structural determinants of antimicrobial and antiplasmodial activity and selectivity in histidine rich amphipathic cationic peptides. J. Biol. Chem. 284: 119–133.

[95] Zhang, T.T., Khang, T.H., Ma, B., Xu, Y., Hung, C.F., Wu, T.C. (2011) LAH4 enhances CD8+ T cell immunity of protein/peptide-based vaccines. Vaccine 30: 784–793.

[96] Kichler, A., Mason, A.J., Bechinger, B. (2006) Cationic amphipathic histidine-rich peptides for gene delivery. Biochimica et Biophysica Acta 1576: 301–307.

[97] Mason, A.J., Leborgne, C., Moulay, G., Martinez, A., Danos, O., Bechinger, B., Kichler, A. (2007) Optimising histidine rich peptides for efficient DNA delivery in the presence of serum. J. Control Release 118: 95–104.

[98] Mason, A.J., Martinez, A., Glaubitz, C., Danos, O., Kichler, A., Bechinger, B. (2006) The antibiotic and DNA-transfecting peptide LAH4 selectively associates with, and disorders, anionic lipids in mixed membranes. FASEB J. 20: 320–322.

[99] Goormaghtigh, E., Raussens, V., Ruysschaert, J.M. (1999) Attenuated total reflection infrared spectroscopy of proteins and lipids in biological membranes [Review]. Biochimica et Biophysica Acta – Reviews on Biomembranes 1422: 105–185.

[100] Escriou, V., Carriere, M., Scherman, D., Wils, P. (2003) NLS bioconjugates for targeting therapeutic genes to the nucleus. Adv. Drug Delivery Rev. 55: 295–306.

[101] Tkachenko, A.G., Xie, H., Coleman, D., Glomm, W., Ryan, J., Anderson, M.F., Franzen, S., Feldheim, D.L. (2003) Multifunctional gold nanoparticle–peptide complexes for nuclear targeting. J. Am. Chem. Soc. 125: 4700–4701.

[102] Imamoto, N. (2000) Diversity in nucleocytoplasmic transport pathways. Cell Struct. Funct. 25: 207–214.

[103] Goncalves, C., Ardourel, M.Y., Decoville, M., Breuzard, G., Midoux, P., Hartmann, B., Pichon, C. (2009) An optimized extended DNA kappa B site that enhances plasmid DNA nuclear import and gene expression. J. Gene Med. 11: 401–411.

[104] Yamada, Y., Akita, H., Kogure, K., Kamiya, H., Harashima, H. (2007) Mitochondrial drug delivery and mitochondrial disease therapy – an approach to liposome-based delivery targeted to mitochondria. Mitochondrion 7: 63–71.

[105] Cheng, R., Feng, F., Meng, F.H., Deng, C., Feijen, J., Zhong, Z.Y. (2011) Glutathione-responsive nano-vehicles as a promising platform for targeted intracellular drug and gene delivery. J. Control Release 152: 2–12.

[106] Meng, F.H., Hennink, W.E., Zhong, Z. (2009) Reduction-sensitive polymers and bioconjugates for biomedical applications. Biomaterials 30: 2180–2198.
[107] Tanaka, K., Kanazawa, T., Ogawa, T., Suda, Y., Takashima, Y., Fukuda, T., Okada, H. (2011) A novel, bio-reducible gene vector containing arginine and histidine enhances gene transfection and expression of plasmid DNA. Chem. Pharm. Bull. 59: 202–207.
[108] Wadia, J.S., Stan, R.V., Dowdy, S.F. (2004) Transducible TAT-HA fusogenic peptide enhances escape of TAT-fusion proteins after lipid raft macropinocytosis. Nat. Med. 10: 310–315.
[109] Michiue, H., Tomizawa, K., Wei, F.Y., Matsushita, M., Lu, Y.F., Ichikawa, T., Tamiya, T., Date, I., Matsui, H. (2005) The NH2 terminus of influenza virus hemagglutinin-2 subunit peptides enhances the antitumor potency of polyarginine-mediated p53 protein transduction. J. Biol. Chem. 280: 8285–8289.
[110] Lundberg, P., El-Andaloussi, S., Sutlu, T., Johansson, H., Langel, U. (2007) Delivery of short interfering RNA using endosomolytic cell-penetrating peptides. Faseb J. 21: 2664–2671.
[111] Plank, C., Oberhauser, B., Mechtler, K., Koch, C., Wagner, E. (1994) The influence of endosome-disruptive peptides on gene transfer using synthetic virus-like gene transfer systems. J. Biol. Chem. 269: 12918–12924.
[112] Wang, Y.H., Mangipudi, S.S., Canine, B.F., Hatefi, A. (2009) A designer biomimetic vector with a chimeric architecture for targeted gene transfer. J. Control Release 137: 46–53.
[113] Akita, H., Kogure, K., Moriguchi, R., Nakamura, Y., Higashi, T., Nakamura, T., Serada, S., Fujimoto, M., Naka, T., Futaki, S., Harashima, H. (2010) Nanoparticles for *ex vivo* siRNA delivery to dendritic cells for cancer vaccines: programmed endosomal escape and dissociation. J. Control Release 143: 311–317.
[114] Kakudo, T., Chaki, S., Futaki, S., Nakase, I., Akaji, K., Kawakami, T., Maruyama, K., Kamiya, H., Harashima, H. (2004) Transferrin-modified liposomes equipped with a pH-sensitive fusogenic peptide: an artificial viral-like delivery system. Biochemistry – US 43: 5618–5628.
[115] Kobayashi, S., Nakase, I., Kawabata, N., Yu, H.H., Pujals, S., Imanishi, M., Giralt, E., Futaki, S. (2009) Cytosolic targeting of macromolecules using a pH-dependent fusogenic peptide in combination with cationic liposomes. Bioconjugate Chemistry 20: 953–959.
[116] Alexis, F., Pridgen, E., Molnar, L.K., Farokhzad, O.C. (2008) Factors affecting the clearance and biodistribution of polymeric nanoparticles. Mol. Pharmaceut. 5: 505–515.
[117] Hatakeyama, H., Ito, E., Akita, H., Oishi, M., Nagasaki, Y., Futaki, S., Harashima, H. (2009) A pH-sensitive fusogenic peptide facilitates endosomal escape and greatly enhances the gene silencing of siRNA-containing nanoparticles *in vitro* and *in vivo*. J. Control Release 139: 127–132.
[118] Buisset J., Seye, AK., Ingrao D., Galy, A., Fenard D. (2012) Hematopoieitic safety evaluation of Vectofusin-1, a new culture additive enhancing lentiviral vector entry into human hematopoietic progenitor cells, Hum. Gene Ther. 23:10, A111.

10

Properties of Disubstituted Ferrocene–Peptide Conjugates: Design and Applications

Sanela Martić, Samaneh Beheshti and Heinz-Bernhard Kraatz

Department of Physical and Environmental Sciences and Department of Chemistry, University of Toronto, Canada

Acronyms

Aib	(*S*)-3-amino-2-methylpropanoic acid
AuNP	Gold nanoparticle
Boc	*N-tert*-butoxycarbonyl
CD	Circular dichroism
Cp	Cylopentadienyl
CWA	Chemical warfare agent
DMSO	Dimethylsulfoxide
DNP	Dinitrophenyl
Fc	Ferrocene
Fca	Ferrocene amino acid
FRET	Fluorescence resonance energy transfer
HIV	Human immunodeficiency virus
IN	Integrase
ITO	Indium tin oxide
NMR	Nuclear magnetic resonance

Peptide Materials: From Nanostructures to Applications, First Edition. Edited by Carlos Alemán, Alberto Bianco and Mariano Venanzi.

PR	Protease
RT	Reverse transcriptase
SWCNRs	Single-wall carbon nanotubes

10.1 Introduction

The design of peptide bioconjugates, as mimics of protein secondary structure, has attracted much attention, since protein folding and conformational changes play a significant role in molecular recognition and function of proteins [1–3]. In this context, molecular scaffolds have been exploited to promote interactions between peptides resulting in specific structural motifs, such as α-helices or β-sheet structures [4]. A number of β-sheet scaffolds rely on strategies, including cyclizations of peptides [5], the coordination to metal ions, and the insertion of β-amino acids into the amino acid sequence [6]. More recently, ferrocene (Fc) has been exploited as a scaffold. The two cyclopentadienyl (Cp) rings separated by 3.3 Å provide an ideal distance for promoting intramolecular hydrogen bonding (H-bonding) between the peptide strands on the two Cp rings. Using Fc scaffolds it is possible to maintain the secondary structural motifs and to influence the optical and electrochemical properties of the conjugate. Fc has long been exploited as a redox active reporter due to its stability and reversible oxidation–reduction properties that are maintained in a biological environment. Not long after its discovery in the 1950s, Fc-labeled amino acids were the first bioconjugates to be studied [7], which paved the way towards other Fc-labeled peptides and proteins [8].

In this chapter, we will explore the structural motifs and inherent properties of 1,*n*′-disubstituted Fc–peptides. Unlike monosubstituted Fc–peptide conjugates, which lack the control over the self-assembly properties, their disubstituted analogues can be tailored with respect to the direction of the assembly and the H-bonding interactions between peptides. A discussion of the structural parameters, as they pertain to protein folding, will focus on the role of Fc as the scaffold directing the intra- and intermolecular H-bonding interactions. For synthetic approaches towards Fc–peptide conjugates, refer to the recent review by Metzler-Nolte [9]. In addition, spectroscopic and electrochemical properties of Fc–peptides will be explored and lead to a discussion of molecular recognition and bioanalytical applications of Fc–peptide conjugates.

10.2 Structural Considerations and Properties

Starting from the three common Fc derivatives, Fc–dicarboxylic acid (**1**), Fc–amino acid (**2**), and Fc-diamine (**3**) (see Figure 10.1), peptide conjugation is readily achieved through standard amide coupling protocols. Each Fc derivative allows for the formation of different H-bonding patterns between peptide strands. For instance, the disubstituted Fc–peptides derived from the Fc–dicarboxylic acid (**1**) and the Fc–diamine (**3**) allow formation of 10- and 14-membered H-bonding rings, while those derived from Fc–amino acid (**2**) impose the 12-membered H-bonding constraints (Figure 10.1). Commonly observed H-bonded rings present in Fc peptide conjugates are 7-, 10-, and 13-membered rings indicative of γ-, β-, and α-turns, respectively [9–12]. More rarely observed are 5-, 8-, and

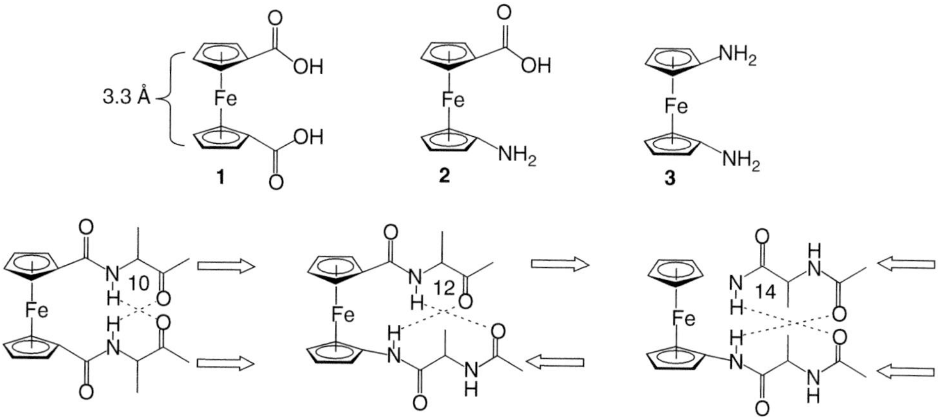

Figure 10.1 *Chemical structures of Fc–dicarboxylic acid (**1**), Fc–amino acid (**2**), and Fc–diamine (**3**). Illustration of the particular intramolecular H-bonding interactions in Fc–peptide conjugates leading to 10-, 12-, and 14-membered H-bonding rings*

11-membered rings, which represent reverse γ-, β-, and α-turns, respectively. Moreover, the directionality of the peptide arms can be tailored through direct N- or C-terminal coupling to the peptides of interest. The influence of the modification site and the directionality of the peptides on secondary structure will be further explored.

The capability of Fc–dicarboxylic acid to act as a scaffold was first exploited by Herrick and coworkers, who prepared a series of Fc–dicarboxylic acid peptide conjugates Fc[CO–Aa–OMe]$_2$ (Aa=Val, Phe, Pro) [13]. Using Fc[CO–Val–OMe]$_2$ (**4**) as a representative example of this group, the podand peptide substituents on the two Cp rings adopt a 1,2′-conformation allowing two equivalent intramolecular hydrogen bonding interactions between the amide NH of one strand and the carbonyl CO of the opposite strand to form a 10-membered H-bonded ring (Figure 10.2). Although theoretical calculations show that 1,2′-conformation or 'Herrick' conformation is energetically favoured [14, 15], there are a number of Fc–conjugates that exhibit alternative conformations. For example, in the solid state structure of Fc[CO–Phe–OMe]$_2$ (**5**), the two peptide substituents also adopt a 1,2′-conformation, but the disposition of the two amide carbonyls allows only the formation of a single cross-strand H-bond. π-Stacking of the aromatic phenyl groups presumably contributes to the stability of the 'van Staveren' conformation in the solid state [16]. However, in solution conjugate **5** presumably adopts the more stable 'Herrick' conformation. Theoretical investigations revealed that the 'Herrick' conformation is ~34 kJ mol^{-1} more stable than the 'van Staveren' conformation [17]. Fc[CO–Cys(OBn)–OMe]$_2$ (**6**) is an example of the 'Xu' conformation in the solid state. The system adopts a 1,3′-conformation and lacks intramolecular H-bonding [10, 18]. Due to the absence of H-bonding in this system, the relative energy is ~64 kJ mol^{-1} higher compared to the 'Herrick' conformation. Investigations into the conformational preferences of Fc–peptides have been extensively explored by Heinze *et al.* and have significantly enhanced our understanding of the factors influencing the structural properties of these conjugates, and provide an elegant and complementary approach to experimental studies [19].

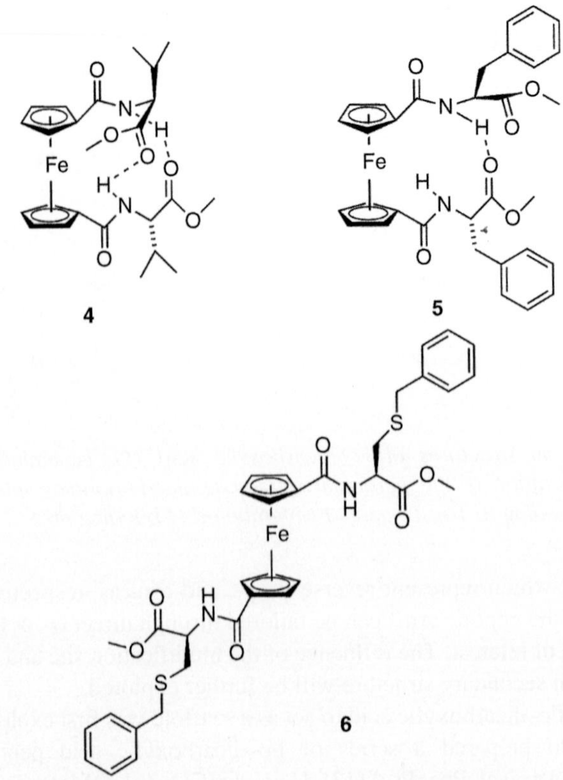

Figure 10.2 *Molecular structures of the Fc–dicarboxylic acid peptide conjugates displaying the 'Herrick' (**4**), 'van Staveren' (**5**), and 'Xu' (**6**) conformations*

Due to the intramolecular H-bonding interactions in these conjugates, rotation about the Cp–Fe–Cp axis is restricted, leading to stable helical isomers, which possess a spectroscopic signature in the Fc region of the circular dichroism (CD) spectrum. Hirao and coworkers first recognized that induced axial chirality results in a distinct pattern for L- and D-amino acid conjugates of Fc. To probe the effects of the amino acid and peptide chirality on the helicity of Fc–peptide conjugates, a variation of the D- and L-amino acid content was systematically explored [20, 21]. Using Fc[CO–L-Ala–Pro–OMe]$_2$ (**7**) (Figure 10.3) as an example, a stable *P*-helical arrangement with two intramolecular H-bonds was observed in the crystal structure and in solution. The corresponding D-Ala (**8**) conjugate is its enantiomer and displays *M*-helicity. In this context, CD spectroscopy is essential for evaluating helical conformations in solution. In general, 1,*n*′-bis(peptides)–Fc conjugates with L-amino acids proximal to the Fc group exhibit a strong positive Cotton effect in the 400–500 nm range of the CD spectrum, indicative of *P*-helicity. The corresponding Fc conjugates with a D-amino acids proximal to Fc display a negative Cotton effect, indicative of *M*-helicity. Fc conjugates with different peptide substituents on the two Cp rings generally adopt the helicity that is determined by the proximal amino acid and follow the rule 'L-amino acids → *P*-helicity; D-amino acids → *M*-helicity'. However, the particular H-bonding pattern may be influenced by steric and other factors and nonclassical H-bonding patterns have been observed in these conjugates [22].

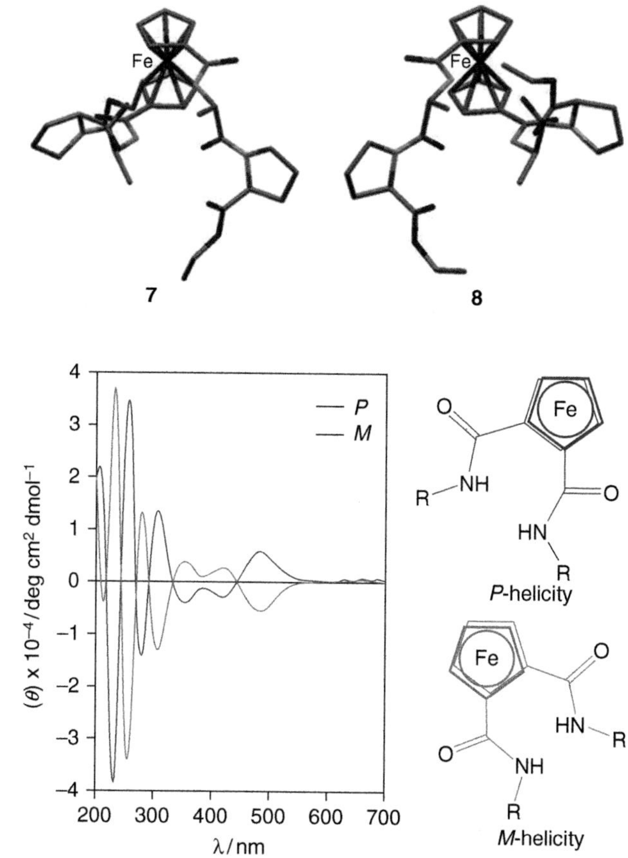

Figure 10.3 *Molecular structures of Fc[CO–L-Ala–Pro–OMe]$_2$ (**7**) and Fc[CO–D-Ala–Pro–OMe]$_2$ (**8**). Circular dichroism spectra of two Fc-–peptide isomers and the representation of the P and M nomenclature used to describe helicity of disubstituted Fc–peptides. Reproduced with permission from reference [20]. Copyright (2010) American Chemical Society*

The substitution at the peptide terminus may also induce conformational changes in Fc–peptides due to the steric effects of the C-terminal modification. For instance, in disubstituted Fc[CO–Asp$_n$]$_2$ dendrimers, the transition from the 'Herrick' to alternative conformations is induced by increasing the size of the dendrimer unit [23]. In addition, this can affect the intra- and intermolecular H-bonding patterns, as is the case for the two Fc–Gly–OR conjugates. Upon ester deprotection of Fc[CO–Gly–OEt]$_2$, the Fc core changes from the 'Xu' to 'Herrick' conformation involving loss of intramolecular H-bonding, while in Fc[CO–Gly–Val–OMe]$_2$, the 'Herrick' pattern is maintained upon methyl ester protection. Extensive intermolecular H-bonding is also established, resulting in the formation of a porous material having hydrophobic channels within the solid state [24].

Recently, examples of the extended β-sheet-like structure were reported for a series of 1,*n*′-Fc–tripeptide conjugates containing hydrophobic amino acids Gly, Val/Leu and Cys(Bn) (**9**) (Figure 10.4). The choice of Gly as the first amino acid provides maximum flexibility to the system and aligns the two peptide chains to engage additional H-bonding

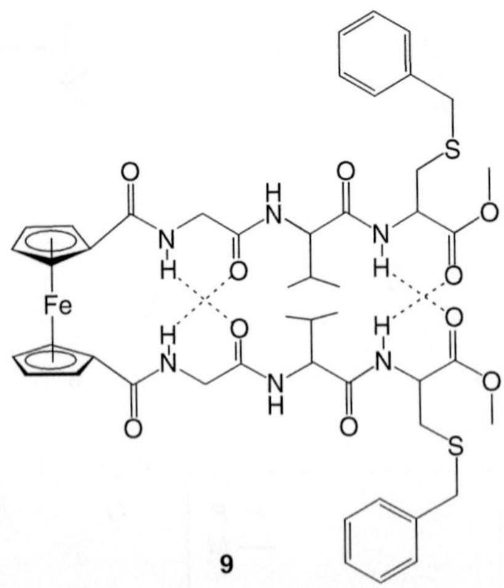

Figure 10.4 Molecular structure of the Fc[CO–Gly–Val–Cys(Bn)OMe]$_2$ (**9**) showing intramolecular H-bonding motifs

interactions akin to those found in sheet-like structures [25]. The choice of Val or Ile, having high β-sheet propensities, is essential in promoting stable peptide alignments leading to intermolecular interactions towards a sheet-like supramolecular assembly. The Fc–dicarboxylic acid structures in the solid state adopt the two 10-membered rings (β-turn) regardless of the number and type of amino acid. Increasing the steric demands in the nonsymmetrically substituted Fc–carboxylic acid conjugates results in the conformational constraints. For example, the Fc[CO–Gly–Ala–Val–NHMe][CO–Gly–Ala–Val–OMe] exhibits a number of NH and three CO groups involved in H-bonding, which may lead to a range of conformations in solution. The longer amino acids induce a conformation with a predominantly 1′-NH···OC intramolecular H-bond. It is not clear if extending the peptide substituents to longer peptides will result in additional intramolecular H-bonding providing access to stable sheet-like systems.

Some of the 1,*n*′-Fc–peptide derivatives discussed to this point form supramolecular aggregates that resemble pleated β-sheets. However, connecting the peptide substituents by a chemical linker removes a degree of freedom that results in curvature of the alignment of the peptide strands. This in turn has interesting consequences for the supramolecular assembly of the Fc–peptides. In the cystamine-linked conjugate Fc[CO–Gly–Val–CSA]$_2$ (**10**), Val residue provides a high β-sheet propensity while the proximal Gly offers flexibility [26]. In this system, the three intramolecular H-bonds (a) and the intermolecular contact in a head-to-tail fashion result in the formation of a 14-membered ring, depicted in Figure 10.5b. The four molecules interact to give a β-barrel (c) and the eight β-strands combine into the pseudo barrel and mimic other eight-stranded barrels found in biological systems [27, 28].

Disubstituted Fc–dicarboxylic acid derivatives are characterized by a stable and reproducible redox signal in the 680–900 mV range versus Ag/AgCl (acetonitrile or dichloromethane). The inherent redox stability of Fc–peptides has been explored for the study of interactions with anions, cations, and other guest molecules and will be discussed in the next section.

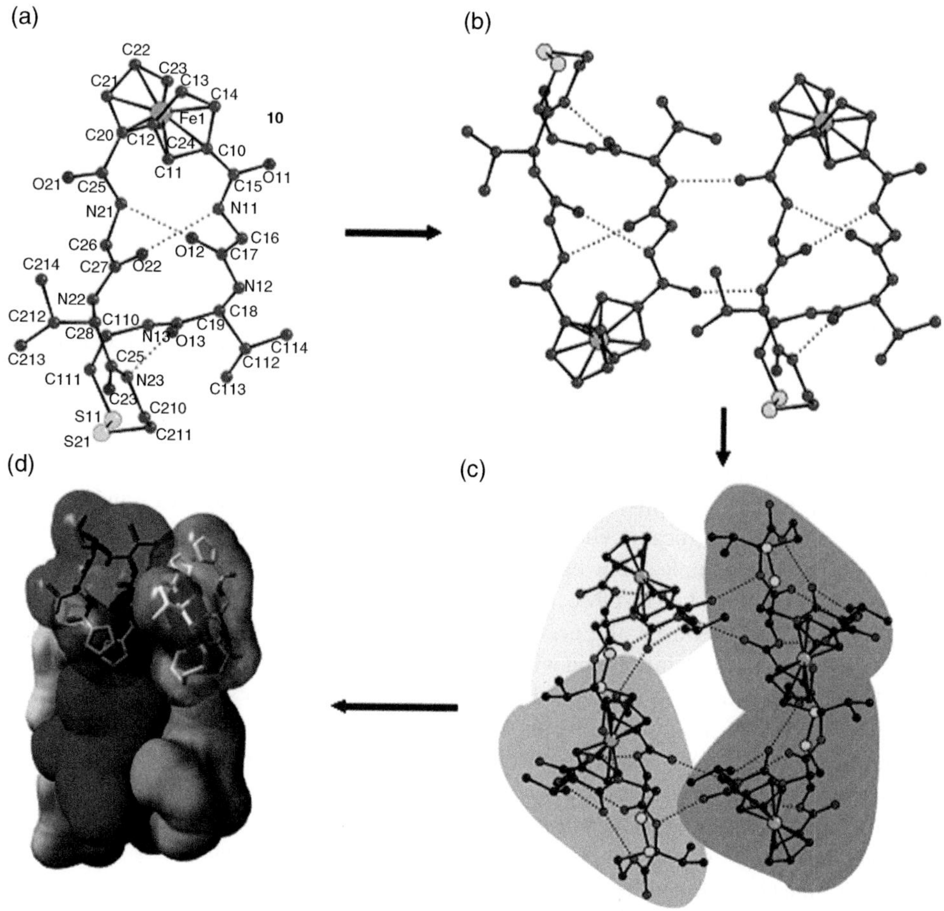

Figure 10.5 *(a) The molecular structure of Fc[CO–Gly–Val–CSA]$_2$ (**10**) showing the intramolecular H-bonding between the peptide strands on the two Cp rings. (b) The formation of an antiparallel β-sheet through intermolecular N(H) ··· O=C H-bonding. Molecules are arranged in a head-to-tail fashion. (c) Curvature of the molecular building blocks promotes H-bonding to form a β-barrel-like structure. (d) Tiling of cyclic ferrocene peptides is illustrated in this molecular surface representation of the barrel viewed along the c-axis. Reproduced with permission from reference [26]. Copyright (2006) Wiley-VCH Verlag GmbH & Co. KGaA (see color plate figure)*

Unlike Fc–dicarboxylic acid, Fc–amino acid (Fca) (**2**) allows for the attachment of two antiparallel peptide chains, which more closely mimics the natural peptide turns. Introduction of an α-(Gly, Ala, Val) and a β-amino acid induces a well-defined secondary structure, giving rise to 9- and 11-membered H-bonded rings [14, 19, 29–32]. The molecular structure of 1′-(*tert*-butoxycarbonyl (Boc))–Ala–Fca–Ala–Ala–OMe reveals an antiparallel arrangement of the two peptide strands, allowing the formation of two intramolecular hydrogen bonds (12-membered rings) [29, 32, 33]. A *P*-helical conformation results in a positive Cotton effect around λ_{max} = 480 nm, due to the Fc chromophore. Extended helical

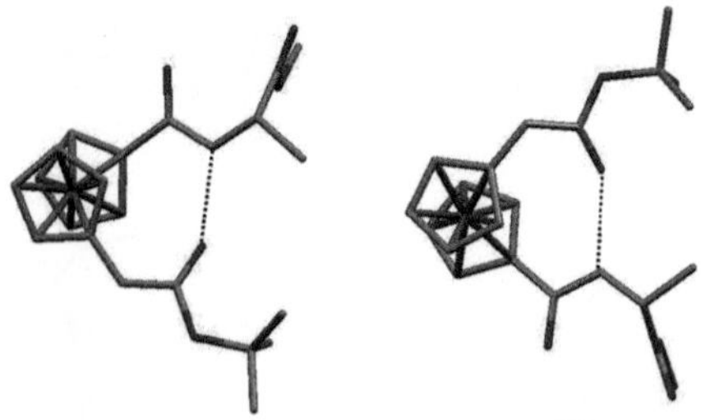

11 12

Figure 10.6 Molecular structures of the Boc–Fca–L-Ala–OMe (11) and Boc–Fca–D-Ala–OMe (12) indicating short intramolecular H-bonding contacts leading to the P- and M-helicity (redrawn from original data and adapted)

foldamers, containing repeating Fca–CO–Ala units, were reported and display a β-helical conformation. In these systems, *P*- and *M*-helicities have been observed for L- and D-Ala, respectively. Unlike Fc–dicaboxylic acid peptide conjugates, the helicity of Fca–peptide conjugates is more complex. For example, *M*- and *P*-helical conformations have been observed for Boc–Fca–L-Ala–OMe (**11**) and Boc–Fca–D-Ala–OMe (**12**), respectively (Figure 10.6) [34]. However, the acetyl (Ac)–Fca–L-Val–L-Ile–OMe conjugate exhibits *P*-helical conformation in solution with a positive CD signal [14]. The interesting differences between Boc- and Ac-terminated Fca conjugates were addressed by performing a systematic study of bisamide X–Fca–Aa–OMe and trisamide X–Aaa–Fc–Aa–OMe (X=Ac, Boc; Aa=Gly, Ala, Val) systems [30]. All trisamides exhibited two intramolecular H-bonds irrespective of the protecting groups and the α-amino acids. Moreover, the trisamides and bisamides are characterized by the positive *P*-helicity. While the effects of the protecting groups on the conformation were indiscernible, the increasing size of the amino acid side chain destabilized certain conformations.

Other Fca derivatives, such as Boc–Fca–Asp(OH)–OH have been explored with Fc[COOMe][CO–Arg–Arg–NH_2] conjugate towards understanding salt bridge-type interactions [35]. In a 1:1 complex, the interaction between diarginine and aspartic acid leads to the formation of the ion-pair without the involvement of the amide backbone. These systems are useful model systems for naturally occurring salt-bridges, which are found in the lysozyme epitope and in dihydroxylate reductase [36–38].

Alternatively, a β-lactam ring may be introduced in order to achieve additional H-bonding interactions and a potential synthetic handle for further modification, since β-lactams are known to serve as intermediates towards α- and β-amino acids [39–42]. The lactam containing Fca conjugates Fc[CO–Ala–Alm]$_2$ and Boc–Fca–Ala–Alm exhibit the intramolecular H-bonds between the two strands and an additional intrachain $NH_{Alm}\cdots CO_{Fc}$ interactions, resulting in the γ-turns and *P*-helicity arrangement [43]. Expanding further on the unsymmetrically substituted Fca, bioconjugates containing (*S*)-3-amino-2-methylpropanoic acid (Aib) were designed to probe the effects of the methylene group insertion on an H-bonding pattern [31]. The comparison between Boc–Ala–Fca–Ala–OMe and Boc–Aib–Fca–Ala–OMe indicates a disordered conformation and preferred *M*-helicity in the latter case. For Boc–Aib–Fca–Aib–OMe, the two intramolecular H-bonds are greatly

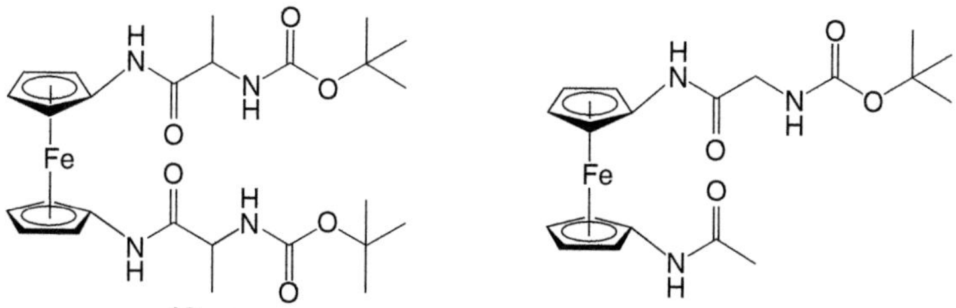

13 14

Figure 10.7 *Molecular structure of Fc–diamine Fc[NH–Ala–Boc]$_2$ (**13**) and Fc[NH–Gly–Boc][NH–Ac] (**14**)*

destabilized, resulting in the inversion of helicity of the Fc group from *P*- to *M*-helicity. Computational studies on these systems revealed that the configuration of the carbon atom of Aa1 in Aa1–Fca–Aa2, for α- and β-amino acids, directs the helicity of the Fc group.

The electrochemical properties of Fca derivatives differ from those of Fc–dicarboxylic acid analogues and display a lower redox potential making it easier to oxidize, reflecting the increased electron density on the Fc core. $E_{1/2}$ values fall in the 300–500 mV range versus Ag/AgCl, due to the amino substitution proximal to the Fc unit. The relatively low redox potential of Fca conjugates makes them useful for probing intermolecular interactions. This will be discussed in more detail in a later section.

The Fc–diamine scaffold allows the attachment of two parallel peptide strands and is complementary to the Fc–dicarboxylic acid scaffold, and provides access to the 14-H-bonded ring. The H-bonded ring size increases from 10- to 12- to 14-membered when going from Fc–dicarboxylic to Fc–amino acid to Fc–diamine conjugates, respectively. The very first examples of the symmetrically substituted Fc–diamine peptides were reported by Kraatz *et al.* [44] and have since been explored for asymmetrically substituted analogues as well [45]. The CD spectra of the two conjugates, containing D- or L- functionalities, reveal that they are mirror images of each other. For example, the Fc[NH–D-Ala–Boc]$_2$ (**13**) in Figure 10.7, containing D-Ala exhibits two intramolecular H-bonds between the amide groups connected to the Cp rings and the adjacent carbonyl of the Boc causing the *M*-helicity. The H-bonding pattern is dramatically influenced by the substitution of Cp rings and the helicity is achieved through the attachment of the first amino acid substituent regardless of the composition of the second modification [46]. In a very recent study, Rapić and coworkers [31] prepared a series of asymmetric 1,1′-diaminoferrocene conjugates of α-amino acids, such as Fc[NH–Aa–Boc][NH–Ac] (**14**) (Aa=Gly, L-Ala, D-Ala, Val) in Figure 10.7.

These systems assemble in weakly coordinating solvents, as well as in DMSO, with intramolecular H-bonds involving the NH groups proximal to the Fc unit. The majority of the low-energy conformations possess a *P*-helical ferrocene chromophore for L-amino acid substituents so that chirality organization is already achieved by attaching one amino acid substituent at the 1,1′-diaminoferrocene central unit. The electrochemical behaviour of Fc–diamine is typically characterized by the quasi-reversible redox potential in the 300–400 mV range.

10.3 Fc–Peptides to Probe Interactions

The ease of synthesis of Fc–peptides combined with their well-established redox properties make these bioconjugates ideal probes, which can be equipped with a receptor region that is able to bind to a specific target molecule or ion. In response to the binding event, the redox properties of the Fc group will be affected and in many cases this allows quantification of specific targets even in complex solutions. Towards this goal, disubstituted Fc–peptides are preferable over their monosubstituted analogues due to the preorganization of the peptide arms on each Cp ring and the ability to introduce a site for surface attachment. The choice of the peptide sequence dictates the type of interactions and binding events with a target molecule. Hence, by carefully choosing the nature of the side arm, from cationic, anionic to chelating, a variety of guests may be probes. Among the 20 common amino acids, the appropriate combination of neutral, charged, or chelating side groups allows for the design of the capture site with high specificity and selectivity. Recent advancements in the field of ion sensing and interactions with biological molecules are presented in the following sections.

10.3.1 Interactions with Ions

There are a range of Fc–peptide conjugates that act as anionic or cationic receptors. Other ion receptor reviews are available that include a comprehensive overview of the chromogenic and fluorogenic chemosensor and reagents for anions [47]. Here, the guest–host interactions including anions and cations will be discussed briefly.

The chromogenic receptor based on Fc–peptide takes advantage of the preorganized peptides substituents and H-bonding ability through thiourea or amide groups. For example, some Fc–dicarboxylic acid peptides (**15** and **16**) can distinguish between *N*-acetyl-L-aspartate and *N*-acetyl-D-aspartate [48]. Fc–carboxylic acid peptides shown in Figure 10.8 are characterized by a strong intramolecular charge transfer absorption, which may be modulated in the presence of anions. The H-bonding with guest anion results in an

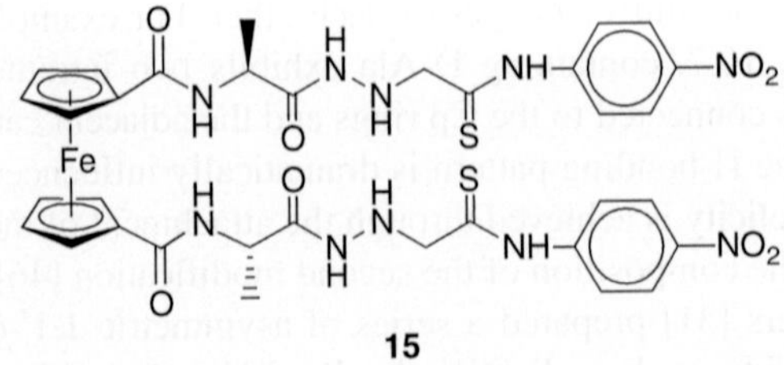

15

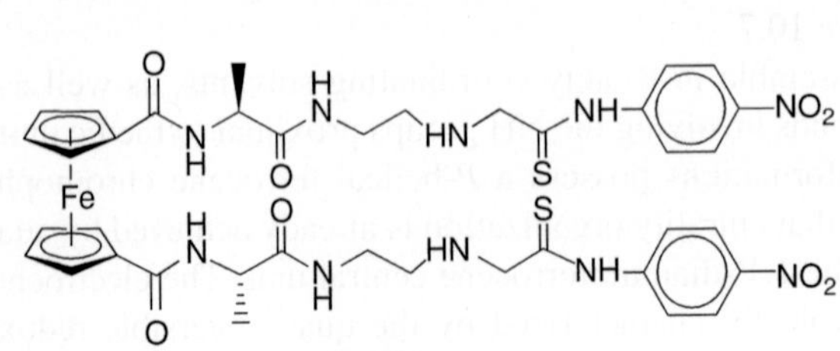

16

Figure 10.8 *Chromogenic Fc–dicarboxylic acid peptides (**15** and **16**) for detection of acetylaspartate*

increased charge transfer and the appearance of new absorptions at lower energies. In a 1:1 complexation with a hydrazine containing Fc–peptide (**15**) greater selectivity was observed with L-amino acids. By switching the hydrazine core to the diethylamine linker (**16**) the enantiomeric selectivity was reversed. These results are rationalized by considering open versus closed conformations in the two Fc–peptides.

In a simple reaction between dihydrazide and dialdehyde, a series of cyclic and acyclic receptors can be designed to produce the ferrocenophanes containing multiple Fc centres [49]. From a series of Fc-based acyclic oligomers (Figure 10.9), a disubstituted

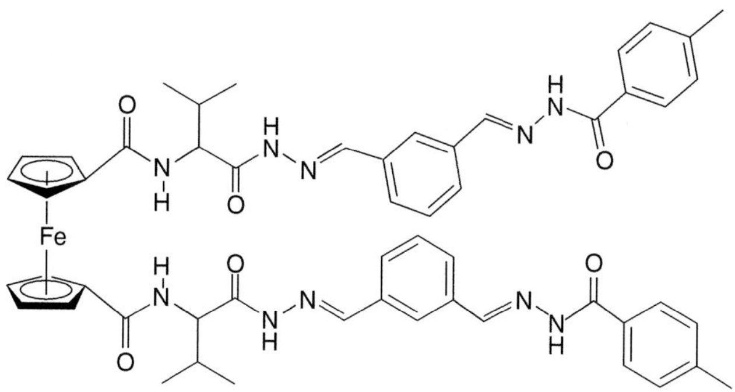

17

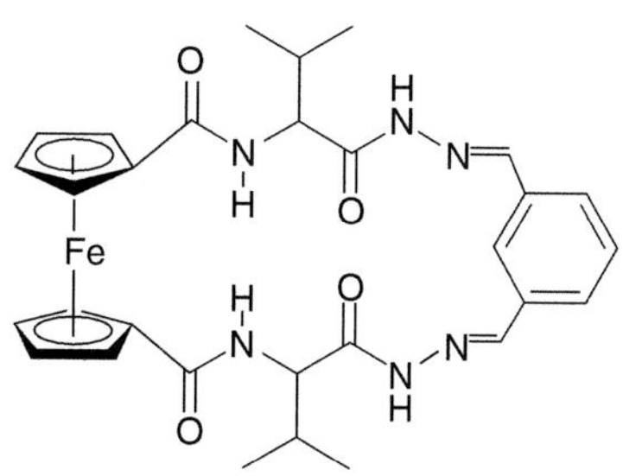

18

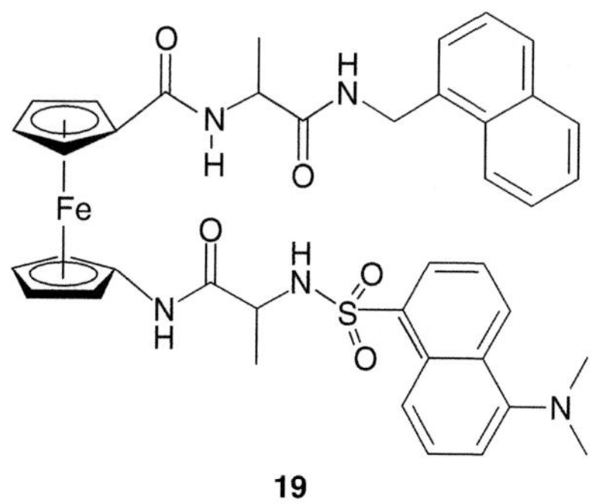

19

Figure 10.9 *Fc–peptides as representative anion probes acyclic Fc–[CO–Val–NHNH$_2$–R]$_2$ (**17**), cyclic Fc–[CO–Val–NHNH$_2$–R]$_2$ (**18**), and DN–Ala–Fca–Ala–NP (**19**)*

ferrocenophane, containing the Fc[CO–Val–NHNH$_2$]$_2$ motif (**17**), serves as a receptor for binding to multiple $H_2PO_4^-$ ions in a cooperative manner [50]. Anion binding constants were determined by monitoring the chemical shift difference in NMR spectra. While a *P*-helical Herrick conformation is present in the CD spectrum of the receptor, in the presence of two anions the helicity is maintained, indicating that the orientation of amino acid proximal to the Fc core does not change dramatically. The receptor efficiency was rationalized by preorganization of the conjugate involving intramolecular hydrogen bonding combined with the Fc group and peptide arm flexibilities.

As the representative example, the macrocycle **18** exhibits strong binding to $H_2PO_4^{3-}$ with a significantly higher binding affinity compared to other anions. NMR studies revealed that the hydrazone NH protons H-bond to the anionic guest molecules but that the amide NH protons are not involved. The macrocycle **18** adopts a van Staveren conformation in solution unlike other macrocycles with multiple Fc groups, which exhibit a *P*-helical arrangement and Herrick conformation. However, the larger macrocycles show little selectivity towards anions, attributed to their flexibility and the presence of multiple binding sites.

Recently, Heinze *et al.* designed an Fca peptide analogue **19** bearing two complementary dyes, a quencher unit, a flexible binding site and chiral groups [29]. The *N*-(1-napthylmethyl) alanyl (NP) at the C-terminus of Fca and a chiral *N*-dansylalanyl (DN) chromophore at the N-site were chosen due to their fluorescence resonant energy transfer (Figure 10.9) [51]. In the absence of anions, the Fca analogue exhibits weak fluorescence due to the inherent quenching by the Fc unit in addition to the NP-to-DN FRET. Two H-bonds between peptide residues hold the two chromophores in close proximity, but anion binding disrupts this folded structure and modifies energy- and electron-transfer pathways. For example, in the presence of $H_2PO_4^-$, HSO_4^-, and L/D-Ac–Ala–O$^-$ ions the increase in fluorescence at λ_{em} = 335 and 506 nm was observed due to the dual emission properties of the host molecule. The overall fluorescence enhancement was attributed to the reduction in the intramolecular luminescence quenching by the Fc unit in the presence of anions. The Fca receptor is characterized by a reversible oxidation at $E_{1/2}$ = 105 mV versus Fc/Fc$^+$. Following the *in situ* oxidation to ferrocenium in the presence of silver tetrafluoroborate, fluorescence anion titrations did not result in a significant enhancement due to the ability of the Fc$^+$ centre to act as an electron sink, preventing efficient FRET.

Bioinorganic chemistry is deeply rooted in understanding the biologically relevance of metal ions [52, 53]. Metal cation–peptide interactions play a critical role in proteins and are responsible for their structural and functional integrity and enzymatic activity [54, 55]. In search of the ‘ideal’ metal ions sensor, ferrocenoyl peptides incorporating thioether functionalities, such as L-methionine, *S*-methyl-L-cysteine, *S*-trityl-L-cysteine, and *S*-benzhydryl-L-cysteine, may preferably bind to soft metals, such as Hg(II). The electrochemical signature signal of the sulfur-containing Fc–peptide **20** (Figure 10.10) may be used to monitor and quantify binding of such metal ions [56, 57]. The interactions of disubstituted Fc derivatives with Hg(II) influences the redox potentials of the Fc group due to through-space and through-bond interactions upon metal ion binding to sulfur. Notably, none of other metal ions tested including Zn(II), Ag(I), and Cd(II) induced any significant response. In a titration experiment, the addition of metal ions to the solution of thioether-containing Fc–dipeptides caused a positive shift in the redox potential ~ 30–70 mV, preferentially for Hg(II) over other ions. Solution NMR studies indicated a 1:1 binding stoichiometry and weak association constants in the 20–25 M^{-1} range. The ^{1}H-NMR shift

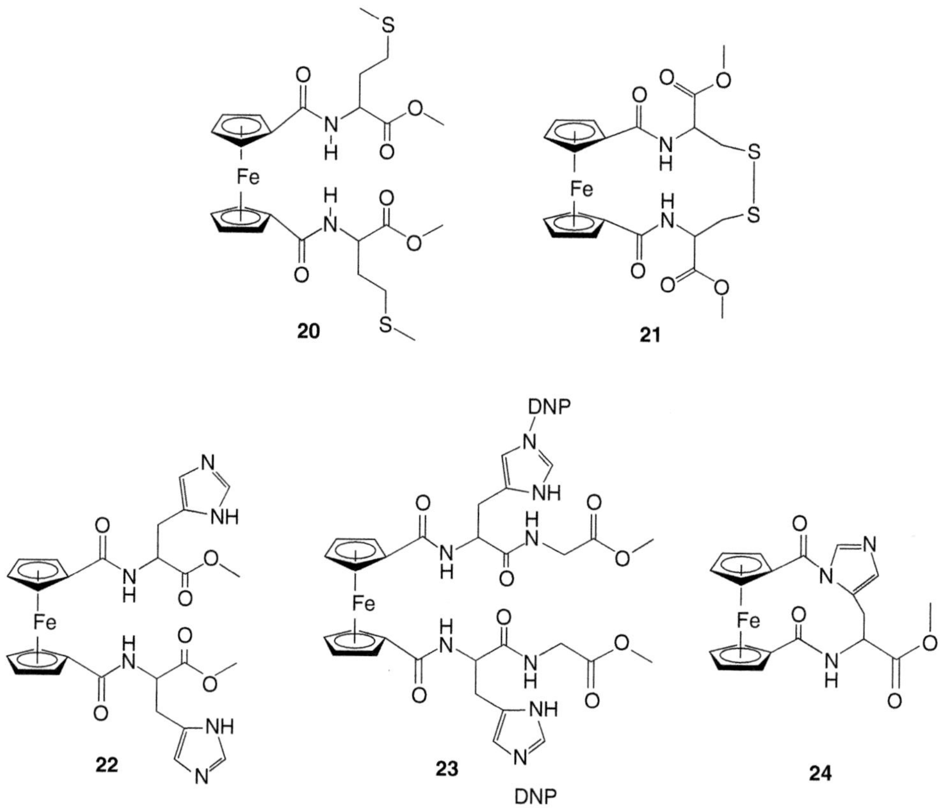

Figure 10.10 *Structures of Fc–peptides for studying interactions with metal ions: Fc–[CO–Cys(Me)–OMe]$_2$ (**20**), cyclic Fc–[CO–Met–OMe]$_2$ (**21**), Fc[CO–His–OMe]$_2$ (**22**), Fc[CO–His(DNP)–Gly–OMe]$_2$ (**23**), and cyclic Fc–peptide (**24**)*

response indicates that binding occurs in the S-containing side chains rather than to the backbone amide groups, which explains the selectivity of the system. The cyclic compound **21** contains a disulfide linkage but is unresponsive towards the addition of transition metal ions, unlike its acyclic counterparts. However, this conjugate was a good host for the Group II metal ions due to the interactions between the metal ions and the cyclic amide moiety.

Similarly, the use of amino acids known to play a critical role for metal complexation in biological systems, such as His, have recently been reported. Conjugate **22** (Figure 10.10) displays *P*-helicity, a 'Herrick' conformation and an $E_{1/2}$=751 mV [58]. This particular peptide is insensitive to the addition of alkali metal ions due to the anticonformation of the two proximal CO groups, which are not involved in metal ion binding. The metal ion interactions are thought to take place through the ester carbonyl group and/or imidazole N atom. Other metal ions have not been explored with this Fc–peptide. To improve the binding ability of the Fc–peptide containing His residues, the dipeptide Fc[CO–His(DNP)–Gly–OMe]$_2$ (**23**) was designed, which possesses more flexibility due to Gly residue proximal to the Fc group (Figure 10.10) [59]. Electrochemical studies show that metal addition to solutions of this conjugate result in cathodic shifts. In some cases, these shifts to lower half-wave

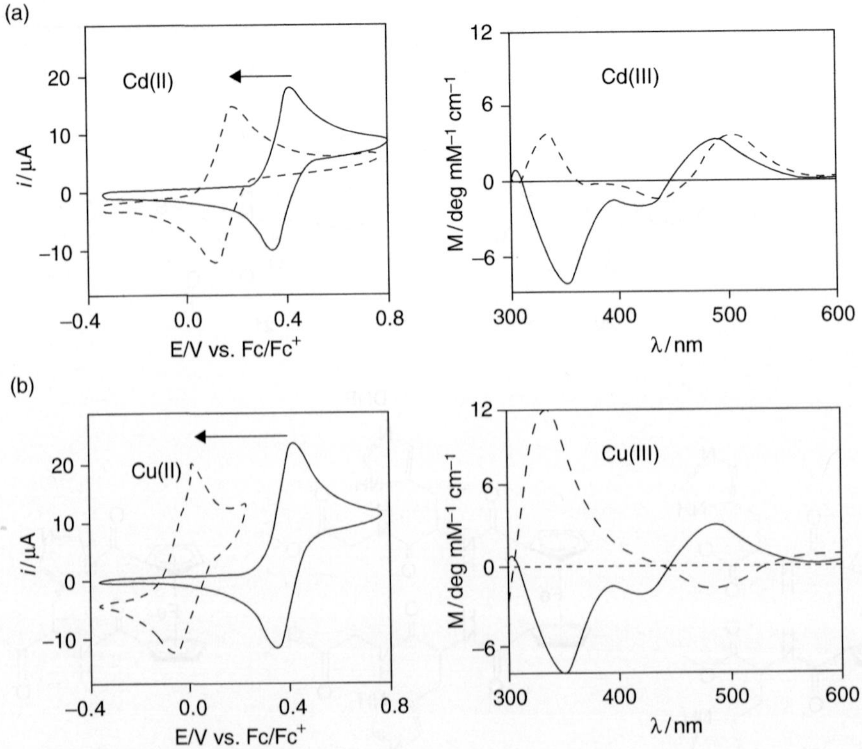

Figure 10.11 *Representative cyclic voltammograms (left) and circular dichroism spectra (right) of Fc–conjugate* ***23*** *before (solid line) and after (dashed line) the addition of (a) Cd(II) and (b) Cu(II) metal ions. Reproduced with permission and modified from reference [59]. Copyright (2010) Wiley-VCH Verlag GmbH & Co. KGaA*

potentials are significant and are a reflection of the interaction and the electronic situation about the Fc core (Figure 10.11). In all cases, a 1:1 complex formation was confirmed by NMR titration experiments. NMR results further indicate that the N atom of the imidazole group is mainly involved in metal ion binding, which leads to an enhanced conformational rigidity of the peptide substituents. However, much more information rich results were obtained from the CD data. The *P*-helicity associated with the Fc group and the orientation of the peptide arms is maintained during the interactions with Mg(II), Na(I), Zn(II), and Cd(II), but *M*-helicity was observed for Fe(II) and Cu(II) (Figure 10.11). The retention of *P*-helicity in CD suggests that the metal ion coordinated in a tetrahedral fashion to the two imidazoles of the Fc–peptide and does not perturb the conformation around the Fc core. In contrast, Fe(II) and Cu(II) ions induce a significant conformational change around the Fc core, leading to *M*-helicity, which was attributed to the particular binding geometry imposed on the complex by the central Cu(II) or Fe(III) ions. In contrast, the cyclic Fc–peptide analogue favourably coordinates to Group I metal ions through the amide carbonyl groups [58], which is a common coordination strategy in other Fc–dipeptides. The cyclic Fc–peptide containing a His residue was investigated first in solution with alkali and alkaline earth ions.

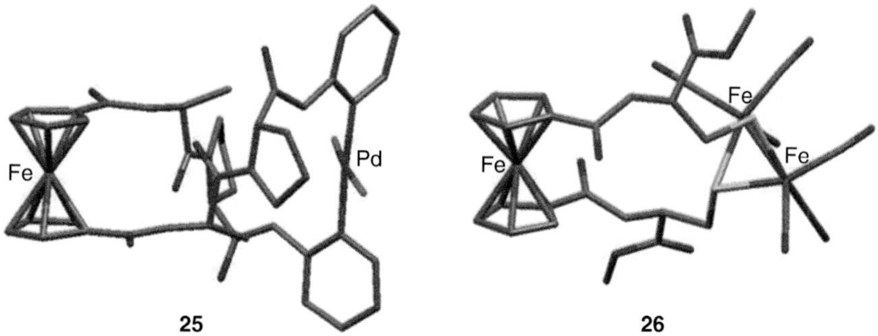

25 26

Figure 10.12 *Molecular structure of Fc[CO–L-Pro–L-Ala–NH-2–Py]$_2$ coordinated to Pd(II) (**25**) and an Fc–hydrogenase mimic (**26**) prepared from the coordination of Fc[CO–Cys–OMe]$_2$ with $Fe_3(CO)_{12}$ (redrawn from original data and adapted)*

The cyclic Fc–conjugate **24** in Figure 10.10 interacts with alkali ions such as Li(I) and Na(I), which induce 46 mV and 100 mV shifts in the half-wave potential from 794 mV, making it easier to oxidize the Fc group. Notably, the presence of the larger alkali ions, such as Cs(I), did not cause any redox shift, which is rationalized by the involvement of the two proximal *syn* carbonyl groups of the Fc group and a relatively small binding cavity.

Recently, metal ion coordination to Fc–peptide conjugates has been explored in more detail due to the potential effects of metal ions on the orientation of the peptide chains. The metallo-peptides may render useful as model systems for biologically relevant proteins. Complexation of Fc[CO–L-Pro–L-Ala–NH-2–Py]$_2$ to the Pd(II) centre resulted in the formation of the *trans* organometallic complex **25**, which stabilized the chirality in solution and the solid state (Figure 10.12) [60]. Metal ion coordination was confirmed by a down-field shift of the Ala N–H proton in the NMR spectrum and by single-crystal X-ray analysis. Some authors have explored extensively the coordination of monosubstituted Fc–peptides to Pd(II) [61, 62], which is not covered in this chapter.

Enzyme mimics involving Fc–peptide conjugates are scarce, but an example exists for modelling Fe–hydrogenase by exploiting Cys–S coordination of Fc–carboxylic acid peptides to a diFe carbonyl fragment, resulting in the formation of the organometallic complex **26** (Figure 10.12) [63]. The complex was characterized by two distinct redox processes involving the two redox components of the complex. The first process is observed at $E_{1/2}$=430 mV and is clearly identified as a one-electron process. The second process is observed at −1.3 V versus Fc/Fc$^+$ and exhibits a sluggish heterogeneous electron-transfer kinetics and a quasi-reversible two electron process, indicating that both Fe centres are reduced simultaneously. While mimicking some of the structural aspects of Fe–hydrogenase, the complex is inactive with respect to the catalytic function of the enzyme.

Next we turn to surface immobilized systems. Unlike Fc–peptides in solution, immobilization of Fc–peptide receptors may allow for a significant amplification of the sensing capabilities and a convenient electrochemical readout of the binding event. Fc–cyclopeptides [64, 65] offer a degree of rigidity and additional H-bonding sites that may render them better binding sites for target molecules or ions. Original work on the interaction of gold surface-bound Fc–peptides with alkali and alkaline earth cations involved the use of the cystamine conjugates Fc[CO–Aa–CSA]$_2$ (**27**) (Aa=Val, Gly, Ala, Leu) and Fc[CO–AAa–CSA]$_2$ (**28**)

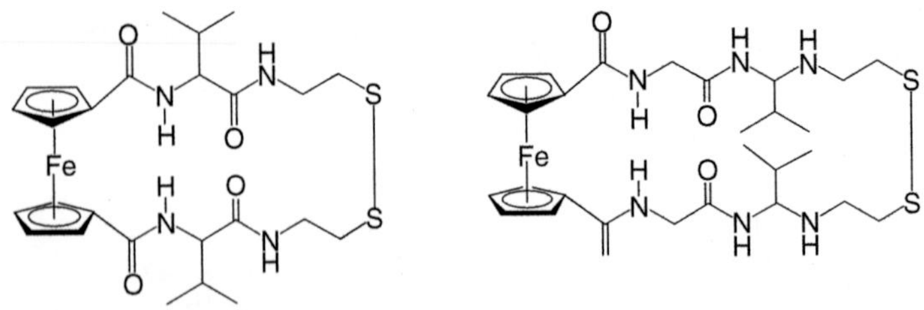

Figure 10.13 *Structure of Fc–conjugate Fc[CO–Val–CSA]$_2$ (**27**) and Fc[CO–Gly–Val–CSA]$_2$ (**28**) as probes for the study of metal ion interactions on surfaces*

(AAa=Gly–Val, Gly–Ile, Ile–Ile) (Figure 10.13) [66]. Unlike monosubstituted Fc derivatives, the corresponding disubstituted Fc–conjugates exhibit restricted mobility on gold surfaces, which greatly influences their interactions with metal ions. The redox potentials of the immobilized Fc–cystamine conjugates appear in the 570–640 mV range versus Ag/AgCl. If the redox activity is related to the reorganization of the Fc–peptide film on the surface, then the potential free dynamics of the rigid disubstituted cyclic Fc–peptides is greatly reduced. Hence, the interactions of the alkali metal ions with immobilized cyclic Fc–peptides were investigated by electrochemical means. Addition of the metal ions resulted in changes of the redox potential, shape of the redox response, as well as in variations of the oxidation and reduction peak separation. The electrochemical changes were ascribed to differences in the electrolyte/cation penetration into the peptide film which affected peptide film organization and electron transfer kinetics [67–69].

The interactions of Fc–peptide **27** with metal ions resulted in a shift in the formal potential of the Fc group (585 mV) to a lower value, following Li(I) (545 mV), Na(I) (570 mV), K(I) (570 mV), Rb(I) (560 mV), and Cs(I) (580 mV). The incorporation of Gly in Fc–conjugate **28** leads to a more complex electrochemical behaviour upon addition of metal ions. The increase from 605 mV was observed in the presence of Na(I) (630 mV), K(I) (660 mV), Rb(I) (635 mV), and Cs(I) (680 mV), but Li(I) resulted in the decrease (570 mV).

10.3.2 Interactions with Other Molecular Targets

For detection of the chemical warfare mimics, such as methylphosphonic acid, diethyl cyanophosphonate, ethyl methylphosphonate, and pinacolyl methylphosphonate, the Boc–Gln–Fca–OMe and Boc–Asn–Fca–OMe conjugates were used in combination with multi-walled carbon nanotubes (MWCNTs) on indium tin oxide (ITO) [70]. The redox properties of the resulting films were highly sensitive to the presence of CWA mimics. Using electrochemical signals associated with Fc, it was possible to detect subnanomolar concentration of these molecules. Presumably, H-bonding interactions between the MWCNT, the Fc–peptide conjugates, and the organophosphonates alters the capacitance within the film, giving rise to significant changes over a wide range of concentrations.

10.3.3 Probing Peptide–Protein Interactions

Unlike the examples provided above, which focus on the solution-based Fc probes, immobilized Fc–peptides offer certain advantages, which result in a lower analyte

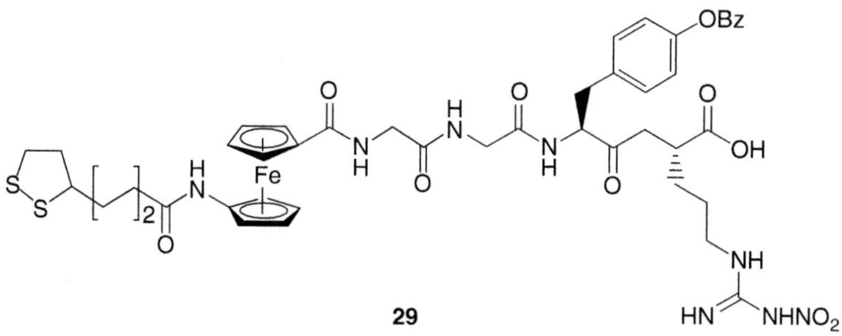

29

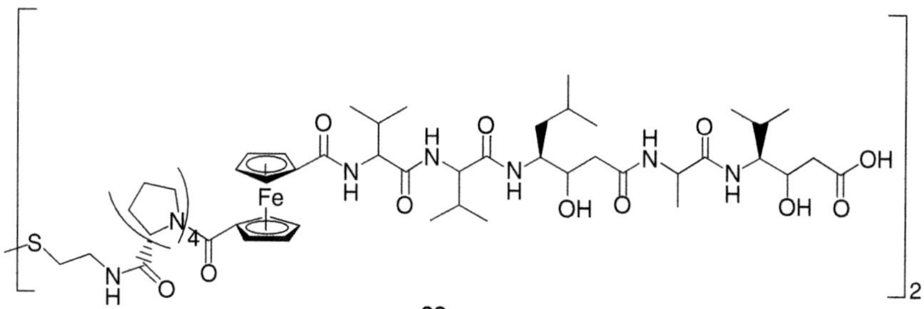

30

Figure 10.14 *The unsymmetrically disubstituted Fca conjugates containing Gly–Gly–Tyr(Bz)–Arg(NO$_2$)–COOH (**29**) recognition peptide and a disulfide containing linker for surface binding. The Fca analogue containing the Val–Val–Sta–Ala–Sta–COOH (**30**) sequence was used for the HIV protein recognition*

consumption, the design of integrated sensors and their miniaturization, and ultimately may allow high-throughput analysis. Examples of Fc conjugates immobilized on a gold surface have been investigated for the purpose of studying peptide–protein interactions and to design protein-specific biosensors. Fca derivatives are preferred over Fc-diacarboxylate derivatives due to their lower reduction potential (~ 250 mV versus ~ 400 mV), which does not overlap with biological oxidation events. Some representative Fca derivatives are shown in Figure 10.14. The disulfide linker is used for the purpose of surface immobilization, and a target peptide is used for studying the peptide–protein interactions, which was in this case the detection of papain enzyme [71, 72]. Originally, the monosubstituted Fc–CO–Gly–Gly–Tyr–Arg–OH conjugate was identified as a competitive papain inhibitor in solution. The study of the Fc–peptide papain interactions was subsequently extended to include surface electrochemistry. The Fc conjugate **29** shown in Figure 10.14 were based on disubstituted Fca and designed to carry the inhibitory peptide sequences Gly–Gly–Tyr–Arg or Gly–Gly–Arg–Tyr that bind to papain, a cystein protease. Lipoic acid was attached to the amino group of this Fca derivative, which allowed for film formation on gold surfaces. The recognition peptides were attached to the carboxyl group of Fca and served as the molecular recognition point for papain. The electrochemical response was associated with the partial shielding of the Fc group by papain, which limits access of anions to the redox group upon oxidation. The effects of ions on the electron-transfer rates and reorganization energies of the Fc–peptide films have been previously explored.

By using a similar surface approach, numerous target strands can be immobilized and tailored towards the analyte of interest if the Fca analogue contains an active ester moiety at its C-terminus. Such methodology was extensively exploited to study peptide interactions

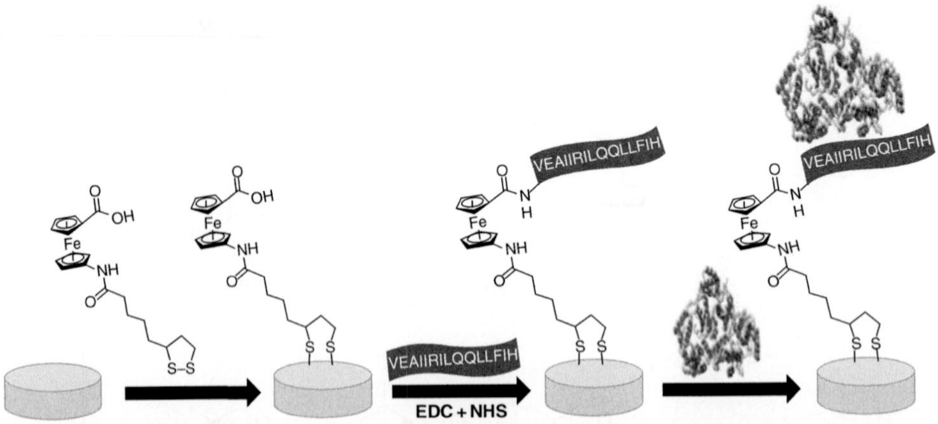

Figure 10.15 *Illustration of HIV-1 protein analysis by using disubstituted Fca conjugate on Au surfaces. Reproduced with permission from reference [75]. Copyright (2011) Elsevier (see color plate figure)*

with HIV-1 related enzymes [73, 74]. For example, the Fc conjugate of pepstatin **30** (Figure 10.14), a pentapeptide Val–Val–Sta–Ala–Sta, with two statin residues (3*S*,4*S*-4-amino-3-hydroxy-6-methylheptanoic acid), was used to study the interaction with HIV-1 protease (PR). Other HIV-1 related proteins, HIV 1 integrase (IN) and HIV 1 reverse transcriptase (RT), were explored with the specific recognition peptides Tyr–Gln–Leu–Leu–Ile–Arg–Met–Ile–Tyr–Lys–Asn–Ile and pepstatin, respectively. As an example, HIV-1 RT is illustrated in Figure 10.15, in a step-by-step build-up of the HIV-1 biosensor and analyte detection [75]. Unlike papain, HIV-1 enzyme concentration caused a shift in the formal potential of the surface-bound Fc bioconjugate to a higher potential as well as a reduction in current intensity. Hence, the oxidation of the Fc group was becoming increasingly difficult, presumably due the encapsulation of the Fc redox centre and its desolvation by the protease, which also influences the access of counterions to the Fc group.

To improve the sensitivity of the HIV-1 biosensor, the Fca–pepstatin conjugate was immobilized on to gold nanoparticles (AuNPs)/gold electrode and on thiolated single-walled carbon nanotubes/gold nanoparticles (SWCNTs/AuNPs) supports [76]. This experimental design allowed for protein detection at the picomolar levels.

Recently, the nonsymmetrically substituted Fc–dicarboxylic acid conjugate **31**, containing the His(DNP)–Gly–OMe sequence and a cystamine moiety, was utilized as an artificial model for carbonic anhydrase (Figure 10.16) [77]. The immobilized Fc conjugate exhibited formal reduction potential at ~ 630 mV versus Ag/AgCl. Zn(II) coordination to N atoms of three His residues results in the formation of a quasi-tetrahedral configuration on the surface, as illustrated in Figure 10.16. The catalytic activity was investigated electrochemically by monitoring the reduction/oxidation waves of the Fc core in the presence of CO_2 gas substrate. Initially, the addition of the substrate gas resulted in an increase in the current density due to the *in situ* formation of HCO_3^- and CO_3^{2-}, which compensate Fc^+.

The catalytic cycle is short-lived, only 5 min, unless the buffered conditions are used to maintain the pH at 7.51. Using the hydrolysis of the *para*-nitrophenyl acetate, which results in the formation of yellow *p*-nitrophenol, the kinetics of this process can

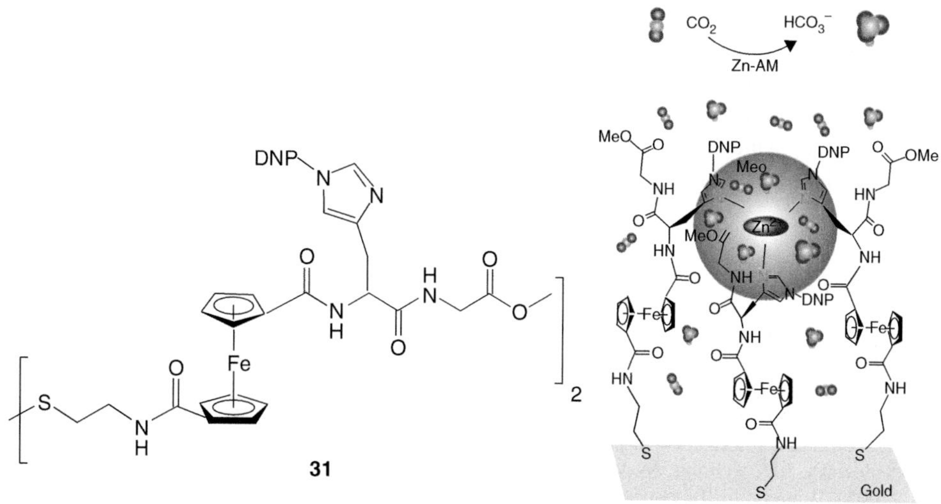

Figure 10.16 *Molecular structure of the artifical carbonic anhydrase model Fc-conjugates **31** and an illustration of the principle used for coordination of Zn(II) and reduction of CO_2 to HCO_3^-. Reproduced with permission from reference [77]. Copyright (2011) The Royal Society of Chemistry (see color plate figure)*

be evaluated by optical spectroscopy. Importantly, the kinetics of this surface-supported Fc–peptide system compared well with the free human carbonic anhydrase (CA-II). However, the catalytic rate constant was larger for the Fc system, indicating a more effective hydrolysis of *para*-nitrophenol. These results suggest that the Fc conjugates may be potentially useful as model systems for CA but also as potential components for the CO_2 sensing, capture, and storage.

10.4 Conclusions

In this short review, we explore the structural diversity of 1,*n*′-disubstituted ferrocene peptide conjugates and highlight major structural motifs and a select number of analytical applications to illustrate the relevance of this class of organometallic bioconjugates. The disubstituted Fc–peptides adopt a wide range of structural motifs that lead to extended H-bonded structures, and together with their electrochemical properties make for ideal bioorganometallic model systems to probe complex structural features in proteins and protein complexes. Intra- and intermolecular H-bonding patterns and interactions and the induced helicity of the Fc core can be conveniently tuned by appropriate peptide substitution, making them useful building blocks, which have shown utility for the design of highly ordered molecular assemblies and biomaterials with interesting optical and electrochemical properties. These properties can be tailored and extended from the molecular to the macroscopic level, aiding in the challenge associated with the design of materials that bridge the gap between the nanoscopic and the macroscopic.

In addition, the peptide component provides an elegant recognition interface that can act as the binding site for ionic and molecular targets, including large proteins. Interactions of such targets with the peptide results in a modulation of the structural and redox properties of the Fc group, which allows the quantification of the target of interest by electrochemical and spectroscopic means. In particular, electrochemical approaches involving such bioconjugates hold great promise for bioanalytical applications.

References

[1] De Vega, M.J., Martin-Martinez, M., Gonzalez-Muniz, R. (2007) Modulation of protein–protein interactions by stabilizing/mimicking protein secondary structure. Curr. Top. Med. Chem. 7: 33–62.
[2] De Grado, W.F., Summa, C.M., Pavone, V., Nastri, F., Lombardi, A. (1999) *De novo* design and structural characterization of proteins and metalloproteins. Annu. Rev. Biochem. 68: 779–819.
[3] Sadowsky, J.D., Schmitt, M.A., Lee, H.-S., Umezawa, N., Wang, S., Tomita, Y., Gellman, S.H. (2005) Chimeric ($\alpha/\beta+\alpha$)-peptide ligands for the BH3-recognition cleft of Bcl-XL: critical role of the molecular scafforld in protein surface recognition. J. Am. Chem. Soc. 127: 11966–11968.
[4] Moriuchi, T., Hirao, T. (2004) Highly ordered structures of peptides by using molecular scaffolds. Chem. Soc. Rev. 33: 294–301.
[5] Li, P., Roller, P.P., Jiecheng Xu, P.P. (2002) Current synthetic approaches to peptide and peptidomimetic cyclization. Curr. Org. Chem. 6: 411–418.
[6] Lelais, G.R., Seebach, D. (2004) β^2-Amino ccids – syntheses, occurrence in natural products, and components of β-peptides. Biopolymers 76: 206–243.
[7] Schlogl, K. (1957) Über Ferrocen-Aminosäuren und Verwandte Verbindungen. Monatash. Chem. 88: 601–603.
[8] van Staveren, D.R., Metzler-Nolte, N. (2004) The bioorganometallic chemistry of ferrocene. Chem. Rev. 104: 5931–5985.
[9] Metzler-Nolte, N. (2010) Biomedical applications of organometal–peptide conjugates. Topics in Organomet. Chem. 32: 195–217.
[10] Appoh, F.E., Sutherland, T.C., Kraatz, H.-B. (2004) Changes in the hydrogen bonding pattern in ferrocene peptides. J. Organomet. Chem. 689: 4669–4677.
[11] Hirao, T. (2009) Control of chirality-organized structures of ferrocene–dipeptide bioconjugates. J. Organomet. Chem. 694: 806–811.
[12] Moriuchi, T., Nagai, T., Hirao, T. (2005) Chirality organization of ferrocenes bearing dipeptide chains of heterochiral sequence. Org. Lett. 7: 5265–5268.
[13] Herrick, R.S., Jarret, R.M., Curran, T.P., Dragoli, D.R., Flaherty, M.B., Lindyberg, S.E., Slate, R.A., Thornton, L.C. (1996) Ordered conformations in bis(amino acid) derivative of 1,1′-ferrocenedicarboxylic acid. Tetrahedron Lett. 37: 5289–5292.
[14] Heinze, K., Beckmann, M. (2005) Conformational analysis of chiral ferrocene peptides. Eur. J. Inorg. Chem. 3450–3457.
[15] Heinze, K., Beckmann, M. (2006) Dynamic behaviour of 1,*n*′-disubsituted ferrocenes. J. Organomet. Chem. 691: 5576–5584.
[16] van Staveren, D.R., Weyhermüller, T., Metzler-Nolte, N. (2003) Organometallic β-turn mimetics. A structural and spectroscopic study of inter-strand hydrogen bonding in ferrocene and cobaltocenium conjugates of amino acids and dipeptides. Dalton Trans. 210–220.
[17] Kirin, S.I., Schatzschneider, U., De Hatten, X., Weyhermuller, T., Mezler-Nolte, N. (2006) 1,*n*′-Disubstituted ferrocenoyl amino acids and dipeptides: conformational analysis by CD spectroscopy, X-ray crystallography and DFT calculations. J. Organomet. Chem. 691: 3451–3457.
[18] Hatten, X.D., Weyhermuller, T., Metzler-Nolte, N. (2004) Ferrocenoyl peptides with sulfur-containing side chains: synthesis, solid state and solution structures. J. Organomet. Chem. 689: 4856–4867.

[19] Heinze, K., Schlenker, M. (2004) Main chain ferrocenyl amides from 1-amino-1′-carboxyferrocene. Eur. J. Inorg. Chem. 2974–2988.
[20] Moriuchi, T., Hirao, T. (2010) Design of ferrocene–dipeptide bioorganometallic conjugates to induce chirality-organized structures. Acc. Chem. Res. 43: 1040–1051.
[21] Nomoto, A., Moriuchi, T., Yamazaki, S., Ogawa, A., Hirao, T. (1998) A highly ordered ferrocene system regulated by podant peptide. Chem. Commun. 1963–1964.
[22] Kirin, S.I., Schatzschneider, U., Koster, S.D., Siebler, D., Metzler-Nolte, N. (2009) A systematic evaluation of different hydorgen bonding patterns in unsymmetrical 1,*n*′-disubstituted ferrocenoyl peptides. Inorg. Chim. Acta 362: 894–906.
[23] Lataifeh, A., Kraatz, H.-B. (2010) Ferrocene–aspartate dendrimers: conformational analysis and electrochemical studies. J. Inorg. Organomet. Polym Mater. 20: 488–502.
[24] Beheshti, S., Lataifeh, A., Kraatz, H.-B. (2011) Hydrogen-bonding interactions in ferrocene peptide conjugates containing valine. J. Organomet. Chem. 696: 1117–1125.
[25] Chowdhury, S., Schatte, G., Kraatz, H.-B. (2008) How useful is ferrocene as a scaffold for the design of β-sheet foldamers? Angew. Chem. Int. Ed. 47: 7056–7059.
[26] Chowdhury, S., Sanders, D.A.R., Schatte, G., Kraatz, H.B. (2006) Discovery of the first pseudo β-barrel: synthesis and formation by tiling of ferrocene-cyclopeptides. Angew. Chem. Int. Ed. 45: 751–754.
[27] Vogt, J., Schultz, G.E. (1999) The structure of the outer membrane protein OmpX from *Escherichia coli* reveals possible mechanisms of virulence. Struc. Folding Design 7: 1301–1309.
[28] Pautsch, A., Schultz, G.E. (1998) Structure of the outer membrane protein. A transmembrane domain. Nat. Struct. Biol. 5: 1013–1017.
[29] Siebler, D., Forster, C., Heinze, K. (2010) Molecular multi-wavelength optical anion sensor. Eur. J. Inorg. Chem. 532–527.
[30] Heinze, K., Wild, U., Beckmann, M. (2007) Solid-phase synthesis of chiral modular ferrocene-based peptides. Eur. J. Inorg. Chem. 617–623.
[31] Čakić Semenčić, M., Siebler, D., Heinze, K., Rapić, V. (2009) Bis- and trisamides derived from 1′-aminoferrocene-1-carboxylic acid and alpha-amino acids: synthesis and conformational analysis. Organometallics 28: 2028–2037.
[32] Čakić Semenčić, M., Heinze, K., Forster, C., Rapić, V. (2010) Bioconjguates of 1′-aminoferrocene-1-carboxylic acid with (*S*)-3-amino-2-methylpropanoic acid and L-alanine. Eur. J. Inorg. Chem. 1089–1097.
[33] Barišić, L., Dropučić, M., Rapić, V., Pritzkow, H., Kirin, S.I., Metzler-Nolte, N. (2004) The first oligopeptide derivative of 1′-aminoferrocene-1-carboxylic acid shows helical chirality with antiparellel strands. Chem. Commun. 2004–2005.
[34] Barišić, L., Čakić, M.K., Mahmoud, A., Liu, Y.-N., Kraatz, H.-B., Pritzkow, H., Kirin, S.I., Metzler-Nolte, N., Rapić, V. (2006) Helically chiral ferrocene peptides containing 1′-aminoferrocene-1-carboxylic acid subunit as turn inducer. Chem. Eur. J. 12: 4965–4980.
[35] Barišić, L., Rapić, V., Metlzer-Nolte, N. (2006) Incorporation of the unnatural organometallic amino acid 1′-aminoferrocene-1-carboxylic acid (Fca) into oligopeptides by a combination of Fmoc and Boc solid phase synthetic methods. Eur. J. Inorg. Chem. 4019–4021.
[36] Lataifeh, A., Bondy, C.R., Kraatz, H.B. (2009) Ferrocene conjugates containing diarginine and aspartic acid: salt bridge interactions in water. Eur. J. Inorg. Chem. 4425–4432.
[37] Somers, W., Ultsch, M., de Vos, A.M., Kossiakoff, A.A. (1994) The X-ray structure of a growth hormone–prolactin receptor complex. Nature 372: 478–481.
[38] Puglisi, J.D., Chen, L., Frankel, A.D., Williamson, J.R. (1993) Role of RNA structure in arginine recognition of TAR RNA. Proc. Natl Acad. Sci. USA 90: 3680–3685.
[39] Maier, T., Podleck, J. (2004) β-Lactam-containing cyclopeptide analogs. Eur. J. Org. Chem. 4379–4386.
[40] Liu, M., Sibi, M.P. (2002) Recent advances in the stereoselective synthesis of β-amino acids. Tetrahedron 58: 7991–8035.
[41] Palomo, C., Aizpurua, J.M., Ganboa, I., Oiarbide, M. (1999) From β-lactams to α- and β-amino acid derived peptides. Amino Acids 16: 321–343.

[42] Ojima, I., Delaloge, F. (1997) Asymmetric synthesis of building-blocks for peptides and peptidomimetics by means of the β-lactam synthon method. Chem. Soc. Rev. 26: 377–386.
[43] Kovač, V., Radolović, K., Habus, I., Siebler, D., Heinze, K., Rapić, V. (2009) Conformational analysis of beta-lactam containing ferrocene peptides. Eur. J. Inorg. Chem. 389–399.
[44] Chowdhury, S., Mahmoud, K.A., Schatte, G., Kraatz, H.-B. (2005) Amino acid conjugates of 1,1′-diaminoferrocene. Synthesis and chiral organization. Org. Biomol. Chem. 3: 3018–3023.
[45] Djaković, S., Siebler, D., Čakić Semenčić, M., Heinze, K., Rapić, V. (2008) Spectroscopic and theoretical study of asymmetric 1,1′-diaminoferrocene conjugates of alpha-amino acids. Organometallics 27: 1447–1453.
[46] Lapić, J., Djaković, S., Kodrin, I., Mihalić, Z., Cetina, M., Rapić, V. (2010) Preparation and conformation analysis of *N*-(ferrocenoyl)dipeptide esters and their 1′-acetyl derivatives. Eur. J. Inorg. Chem. 2512–2524.
[47] Moragues, M.E., Martinez-Manez, R., Sancenon, F. (2011) Chromogenic and fluorogenic chemosensors and reagents for anions. Chem. Soc. Rev. 40: 2593–2643.
[48] Qing, G.Y., Sun, T.L., Wang, F., He, Y.B., Yang, X. (2009) Chromogenic chemosensors for *N*-acetylaspartate based on chiral ferrocene-bearing thiourea derivatives. Eur. J. Org. Chem. 841–849.
[49] Beeren, S.R., Sanders, J.K.M. (2011) Ferrocene–amino acid and macrocycles as hydrazone-based receptors for anions. Chem. Sci. 2: 1560–1567.
[50] Beeren, S.R., Sanders, J.K.M. (2011) Descovery of linear receptor for multiple dihydrogen phosphate ions using dynamic combinatorial chemistry. J. Am. Chem. Soc. 133: 3804–3807.
[51] Wang, Z., Palacios, M.A., Zyryanov, G., Anzenbacher, P. (2008) Harnessing a ratiometric fluorescence output from a sensor array. Chem. Eur. J. 14: 8540–8546.
[52] Kaim, W., Schwederski, B. (1994) Bioinorganic Chemistry: Inorganic Elements in the Chemistry of Life. Wiley, New York.
[53] Ghadiri, M.R., Choi, C. (1990) Secondary structure nucleation in peptides. Transition metal ion stabilized alpha-helices. J. Am. Chem. Soc. 112: 1630–1632.
[54] Shimazaki, Y., Takani, M., Yamauchi, O. (2009) Metal complexes of amino acids and amino acids side chain groups. Structures and properties. Dalton Trans. 7854–7869.
[55] Kerman, K., Kraatz, H.B. (2008) Metals coordinate protein–protein interactions. Angew. Chem. Int. Ed. 47: 6522–6524.
[56] Scully, C.C.G., Lau, Y.H., Jensen, P., Rutledge, P.J. (2011) Synthesis, electrochemistry and metal binding properties of monosubstituted ferrocenoyl peptides with thioether-containing sidechains. J. Organomet. Chem. 696: 715–721.
[57] Scully, C.C.G., Rutledge, P.J. (2010) Synthesis and electrochemical studies of disubsituted ferrocene–dipeptide conjugates with sulfur-containing side chains. Tetrahedron 66: 5653–5659.
[58] Chowdhury, S., Schatte, G., Kraatz, H.B. (2006) A cyclic Fc histidine conjugate: synthesis and properties. Interactions with alkali metal ions. Eur. J. Inorg. Chem. 988–993.
[59] Cheng, L., Long, Y-T., Tian, H., Kraatz, H.B. (2010) Spectroscopic and electrochemical investigations into the interactions of metal ions with a ferrocene–histidine peptide conjugate. Eur. J. Inorg. Chem. 5231–5238.
[60] Moriuchi, T., Yoshida, K., Hirao, T. (2001) β-Turn-structure-assembled palladium complexes by complexation-induced self-organization of ferrocene–dipeptide conjugates. Organometallics 20: 3101–3105.
[61] Moriuchi, T., Fujiwara, T., Hirao, T. (2009) β-Turn-structure-assembled palladium complexes by complexation-induced self-organization of ferrocene–dipeptide conjugates. Dalton Trans. 4286–4288.
[62] Moriuchi, T., Yoshida, K., Hirao, T. (2001) Structural characterization and complexation behaviour of ferrocene bearing dipeptide chain (L-Ala–L-Pro–NHPy). J. Organomet. Chem. 637–639: 75–79.
[63] De Hatten, X., Bothe, E., Merz, K., Huc, I., Metzler-Nolte, N. (2008) A ferrocene–peptide conjugate as a hydrogenase model system. Eur. J. Inorg. Chem. 4530–4537.
[64] Orlowski, G.A., Chowdhury, S., Kraatz, H.B. (2007) The effects of alkali metal ions on the electrochemical behaviour of ferrocene–peptide conjugates immobilized on gold surfaces. Electrochim. Acta 31: 2034–2039.

[65] Curran, T.P., Handy, E.L. (2009) A 310-helix single turn enforced by crosslinking of lysine with 1,1′-ferrocenedicarboxylic acid. J. Organomet. Chem. 694: 902–907.
[66] Chowdhury, S., Schatte, G., Kraatz, H.B. (2004) Synthesis, structure and electrochemistry of ferrocene–peptide macrocycles. Dalton Trans. 1726–1730.
[67] Lee, L.Y.S., Sutherland, T.C., Rucareanu, S., Lennox, R.B. (2006) Ferrocenylalkylthiolates as a probe of heterogeneity in binary self-assembled monolayers on gold. Langmuir 22: 4438–4444.
[68] Lee, L.Y.S., Lennox, R.B. (2007) Electrochemical desorption of *n*-alkylthiol SAMs on polycrystalline gold: studies using a ferrocenylalkylthiol probe. Langmuir 23: 292–296.
[69] Chidsey, C.E.D., Bertozzi, C.R., Putvisnki, T.M., Mujsce, A.M. (1990) Coadsorption of ferrocene-terminated and unsubstituted alkanethiols on gold: electroactive self-assembled monolayers. J. Am. Chem. Soc. 112: 4301–4306.
[70] Khan, M.A.K., Kerman, K., Petryk, M., Kraatz, H.–B. (2008) Non-covalent modification of carbon nanotubes wiht ferrocene–amino acid conjugates for electrochemical sensing of chemical warfare agent mimics. Anal. Chem. 80: 2574–2582.
[71] Mahmoud, K.A., Hrapović, S., Luong, J.H.T. (2008) Picomolar detection of protease using peptide/single walled carbon nanotube/gold nanoparticle-modified Electrode. ACS Nano 2: 1051–1057.
[72] Plumb, K., Kraatz, H.-B. (2003) Interactions of a ferrocenoyl-modified peptide with papain: toward protein-sensitive electrochemical probes. Bioconjugate Chem. 14: 601–606.
[73] Kerman, K., Mahmoud, K.A., Kraatz, H.–B. (2007) An electrochemical approach for the detection of HIV-1 protease. Chem. Commun. 37: 3829–3831.
[74] Kerman, K., Kraatz, H.-B. (2009) Electrochemical probing of HIV enzymes using ferrocene-conjugated peptides on surfaces. Analyst 134: 2400–2404.
[75] Labib, M., Shipman, P.O., Martić, S., Kraatz, H.-B. (2011) A bioorganometallic approach for rapid electrochemical analysis of HIV-1 reverse transcriptase serum Levels. Electrochim. Acta 56: 5122–5128.
[76] Mahmoud, K.A., Kraatz, H.-B. (2007) A bioorganometallic approach for the electrochemical detection of proteins: a study of the interaction of ferrocene–peptide conjugates with papain in solution and on Au surfaces. Chem. Eur. J. 13: 5885–5895.
[77] Cheng, L., Long, Y.T., Kraatz, H.-B., Tian, H. (2011) Evaluation of an immobilized artificial carbonic anhydrase model for CO_2 sequestration. Chem. Sci. 2: 1515–1518.

11

Mechanisms of Adsorption of Short Peptides on Metal and Oxide Surfaces

Vincent Humblot, Jessem Landoulsi and Claire-Marie Pradier

Laboratoire de Réactivité de Surface, UMR CNRS 7197, Université Pierre et Marie Curie, France

Acronyms

AA	Amino acid
AFM	Atomic force microscopy
DFT	Density functional theory
FP	First principle
GSH	Glutathione (Glu–Cys–Gly)
HOPG	Highly oriented pyrolytic graphite
IGF	Insulin growth factor
IR-ATR	Infrared attenuated total reflection spectroscopy
KPL	Ketopantolactone
LEED	Low energy electron diffraction
MD	Molecular dynamic
NEXAFS	Near edge X-ray absorption Fine structure spectroscopy
PhD	Photoemission diffraction
PM-RAIRS	Polarization modulation reflection absorption infrared spectroscopy
RAIRS	Reflection absorption infrared spectroscopy

Peptide Materials: From Nanostructures to Applications, First Edition. Edited by Carlos Alemán, Alberto Bianco and Mariano Venanzi.

RAS	Reflection anisotropy spectroscopy
RGD	Arg–Gly–Asp
SAM	Self-assembled monolayer
STM	Scanning tunneling microscopy
TPA	Terephtaldehic ccid
TPD	Temperature programmed desorption
UHV	Ultra-high vacuum
XPD	X-ray photoelectron diffraction
XPS	X-ray photoelectron spectroscopy

11.1 Introduction

Interfaces between protein molecules and inorganic materials are now among one of the hottest research topics in various fields, such as biomedicine, biochemistry and biotechnologies. For instance, molecular recognition events resulting in the formation of elegant structures, often found in hard biomineralized tissues, are mediated by proteins and other biomolecules [1]. In another context, the performances of advanced implants and therapeutic devices are strongly related to the biocompatibility of materials, which directly depends on the protein/material interface [2, 3]. Recent progresses have been gained in the nanomanipulation of proteins at interfaces. In particular, a great advance in combinatorial biology (e.g. the phage display method and other related artificial techniques) provides amino acid sequences, usually called aptamers or binders, possessing specific affinities to their targeted inorganic materials, ranging over metals and oxides [4]. These findings make peptides good candidates to mimic and control biointeraction events.

Among various relevant applications, peptides provide new routes for surface modification, replacing common functionalization procedures (e.g. self-assembled monolayers, or SAMs), as these molecules may be anchored directly on the substrate with specific affinity [5]. Despite these major advances, several important and basic issues regarding the interaction of peptides with metal surfaces remain to be addressed.

This chapter aims to point out particular aspects, either experimental or conceptual, that are of primary importance to understand and anticipate the behavior of peptides on metal and oxide surfaces. For this purpose, a description of how short, di- or tripeptides interact with a variety of metals and oxides in aqueous solutions or in ultra-high vacuum conditions is given. Examples have been selected to illustrate some key considerations: (i) physicochemical features of metal and oxide surfaces and (ii) parameters influencing peptide adsorption, including intrinsic properties of peptides and external parameters related to the condition under which the adsorption is performed. Considering these aspects is essential to make progress in deciphering interfacial mechanisms involved in many relevant biological events, such as molecular recognition and cell adhesion.

Under vacuum conditions, i.e. on model metal surfaces, adsorption of short peptides, tripeptides at a maximum, has been investigated by evaporating the molecules from powder, giving the opportunity to identify simultaneously the chemical state of the molecule, its orientation and, more important, the chemical sites interacting with the surface. The main utilized techniques are reflection absorption infrared spectroscopy (RAIRS), which provides *in situ* (under low pressure of peptide) surface characterization, and *ex situ* X-ray photoelectron spectroscopy (XPS) and near edge X-ray absorption fine structure spectroscopy (NEXAFS) analyses. Scanning tunneling microscopy (STM) and atomic force microscopy (AFM) have also been applied to the characterization of peptide layers on metal surfaces. These 'model' experimental investigations sometimes benefit from parallel theoretical studies for refining spectral interpretations.

With the aim of approaching more realistic conditions, a number, still limited, of adsorption measurements have been performed in the liquid phase, taking advantage of electrochemical techniques, like capacitance measurements. Other works rely on *in situ* ATR-IR spectroscopy, or 'post-mortem' characterization of the peptide layers formed in liquid conditions. Several examples will be detailed in this chapter, demonstrating that results at a molecular lever could still be deduced from these experiments! Last but not least, advances in the description of complex peptide/oxide interfaces have been permitted by the use of molecular dynamics calculation; Arg–Gly–Asp (RGD)/TiO_2 is byfar the most studied system because of its relevance for biocompatibility.

The first part (Sections 11.2 to 11.4) of this chapter concentrates on the reasons motivating such investigations, the description of metal and oxide surface chemistry, as well as on some thermodynamic principles at the origin of peptide adsorption, These considerations will then be illustrated in two main sections (Sections 11.5 and 11.6), one devoted to the adsorption of short peptides on metal and oxide surface in vacuum and the second in the liquid phase. It is important to make clear that this chapter does not aim to review all the works of the literature on this topic; our objective is, by selecting the most demonstrative examples, to uncover or confirm great lines governing peptide/solid interactions.

11.2 Why Studying the Interaction of Short Peptides with Solid Surfaces?

During the last decades, the self-assembly of peptides in solution has attracted growing attention to synthesis of new materials with biological interest [6]. Many results have shown that short peptides may organize at the nanoscale and offer a fascinating and rich repertoire of shapes and nanostructures. By contrast, little work has been devoted to the self-assembly, and even to the adsorption of short peptides, at the solid/liquid interface. In contrast to bulk solution, interfaces exhibit different energetic balances, leading to conformational change of peptides upon adsorption. The most outstanding phenomenon for most short peptides is their ability to self-assemble [7] and form well-ordered nanostructures of high interest for many technological applications; this is due to structural complementarities and intermolecular coupling via hydrogen bonding, electrostatic interaction, hydrophobic affinity, etc [8].

Controlling peptide/solid interactions and tailoring the structural and chemical properties of solid material with these small molecules has numerous applications in the field of biocompatibility, biodetection and enantioselective catalysis. Note also the use of peptides in metal nanoparticle synthesis, where the composition, sequence and interaction capability were shown greatly to influence nanoparticle shape and size [9]. Recently, short peptides were used for the elaboration of highly sensitive metal ion sensors through the modification of nanostructured materials [10]. Indeed, when adsorbed on various solid surfaces, tripeptides have remarkable properties of metal ion complexation, highly dependent on their nature, sequence and immobilization mode. Peptides containing a histidine residue are particularly good candidates for cation complexation thanks to their donor nitrogen atom [11]. A recent example even showed that Gly–Gly–His and His–Gly–Gly tripeptides, immobilized on aldehyde-decorated silicon surfaces, display different Cu^{2+} complexation properties, most likely due to different binding modes and consequent crowdedness on the surface [12]. Interesting to our view is to retain from this example the importance of understanding and controlling the interaction mode of peptides with solid surfaces.

Chirality is another fascinating property of peptides or other biomolecules, which has governed the living world for ages. Chiral induction has been well studied and established by the Raval group in their studies of amino acid adsorption on metal surfaces [13–15]. This property may induce several distinct geometries of adsorption for the same chemical species and, of course, influences the interaction and recognition of molecules in a further step. Except glycine, all amino acids have a chiral center at the origin of a chiral 2D arrangement on a surface [13–15] and of chiral recognition. Besides, by using chiral inductors, solid catalyst surfaces may be modified and favor enantioselective catalysis, an applied field of great added value [16]. In this context amino acids and short peptides are natural, relevant candidates as chiral inducers, thanks to their easy binding to a metal or oxide support and their ability to form 2D or 3D arrangements that will provide chiral adsorption sites and docking structures for prochiral reactants. Peptides, because of their more complex structure compared to amino acids, have been very seldom used as chiral inductors. Recent examples will be detailed hereafter, illustrating new promising applications of peptide immobilization.

11.3 Metal and Oxide Surfaces

The interaction of peptide molecules with metal and oxide surfaces strongly depends on the physicochemical properties of surfaces, which govern interfacial forces. Although these properties may differ from one metal to another, some key features can be depicted in a common description.

Metal and alloys are generally unstable thermodynamically in aqueous media, except for some noble ones, such as Au and Pt [17]. They are, therefore, subjected to anodic dissolution according to

$$M \rightarrow M^{z+}_{(aq)} + ze^- \tag{11.1}$$

In many cases, depending on the composition of the medium, the active state, described in Equation (11.1), is followed by the formation of an insulating thin oxide layer, of typically a few nanometers thick, to reach a passive state. It may form according to the anodic reaction described by

$$M + \frac{z}{2} H_2O \rightarrow MO_{z/2} + zH^+ + ze^- \tag{11.2}$$

This event appreciably slows down ionic transports and thus also the anodic dissolution rate. The driving force for the oxide layer growth and stability is the potential gap between the metal and the solution, inducing a high electrical field (up to 10^6 V/cm). The oxide film may form under electrochemical polarization control or spontaneously (ageing) in the presence of oxidant agent (electron acceptor). The properties of the oxide layer strongly depend on the material itself (composition, microstructure, etc.), and on the conditions in which the oxide is formed (pH, ionic strength, dissolved oxygen, etc.). The composition of an oxide layer that forms on metal or alloy substrates is difficult to determine with accuracy; it always tends to be more hydrated in the vicinity of the solution. Upon adsorption of peptides, the outer part of oxides/hydroxide layer is involved in molecule–surface interaction; accordingly, particular attention should be dedicated to its composition, its acid–base properties and electrochemical behavior, when studying adsorption mechanisms.

In liquid, metal and oxide surface charges induce electrostatic forces, which are determined in the adsorption of charged peptide molecules. Interaction of charged species and oriented dipoles (from the bulk) with the surface result in the formation of the so-called electric double layer [18]. The inner layer (close to the surface, the 'Stern layer') is constituted of solvent molecules and other ions or molecules that adsorb specifically on the surface. The external side of the double layer (diffuse layer) includes solvated ions that interact with the surface through long-range electrostatic forces, and is considered to comprise several layers.

In many situations of practical relevance, the surface of metals and oxides is covered by an organic adlayer originating from ambient atmosphere, formed during surface preparation, analysis or conditioning. This is related to the strong forces that are unbalanced at the surface and responsible for a high excess of free energy (surface energy) [19, 20]. Indeed, XPS analyses of metal or oxide surfaces generally reveal the presence of carbon in a significant amount, independently of the surface preparation [21, 22]. Actually, wet cleaning does not eliminate, but essentially standardizes, surface contamination. More severe cleaning procedures result in a surface with a high excess of potential energy, which quickly adsorbs significant amounts of contaminants upon further treatments [21].

11.4 Factors Influencing Peptide Adsorption

11.4.1 Driving Force

In contrast to proteins or high molecular weight polypeptides, which may be subjected to significant structural changes in the adsorbed state, short peptides may be considered as small molecules that behave like rigid particles upon adsorption on solid surfaces. To better understand the difference between the adsorption mechanism of short peptides and of proteins at the molecular level, enthalpy and entropy contributions have to be considered in the variation of the free energy. On the one hand, adsorption is generally an exothermic process, as it originally tends to minimize the free enthalpy of the system. On the other hand, the reduction of molecule mobility (adsorption on the surface) leads to a decrease of the system entropy; this illustrates the compensation between ΔH and $T\Delta S$ terms in the adsorption process. For a similar total adsorbed mass, the entropy decrease is less pronounced for long polypeptides compared to several short peptides. Accordingly, polypeptides have a

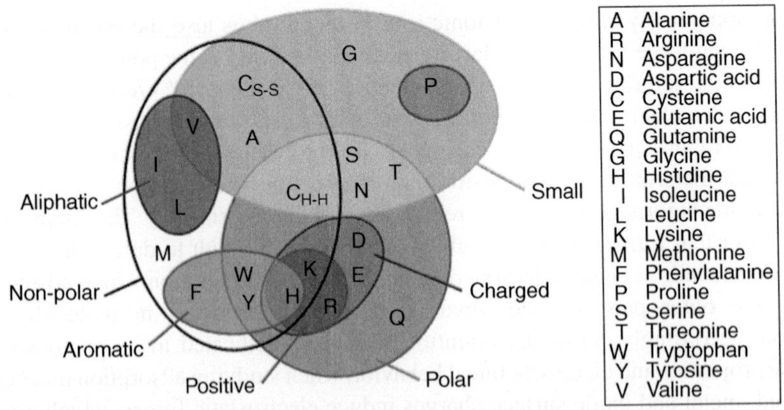

Figure 11.1 *The main properties of the 20 natural amino acids*

more pronounced tendency to adsorb compared to short peptides (with the same amino acid sequence).

The main interactions involved in the driving force of short peptide adsorption can be summarized as follows:

- Electrostatic interaction between the electric double layer of metal and oxide surfaces, and charged groups of peptide molecules
- Van der Waals (London) forces
- Solvation forces (positive contribution in ΔS).

11.4.2 Influence of Intrinsic Properties

The diversity of peptides is related to the existence of 20 natural amino acids, the basic building blocks [23]. All the amino acids bear a variable group that governs the essential properties of amino acids. The latter can thus be charged, polar, nonpolar and aliphatic (Figure 11.1). The synthesis of dipeptide molecules results from the removal of a water molecule between two amino acids and the formation of one amide bond. In practice, the synthesis of peptides can now be readily carried out using the solid phase Fmoc reaction, providing ample opportunities for peptide design and the subsequent exploration of the relationship between molecular architecture and peptide physiochemical properties. Intermolecular organization of short peptides, simpler than proteins, mainly lead to secondary structures, mainly alpha helices.

11.4.3 Influence of External Parameters

Adsorption of peptides on solid surfaces is strongly influenced by experimental conditions under which the adsorption is carried out. The main parameters to be considered are pH, ionic strength, composition of the solution and temperature.

The pH influences the electrostatic state of both the peptide and the surface, in direct correlation with the electric double layer [18]. The surface charge of metal oxides/

hydroxides is mainly governed by their ability to gain or lose protons in aqueous media and the pH at which the numbers of negative and positive charges are equal, resulting in a net neutral molecule, is called the isoelectric point (iep). Although the calculation of surface charge using the ζ potential remains difficult [24], estimated iep values have been reported for many oxides/hydroxides. Note that a theoretical iep value of peptides can be computed using the amino acid sequence of the peptide on the basis of the ProtParam tool available on the ExPASy molecular biology server [25].

The concentration of dissolved ions, expressed by the ionic strength term, is another parameter influencing adsorption. On the one hand, the ionic strength determines the properties of the double layer of the metal or oxide surface, its thickness and electric charge. On the other hand, it determines the Debye length, due to the damping of the electric potential of a fixed charge in an electrolyte. An increase in the ionic strength leads to a decrease of electrostatic interactions between charged entities, including peptide–peptide and peptide–surface interactions.

The type of salt ion also influences the adsorption processes. At high concentration, ions may promote (e.g. SO_4^{2-}, Mg^{2+}) or reduce (e.g. NH_4^+) peptide precipitation in solution [26, 27]. Ca^{2+}, among other ions, is reported to act as a bridge and promote peptide–surface bonds [28].

Temperature may influence the equilibrium state related to the adsorption process, as described above, as well as the kinetics of peptide adsorption. An increase of the temperature is expected to accelerate (i) the diffusivity of molecules toward the surface and (ii) the release of adsorbed water molecules and salt ion (entropy gain).

In UHV conditions, adsorption conditions and thus variable parameters are obviously different from those used in the liquid phase. The temperature and the exposure time basically constitute the external parameters influencing the adsorption processes. Examples of these complex issues are detailed below.

11.5 Adsorption at the Solid/Gas Interface

11.5.1 Adsorption of Dipeptides

Interpretation of peptide adsorption is often based on data previously obtained with single amino acids interacting with a solid surface. This is the case for the following set of data, describing the adsorption of glycine (Gly) and alanine (Ala), followed by that of Gly-based and Ala-based (homo)–dipeptides on Cu(110), which demonstrates that, though displaying similar reactive groups (carboxylic acids, amino groups), all susceptible to interact with metal or oxide surfaces, the interaction of a peptide is not as simple as the sum of the interactions of each of its fragments.

Briefly, Gly and Ala amino acids adsorption on Cu(110) has been studied using a broad range of surface science techniques (RAIRS, LEED, STM, TPD, XPS, NEXAFS, XPD, PhD; for more details see the review by Barlow and Raval [29] and references therein). These data point out that glycine is adsorbed in its anionic form up to a saturated monolayer. The binding and orientation of the molecules vary with surface coverage and temperature; at low coverage, glycine is anchored via both O atoms of the deprotonated carboxylate group, with the amino group pointing away from the surface, while at higher coverage the molecules bind through one O atom of the tilted COO^- group, the N atom of the amino group being now closer to the surface [30]. Importantly, above room temperature, there is no possibility of bi- or multilayer formation, as there are no compatible available moieties for H-bond interactions (e.g. COO^- and NH_3^+). The data obtained for alanine end up with the same conclusions (anionic form with two different anchoring modes), but with possible multilayer growth, made possible by the presence of alanine molecules under their zwitterionic chemical form [31].

As a first example of peptide, the simplest homo–dipeptide, composed of two glycine molecules, Gly–Gly, has been adsorbed and characterized under UHV by XPS, NEXAFS, together with DFT calculations [32, 33]. UHV deposition results from evaporation of the powder placed in a Knudsen cell. XPS analysis after adsorption at room temperature indicates the presence of molecules adsorbed under either zwitterionic or anionic form; after annealing at 420 K the anionic form dominates.

Adsorption geometry of the Gly–Gly molecules was proposed after NEXAFS data analysis and DFT calculations for spectral simulation and structure optimization. Gly–Gly, under its anionic form, is anchored to the Cu(110) surface via a three-point contact mode:

- the two O atoms of the COO^- groups (bidentate conformation) and the amino *nitrogen* atom for the monolayer regime (Figure 11.2a);
- the two O atoms of the COO^- groups and one amino *hydrogen* atom in the submonolayer regime (Figure 11.2b).

Note that, at very low coverage, zwitterionic molecules are present, anchored only via the carboxylate groups in a monodentate mode, with the NH_3^+ group pointing towards the surface (Figure 11.2c). Eventually, when the copper surface was precovered with a fraction of oxygen monolayer ($p(2\times1)$ structure), anionic dipeptide species were formed immediately from the submonolayer regime. The adlayer is stabilized when increasing the coverage, and the molecule–surface interactions proceed via the oxygen atoms of COO^- groups.

Another, more complex, homodipeptide, the diphenylalanine (Phe–Phe), was adsorbed on Cu(110) under UHV conditions; its self-assembly has been characterized by XPS, NEXAFS [34], scanning tunneling microscopy (STM) and modeling by first principle (FP) and molecular dynamic (MD) calculations [35]. This example demonstrates the importance of the fragment structure, more precisely flexibility, upon the supramolecular arrangement of peptides in an adsorbed layer. As a matter of fact, on each part of the peptidic bond, the two end-phenyl groups are rather mobile, which provides to the dipeptide molecules some

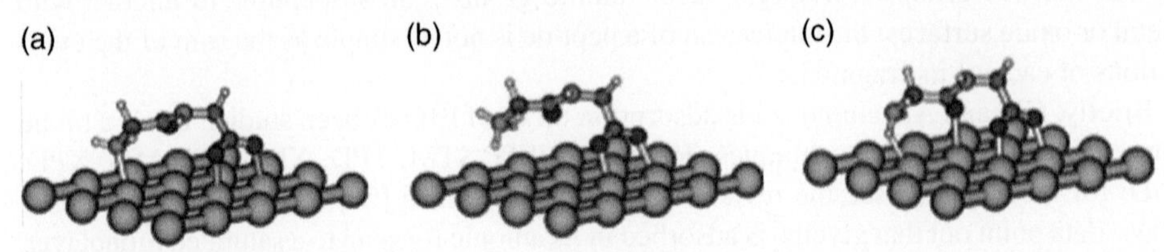

Figure 11.2 Adsorption model for the Gly–Gly molecules adsorbed on Cu(110) as in anionic bidentate fashion in (a) and (b), and in zwitterionic monodentate fashion in (c). Reproduced with permission from reference [33]. Copyright (2010) American Chemical Society

flexibility. Two distinct adsorption modes, characterized by Phe–Phe different chemical forms and 2D arrangements, have been identified on Cu(110), depending on the coverage. At low coverage, the adsorbed dipeptides are isolated, as neutral molecules (Figure 11.3a) [35]. When the coverage is increased, molecules become zwitterionic, as a result of a proton transfer from the carboxylic acid group towards the amine group [34]. In addition, STM images reveal a peptide arrangement in linear chains, molecules forming pairs thanks to H-bonds, to increase their adsorption energy. When Phe–Phe molecules are adsorbed as single units (under the neutral chemical form), they anchor via their amino N atom and the oxygen atoms from the carboxyl and peptidic groups, while when adsorbed under a chain configuration (zwitterionic chemical form), they bind to the surface via the amino N atom and the two carboxylate O atoms. In the 2D chain assembly, Phe–Phe molecules are tilted, compared to isolated dipeptides, with the amino group coming above the surface to facilitate H-bonds in between neighbouring molecules; the CONH plane thus passes from perpendicular to parallel to the surface. Importantly, the assembly cannot go for chains longer than nine pairs; in order to create long-range arrays, the chains need to be stabilized by coadsorbed molecules, such as terephtaldehic acid (TPA) [36], resulting in chain lengths up to several tenths of nanometers on the same terrace (Figure 11.3b); the extent of the supramolecular assembly is only limited by the size of the atomic terrace.

As an illustration of the importance of H-bonds in the creation of a long-range supramolecular arrays, it is worth mentioning the adsorption of a close Phe–Phe look-a-like molecule: Ala–Ala. Dialanine peptide is the simplest chiral homopeptide, differing mainly from Phe–Phe by the absence of phenyl rings at its extremities. Stengaard reports its adsorption on Cu(110) surface as a function of adsorption temperature [37]. Based on the adsorption of single Ala and tri-Ala molecules on a Cu(110) surface, the author was able to propose (i) the chemical anionic form of the adsorbed molecules and (ii) an adsorption geometry model where the di-Ala peptide is anchored in a three-point mode to the surface via both oxygen of a carboxylate group and the N atom of the amine group at the other end of the molecules. The absence of phenyl rings present in Phe–Phe allows a rotation freedom for the Ala–Ala

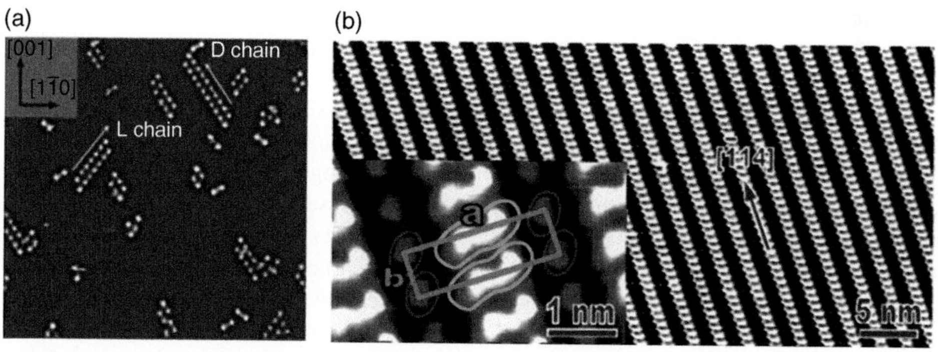

Figure 11.3 *STM images of Phe–Phe dipeptide molecules adsorbed on Cu(110). In (a) single molecules (two white blobs) and chains of molecules [35]; in (b) extended 2D arrays of Phe–Phe molecules stabilized by coadsorption of TPA molecules [36]. Reproduced with permission from references [35] and [36]. Copyright (2007) John Wiley & Sons, Ltd and (2007) American Chemical Society (see color plate figure)*

molecules, resulting in H-binding between terminal O and N atoms of adjacent molecules. As a direct consequence, the 2D organization starts in the same way as for a submonolayer, by small chains of identical molecules. However, when the coverage is increased, the possibility of H-bonds in between adjacent molecules enables the creation of dimer chains up to large islands of chains. If the coverage is increased, the entire surface can be covered without stabilizing molecules such as TPA in the previous Phe–Phe example.

Having seen the role of the amino acid nature and flexibility upon a dipeptide adsorption, we now focus on the possible influence of the substrate. Let us take the example of Gly–Pro, whose adsorption has been investigated, under identical low pressure conditions, on two different surfaces, namely Au(110) [38] and Cu(110) [39]. The surfaces were characterized *in situ* by PM-RAIRS and XPS spectroscopies. Interpretation of Gly–Pro adsorption was achieved by comparing the data with those of glycine [30] and proline [40] on copper or gold surfaces.

On Au(110), at low and intermediate Gly–Pro coverage (15 and 30 minutes of exposure), the PM-RAIRS analysis shows that carboxylic acids and carboxylate groups are simultaneously present; Au(110) thus stabilizes neutral and zwitterionic adsorbed species. XPS data recorded for the C 1 s and N 1 s region confirm the PM-RAIRS conclusions, with the obvious presence, on the carbon peak, of two components at high binding energy [38], for carbon in COOH and COO^- respectively; the nitrogen peak also shows distinct contributions for protonated and standard NH_2 groups respectively (Figure 11.4b). XPS data are quantitative, leading to a zwitterionic-over-neutral chemical form ratio equal to 1:2. When the coverage increases and multilayers start to grow on the surface, the fraction of zwitterionic molecules increases, and become prominent.

The same experiments were conducted on Cu(110). This comparative study was motivated by the already observed different binding modes of simple amino acids on these two surfaces: on Au surfaces, AA mainly adsorbs under their zwitterionic chemical forms, while these are known to preferentially bind in the anionic form on copper [30, 31, 40–42]. Passing from gold to copper also changes the way dipeptides interact with the metal surface. Gly–Pro on Cu(110) data highlight the presence of carboxylate groups and no protonated acid groups (no signal at ~1720 cm^{-1}). Additional XPS data in the nitrogen region reveal the presence of one main peak at 400.3 eV, assigned to amide or amine groups, and no contribution at higher binding energy, ~401.8 eV, the expected signature of protonated NH_3^+ groups observed in the Gly-Pro powder (Figure 11.4a); Gly–Pro molecules are thus adsorbed under their anionic chemical form. The N 1 s peak makes clear an additional component at lower binding energy (Figure 11.4c), 398.6 eV, which has been assigned to a nitrogen atom in strong interaction with the copper substrate. Since this XPS feature was not present in the adsorption of simple proline [43, 44], the authors deduced that Gly–Pro interacts with copper by the amino N atoms from the terminal NH_2 groups, in addition to the carboxylate O atoms. Very importantly, this adsorption saturates when the coverage reaches one monolayer, likely because the absence of protonated COOH groups impedes further intermolecular H-bonds.

The example of Gly–Pro, and this is rare in the literature, demonstrates the role of the metal nature: copper favouring deprotonation, compared to gold, and direct interaction with nitrogen atoms. A consequence of the absence of zwitterionic molecules is the impossible intermolecular H-bond interactions, which are at the origin of peptide multilayers on a surface.

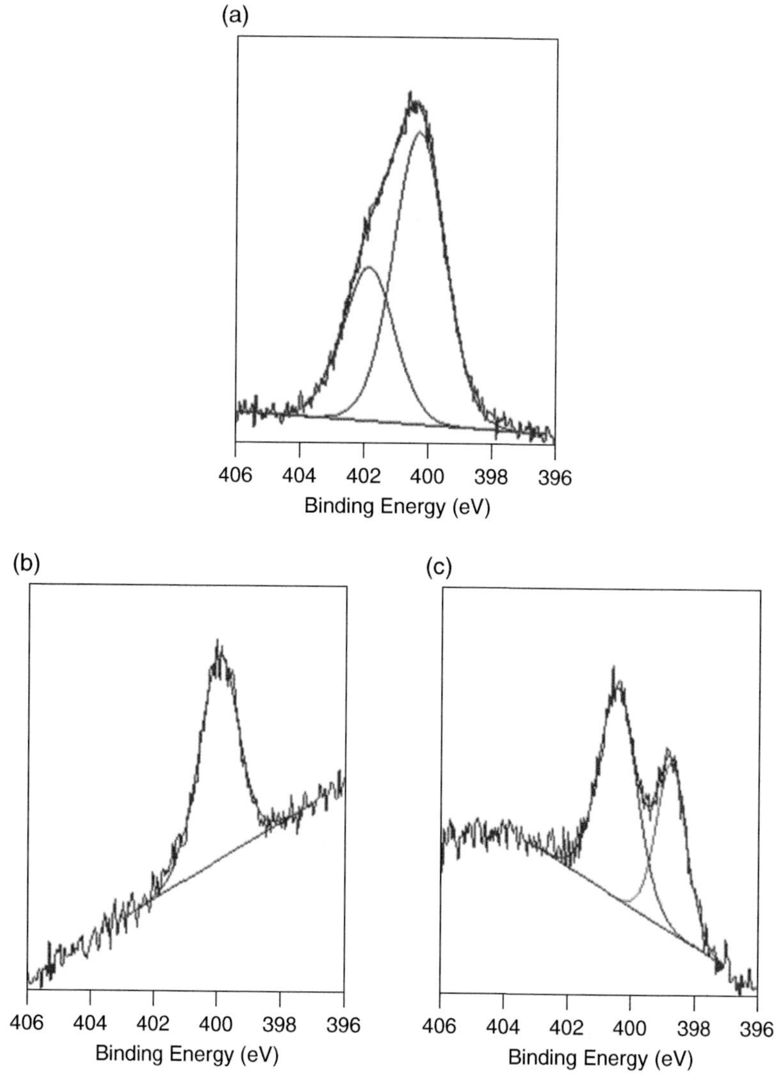

Figure 11.4 *XPS spectra of Gly–Pro in the crystalline state in (a), adsorbed on Au(110) in (b) and on Cu(110) in (c) at a low coverage phase*

11.5.2 Adsorption of Tripeptides

The simplest tripeptides are of course those constituted of three identical amino acids; two of them, tri-L-alanine and tri-L-leucine, have been adsorbed on Cu (110) and characterized by RAIRS under UHV conditions [45]. Peptides in the powder form were evaporated in vacuum, supposed to be in a zwitterionic form. Tripeptides can arrange in parallel or antiparallel β-sheets form [46], depending on the network of intermolecular H-bonds. On Cu(110), trialanine is in an anionic form after COOH deprotonation and displays various orientations depending on the surface coverage; at low coverage, the peptide layer is disordered whereas at a full monolayer, shifts in the C=O and NH RAIRS signals suggest that the molecules are lying parallel to the surface and in interaction with each other via H-bond, forming a β-antiparallel-like arrangement. As for trileucine, differing from trialanine by an additional isopropyl group on its side chain, it also binds to the surface via its COO^- group and forms a disordered phase at low coverage, followed by a monolayer of H-bond interacting molecules when the coverage increases to one monolayer. The interesting difference between trialanine and trileucine appears at higher exposure, when multilayers are expected to form, but do not in the second case, likely due to the bulky side groups, which hinder H-binding between C=O and NH groups of two different molecules. This study is a rare one where the tridimensional structure of short peptides, in an adsorbed phase, is investigated, thus demonstrating the role of side groups on the growth and arrangement of peptides on a solid surface. We will see another example below.

Because it shows the determining influence of an additional fragment to a peptide, let us examine the adsorption on copper and gold surfaces of the Gly–Pro dipeptide and of the Gly–Pro–Glu (IGF) tripeptide, which is nearly the same as Gly–Pro with a glutamate residue instead of its acid terminal group. As described above, adsorption of the insulin-like growth factor tripeptide, IGF, was thus performed under low pressure on the two well-defined surfaces, Au(110) and Cu(110). This set of data offers a direct comparison and enables some conclusions to be deduced concerning the effect of the peptide size and of the nature of the metal upon adsorption.

We have just seen that the Gly–Pro dipeptide adsorbs intact on both copper and gold surfaces, under a neutral form with a fraction of zwitterionic species. IGF, only differing by an additional glutamate residue, was allowed to interact with Au(110) under the same conditions, namely 10^{-9} torr. First, IGF is mainly adsorbed under a neutral form; the RAIRS spectra of adsorbed IGF is indeed dominated by the C=O stretch vibration, at ca. 1743 cm^{-1}, showing that a major fraction of the IGF molecules keeps their protonated COOH groups (Figure 11.5). Second, the kinetics of adsorption is much more rapid for IGF than for Gly–Pro; this was evaluated from the measured average thickness of the organic layer on gold, which is more than twice as high as that of Gly–Pro after identical doses; this indicates that an additional residue also implies stronger adsorption, likely due to one or several additional binding sites; the additional interaction with the surface cannot be ascribed to COO^- interacting with gold.

Moreover, besides the expected contribution attributed to nitrogen in NH_2 or NH_3^+ groups, the XPS spectra of IGF made clear an unexpected N 1 s contribution at low binding energy (398.7 eV), which was not present on the Gly–Pro XPS analysis on the same surface; this contribution is associated with nitrogen atoms in interaction with gold, thus constituting the main difference with Gly–Pro in their interaction with gold. These data lead to the conclusion that a longer peptide interacts more strongly than a shorter one just because it has more binding sites. At higher exposure, IGF continues to adsorb and form 3D aggregates, as suggested by strong charge effect observed during XPS analyses.

The same comparison, Gly–Pro versus IGF, was done on Cu(110) and results were slightly different. The tripeptide is adsorbed under anionic form. As seen before, Gly–Pro interacts with copper via its nitrogen atom and its COO^- groups (Figure 11.6a), with a maximum coverage attained for one monolayer (saturation). Addition of a glutamate residue induces slight changes in the interaction, but significant ones in the growth mode of the peptide. The IGF molecule binds to copper by nitrogen from the initial stage of

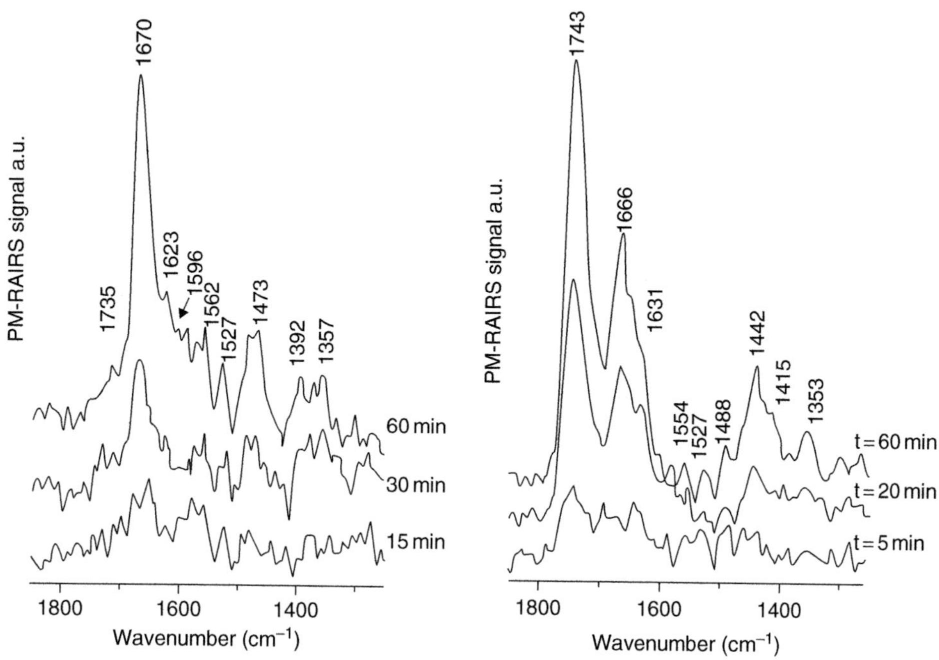

Figure 11.5 *PM-RAIRS spectra recorded on Au(110) during exposure to Gly–Pro (left) and IGF (right). Reproduced with permission from reference [38]. Copyright (2009) American Chemical Society*

adsorption and may also be by deprotonated COO^- groups (Figure 11.6b); like on gold, the adsorption does not seem to saturate at one monolayer and IR spectra show the existence of protonated acid end groups, which make intermolecular interactions possible and explains the progressive, and apparently nonlimited, growth of multilayers (see Figure 11.5).

Adsorption of another tripeptide, glutathione (GSH) or Glu–Cys–Gly, currently studied for its role in human metabolism, was also characterized on the Au(110) surface. GSH differs from IGF by its central fragment, cystein instead of proline for the latter. Like IGF, GSH strongly interacts with gold, but via its S atom at very low coverage. This was expected from the well-known affinity of sulfur for gold. More surprising was the growth mode of the GSH layer; molecules come in interaction with those that have been adsorbed, forming interacid group H-bonds – just like IGF!

These comparisons, Gly–Pro versus IGF, Au versus Cu, give us a few lessons to keep in mind: peptides bind to metal surfaces via their nitrogen or the deprotonated COO^- groups, likely depending on the conditions of adsorption and the ability of the metal to permit electron donation to the metal via the H atom, which indicates the possible influence of the nature of the metal. Eventually, we observed an increased rate of adsorption with a tripeptide, when the latter displays more than one site of interaction, and possible intermolecular H-binding.

Though this chapter is focused on experimental data describing the interaction of short peptides with metal or oxide surfaces, we will make an exception for the Arg–Gly–Asp tripeptide (RGD), which raised a large amount of interest in the community of theoreticians.

(a)

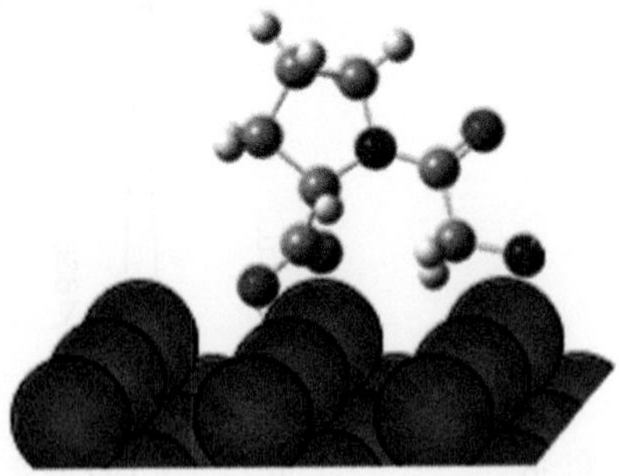

(b)

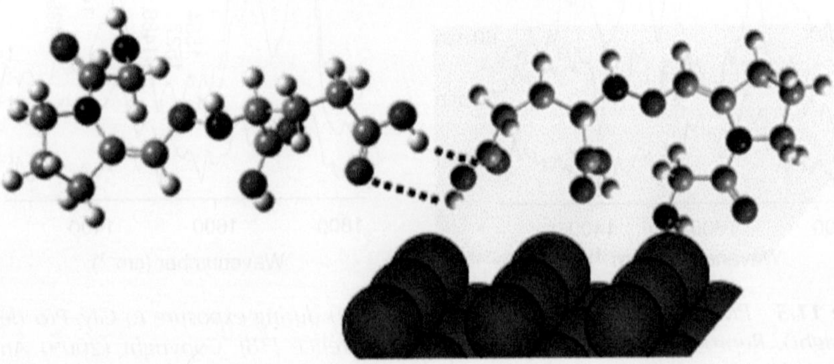

Figure 11.6 Geometry and growth mode of (a) Gly–Pro and (b) Gly–Pro–Glu on Cu(110). Reproduced with permission from reference [39]. Copyright (2011) American Chemical Society

The reason for that is, first, the crucial and well-established role of RGD in cell adhesion [47] and, second, the size of the molecule, only three amino acids, which permits reliable modeling studies of its interaction with the TiO_2 surface, representative of implant biocompatible material. TiO_2 can exist either on the anatase, rutile or brookite form, and various crystallographic orientations can be considered. On anatase, the strongest interaction of RGD was calculated on the (001) face, while it was on the (010) surface of rutile; on both surfaces, RGD adopts the same conformation, in order to facilitate H-bindings with the Ti cations on the surface (Figure 11.7) [48].

Another molecular dynamics simulation study demonstrates that, in aqueous solution, the amino and carboxyl groups of RGD bind to the hydrophilic TiO_2 surface via electrostatic and van der Waals interactions. Water molecules form an intermediate layer that also interacts directly with RGD groups. Stronger adsorption of RGD is observed on a TiO_2 surface containing structural defects than on a perfect TiO_2 (110) surface; this is related to the stabilization of low coordinated titanium atoms by water molecules, which in turn interact with the peptide carboxyl groups [49].

To make it closer to real physiological conditions, not only water molecules but cations have to be taken into account; this was done in a recent simulation investigation [50]. The

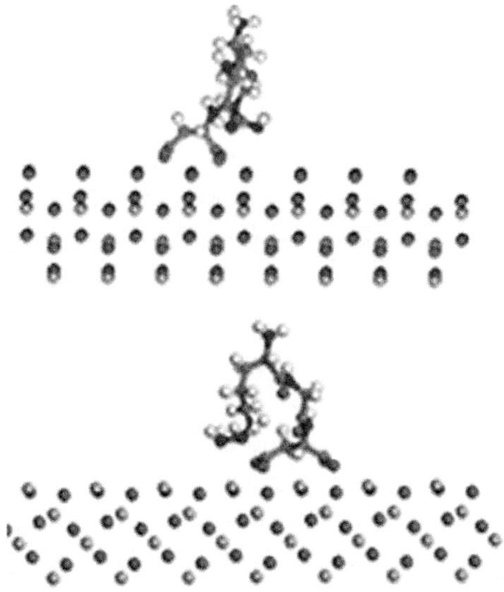

Figure 11.7 Simulated RGD peptide configuration on the anatase (001) and rutile (010) surfaces. Reproduced with permission from reference [48]. Copyright (2008) Springer Science

role of Na^+ cations in the adsorption of RGD on negatively charged hydroxylated rutile surfaces was demonstrated. As written above, RGD indeed reacts via its NH_2 and COO^- side groups; adsorption of Na^+ cations competes with that of the positively charged Arg side of the peptides in such a way that the latter detaches from the surface, inducing a reorientation of the peptide. Na^+ may also stabilize the interaction of COO^- groups with surface hydroxyls and this can be, at least partially, controlled by modifying the ionic strength of solutions.

Also by molecular dynamics simulation, the adsorption mode of the Lys–Glu–Lys tripeptide was studied on three different types of surfaces, TiO_2 rutile (110), muscovite mica and oriented graphite (HOPG) [51]. Remarkable is the good agreement between this recent study and previous experimental investigations, which showed adsorption strengths in the following order, graphite < mica ≈ titanium oxide. On the latter surfaces, peptides are stabilized by the side groups of both Lys and Glu amino acids; Ca^{2+} cations appear to act as linkers between the peptide COO^- groups and the surface.

11.6 Adsorption at the Solid/Liquid Interface

Despite the huge interest to understand the adsorption behavior of short peptides in the liquid phase, only a few papers dealing with adsorption on bare metal and oxide surfaces are reported in the literature. In most cases, short peptides are immobilized on modified surfaces or conjugated to other polymers or polypeptides exhibiting chemical functional groups, as broadly reported in the case of RGD peptides [52, 53]. In the following, on the basis of examples from the literature, we describe some important features with fundamental and practical relevance for the adsorption of short peptides on metals and oxides in aqueous solution.

Feyer *et al.* studied the adsorption of a dipeptide composed of histidine and glycine, Gly–His, from an aqueous solution [32]. The adlayer was prepared by deposition of a drop of a Gly–His saturated aqueous solution, under N_2 atmosphere on a clean Au surface for 5 minutes; samples were then heavily rinsed and annealed at 420 K in vacuum to remove weakly adsorbed molecules. The Gly–His adlayer was then characterized by combining synchrotron radiation–XPS and NEXAFS techniques. This study is of particular interest if one has in mind the controversy raised by the adsorption of the histidine, one of this peptide's fragments, on the same Au(111) surface, in particular, the mode of interaction of carboxylate groups with the surface (one or two points) [54] or that of Matéo Marti *et al.* [55, 56].

As for Gly–His, the C 1 s and O 1 s spectra are very similar to those obtained for His alone by the same authors, Feyer *et al.* [33], leading to the same conclusions concerning the presence of carboxylates. As for the N 1 s peak, expected to be different from the single AA one, due to the presence of a peptidic bond, it is indeed the case with an additional peak at 399.8 eV, assigned to the nitrogen atom in CONH peptidic moieties. NEXAFS studies allowed the authors to determine the orientation of the molecules of Gly–His with respect to the surface and propose an adsorption model: the Gly–His molecules bind to the surface, under their anionic forms, via the imino nitrogen and the two oxygen atoms of deprotonated carboxylic groups from the histidinate moieties. Molecules are tilted compared to the His AA, with no interaction between the peptide group and the surface; the latter stands roughly perpendicular to the surface above the imidazole group, as shown in Figure 11.8.

Other tripeptides, Gly–His–Gly and Gly–Gly–His, have been adsorbed on Au(111) and compared to their dipeptidic fragments in the X-ray photoelectron spectroscopy and near edge absorption spectroscopy (NEXAFS) study reported by Feyer *et al.* [33]. As soon as histidine is present in the peptide sequence, one may expect an interaction via the nitrogen atom of the ring and the carboxylate group. An additional glycine residue does not imply that the second carboxylate binds with the surface. It may, however, change the conformation of the tripeptide, as schematized in Figure 11.8. XPS data suggest an adsorption mode with two anchoring points and the peptide chain almost perpendicular to the surface. This

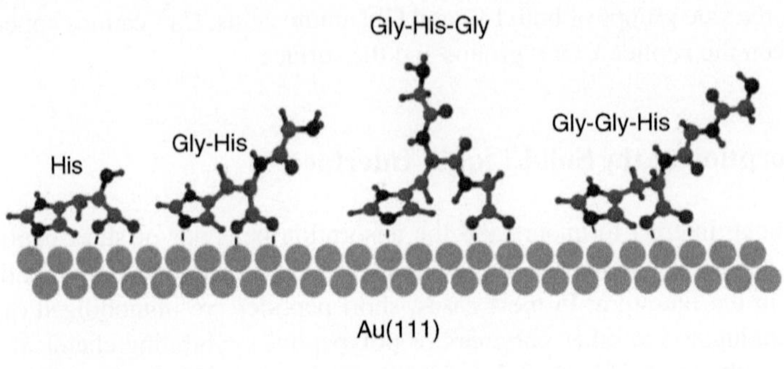

Figure 11.8 *Adsorption model for histidine and histidine-derived peptides on Au(111). Reproduced with permission from reference [32]. Copyright (2010) American Chemical Society*

is an example where the binding does not imply all carboxylate groups in interaction with the surface.

The metal surfaces may also be studied under electrochemically controlled conditions. By varying the applied potential of the Au electrode, some authors have evidenced drastic changes at the Au/electrolyte interface upon the adsorption of the Cys–Trp peptide, characterized by means of reflection anisotropy spectroscopy. At a potential value equal to –0.6 V/SCE, the adsorption of peptides occurs mainly through the formation of Au–S bonds, with an additional link involving the NH_2 group of the N-terminal. When the applied potential increases to more positive values, the NH_2 groups detach from the surface, leading to conformational/orientation change that authorizes adjacent peptide molecules to interact with the Au surface through a disulfide bridge. One point of primary importance is related to the fact that the orientation of Trp moieties is very sensitive to the applied potential. Moreover, by performing subsequent cycling of the applied potential, it is shown that the described events are not perfectly reversible. These findings open new ways to take electrochemical considerations into account in the study of peptide adsorption on metals.

Adsorption of lysine-based peptides, of different molecular weights, on the TiO_2 surface was investigated by *in situ* IR-ATR measurements in aqueous medium (pH ~ 7.4). To this end, the authors used dilysine, trilysine, tetralysine, pentalysine and polylysine (169 lysine units). In these conditions, adsorption occurs in all cases, as expected, due to electrostatic interactions between a negatively charged TiO_2 surface and positively charged peptides. This result was confirmed by varying the pH of the medium from 2.4 to 12.2. As shown in Figure 11.9, differences in the IR spectrum of adsorbed tetralysine are noticeable compared to the spectrum of the same molecules in solution. Indeed, the evolution of the band at 1637 cm^{-1} could be attributed to the fact that carboxylate functions are likely to form coordinative bonds to surface Ti^{4+}. Based on the determination of binding constant (K), it is evidenced that K values increase with the number of lysine units, in agreement with thermodynamic expectations detailed above. The change of the secondary structure of

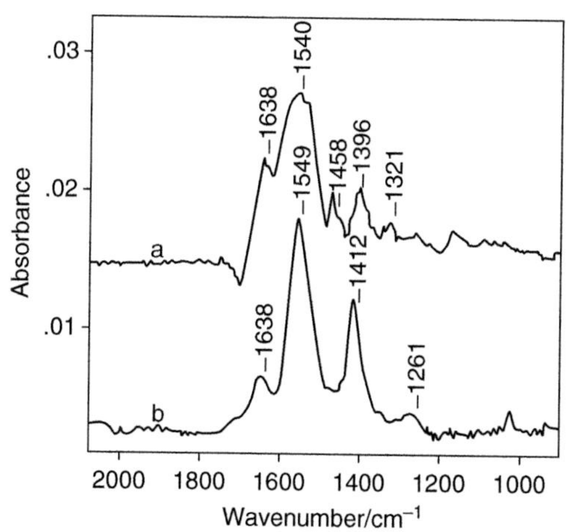

Figure 11.9 *IR spectra of (a) tetralysine adsorbed to TiO_2 and (b) tetralysine on a bare ZnSe prism. Reproduced with permission from reference [57]. Copyright (1999) Academic Press*

adsorbed peptides was only observed on polylysine ($n=169$). However, it must be kept in mind that such conformational change is difficult to observe with IR spectroscopic analyses for short peptides.

Glutathione (Glu–Cys–Gly, or GSH) was adsorbed on gold in the liquid phase and characterized by ATR-IR, combined with modulation excitation spectroscopy to monitor protonation/deprotonation induced by changing the solvent acidity [58]. On such a metal, GSH forms a self-assembled monolayer strongly anchored via the S atom, with a possible additional binding site via one of the deprotonated carboxylic groups. The molecule may be reprotonated and rearrange on the surface in order to optimize H-bond interactions with the surface and possibly between adjacent molecules. This *in situ* IR study shows how the ionic state of the peptide is important for its conformation and interaction mode with the surface.

Adsorption of another tripeptide on platinum was investigated using continuous capacitance and ellipsometry measurements [59]. Five derivatives of the Phe–Val–Arg tripeptide have been synthesized and adsorbed on platinum electrodes. This study was motivated by the current use of these peptides to measure serine protease activities, the question being how to improve their immobilization on the electrode. The adsorption properties of peptides strongly depend on the number of aromatic rings, which are strong binding groups with platinum, whereas the valine residue has no influence on the adsorption. Interestingly, the authors observed the same occupied surface areas for all peptides at a given concentration, which is explained by peptide–peptide lateral interactions in the adsorbed phase. These interactions may be hydrophobic between nonpolar amino acids (Phe or Val) or electrostatic (attractive or repulsive) between charged groups; they induce conformational changes along with coverage increases.

In another context related to the use of peptides for biocompatibility, some authors have used short peptide sequences to mimic the EAK 16 oligopeptides [60–62]. For this purpose, Ala–Glu and Ala–Lys were adsorbed on the TiO_2 surface in aqueous medium. Molecular dynamic simulations have allowed the most probable contact points between peptides and the surface to be identified. Based on a comparison between experimental and *ab initio* simulated XPS spectra, the authors evidenced the presence of multiple coordination points in the peptide/surface binding [62]. For Ala–Glu, in addition to the expected presence of coordinative bounded carboxylate species, hydrogen binding between the NH_2 groups and oxygen, from the TiO_2 surface, is observed. In the case of Ala–Lys, two carbonyl oxygen and one hydrogen-bound NH_2 are likely to be involved in the adsorption of the peptide on the TiO_2 surface.

For another promising application, namely chiral induction, Mondelli *et al*. synthesized short-chain peptides, such as L-Trp–Ala, L-Trp–Trp and L-Trp–Gly–Gly, and adsorb these peptides on Pt(111) surfaces [62]. This procedure aims at promoting an enantioselective catalysis reaction, namely the hydrogenation of ketopantolactone (KPL). First principle computational modeling was also used to describe the chiral active sites and their environment. By testing a huge set of possible geometries and considering measured enantiomeric excesses, the authors came to important conclusions: the geometry of these molecules on a surface, and thus the shape of the chiral site, is strongly governed by the adsorption mode of the indole ring, flat, upside or tilted with regard to the surface; the preferred adsorption is with the indole moiety parallel to the surface and the adsorption energy is mostly due to this interaction. This work also shows the crucial role of Trp- containing peptides to generate docking chiral surfaces; best enantiomeric excesses were obtained with Trp–Gly, Trp–Trp dipeptides and even better with Trp–Gly–Gly or Trp–Gly–Tyr tripeptides, where the amino acid chain is more elongated and thus stabilizes the chiral active sites. This study opens a new, and very promising, field of investigation applying controlled peptide/solid interfaces for their chiral properties.

11.7 Conclusions and Guidelines for the Future

Through the different examples described in this chapter, the interest of fundamental data gathered from peptide adsorption in UHV conditions is clearly demonstrated. Some recent studies describing the interaction of short peptides with metal or oxide surfaces, reported in the literature, are cited in Table 11.1.

Indeed, relevant events regarding the mechanism of interaction of short peptides with metal surfaces are acquired, revealing various interfacial processes: peptide–surface interactions presumably through coordinative bonds, supramolecular organization and

Table 11.1 *Examples of peptide adsorption on metal or oxide surfaces*

Peptide sequence	Surface	Adsorption medium	Surface characterization techniques	Reference
Lys–Lys Lys–Lys–Lys	TiO_2	HEPES buffer (pH ~ 7)	IR-ATR	[57]
Cys–Trp	Au(110)	Phosphate buffer	RAS	[63]
Arg–Cys	Au	Ultrapure solution	XPS; RAIRS	[64]
Ala–Glu Ala–Lys	TiO_2	Solution of NaCl containing HCl (pH ~ 4)	XPS; NEXAFS	[61]
Ala–Glu Ala–Lys	TiO_2	Solution of NaCl containing HCl (pH ~ 4)	XPS	[60]
Gly–His Gly–His–Gly Gly–Gly–His	Au(111)	Aqueous solution UHV	XPS; NEXAFS	[32]
Ala–Ala–Ala	Cu(110)	UHV	RAIRS	[45]
Leu–Leu–Leu	Cu(110)	UHV	RAIRS	[45]
Ala–Ala	Cu(110)	UHV	STM	[37]
Phe–Phe	Cu(110)	UHV	STM	[35] [36] [34]
Gly–Gly	Cu(110)	UHV	XPS; NEXAFS	[33]
Gly–Pro	Au(110)	UHV	XPS; RAIRS	[38]
Gly–Pro	Cu(110)	UHV	XPS; RAIRS	[39]
Gly–Pro–Glu	Au(110)	UHV	XPS; RAIRS	[38]
Gly–Pro–Glu	Cu(110)	UHV	XPS; RAIRS	[39]
Gly–Cys–Glu	Au(111)	UHV	XPS; RAIRS	[65]

multilayer growth via intermolecular interactions. On another side, investigating the adsorption of peptides on oxidized surfaces provides a unique insight into how peptides interact with 'real surfaces' involved for the different applications cited above.

Making a direct connection between adsorption in UHV conditions and in the liquid phase remains not obvious, owing to the state of the interface and the behavior of peptides in the gas phase versus aqueous solutions. Interactions taking place between peptides and a metallic surface in the liquid phase can be in some way predicted from data obtained under clean and model UHV conditions. However, some differences inherent to the liquid phase are not predictable from model experiments, such as the role of water, the chemical form of the molecules dictated by the pH of the media or even some oxidation and/or dissolution of the metallic substrate upon peptide adsorption/media contact.

The *properties of modified substrates* are very much linked to the *interactions* with *admolecules*. It is thus essential to know what kind of binding and interactions occur in order to design a wished surface, for tailoring catalytic surfaces, for biorecognition or chiral discrimination systems.

Understanding and controlling the peptide–surface interactions is an obvious key step towards 'greener' processes, biomimetic synthesis of nanoparticles, enantioselective catalysis or biodetection chips. Nevertheless, there is a lot left to study, especially regarding the role of water during the adsorption process, as little is known about the crosslinking of water in the building of two-dimensional supramolecular assemblies, or even in the building of three-dimensional structures, and, more importantly, how the interactions induced by water molecules in a 2D or 3D building block will evolve upon temperature, concentration or media changes.

References

[1] Sarikaya, M., Tamerler, C., Jen, A.K.Y., Schulten, K., Baneyx, F. (2003) Molecular biomimetics: nanotechnology through biology. Nature Mater. 2: 577–585.

[2] Ratner, B.D. (2004) Biomaterials Science: An Introduction to Material in Medicine. 2nd edition. Elsevier Academic Press, USA.

[3] Horbett, T.A., Brash, J.L. (1995) Proteins at Interfaces II: Fundamentals and Applications. American Chemical Society, Washington.

[4] Sarikaya, M., Tamerler, C., Schwartz, D.T., Baneyx, F. (2004) Materials assembly and formation using engineering polypeptides. Annu. Rev. Mater. Res. 34: 373–408.

[5] Tamerler, C., Sarikaya, M. (2009) Genetically designed peptide-based molecular materials. ACS Nano 3: 1606–1615.

[6] Zhang, S. (2003) Fabrication of novel biomaterials through molecular self-assembly. Nat. Biotech. 21: 1171–1178.

[7] Baumann, M.K., Textor, M., Reimhult, E. (2008) Understanding self-assembled amphiphilic peptide supramolecular structures from primary structure helix propensity. Langmuir 24: 7645–7647.

[8] Ghosh, S., Reches, M., Gazit, E., Verma, S. (2007) Bioinspired design of nanocages by self-assembling triskelion peptide elements. Ang. Chem. Int. Ed. 46: 2002–2004.

[9] Tan, Y.N., Lee, J.Y., Wang, D.I.C. (2010) Uncovering the design rules for peptide synthesis of metal nanoparticles. J. Am. Chem. Soc. 132: 5677–5686.

[10] Bi, X., Agarwal, A., Yang, K.-L. (2009) Oligopeptide-modified silicon nanowire arrays as multichannel metal ion sensors. Biosens. Bioelectr. 24: 3248–3251.

[11] Deschamps, P., Kulkarni, P.P., Gautam-Basak, M., Sarkar, B. (2005) The saga of copper(II)-L-histidine. Coord. Chem. Rev. 249: 895–909.
[12] Bi, X., Yang, K.-L. (2007) Complexation of copper ions with histidine-containing tripeptides immobilized on solid surfaces. Langmuir 23: 11067–11073.
[13] Barlow, S.M., Raval, R. (2008) Nanoscale insights in the creation and transfer of chirality in amino acids monolayers at defined metal surfaces. Curr. Op. Coll. Interf. Sci. 13: 65–73.
[14] Raval, R. (2009) Nanoscale insights in the creation of chiral surfaces. J. Mol. Catal. A 305: 112–116.
[15] Raval, R. (2009) Chiral expression from molecular assemblies at metal surfaces: insights from surface science techniques. Chem. Soc. Rev. 38: 707–721.
[16] Mallat, T., Orglmeister, E., Baiker, A. (2007) Asymmetric catalysis at chiral metal Surfaces. Chem. Rev. 107: 4863–4890.
[17] Pourbaix, M. (1963) Atals d'Equilibres Electrochimiques. Gauthier-Villars, Paris.
[18] Bockris, J.O.M., Reddy, A.K.N. (1970) Modern Electrochemistry: A Molecular Level Approach. Plenum Press, New York.
[19] Kinloch, A. J. (1990) Adhesion and Adhesive, Vol. 32. Chapman & Hall, London.
[20] Mantel, M., Rabinovich, Y.I., Wightman, J.P., Yoon, R.H. (1995) A study of hydrophobic interactions between stainless steel and silanated glass surface using atomic force microscopy. J. Coll. Interf. Sci. 170: 203–214.
[21] Caillou, S., Gerin, P.A., Nonckreman, C.J., Fleith, S., Dupont-Gillain, C.C., Landoulsi, J., Pancera, S.M., Genet, M.J., Rouxhet, P.G. (2008) Enzymes at solid surfaces: nature of the interfaces and physico-chemical processes. Electroch. Acta 54: 116–122.
[22] Landoulsi, J., Genet, M.J., Richard, C., El Kirat, K., Pulvin, S., Rouxhet, P.G. (2008) Evolution of the passive film and organic constituents at the surface of stainless steel immersed in fresh water. J. Coll. Interf. Sci. 318: 278–289.
[23] Nelson, D.L., Cox, M.M. (2005) Lehninger Principles of Biochemistry. 4th edition. W.H. Freeman and Company, New York.
[24] Lefèvre G., D.M., Fédoroff M., Johannes L. (2006) Accuracy in the determination of acid-base properties of metal oxides surfaces. In: Surface Complexation Modelling. Elsevier, pp. 35–66.
[25] ExPASy, http://web.expasy.org/.
[26] Léoonil, J., Mollé, D., Bouhallab, S., Henry, G. (1994) Precipitation of hydrophobic peptides from tryptic casein hydrolysate by salt and pH. Enz. Microb. Tech. 16: 591–595.
[27] Sengupta, P., Garai, K., Sahoo, B., Shi, Y., Callaway, D.J.E., Maiti, S. (2003) The Amyloid b peptide (Ab1-40) is thermodynamically soluble at physiological concentrations. Biochem. 42: 10506–10513.
[28] Pradier, C.-M., Humblot, V., Stievano, L., Méthivier, C., Lambert, J.F. (2007) Salt concentration and pH-dependent adsorption of two polypeptides on planar and divided alumina surfaces. *In situ* IR investigations. Langmuir 23: 2463–2471.
[29] Barlow, S.M., Raval, R. (2003) Complex organic molecules at metal surfaces: bonding, organisation and chirality. Surf. Sci. Rep. 50: 201–341.
[30] Barlow, S.M., Kitching, K.J., Haq, S., Richardson, N.V. (1998) A study of glycine adsorption on a Cu(110) surface using reflection absorption infrared spectroscopy. Surf. Sci. 401: 322–335.
[31] Barlow, S.M., Louafi, S., Le Roux, D., Williams, J., Muryn, C., Haq, S., Raval, R. (2005) Polymorphism in supramolecular chiral structures of *R*- and *S*-alanine on Cu(110). Surf. Sci. 590: 243–263.
[32] Feyer, V., Plekan, O., Tsud, N., Chab, V., Matolin, V., Prince, K. C. (2010) Adsorption of histidine and histidine-containing peptides on Au(111). Langmuir 26: 8606–8613.
[33] Feyer, V., Plekan, O., Tsud, N., Lyamayev, V., Chab, V., Matolin, V., Prince, K.C., Carravetta, V. (2010) Adsorption structure of glycyl–glycine on Cu(110). J. Phys. Chem. C 114: 10922–10931.
[34] Tomba, G., Lingenfelder, M., Costantini, G., Kern, K., Klappenberger, F., Barth, J.V., Ciacchi, L.C., De Vita, A. (2007) Structure and energetics of diphenylalanine self-assembling on Cu(110). J. Phys. Chem. A 111: 12740–12748.

[35] Lingenfelder, M., Tomba, G., Costantini, G., Colombi Ciacchi, L., De Vita, A., Kern, K. (2007) Tracking the chiral recognition of adsorbed dipeptides at the single-molecule level. Ang. Chem. Int. Ed. 46: 4492–4495.
[36] Wang, Y., Lingenfelder, M., Classen, T., Costantini, G., Kern, K. (2007) Ordering of dipeptide chains on Cu surfaces through 2D cocrystallization. J. Am. Chem. Soc. 129: 15742–15743.
[37] Stensgaard, I. (2003) Adsorption of di-L-alanine on Cu(100) investigated with scanning tunneling microscopy. Surf. Sci. 545: L747–L752.
[38] Vallée, A., Humblot, V., Méthivier, C., Pradier, C.-M. (2009) Adsorption of di- and tripeptides on Au(110) under ultrahigh vacuum conditions. 1. Polarization modulation reflection absorption infrared spectroscopy and X-ray photoelectron spectroscopy characterization. J. Phys. Chem. C 113: 9336–9344.
[39] Méthivier, C., Lebec, V., Landoulsi, J., Pradier, C.-M. (2011) Probing the binding mechanism of peptides on a copper surface: multilayer self-assembly promoted by glutamate residues. J. Phys. Chem. C 115: 4041–4046.
[40] Matéo Marti, E., Barlow, S.M., Haq, S., Raval, R. (2002) Bonding and assembly of the chiral amino acid *S*-proline on Cu(110): the influence of structural rigidity. Surf. Sci. 501: 191–202.
[41] Humblot, V., Méthivier, C., Raval, R., Pradier, C.-M. (2007) Amino acid and peptides on Cu(110) surfaces: chemical and structural analyses of l-lysine. Surf. Sci. 601: 4189–4194.
[42] Jones, T.E., Baddeley, C.J. (2002) A RAIRS, STM and TPD study of the Ni(111)/*R,R*-tartaric acid system: modelling the chiral modification of Ni nanoparticles. Surf. Sci. 513: 453–467.
[43] Fleming, G.J., Adib, K., Rodriguez, J.A., Barteau, M.A., Idriss, H. (2007) Proline adsorption on TiO_2(110) single crystal surface: a study by high resolution photoelectron spectroscopy. Surf. Sci. 601: 5726–5731.
[44] Fleming, G.J., Adib, K., Rodriguez, J.A., Barteau, M.A., White, J.M., Idriss, H. (2008) The adsorption and reactions of the amino acid proline on rutile TiO_2(110) surfaces. Surf. Sci. 602: 2029–2038.
[45] Barlow, S.M., Haq, S., Raval, R. (2001) Bonding, organization, and dynamical growth behavior of tripeptides on a defined metal surface: tri-L-alanine and tri-L-leucine on Cu(110). Langmuir 17: 3292–3300.
[46] Hempel, A., Camerman, N., Camerman, A. (1991) L-Alanyl-L-alanyl-L-alanine: parallel pleated sheet arrangement in unhydrated crystal structure, and comparisons with the antiparallel sheet structure. Biopolymers 31: 187–192.
[47] Singhvi, R., Stephanopoulos, G., Wang, D.I.C. (1994) Effects of substratum morphology on cell physiology. Biotech. Bioeng. 43: 764–771.
[48] Zhang, H.P., Lu, X., Fang, L.M., Weng, J., Huang, N., Leng, Y. (2008) Molecular dynamics simulation of RGD peptide adsorption on titanium oxide surfaces. J. Mater. Sci. Mater. Med. 19: 3437–3441.
[49] Song, D.P., Chen, M.J., Liang, Y.C., Bai, Q.S., Chen, J.X., Zheng, X.F. (2010) Adsorption of tripeptide RGD on rutile TiO_2 nanotopography surface in aqueous solution. Acta Biomater. 6: 684–694.
[50] Wu, C., Skelton, A.A., Chen, M., Vlček, L., Cummings, P.T. (2011) Modeling the interaction between integrin-binding peptide (RGD) and rutile surface: the effect of Na^+ on peptide adsorption. J. Phys. Chem. C 115: 22375–22386.
[51] Monti, S., Alderighi, M., Duce, C., Solaro, R., Tiné, M.R. (2009) Adsorption of ionic peptides on inorganic supports. J. Phys. Chem. C 113: 2433–2442.
[52] Davis, D.H., Giannoulis, C.S., Johnson, R.W., Desai, T.A. (2002) Immobilization of RGD to <111>silicon surfaces for enhanced cell adhesion and proliferation. Biomaterials 23: 4019–4027.
[53] Falconnet, D., Csucs, G., Michelle Grandin, H., Textor, M. (2006) Surface engineering approaches to micropattern surfaces for cell-based assays. Biomaterials 27: 3044–3063.
[54] Zubavichus, Y., Zharnikov, M., Yang, Y., Fuchs, O., Heske, C., Umbach, E., Tzvetkov, G., Netzer, F.P., Grunze, M. (2005) Surface chemistry of ultrathin films of histidine on gold as probed by high-resolution synchrotron photoemission. J. Phys. Chem. B 109: 884–891.
[55] Matéo Marti, E., Méthivier, C., Dubot, P., Pradier, C.-M. (2003) Adsorption of (*S*)-histidine on Cu(110) and oxygen-covered Cu(110), a combined Fourier transform reflection absorption infrared spectroscopy and force field calculation study. J. Phys. Chem. B 107: 10785–10792.

[56] Matéo Marti, E., Quash, A., Méthivier, C., Dubot, P., Pradier, C.-M. (2004) Interaction of *S*-histidine, an amino acid, with copper and gold surfaces, a comparison based on RAIRS analyses. Coll. Surf. A 249: 85–89.

[57] Roddick-Lanzilotta, A.D., McQuillan, A.J. (1999) An *in situ* infrared spectroscopic investigation of lysine peptide and polylysine adsorption to TiO_2 from aqueous solutions. J. Coll. Interf. Sci. 217: 194–202.

[58] Bieri, M., Bürgi, T. (2005) Probing enantiospecific interactions between proline and an L-glutathione self-assembled monolayer by modulation excitation ATR-IR spectroscopy. J. Phys. Chem. B 109: 10243–10250.

[59] Arwin, H., Lundstroem, I., Arielly, S., Claeson, G. (1990) Orientation of a tripeptide on platinum. Langmuir 6: 1551–1557.

[60] Monti, S., Carravetta, V., Battocchio, C., Iucci, G., Polzonetti, G. (2008) Peptide/TiO_2 surface interaction: a theoretical and experimental study on the structure of adsorbed Ala–Glu and Ala–Lys. Langmuir 24: 3205–3214.

[61] Polzonetti, G., Battocchio, C., Dettin, M., Gambaretto, R., Di Bello, C., Carravetta, V., Monti, S., Iucci, G. (2008) Self-assembling peptides: a combined XPS and NEXAFS investigation on the structure of two dipeptides Ala–Glu, Ala–Lys. Mat. Sci. Engin. C 28: 309–315.

[62] Mondelli, C., Vargas, A., Santarossa, G., Baiker, A. (2009) Fundamental aspects of the chiral modification of platinum with peptides: asymmetric induction in hydrogenation of activated ketones. J. Phys. Chem. C 113: 15246–15259.

[63] Morozzo della Rocca, B., Smith, C. I., Tesauro, C., Desideri, A., Weightman, P. (2010) Adsorption of the cysteine–tryptophan dipeptide at the Au(110)/liquid interface studied using reflection anisotropy spectroscopy. Surf. Sci. 604: 2170–2176.

[64] Petoral Jr, R.M., Uvdal, K. (2002) Arg–Cys and Arg–cysteamine adsorbed on gold and the G-protein–adsorbate interaction. Coll. Surf. B 25: 335–346.

[65] Vallée, A., Humblot, V., Méthivier, C., Pradier, C.-M. (2008) Adsorption of a tripeptide, GSH, on Au(111) under UHV conditions; PM-RAIRS and low T-XPS characterisation. Surf. Sci. 602: 2256–2263.

Part IV

Applications of Peptide Materials

12

Bioactive Rosette Nanotubes for Bone Tissue Engineering and Drug Delivery

Rachel L. Beingessner[1], *Alaaeddin Alsbaiee*[2], *Baljit Singh*[3], *Thomas J. Webster*[4] *and Hicham Fenniri*[2]

[1] *National Institute for Nanotechnology, Canada*
[2]*Department of Chemistry, National Institute for Nanotechnology and University of Alberta, Canada*
[3]*Department of Veterinary Biomedical Sciences, Western College of Veterinary Medicine, University of Saskatchewan, Canada*
[4]*Division of Engineering and Institute for Molecular and Nanoscale Innovation, Brown University, USA*

Acronyms

AFM	Atomic force microscopy
BMPs	Bone morphogenetic proteins
BRD	Bovine respiratory disease
CD	Circular dichroism
COSY	Correlation spectroscopy
DDA	Donor–donor–acceptor
ELISA	Enzyme-linked immunosorbent assay
EMAP-II	Endothelial–monocyte activating polypeptide
ESI-MS	Electrospray ionization mass spectrometry
G∧C	Guanine–cytosine motif
HA	Hydroxyapatite

Peptide Materials: From Nanostructures to Applications, First Edition. Edited by Carlos Alemán, Alberto Bianco and Mariano Venanzi.

KRSR	Lysine–arginine–serine–arginine
LPS	Lipopolysaccharides
MAG_3	Mercaptoacetyltriglycine
NOESY	Nuclear Overhauser effect spectroscopy
NPs	Nanoparticles
OBs	Osteoblasts
PMN	Polymorphonuclear cell
QM	Quartz microparticles
RGD	Arginine–glycine–aspartic acid
RNTs	Rosette nanotubes
ROESY	Rotation-frame Overhauser effect spectroscopy
SEM	Scanning electron microscopy
S_NAr	Nucleophilic aromatic substitution
STM	Scanning tunneling microscopy
TEM	Transmission electron microscopy
TFA	Trifluoroacetic acid
TNF-α	Tumor necrosis factor α
TOCSY	Total correlation spectroscopy

12.1 Introduction

Injectable, biocompatible and biodegradable polymers [1–3] have been intensely investigated for drug delivery applications. Likewise, self-assembled nanoscale materials [4] have also received extensive interest, because of their highly tunable chemical, physical and mechanical properties and impressive engineering control. Their potential for specific morphologies, shapes, sizes and composition [4] can create a specific surface area, charge and topography that mimic the dimensions of constituent components of natural tissues. In this respect, self-assembled materials can be ideal for regenerative tissue engineering [5–7], by serving as a biocompatible scaffold that encourages the controlled delivery of mammalian cells to an injured tissue site, while displaying or releasing therapeutic factors that promote the healing process [4].

In this chapter, the design, properties, toxicology and biomedical applications of a class of biologically inspired, biocompatible, nanotubular architectures called rosette nanotubes (RNTs) are described [8]. These materials have a facile synthetic accessibility, chemical tunability and innate biomimetic nanostructured features that are ideal for the display and delivery of therapeutic factors. As a self-assembled scaffold, RNTs offer great advantages since they can covalently express a variety of amino acid residues and proteins on their outer surface and also incorporate hydrophobic drug molecules [9] via noncovalent mechanisms. In addition, RNTs mimic the natural nanoscale structure and features of the extracellular matrices in bone and cartilage tissues and thus, in conjunction with delivering therapeutic molecules, they can also create a favorable surface environment on implant materials for enhanced cellular adhesion and function.

12.2 Rosette Nanotubes (RNTs)

12.2.1 RNT Design

The self-assembling unit of the RNTs is the guanine–cytosine motif (G∧C) [10, 11] and was designed to be hydrophobic and possess the Watson–Crick donor–donor–acceptor (DDA) H-bond array of guanine (G) and acceptor–acceptor–donor (AAD) H-bond array of cytosine (C) [8]. Figure 12.1 shows the structure of this bicyclic molecule, along with a

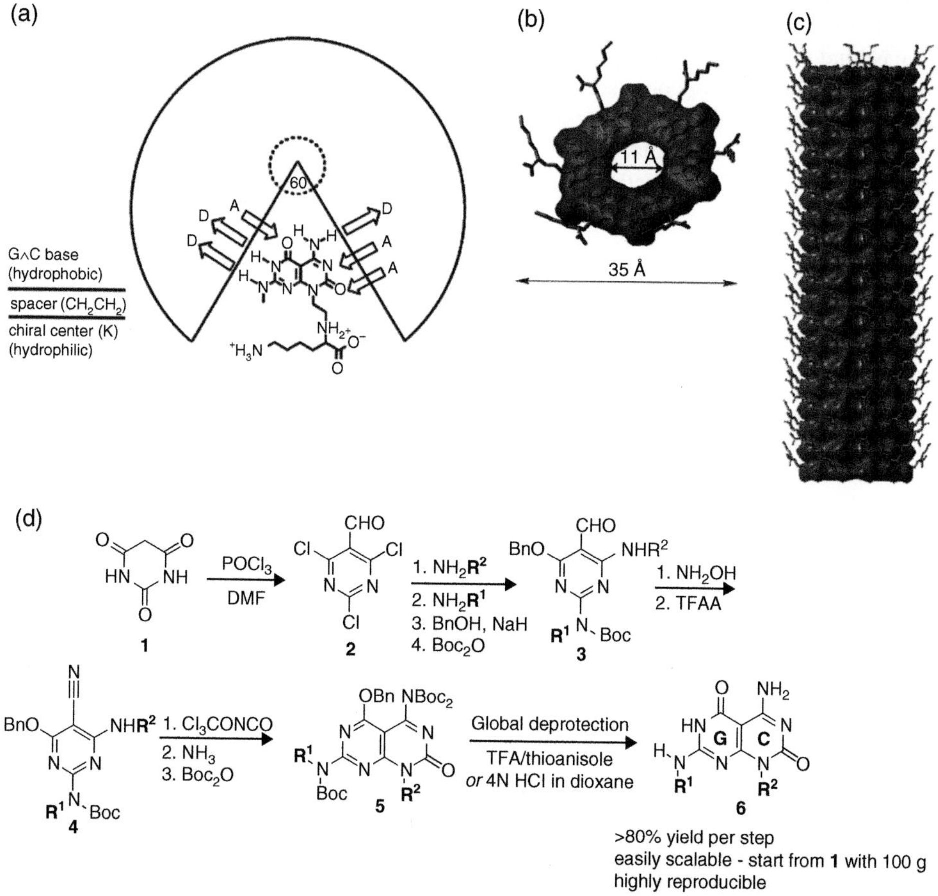

Figure 12.1 *(a) Bicyclic G∧C design, where 'A' and 'D' denote a hydrogen bond acceptor and hydrogen bond donor, respectively. The lysine group on the motif imparts water solubility and supramolecular chirality on the RNTs. (b) Six G∧C motifs spontaneously organize through H-bonding interactions to form a hexameric rosette. (c) Stacking of several rosettes generates the RNT having an inner channel diameter of 1.1 nm and length up to several millimeters. (d) General synthesis of the bicyclic G∧C core (**6**) from barbituric acid (**1**). Substituents R^2 or R^1 on compound **5** can undergo further modification prior to the deprotection to generate compound **6***

high yielding and readily scalable (>100 g) synthetic scheme for its preparation. Upon dissolution and self-assembly in water, the spatial arrangement of the guanine–cytosine H-bonding array strains the G∧C motif to form a six-membered supermacrocycle (rosette) maintained by 18 H-bonds. This organization has been corroborated by a variety of nuclear magnetic resonance spectroscopy techniques such as correlation spectroscopy (COSY), rotating-frame nuclear Overhauser enhancement spectroscopy (ROESY), nuclear Overhauser effect spectroscopy (NOESY) and total correlation spectroscopy (TOCSY). Electrospray ionization mass spectrometry (ESI-MS) has also been utilized to establish the formation of the hexameric rosette [8]. These rosettes then organize through π-stacking, van der Waals interactions and solvophobic effects to form the RNT architectures [12]. The drive behind this self-assembly process is entropic in nature, whereby water molecules are released from the hydrophobic surfaces of the motifs located at the ends of the tubes, as new rosettes are being recruited to extend the RNT [12].

Molecular modeling calculations have predicted that water contributes to the stability of the RNTs by mediating effective interactions between the rosettes and determining the

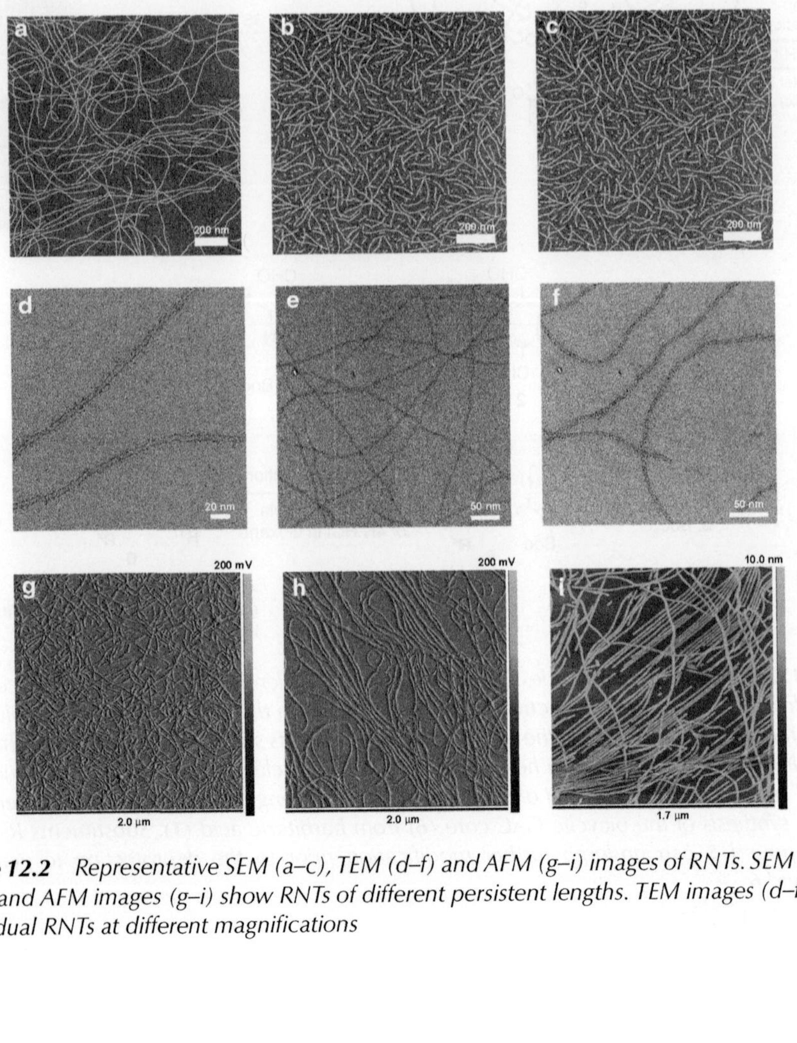

Figure 12.2 *Representative SEM (a–c), TEM (d–f) and AFM (g–i) images of RNTs. SEM images (a–c) and AFM images (g–i) show RNTs of different persistent lengths. TEM images (d–f) show individual RNTs at different magnifications*

RNTs' formation pathway [13]. These solvent molecules are likely to form a wetting monolayer in the RNT channel, penetrate the pockets on the RNT outer surface and bridge the adjacent π-stacked rosettes. Figure 12.2 shows representative scanning electron microscopy (SEM, a–c), transmission electron microscopy (TEM, d–f) and atomic force microscopy (AFM, g–i) images as visual evidence for the formation of the nanotubular assemblies. These materials can also undergo long-range organization on Si/SiO_x and poly(methyl methacrylate)-coated Si/SiO_x substrates using directional flow techniques [14, 15]. As well, further hierarchical organization can lead to bundles [8], sheets [12] and superhelices [16] in water, as well as prolate nanospheroids in organic solvents [17].

With regards to their physical dimensions, the RNT length can be tuned from several hundred nanometers up to several millimeters. High temperatures and longer aging times, for example, can promote the formation of longer nanotubes and higher aggregation states [12]. The inner channel diameters and hydrophobic properties are alternatively, dependent on the size of the G∧C scaffold and can also be tuned accordingly [18, 19]. This is an important feature of the RNT design for drug incorporation. The bicyclic G∧C motif, for example, yields a RNT with a channel diameter of 1.1 nm, as shown in Figure 12.3a. The inner diameter can be further increased to 1.4 nm by using the more hydrophobic tricyclic motif, which has the same G and C H-bonding arrays but is separated by a pyridine ring (Figure 12.3b) [18]. Interestingly, extending this core by the incorporation of the pyridine

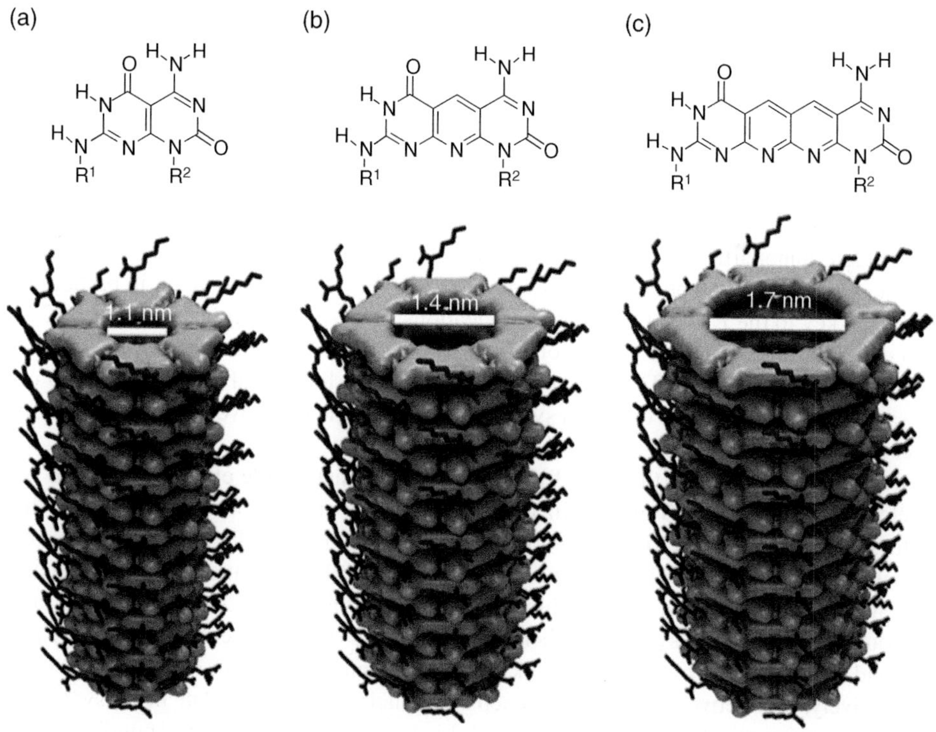

Figure 12.3 *Self-complementary G∧C motifs: (a) bicycle, (b) tricycle and (c) tetracycle and their self-assembly into RNTs*

ring also leads to unique J-type properties [20], indicative of stronger dipole transitions between adjacent molecules in the RNT. It was proposed that the increased hydrophobic/amphiphilic character of the tricyclic core, compared to the bicyclic core, results in stronger assemblies in water and polar solvents, promotes stronger and larger π–π interactions – which are favorable for establishing optimal interchromophoric distances and geometries for exciton coupling – and increases the electronic delocalization [20]. Along with this tricyclic motif, a tetracyclic motif has also been developed in order to generate RNTs with an even larger inner diameter of 1.7 nm, such as shown in Figure 12.3c [19].

12.2.2 RNT Functionalization

An important element of the RNTs, which sets them apart from other self-assembled materials, is their ability to express a variety of R^1 and R^2 functional groups (Figure 12.1) on their outer surface. This is accomplished through a covalent link of R^1 and R^2 to the exocyclic nitrogen atom of the G-face and the endocyclic amine on the C-face, respectively. This provides a facile strategy in which to dictate the properties of RNT depending on the biochemical or physical characteristics of these functional groups. We first demonstrated this concept with the synthesis of RNTs expressing either D- or L-lysine surface groups [8]. Using circular dichroism (CD) spectroscopy, the RNT assemblies were shown to display opposite supramolecular chiralities dictated by the *covalently* attached chiral amino acid residues. Shortly thereafter, it was found that amino acid residues, which are electrostatically bound to crown ether anchor points on the RNT surface, could similarly dictate the RNT chiroptical properties [21]. This further led to solvent-induced chirality inversion (water versus methanol) of RNTs generated from a single chiral G∧C motif, the results of which were attributed to thermodynamic and kinetic self-assembly processes involving different conformational isomers (called *chiromers*) [22].

Among the hundreds of different functional groups that have been expressed on the surface of the RNTs, some of these have included crown ethers [21], long and short alkyl chains [23], nanoparticles (NPs) [24], inorganic complexes [25], as well as amino acid residues and bioactive peptides (up to 15 amino acids) [8, 12, 26]. The latter, in particular, have been key to the RNTs' biomedical applications, as described further in Section 12.3. Two different functionalized G∧C motifs can also be co-assembled together, in order to obtain multi-functional RNTs that express more than one biological functional group on the same scaffold [26]. This creates a synergetic effect of the bioactive components on the outer surface.

Given the importance of the RNT functionalization, considerable effort has focused on developing and optimizing functionalization strategies that proceed both early or late in the synthesis of the G∧C core (compound **6**, Figure 12.1). For example, the introduction of amines NH_2R^2 and NH_2R^1 via consecutive S_NAr (nucleophilic aromatic substitution) reactions at positions 2 and 4 on 2,4,6-trichloropyrimidine carbaldehyde (**2**), installs the R^2 and R^1 groups early on in the synthetic scheme (Figure 12.1). This is an excellent strategy for acid and base stable groups such as alkyl chains, aromatics and unsaturated moieties. Alternatively, a reductive amination approach is routinely implemented late in the synthetic scheme, whereby motif **5a** containing R^2=allyl is transformed to the corresponding aldehyde via an oxidative cleavage reaction (Figure 12.4). Following a reductive amination of the aldehyde with NH_2R^3, the protected functionalized G∧C motif **5aa** is obtained. Using

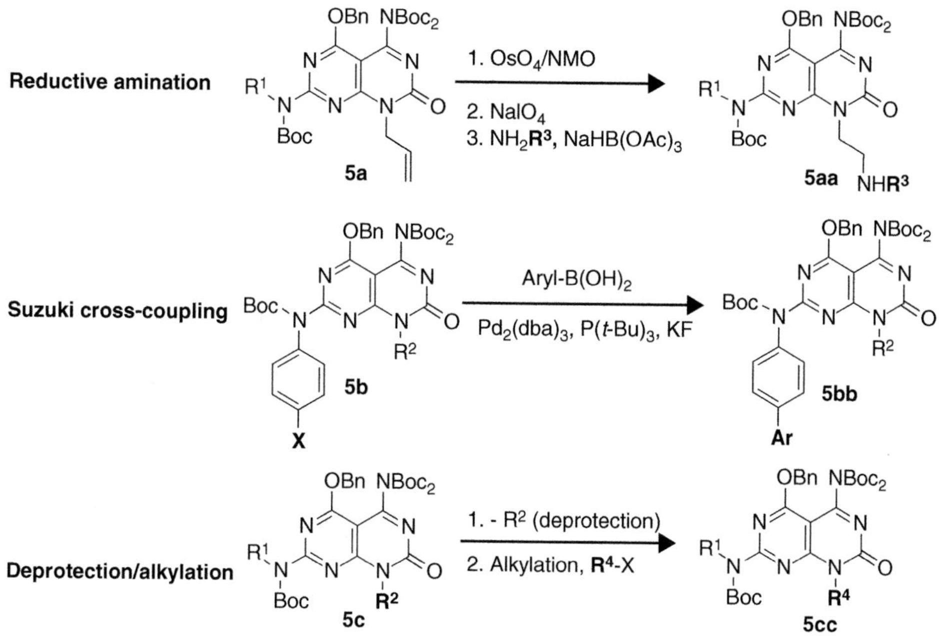

Figure 12.4 *Reductive amination, Suzuki cross-coupling and deprotection/alkylation strategies for the functionalization of the G∧C motif*

this particular approach, G∧C motifs expressing many different motieties have been prepared for both water-soluble and organic-soluble RNTs. The facile expression of different peptides, for example, the tubular renal imaging mercaptoacetyl triglycine (MAG_3) ligand, has been paramount in tuning the properties and applications of these nanotubes [25]. This conjugated MAG_3 group has been used for the development of an RNT radiolabeling strategy [25] by taking advantage of its strong chelation to the $^{186/188}Re$ or Tc-99m [27–30] radionuclides.

In another functionalization approach, the application of cross-coupling reactions, namely the Suzuki cross-coupling, has been applied in the functionalization of motif **5b** in order to prepare electronically and structurally unique aromatic motifs **5bb** in excellent yields [31]. A streamlined approach to RNT synthesis [32] has also been developed, which relies on the selective removal of the R^2 protecting group of **5c** late in the synthesis, followed by the alkylation of the free nitrogen atom with an electrophilic moiety, such as a carbohydrate.

While expressing amino acid residues and proteins on the outer surface of the RNTs is an important component for their biomedical applications, as described in Section 12.3, lysine-functionalized RNTs have also been used for the one-pot nucleation, growth, morphogenesis and passivation of 1.4 nm Au NPs [24]. As illustrated in Figure 12.5, the lysine side chains can be used to coordinate negatively charged tetrachloroaurate ($AuCl_4^-$). Upon reduction with hydrazine, evenly distributed nucleation sites on the surface of the RNTs lead to the formation of nearly monodisperse Au NPs. This ability to express metal nanoparticles on the surface of the RNTs could have interesting applications in molecular electronics, catalysis and nanomedicine [33, 34].

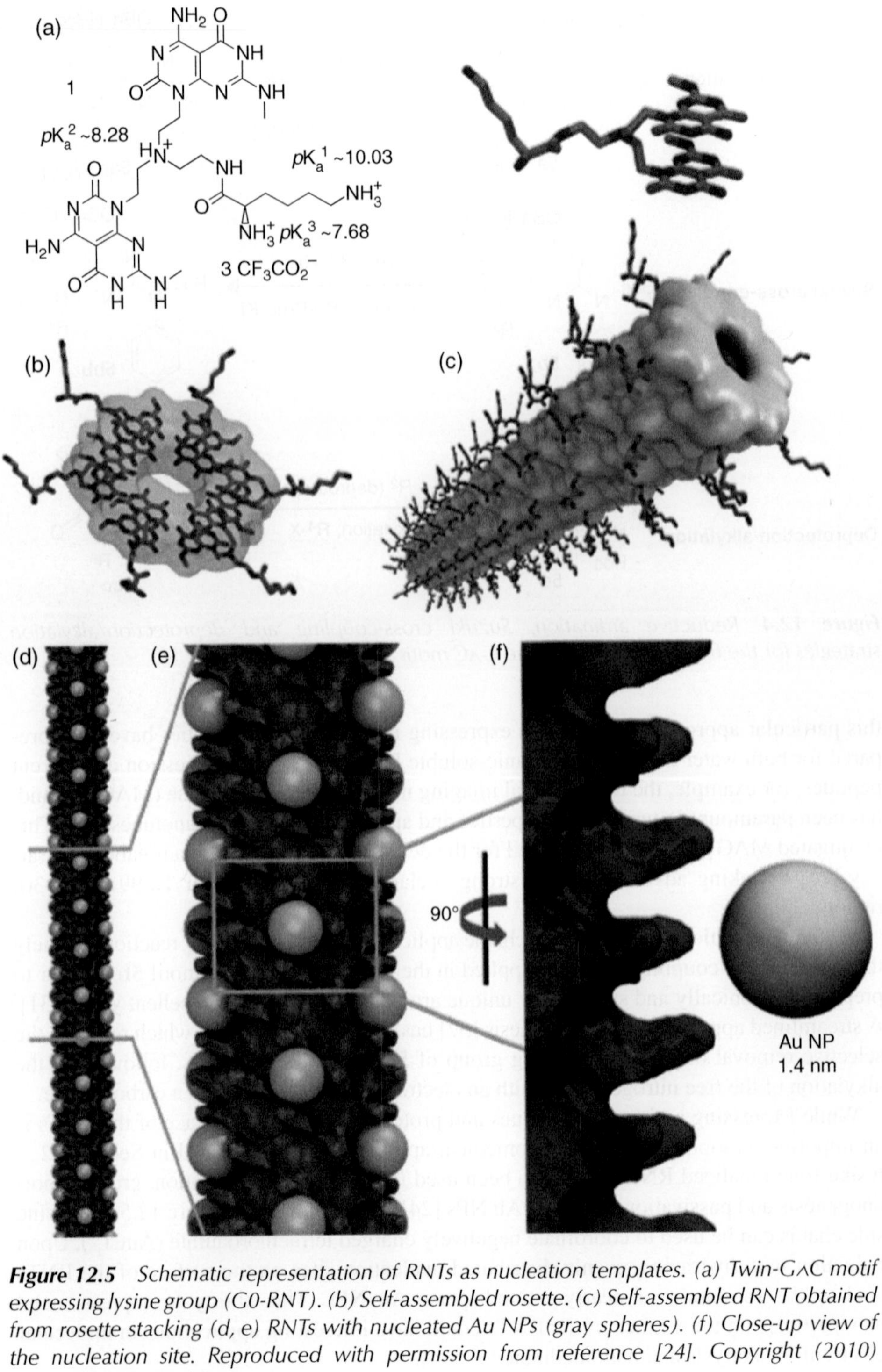

Figure 12.5 *Schematic representation of RNTs as nucleation templates. (a) Twin-G∧C motif expressing lysine group (G0-RNT). (b) Self-assembled rosette. (c) Self-assembled RNT obtained from rosette stacking (d, e) RNTs with nucleated Au NPs (gray spheres). (f) Close-up view of the nucleation site. Reproduced with permission from reference [24]. Copyright (2010) American Chemical Society*

12.2.3 RNT Stability

As a novel scaffold for biomedical applications, including drug delivery, RNTs have the advantage of extensive functional group tolerance (stability) and exceptional control over the self-organization process. This is evidenced by their stability over a broad range of pH values and temperatures and also their dissociation upon high dilution, application of heat (>85 °C) or treatment with acid (TFA (trifluoroacetic acid), HCl), as studied spectroscopically (circular dichroism, UV–visible spectroscopy) and microscopically (SEM, TEM, scanning tunneling microscopy (STM), AFM) [8].

Given their noncovalent mechanism of formation, the stability of the RNT assemblies depends mainly on the density (sterics) and net charge (electrostatics) [35] of the surface functional groups. That being said, in order to accommodate larger charged bioactive molecules (i.e. peptides) on the RNTs without compromising their self-assembly and stability, the twin version of the G∧C motif shown in Figure 12.6 was developed. This molecule features two bicyclic G∧C motifs linked by ethylene spacers. When self-assembled, the twin-RNT construct is substantially more stable than the single-RNT construct (prepared from the single G∧C motif) when identically functionalized and of equal length. This is because (a) the functional group density and net charge are reduced by a factor of 2; (b) the thermal stability of the nanotubes is enhanced as a result of preorganization, increased amphiphilic character and a greater number of H-bonds per module (12 instead of 6); (c) the twin rosettes are maintained by 36-H bonds instead of 18; and (d) the resulting RNT is less sterically congested and experiences reduced electrostatic repulsion on its surface [35].

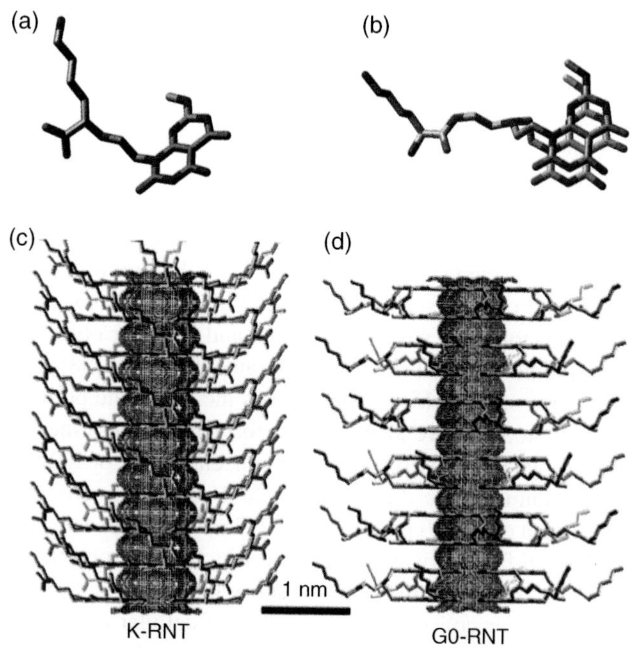

Figure 12.6 Motif **a** self-assembles into K-RNT (**c**). Motif **b** self-assembles into G0-RNT (**d**). Reproduced with permission from reference [35]. Copyright (2005) American Chemical Society

Molecular modeling calculations [36] revealed that the association free energy for the assembly of six molecules of motif **a** (Figure 12.6) into rosettes in water and the subsequent stacking of two of these rosettes is 101 kcal/mol and 168 kcal/mol, respectively [35]. For further reference, the RNT formed from motif **a** is called K-RNT. By comparison, the association energy of six hydrated twin motifs of **b** into a twin rosette (called G0-RNT herein) and the stacking of two of these rosettes is substantially greater at 770 kcal/mol and 1890 kcal/mol, respectively. The free energy of adding a new rosette to the nanotube increases nearly linearly by 206 kcal/mol for each new single rosette for **a** and by 1890 kcal/mol for each new twin rosette for **b**. This linear increase is indicative of an effective electrostatic field that accumulates with the nanotube length and therefore is suggestive of macrodipole formation. Overall, it was predicted that the association free energy of a twin motif is approximately twice that of a single G∧C motif at the same given length [35]. This enhanced stability and functional group tolerance of the twin RNTs has allowed for bioactive peptides to be displayed on the RNT surface including decapeptides derived from bone morphogenetic proteins (BMPs) for bone tissue engineering applications (see Section 12.3) [37].

12.2.4 RNT Toxicity and Biocompatibility

As discussed in Sections 12.2.1 to 12.2.3, RNTs have many desirable structural features that make them suitable for biomedical applications, including their bioinspired design, water solubility and their high tolerance for functionalization with bioactive molecules such as amino acid residues and peptides. Several *in vitro* and *in vivo* toxicology studies have also revealed their biocompatibility and low toxicity [38–40]. Journeay *et al.*, for example, demonstrated that pulmonary exposure of C57/BL mice to G0-RNT (see Section 12.2.3) induces only a mild and transient inflammatory lung response [38].

As illustrated in Figure 12.7a, for example, mice treated with 5 µg of G0-RNT in nanopure water had a bronchoalveolar lavage fluid (BALF) total inflammatory cell count similar to the low levels observed in the control groups (lysine or nanopure water) after 24 h and 7 d. Only the 25 µg and 50 µg groups showed increased total cell counts compared to the control and 5 µg treatment groups after 24 h [38]. Because lysine did not trigger a similar influx of inflammatory cells, the observed inflammation was attributed to the nanotube structure [38]. Most significantly, this acute inflammatory response in the 50 µg treatment group resolved after 7 d, as evidenced by a substantial reduction in the total cell number compared to the 24 h time-point. As also illustrated in Figure 12.7b to e, these results were well paralleled by the histological study of the lung and by other observations, including the absolute polymorphonuclear cell (PMN) and monocyte counts [38].

Along with this *in vivo* study, which demonstrates the low acute pulmonary toxicity of the RNTs, *in vitro* cell studies have also been performed to assess the cell viability and inflammatory response when exposed to varying concentrations of K-RNT. The two cell lines that have been examined are the human Calu-3 epithelial cells (human bronchial epithelial adenocarcinoma cell) [39] and human U937 macrophage cells [40]. Whereas the former cell line is derived from airway epithelium, which is an important defense against inhaled pollutants and acts by secreting inflammatory mediators and expressing adhesion molecules [41–45], the latter constitute the first line of cellular defense against harmful external agents by engulfing foreign particulates before being transported via the mucociliary escalator towards the pharynx [46, 47].

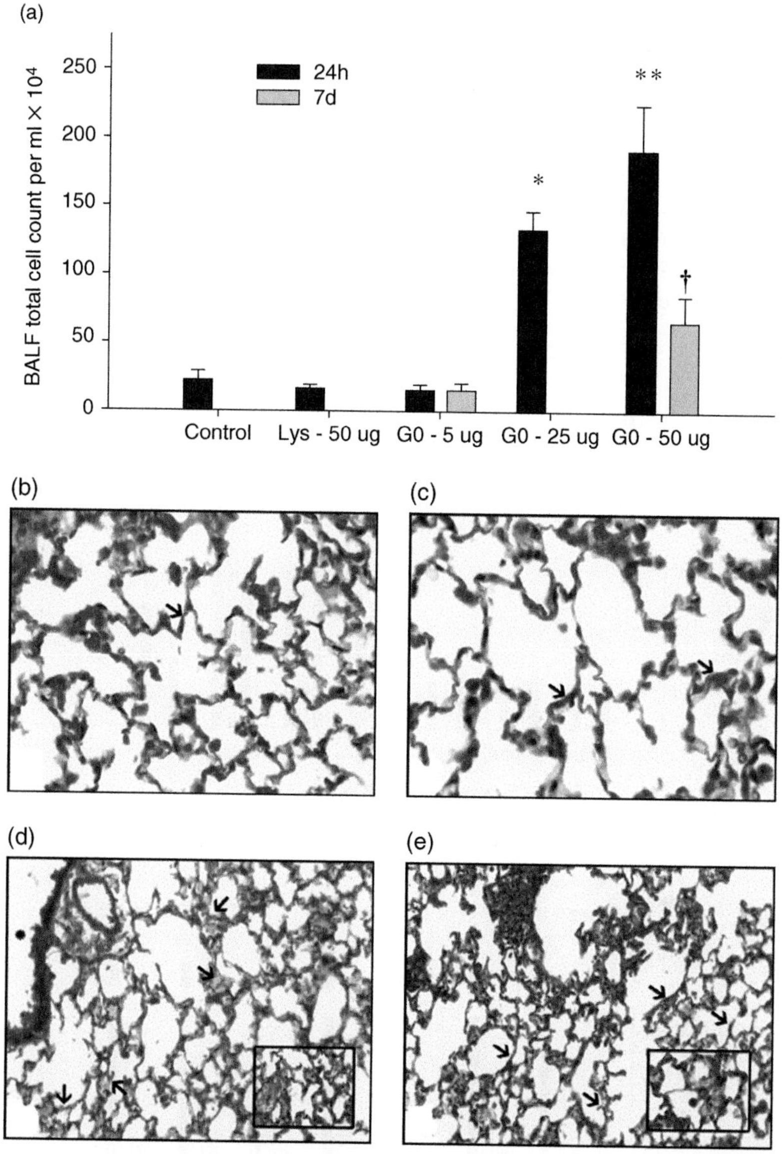

Figure 12.7 *(a) BALF total cell count from mice at 24 h and 7 d. Values represent means ± SD. Note that * shows a significant difference from control, lysine and 5 µg groups at 24 h ($p < 0.05$); ** shows a significant difference from all other groups at 24 h; † shows a difference from the same treatment dose at 24 h ($p < 0.05$). (b–e) Hematoxylin and eosin-stained lung sections. Lung sections from (b) for control mice and (c) for 5 µg G0-RNT treated mice show normal alveolar septa (arrows) while those from (d) for 50 µg G0-RNT treated mice at 24 h show septal congestion (inset), septal thickening and edema (arrows). (e) Section of a lung collected from 50 µg G0-RNT treated mice after 7 d show nearly normal alveolar septa (arrows) and some congestion (arrows). Original magnification: ×40. Reproduced with permission from reference [38]. Copyright (2008) Dove Medical Press Ltd*

In the study involving human Calu-3 epithelial cells [39], the cell viability when exposed to different concentrations of K-RNT (1, 5 and 50 µg/mL), was compared to exposure with media only (control), lysine (50 µg/mL), Min-U-Sil 5 quartz microparticles (QM) (80 µg/mL) and lipopolysaccharides (LPS) (1 µg/mL) after 1 h, 6 h, and 24 h. As presented in the Trypan blue assay in Figure 12.8a, K-RNT doses of 1 µg/mL and 5 µg/mL were tolerated well by the cells. Although cell viability was reduced in the 50 µg/mL RNT-treated group, this was less than the QM group after 24 h of exposure. Because the Trypan blue method is not specific to other forms of cell death or injury such as apoptosis, these results were interpreted cautiously, yet provided insight into the nontoxic dosage range of K-RNT for future cellular studies [39].

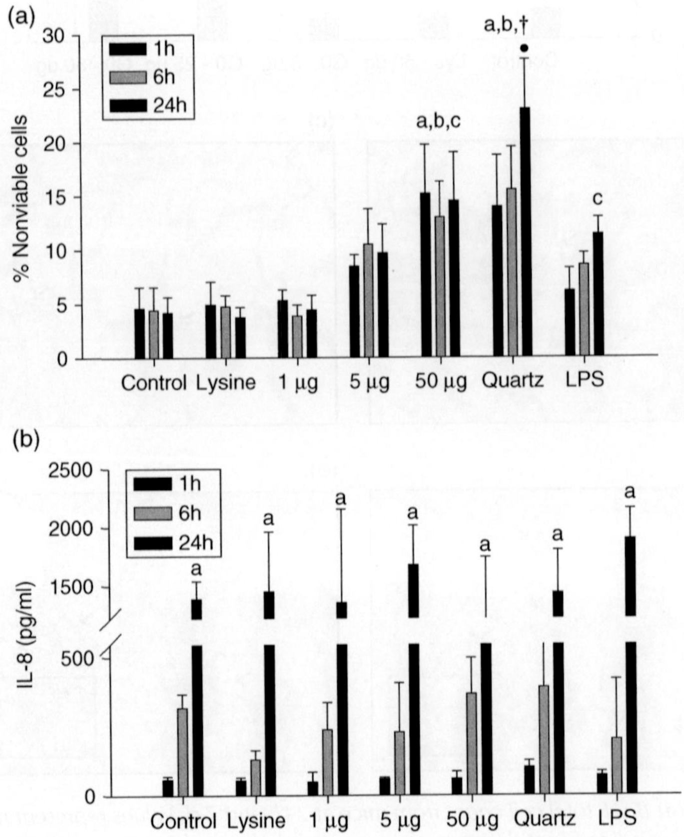

Figure 12.8 *(a) Trypan blue cell viability assay. Values plotted are a mean percentage of nonviable cells ± SD (standard deviation) of triplicate exposures. Letters a, b and c denote significantly greater than control, lysine (50 µg/mL) and LPS (1 µg/mL) groups at 1, 6 and 24 h, respectively. Note that * shows greater than 1 h and 6 h within the QM treatment ($p<0.05$) and † shows greater than all treatment groups at 24 h ($p<0.05$). (b) ELISA for IL-8 on cell supernatant. Letter 'a' denotes significantly greater than 1 h and 6 h values within the same treatment group ($p<0.05$). Values represent mean ± SD. Reproduced with permission from reference [39]. Copyright (2008) John Wiley & Sons, Ltd*

Enzyme-linked immunosorbent assay (ELISA) analysis was also performed on the supernatants in this study and revealed that the proinflammatory mediator tumor necrosis factor alpha (TNF-α) and endothelial–monocyte activating polypeptide (EMAP-II), which initiate and amplify inflammation, were below the detection limit for all treatment groups. This is perhaps because of their dilution in the supernatant or lack of release from intracellular sites [39]. The interleukin-8 (IL-8) levels alternatively all increased significantly within the same treatment group after 24 h, as illustrated in Figure 12.8b. Interestingly, a quantitative real-time reverse transcriptase polymerase chain reaction (qRT-PCR) of IL-8 mRNA showed increased expression in the high-dose RNT-treated and QM groups at both 1 h and 6 h. LPS-treated cells also showed a significant increase by 6 h. Importantly, all treatment groups, with the exception of lysine-treated cells, presented an increase in the IL-8 mRNA over the controls [39]. This suggested that a 1–6 h timeframe is needed in some of the treatments for a significant transcriptional response. Although this mRNA data did not coincide with the observed protein secretion behavior in the supernatant (Figure 12.8b), it was speculated that transcription activation of the IL-8 gene may not necessarily lead to greater levels of protein secretion, because of post-translational blocks or an intracellular accumulation of the IL-8 protein [39].

A comparison of this study with that involving U937 macrophage cells again revealed that K-RNT doses of 1 μg/mL and 5 μg/mL were tolerated well by the cells, similar to the lysine and control samples (Figure 12.9) [40]. Although cells exposed to the highest dose of K-RNT (50 μg/mL) experienced reduced cell viability after 24 h compared to the 1 h and 6 h groups, this was much less significant than for quartz-treated cells, which after 24 h showed reduced viability compared to all other treatments [40]. In addition, while K-RNT did not induce EMAP-II expression, the transcription of IL-8 and TNF-α were activated within 1 h, but this was not accompanied by protein secretion into the supernatant.

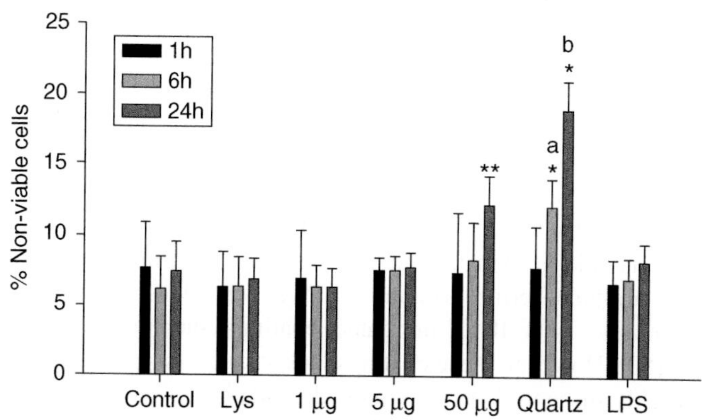

Figure 12.9 *Trypan blue cell viability assay. Values are the mean percentage of nonviable cells ± the standard deviation (SD) of triplicate exposures. Note that * shows a difference between other time points within the quartz treatment (p<0.05) and ** shows a significant difference from 1 h and 6 h time-points within 50 μg/mL K-RNT treated cells. Letter 'a' denotes reduced viability as compared to the control, lysine and K-RNT treated cells (1 and 5 μg/mL) at 6 h and 'b' denotes a significant difference from all other treatments at 24 h. Reproduced with permission from reference [40]. Copyright (2009) John Wiley & Sons, Ltd*

In summary, these *in vivo* and *in vitro* studies of K-RNT and G0-RNT showed that they do not induce a robust inflammatory response or cytotoxicity. This is likely to be due to their biologically inspired and metal-free, self-assembled design.

12.3 Applications of RNTs in Bone Tissue Engineering

12.3.1 Introduction

Tissue engineering is an interdisciplinary field of research that relies on the principles of chemistry, biology, engineering, material science and medicine in order to design novel materials that can repair or replace portions of damaged tissues. There have been great advances in the design of smart synthetic biomaterials that can interact with cells and mimic the nanoscale features of soft and hard tissues including bone, cartilage, blood vessels, brain, skin and liver [48, 49]. Bone and cartilage tissue engineering, in particular, have received tremendous attention over the past two decades because of the wide range of costly bone complications resulting from trauma or sport injuries, deterioration with age and osteoporosis. In 2000, for example, there were 9 million new osteoporotic fractures in the hip, forearm and the vertebral region in Europe and the United States [50]. Furthermore, it is estimated that 1 in 3 women and 1 in 5 men over 50 will experience osteoporotic fractures in their lifetime [50].

Traditionally, the treatment of bone fractures is clinically performed using metallic implant materials such as Ti metal and its alloys. These have several advantages including high mechanical strength, durability, chemical stability, corrosion resistance and low toxicity [51]. There are still significant and unresolved problems associated with these implants, however, such as their low biocompatibility, poor long-term stability, slow rate of osseointegration and low healing response [52]. Unfortunately, for many patients these deficits often result in revision surgeries, thereby causing tremendous pain and inconvenience as well as being financially taxing for the patient and the health system.

The surface topography and properties of Ti have a crucial impact on the interface between the implanted material and the tissue and eventually on the success of the implant after the surgery. Several natural as well as synthetic materials, which mimic the structure and function of the extracellular matrix of bone, and thus increase the rate of osseointegration around the implant, have been used as coatings on Ti implants prior to the surgical installation. These include growth factors such as BMP, arginine–glycine–aspartic acid (RGD) and lysine–arginine–serine–arginine (KRSR) cell adhesive peptides, DNA coatings, enzyme coatings, as well as inorganic coatings such as hydroxyapatite (HA, $Ca_{10}(PO_4)_6(OH)_2$) [53]. However, these coatings lack strong adsorption and stability on the implant and are highly susceptible to enzymatic degradation.

Self-assembled nanomaterials in particular have received significant interest over the past decade as coating materials on Ti implants, due to their robust nature, large surface area, high biocompatibility, capacity for covalent functionalization with osteogenic factors and their strong adsorption and stability on the surface. Furthermore, their nanostructural features can mimic the extracellular matrix of bone, in order significantly to promote selective protein adsorption and efficiently to stimulate new bone formation on the implant compared with conventional materials. Among the various self-assembled nanomaterials used for this purpose, the most widely explored have included peptide nanomaterials such as self-complementary ionic peptides [49] and surfactant-like peptides [48], as well as bioinspired nanomaterials such as RNTs [8]. The following sections will discuss the most recent advances in the applications of RNTs as surface coatings for Ti implants and as functional fillings in hydrogels that are used as injectable therapeutics.

12.3.2 RNTs as 2D Coatings on Ti Implants

12.3.2.1 Lysine-Functionalized RNTs (K-RNT)

An ideal coating material for Ti implants must possess good adhesion and stability on the Ti surface and have a nanoarchitecture that enables sufficient interactions with the bone cells for enhanced attachment. With these features in mind, RNTs were designed to express positively charged amino acids such as lysine (K-RNT) or arginine (R-RNT) on their outer surface, in order to strongly adhere to the negatively charged TiO_2 layer and negatively charged membranes of bone cells via electrostatic interactions [54]. This strategy proved very effective, as solutions of K-RNT (Figure 12.10) or R-RNT (1, 0.5, 0.125 and 0.005 mg/mL) coated on Ti surfaces by simple physical adsorption showed approximately 50% more cell adhesion of human osteoblasts (OB) (bone-forming cells) compared to the uncoated Ti substrates. Notably, this OB adhesion enhancement is effective at very low RNT concentrations (5 μg/mL), which is significant since less material is required to achieve excellent cell adhesion (Figure 12.10) [54].

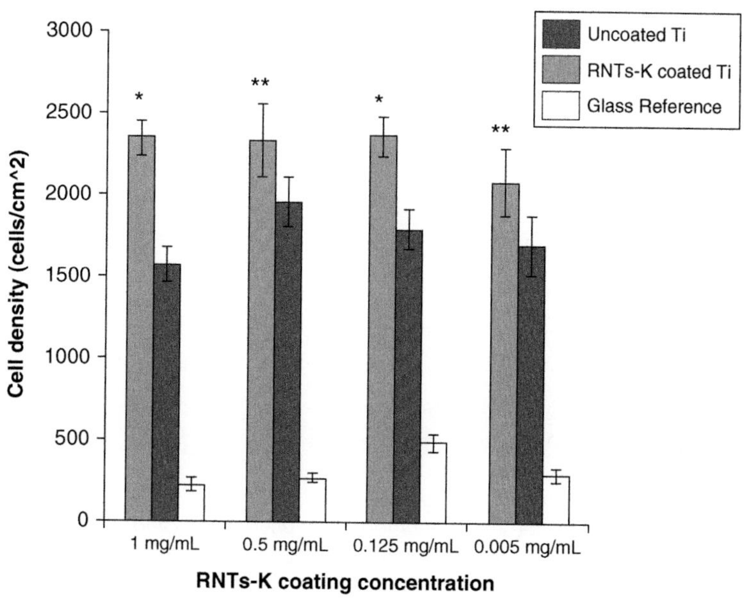

Figure 12.10 *OB adhesion on K-RNT-coated Ti (1, 0.5, 0.125 and 0.005 mg/mL), uncoated Ti and glass substrates. The data are mean ± SEM, n = 3; *p < 0.01, **p < 0.10 when compared to uncoated Ti. Reproduced with permission from reference [54]. Copyright (2004) IOP Publishing Ltd*

Because RNTs undergo extensive self-assembly at higher temperatures, resulting in long nanotubes and higher aggregation states as shown in the TEM images in Figure 12.11b and c, the effect of heated (+T) versus unheated (–T) K-RNT on OB adhesion in the presence (+S) and absence (–S) of serum proteins has also been examined [55]. In general, serum proteins [56–59] are known to enhance OB adhesion on nanophase alumina, titania and HA [60–62]. Figure 12.11a shows that (a) under –S and + S conditions, +T and –T K-RNT perform better than Ti and glass controls; (b) OB adhesion is enhanced for –T K-RNT regardless of whether proteins are present or absent; (c) + S conditions significantly enhance OB adhesion compared to –S for + T K-RNT; (d) under –S conditions –T K-RNT display better

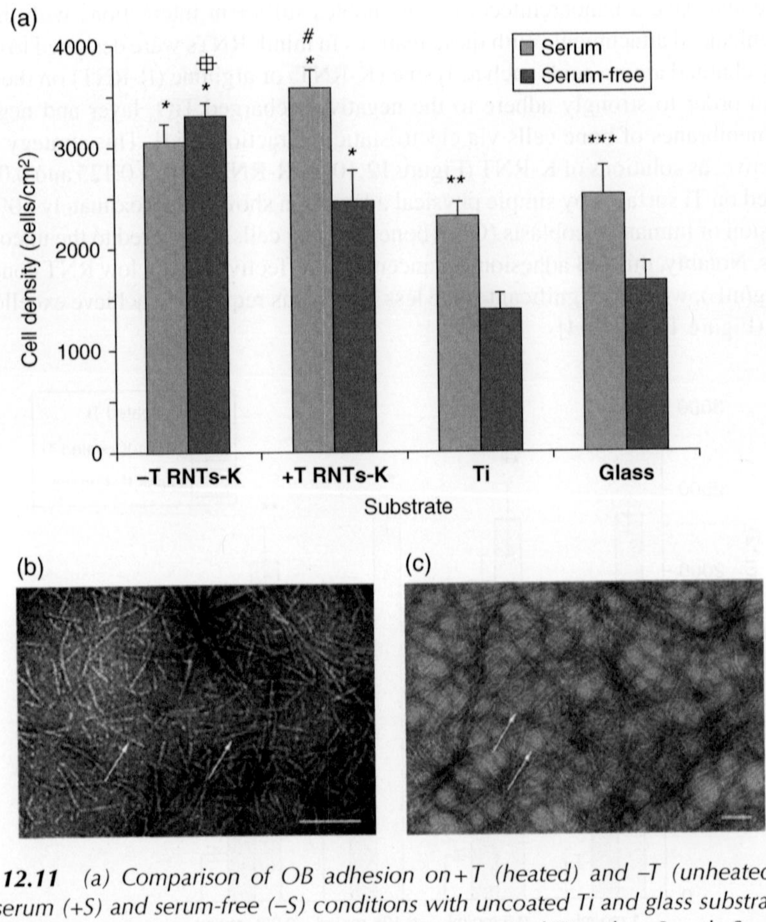

Figure 12.11 (a) Comparison of OB adhesion on + T (heated) and –T (unheated) K-RNT under serum (+S) and serum-free (–S) conditions with uncoated Ti and glass substrates. Data are mean ± SEM, n = 3; *p < 0.01 when compared to uncoated Ti under + S and –S conditions, respectively; #p < 0.01 when compared to + T K-RNT under the –S condition; ⊕p < 0.05 when compared to + T K-RNT under the –S condition; **p < 0.01 when compared to uncoated Ti under the –S condition; ***p < 0.01 when compared to glass under the –S condition. (b) TEM image of –T (unheated) K-RNT, showing short RNTs (arrows) and (c) TEM image of + T (heated) K-RNT, showing long RNTs (arrows). Scale bar = 50 nm in both B and C. Reproduced with permission from reference [55]. Copyright (2005) Elsevier

adhesion than +T K-RNT, while under + S conditions the difference is statistically insignificant ($p < 0.1$); (e) uncoated Ti (negative control) and glass (reference) display better OB adhesion under + S versus –S conditions [55]. Although the reasoning behind this enhanced OB adhesion and the relationship to the absence or presence of proteins is not well understood, it is reasonable to assume that the lysine distribution on the RNT surface plays an important role [55]. Because + T K-RNT coated Ti under –S conditions has the same level of OB adhesion as uncoated Ti in the presence of proteins, this suggests that the RNTs have signaling properties that are similar to the proteins that enhance OB adhesion or that they work by inducing a different signaling mechanism [55]. The lysine residues may perhaps act as lysine-rich bovine proteins, which are known to promote OB adhesion, proliferation and differentiation [63]. In any case, we have confirmed from subsequent studies that this enhancement effect is not merely the result of the lysine itself. Ti-coated with K-RNT at a concentration of 0.5 mg/mL, for example, showed comparable OB attachment to Ti coated with poly-K when the concentration was 0.025 mg/mL [64]. Poly-K [65] is known to adsorb strongly on surfaces and promote cell adhesion. Considering the lysine content of the RNTs and that the nanotubes lie flat on the Ti surface, the upper limit of the accessible lysine on K-RNT would not be likely to exceed 0.0005 mg/mL for the 0.5 mg/mL solution [64]. Using the same argument, there is only 0.0125 mg/mL of accessible lysine in the case of the 0.025 mg/mL poly-K. Given the discrepancies in these concentrations, it is clear that the RNTs, in combination with the expressed lysine resides, are key for OB activity [64].

In addition to improving OB cell attachment, K-RNTs have also been explored to enhance endothelial cell attachment and proliferation on Ti implants used for vascular stents [66]. Vascular stents are artificial metallic tubes inserted into partially clogged arteries to prevent blood flow restriction and myocardial infarctions. Due to the high demand for these implants, the market for vascular stents reached $918 million in 2005 in the United States alone [67]. Unfortunately, there have been several unresolved issues associated with these materials due to their lack of interaction with the endothelial cells [68, 69]. As a result, an inflammatory response is often triggered, leading to scar tissue formation, known as restenosis, which constricts blood flow and defeats the original purpose of the stent [68]. To minimize this failure, drug-eluting stents are often prescribed to counteract such events by releasing immunosuppressants. However, these can also suppress desirable endothelial cell growth and sometimes lead to late-stage thrombosis [69]. Thus, the ideal solution is to create a biocompatible coating on the implant surface that promotes endothelial cell attachment in order to minimize the inflammatory response.

In this regard, K-RNT-coated Ti has been found to greatly improve endothelial cell functions compared to uncoated Ti [66]. At concentrations of 0.01 mg/mL and 0.001 mg/mL, K-RNT-coated Ti, for instance, increases the endothelial cell density by 52% and 21%, respectively, compared to uncoated Ti after three days of culture. Based on these observations, higher concentrations of RNTs seem to promote endothelial cell coverage better, possibly by displaying more lysine side chains per unit of surface area and/or by more closely resembling the natural structure of vascular tissue when coated [66]. As shown in Figure 12.12a, by day 5 the RNT-coated Ti still had a higher average endothelial cell density, although the differences were not statistically significant. However, analysis of the endothelial cell numbers from day 1 to day 5 using fluorescence microscopy, revealed that the cells reached confluency faster on the K-RNT-coated Ti compared to the uncoated Ti (Figure 12.12b) [66].

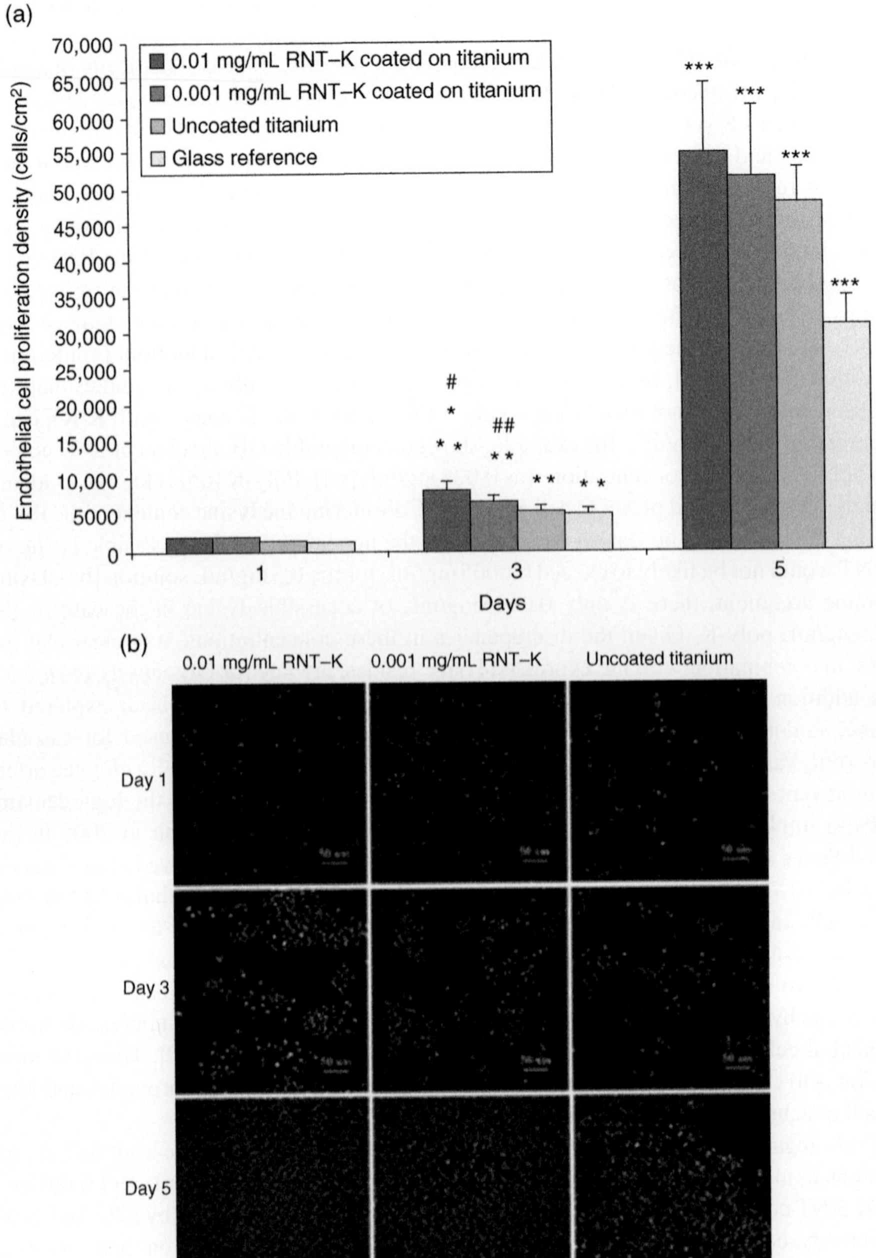

Figure 12.12 *(a) Endothelial cell density on 0.01 mg/mL and 0.001 mg/mL K-RNT-coated Ti, uncoated Ti and glass substrates after 1, 3 and 5 days. Data are mean values ± SEM; n = 3; #p < 0.01 and ##p < 0.05 compared to uncoated Ti on day 3; *p < 0.05 compared to 0.001 mg/mL K-RNT-coated Ti on day 3; **p < 0.001 compared to respective substrates on day 1; ***p < 0.001 compared to respective substrates on day 3. (b) Fluorescence microscopy images showing increased endothelial cell densities (4′,6-diamidino-2-phenylindole stain) on 0.01 mg/mL and 0.001 mg/mL K-RNT-coated Ti compared to uncoated Ti after 1, 3 and 5 days of culture. Reproduced with permission from reference [66]. Copyright (2009) Dove Medical Press Ltd*

The demonstrated potential of lysine functionalized RNTs as 2D coatings on Ti implants can be attributed to their nanoscale architecture mimicking the dimensions and structure of collagen nanofibers (the major organic component in bone) in the extracellular matrix and also their ability to express bioactive amino acids on the outer surface, which further promote cell attachment. Along with collagen nanofibers, another major component in the extracellular matrix in bone is HA. This is the inorganic portion of bone that strengthens the collagen fibers and enhances their mechanical properties. Although HA has been used for decades as a filler and as a coating material for orthopedic and dental applications [70–72], current HA coatings still have significant shortcomings such as lack of rapid osseointegration, a low degree of crystallinity and weak interactions with Ti.

Therefore, in order to take advantage of the promising attributes of both RNTs and nanocrystalline HA for enhanced OB adhesion, nanocomposite materials consisting of K-RNT and different grain sizes of nanocrystalline HA (small ca. 80 nm, medium ca. 90 nm and large ca. 130 nm in length) have been prepared and coated on a Ti surface by simple adsorption [73]. As shown in the TEM image in Figure 12.13, K-RNT and nanocrystalline HA have a good affinity for each other and, moreover, the RNTs appear oriented on the HA surface.

OB adhesion studies revealed that the nanocomposites containing the small- and medium-sized HA grains enhanced OB adhesion more than those containing the large-sized HA grain (Figure 12.14) [73]. This is in agreement with our earlier work, which has shown that nanophase materials (67 nm HA) improve specific protein (such as fibronectin and vitronectin) interactions compared with large grain size (179 nm HA) conventional materials and thus contribute to enhanced OB functions [60].

It has also been demonstrated that RNTs can act as a biomimetic template for HA deposition [73]. As illustrated in Figure 12.15a–c, $CaCl_2$ and Na_2HPO_4 solutions rapidly form HA crystals on RNTs in a pattern similar to HA/collagen fibers in the natural bone.

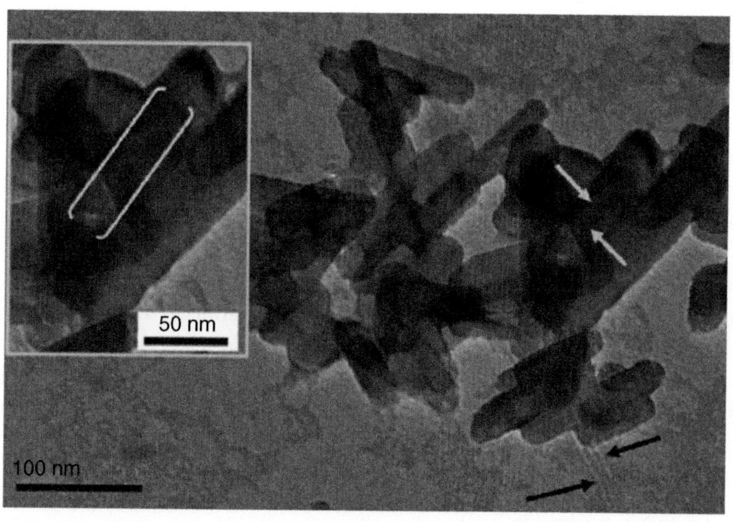

Figure 12.13 TEM images of 0.01 mg/mL K-RNT with small grain size nanocrystalline HA. Inset shows close K-RNT/HA interaction. Reproduced with permission from reference [73]. Copyright (2008) Dove Medical Press Ltd

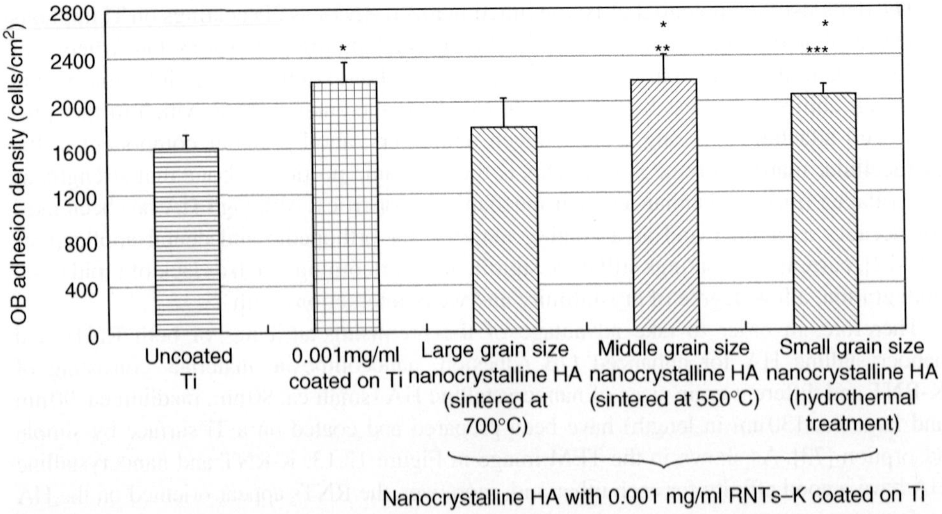

Figure 12.14 *OB adhesion on K-RNT/nano-HA composites coated with Ti, K-RNT-coated Ti and uncoated Ti. Data are mean ± SEM; n = 9; *p < 0.05 when compared to uncoated Ti; **p < 0.01 and ***p < 0.1 when compared with large grain size nanocrystalline HA with 0.001 mg/mL K-RNT coated on Ti. Reproduced with permission from reference [73]. Copyright (2008) Dove Medical Press Ltd*

In comparison, Figure 12.15d shows randomly agglomerated HA crystals in the absence of RNTs. Since the deposition, nucleation and growth of HA in pores of the extracellular collagen matrix are key steps for *in vivo* bone remodeling, nanocomposites containing K-RNT/nano-HA are anticipated to mimic the composition and morphology of bone tissue formation and possibly lower implant failure rates [73].

12.3.2.2 RNTs Functionalized with Bioactive Peptides

In the previous Section 12.3.2.1, K-RNTs were shown to enhance OB attachment on a Ti surface by approximately 50% compared to uncoated Ti [54]. Given these encouraging results, it was anticipated that OB adhesion could be further increased by conjugating RNTs with short bioactive peptide sequences that are known to enhance OB adhesion and function on Ti surfaces. Among these peptides is the adhesive arginine–glycine–aspartic acid (RGD) peptide. It is found in several natural proteins (e.g. fibronectin and collagen type I) in the extracellular matrix and is known to activate OB membrane integrin receptors [74–76]. This peptide and its derivatives have been used as a coating material on Ti implants both *in vitro* and *in vivo* and have enhanced cell attachment on the surface [77–81]. Another important cell adhesive peptide sequence is lysine–arginine–serine–arginine (KRSR). This is a heparin-binding domain in many extracellular matrix proteins with selectivity for OB cell adhesion versus fibroblasts or endothelial cells [82]. Fibroblasts, in particular, are undesirable cells for orthopedic applications because of their tendency to synthesize fibrous soft tissue around implants, which weakens the connection between the implant and the bone tissue.

In order to combine the positive attributes of the RNTs' nanostructured scaffold and the bioactive peptides, two G∧C motifs functionalized with RGD and KRSR have been synthesized using a solid phase peptide synthesis strategy [83]. These two motifs readily

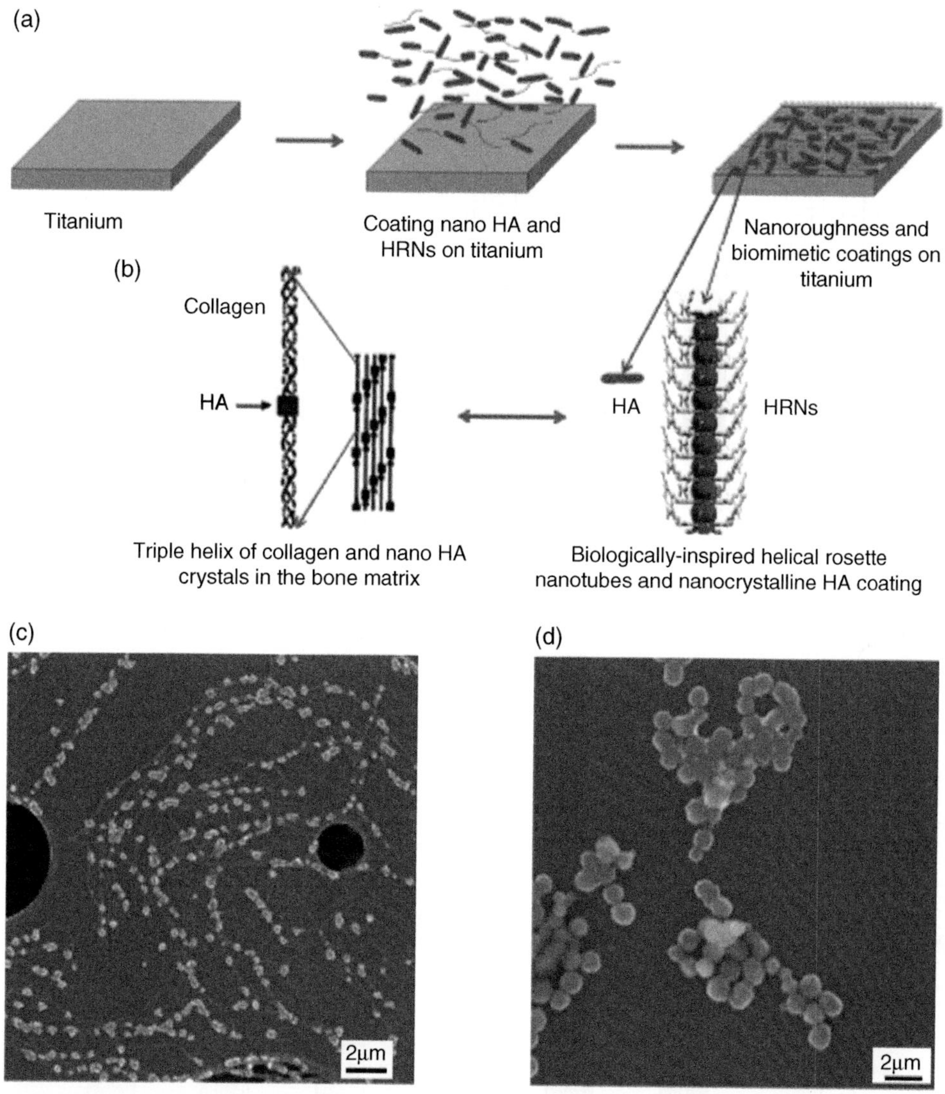

Figure 12.15 *(a) Schematic representation of the K-RNT/nano-HA coating on Ti. (b) The K-RNT/nano-HA coating mimics the natural bone extracellular matrix, which consists of triple helices of collagen assembled with nanocrystalline HA. (c) SEM images of the K-RNT/nano-HA composite on a porous carbon TEM grid and (d) only an HA coating on the TEM grid. Reproduced with permission from reference [73]. Copyright (2008) Dove Medical Press Ltd*

self-assemble into RNTs in aqueous solution, with the respective peptide sequences displayed on their periphery. RGD-RNT and K-RNT have also been co-assembled to create a synergetic effect of the bioactive components on the outer surface. As presented in Figure 12.16a, compared to uncoated Ti, KRSR-RNT and co-assembled K/RGD-RNT (95:5 molar ratio) significantly improve OB adhesion when coated on Ti, by 122% and 124% respectively [83]. Furthermore, OB adhesion on Ti coated with KRSR-RNTs is 84% higher than Ti coated with the peptide KRSR alone, which establishes the nanostructured

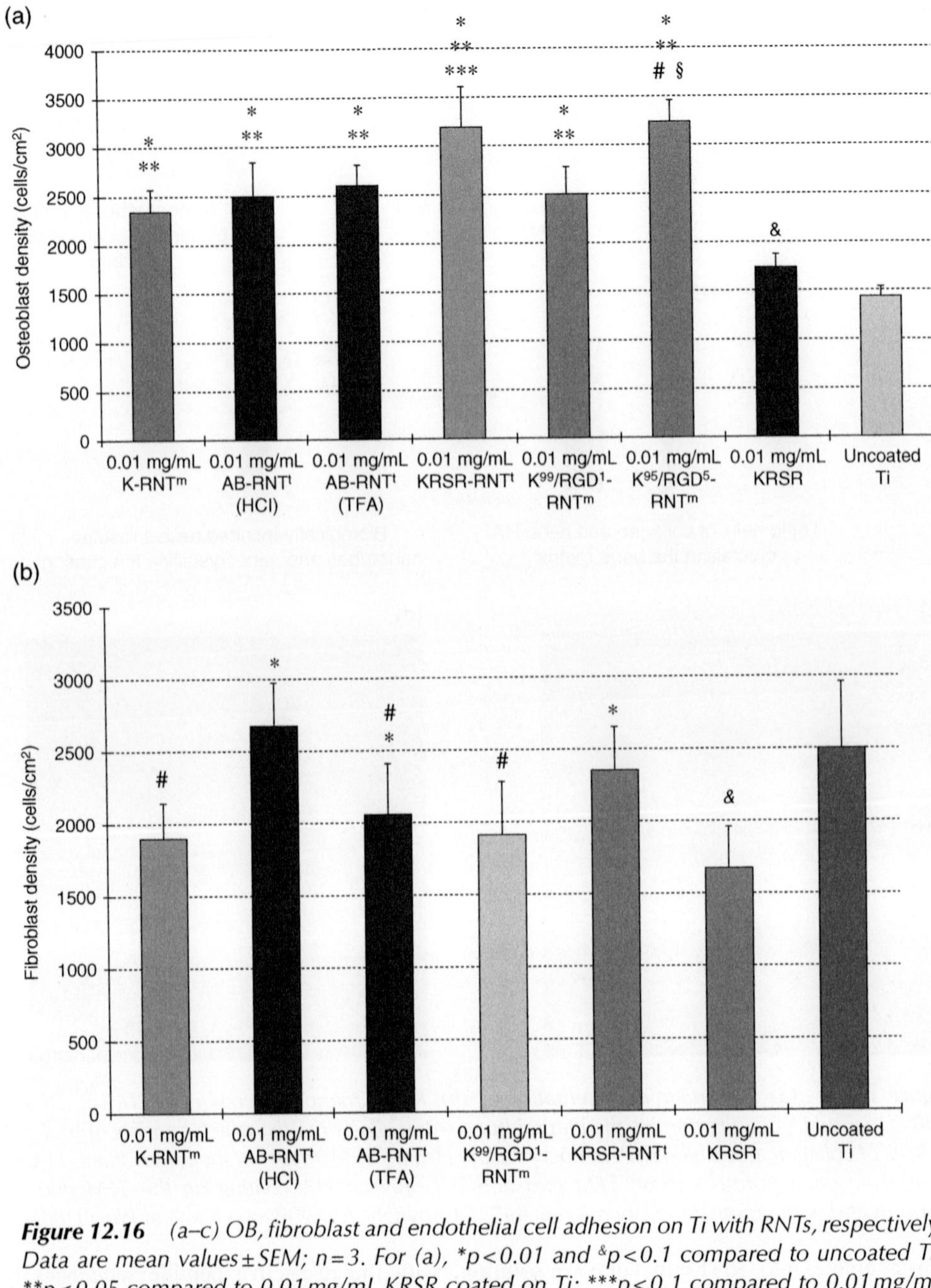

Figure 12.16 *(a–c) OB, fibroblast and endothelial cell adhesion on Ti with RNTs, respectively. Data are mean values ± SEM; n = 3. For (a), *p < 0.01 and &p < 0.1 compared to uncoated Ti; **p < 0.05 compared to 0.01 mg/mL KRSR coated on Ti; ***p < 0.1 compared to 0.01 mg/mL K-RNT^m, K/RGD-RNT^m (99:1, m = mono G∧C motif), AB-RNT^t (AB = amino butane, t = twin G∧C motif) (HCl) coated on Ti; #p < 0.05 compared to 0.01 mg/mL K-RNT^m coated on Ti; §p < 0.1 compared to K/RGD-RNT^m (99:1), AB-RNT^t (HCl or TFA salts) coated on Ti. For (b), *p < 0.05 compared to 0.01 mg/mL KRSR coated on Ti; #p < 0.05 compared to 0.01 mg/mL AB-RNT^t (HCl) coated on Ti; &p < 0.05 compared to uncoated Ti.*

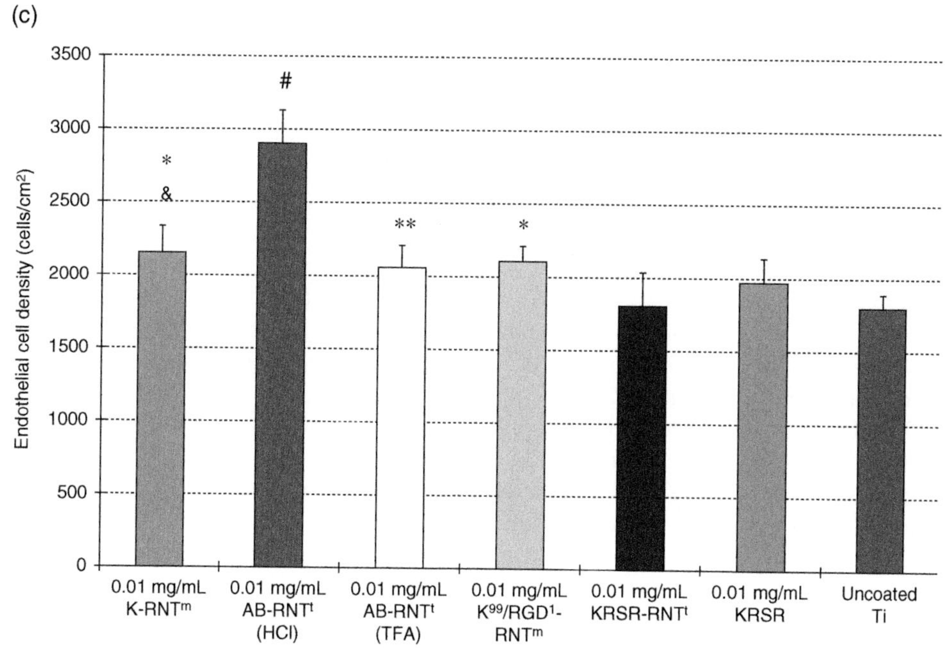

Figure 12.16 (cont'd) *For (c), *$p<0.05$ and **$p<0.1$ compared to uncoated Ti; #$p<0.05$ compared to all other substrates; §$p<0.1$ compared to 0.01 mg/mL KRSR-RNT^t coated on Ti. Reproduced with permission from reference [83]. Copyright (2010) John Wiley & Sons, Ltd*

role of RNTs in delivering the peptide on the Ti surface. Notably, K-RNT, KRSR-RNT and K/RGD-RNT also do not enhance the adhesion of fibroblast cells (Figure 12.16b) compared to uncoated Ti. Although Ti coated with K-RNT and K/RGD-RNT do enhance endothelial cell adhesion (relative to uncoated Ti), KRSR-RNT does not (Figure 12.16c). Thus, it is possible selectively to improve OB attachment on Ti relative to fibroblasts and endothelial cells by using KRSR-functionalized RNTs [83].

Adequate cell adhesion on a Ti surface is essential for improving its integration with the surrounding tissue and reducing inflammation. For successful long-term treatment and rapid healing, however, the implant surface needs to promote facile bone generation and growth as well as calcium deposition. This can be achieved by coating the implant surface with osteogenic materials that promote bone cell proliferation and differentiation functions. Among the different osteogenic materials for this purpose are natural growth factors produced by bone cells and stored in the bone matrix, such as the transforming growth factor β (TGF-β) and the BMPs [84–87]. In particular, BMP-2, BMP-7 and TGF-β1 are promising growth factors for enhanced bone formation around the implant [88, 89]. These proteins serve as signaling agents for cells to promote replication, differentiation, protein synthesis and/or migration of proper cell types. Unfortunately, immobilizing these growth factors on the implant by physical adsorption is not without its challenges, since these proteins lack stability and sustainability on the surface and are highly susceptible to enzymatic degradation. Thus, for the success of these growth factors, they need to be delivered at an appropriate rate and under controlled release conditions. Nanostructured materials that incorporate these proteins offer great promise in this regard because of their potentially better interaction and accessibility to bone cells and stability on the implant surface. However, because of the proteins' large size, several studies have aimed to identify short peptide sequences that mimic their bioactivity. Saito *et al.*, for example, reported a 20 amino acid peptide sequence (KIPKASSVPTELSAISTLYL) derived from the 'knuckle' region (bioactive region) of BMP-2 that promotes alkaline phosphatase activity in OB cells [90]. Kirkwood *et al.* also indentified a 20 amino acid long (AISVLYFDDSSNVILKKYRN) peptide sequence in BMP-7 that promotes calcium deposition and mineralization processes in OB cells [91]. More recently, we have described three peptides taken from the knuckle (bioactive) region of BMP-7, namely, **A** (SNVILKKYRN), **B** (KPSSAPTQLN) and **C** (KAISVLYFDDS) [92]. Peptide B and a combination of A and B have been found to increase OB proliferation, while peptide C and its combinations AC, BC and ABC were found to increase calcium deposition of OB cells (Figure 12.17).

To utilize these peptides in promoting bone cell functioning with the nanostructured features of the RNTs, three twin G∧C motifs functionalized with A, B and C have been recently synthesized using a solid-phase peptide synthesis strategy [37]. In the design, a lysine amino acid spacer was incorporated between the twin G∧C motif and the bioactive peptides so that they are more accessible to the bone cells (Figure 12.18). Thus far, upon

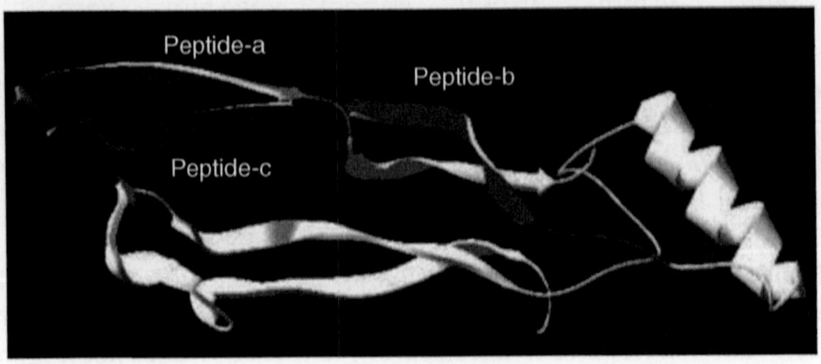

Figure 12.17 *Structure of bone morphogenetic protein-7 (BMP-7). Reproduced with permission from reference [92]. Copyright (2009) John Wiley & Sons, Ltd*

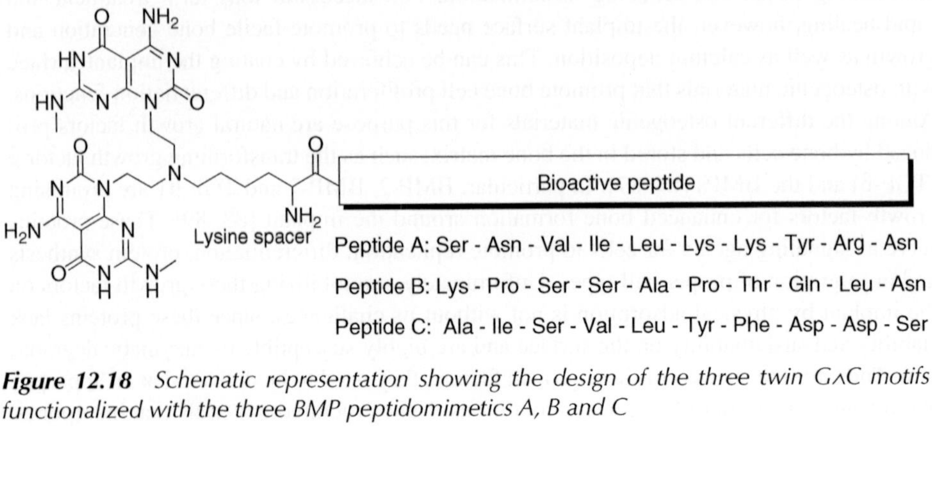

Figure 12.18 *Schematic representation showing the design of the three twin G∧C motifs functionalized with the three BMP peptidomimetics A, B and C*

dissolution in water under physiological conditions, motifs A-G∧C and B-G∧C have been found to undergo self-assembly into RNTs according to SEM imaging [37]. This demonstrates the high tolerance of the RNTs in expressing multicharged and structurally unique peptides on their outer surface. Having optimized the self-assembly process, the materials will next be coated on Ti and then studied for OB proliferation and functioning. The three motifs will also be co-assembled with K-RNT in order to create multifunctional RNTs.

12.3.3 RNTs Embedded in Hydrogels

RNTs not only show promising applications as 2D coatings on Ti implants used as bone substituents or fixation devices but also as injectable therapeutic agents to heal bone or cartilage fractures *in situ*. This is because RNTs undergo a phase transition from a liquid to a viscous gel when heated to 60 °C or when added to serum-supplemented media at body temperature [55]. Having said this, the mechanical properties of the viscous gel formed by RNTs can be further enhanced by combining with hydrogels that are currently used as injectable therapeutics. pHEMA (poly(2-hydroxyethyl methacrylate)) for example, is one of the most widely used hydrogels for bone tissue repair applications [93, 94] because it is biocompatible and has mechanical properties that can be easily modified. For instance, pHEMA can be used as a coating on orthopedic implant materials, cast into desirable shapes before implantation or used as an injectable biomaterial at a fracture site. In addition, when pHEMA is prepared by bulk polymerization, it has hardness properties that are similar to natural bone [95].

Given these favorable properties, composite materials consisting of K-RNT embedded in or coated on pHEMA hydrogels have been prepared [96]. In the first case, the composites were synthesized by polymerizing the HEMA monomer in two pre-prepared K-RNT solutions (0.01 mg/mL and 0.001 mg/mL). In the second case, the composites were generated by coating the pHEMA hydrogel with either 0.01 mg/mL or 0.001 mg/mL K-RNT by simple adsorption at room temperature. TEM imaging of the RNTs embedded within the hydrogel scaffolds showed a dense network of long RNTs (up to 3 μm in length), which indicated that the nanostructured properties of the RNTs were preserved in the hydrogel during the polymerization process [96]. Furthermore, gel condition studies revealed that significantly less polymerization time is required when K-RNTs are embedded within hydrogels, compared to hydrogel controls. For example, viscous constructs were made in 150 min at 37 °C for K-RNT embedded in pHEMA. Hydrogel controls (e.g. without K-RNT) alternatively required over 290 min at the same temperature [96]. This decrease in polymerization time offers significant advantages if used as an orthopedic material prior to implantation or as an injectable agent. In addition to this, hydrogels coated with K-RNT (0.01 mg/mL) were found to increase OB adhesion by 129.5% compared to the uncoated hydrogel control [96]. Hydrogels coated with 0.01 mg/mL K-RNT also increased OB adhesion by 75.1% compared to 0.001 mg/mL K-RNT, thereby demonstrating a substantial concentration effect. Similar observations were found for K-RNT embedded within the hydrogel.

Given that the pHEMA biocompatibility (OB adhesion) and its gelation properties are improved by using K-RNT, a subsequent study was performed using K-RNT co-assembled with 1%, 5% and 10% molar ratios of RGD functionalized RNTs (RGD-RNT) [97]. As mentioned previously, RGD is found in several natural proteins (e.g. fibronectin and collagen type I) in the extracellular matrix and is known to activate OB membrane integrin receptors [74–76]. Cell studies of these solutions (0.01 mg/mL) embedded in or coated on pHEMA hydrogels revealed an increasing trend in OB adhesion with increasing RGD molar ratios and an improvement in the cytocompatibilty properties compared to the K-RNT hydrogels. For example, K/RGD-RNT (90:10) (coated) enhanced OB attachment by 209% compared to the hydrogel control [97]. As well, K/RGD-RNT (90:10) (embedded) improved OB adhesion by 74.1% compared to K-RNT (embedded) (Figure 12.19a). Interestingly, these studies also showed that the RNT hydrogels (coated) greatly improved OB attachment compared to poly-L-lysine and collagen-coated hydrogels at the same concentration (Figure 12.19b). Type I collagen is one of the main extracellular matrix components in bone that contains cell-adhesive RGD peptides [97]. Although a detailed understanding of the mechanisms behind the enhanced OB attachment is yet to be established, we suspect that the greater fibronectin [98–100] adsorption observed in the case of K/RGD-RNT (99:1), K/RGD-RNT (95:5) and K-RNT-coated hydrogels, when compared to the uncoated hydrogels, may be partly responsible [97].

As mentioned in Section 12.3.2.1, nanocrystalline HA is the inorganic portion of the extracellular matrix in bone, which fills the gaps in the organic matrix (collagen fibers) and strengthens the whole tissue. To further increase the mechanical properties of the hydrogels and explore the effect on OB adhesion, composites consisting of embedded K-RNT (0.01 mg/mL) with dispersed 2% and 10% weight HA have been prepared [101]. As anticipated, the presence of nanocrystalline HA in the K-RNT-modified hydrogel supported greater tensile stresses and had greater compressive and tensile moduli compared to hydrogel controls.

Figure 12.20 shows that 2% (w/w) nanocrystalline HA (prepared by sintering) in a hydrogel containing 0.01 mg/mL K-RNT resulted in the greatest OB densities, which were 235% and 99% higher than hydrogel controls and 2% (w/w) nanocrystalline HA (prepared by sintering) containing hydrogels, respectively. Thus, K-RNT plays an important role in improving the cytocompatibility properties of the nanocomposite [101]. As well, there were no significant differences in the cell adhesion between the 2% (w/w) nanocrystalline HA prepared by hydrothermal treatment or sintering at 550 °C (in the absence of RNTs). However, when combined with the K-RNT hydrogel, the 2% nanocrystalline HA prepared by sintering attracted a greater number of OB cells than when hydrothermally treated. It is speculated that the HA prepared by sintering has a different surface chemistry that is more favorable for cell adsorption/interaction compared to the rod-like hydrothermally treated HA [101]. As also shown in Figure 12.20, increasing the HA weight ratio from 2% to 10% did not increase the OB density, possibly because the favorable K-RNT surface coverage on the hydrogel composites was minimized [101].

12.4 RNTs for Drug Delivery

As a novel platform for drug delivery, RNTs have unique chemical and physical properties that make them particularly attractive. For example, they can be chemically modified on their outer surface with bioactive molecules and be structurally tuned to have dimensions (lengths, diameters) and properties that are suitable for incorporating hydrophobic drug molecules via non-covalent interactions. For tissue engineering applications, RNTs have the potential to be loaded with a variety of hydrophobic drugs, including anticancer drugs for site-specific delivery at the surface of the implant materials. As a result, RNTs may serve as a multifunctional scaffold that delivers osteogenic peptides and creates a biocompatible surface to promote healthy cell and tissue growth and also slowly release encapsulated drugs into the targeted areas.

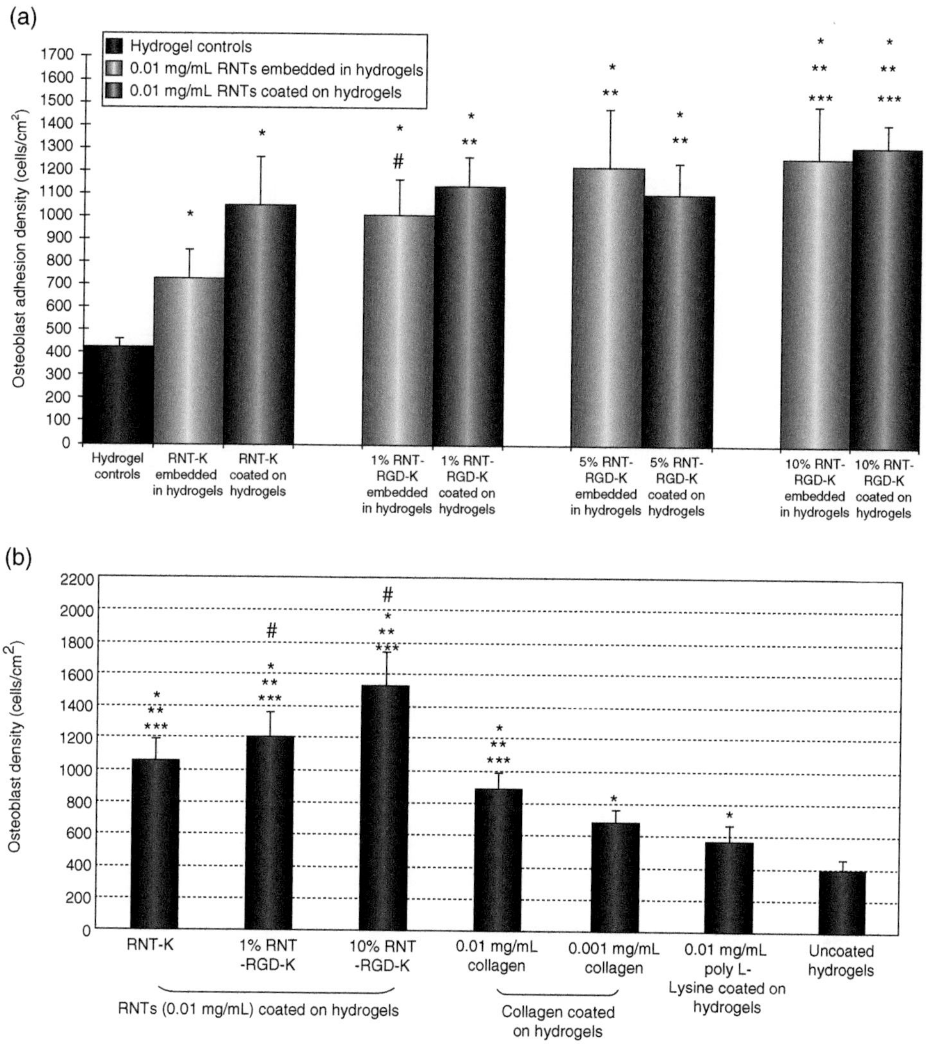

Figure 12.19 *(a) Osteoblast adhesion on K-RNT or K/RGD-RNT (0.01 mg/mL) embedded in or coated on the pHEMA hydrogel after 4 h. Data are mean values ± SEM; n = 3; *p < 0.05 compared to uncoated hydrogels; **p < 0.05 and #p < 0.1 compared to 0.01 mg/mL K-RNT embedded in hydrogels; ***p < 0.1 compared to respective 0.01 mg/mL K/RGD-RNT (99:1) coated on and embedded hydrogels. (b) OB density on K/RGD-RNT-coated hydrogels, type I collagen-coated hydrogel and poly-L-lysine-coated hydrogel. Data are mean values ± SEM; n = 3; *p < 0.05 compared to uncoated hydrogels; **p < 0.05 compared to 0.001 mg/mL collagen-coated hydrogels; ***p < 0.01 compared to 0.01 mg/mL poly-L-lysine-coated hydrogels; #p < 0.05 compared to 0.01 mg/mL collagen-coated hydrogels. Reproduced with permission from reference [97]. Copyright (2009) Elsevier*

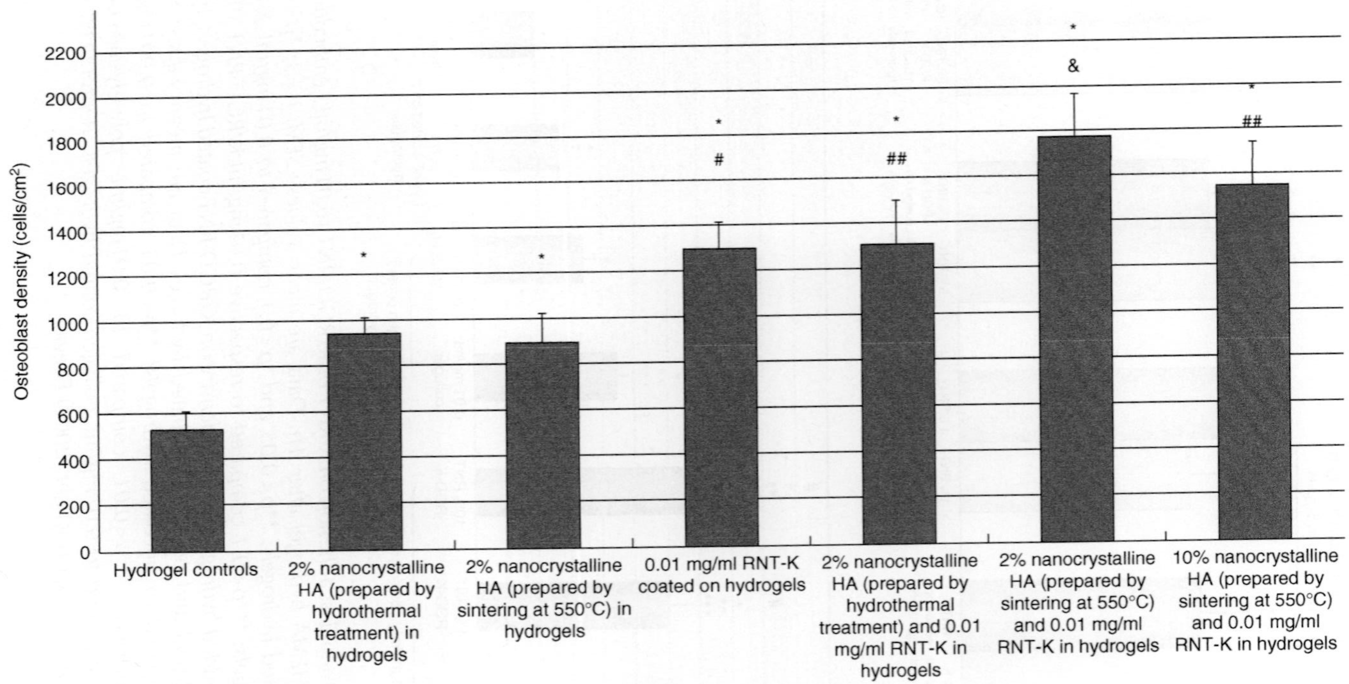

Figure 12.20 *Osteoblast adhesion on nano-HA/K-RNT/pHEMA hydrogel composite materials compared with controls. Data are mean ± SEM; $n = 3$; *$p < 0.005$ compared to hydrogel controls; #$p < 0.005$ and ##$p < 0.05$ compared to 2% (w/w) nanocrystalline HA (prepared by hydrothermal or sintering methods) in hydrogels; &$p < 0.05$ compared to all other scaffolds except 10% (w/w) nanocrystalline HA in K-RNT hydrogels. Reproduced with permission from reference [101]. Copyright (2009) IOP Publishing Ltd*

As a demonstration of the encapsulation potential of the RNTs, the loading of a water-insoluble anticancer drug tamoxifen (TAM) [102] into K-RNT and TBL was studied using ^{1}H-NMR, diffusion ordered spectroscopy (DOSY) and AFM, as well as UV–Vis spectroscopy [103]. TBL are RNTs generated from a twin G∧C motif functionalized with an amino butyl group (Figure 12.21). ^{1}H-NMR studies performed in CD_3OD showed that the incorporation of TAM reached an average maximum of 22% within one day at a TAM:K-RNT molar ratio of 5:3 [103], with the more hydrophobic and stable TBL alternatively incorporated 11.3%±2.1% more TAM than K-RNT. Successful TAM loading was also confirmed by DOSY NMR, which showed a reduction in the diffusion coefficient of TAM in the presence of the RNTs.

UV–Vis spectra revealed that the profiles of TAM-K-RNT and TAM-TBL were significantly different than those from the free RNTs. For example, a hypochromic effect was observed in the UV–Vis spectra of TAM-TBL in MeOH, suggesting that TAM molecules may be intercalating between the G∧C motifs (Figure 12.21) [103]. AFM height profiles of TAM-K-RNT and TAM-TBL also showed a dramatic increase in the RNT heights, thereby further supporting an interaction with TAM. Most importantly, the integrity of the RNTs was maintained in the presence of the drug, which is important for the delivery applications.

In a similar study, the incorporation of the widely used dexamethasone (DEX) drug into RNTs was also investigated [104]. Although DEX enhances osteoconductivity [105, 106], it also interferes with the differentiation of bone marrow-derived stem cells into muscle, cartilage, fat and other tissue-forming cells [107]. Thus, it is very important to deliver DEX

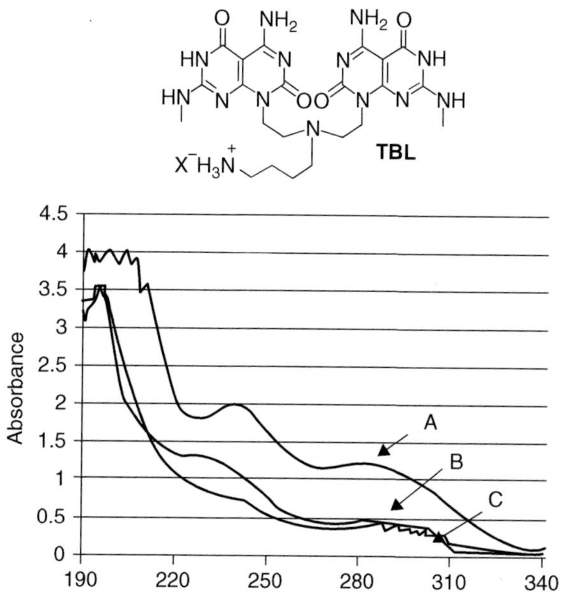

Figure 12.21 *UV–Vis spectra recorded in MeOH. (A) TAM (5 µg/mL), (B) TBL (25 µg/mL) and (C) TAM-TBL (25 µg/mL). Reproduced with permission from reference [103]. Copyright (2011) Dove Medical Press Ltd*

to the desired site of new bone growth, in order to achieve the optimum therapeutic activity [104]. As in the previous study, ^{1}H-NMR spectroscopy showed that at the molar ratio of 5:3 for DEX:K-RNT, approximately 32% of the drug was incorporated [104]. In agreement with this observation was DOSY, AFM and UV–Vis data, which supported an interaction between the two species and suggested that this was via intercalation/stacking.

To gain some insight into the loading and release of DEX from the RNTs, the DEX-K-RNT was prepared using two different methods. In the first method, K-RNT was adsorbed on to a glass slide and then dipped into a water-soluble DEX solution. In the second method, K-RNT was assembled in a water-soluble DEX solution and then adsorbed on to a glass slide [104]. Subsequent release experiments performed under physiological conditions (water buffer at 37 °C) showed that the first approach retained approximately 24.8% more DEX than control samples and released the drug for up to 6 days. The second method alternatively had 42% more DEX loading compared to the control samples [104].

After establishing that K-RNT could incorporate and slowly release DEX, the bioactivity of the drug was determined. As presented in Figure 12.22, DEX-K-RNT-coated glass slides not only exerted the same biological effect as the drug released from the DEX-coated glass slides [104] but also promoted a higher OB density, which is in agreement with other studies described in this chapter [54, 55, 73].

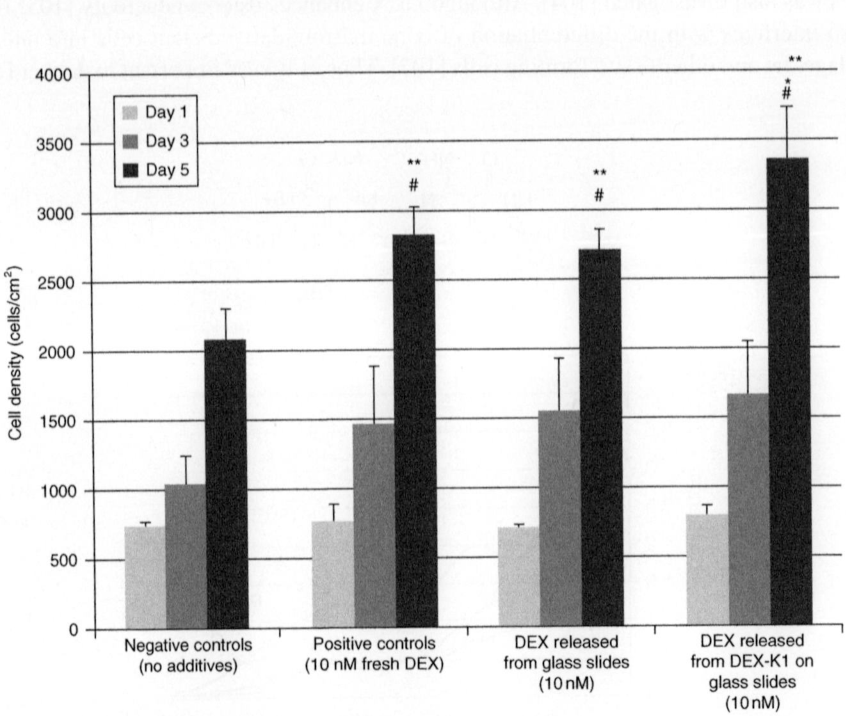

Figure 12.22 OB density cultured with released DEX. Data are mean ± SEM; $n = 9$; $^{*}p < 0.05$ compared to both controls and DEX released from glass slides at day 5; $^{**}p < 0.05$ compared to negative controls (no additives) at day 5; $^{\#}p < 0.05$ compared to day 1 and day 3 results on respective samples. Reproduced with permission from reference [104]. Copyright (2011) Dove Medical Press Ltd

While the incorporation of DEX and TAM into the RNTs and release into OB cells is an excellent demonstration of the drug delivery potential of the nanotubes, our longer-term goal is to target specific cells and not bystander cells. As described in Section 12.2.4, K-RNT causes only a transient inflammation (no signs of cell death) in Calu-3 cells [39]. By co-assembling K-RNT with RGDSK-RNT (Arg–Gly–Asp–Ser–Lys) in a 10:1 molar ratio, however, we aimed to specifically target integrins (e.g. αvβ3) within this same cell line, in order to initiate an inflammatory and apoptosis response [26]. Integrins are heterodimeric transmembrane proteins that recognize the RGD [108] peptide sequence and play a key role in cell signaling, migration of leukocytes [109], angiogenesis [110] and cell–cell and cell–extracellular matrix adhesion [111]. Notably, the αvβ3 integrin is generally not expressed in normal or resting enthothelial cells, but is present in proliferating endothelial cells in tumors, and thus can be used for targeted drug delivery in cancer treatment [112, 113].

To indicate whether this modified K/RGDSK-RNT (10:1) construct could indeed target integrins, the activation of p38 mitrogen-activated protein kinase (MAPK) [114] and the resulting triggering of cell signaling in Calu-3 cells was investigated. In general, p38 MAPK is activated by the phosphorylation of tyrosine and threonine and is involved in regulating inflammation [115–117] and apoptosis [118–120]. When examined, the RNTs were in fact able to induce a rapid, but transient, activation of p38 MAPK [26]. More specifically, K/RGDSK-RNT (10:1) induced an increase in the phosphorylation of the ~40 kDa band of p38 MAPK within 2 min, reached a maximum at 15 min and then declined to baseline at 25 min. The nonphosphorylated p38 MAPK loading control alternatively remained unchanged following exposure to the RNTs [26]. As well, phosphorylation of p38 MAPK was successfully inhibited by the potent cell permeable inhibitor, SB239063.

Given the role of p38 MAPK in inflammation [115–117] and apoptosis [118–120], the ability of K/RGDSK-RNT (10:1) to induce the expression of TNF-α and ICAM-1 [121–123] as evidence for an inflammatory response, along with caspase-3, which is a death protease that catalyzes the cleavage of cellular regulatory proteins resulting in DNA fragmentation, were investigated. In this regard, a significant dose-dependent increase in TNF-α release into the media was observed after treatment with the RNTs. The highest concentration of K/RGDSK-RNT (400:40 μM) induced an expression of TNF-α comparable to LPS, which is a known inflammatory agent. Control studies involving dimethyl sulfoxide (DMSO), lysine, RGDSK and RSDGK peptides alternatively did not promote its secretion from the Calu-3 cells [26]. Importantly, blocking the phosphorylation of p38 MAPK by incubating cells with SB239063 prior to exposure with the RNTs inhibited the release of TNF-α [26]. These studies therefore revealed that K/RGDSK-RNT (10:1) induces the expression of TNF-α through the p38 MAPK pathway. In a similar study, Calu-3 cells were also determined to have a concentration-dependent increase in the release of ICAM-1 upon exposure to RNTs-K/RGDSK. However, the release occurred in a p38 MAPK-independent manner [26]. Finally, K/RGDSK-RNT (10:1) induced caspase-3 activation and DNA fragmentation (indicator of apoptosis) upstream through the p38 MAPK cascade.

Based on these results, it is clear that unlike K-RNT [26], the co-assembled K/RGDSK-RNT (10:1) induces inflammation and apoptosis in Calu-3 cells. The proposed mode of action is summarized in Figure 12.23. It is postulated that the RNTs interact with the αvβ3 integrin (the expression of the αvβ3 integrin was verified) and this leads to the activation of the p38 MAPK. As shown in the figure, it is possible that apoptosis is partially caused by an increased expression of TNF-α, since immobilized RGD has been shown to promote

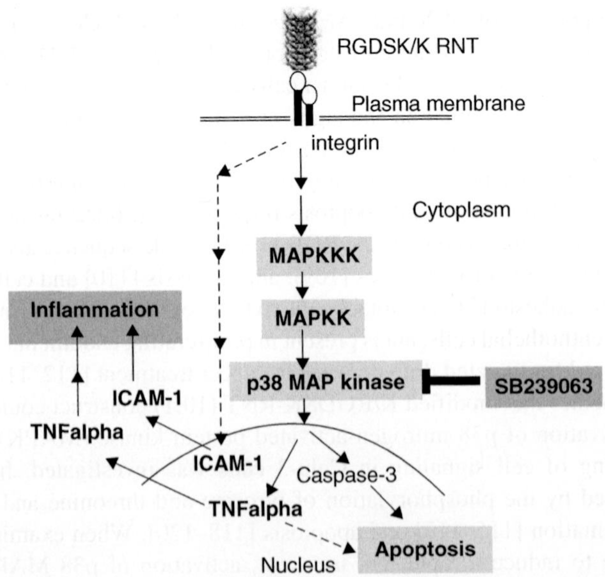

Figure 12.23 *Proposed action of Calu-3 cells treated with K/RGDSK-RNT (10:1). Reproduced with permission from reference [26]. Copyright (2009) Elsevier*

TNF-α-dependent apoptosis in mesangial cells through a MAPK-independent mechanism [124, 125]. Moreover, TNF-α is also known to regulate apoptosis in lung adenocarcinoma cells [126–128].

Overall, it is evident that RNTs are able to induce different cellular responses based on their strategic expression of bioactive sequences. In particular, K/RGDSK-RNT (10:1) has been shown to be a potentially excellent tool for targeting cellular functions such as cell signaling, inflammatory gene expression and apoptosis. This study is very significant because it highlights the therapeutic potential of the RNTs as a new class of nanomaterials for targeted drug delivery [26].

In a related study, the effects of K/RGDSK-RNT (95:5) on bovine neutrophil chemotaxis, cell signaling and apoptosis were investigated *in vitro* [129]. Neutrophils in general are the first line of defense against bacterial infections. However, intensive sequestration of activated neutrophils in injured lungs can cause excessive inflammation [130]. Their activation also suppresses apoptosis and therefore clearance by macrophages, resulting in a longer lifespan and tissue damage [131, 132]. Bovine respiratory disease (BRD) is one of the most common and costly inflammatory diseases in feedlot cattle [133] in terms of average daily gain, treatment and mortality [134]. Despite best efforts, including vaccination, antibiotics, genetic selection and antiinflammatory drugs, BRD continues to be a significant problem [133]. Thus, understanding the mechanisms to control the migration and lifespan of neutrophils in lung inflammation is of great need.

Neutrophil migration is regulated by chemoattractants and adhesive proteins including selectins and integrins such as β2 and possibly αvβ3 [135, 136]. When an RGD-ligand is engaged with integrin, this leads to intracellular signals including MAPK pathways that regulate cell migration via activation of focal adhesion kinases [137, 138]. In turn, activity of the extracellular signal regulated protein kinase (ERK) is important for focal adhesion disassembly and direct phosphorylation of myosin light chain kinase, which leads to cell migration [139, 140]. Although ERK enables direct and rapid regulation of cytoskeletal dynamics in migrating cells, the integrin ligation is necessary for the induction as well as the persistence of cell migration with regard to MAPK regulation [129, 141, 142].

Based on these cascading cellular events, it was reasoned that K/RGDSK-RNT (95:5) could be used to target the neutrophil $\alpha v\beta 3$ integrin and in doing so inhibit the chemotaxis that leads to excessive inflammation. This hypothesis was verified experimentally, as the RNTs did indeed inhibit neutrophil migration in response to the strong chemoattractant *N*-formyl-methionine–leucine–phenylalanine (fMLP). Moreover, since the RNTs had no

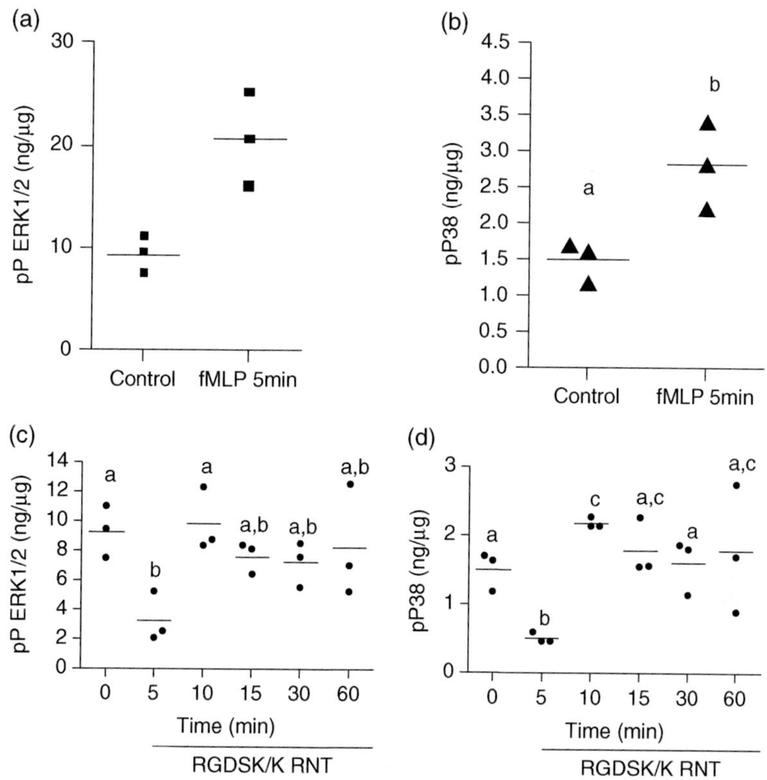

Figure 12.24 *Phosphorylation (pP) of ERK1/2 (a, c) and p38 (b, d) MAPK in bovine neutrophils. Results are of three independent experiments and are mean ± SEM. Significant differences between the treatment groups are expressed by different letters (a–c) above the bars ($p<0.001$ and $p<0.01$ for ERK and p38, respectively). fMLP induced significant phosphorylation of ERK1/2 (a) and p38 (b) MAPK within 5 min of exposure. RNTs-K/RGDSK significantly suppressed phosphorylation of ERK1/2 (c) and p38 (d) MAPK within 5 min of treatment. The phosphorylation of ERK1/2 (c) and p38 (MAPK) returned to control values at 10 min. Reproduced with permission from reference [129]. Copyright (2010) EDP Sciences. http://www.vetres.org/index.php?option=com_issues&task=all&Itemid=39&lang=en_GB.utf8%2C+en_GB.UT*

effect on neutrophil chemotaxis pre-treated with an anti-αvβ3 antibody, this is strong evidence for their binding to αvβ3 [129].

To understand the molecular effects on neutrophil migration, cells were exposed to K/RGDSK-RNT (95:5) with or without fMLP, followed by quantification of the phosphorylated ERK1/2 and p38 MAPK [129]. As shown in Figure 12.24a and b, exposure to fMLP resulted in an increase in the phosphorylation of both kinases relative to the control. However, treatment with K/RGDSK-RNT (95:5) significantly suppressed the phosphorylation and therefore activation of both ERK1/2 and MAPK (Figure 12.24c and d) within 5 min of treatment. The impact of this suppression was evident by the fact that the chemical inhibitors UO126 (ERK1/2 inhibitor) and SB239063 (p38 MAPK inhibitor) were similarly potent in inhibiting neutrophil chemotaxis [129].

Because excessive tissue damage [131, 132] is believed to result from the prolonged lifespan of activated neutrophils, the effect of K/RGDSK-RNT (95:5) and K/RGDSK-RNT (90:10) (in the presence or absence of LPS) on their lifespans was measured by the caspase-3 expression at 0, 18, 24 and 48 h (Figure 12.25). Most notably, whereas LPS

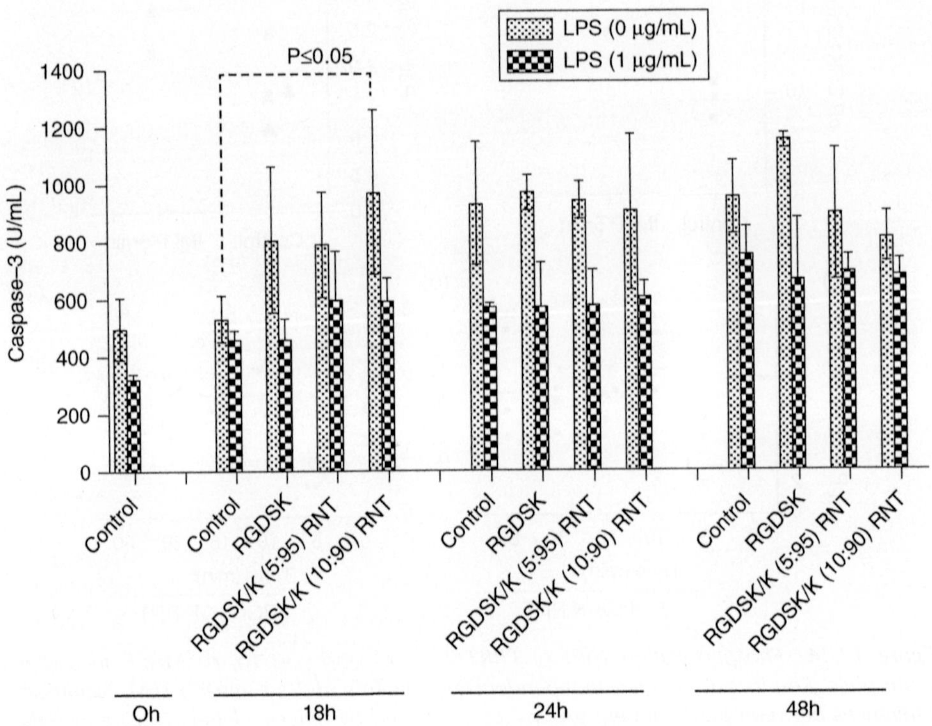

Figure 12.25 *Effect of K/RGDSK-RNT on caspase-3 activity. Results of three different experiments are displayed as mean ± SEM; p<0.001 when compared to LPS and non-LPS treated groups. Neutrophils with or without pre-exposure to LPS (1 μg/mL) were treated with RGDSK, K/RGDSK-RNT (95:5) or K/RGDSK-RNT (90:10) for 0, 18, 24 and 48 h. Cell lysates were used for measurement of caspase-3 activity. Reproduced with permission from reference [129]. Copyright (2010) EDP Sciences. http://www.vetres.org/index.php?option=com_issues&task=all&Itemid=39&lang=en_GB.utf8%2C+en_GB.UT*

caused significant suppression of caspase-3 activity in all treatments at all time-points ($p < 0.001$), the RNTs did not induce an increase in caspase-3 activity except for the significant induction by K/RGDSK-RNT (90:10) at 18 h compared to the control ($p < 0.05$) [129].

Further assessment of the neutrophil apoptosis at 24 h revealed that the LPS treatment significantly inhibited apoptosis, whereas the RGDSK peptide, K/RGDSK-RNT (95:5) or K/RGDSK-RNT (90:10) had no effect compared to the control. Interestingly, this is in contrast to the study described above, which showed that K/RGDSK-RNT (10:1) induces apoptosis in the Calu-3 cells [26]. It is possible that this is due to the fact that the neutrophil is a terminally differentiated cell, whereas the adenocarcinoma is a dividing cell line, thereby causing differences in the actions of the RNTs. As well, because the RNTs increased caspase-3 expression at 18 h, perhaps a higher concentration of the RNTs, a higher proportion of the K-RNT component or prolonged treatment is needed to induce caspase-3 activity and apoptosis in activated neutrophils [129].

Nevertheless, this study provides the first evidence that K/RGDSK-RNT inhibits the phosphorylation of the ERK1/2 and p38 MAPK and chemotaxis of bovine neutrophils. This is a further demonstration of the RNTs' strategic expression of bioactive sequences in order to target cells and initiate a powerful cellular response.

12.5 Conclusions

As presented in this chapter, rosette nanotubes (RNTs) are a novel organic-based nanomaterial formed by the self-assembly of a guanine–cytosine (G∧C) hybrid motif in aqueous solution [8]. In addition to featuring tunable lengths [12] and inner channel diameters [18, 19], RNTs can display a variety of electronically and structurally diverse functional groups on their outer surface. This is achieved by covalently linking these groups to the G∧C motif using reductive amination [8], Suzuki cross-coupling [31] or deprotection/alkylation strategies [32]. Importantly, this facile functionalization enables the RNTs to have unique physical and chemical properties for a variety of biomedical and other materials-based applications.

Along with having highly tailored properties, RNTs possess high biocompatibility and low toxicity both *in vivo* and *in vitro* [38–40]. This can be attributed to their water solubility, biomimetic design and their nanometric dimensions. As such, these architectures have promising applications in biomedicine, especially as delivery scaffolds for amino acids and bioactive peptides for bone tissue engineering. In Section 12.3.2, for example, RNTs were shown to be a very promising 2D coating for Ti implant materials that are currently used to treat bone complications resulting from traumatic injuries or osteoporosis. In particular, RNTs functionalized with lysine and arginine [54, 55, 66, 73] amino acids, as well as KRSR and RGD cell adhesive peptides [83], dramatically increase OB cell attachment when coated on Ti, compared to uncoated Ti and other controls. This enhancement is not only due to the presence of the amino acids and bioactive peptides but also to the innate ability of the RNTs to mimic the organic components of the extracellular matrix in bone and effectively deliver these functional motifs on to the Ti surface. RNTs have also been found to improve the gelation properties and OB cell adhesion when they are coated on or embedded within pHEMA hydrogels [96, 97, 101]. Moreover, although not described in this chapter, RNTs have been explored as alternative cartilage regenerative materials and again have improved cellular functioning [143].

While covalently functionalizing the outer surface is one method that the RNTs can use to deliver bioactive moieties, these self-assembled constructs can also be loaded with drug molecules (e.g. tamoxifen and dexamethasone) via noncovalent interactions and slowly release the cargo into their physiological media [103, 104]. In this manner, RNTs can serve as a biocompatible and multifunctional matrix that delivers osteogenic peptides and slowly releases the encapsulated drug molecules to help generate healthy tissue around the implant. RNTs featuring integrin cell targeting peptide sequences such as RGDSK have also been used to specifically target and initiate inflammation and apoptosis events in human adenocarcinoma cells [26] and to inhibit chemotaxis of bovine neutrophil cells [129]. In combination, these findings set the stage for the use of RNTs as drug carriers for site-specific drug delivery applications in order to treat cancer and other chronic diseases.

References

[1] Grund, S., Bauer, B., Fischer D. (2011) Polymers in drug delivery – state of the art and future trends. Adv. Biomater. 13: B61–B87.

[2] Liechty, W.B., Kryscio, D.R., Slaughter, B.V., Peppas, N.A. (2010) Polymers for drug delivery systems. Annu. Rev. Chem. Biomol. Eng. 1: 149–173.

[3] Amsden, B.G. (2010) Liquid, injectable, hydrophobic and biodegradable polymers as drug delivery vehicles. Macromol. Biosci. 10: 825–835.

[4] Goldberg, M., Langer, R., Jia, X. (2007) Nanostructured materials for applications in drug delivery and tissue engineering. J. Biomater. Sci. Polymer Edn. 18: 241–268.

[5] Balasundaram, G., Webster, T.J. (2006) Nanotechnology and biomaterials for orthopedic medical applications. Nanomedicine (Lond.) 1: 169–176.

[6] Balasundaram, G., Webster, T.G. (2007) An overview of nano-polymers for orthopedic applications. Macromol. Biosci. 7: 635–642.

[7] Bozic, K.J., Kurtz, S.M., Lau, E., Ong, K., Vail, T.P., Berry, D.J. (2009) The epidemiology of revision total hip anthroplasty in the United States. J. Bone. Joint Surg. (Am.) 91: 128–133.

[8] Fenniri, H., Mathivanan, P., Vidale, K.L., Sherman, D.M., Halenga, K., Wood, K.V., Stowell, J.G. (2001) Helical rosette nanotubes: design, self-assembly and characterization. J. Am. Chem. Soc. 123: 3854–3855.

[9] Song, S., Chen, Y., Yan, Z., Fenniri, H., Webster, T.J. (2011) Self-assembled rosette nanotubes for incorporating hydrophobic drugs in physiological conditions. Int. J. Nanomed. 6: 101–107.

[10] Marsh, A., Silvestri, M., Lehn, J.-M. (1996) Self-complementary hydrogen bonding heterocycles designed for the enforced self-assembly into supramolecular macrocycles. Chem. Comm. 1527–1528.

[11] Mascal, M., Hext, N.M., Warmuth, R., Moore, M.H., Turkenburg, J.P. (1996) Programming a hydrogen-bonding code for the specific generation of a supermacrocycle. Angew. Chem. Int. Ed. 35: 2204–2206.

[12] Fenniri, H., Deng, B.-L., Ribbe, A.E., Hallenga, K., Jacob, J., Thiyagarajan, P. (2002) Entropically driven self-assembly of multichannel rosette nanotubes. Proc. Natl Acad. Sci. 99: 6487–6492.

[13] Yamazaki, T., Fenniri, H., Kovalenko, A. (2010) Structural water drives self-assembly of organic rosette nanotubes and holds host atoms in the channel. Chem. Phys. Chem. 1: 361–367.

[14] Maubach, G., Fritzsche, W. (2004) Precise positioning of individual DNA structures in electrode gaps by self-organization onto guiding microstructures. Nano Lett. 4: 607–611.

[15] Raez, J., Moralez, J.G., Fenniri, H. (2004) Long-range flow-induced alignment of self-assembled rosette nanotubes on Si/SiO$_x$ and poly(methyl methacrylate)-coated Si/SiO$_x$. J. Am. Chem. Soc. 126: 16298–16299.

[16] Moralez, J.G., Raez, J., Yamazaki, T., Motkuri, R.K., Kovalenko, A., Fenniri, H. (2005) Helical rosette nanotubes with tunable stability and hierarchy. J. Am. Chem. Soc. 127: 8307–8309.

[17] Tikhomirov, G., Yamazaki, T., Kovalenko, A., Fenniri, H. (2008) Hierarchical self-assembly of organic prolate nanospheroids from hydrophobic rosette nanotubes. Langmuir 24: 4447–4450.

[18] Borzsonyi, G., Johnson, R.S., Myles, A.J., Cho, J.-Y., Yamazaki, T., Beingessner, R. L., Kovalenko, A. Fenniri, H. (2010) Rosette nanotubes with 1.4 nm inner diameter from a tricyclic variant of the Lehn–Mascal G∧C base. Chem. Comm. 46: 6527–6529.

[19] Borzsonyi, G., Alsbaiee, A., Beingessner, R.L., Fenniri, H. (2010) Synthesis of a tetracyclic G∧C scaffold for the assembly of rosette nanotubes with 1.7 nm inner diameter. J. Org. Chem. 75: 7233–7239.

[20] Borzsonyi, G., Beingessner, R.L., Yamazaki, T., Cho, J.-Y., Myles, A.J., Malac, M., Egerton, R., Kawasaki, M., Ishizuka, K., Kovalenko, A., Fenniri, H. (2010) Water-soluble J-type rosette nanotubes with giant molar ellipticity. J. Am. Chem. Soc. 132: 15136–15139.

[21] Fenniri, H., Deng, B.-L., Ribbe, A.E. (2002) Helical rosette nanotubes with tunable chiroptical properties. J. Am. Chem. Soc. 124: 11064–11072.

[22] Johnson, R.S., Yamazaki, T., Kovalenko, A., Fenniri, H. (2007) Molecular basis for water-promoted supramolecular chirality inversion in helical rosette nanotubes. J. Am. Chem. Soc. 129: 5735–5743.

[23] Tikhomirov, G., Oderinde, M., Makeiff, D., Mansouri, A., Lu, W., Heirtzler, F., Kwok, D.Y., Fenniri, H. (2008) Synthesis of hydrophobic derivatives of the G∧C base for rosette nanotube self-assembly in apolar media. J. Org. Chem. 73: 4248–4251.

[24] Chhabra, R., Moralez, J.G., Raez, J., Yamazaki, T., Cho, J.-Y., Myles, A.J., Kovalenko, A., Fenniri, H. (2010) One-pot nucleation, growth, morphogenesis and passivation of 1.4 nm Au nanoparticles on self-assembled rosette nanotubes. J. Am. Chem. Soc. 132: 32–33.

[25] Alsbaiee, A., St Jules, M., Beingessner, R.L., Cho, J.-Y., Yamazaki, T., Fenniri, H. (2012) Synthesis of rhenium chelated MAG$_3$ functionalized rosette nanotubes. Tetrahedron Lett. 53: 1645–1651.

[26] Suri, S.S., Rakotondradany, F., Myles, A.J., Fenniri, H., Singh, B. (2009) The role of RGD-tagged helical rosette nanotubes in the induction of inflammation and apoptosis in human lung adenocarcinoma cells through the P38 MAPK pathway. Biomaterials 30: 3084–3090.

[27] Fritzberg, A.R., Kasina, S., Eshima, D., Johnson, D.L. (1986) Synthesis and biological evaluation of technetium-99 m MAG$_3$ as a hippuran replacement. J. Nucl. Med. 27: 111–116.

[28] Roberts, J., Chen, B., Curtis, L.M., Agarwal, A., Sanders, P.W., Zinn, K.R. (2007) Detection of early changes in renal function using ^{99m}Tc-MAG$_3$ imaging in a murine model of ischemia-reperfusion injury. Am. J. Physiol. Renal. Physiol. 293: F1408–F1412.

[29] Reyes, L., Navarrete, M. (2002) Synthesis and preliminary study of ^{99m}Tc-mercaptoacetyl-triglycine (^{99m}Tc-MAG$_3$) for radiolabeling of immune globulins. J. Radioanal. Nucl. Ch. 251: 245–248.

[30] Wang, Y., Liu, X., Hnatowich, D.J. (2007) An improved synthesis of NHS-MAG$_3$ for conjugation and radiolabeling of biomolecules with ^{99m}Tc at room temperature. Nat. Protocols 2: 972–978.

[31] Beingessner, R.L., Deng, B.-L., Fanwick, P.E., Fenniri, H. (2008) A regioselective approach to trisubstituted 2(or 6)-arylaminopyrimidine-5-carbaldehydes and their application in the synthesis of structurally and electronically unique G∧C base precursors. J. Org. Chem. 73: 931–939.

[32] Beingessner, R.L., Diaz, J.A., Hemraz, U.D., Fenniri, H. (2011) Synthesis of a β-glycoside functionalized G∧C motif for self-assembly into rosette nanotubes with predefined length. Tetrahed. Lett. 52: 661–664.

[33] Daniel, M.-C., Astruc, D. (2004) Gold nanoparticles: assembly, supramolecular chemistry, quantum-size-related properties, and applications toward biology, catalysis, and nanotechnology. Chem. Rev. 104: 293–346.

[34] Rosi, N.L., Mirkin, C.A. (2005) Nanostructures in biodiagnostics, Chem. Rev. 105: 1547–1562.

[35] Moralez, J.G., Raez, J., Yamazaki, T., Motkuri, R.K., Kovalenko, A., Fenniri, H. (2005) Helical rosette nanotubes with tunable stability and hierarchy. J. Am. Chem. Soc. 127: 8307–8309.

[36] Kovalenko, A. (2003) Three-dimensional RISM theory for molecular liquids and solid-liquid interfaces. In: Molecular Theory of Solvation, Vol. 24. Understanding Chemical Reactivity Series. Kluwer Academic Publishers, Dordrecht.

[37] Alsbaiee, A., El-Bakkari, M., Fenniri, H. (2011) The synthesis and self-assembly studies of two bioactive BMP-7 short peptides modified rosette nanotubes for bone tissue engineering. Mater. Res. Soc. Symp. Proc. 1316: mrsf10-1316-qq12–18 (6 pages).

[38] Journeay, W.S., Suri, S.S., Moralez, J.G., Fenniri, H., Singh, B. (2008) Rosette nanotubes show low acute pulmonary toxicity *in vivo*. Int. J. Nanomed. 3: 373–383.

[39] Journeay, W.S., Suri, S.S., Moralez, J.G., Fenniri, H., Singh, B. (2008) Low inflammatory activation by self-assembling rosette nanotubes in human Calu-3 pulmonary epithelial cells. Small 4: 817–823.

[40] Journeay, W.S., Suri, S.S., Moralez, J.G., Fenniri, H., Singh, B. (2009) Macrophage inflammatory response to self-assembling rosette nanotubes. Small 5: 1446–1452.

[41] Gurr, J.R., Wang, A.S.S., Chen, C.H., Jan, K.Y. (2005) Ultrafine titanium dioxide particles in the absence of photoactivation can induce oxidative damage to human bronchial epithelial cells. Toxicology 213: 66–73.

[42] Ning, Y., Tao, F., Qin, G., Imrich, A., Goldsmith, C., Yang, Z., Kobzik, L. (2004) Particle–epithelial interaction: effect of priming and bystander neutrophils on interleukin-8 release. Am. J. Respir. Cell. Mol. Biol. 30: 744–750.

[43] Bergamaschi, E., Bussolati, O., Magrini, A., Bottini, M., Migliore, L., Bellucci, S., Iavicoli, I., Bergamaschi A. (2006) Nanomaterials and lung toxicity: interactions with airways cells and relevance for occupational health risk assessment. Int. J. Immunopathol. Pharmacol. 19: 3–10.

[44] Auger, F., Gendron, M.C., Chamot, C., Marano, F., Dazy, A.C. (2006) Responses of well-differentiated nasal epithelial cells exposed to particles: role of the epithelium in airway inflammation. (2006) Toxicol. Appl. Pharmacol. 215: 285–294.

[45] Takizawa, H. (1998) Airway epithelial cells as regulators of airway inflammation. Int. J. Mol. Med. 1: 367–378.

[46] Lehnert B.E. (1992) Pulmonary and thoracic macrophage subpopulations and clearance of particles from the lung. Environ. Health. Perspect. 97: 17–46.

[47] Duffield, J.S. (2003) The inflammatory macrophage: a story of Jekyll and Hyde. Clin. Sci. 104: 27–38.

[48] Cui, H., Webber, M.J., Stupp, S.I. (2010) Self-assembly of peptide amphiphiles: from molecules to nanostructures to biomaterials. Biopolymers 94: 1–18.

[49] Yang, Y., Khoe, U., Wang, X., Horii, A., Yokoi, H., Zhang, S. (2009) Designer self-assembling peptide nanomaterials. Nanotoday 4: 193–210.

[50] Johnell, O., Kanis, J.A. (2006) An estimate of the worldwide prevalence and disability associated with osteoporotic fractures. Osteoporos. Int. 17: 1726–1733.

[51] Rautray, T.R., Narayanan, R., Kwon, T.-Y., Kim, K.-H. (2010) Surface modification of titanium and titanium alloys by ion implantation. J. Biomed. Mater. Res. Part B 93: 581–591.

[52] Molenberg, A., Schwarz, F., Herten, M., Berner, S., de Wild, M., Wieland, M. (2009) Improved osseointegration of a novel, hydrophilic Ti surface. Mat.-wiss. u. Werkstofftech. 40: 31–35.
[53] de Jonge, L.T., Leeuwenburgh, S.C.G., Wolke, J.G.C., Jansen, J.A. (2008) Organic–inorganic surface modifications for titanium implant surfaces. Pharmaceut. Res. (2008) 25: 2357–2369.
[54] Chun, A.L., Moralez, J.G., Fenniri, H., Webster, T.J. (2004) Helical rosette nanotubes: a more effective orthopaedic implant material. Nanotechnology 15: S234–S239.
[55] Chun, A.L., Moralez, J.G., Webster, T.J., Fenniri, H. (2005) Helical rosette nanotubes: a biomimetic coating for orthopedics? Biomaterials 26: 7304–7309.
[56] Chapman, R.G., Otsuni, E., Takayama, S., Holmlin, R.E., Yan, L., Whitesides, G.M. (2000) Surveying for surfaces that resist the adsorption of proteins. J. Am. Chem. Soc. 122: 8303–8304.
[57] Tirrell, M., Kokkoli, E., Biesalski, M. (2002) The role of surface science in bioengineered materials. Surf. Sci. 500: 61–83.
[58] Webster. T.J. (2004) Proteins: structure and interaction patterns to solid surfaces. In: Encyclopedia of Nanoscience and Nanotechnology, Vols 1–16. Schwarz, J.A., Contescu, C., Putyera, K. (Eds). Marcel Dekker, New York.
[59] Puleo, D.A., Nanci, A. (1999) Understanding and controlling the bone implant interface. Biomaterials 20: 2311–2321.
[60] Webster, T.J., Siegel, R.W., Bizios, R. (2000) Specific proteins mediate osteoblast adhesion on nanophase ceramics. J. Biomed. Mater. Res. 51: 475–483.
[61] Webster, T.J., Schadler, L.S., Siegel, R.W., Bizios, R. (2001) Mechanisms of enhanced osteoblast adhesion on nanophase alumina involve vitronectin. Tissue. Eng. 7: 291–301.
[62] Kyung, M.W., Chen, V.J., Ma, P.X. (2003) Nano-fibrous scaffolding architecture selectively enhances protein adsorption contributing to cell attachment. J. Biomed. Mater. Res. 67A: 531–537.
[63] Zhou, H.Y., Ohnuma, Y., Takita, H., Fujisawa, R., Mizuno, M., Kuboi, Y. (1992) Effects of a bone lysine-rich 18 kDa protein on osteoblast-like MC3T3-E1 cells. Biochem. Biophys. Res. Commun. 186: 1288–1293.
[64] Chun, A.L., Webster, T.J., Fenniri, H. (2006) Helical rosette nanotubes: topographical and chemical roles in osteoblast attachment. Adv. Sci. Technol. 53: 1–8.
[65] Mazia, D., Schatten, G., Sale, W. (1975) Adhesion of cells to surfaces coated with polylysine. Applications to electron microscopy. J. Cell. Biol. 66: 198–200.
[66] Fine, E., Zhang, L., Fenniri, H., Webster, T.J. (2009) Enhanced endothelial cell functions on rosette nanotube-coated titanium vascular stents. Int. J. Nanomed. 4: 91–97.
[67] Medtechinsight.com. US Markets for Peripheral Vascular Stents, Irvine: Medtech Insight. Available from: http://www. medtechinsight.com/ReportA250.html.
[68] Hoffman, G.S., Weyand, C.M. (2002) Inflammatory Diseases of Blood Vessels. Marcel Dekker, New York.
[69] White, C.J. (2005) Drug-eluting Stents: Advanced Applications for the Management of Coronary Disease. Taylor & Francis, New York.
[70] Hoexter, D.L. (2002) Bone regeneration graft materials. J. Oral. Implantol. 28: 290–294.
[71] Sammarco, V.J., Chang, L. (2002) Modern issues in bone graft substitutes and advances in bone tissue technology. Foot Ankle Clin. 7: 19–41.
[72] Chris Arts, J.J., Verdonschot, N., Schreurs, B.W., Buma, P. (2006) The use of a bioresorbable nano-crystalline hydroxyapatite paste in acetabular bone impaction grafting. Biomaterials 27: 1110–1118.
[73] Zhang, L., Chen, L.Y., Rodriguez, J., Fenniri, H., Webster, T.J. (2008) Biomimetic helical rosette nanotubes and nanocrystalline hydroxyapatite coatings on titanium for improving orthopedic implants. Int. J. Nanomed. 3: 323–333.

[74] Puleo, D.A., Bizios, R. (1992) Mechanisms of fibronectin mediated attachment of osteoblasts to substrates in vitro. Bone Miner. 18: 215–226.

[75] Hasenbein, M.E., Andersen, T.T., Bizios, R. (2002) Micropatterned surfaces modified with selected peptides promote exclusive interactions with osteoblasts. Biomaterials 19: 3937–3942.

[76] Webster, T.J., Schadler, L.S., Siegel, R.W., Bizios, R. (2001) Mechanisms of enhanced osteoblast adhesion on nanophase alumina involve vitronectin. Tissue Eng. 7: 291–301.

[77] Kroese-Deutman, H.C., Van Den Dolder, J., Spauwen, P.H.M., Jansen, J.A. (2005) Influence of RGD-loaded titanium implants on bone formation *in vivo*. Tissue Eng. 11: 1867–1875.

[78] Schliephake, H., Scharnweber, D., Dard, M., Rössler, S., Sewing, A., Meyer, J., Hoogestraat, D. (2002) Effect of RGD peptide coating of titanium implants on periimplant bone formation in the alveolar crest. An experimental pilot study in dogs. Clin. Oral Implants Res. 13: 312–319.

[79] Chua, P.H., Neoh, K.G., Kang, E.T., Wang, W. (2008) Surface functionalization of titanium with hyaluronic acid/chitosan polyelectrolyte multilayers and RGD for promoting osteoblast functions and inhibiting bacterial adhesion. Biomaterials 29: 1412–1421.

[80] Masuko, T., Iwasaki, N., Yamane, S., Funakoshi, T., Majima, T., Minami, A., Ohsuga, N., Ohta, T., Nishimura, S.I. (2005) Chitosan–RGDSGGC conjugate as a scaffold material for musculoskeletal tissue engineering. Biomaterials 26: 5339–5347.

[81] Zreiqat, H., Akin, F.A., Howlett, C.R., Markovic, B., Haynes, D., Lateef, S.S., Hanley, L. (2003) Differentiation of human bone-derived cells grown on GRGDSP–peptide bound titanium surfaces. J. Biomed. Mater. Res. Part A 64: 105–113.

[82] Dee, K.C., Andersen, T.T., Bizios, R. (1998) Design and function of novel osteoblast-adhesive peptides for chemical modification of biomaterials. J. Biomed. Mater. Res. 40: 371–377.

[83] Zhang, L., Hemraz, U.D., Fenniri, H., Webster, T.J. (2010) Tuning cell adhesion on titanium with osteogenic rosette nanotubes. J. Biomed. Mater. Res. Part A 95A: 550–563.

[84] Babensee, J.E., McIntire, L.V., Mikos, A.G. (2000) Growth factor delivery for tissue engineering. Pharm. Res. 17: 497–504.

[85] Solheim, E. (1998) Growth factors in bone. Int. Orthop. 22: 410–416.

[86] Giannobile, W.V. (1996) Periodontal tissue engineering by growth factors. Bone 19: 23s–37s.

[87] Lind, M., Deleuran, B., Thestrup-Pedersen, K., Søballe, K., Eriksen, E.F., Bünger, C. (1995) Chemotaxis of human osteoblasts. Effects of osteotropic growth factors. APMIS 103: 140–146.

[88] Reddi, A.H. and Cunningham, N.S. (1993) Initiation and promotion of bone differentiation by bone morphogenetic proteins. J. Bone Mineral Res. 8: s499–s502.

[89] Friedlaender, G.E., Perry, C.R., Cole, J.D., Cook, S.D., Cierny, G., Muschler, G.F., Zych, G.A. Calhoun, J.H., LaForte, A.J., Yin, S. (2001), Osteogenic protein-1 (bone morphogenetic protein-7) in the treatment of tibial nonunions. J. Bone Joint Surg. Am. 83: 151–158.

[90] Saito, A., Suzuki, Y., Ogata, S., Ohtsuki, C., Tanihara, M. (2003) Activation of osteo-progenitor cells by a novel synthetic peptide derived from the bone morphogenetic protein-2 knuckle epitope. Biochim. Biophys. Acta 1651: 60–67.

[91] Kirkwood, K., Rheude, B., Kim, Y.J., White, K. and Dee, K.C. (2000) *In vitro* mineralization studies with substrate-immobilized bone morphogenetic protein peptides. J. Oral. Implantol. 29: 57–65.

[92] Chen, Y., Webster, T.J. (2009) Increased osteoblast functions in the presence of BMP-7 short peptides for nanostructured biomaterial applications. J. Biomed. Mater. Res. Part A. 91: 296–304.

[93] Lee, K.Y. Mooney, D.J. (2001) Hydrogels for tissue engineering. Chem. Rev. 101: 1869–1879.

[94] Hoffmann, A.S. (2002) Hydrogels for biomedical applications. Adv. Drug Delivery Rev. 54: 3–12.

[95] Filmon, R., Grizon, F., Baslé, M.F., Chappard, D. (2002) Effects of negatively charged groups (carboxymethyl) on the calcification of poly(2-hydroxyethylmethacrylate). Biomaterials 23: 3053–3059.

[96] Zhang, L., Ramsaywack, S., Fenniri, H., Webster, T.J. (2008) Enhanced osteoblast adhesion on self-assembled nanostructured hydrogel scaffolds. Tissue Eng. A 14: 1353–1364.

[97] Zhang, L., Rakotondradany, F., Myles, A.J., Fenniri, H., Webster, T.J. (2009) Arginine–glycine–aspartic acid modified rosette nanotube–hydrogel composites for bone tissue engineering. Biomaterials 30: 1309–1320.

[98] Tirrell, M., Kokkoli, E., Biesalski, M. (2002) The role of surface science in bioengineered materials. Surf. Sci. 500: 61–83.

[99] Webster, T.J., Ergun, C., Doremus, R.H., Siegel, R.W., Bizios, R. (2000) Specific proteins mediate enhanced osteoblast adhesion on nanophase ceramics. J. Biomed. Mater. Res. 51: 475–483.

[100] Webster, T.J., Schadler, L.S., Siegel, R.W., Bizios, R. (2001) Mechanisms of enhanced osteoblast adhesion on nanophase alumina involve vitronectin. Tissue Eng. 7: 291–301.

[101] Zhang, L., Rodriguez, J., Raez, J., Myles, A.J, Fenniri H., Webster, T.J. (2009) Biologically inspired rosette nanotubes and nanocrystalline hydroxyapatite hydrogel nanocomposites as improved bone substitutes. Nanotechnology 20: 1–12.

[102] Macgregor, J., Jordan, V.C. (1998) Basic guide to the mechanisms of antiestrogen action. Pharmacol. Rev. 50: 151–196.

[103] Song, S., Chen, Y., Yan, Z., Fenniri, H., Webster, T.J. (2011) Self-assembled rosette nanotubes for incorporating hydrophobic drugs in physiological environments. Int. J. Nanomed. 6: 101–107.

[104] Chen, Y., Song, S., Yan, Z., Fenniri, H., Webster, T.J. (2011) Self-assembled rosette nanotubes encapsulate and slowly release dexamethasone. Int. J. Nanomed. 6: 1035–1044.

[105] Guzmán-Morales, J., El-Gabalawy, H., MPham, H., Tran-Khanh, N., McKee, M.D., Wu, W., Centola, M., Hoemann, C.D. (2009) Effect of chitosan particles and dexamethasone on human bone marrow stromal cell osteogenesis and angiogenic factor secretion. Bone 45, 617–626.

[106] Beule, A.G., Steinmeier, E., Kaftan, H., Biebler, K.E., Göpferich, A., Wolf, E., Hosemann, W. (2009) Effects of a dexamethasone-releasing stent on osteoneogenesis in a rabbit model. Am. J. Rhinol Allerg. 23: 433–436.

[107] Grigoriadis, A.E., Heersche, J.N., Aubin, J.E. (1988) Differentiation of muscle, fat, cartilage, and bone from progenitor cells present in a bone-derived clonal cell population: effect of dexamethasone. J. Cell Biol. 106: 2139–2151.

[108] Humphries J.D., Byron, A., Humphries, M.J. (2006) Integrin ligands at a glance. J. Cell Sci. 119: 3901–3903.

[109] Lämmermann, T., Bader, B.L., Monkley, S.J., Worbs, T., Wedlich-Söldner, R., Hirsch, K., Keller, M., Förester, R., Critchley, D.R., Fässler, R., Sixt, M. (2008) Rapid leukocyte migration by integrin-independent flowing and squeezing. Nature 453: 51–55.

[110] Hodivala-Dilke, K. (2008) Alphavbeta3 integrin and angiogenesis: a moody integrin in a changing environment. Curr. Opin. Cell Biol. 20: 514–519.

[111] Nemeth, J.A., Nakada, M.T., Trikha, M., Lang, Z., Gordon, M.S., Jayson, G.C., Corringham, R., Prabhakar, U., Davis, H.M., Beckman, R.A. (2007) Alpha-V integrins as therapeutic targets in oncology. Cancer Invest. 25: 632–646.

[112] Hu, Z., Luo, F., Pan, Y., Hou, C., Ren, L., Chen, J., Wang, J., Zhang, Y. (2008) Arg–Gly–Asp (RGD) peptide conjugated poly(lactic acid)–poly(ethylene oxide) micelle for targeted drug delivery. J. Biomed. Mater. Res. Part A 85: 797–807.

[113] Dunehoo, A.L., Anderson, M., Majumdar, S., Kobayashi, N., Berkland, C., Siahaan, T.J. (2006) Cell adhesion molecules for targeted drug delivery. J. Pharm. Sci. 95: 1856–1872.

[114] Yee, K.L., Weaver, V.M., Hammer, D.A. (2008) Integrin-mediated signalling through the MAP–kinase pathway. IET Syst. Biol. 2: 8–15.

[115] Kumar, S., Boehm, J., Lee, J.C. (2003) p38 MAP kinases: key signalling molecules as therapeutic targets for inflammatory diseases. Nat. Rev. Drug Discov. 2: 717–726.

[116] Wang, L., Luo, J., Fu, Y., He, S. (2006) Induction of interleukin-8 secretion and activation of ERK1/2, p38 MAPK signaling pathways by thrombin in dermal fibroblasts. Int. J. Biochem. Cell Biol. 38: 1571–1583.

[117] Strassburger, M., Braun, H., Reymann, K.G. (2008) Anti-inflammatory treatment with the p38 mitogen-activated protein kinase inhibitor SB239063 is neuroprotective, decreases the number of activated microglia and facilitates neurogenesis in oxygen–glucose-deprived hippocampal slice cultures. Eur. J. Pharmacol. 592: 55–61.

[118] Reddy, K.B., Nabha, S.N., Atanaskova, N. (2003) Role of MAP kinase in tumor progression and invasion. Cancer Metastasis Rev. 22: 395–403.

[119] Sun, Y., Sinicrope, F.A. (2005) Selective inhibitors of MEK1/ERK44/42 and p38 mitogenactivated protein kinases potentiate apoptosis induction by sulindac sulfide in human colon carcinoma cells. Mol. Cancer Ther. 4: 51–59.

[120] Wang, X.B., Gao, H.Y., Hou, B.L., Huang, J., Xi, R.G., Wu, L.J. (2007) Nanoparticle realgar powders induce apoptosis in U937 cells through caspase MAPK and mitochondrial pathways. Arch. Pharm. Res. 30: 653–658.

[121] Basit, A., Reutershan, J., Morris, M.A., Solga, M., Rose Jr, C.E., Ley, K. (2006) ICAM-1 and LFA-1 play critical roles in LPS-induced neutrophil recruitment into the alveolar space. Am. J. Physiol. Lung Cell Mol. Physiol. 291: L200–L207.

[122] Chan, S.C., Shum, D.K., Tipoe, G.L., Mak, J.C., Leung, E.T., Ip, M.S. (2008) Upregulation of ICAM-1 expression in bronchial epithelial cells by airway secretions in bronchiectasis. Respir. Med. 102: 287–298.

[123] Hu, G., Vogel, S.M., Schwartz, D.E., Malik, A.B., Minshall, R.D. (2008) Intercellular adhesion molecule-1-dependent neutrophil adhesion to endothelial cells induces caveolae-mediated pulmonary vascular hyperpermeability. Circ. Res. 102: e120–e131.

[124] Moreno-Manzano, V., Lucio-Cazana, J., Konta, T., Nakayama, K., Kitamura, M. (2000) Enhancement of TNF-alpha-induced apoptosis by immobilized arginineglycine–aspartate: involvement of a tyrosine kinase-dependent, MAP kinase independent mechanism. Biochem. Biophys. Res. Commun. 277: 293–298.

[125] Furusu, A., Nakayama, K., Xu, Q., Konta, T., Kitamura, M. (2007) MAP kinase-dependent, NF-kappaB-independent regulation of inhibitor of apoptosis protein genes by TNF-alpha. J. Cell Physiol. 210: 703–710.

[126] Wang, R., Alam, G., Zagariya, A., Gidea, C., Pinillos, H., Lalude, O., Choudhary, G., Oezatalay, D., Uhal, B.D. (2000) Apoptosis of lung epithelial cells in response to TNF-alpha requires angiotensin II generation *de novo*. J. Cell. Physiol. 185: 253–259.

[127] Pantano, C., Anathy, V., Ranjan, P., Heintz, N.H., Janssen-Heininger, Y.M. (2007) Nonphagocytic oxidase 1 causes death in lung epithelial cells via a TNF-RI-JNK signaling axis. Am. J. Respir. Cell Mol. Biol. 36: 473–479.

[128] Jin, C.Y., Moon, D.O., Lee, J.D., Heo, M.S., Choi, Y.H., Lee, C.M., Park, Y.M., Kim, G.Y. (2007) Sulforaphane sensitizes tumor necrosis factor-related apoptosis-inducing ligand-mediated apoptosis through downregulation of ERK and Akt in lung adenocarcinoma A549 cells. Carcinogenesis 28: 1058–1066.

[129] Anh Le, M.H., Suri, S.S., Rakotondradany, F., Fenniri, H., Singh, B. (2010) Rosette nanotubes inhibit bovine neutrophil chemotaxis. Vet. Res. 41: 75 (11 pages).

[130] Ackermann, M.R., Brogden, K.A. (2000) Response of the ruminant respiratory tract to *Mannheimia* (*Pasteurella*) *haemolytica*. Microbes Infect. 2: 1079–1088.

[131] Meszaros, A.J., Reichner, J.S., Albina, J.E. (2000) Macrophage-induced neutrophil apoptosis. J. Immunol. 165: 435–441.
[132] Vucic, D., Fairbrother, W.J. (2007) The inhibitor of apoptosis proteins as therapeutic targets in cancer. Clin. Cancer Res. 13: 5995–6000.
[133] Lekeux P. (2006) BRDC and the modulation of lung inflammation. Vet. J. 171: 14–15.
[134] Snowder, G.D., Van Vleck, L.D., Cundiff, L.V., Bennett, G.L. (2006) Bovine respiratory disease in feedlot cattle: environmental, genetic, and economic factors. J. Anim. Sci. 84: 1999–2008.
[135] Lawson, M.A., Maxfield, F.R. (1995) Ca(2+)- and calcineurin-dependent recycling of an integrin to the front of migrating neutrophils. Nature 377: 75–79.
[136] Berton, G., Lowell, C.A. (1999) Integrin signalling in neutrophils and macrophages. Cell Signal 11: 621–635.
[137] Giancotti, F.G., Ruoslahti, E. (1999) Integrin signaling. Science 285: 1028–1032.
[138] Huang, C., Jacobson, K., Schaller, M.D. (2004) MAP kinases and cell migration. J. Cell Sci. 117: 4619–4628.
[139] Klemke, R.L., Cai S., Giannini, A.L., Gallagher, P.J., de Lanerolle, P., Cheresh, D.A. (1997) Regulation of cell motility by mitogen-activated protein kinase. J. Cell. Biol. 137: 481–492.
[140] Pullikuth, A.K., Catling, A.D. (2007) Scaffold mediated regulation of MAPK signaling and cytoskeletal dynamics: a perspective. Cell Signal 19: 1621–1632.
[141] Juliano, R.L., Reddig, P., Alahari, S., Edin, M., Howe, A., Aplin, A. (2004) Integrin regulation of cell signalling and motility. Biochem. Soc. Trans. 32: 443–446.
[142] Wu, W.S., Wu, J.R. Hu, C.T. (2008) Signal cross talks for sustained MAPK activation and cell migration: the potential role of reactive oxygen species. Cancer Metastasis Rev. 27: 303–314.
[143] Chen, Y., Bilgen, B., Pareta, R.A., Myles, A.J., Fenniri, H., Ciombor, D. McK., Aaron, R.K., Webster, T.J. (2010) Self-assembled rosette nanotube/hydrogel composites for cartilage tissue engineering. Tissue Eng. C 16: 1233–1243.

13

Peptide Secondary Structures as Molecular Switches

Fernando Formaggio, Alessandro Moretto, Marco Crisma and Claudio Toniolo

Institute of Biomolecular Chemistry, Padova Unit, CNR, Department of Chemistry, University of Padova, Italy

Acronyms

Ac	Acetyl
Aib	α-Aminoisobutyric acid or $C^{\alpha,\alpha}$-dimethylglycine
Ala	Alanine
Beg	C^{α}-*n*-butyl, C^{α}-ethylglycine
CD	Circular dichroism
Deg	$C^{\alpha,\alpha}$-diethylglycine
Epg	C^{α}-ethyl, C^{α}-*n*-pentylglycine
Gly	Glycine
HFA	Hexafluoroacetone
HFIP	1,1,1,3,3,3-Hexafluoroisopropanol
Lys	Lysine
MeOH	Methanol
(αMe)Val	C^{α}-methylvaline
NH*i*Pr	Isopropylamino
NHMe	Methylamino
NMR	Nuclear magnetic resonance
OMe	Methoxy
Pro	Proline
Tfa	Trifluoroacetyl
TFE	2,2,2-Trifluoroethanol

Peptide Materials: From Nanostructures to Applications, First Edition. Edited by Carlos Alemán, Alberto Bianco and Mariano Venanzi.

13.1 Introduction

Correct 3D structural disposition and relative orientation of functional groups in materials systems is essential to their performance. For example, an area where this dogma is inescapable is electron/ energy transfer in which spatial control of donor and acceptor chromophores must be precisely achieved [1]. In this framework, the design of molecular switches offers an exceptional challenge to nanochemists on the road toward the development of novel materials [2–9]. Here, use of peptide molecules provides great advantages, such as a relatively easy synthesis, the chance to shape them into the desired 3D structures, the fine tuning of a large variety of spectroscopic and physical properties by conformational changes, and the characterization of each of the single isolated forms to study basic issues [10–19]. The fundamental property for any type of molecular switch is bistability, namely the occurrence of an equilibrium between two different molecular conformers, which can be shifted toward either direction by means of an external stimulus or an environmental modification.

To date, almost exclusively the four most popular polypeptide conformations [20] have been utilized in materials chemistry. They are: (a) the α-helix, with its related α-helix coiled coil superstructure; (b) the extended, pleated sheet β-structure, with its intra-/intermolecular and parallel/antiparallel variants, which include the well-known intramolecular/antiparallel double-stranded β-hairpin conformation and the multistranded cross-β (meander) structure; (c) the very specific type-I/type-II poly-(L-Pro)$_n$ structures; and (d) the β-turn conformations. These structures have inspired materials chemists to a large extent because they are the fundamental biomolecular motifs characterizing the most common, natural fibrous proteins: (a) the α-helical keratin; (b) the β-sheet forming silk fibroin; (c) the triple helical collagen molecule, each component of which is based on the type-II poly-(L-Pro)$_n$ structure; and (d) the multiple, but not strictly consecutive, type-II β-turn conformations, generating the typical properties of elastin.

In the present review article, in addition to spring systems involving the *classical* α-helix, β-sheet and poly-(L-Pro)$_n$ conformations, we also considered two *recent* developments in the area of peptide secondary structures as potential components of novel types of molecular switches. Both switches imply the participation of the 3_{10}-helical structure, which, depending upon the chemical characteristics of the polypeptide molecule under investigation (such as main-chain length or nature of amino acid C^{α}-alkylation), may easily undergo transitions either to the α-helix or to the much less studied, multiple, consecutive, fully extended (2.0_5-helix) conformation.

13.2 Classical Secondary Structure Switches

13.2.1 α-Helix/β-Pleated Sheet Switch

In the late 1960s the first transition between peptide secondary structures (α-helix/β-pleated sheet), the average parameters of which are listed in Table 13.1, was investigated in detail both experimentally and theoretically using a chirospectroscopic technique (CD) still extensively exploited nowadays in studies of polypeptide and protein conformational changes [21, 22]. It involved the high-molecular weight poly-(L-Lys)$_n$ (Figure 13.1). This

Table 13.1 *Average parameters for peptide helices based on L-α-amino acids*

Parameter	α-Helix	3_{10}-Helix	β-Pleated sheet		2.0_5-Helix	Poly-(L-Pro)$_n$ helix	
			Parallel	Antiparallel		Type I[a]	Type II[a]
ϕ (°)[b]	−63	−57	−120	−140	180	−70	−70
ψ (°)[c]	−42	−30	115	135	180	160	145
n[d]	3.63	3.24	2.00	2.00	2.00	3.30	−3.00
d (Å)[e]	1.56	1.94	3.25	3.47	3.70	2.22	3.12
p (Å)[f]	5.66	6.29	6.50	6.94	7.40	7.33	9.36

[a] In the poly-(Pro)$_n$ type I helix all ω (C^{α}_i–C'_i–N_{i+1}–C^{α}_{i+1}) torsion angles are *cis* (0°), whereas in poly-(Pro)$_n$ type-II they are all *trans* (180°).
[b] The C'_{i-1}–N_i–C^{α}_i–C'_i torsion angle.
[c] The N_i–C^{α}_i–C'_i–N_{i+1} torsion angle.
[d] Number of amino acid residues per helical turn (positive values refer to right-handed helices, whereas negative values refer to left-handed helices).
[e] Axial translation per residue.
[f] Pitch or axial translation per helical turn.

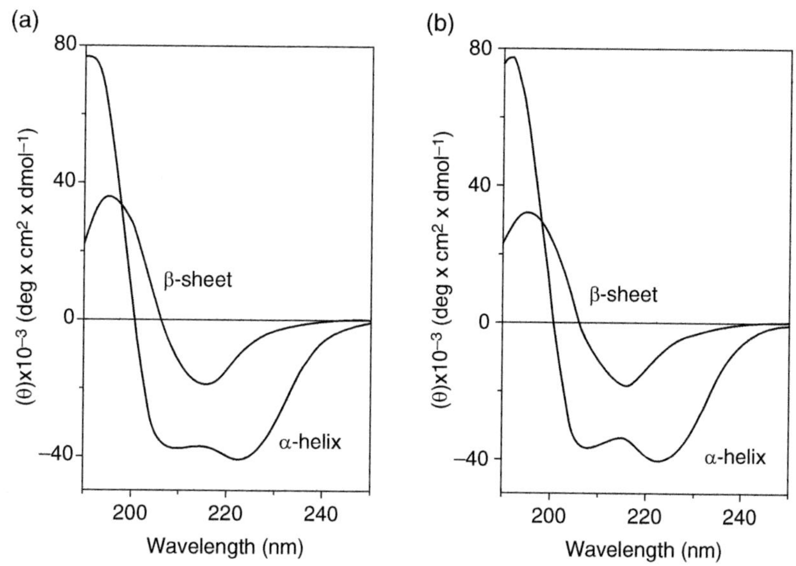

Figure 13.1 *(a) Experimental and (b) computed CD spectra of poly-(L-Lys)$_n$ in the α-helix and β-sheet conformations. Adapted with permission from reference [22]. Copyright (1969) American Chemical Society*

homo-polymer is a good model considering the remarkably cooperative aspects of the formation of its secondary structures. However, it is doubtful that its CD spectra represent an acceptable approximation of those of short segments of a protein composed of different amino acids (some of them even endowed with strong, induced CD contributions in the far-UV region associated with their side-chain aromatic chromophores) in the α-helical or β-sheet conformations. The α-to-β 3D structural transition takes place in water at pH 11.0–11.1 (where the Lys side-chain amino groups are not protonated) and at millimolar concentration upon heating the solution at about 50 °C for 10–15 minutes and cooling it back to room temperature. The CD curve of the right-handed α-helix of poly-(L-Lys)$_n$ exhibits two very intense, negative Cotton effects at 222 nm and 208 nm, respectively, while that of the β-sheet conformation is characterized by low ellipticity values at those wavelengths but by a relatively strong, negative CD band at 217 nm.

In a meeting on poly-α-amino acids held in Israel in 1974, which one of the authors (C.T.) of this contribution attended, Chou and Fasman were the first to present an intriguing list of segments of peptide hormones and proteins with potential for conformational changes in response to variations in the experimental conditions [23]. Many regions of bioactive polypeptides have both α-helical and β-sheet forming tendencies. One of the two conformations is typically preferred, depending on the environmental factors. The authors' predictive model for the secondary structure of polypeptide molecules applied to the 29-mer hormone glucagon well illustrates the delicate balance of this bioactive polypeptide between its two major conformational states. The 3D structural sensitivity of glucagon is limited to the C-terminal residue stretch 19–27, which has almost equally high α-helix and β-sheet computed potentials. In the predicted conformation (a), this nonapeptide forms an α-helical region, while in the predicted conformation (b) this region is β-sheet (Figure 13.2).

These two conformations were identified experimentally for this hormone by CD and IR absorption spectroscopies. This conformational change can be brought about by a change in peptide concentration. In particular, the β-sheet rich conformation prevails at very high peptide concentrations, eventually forming gels and fibrils. Interestingly, recently it was shown that glucagon forms different types of fibrils depending on temperature, pH, ionic strength, concentration, and mechanical stress [24].

Understanding the factors that govern the relative thermodynamic stabilities of α-helix and β-sheet structures in polypeptides is of paramount importance. In particular, the possibility of their interconversion is a relevant characteristic of novel biomaterials [25] and

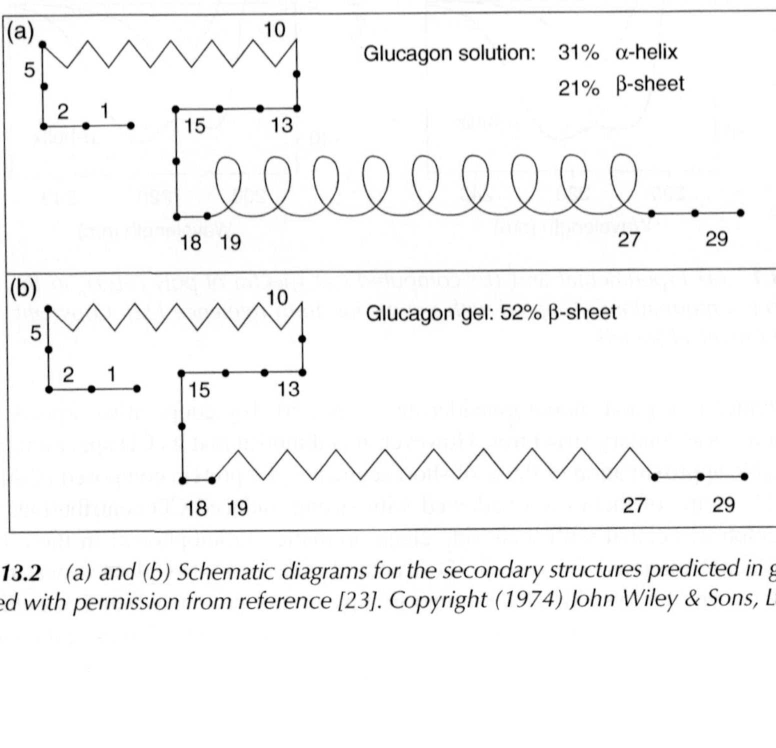

Figure 13.2 *(a) and (b) Schematic diagrams for the secondary structures predicted in glucagon. Adapted with permission from reference [23]. Copyright (1974) John Wiley & Sons, Ltd*

also has direct implications for extremely serious human diseases (Alzheimer, Parkinson, prion, type-II diabetes) [26]. In this framework, it was shown that α-helix/β-sheet transitions are main-chain length dependent [27]. In conclusion, although numerous experimental and theoretical studies, particularly on model compounds [28], peptides (β-amyloid) [29], and proteins (α-synuclein) [30, 31] have recently been published on this crucial topic, 3D structural details and the various types of mechanisms of this phenomenon are not yet well understood nor quantitatively described.

13.2.2 Type-I Poly-(L-Pro)$_n$/Type-II Poly-(L-Pro)$_n$ Switch

Two types of poly-(L-Pro)$_n$ helices are known, termed type-I (right-handed) and type-II (left-handed) (Table 13.1 and Fig. 13.3). This 3D structural phenomenon is called dimorphism. Their sets of ϕ, ψ backbone torsion angles are quite close (in particular, ϕ is fixed to about –70° by the constraints imposed by the five-membered pyrrolidine cyclic structure). Nonetheless, these two 3D structures differ substantially because form I has all tertiary peptide bonds in the *cis* conformation (their ω torsion angle is 0°), while in form II these bonds are all in the *trans* conformation (ω is 180°). Therefore, the less stable form I is remarkably more compact than the more stable form II (the latter also occurs in the collagen triple helix).

Because all peptide bonds are tertiary (no NH groups are present), these two helices are not stabilized by any intramolecular N–H···O=C H-bond. Their interconversion, termed 'mutarotation' because it was originally followed by looking at variations of the polypeptide polarimetric value with time, are governed by the nature of the solvent. Type-II poly-(L-Pro)$_n$ is the dimorph overwhelmingly prevailing in polar solvents. For a complete overview of the literature on this subject to 1966, the reader is referred to the monumental article by Mandelkern [32].

In 1967 the CD spectra showing the slow kinetics of the type-I → type-II transformation of poly-(L-Pro)$_n$ were first reported by Bovey and Hood [33] (Figure 13.4). The solvent is TFE. At zero time, a very weak negative CD band is seen at ca. 232 nm, followed by a large positive band near 215 nm along with a weaker negative band at 199 nm. As the isomerization proceeds, the positive band becomes weaker and the initial negative 199 nm band is displaced by a somewhat stronger negative band at 206 nm. Even after almost one day of equilibration, a weak positive band, centered at approximately 226 nm, is still apparent. The initial curves a and a′ correspond to the poly-(L-Pro)$_n$ type-I conformation, curve c to type-II, and curve b (1 h after dissolution) to a mixture of the two concomitantly occurring dimorphs.

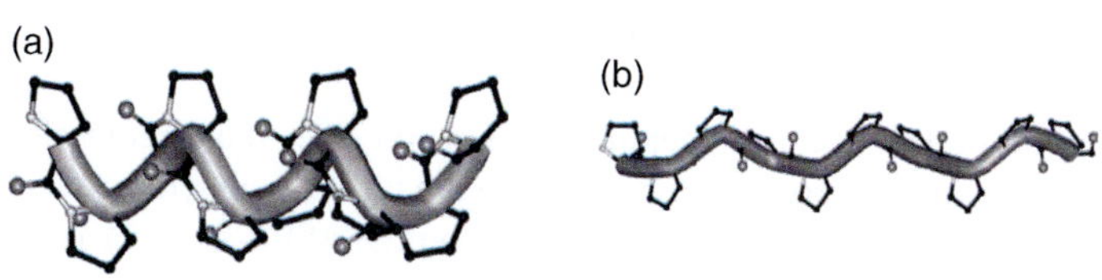

Figure 13.3 *An (L-Pro)$_{10}$ stretch in the type-I (a) and type-II (b) poly-(L-Pro)$_n$ helix conformations*

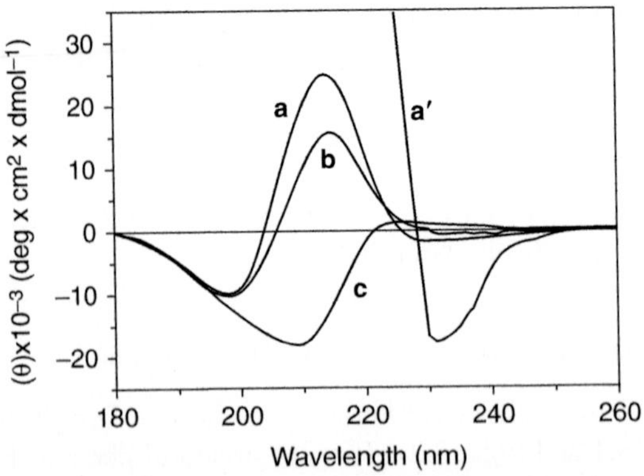

Figure 13.4 *CD spectra of poly-(L-Pro)$_n$ in TFE. The initial curve is shown in (a) (a' is the same curve after appropriate expansion above 225 nm). Curves (b) and (c) are taken after 1 h and 19 h, respectively. Adapted with permission from reference [33]. Copyright (1967) John Wiley & Sons, Ltd*

The first calculated CD spectra for type-I and type-II poly-(L-Pro)$_n$ structures were reported by Pysh [34] in the same year. CD spectra for the two poly-(L-Pro)$_n$ dimorphs (type-I in a 9:1 1-propanol–water solvent mixture and type-II in water) were subsequently reported by Giralt and coworkers [35].

The use of homo-oligomeric model compounds proved to be instrumental to obtain further information on the (L-Pro)$_n$ switch. The CD spectra of unprotected (L-Pro)$_n$ (n=2–40) show the typical features of each form of the poly-(L-Pro)$_n$ structures from the trimer stage, although the ellipticities of the Cotton effects increase with main-chain length and reach the characteristic values of the homo-polymer from the 20-mer upward [36, 37]. However, it is worth noting that the corresponding N^{α}-protected homo-oligomers do not show any mutarotation on changing the solvent from $CHCl_3$ to 1-propanol (only the type-II form seems to be present). The kinetics of the conformational switch for the unprotected (L-Pro)$_{13}$ oligomer was recently investigated in detail in various alcohols [38]. From the results in solution discussed above, and other data in the solid state (summarized in reference [39]), it turns out that mixtures of the *cis* and *trans* peptide bonds (and of type-I and type-II (L-Pro)$_n$ isomeric stretches as well) concomitantly take place in all physical states investigated. On these bases, it is clear that the exploitation of (L-Pro)$_n$ peptide segments as rigid spacers in physicochemical and biochemical studies should be handled with extreme caution.

A very promising, unexpected conformational switch was recently observed for the unprotected (2*S*,4*S*)-4-aminoproline nonamer [40]. In TFE, this homo-oligomer surprisingly adopts an antiparallel β-sheet structure, stabilized by enhancing peptide concentration, i.e. by increasing the extent of intermolecular (side chain) N–H···O=C (main chain) H-bonds (Figure 13.5).

Interestingly, a closely related peptide, based on the diastereomeric (2*S*,4*R*)-4-aminoproline in the same fluoroalcohol is folded in the type-II poly-(L-Pro)$_n$ conformation. The H-bonding scheme of the β-sheet structure of the (2*S*,4*S*)-4-aminoproline nonamer is disrupted in water and rearranged to an intramolecular H-bonding motif that favors the type-II poly-(L-Pro)$_n$ conformation via enrichment of the *trans* tertiary peptide bond geometry. This is the first example of the β-sheet structure in poly-(L-Pro)$_n$-like oligomers which

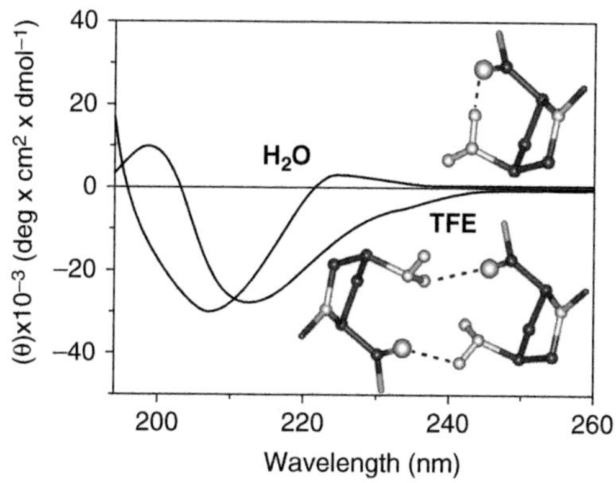

Figure 13.5 *CD spectra of a peptide based on a stretch of nine consecutive (2S,4S)-aminoproline residues which forms an unusual β-sheet structure in TFE that switches to type-II poly-(L-Pro)$_n$ in an aqueous medium. The intra- and intermolecularly H-bonded basic units of the two structures are also represented. Adapted with permission from reference [40]. Copyright (2010) American Chemical Society*

adds a new design principle to the continuously growing repertoire of strategies for an engineering peptide secondary structural switch. However, this β-sheet structure is expected to be largely distorted (hence, we prefer to classify it as a 'pseudo β-sheet structure') since the blocked ϕ value for an L-Pro residue is typically −65 ± 5°, whereas that of a classical β-sheet structure is ca. −140°. Moreover, in this case the intermolecular H-bonding donor N–H groups are of the amino (not amide) type and are located in the amino acid side chain (not in the peptide backbone, as usual).

13.3 Recently Discovered Secondary Structure Switches

13.3.1 The 3_{10}-Helix/α-Helix Switch

As a significant part of the stabilization energy of peptide helices, except for the poly-(L-Pro)$_n$ helices, arises from intramolecular (backbone) C=O···H–N (backbone) H-bonds, their nomenclature is usually described as n_m, where n is the number of amino acid residues per helical turn and m is the number of atoms involved in the (pseudo)cycle generated by the intramolecular H-bond. The most commonly observed 3D structures in a system of five linked peptide units are the $i \leftarrow i+3$ and $i \leftarrow i+4$ intramolecularly H-bonded conformations schematically depicted in Figure 13.6.

As they are (pseudo)*cyclic* forms and have different m values, these structures are also called C_{10} and C_{13} conformations, respectively. An alternative nomenclature in common use for these structures is β-turn [41, 42] and α-turn [43], respectively (Figure 13.7). Helical structures originated from consecutively repeating these two intramolecularly H-bonded conformations are called 3.0_{10}- (or more simply 3_{10}-)helix, and 3.6_{13}- (or α-)helix [44–46], respectively (Figure 13.7), where the values 3.0 and 3.6 correspond to their n parameter. These turns and helices are all characterized by a right-handed screw sense if entirely based on C^{α}-trisubstituted α-amino acids with the usual L-configuration at the α-carbon.

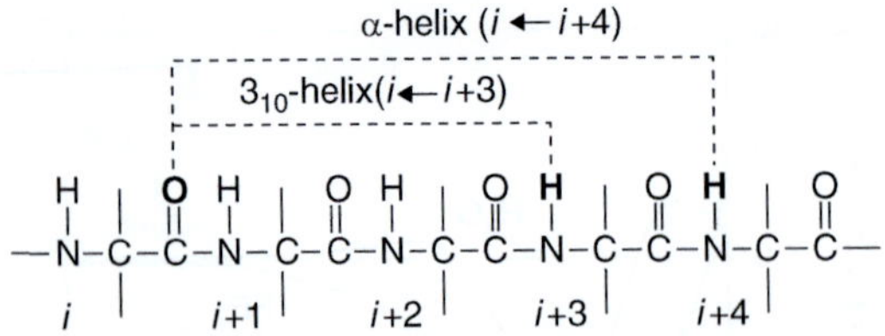

Figure 13.6 *Intramolecular (backbone) C=O···H–N (backbone) H-bonds: i ← i+3 for the 3_{10}-helix and i ← i+4 for the α-helix*

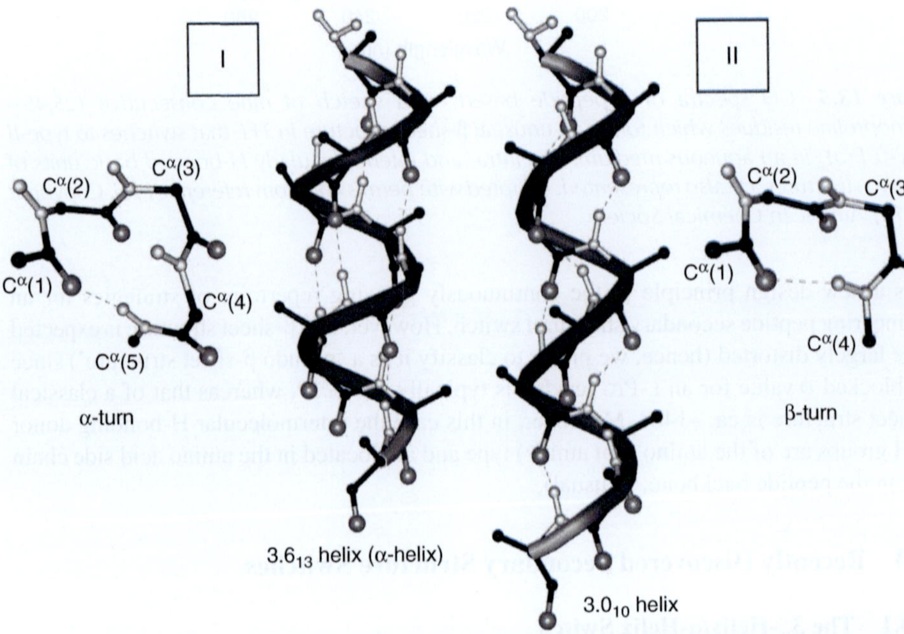

Figure 13.7 *(I) The 3.6_{13}-helix (α-helix) and its basic unit, the helical α-turn (or C_{13}-conformation). (II) The 3.0_{10}-helix and its basic unit, the helical type-III β-turn (or C_{10}-conformation)*

The classical α-helix is the most common regular secondary structure of peptides and proteins and, not surprisingly, the first to be experimentally authenticated (by X-ray diffraction analysis). Another long-range helical structure that occurs in peptides and in proteins is the 3_{10}-helix. This ternary helix is more tightly bound and more elongated than the α-helix (Table 13.1). The sets of ϕ, ψ torsion angles of the α- and 3_{10}-helices do not differ significantly, falling within the same region of the Ramachandran map (termed the 'helical' region). However, as mentioned above, their intramolecular C=O···H–NH-bonding schemes are remarkably distinct.

In a long polypeptide chain based on C^{α}-trisubstituted (e.g. protein) α-amino acid residues, the 3_{10}-helix is less stable than the α-helix. In particular, its van der Waals energy is less favorable (it has several close, although not forbidden, short contacts) and the H-bond

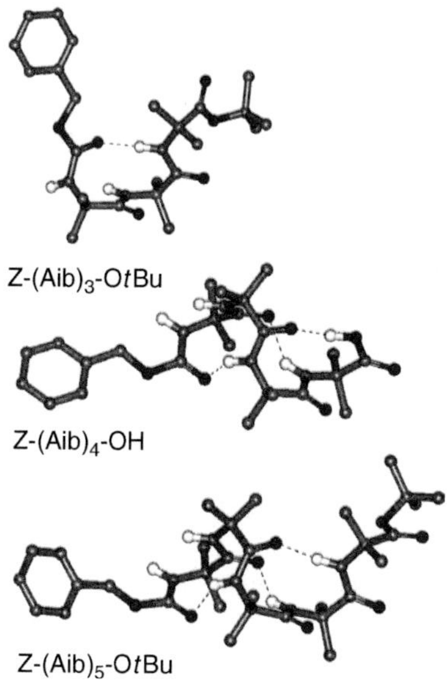

Figure 13.8 *The X-ray diffraction structures of the N^{α}-acylated Aib homo-trimer, tetramer and pentamer. Adapted with permission from reference [51]. Copyright (1982) American Chemical Society*

geometry is less appropriate. Thus, for many years it was considered unlikely that long stretches of 3_{10}-helix would be observed. However, there is no disallowed region of the ϕ, ψ space completely separating these two regularly folded secondary structures. Therefore, the α-helix may be gradually transformed into a 3_{10}-helix (and vice versa), maintaining a nearly helical conformation of the chain throughout. Furthermore, if one of the conformations should turn out to be impossible (say, as a result of markedly unfavorable side-chain interactions), the main chain may slip into the other conformation. Actually, the 3_{10}-helix appears to derive its structural relevance mainly from its proximity in the conformational energy map to the more stable α-helix. Not surprisingly, a role of the 3_{10}-helix as an important intermediate in the mechanism of folding of α-helical proteins has been envisaged [47].

Forty years ago, Marshall [48] used energy calculations of sterically allowed conformations to show that Aib, the prototype of achiral C^{α}-tetrasubstituted α-amino acids, can promote the onset of peptide helices very efficiently, due to steric interactions involving the *gem*-methyl groups linked to the α-carbon. Since 1978 [49], by taking advantage of the extremely high crystallinity of peptides rich in C^{α}-tetrasubstituted α-amino acids, we and others have solved the X-ray diffraction structures of numerous Aib homo-oligo- and co-oligo-peptides, forming 3_{10}-helical structures [44–46, 50]. Interestingly, we found that there is no critical main-chain length dependence for 3_{10}-helix formation in the crystal state, i.e. incipient 3_{10}-helices are formed at the lowest possible level (of an N^{α}-acylated tripeptide) (Figure 13.8) [51].

The two longest, regular 3_{10}-helices reported so far are those of the N^{α}-acylated Aib homo-decamer and undecamer (Figure 13.9) [52–54]. Moreover, a fully developed, stable 3_{10}-helix in a structure supporting solvent requires only about eight C^{α}-tetrasubstituted α-amino acids (e.g. Aib residues) [55].

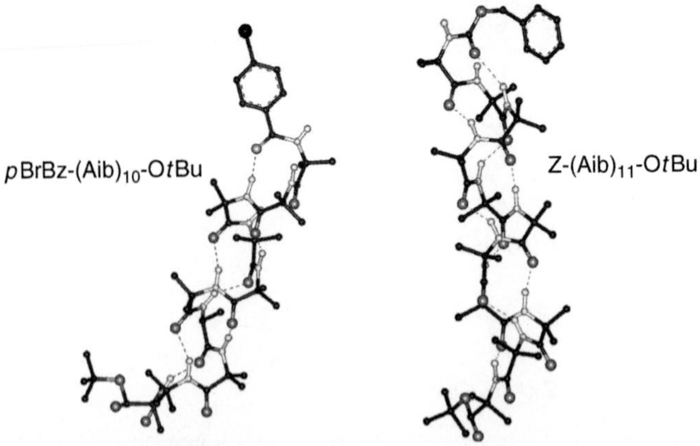

Figure 13.9 The X-ray diffraction structures of the N^{α}-acylated Aib homo-decamer and undecamer. Adapted with permission from references [52], [53], and [54]. Copyright (1990) Elsevier, (1991) and (2003) John Wiley & Sons, Ltd

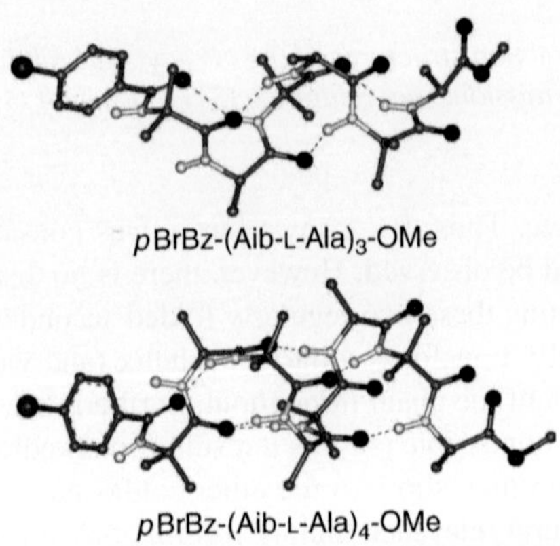

Figure 13.10 The X-ray diffraction structures of the N^{α}-acylated (Aib–L-Ala)$_n$ sequential hexamer and octamer. Adapted with permission from reference [56]. Copyright (1990) Taylor & Francis

To check experimentally the details of the switch between the peptide 3_{10}-helix and the α-helix, we synthesized a complete series of terminally protected, sequential (Aib–L-Ala)$_n$ oligomers to $n=6$. By X-ray diffraction analysis, we showed that the type of helix adopted by these Aib-rich peptides in the crystal state depends on the length of the peptide main chain [56]. The 6-mer ($n=3$) folds into a perfect (right-handed) 3_{10}-helix (Figure 13.10). A dramatic conformational transition was observed between the 6-mer and the 8-mer ($n=4$) by changing only the length of the peptide backbone, not the amino acid sequence. The core of the 8-mer is essentially α-helical, but with two β-turns at the N-terminus (Figure 13.10). The transition is unveiled by formation of a three-center H-bond involving the C(1)=O(1) acceptor with the N(4)H and N(5)H donors.

The two higher oligomers (with $n=5$ and 6) are also overwhelmingly, although not fully, α-helical [57]. It was subsequently found that in the X-ray diffraction structures of peptides and proteins one or two 3_{10}-helical residues at either end of an α-helix stretch are rather common (the short bits of the 3_{10}-helix tighten up the ends of the α-helix by moving the terminal peptide groups near the axis) [46, 50]. These results strongly support the view that an efficient cooperativity effect is needed for the onset of a stable α-helix. It was also demonstrated that, besides the main-chain length, additional parameters, such as percentage of Aib residues in the amino acid composition [44] and sequence permutation [58], have a direct influence on the exact value of the critical main-chain length for the 3_{10}- to α-helix switch in the crystal state. Clearly, a well-defined range of values for this main-chain dependent 3D structural switch is of great interest to theoretical peptide and protein biochemists.

Moreover, FT-IR absorption and NMR studies support the conclusion that in $CDCl_3$, a solvent of low polarity, the pentamers and heptamers –L-Ala–(Aib–L-Ala)$_n$– (n=2, 3) are essentially 3_{10}-helical [59–61]. However, partially conflicting results were reported for the longer oligomer –(Aib–L-Ala)$_5$– in this solvent, in the sense that the NMR data are consistent with a 3_{10}-helicity while FT-IR absorption favors the concomitant occurrence of both 3_{10}- and α-helices. On the other hand, based on an NMR analysis in CD_3CN, it was concluded that the octamer (L-Ala–Aib)$_4$ is folded in the α-helix structure [58]. To complete the conformational picture in organic solvents, it was shown that in the polar, H-bonding donor, alcohols the characteristic features of right-handed 3_{10}- and α-helical conformations are partially developed at the hexa- and octamer levels, respectively, for the (Aib–L-Ala)$_n$ series [62–73].

Recently, to further use these sequential peptides as molecular electronic elements [58, 64–72], we decided to prepare the (Aib–L-Ala)$_n$ and L-Ala–(Aib–L-Ala)$_n$ series to the nonamer stage with the N^{α}-Ac blocking group replacing the usual aromatic benzyloxycarbonyl or *para*-bromobenzoyl moieties. During the step-by-step synthesis in solution, we serendipitously discovered that a remarkable amount of each peptide was lost during the purification procedures (extraction with water from an ethyl acetate or dichloromethane solution) [74]. The unexpected aqueous solubility of this family of peptides based exclusively on aliphatic residues, even in the absence of a covalently bound, strongly water-solubilizing polymer [62], prompted us to exploit CD spectroscopy in the far-UV region to assess the critical main-chain lengths for helix formation in water for the first time. We found that six and eight residues are required, respectively, for the onset of 3_{10}- and α-helices, although both secondary structures were still being formed to a limited extent (Figure 13.11). The CD fingerprints of both types of right-handed helices are characterized by two negative maxima at about 222 and 205 nm, but the former band is relatively much stronger in the α-helix.

In search of a *reversible* 3_{10}-helix/α-helix transition and to drastically simplify the complex interplay of peptide sequence and length, we decided to focus our attention on homopeptides based on chiral, C^{α}-methylated α-amino acids, known to be effective helicogenic building blocks [75]. In this connection, L-(αMe)Val is an excellent right-handed helix forming residue [76], despite the fact that its unmethylated counterpart, Val, is a poor helicogenic amino acid [77–80]. We studied by CD homo-oligomers of different main-chain lengths and found that four of the peptides examined, namely Ac–[L-(αMe)Val]$_n$–OMe (n=6–8) and Ac–[L-(αMe)Val]$_7$–NH*i*Pr afford reversible, solvent-driven, 3_{10}-helix/α-helix transitions from MeOH to HFIP or HFA trihydrate [81]. Figure 13.12 shows this conformational switch for the homo-hexamer.

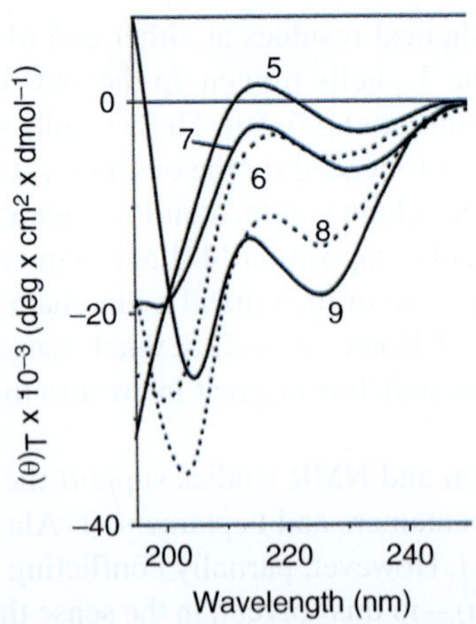

Figure 13.11 *Far-UV CD spectra of Ac–L-Ala–(Aib–L-Ala)$_n$–OMe (n=2,* ***5****; n=3,* ***7****; n=4,* ***9****) and Ac–(Aib–L-Ala)$_n$–OMe (n=3,* ***6****; n=4,* ***8****) in water at 20 °C. Adapted with permission from reference [74]. Copyright (2011) John Wiley & Sons, Ltd*

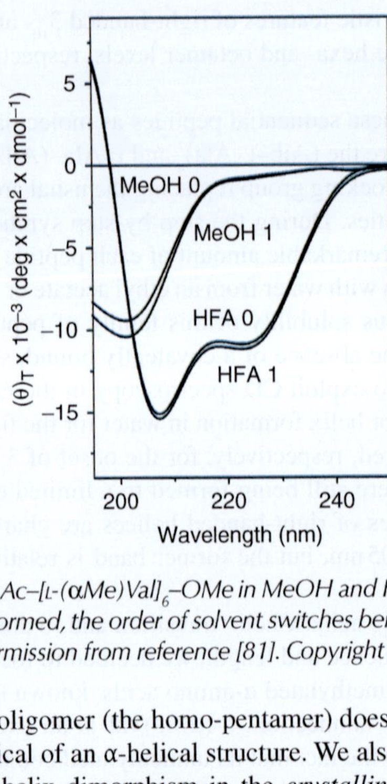

Figure 13.12 *CD spectra of Ac–[L-(αMe)Val]$_6$–OMe in MeOH and HFA solutions. Repeated cycles of helix conversion were performed, the order of solvent switches being MeOH 0, HFA 0, MeOH 1, and HFA 1. Adapted with permission from reference [81]. Copyright (2011) John Wiley & Sons, Ltd*

Interestingly, the lower oligomer (the homo-pentamer) does not give any clear-cut indication of a CD pattern typical of an α-helical structure. We also described an example of a solvent-driven 3_{10}-helix/α-helix dimorphism in the *crystalline* state [82]. The (αMe)Val homo-heptamer Ac–[L-(αMe)Val]$_7$–NH*i*Pr is completely 3_{10}-helical when its crystals are grown from an MeOH solution (Figure 13.13). By contrast, it is folded in the α-helical conformation when crystallized from HFIP, an alcohol of higher polarity.

The 3D structures of this peptide in the two crystals mirror those discussed above occurring in the two solvents. These studies from our group clearly suggest that the

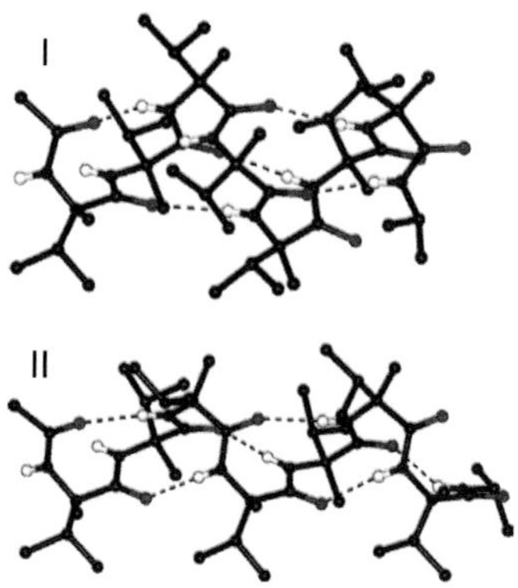

Figure 13.13 The X-ray diffraction structures of Ac–[L-(αMe)Val]$_7$–NHiPr: (I) crystals grown from an MeOH solution (molecule A in the asymmetric unit) and (II) crystals grown from an HFIP solution (molecule A in the asymmetric unit). Adapted with permission from reference [82]. Copyright (2007) American Chemical Society

interconversion between 3_{10}- and α-helices is allowed even in peptides exclusively composed of C$^{\alpha}$-tetrasubstituted α-amino acids. Remarkably, the X-ray diffraction structures of two peptides with the same $C^{\alpha,\alpha}$-di-*n*-propylglycine containing sequence (but having different N-terminal groups) are 3_{10}- and α-helical, respectively, despite being crystallized from the same solvent [83]. A limited number of earlier attempts was reported where a few peptides of markedly different lengths, each based on both C^{α}-trisubstituted and C$^{\alpha}$-tetrasubstituted α-amino acids, were shown to undergo this reversible process efficiently, induced either by heating or changing the solvent properties [84–86].

In summary, we and others developed peptides able to generate molecular spring systems upon modifying the nature of their helical structure. All peptide sequences utilized are characterized by at least one helicogenic C$^{\alpha}$-tetrasubstituted α-amino acid. It is worth pointing out that the distance between the N(1) and N(8) atoms in Ac–[L-(αMe)Val]$_7$–NH*i*Pr significantly changes from 13.58 Å, when the N$^{\alpha}$-acylated heptapeptide alkylamide adopts the more elongated 3_{10}-helix, to 10.92 Å, when it is folded in the more compressed α-helix. The Δ[N(1)···N(8)] separation is 2.66 Å (0.38 Å per residue). Interestingly, a 3_{10}-helix to α-helix conversion occurring in nature has been recently experimentally discovered. It involves the major histocompatibility complex II protein and drives the interaction of this complex with HLS-DM, the partner protein [87]. For selected theoretical studies on cooperativity in 3_{10}-/α-helix formation, modeling, topological analysis of the electron density in H-bonds, and the transition pathway between these two peptide ordered secondary structures, references [88] to [95] will be beneficial to the reader.

13.3.2 The 2.0_5-Helix/3_{10}-Helix Switch

The most recently investigated conformational conversion between ordered polypeptide secondary structures involves the 2.0_5- and 3_{10}-helices (for details on the latter conformation, see the preceding section). As already stated for the 3_{10}-helix, the 2.0_5-helix is a structure proposed a long time ago [96], which has been experimentally studied only relatively recently (Table 13.1).

The 3D structural propensities of C$^{\alpha}$-tetrasubstituted α-amino acids are quite different from 3_{10}- and α-helices, namely those overwhelmingly preferred by the Aib and (αMe)Val

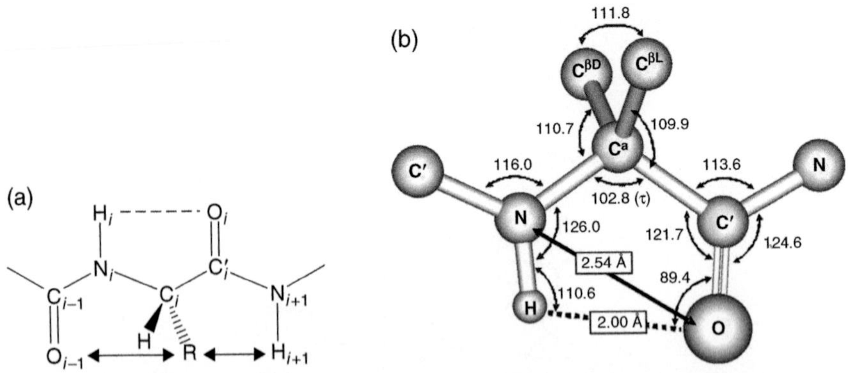

Figure 13.14 The fully extended peptide conformation. (a) chemical formula. (b) lengths and angles for covalent bonds and the H-bond from a statistical analysis

residues discussed above, when both side chains are identical, longer than methyl groups, and not interconnected in a cyclic system. In this class of residues, those examined to date in detail are Deg [46, 50, 97–107], $C^{\alpha,\alpha}$-di-*n*-propylglycine [46, 50, 108, 109], $C^{\alpha,\alpha}$-diphenylglycine [50, 110], and $C^{\alpha,\alpha}$-dibenzylglycine [50, 110]. All of them strongly favor the stabilization of the multiple, consecutive, fully extended (C_5) conformation, termed the 2.0_5-helix, which is characterized by two amino acids per turn and five atoms in the pentagonal (pseudo)ring closed by the intramolecular N–H···O=C H-bond. The repeating motif of the fully extended polypeptide conformation is the $i \rightarrow i$ intramolecularly H-bonded form (Figure 13.14a). The relative spatial disposition of the two dipoles, N_i–H_i and C'_i=O_i, is such that there is obviously some interaction between them.

The C_5 structure was taken into consideration (and often overemphasized) in a variety of conformational energy calculations [42, 97, 100–103]. Interestingly, however, it was reported that the C_5 conformation of Ac–Deg–NHMe becomes more stable than the helical structure when the critical bond angle N–C^α–C′ (τ) is lower than 107° (Figure 13.15) [97, 100]. Its earlier experimentally authenticated occurrence was suggested in noninteracting solvents of low polarity using mainly IR absorption and NMR measurements of model peptides based on protein amino acids [42, 98, 109, 111]. It was found that Gly derivatives exhibit the highest population of C_5 structure when compared to the derivatives of residues carrying a side chain.

Unambiguous validation of the presence of the C_5 form was obtained in the crystal state by the X-ray diffraction analyses of a few, favorable compounds, i.e. Gly-rich short peptides [42, 112]. In globular proteins, a repeating C_5 motif has so far only been assessed in the X-ray diffraction analysis of the –(Gly)$_4$– stretch of His-*t*RNA synthetase [113]. Moreover, a recent statistical analysis of proteins demonstrated that more than 99% of the residues populating the C_5 region (ϕ, ψ = 180°, 180°) of the Ramachandran map are Gly [114]. As determined by a wealth of X-ray diffraction analyses on peptides rich in Deg and related residues, in the intraresidue H-bonded C_5 conformation the internal τ bond angle (≈ 103°) is much smaller than the sp^3 tetrahedral angle (≈ 109°) and the N···O distance (2.54 Å) is much shorter than that commonly seen in helical peptides (2.8–3.0 Å) (Figure 13.14b).

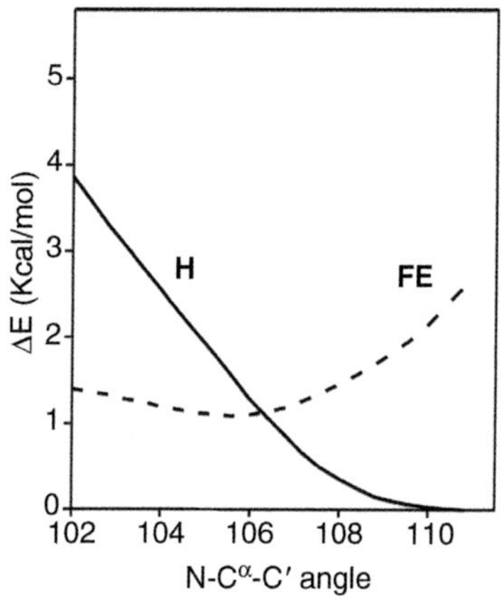

Figure 13.15 *Conformational energy calculations for Ac–Deg–NHMe in the helical (H) and fully extended (FE) conformation as a function of the N–Cα–C′ (τ) bond angle. Adapted with permission from references [97] and [100]. Copyright (1985) and (1988) John Wiley & Sons, Ltd*

On the basis of the aforementioned results, for many years the general view held by structural peptide chemists was that many *achiral* C^{α}-tetrasubstituted α-amino acids could represent useful building blocks for the production of the 2.0_5-helix structure. However, more recently, in a series of papers published by Tanaka, Suemune, and their co-workers [115–117], it was clearly shown that the essential prerequisite for the peptide C_5-structure formation is not side-chain symmetry (i.e. amino acid achirality) but rather any kind of substitution at both C^{β} atoms (i.e. even *chiral*, acyclic C^{α}-tetrasubstituted α-amino acids can be involved). In particular, their most exciting result was the X-ray diffraction structure of the 2.0_5-helical, terminally protected, homo-tetramer from (*S*)-Beg. This finding changed the picture of the conformational preferences of C^{α}-tetrasubstituted α-amino acids to a relevant extent, in the sense that it paved the way to the current belief that a peptide sequence of acyclic α-amino acid residues, each of them with both side chains longer than a methyl group, might result in a marked stabilization of the otherwise unstable 2.0_5-helix. In 2011, we extended this research to homo-peptide esters from (*S*)-Epg [118]. In agreement with the conclusions of the Japanese authors [115–117], by use of X-ray diffraction (Figure 13.16b), FT-IR absorption and NMR techniques we found that these oligomers extensively adopt the 2.0_5-helix, irrespective of their main-chain length, in the crystal state and in solvents of low polarity (Figure 13.16a also shows the X-ray diffraction structures of the 2.0_5-helix forming $(Deg)_{3\text{-}5}$ homo-oligomers).

Attempts to unambiguously define the CD spectrum characteristic of the 2.0_5-helix, a potentially very useful spectroscopic parameter for a rapid assignment of this 3D structure, are currently underway in our laboratory.

Usually, in the solution phase the determination of the conformation for homo-peptides rich in C^{α}-tetrasubstituted α-amino acids is based on the ^{1}H NMR and FT-IR absorption spectroscopies [98, 103, 105, 109, 115, 116, 118]. In particular, the dependence of amide NH proton chemical shifts on temperature and addition of external perturbing agents, e.g. paramagnetic compounds or polar solvents, highlights the solvent accessibility of these

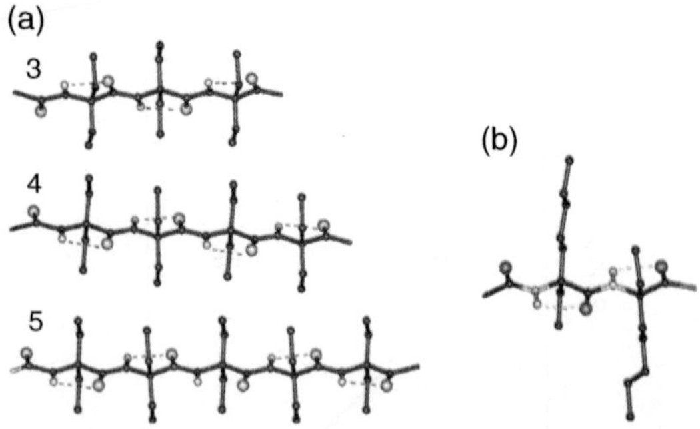

Figure 13.16 *The X-ray diffraction structures of: (a) the three N^{α}-acylated/esters from the $(Deg)_n$ (n=3, 4, 5) homo-oligomers; (b) the N^{α}-acylated ester from the $(Epg)_2$ homo-dimer. Adapted with permission from reference [118]. Copyright (2011) Springer Science*

groups, which in turn provides information on their participation in N–H···O=C H-bonding. Similarly, the frequency of the N–H stretching bands in the FT-IR absorption spectra are sensitive to this type of H-bonds. A more direct method was recently established to determine the structure of Deg homo-peptides in the solution phase [104]. It involves the study of the amide I and amide II absorptions, which are strictly related to conformational parameters such as distances and angles between different modes. In particular, Ge and collaborators determined the 3D-structure of a series of Deg homo-peptides to the pentamer level in $CDCl_3$ solution by linear and 2D FT-IR absorption spectroscopies [104]. These authors were able to characterize the amide I and II spectral signatures for the 2.0_5-helix, thus facilitating the conformational analysis of model oligopeptides that contain this particular 3D structural motif.

The 3D properties of the 2.0_5-helix discussed above make it a very appealing peptide secondary structure. However, this helix is still only poorly understood and utilized, in part because it was reported that it is constitutionally fragile [103, 105–107]. The α-amino acid residue commercially available, easiest to handle, and therefore highly recommended for building up this unusual 3D structure is Deg. Recent evidence from our and Tanaka/Suemune's laboratories indicates that the most stabilizing, terminal protecting (or blocking) groups for the peptide main chain are Tfa at the N-terminus and an alkyl ester at the C-terminus [97, 98, 103, 105]. In the Tfa-protected Deg homopeptides, the N-terminal residue is held tightly in the C_5 conformation by virtue of a stabilizing, multiple H-bonding of the F···H–N(2)···O(2)=C′(2) type. In the absence of the fluorine atoms, the same peptide with the closely related Ac group at the N-terminus is folded in the more compressed 3_{10}-helical conformation (Figure 13.17) [119].

C-terminal primary and secondary amides, at variance with a tertiary amide, are not tolerated because their unsatisfied –NH– function can provide the H-bonding donor responsible for the switch to the 3_{10}-helix (Figure 13.18) [105].

Moreover, we made an attempt to exploit the 2.0_5-helix conformation as a rigid molecular ruler using (static) fluorescence spectroscopy [107]. However, this physicochemical technique implies the introduction of aromatic moieties as a fluorophore and a quencher. Our X-ray diffraction and spectroscopic results forced us to conclude that not only in the crystal state but in chloroform as well, in the Deg homo-peptides investigated in

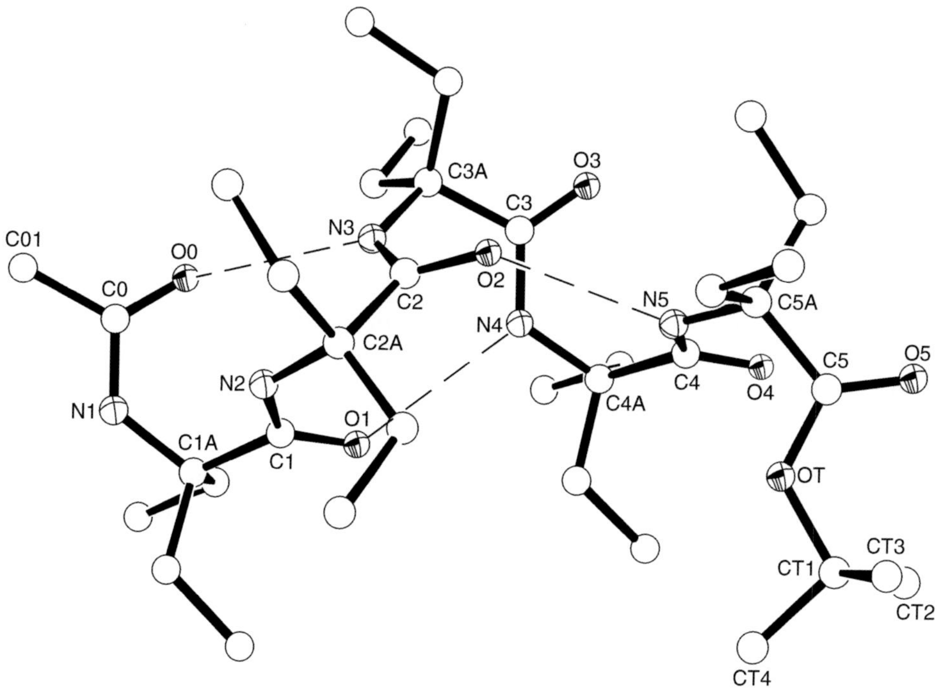

Figure 13.17 *The X-ray diffraction structure of Ac–(Deg)$_5$–OtBu [119]*

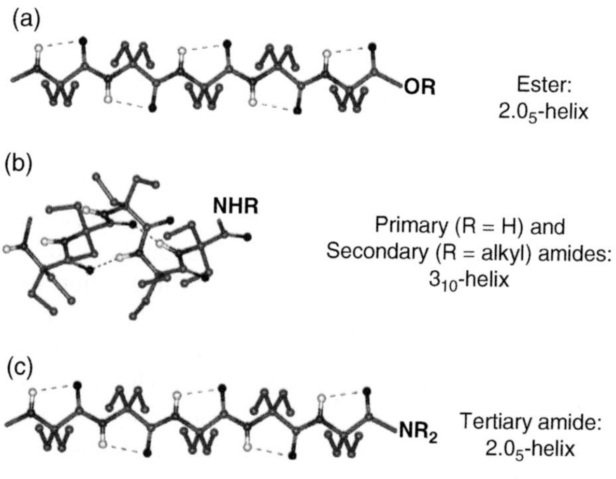

Figure 13.18 *Conformation (2.0$_5$-helical or 3$_{10}$-helical) of an N$^{\alpha}$-acylated (Deg)$_4$ (R= R=–CH$_2$–CH$_3$) homo-oligomer controlled by the chemical nature of the C-terminal moiety*

this work the 2.0$_5$-helix conformation and the 3$_{10}$-helix coexist to different extents. Clearly, terminal aromatic groups are not recommended for the fabrication of a stable 2.0$_5$-helix conformation. Also, to avoid destabilization of this conformation, the choice of solvent is critical (chloroform turned out to be the most appropriate).

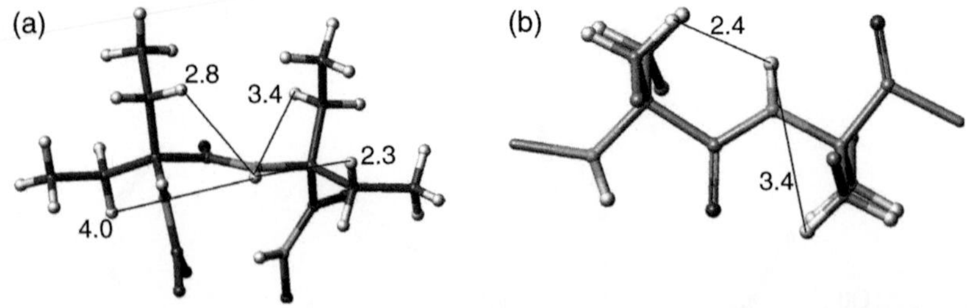

Figure 13.19 Intramolecular proton···proton distances in a Deg homo-dipeptide model: (a) right-handed 3_{10}-helical conformation (ϕ, ψ = −57°, −30°), with the pro-S and pro-R side chains in the trans and g^+ dispositions, respectively, as commonly observed in the crystal state. (b) the 2.0_5-helix conformation (ϕ, ψ = 180°, 180°), with the pro-S and pro-R side chains in the g^+ and g^- disposition, respectively. Adapted with permission from reference [106]. Copyright (2012) Elsevier

Intermolecular H-bonds arising from solvents of higher polarity [106] and packing forces occurring in crystals [103] may compete with the relatively weak intraresidue H-bond of the C_5 basic unit, thus switching the conformation to the 3_{10}-helix.

To discriminate more effectively between the 2.0_5-helix and the 3_{10}-helix conformations for the Deg homo-peptides in solution, we recently introduced a 2D NMR method [106]. It is based on the observation of divergent cross-peak intensities in the NOESY spectra. In particular, any $\beta CH_2(i\text{-}1) \rightarrow NH(i)$ cross peak is more intense than the intraresidue $\beta CH_2(i) \rightarrow NH(i)$ cross peak when the peptide adopts the 2.0_5-helix conformation. In this 3D structure a marked splitting of the chemical shifts of the two nonequivalent βCH_2 protons is also apparent. In contrast, an opposite trend of intensities of the same NOE cross peaks indicates the occurrence of a 3_{10}-helical conformation. The related intramolecular proton···proton separations in a $(Deg)_2$ peptide model in the two types of helical structures are given in Figure 13.19. This 3D structural shift is induced by a change in the nature of the solvent.

In conclusion, our and Tanaka/Suemune groups have expanded the set of available helical peptide spring systems (to the 2.0_5-helix/3_{10}-helix switch) simply by replacing the introduction of a C^α-methyl substituent, as in Aib and (αMe)Val, to that of a C^α-ethyl substituent, as in Deg, Beg, and Epg, in C^α-trisubstituted α-amino acids. The 2.0_5-helix, a unique type of 3D structure, extremely rare with protein amino acids, may represent an excellent molecular spacer in long-range probe···probe interaction studies, as it is the longest per residue conformation any peptide can adopt. Interestingly, the shortening in length of a Deg pentapeptide switching from the 2.0_5- to the 3_{10}-helix is as high as 45%. The N(1) to N(5) distance is 7.86 Å when it is folded in the shorter 3_{10}-helix, while it is as long as 14.35 Å (6.49 Å longer, 1.62 Å per residue) when it adopts the more stretched 2.0_5-helix (Figure 13.20). We envisage that this attractive 'spring-like' behavior could be exploited in further studies aimed at developing peptide-based molecular devices.

13.4 Conclusions

Switch peptides are amino acid sequences that are able to alter their conformation radically in response to variations in the surrounding environment (solvent polarity, pH, temperature, concentration) or under the action of an external stimulus. In this review article, we

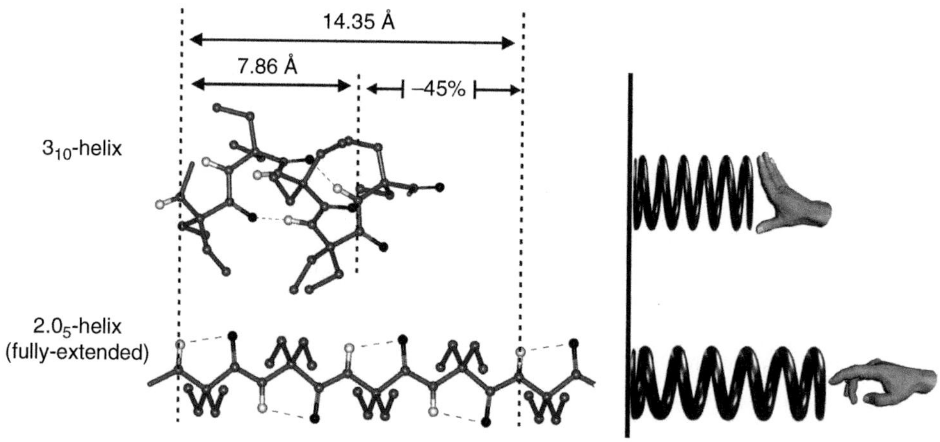

Figure 13.20 The X-ray diffraction structures of two N^{α}-acylated $(Deg)_5$ homo-oligomer esters in the 3_{10}-helix and 2.0_5-helix conformation, which underscore the dramatic change in the distance between the N(1) and N(5) atoms

focused only on conformational changes from an ordered peptide secondary structure to another (in other words, those involving unordered conformations were not taken into consideration).

Four types of such molecular switches are well documented and are discussed here. Two of them, those we termed *classical switches*, were known more than fifty years ago. The α-helix/β-pleated sheet switch implies a large, concomitant variation of both ϕ, ψ backbone torsion angles, the latter conformation being significantly more elongated. In contrast, in the type-I poly-(L-Pro)$_n$/type-II poly-(L-Pro)$_n$ switch the ϕ, ψ torsion angles vary only slightly. However, the value of the third backbone torsion angle (ω), involving the *cis*/*trans* isomerization of the peptide bond, changes dramatically, from 0° (*cis*) to 180° (*trans*) and vice versa, with a resulting, significant stretching or compression of the peptide length.

We called the two other peptide molecular switches *recently discovered*, since both deal with the 3_{10}-helix, an ordered secondary structure first authenticated experimentally only in the late 1970s [49]. The 3_{10}-helix/α-helix switch is that in which the peptide main chain undergoes the shortest elongation/reduction, in that the ϕ, ψ torsion angles of the more compressed α-helix differ by 6–12° only from those of the 3_{10}-helix (Table 13.1). The newest spring system involves the 2.0_5-helix and the 3_{10}-helix structures. Here, too, the torsion angles involved are ϕ, ψ, but the elongation to or shortening from the recently rediscovered 2.0_5-helix is radical as this 3D structure is the longest possible peptide conformation (both ϕ and ψ are in the *trans* disposition).

In conclusion, considering the rather limited number of studies in this area, there is ample space for the design of artificial molecular switches and their application to the preparation of peptide materials with novel, exciting properties. Peptide science is a typical interdisciplinary subject, which is expected to contribute heavily to narrow the gap and strengthen the ties between synthetic organic chemistry on one side and applications to nanoscience on the other side. In this connection, the unique modular properties of peptides, combined with the enormous chemical diversity of coded and noncoded amino acid building blocks, offer huge advantages to materials chemists who are limited only by their imagination in their research projects.

Finally, it is worth mentioning here that, in addition to amino acid C^{α}-alkylation (or C^{α}-arylation) extensively discussed above, another backbone chemical modification was recently investigated with the purpose of triggering a drastic conformational change in a polypeptide molecule. This noninvasive approach is the quasi-isosteric substitution of the carbonyl oxygen atom of one peptide group with a sulfur atom [120–122]. The resulting thiopeptide unit, which is predominantly in the *trans* conformation in thermal equilibrium, can be selectively excited by a (260 nm) UV pulse. Thus, a *trans* → *cis* isomerization of the ω torsion angle of the –C(=S)–NH bond will efficiently take place, which, although site specific, may result in a huge variation in the overall peptide molecular conformation and end-to-end distance. After photoisomerization, the metastable *cis* conformation in the electronic ground state relaxes back to *trans* with a time constant that ranges from a few seconds to many minutes.

References

[1] Polese, A., Mondini, S., Bianco, A., Toniolo, C., Scorrano, G., Guldi, D.M., Maggini, M. (1999) Solvent-dependent intramolecular electron transfer in a peptide-linked $[Ru(bpy)_3]^{2+}$– C_{60} dyad. J. Am. Chem. Soc. 121: 3446–3452.

[2] Feringa, B.L., Huck, N.P.M., Schoevaars, A.M. (1996) Chiroptical molecular switches. Adv. Mater. 8: 681–684.

[3] Berkovic, G., Kronganz, V., Weiss, V. (2000) Spiropyrans and spirooxazines for memories and switches. Chem. Rev. 100: 1741–1753.

[4] Pease, A.R., Jeppesen, J.O., Stoddart, J.F., Luo, Y., Collier, C.P., Heath, J.R. (2001) Switching devices based on interlocked molecules. Acc. Chem. Res. 34: 433–444.

[5] Balzani, V., Credi, A., Venturi, M. (2002) The bottom-up approach to molecular level devices and machines. Chem. Eur. J. 8: 5524–5532.

[6] Raymo, F.M., Tomasulo, M. (2005) Fluorescence modulation with photochromic switches. J. Phys. Chem. A 109: 7343–7352.

[7] Morimoto, M., Irie, M. (2005) Photochromism of diarylethene single crystals: crystal structures and photochromic performances. J. Chem. Soc., Chem. Commun. 3895–3905.

[8] Kottas, G.S., Clarke, L.I., Horinek, D., Michl, J. (2005) Artificial molecular rotors. Chem. Rev. 105: 1281–1376.

[9] Kinbara, K., Aida, T. (2005) Toward intelligent molecular machines: directed motions of biological and artificial molecules and assemblies. Chem. Rev. 105: 1377–1400.

[10] Cerpa, R., Cohen, F.E., Kuntz, I.D. (1996) Conformational switching in designed peptides: the helix/sheet transition. Folding and Design 1: 91–101.

[11] Idiris, A., Alam, M.T., Ikai, A. (2000) Spring mechanics of an α-helical polypeptide. Protein Eng. 13: 763–770.

[12] Reiersen, H., Rees, A.A. (2001) The hunchback and its neighbours: proline as an environmental modulator. Trends Biochem. Sci. 26: 679–684.

[13] Ma, K., Kan, L.-S., Wang, K. (2001) Polyproline II helix is a key structural motif of the elastic PEVK segment of titin. Biochemistry 40: 3427–3438.

[14] Pieroni, O., Fissi, A., Angelini, N., Lenci, F. (2001) Photoresponsive polypeptides. Acc. Chem. Res. 34: 9–17.

[15] Li, B., Alonso, D.O.V., Daggett, V. (2002) Stabilization of globular proteins via introduction of temperature-activated, elastin-based switches. Structure 10: 989–998.

[16] Oda, M.N., Forte, T.M., Ryan, R.O., Voss, J.C. (2003) The C-terminal domain of apolipoprotein A-I contains a lipid-sensitive conformational trigger. Nat. Struct. Biol. 10: 455–459.

[17] Gehenn, K., Pipkorn, R., Reed, J. (2004) Successful design and synthesis of a polarity-triggered $\beta \rightarrow \alpha$ conformational switch using the side-chain interaction index (SCII) as a measure of local structural stability. Biochemistry 43: 607–612.
[18] Pagel, K., Vagt, T., Köksch, B. (2005) Directing the secondary structure of polypeptides at will: from helices to amyloids and back again. Org. Biomol. Chem. 3: 3843–3850.
[19] Cerasoli, E., Sharpe, B.K., Woolfson, D.K. (2005) ZiCo: a peptide designed to switch folded state upon binding zinc. J. Am. Chem. Soc. 127: 15008–15009.
[20] Crisma, M., Formaggio, F., Moretto, A., Toniolo, C. (2006) Peptide helices based on α-amino acids. Biopolymers (Pept. Sci.) 84: 3–12.
[21] Sarkar, P.K., Doty, P. (1966) The optical rotatory properties of the β-configuration in polypeptides and proteins. Proc. Natl Acad. Sci. USA 55: 981–989.
[22] Greenfield, N., Fasman, G.D. (1969) Computed circular dichroism spectra for the evaluation of protein conformation. Biochemistry 8: 4108–4116.
[23] Fasman G.D., Chou, P.Y. (1974) Prediction of protein conformation: consequences and aspirations. In: Peptides, Polypeptides and Proteins. Blout, E.R., Bovey, F.A., Goodman, M., Lotan, N. (Eds). John Wiley & Sons, Inc., New York, pp. 114–125.
[24] Macchi, F., Hoffmann, S.V., Carlsen, M., Vad, B., Imparato, A., Rischel, C., Otzen, D.E. (2011) Mechanical stress affects glucagon fibrillation kinetics and fibril structure. Langmuir 27: 12539–12549.
[25] Buehler, M.J., Yung, Y.C. (2009) Deformation and failure of protein materials in physiologically extreme conditions and disease. Nat. Mater. 8: 175–188.
[26] Chiti, F., Dobson, C.M. (2006) Protein misfolding, functional amyloid, and human disease. Annu. Rev. Biochem. 75: 333–366.
[27] Ricchiuto, P., Brukhno, A.V., Paci, E., Auer, S. (2011) Conformation state diagram of polypeptides: a chain-length induced α–β transition. J. Chem. Phys. 135: 061101.
[28] Hoernke, M., Falenski, J.A., Schwieger, C., Koksch, B., Brezesinki, G. (2011) Triggers for β-sheet formation at the hydrophobic–hydrophilic interface: high concentration, in-plane orientational order, and metal ion complexation. Langmuir 27: 14218–14231.
[29] Rojas, A.V., Liwo, A., Scheraga, H.A. (2011) A study of the α-helical intermediate preceding the aggregation of the amino-terminal fragment of the β-amyloid peptide ($A\beta_{1\text{-}28}$). J. Phys. Chem. B 115: 12978–12983.
[30] Wang, W., Perovic, I., Chittuluru, J., Kaganovich, A., Nguyen, L.T.T., Liao, J., Auclair, J.R., Johnson, D., Landeru, A., Simorellis, A.K., Ju, S., Cookson, M.R., Asturias, F.J., Agar, J.N., Webb, B.N., Kang, C.H., Ringe, D., Petsko, G.A., Pochepsky, T.C., Hoang, Q.Q. (2011) A soluble synuclein construct forms a dynamic tetramer. Proc. Natl Acad. Sci. USA 108: 17797–17802.
[31] Ullman, O., Fisher, C.K., Stultz, C.M. (2011) Explaining the structural plasticity of α-synuclein. J. Am. Chem. Soc. 133: 19536–19546.
[32] Mandelkern, L. (1967) Poly-L-proline. In: Poly-α-Amino Acids: Protein Models for Conformational Studies. Fasman, G.D. (Ed.). Marcel Dekker, New York, pp. 675–724.
[33] Bovey, F.A., Hood, F.P. (1967) Circular dichroism of poly-L-proline. Biopolymers 5: 325–326.
[34] Pysh, E.S. (1967) The calculated ultraviolet optical properties of poly-L-proline I and II. J. Mol. Biol. 23: 587–599.
[35] Rabanal, F., Ludevid, M.D., Pons, M., Giralt, E. (1993) CD of proline-rich polypeptides: application to the study of the repetitive domain of maize glutelin-2. Biopolymers 33: 1019–1028.
[36] Okabayashi, H., Isemura, T., Sakakibara, S. (1968) Steric structure of L-proline oligopeptides. II. Far-UV absorption spectra and optical rotations of L-proline oligopeptides. Biopolymers 6: 323–330.
[37] Rothe, M., Rott, H., Mazanek, J. (1976) Solid phase synthesis and conformation of monodisperse, high molecular weight oligo-L-proline. In: Peptides 1976. Loffet, A. (Ed.). Editions de l'Université de Bruxelles, Bruxelles, Belgium, pp. 309–318.
[38] Yuge, M., Kakinoki, S., Hirano, Y., Oka, M. (2006) Conformational transition of poly(proline) in various solvents. In: Peptide Science 2006. Ishida, H., Mihara, H. (Eds). The Japanese Peptide Society, Osaka, Japan, pp. 84–85.

[39] Benedetti, E., Bavoso, A., Di Blasio, B., Pavone, V., Pedone, C., Toniolo, C., Bonora, G.M. (1983) Solid-state geometry and conformation of linear, diastereomeric oligoprolines. Biopolymers 22: 305–317.

[40] Sonar, M.V., Ganesh, K.N. (2010) Water-induced switching of β-structure to polyproline II conformation in the 4*S*-aminoproline polypeptide via H-bond rearrangement. Org. Lett. 12: 5390–5393.

[41] Venkatachalam, C.M. (1968) Stereochemical criteria for polypeptides and proteins. V. Conformation of a system of three linked peptide units. Biopolymers 6: 1425–1436.

[42] Toniolo, C. (1980) Intramolecularly hydrogen-bonded peptide conformations. CRC Crit. Rev. Biochem. 9: 1–44.

[43] Pavone, V., Gaeta, G., Lombardi, A., Nastri, F., Maglio, O., Isernia, C., Saviano, M. (1996) Discovering protein secondary structures: classification and description of isolated α-turns. Biopolymers 38: 705–721.

[44] Karle, I.L., Balaram, P. (1990) Structural characteristics of α-helical peptide molecules containing Aib residues. Biochemistry 29: 6747–6756.

[45] Toniolo, C., Benedetti, E. (1991) The polypeptide 3_{10}-helix. Trends Biochem. Sci. 16: 350–353.

[46] Benedetti, E., Di Blasio, B., Pavone, V., Pedone, C., Toniolo, C., Crisma, M. (1992) Characterization at atomic resolution of peptide helical structures. Biopolymers 32: 453–456.

[47] Millhauser, G.L. (1995) Views of helical peptides: a proposal for the position of 3_{10}-helix along the thermodynamics folding pathways. Biochemistry 34: 3873–3877.

[48] Marshall, G.R. (1971) Studies of the biologically active conformations of angiotensin. In: Intra-Science Chemistry Reports, Kharasch, N. (Ed.), Vol. 5. Gordon and Breach, New York, pp. 305–316.

[49] Shamala, N., Nagaraj, R., Balaram, P. (1978) The 3_{10}-helical conformation of a pentapeptide containing α-aminoisobutyric acid (Aib): X-ray crystal structure of Tos–$(Aib)_5$–OMe. Chem. Commun. 996–997.

[50] Toniolo, C., Crisma, M., Formaggio, F., Peggion, C. (2001) Control of peptide conformation by the Thorpe-Ingold effect (C^{α}-tetrasubstitution). Biopolymers (Pept. Sci.) 60: 396–419.

[51] Benedetti, E., Bavoso, A., Di Blasio, B., Pavone, V., Pedone, C., Crisma, M., Bonora, G.M., Toniolo, C. (1982) Solid-state and solution conformation of homo-oligo (α-aminoisobutyric acids) from tripeptide to pentapeptide: evidence for a 3_{10}-helix. J. Am. Chem. Soc. 104: 2437–2444.

[52] Pavone, V., Di Blasio, B., Santini, A., Benedetti, E., Pedone, C., Toniolo, C., Crisma, M. (1990) The longest, regular polypeptide 3_{10}-helix at atomic resolution. J. Mol. Biol. 214: 633–635.

[53] Toniolo, C., Crisma, M., Bonora, G.M., Benedetti, E., Di Blasio, B., Pavone, V., Pedone, C., Santini, A. (1991) Preferred conformation of the terminally blocked $(Aib)_{10}$ homo-oligopeptide: a long, regular 3_{10}-helix. Biopolymers 31: 129–138.

[54] Gessmann, R., Brückner, H., Petratos, K. (2003) Three complete turns of a 3_{10}-helix at atomic resolution: the crystal structure of Z–$(Aib)_{11}$–O*t*Bu. J. Pept. Sci. 9: 753–762.

[55] Toniolo, C., Bonora, G.M., Barone, V., Bavoso, A., Benedetti, E., Di Blasio, B., Grimaldi, P., Lelj, F., Pavone, V., Pedone, C. (1985) Conformation of pleionomers of α-aminoisobutyric acid. Macromolecules 18: 895–902.

[56] Pavone, V., Benedetti, E., Di Blasio, B., Pedone, C., Santini, A., Bavoso, A., Toniolo, C., Crisma, M., Sartore, L. (1990) Critical main-chain length for conformational conversion from 3_{10}-helix to α-helix in polypeptides. J. Biomol. Struct. Dyn. 7: 1321–1331.

[57] Benedetti, E., Di Blasio, B., Pavone, V., Pedone, C., Santini, A., Bavoso, A., Toniolo, C., Crisma, M., Sartore, L. (1990) Linear oligopeptides. Part 227. X-ray crystal and molecular structures of two α-helix forming (Aib–L-Ala) sequential oligopeptides, *p*BrBz–$(Aib–L-Ala)_5$–OMe and *p*BrBz–$(Aib–L-Ala)_6$–OMe. J. Chem. Soc., Perkin Trans. 2: 1829–1837.

[58] Otoda, K., Kitagawa, Y., Kimura, S., Imanishi, Y. (1993) Chain length dependent transition of 3_{10}- to α-helix of Boc–$(Ala–Aib)_n$–OMe. Biopolymers 33: 1337–1345.

[59] Kennedy, D.F., Crisma, M., Toniolo, C., Chapman, D. (1991) Studies of peptides forming 3_{10}- and α-helices and β-bend ribbon structures in organic solution and in model membranes by Fourier transform infrared spectroscopy. Biochemistry 30: 6541–6548.

[60] Vijayakumar, E.K.S., Balaram, P. (1983) Stereochemistry of α-aminoisobutyric acid peptides in solution: helical conformations of protected decapeptides with repeating Aib–L-Ala and Aib–L-Val sequences. Biopolymers 22: 2133–2140.
[61] Vijayakumar, E.K.S., Balaram, P. (1983) Solution conformations of penta- and heptapeptides containing repetitive α-aminoisobutyryl-L-alanyl and α-aminoisobutyryl-L-valyl sequences. Tetrahedron 39: 2725–2731.
[62] Mayr, W., Oekonomopulos, R., Jung, G. (1979) Synthesis and conformation of a polyoxyethylene-bound undecapeptide of the alamethicin helix and (2-methylalanyl-L-alanine)$_{1-7}$. Biopolymers 18: 425–450.
[63] Oekonomopulos, R., Jung, G. (1980) Circular dichroism and conformational analysis of the membrane-modifying peptide *N-t*-Boc–(Aib–Ala)$_5$–Gly–Ala–Aib–Pro–Ala–Aib–Aib–Glu(OBzl)–Gln–OMe with respect to alamethicin. Biopolymers 19: 203–214.
[64] Otoda, K., Kimura, S., Imanishi, Y. (1993) Orientation and aggregation of hydrophobic helical peptides in phospholipid bilayer membrane. Biochim. Biophys. Acta 1150: 1–8.
[65] Kitagawa, K., Morita, T., Kimura, S. (2005) A helical molecule that exhibits two lengths in response to an applied potential. Angew. Chem. Int. Edit. 44: 6330–6333.
[66] Ishikawa, T., Morita, T., Kimura, S. (2007) Unique helical triangle molecular geometry induced by dipole–dipole interactions. Bull. Chem. Soc. Jpn 80: 1483–1491.
[67] Kai, M., Takeda, K., Morita, T., Kimura, S. (2008) Distance dependence of long-range electron transfer through helical peptides. J. Pept. Sci. 14: 192–202.
[68] Takeda, K., Morita, T., Kimura, S. (2008) Effect of monolayer structures on long-range electron transfer in helical peptide monolayer. J. Phys. Chem. B 112: 12840–12850.
[69] Okamoto, S., Morita, T., Kimura, S. (2009) Electron transfer through a self-assembled monolayer of a double-helix peptide with linking the terminals by ferrocene. Langmuir 25: 3297–3304.
[70] Nakayama, H., Morita, T., Kimura, S. (2010) Electric field effect of helical peptide dipole in self-assembled monolayers on electronic structure of oligo(phenylene–ethynylene). J. Phys. Chem. C 114: 4669–4674.
[71] Arikuma, Y., Nakayama, H., Morita, T., Kimura, S. (2010) Electron hopping over 100Å along an α-helix. Angew. Chem. Int. Edit. 49: 1800–1804.
[72] Arikuma, Y., Nakayama, H., Morita, T., Kimura, S. (2011) Ultra-long range electron transfer through a self-assembled monolayer of gold composed of 120Å long α-helices. Langmuir 27: 1530–1535.
[73] Toniolo, C., Formaggio, F., Crisma, M., Schoemaker, H.E., Kamphuis, J. (1994) The *p*-bromobenzamido chromophore as a circular dichroic probe for the assignment of the screw sense of helical peptides. Tetrahedron: Asymmetry 5: 507–510.
[74] Longo, E., Moretto, A., Formaggio, F., Toniolo, C. (2011) The critical main-chain length for helix formation in water determined in a peptide series with alternating Aib and Ala residues exclusively and detected with ECD spectroscopy. Chirality 23: 756–760.
[75] Toniolo, C., Crisma, M., Formaggio, F., Valle, G., Cavicchioni, G., Précigoux, G., Aubry, A., Kamphuis, J. (1993) Structures of peptides from α-amino acids methylated at the α-carbon. Biopolymers 33: 1061–1072.
[76] Polese, A., Formaggio, F., Crisma, M., Valle, G., Toniolo, C., Bonora, G.M., Broxterman, Q.B., Kamphuis, J. (1996) Peptide helices as rigid molecular rulers: a conformational study of isotactic homo-peptides from C^{α}-methyl,C^{α}-isopropylglycine [L-(αMe)Val]. Chem. Eur. J. 2: 1104–1111.
[77] Toniolo, C., Bonora, G.M., Fontana, A. (1974) Three-dimensional architecture of monodisperse β-branched linear homo-oligopeptides. Int. J. Pept. Protein Res. 6: 371–380.
[78] Baron, M.H., de Lozé, C., Toniolo, C., Fasman, G.D. (1978) Structure in solution of protected homo-oligopeptides of L-valine, L-isoleucine, and L-phenylalanine: an infrared absorption study. Biopolymers 17: 2225–2239.
[79] Toniolo, C., Bonora, G.M., Mutter, M. (1979) Conformation of poly(ethyleneglycol)-bound homo-oligo-L-alanines and L-valines in aqueous solution. J. Am. Chem. Soc. 101: 450–454.
[80] Moretto, V., Crisma, M., Bonora, G.M., Toniolo, C., Balaram, H., Balaram, P. (1989) Comparison of the effect of five guest residues on the β-sheet conformation of host (L-Val)$_n$ oligopeptides. Macromolecules 22: 2939–2944.

[81] Moretto, A., Formaggio, F., Kaptein, B., Broxterman, Q.B., Wu, L., Keiderling, T.A., Toniolo, C. (2008) First homo-peptides undergoing a reversible 3_{10}-helix/α-helix transition: critical main-chain length. Biopolymers (Pept. Sci.) 90: 567–574.

[82] Crisma, M., Saviano, M., Moretto, A., Broxterman, Q.B., Kaptein, B., Toniolo, C. (2007) Peptide $\alpha/3_{10}$-helix dimorphism in the crystal state. J. Am. Chem. Soc. 129: 15471–15473.

[83] Vijayalakshmi, S., Balaji Rao, R., Karle, I.L., Balaram, P. (2000) Comparison of helix-stabilizing effects of α,α-dialkyl glycines with linear and cycloalkyl side chains. Biopolymers 53: 84–98.

[84] Hungerford, G., Martinez-Insua, M., Birch, D.J.S., Moore, B.D. (1996) A reversible transition between an α-helix and a 3_{10}-helix in a fluorescence-labeled peptide. Angew. Chem. Int. Edit. Engl. 35: 326–329.

[85] Pengo, P., Pasquato, L., Moro, S., Brigo, A., Fogolari, F., Broxterman, Q.B., Kaptein, B., Scrimin, P. (2003) Quantitative correlation of solvent polarity with the α-/3_{10}-helix equilibrium: a heptapeptide behaves as a solvent-driven molecular spring. Angew. Chem. Int. Edit. 42: 3388–3392.

[86] Bellanda, M., Mammi, S., Geremia, S., Demitri, N., Randaccio, L., Broxterman, Q.B., Kaptein, B., Pengo, P., Pasquato, L., Scrimin, P. (2007) Solvent polarity controls the helical conformation of short peptide rich in C^{α}-tetrasubstituted amino acids. Chem. Eur. J. 13: 407–416.

[87] Painter, C.A., Negroni, M.P., Kellersberger, K.A., Zavala-Ruiz, Z., Evans, J.E., Stern, L.J. (2011) Conformational lability in the class II MHC 3_{10}-helix and adjacent extended strand dictate HLA-DM susceptibility and peptide exchange. Proc. Natl Acad. Sci. USA 108: 19329–19334.

[88] Basu, G., Kuki, A. (1992) Conformational preferences of oligopeptides rich in α-aminoisobutyric acid. II. A model for 3_{10}-/α-helix transition with composition and sequence sensitivity. Biopolymers 32: 61–71.

[89] Huston, S.E., Marshall, G.R. (1994) $\alpha/3_{10}$-Helix transition in α-methylalanine homo-peptides: conformational transition pathway and potential of mean force. Biopolymers 34: 75–90.

[90] Zhang, L., Hermans, J. (1994) 3_{10}-Helix versus α-helix: a molecular and dynamics study of conformational preferences of Aib and alanine. J. Am. Chem. Soc. 116: 11915–11921.

[91] Sheinerman, F.B., Brooks, C.L. (1995) 3_{10}-Helices in peptides and proteins as studied by modified Zimm–Bragg theory. J. Am. Chem. Soc. 117: 10098–10103.

[92] Rohl, C.A., Doig, A.J. (1996) Models for the 3_{10}-helix/coil, π-helix/coil, and α-helix/3_{10}-helix/coil transitions in isolated peptides. Protein Sci. 5: 1687–1696.

[93] Topol, I.A., Burt, S.K., Deretey, E., Tang, T.-H., Perczel, A., Rashin, A., Csizmadia, I.G. (2001) α-/3_{10}-Helix interconversion: a quantum-chemical study on polyalanine systems in the gas phase and in aqueous solvent. J. Am. Chem. Soc. 123: 6054–6060.

[94] Wu, Y.-D., Zhao, Y.-L. (2001) A theoretical study on the origin of cooperativity in the formation of 3_{10}- and α-helices. J. Am. Chem. Soc. 123: 5313–5319.

[95] Liebschner, D., Jelsch, C., Espinosa, E., Lecomte, C., Chabrière, E., Guillot, B. (2011) Topological analysis of hydrogen bonds and weak interactions in protein helices via transferred experimental charge density parameters. J. Phys. Chem. A 115: 12895–12904.

[96] Pauling, L., Corey, R.B. (1951) Configuration of polypeptide chains with favored orientations around single bonds: two new pleated sheets. Proc. Natl Acad. Sci. USA 37: 729–740.

[97] Benedetti, E., Barone, V., Bavoso, A., Di Blasio, B., Lelj, F., Pavone, V., Pedone, C., Bonora, G.M., Toniolo, C., Leplawy, M.T., Kaczmarek, K., Redlinski, A. (1988) Structural versatility of peptides from $C^{\alpha,\alpha}$-dialkylated glycines. I. A conformational energy computation and X-ray diffraction study of homo-peptides from $C^{\alpha,\alpha}$-diethylglycine. Biopolymers 27: 357–371.

[98] Toniolo, C., Bonora, G.M., Bavoso, A., Benedetti, E., Di Blasio, B., Pavone, V., Pedone, C., Barone, V., Lelj, F., Leplawy, M.T., Kaczmarek, K., Redlinski, A. (1988) Structural versatility of peptides from $C^{\alpha,\alpha}$-dialkylated glycines. II. An IR absorption and ^{1}H NMR study of homo-peptides from $C^{\alpha,\alpha}$-diethylglycine. Biopolymers 27: 373–379.

[99] Toniolo, C., Benedetti, E. (1991) The fully extended polypeptide conformation. In: Molecular Conformation and Biological Interactions. Balaram, P., Ramaseshan, S. (Eds). Indian Academy of Sciences, Bangalore, India, pp. 511–521.

[100] Barone, V., Lelj, F., Bavoso, A., Di Blasio, B., Grimaldi, P., Pavone, V., Pedone, C. (1985) Conformational behavior of α,α-dialkylated peptides. Biopolymers 24: 1759–1767.
[101] Ramnarayan, K., Chan, M.F., Balaji, V.N., Profeta Jr, S., Rao, S.N. (1995) Conformational studies on model dipeptides of Gly, L-Ala and their C^{α}-substituted analogs. Int. J. Pept. Protein Res. 45: 366–376.
[102] Cirilli, M., Coiro, V.M., Di Nola, A., Mazza, F. (1998) Relationship between conformation and geometry as evidenced by molecular dynamics simulation of $C^{\alpha,\alpha}$-dialkylated glycines. Biopolymers 46: 239–244.
[103] Tanaka, M., Imawaka, N., Kurihara, M., Suemune, H. (1999) Helical versus planar conformation of homo-oligopeptides prepared from diethylglycine (= 2-amino-2-ethylbutanoic acid). Helv. Chim. Acta 82: 494–510.
[104] Maekawa, H., Ballano, G., Toniolo, C., Ge, N.-H. (2011) Linear and two-dimensional infrared spectroscopic study of the amide I and II modes in fully-extended peptide chains. J. Phys. Chem. B 115: 5168–5182.
[105] Formaggio, F., Crisma, M., Peggion, C., Moretto, A., Venanzi, M., Toniolo, C. (2012) Looking for a robust, fully-extended (2.0_5-helical) peptide structure. Effect of terminal groups. Eur. J. Org. Chem. 2012: 167–174.
[106] Peggion, C., Crisma, M., Toniolo, C., Formaggio, F. (2012) A solvent-dependent peptide spring unreaveled by 2D-NMR. Tetrahedron 68: 4429–4433.
[107] Formaggio, F., Crisma, M., Ballano, G., Peggion, C., Venanzi, M., Toniolo, C. (2012) Novel peptide foldameric motifs: a step forward in our understanding of the fully-extended conformation/3_{10}-helix coexistence. Org. Biomol. Chem. 10: 2413–2421.
[108] Benedetti, E., Toniolo, C., Hardy, P., Barone, V., Bavoso, A., Di Blasio, B., Grimaldi, P., Lelj, F., Pavone, V., Pedone, C., Bonora, G.M., Lingham, I. (1984) Folded and extended structures of homo-oligopeptides from α,α-dialkylated glycines. A conformational energy computation and X-ray diffraction study. J. Am. Chem. Soc. 106: 8146–8152.
[109] Bonora, G.M., Toniolo, C., Di Blasio, B., Pavone, V., Pedone, C., Benedetti, E., Lingham, I., Hardy, P. (1984) Folded and extended structures of homo-oligopeptides from α,α-dialkylated α-amino acids. An infrared absorption and ^{1}H nuclear magnetic resonance study. J. Am. Chem. Soc. 106: 8152–8156.
[110] Crisma, M., Valle, G., Bonora, G.M., Toniolo, C., Lelj, F., Barone, V., Fraternali, F., Hardy, P.M., Maia, H.L.S. (1991) Preferred conformation of peptides from $C^{\alpha,\alpha}$-symmetrically disubstituted glycines: aromatic residues. Biopolymers 31: 637–641.
[111] Cung, M.T., Marraud, M., Néel, J. (1972) Étude expérimentale de la conformation dipeptidique. Comparaison avec les previsions théoriques. Ann. Chim. (Paris) 7: 183–209.
[112] Birkedal, H., Schwarzenbach, D., Pattison, P. (2002) A fully-extended tetrapeptide consisting of natural amino acids. Chem. Commun. 2812–2813.
[113] Åberg, A., Yaremchuk, A., Tukalo, M., Rasmussen, B., Cusack, S. (1997) Crystal structure analysis of the activation of histidine by *Thermus thermophilus* histidyl-tRNA synthetase. Biochemistry 36: 3084–3094.
[114] Hovmöller, S., Zhou, T., Ohlson, T. (2002) Conformation of amino acids in proteins. Acta Crystallogr. D 58: 768–776.
[115] Imawaka, N., Tanaka, M., Suemune, H. (2000) The first fully planar C_5 conformation of homo-oligopeptides prepared from a chiral α-ethylated α,α-disubstituted amino acid: (*S*)-butylethylglycine (= (2*S*)-2-amino-2-ethylhexanoic acid). Helv. Chim. Acta, 83: 2823–2825.
[116] Oba, M., Tanaka, M., Kurihara, M., Suemune, H. (2002) Conformation of peptides containing a chiral α-ethylated α,α-disubstituted α-amino acid: (*S*)-α-ethylleucine (=(2*S*)-2-amino-2-ethyl-4-methylpentanoic acid) within sequences of dimethylglycine and diethylglycine residues. Helv. Chim. Acta 85: 3197–3217.
[117] Tanaka, M. (2007) Design and synthesis of chiral α,α-disubstituted amino acids and conformational study of their oligopeptides. Chem. Pharm. Bull. (Tokyo) 55: 349–358.
[118] Crisma, M., Moretto, A., Peggion, C., Panella, L., Kaptein, B., Broxterman, Q.B., Formaggio, F., Toniolo, C. (2011) Chiral, fully-extended helical peptides. Amino Acids 41: 629–641.
[119] Crisma, M., Formaggio F., Toniolo, C. Unpublished results.

[120] Wildemann, D.H., Schiene-Fischer, C., Aumüller, T., Bachmann, A., Kiefhaber, T., Lücke, C., Fischer, G. (2007) A new isosteric photosensitive amide-backbone substitution allows enzyme activity switching in Ribonuclease S. J. Am. Chem. Soc. 129: 4910–4918.
[121] Huang, Y., Cong, Z., Yang, L., Dong, S. (2008) A photoswitchable thioxopeptide bond facilitates the conformation-activity correlation study of insect kinin. J. Pept. Sci. 14: 1062–1068.
[122] Bregy, H., Heimgartner, H., Helbing, J. (2009) A time-resolved spectroscopic comparison of the photoisomerization of small β-turn forming thioxopeptides. J. Phys. Chem. B 113: 1756–1762.

14

Peptide Nanostructured Conjugates for Therapeutics: The Example of P140 Peptide for the Treatment of Systemic Lupus Erythematosus

Yves Frère[1], *Louis Danicher*[1] *and Sylviane Muller*[2]

[1] *CNRS UPR22, Institut Charles Sadron, France*

[2] *CNRS UPR9021, Institut de Biologie Moléculaire et Cellulaire, Immunologie et Chimie Thérapeutiques, France*

Acronyms

AFM	Atomic force microscopy
ALG	Alginate
APL	Altered peptide ligand
APCs	Antigen-presenting cells
CFA	Complete Freund's adjuvant
CS	Chitosan
CT	Cholera toxin
EE	Encapsulation efficiency
FDA	Food and Drug Administration
HA	Hyaluronic acid
LCs	Langerhans cells
LPS	Lipopolysaccharide

Peptide Materials: From Nanostructures to Applications, First Edition. Edited by Carlos Alemán, Alberto Bianco and Mariano Venanzi.

LNCs	Lymph node cells
LT	Heat-labile enterotoxin
MAP	Multiple antigen peptide
MHCII	Major histocompatibility complex class II molecules
PCA	Poly(alkylcyanoacrylate)
PCL	Poly(ε-caprolactone)
PLGA	Poly(lactide-co-glycolide)
s.c.	Subcutaneous
SLE	Systemic lupus erythematosus
SOC	Sequential oligopeptide carrier
TASP	Template assembled synthetic protein
TCR	T cell receptor
TEM	Transmission electron microscopy
TLR	Toll-like receptor

14.1 Introduction

The introduction in 1963 of solid-phase peptide synthesis by Bruce Merrifield has revolutionized the synthesis of peptides and allowed the rapid production of synthetic antigens, biologically active peptides, artificial proteins, active enzymes, and peptide libraries. The therapeutic peptides market emerged in the 1970s, when Lypressin, a vasopressin analogue was launched. Some 40 peptide drugs are currently on the market and hundreds are in the pipeline [1, 2]. Therapeutic peptides increase in length and complexity, most of them having more than 30 residues in their sequences. The development of higher-quality resins for solid-phase synthesis and price reductions for reagents and amino acid derivatives used in peptide synthesis contribute to lowering the cost of synthesizing such peptides under Good Manufacturing Practice conditions.

Beside numerous advantages peptides have over other drug candidates such as antibodies and fusion proteins (higher activity per unit mass, increased selectivity and specificity, greater stability at storage, weaker immunogenicity, better organ or tumor penetration), some limitations require attention. These limitations concern in particular their bioavailability and biodistribution. The extent of these drawbacks will depend on their sequence and structure. Some peptides display a rapid hepatic or renal clearance; some are rapidly degraded by proteolytic enzymes. A number of original strategies have been developed to overcome these limitations. Introducing nonnatural residues or modifying peptide bonds in generating pseudopeptides has been exploited with success to increase peptide stability without altering their bioactivity [3, 4]. Lipopolysaccharides, fatty acids, and polyethylene glycol conjugates were used to enhance their bioavailability and membrane permeability. The intracellular delivery of peptides using membrane permeable carrier peptide vectors also showed some promise. Finally, their formulation with nanocarriers, such as liposomes, nanoparticles, or nanocapsules, was found to be advantageous.

The route of peptide administration is also central. The way a peptide is given can dramatically modify its bioactivity. Among noninvasive delivery routes, oral delivery of therapeutic peptides remains problematic. Except for cyclosporine, very limited success has been achieved using this route of administration, the pulmonary and transcutaneous routes providing so far the most encouraging data for noninvasive peptide delivery [5, 6].

14.2 Noninvasive Routes of Peptide Administration

14.2.1 The Transcutaneous Route

The skin is the outermost layer of the body, which acts as a mechanical, physicochemical, and immunological barrier. Like the mucosa, the skin-associated lymphoid tissue is rich in antigen-presenting cells (APCs) and functions as an immune barrier. The Langerhans cells (LCs) of the epidermis are powerful APCs that can initiate immune responses by picking up the antigen, processing it, and presenting it to naive T cells in the draining lymph nodes. The keratinocytes of the epidermis are also involved in the structuration of the skin, and are immunologically active cells by producing proinflammatory cytokines like IL-1, TNF-α in response to danger signals, and cutaneous barrier perturbation. This leads to the activation, maturation, and migration of the LCs, and attraction of effector cells at the site of inflammation. As the cutaneous microenvironment allows efficient uptake of antigen by APCs, this route appears to be advantageous for vaccination against infectious agents and potentially also for administrating therapeutics such as therapeutic peptides [6].

Peptides that normally have low molecular weight can easily bridge the skin barrier. Thus, it has been demonstrated that peptide immunogens representing T-helper cell epitopes from influenza virus haemaglutinin or tetanus toxoid can elicit peptide- and antigen-specific $CD4^+$ T cells when they are co-applied onto bare skin with ADP-ribosylating exotoxins, cholera toxin (CT), and heat-labile enterotoxin (LT) of *Escherichia coli* as adjuvants [7]. This is particularly important in the context of vaccine design and delivery since $CD4^+$ T cells produce cytokines that have a direct effect on viruses, help B cells to produce neutralizing and protective antivirus antibodies, and enhance the magnitude of cytotoxic T cell responses to clear virus-infected cells. Safe application of a patch containing LT to humans results in robust LT-antibody responses [8], showing that this route of administration is feasible for human immunization. Apart from CT, LT, and effective LT mutants, several other adjuvants have been tested in experimental models for skin immunization, such as oligodinucleotides containing CpG motifs, lipopolysaccharide (LPS), muramyl dipeptide, alum, IL-2, and IL-12 [7]. Among the Toll-like receptor (TLR) agonists that were evaluated, Imiquimod, which activates immune cells through the TLR7, associated to synthetic peptides, was found to generate strong T cell responses that exhibit effective antitumor effects in a murine melanoma model system [9].

14.2.2 The Transmucosal Routes for Peptide Delivery

The successful delivery of drugs across the oral, nasal, and ocular mucosa represents a continuing challenge [10]. These mucosa are attractive sites for drug delivery thanks to the ease of administration.

Buccal and sublingual [11], nasal [12], and ocular [13] routes are tempting since the enzymatic activity is relatively reduced and the active molecules are only transiently subjected to degradation. However, the surface of absorption is relatively small and bioactive molecules remain during a short time only in contact. The pulmonary route offers a vast surface of contact ($75\,m^2$) thanks to air cells [14]. The walls of these cells are fine and well vascularized, allowing the therapeutic effect to be fast. However, the effective absorbed dose is difficult to control and can vary from a drug-taking one to another one.

The rectal route is also interesting since the human rectum represents a body cavity in which drugs can be easily introduced and retained. Epithelial cells of the colon present a brushy border, which contains a small quantity only of proteolytic enzymes [15]. As the epithelium is not very permeable to peptides and proteins, this route requires the use of absorption promoters. Drawbacks of rectal drug administration, however, include the interruption of absorption by defecation and lack of patient acceptability.

The oral route is the route of administration that is the most physiological, the most comfortable, and the best accepted by patients. However, active molecules administered orally have only a very weak bioavailability (in the order of 3%) [16]. Until now, solutions allowing increasing this bioavailability to a satisfactory value are weak. Use of the oral route for the administration of peptide or protein drugs thus remains a challenge of the twenty-first century.

14.2.3 The Oral Route

14.2.3.1 The Difficulties Met with Oral Administration

Oral administration obliges peptides or proteins to cross two barriers and meet adverse conditions: first the gastrointestinal tract, by the pH variations [17] and the presence of proteolytic enzymes [18], then the intestinal epithelium crossing, by its hydrophobicity and its impermeability to high molar mass molecules [19]. The development of a peptide or a protein oral formulation requires finding a solution for each of these two natural barriers.

Gastrointestinal Tract Barrier. Peptides and proteins administered via the oral route meet important pH variations during their passing through the gastrointestinal tract. Once ingested, they are confronted with the very acidic environment of the stomach (pH 2). In acid media, the majority of peptides and proteins precipitates and can suffer from a limited hydrolysis [20]. Generally a peptide or a protein does not cross the chemical barrier of the stomach.

The presence of numerous proteolytic enzymes in the digestive tract leads also to an important degradation of proteins and peptides. According to the enzymes, peptides and proteins meet, they are split in various sites, and it is in the form of amino acid residues or small peptides that they are absorbed at the level of the intestinal epithelium. After endocytosis, small peptides are hydrolyzed within enterocytes and finally amino acid residues only reach the blood circulation and are recycled [21].

Intestinal Epithelial Barrier. The small intestine, which represents the widest interface between the outside environment and the body (more than 200 m^2), is the target of choice for drugs liberation [22]. The role of intestinal epithelium is to absorb nutrients while it prevents the penetration of antigens and other micro-organisms (viruses, toxin, and allergens) into the blood circulation. It is formed by a waterproof monolayer of various cells that remain joined by tightly closed junctions (Figure 14.1).

These junctions are dynamic and complex structures, which regulate the permeability of the epithelium. This layer of epithelial cells recovers a segment of intestine, which is rich in blood and lymphatic vessels. Enterocytes, the most numerous cells of the intestine, act by absorbing nutriments and assuring them passage through in the blood circulation. Caliciform cells that accumulate mucinogen-containing secretory granules at their apical

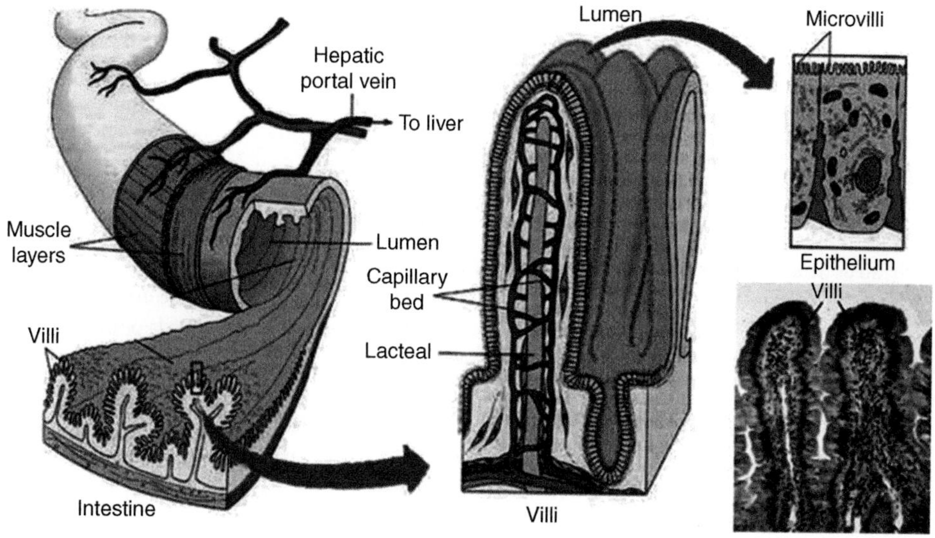

Figure 14.1 Schema of intestinal epithelium, villi, and microvilli. From http://www.daviddarling.info/encyclopedia/S/small_intestine.html (see color plate figure)

end secrete this mucus, a viscoelastic gel present on the surface of epithelium that protects cells from luminal contents (bacteria, pathogenic agents) and facilitates the adsorption of nutriments. M cells (or microfold cells) are found in the follicle-associated epithelium of Peyer's patches. Smaller and far less abundant than enterocytes, these cells, which possess irregular microvillosities and do not secrete mucus or digestive enzymes, are readily accessible to antigens and transport them towards underlying lymphoid follicles. They have a central role in mucosal immunity.

14.2.3.2 Bioavailability Improvement of Peptides or Proteins Administered by the Oral Route

To improve the peptide or protein bioavailability, it is necessary either to decrease the proteolytic activity in the gastrointestinal tract by using protease inhibitors [23] or to increase the intestine permeability by using absorption promoters [24].

Protease Inhibitors. To decrease proteolytic activity, active molecules can be co-administered with enzyme inhibitors. To be effective, the latter has to be chosen according to the peptide or protein structure that needs to be protected and in sufficient amount that has to be calibrated.

Absorption Promoters. Their role is to increase intestine permeability either by making permeable the lipid membranes of epithelial cells or by disrupting the tight junctions. By modulating this permeability, it is possible to increase the paracellular transport of active molecules. These adsorption promoters can be fatty acid, chelators [24], or surfactants [20].

Alternative options consist of increasing the stability of peptides (or proteins) by chemical modification [25] or of protecting them, for example, by encapsulation [26].

Chemical Modification. An abundant literature describes the chemical synthesis of peptide analogues resistant to degradation by proteases present in the digestive tract. A first approach consists in coupling hydrophobic groups or macromolecular vectors to peptides. These changes alter the lipophilic character of bioactive molecules, their charge, molar mass, solubility, conformation, or isoelectric point, leading to an improvement of their stability and absorption. A second approach aims at synthesizing peptide analogues by replacing natural –CO–NH– peptide bond(s) by more stable bond(s) (e.g. reduced peptide bonds $-CH_2-NH-$, –NH–CO– peptide bonds in retro-inverso peptide analogues). Such structural modifications (changes in flexibility and hydrogen-bonding character) can deeply influence the resistance to enzymatic degradation of bioactive molecules and affect their biological activity. The challenge is therefore to generate pseudopeptides that retain or enhance the specificity and/or potency of nominal peptides. Multipresentation of bioactive peptides, for example as dendrimers, multiple antigen peptide (MAP), template assembled synthetic protein (TASP), or sequential oligopeptide carrier (SOC), is also a very promising way to improve their stability and potency [3]. Biologically active peptides can also be linked to an inert surface, such as carbon nanotubes [27]. This kind of approach is being actively investigated.

Encapsulation. Protection of biologically active peptides within capsules or particles brings a global solution to problems raised by oral administration. These vectors protect peptides from chemical and biochemical degradation that occurs along the gastrointestinal tract. With adequate characteristics in terms of size, charge, or hydrophobicity, they can facilitate their transport through biological barriers as intestinal epithelium. This aspect will be described in further detail in the following paragraphs.

14.3 Encapsulation of Peptides and Proteins for Oral Delivery

With the aim of serving as pharmaceutical vectors, numerous galenic forms were developed. They are classified into two main categories, which are lipidic and polymeric vectors.

14.3.1 Lipidic Vectors

Lipidic vectors (Figure 14.2) are mainly liposomes [28], solid lipidic nanoparticles [29], and multiple emulsions [30].

14.3.1.1 Liposomes

The purpose of using liposomes is to increase the biological efficacy of certain drugs and to decrease their toxicity. Liposomes are spherical vesicles of variable size (10 to 1000 nm) constituted by phospholipidic bilayers, which are separated by aqueous compartments. Their mono- or multilamellar structures allow them to transport a wide range of active molecules, a hydrophilic one in the aqueous compartments and a lipophilic one in the heart of phospholipidic bilayers. Liposomes present a weak efficiency of encapsulation (in the order of 10%) and short lifetimes in the gastrointestinal tract and peripheral blood. Today, however, a few formulations only have obtained marketing authorization for human administration.

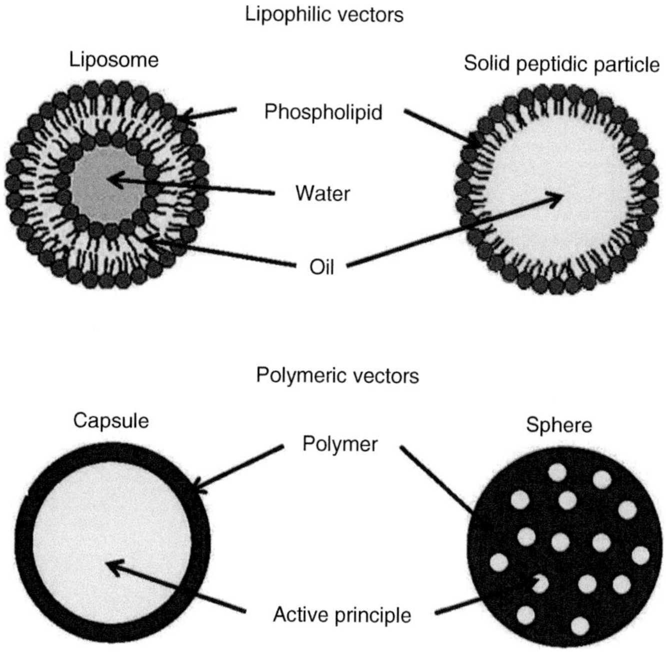

Figure 14.2 Schematic representation of lipidic and polymeric vectors (see color plate figure)

14.3.1.2 Solid Lipidic Nanoparticles

These are nanoparticles with an oily heart and with a size between 50 and 1000 nm. As lipid degradation in the gastrointestinal tract is fast and encapsulation of hydrophilic active molecules such as peptides and proteins are not possible, they are not adapted to oral administration.

14.3.1.3 Multiple Emulsions

These emulsions, in which one of two phases are dispersed in a third phase, are either water in oil in water (w/o/w) or oil in water in oil (o/w/o). Liberation of the bioactive molecule is made by diffusion through the lipidic phase and at the same time by digestion of this phase. The bioactive molecule is dissolved in micelles and vesicles stemming from products of lipidic phase degradation and then absorbed at the level of the intestinal membrane.

These three types of lipidic vectors present the same drawback, namely a weak stability in the gastrointestinal tract.

14.3.2 Polymeric Vectors

Vectors based on biocompatible, biodegradable and nontoxic polymers are much more stable. According to their size their appellation changes; they are either microparticles (1 to 1000 μm) or nanoparticles (1 to 1000 nm). Their structure depends on their formation method and on the active molecule solubility. They are under the shape of either a

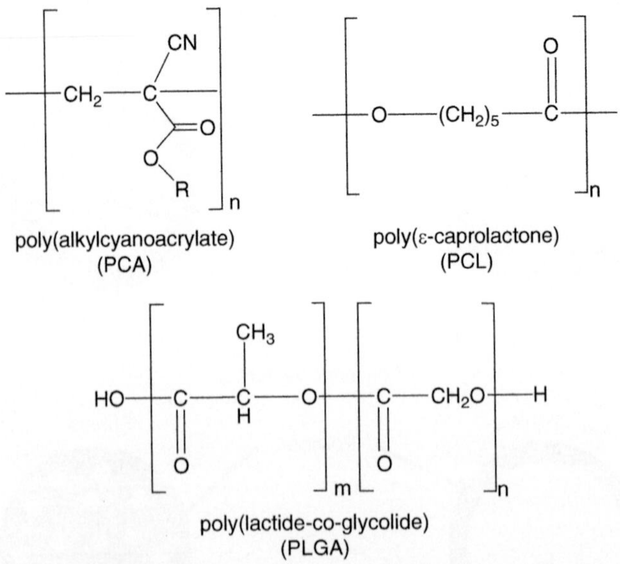

Figure 14.3 *Formula of synthetic polymers*

sphere, a continuous matrix of polymer in which the active molecule is uniformly scattered, or a capsule, i.e. a tank system with an aqueous heart (hydrophilic active molecule) or an oily one (hydrophobic active molecule) surrounded by a polymer membrane (Figure 14.2). In both cases, polymers arise either from the chemical synthesis or from a natural origin.

14.3.2.1 Synthetic Polymers

To improve peptide or protein bioavailability when they are administered by the oral route, numerous synthetic polymers were developed. The most widely studied polymers are poly(lactide-co-glycolide) (PLGA), poly(ε-caprolactone) (PCL), and poly(alkylcyanoacrylate) (PCA) (Figure 14.3).

PLGA. As the human use of PLGA is approved by the Food and Drug Administration (FDA), it is largely employed for peptide or protein vectorization. In the biomedical field, it is present in absorbable sutures, osseous implants, or grafting material for artificial organs. The auto-catalyzed hydrolysis of ester bonds transforms PLGA into lactic acid and glycolic acid. In the human body, these compounds are eliminated via the Krebs cycle [31]. The degradation kinetics depend on copolymer composition – the more the proportion of poly(glycolic)acid is weak, the more the degradation of PLGA is slow.

PLGA vectors are obtained either under the form of spheres by emulsion/evaporation of solvent (lipophilic active molecules) or under the form of capsules by double emulsion/ evaporation of solvent (hydrophilic active molecules) [32]. If PLGA vectors increase the bioavailability of the encapsulated active molecules and prolong the duration of their action in the body, they present numerous disadvantages. They are very hydrophobic, immunogenic, and can produce nonspecific inflammatory responses. The conditions of nanocapsule synthesis are drastic: the emulsions are carried out by homogenization using a stirrer at either high speed or using ultrasound. For an *in vivo* application, the presence of organic solvents raises toxicity problems. The degradation kinetics of PLGA is slow (two weeks to some months) and polymer can accumulate in the body while repeated administrations are given. Furthermore, the acidic character of degradation compounds can destabilize encapsulated peptides or proteins.

PCL. The ester functions of this semi-crystalline aliphatic polyester are degraded by hydrolysis in physiological media. If, contrary to PLGA, degradation products are not acidic, degradation kinetics however are slower and can extend over more than one year. PCL is more interesting for drug delivery in the long term than for repeated administrations of an active molecule that displays a short release time.

Numerous techniques were developed to obtain particles [33]. The emulsion oil in water or the double emulsion (water in oil) in water [34] allows encapsulation of hydrophilic or lipophilic bioactive molecules with a good efficacy. Nanospheres obtained by nanoprecipitation have a size comprising between 200 and 280 nm [35].

PCA. Initially used in surgery as tissue glue, PCA is also used to generate biodegradable nanoparticles [36]. PCA is hydrolyzed under the effect of esterases. The lateral chains of polymer are transformed into soluble poly(cyanoacrylic) acid. In physiological media, degradation occurs in some hours and depends on the length of the lateral alkyl chain: the longer this one is, the slower the degradation kinetics.

14.3.2.2 Natural Polymers

Natural proteins, as gelatin [37], albumin [38], or collagen [39], are widely used for the formulation of pharmaceutical vectors. However, they are sensitive to enzymatic degradation and cannot ensure the protection of bioactive peptides and proteins in the digestive tract. Polysaccharides, which are semi-flexible hydrophilic polymers often having the capacity to form hydrogels, are also used for the vectorization of bioactive molecules. The most frequently studied polysaccharides are chitosan (CS), alginate (ALG), and hyaluronic acid (HA) (Figure 14.4).

CS. Chitin, the polymer that is the second most abundant in the natural world after cellulose, is present in invertebrates, shellfish shells, and cellular membranes in the majority of mushrooms and algae or in certain yeasts and bacteria. Its structure, formed by a chain of residues, *N*-acetyl-β-D-glucosamine bounded by glycosidic connections, is similar to the cellulose structure, except that acetamido groups replace the hydroxyl groups situated in the C2 position. Because of its weak solubility in a large number of solvents, numerous soluble derivatives of chitin were synthesized. The most known and used derivative is CS, which is obtained by partial or total deacetylation of chitin in alkaline media.

CS is formed by a statistical alternation of *N*-acetyl-D-glucosamine and D-glucosamine units. It is characterized by its degree of acetylation, which represents the fraction of *N*-acetyl-D-glucosamine units with regard to the total number of *N*-acetyl-D-glucosamine and D-glucosamine units. A characteristic of CS is its content of the deacetylated amine functions in the C2 position, which is higher than 50%. CS is a weak base having an intrinsic pK_a of 6.0; it is soluble only in acidic pH in which its amine functions are ionized.

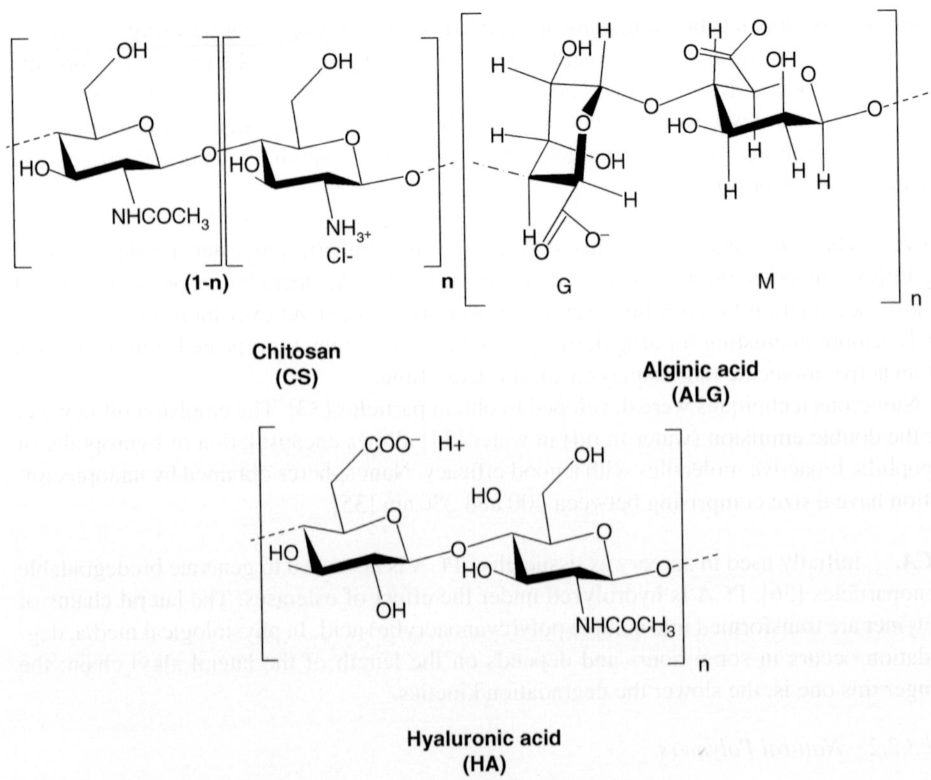

Figure 14.4 Formula of natural polymers

It behaves then as a cationic polyelectrolyte with a strong density of charges. Its toxicity is very weak, close to that of some sugars and salts (16 g/kg in mice). Its hemostatic properties and its capacity to accelerate healing are exploited in bandages or absorbable stitches. CS allows a reduction in the cholesterol rate in the blood by trapping lipids and making them indigestible. Finally, its mucoadhesive character and its capacity to promote drug absorption are very popular characteristics for drug delivery [40].

Numerous methods have been developed to obtain microparticles. The most usual one consists in forming an emulsion of a CS aqueous solution in oil, which is then stabilized by adding a cross-linker. The main drawback of this method is the use of cross-linkers that are often toxic and can inactivate bioactive molecules. Gentle methods can be used. Thus, a simple coacervation approach consists of vaporizing an acidic solution of CS in a basic solution of sodium hydroxide or of a mixture of sodium hydroxide–methanol. In these media, CS is insoluble and precipitates. The complex coacervation allows the polyelectrolyte complexes formation between CS (positively charged polyelectrolyte in acidic pH) and a polyanion.

ALG. This anionic polysaccharide is isolated from various types of brown algae *Laminaria hyperborean, Ascophyllum nodosum, Macrocystis pyrifera*) or certain bacteria (*Azotobacter vinelandii, Pseudomonas*). Its linear structure consists of two residues, mannuronic acid (M; pK_a 3.38) and glucuronic acid (G; pK_a 3.65) bounded by a 1.4 connection. ALG has the structure of a copolymer in blocks, with alternating GG, GM, and MM blocks. In the chain, the G/M proportion and the block sequence depend essentially on the ALG origin. Divalent cations (typically calcium, strontium, or barium) can take place in the cavities formed by ALG chains (Figure 14.5).

In GG sequences, a particularly stable cross-linked gel is formed according to the model of the 'egg-box' [41]. In MM sequences, the interactions of polymer/cation are much weaker [42]. These interchain associations can be permanent or temporary according to the quantity of cations present in the gel. Structure and molar mass of ALG as well as the nature of cations have an influence on the obtained gel characteristics.

The most usual way to obtain ALG particles is the drip addition of an ALG solution containing the active molecule in a calcium chloride solution. Calcium ions diffuse inside the drop, leading to the formation of a gel, which finally solidifies. Because ALG is protonated in gastric medium (acidic medium), gel particles contract and the size of gel pores reduces. This brings protection to the bioactive molecule from the gastric medium, and its release in the stomach becomes impossible. In the intestinal medium (weakly basic medium), gel particles swell and slowly dissolve. Bioactive molecules are then released by diffusion and by degradation of the ALG matrix. This behavior explains the large interest in ALG particles for oral administration of fragile bioactive molecules. The very fast dissolution of gel particles in intestinal medium provokes an immediate and complete release of active molecules in the intestinal lumen. To obtain a profile of the extended release, it is possible to cross-link ALG particles with aldehydes or to decrease their porosity by adding a layer of CS or poly-L-lysine on their surface.

HA. Hyaluronan, the other name for HA, is present in numerous animal tissues. Commercially available HA is mainly extracted from umbilical cords, the crests of cocks, or from certain strains of Streptococci (*zooepidemicus* or *equi*). It is a hydrosoluble polymer of high molar mass (from several hundred to several thousand million Da) having a glycosaminoglycan structure [43]. It is constituted by a disaccharides linear chain formed from D-glucuronic acid and *N*-acetyl-D-glucosamine bounded alternately by beta-1,4 and beta-1,3 connections. This configuration, which places all blocked groups in the equatorial position and hydrogen atoms in the axial position, makes it very stable. Axial hydrogen atoms form a relatively apolar hydrophobic face and equatorial groups form a more polar and hydrophilic face. In diluted solution, chains adopt a helical configuration. With an intrinsic pK_a of 2.9, HA behaves in physiological pH as a negatively charged polyelectrolyte. It is constantly synthesized under the control of hyaluronan synthetase and metabolized in epithelial cells of liver or lymph nodes. Hyaluronidase quickly degrades HA, thus avoiding any possibility of accumulation in the case of repeated administration. Its unique viscoelasticity properties are used in ophthalmology, in rheumatology, and in cosmetic medicine.

Compared to CS and ALG, HA is more seldom used as a vector of bioactive molecules. However, its nature of polyanion allows the generation of microspheres by mixing HA with polycations such as CS or proteins [44]. Thus, films or particles with extended release properties are obtained by cross-linking HA. The unique bioadhesive properties of HA make HA a component of choice for covering particles used in the administration of bioactive molecules via a transmucosal route.

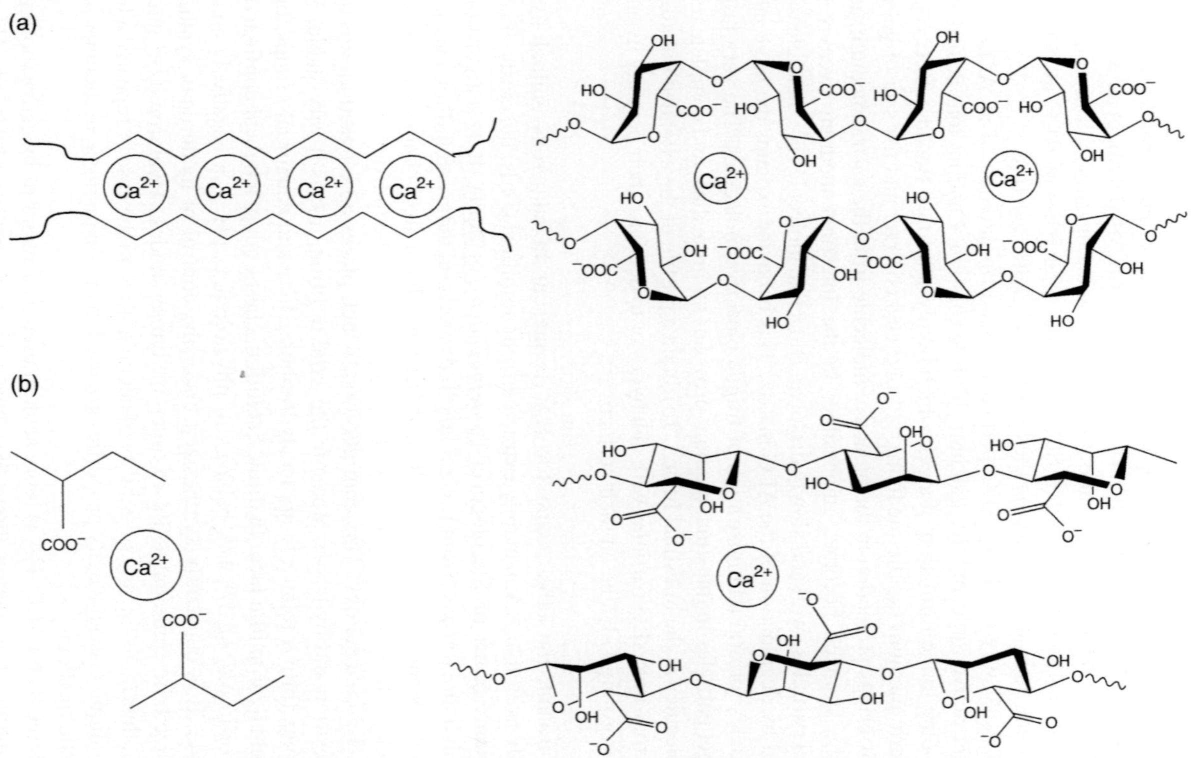

Figure 14.5 *Interactions between alginic acid and calcium ions: (a) GG segments (egg-box) and (b) MM segments*

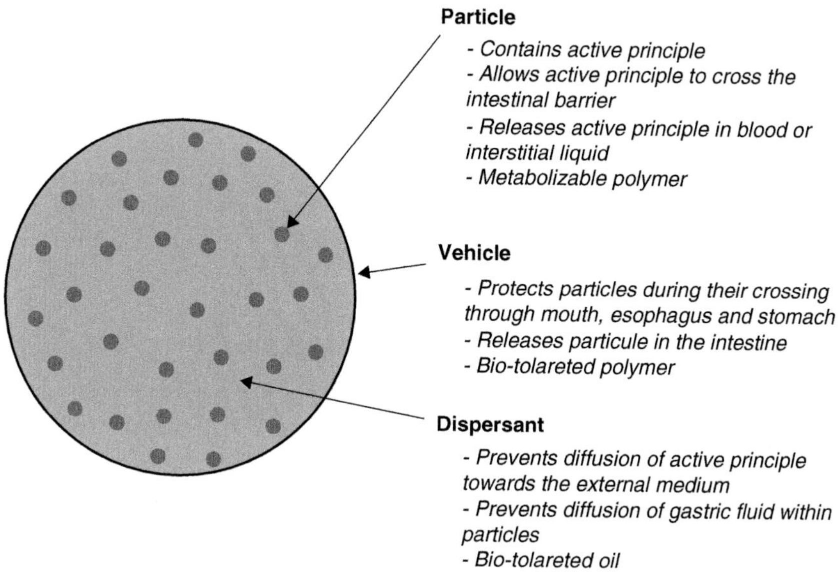

Figure 14.6 *Schematic representation of the pharmaceutical complex vector*

14.3.3 The Vector for the Oral Route

Encapsulation is presented as the best solution to protect the integrity of a bioactive molecule during its passage through the digestive tract. While specific solutions exist to solve each problem met during the various stages of this passage (gastric acidity, enzymes), no global solution is totally satisfactory. Administered in equivalent amounts, an active molecule given orally or subcutaneously can have a therapeutic effect sharply weaker in the former conditions. This decrease of activity is mainly accredited to the weak rate of passage through the intestinal epithelium.

Results obtained with a pharmaceutical vector designed in our laboratory [45] are described below. This vector is constituted by three different biocompatible parts, namely nanoparticles integrating the bioactive molecule, a lipophilic dispersant, and a vehicle (Figure 14.6).

Each of these constituents brings an answer to the problems met by peptides or proteins during oral administration. This pharmaceutical vector provides a protection against digestive tract attacks, a more efficient passing through the intestinal epithelium, and total release in blood (Figure 14.6). It greatly increases the bioavailability of peptides or proteins.

14.3.3.1 The Nanoparticles

Particles obtained by complex coacervation between two polyelectrolytes of opposite charges are polyelectrolyte complexes [46]. They are constituted by a peptide or a protein, which appears in the shape of a negatively or positively charged polyelectrolyte (in function having the nature of the peptide or protein) and a polyelectrolyte of the opposite charge, for example CS (positive charge) or HA (negative charge). Stable in the intestinal lumen, this complex vector protects the peptide or protein from enzymatic attacks. Its small

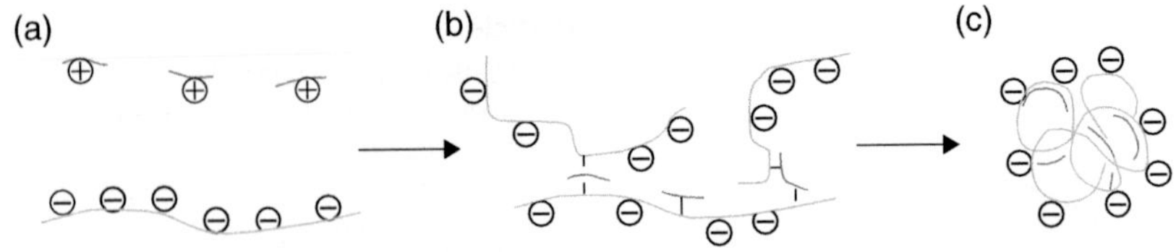

Figure 14.7 Formation of a polyelectrolyte complex: (a) addition of a short-chain cationic polyelectrolyte (grey) to a long-chain anionic polyelectrolyte (black); (b) polyelectrolyte chains cross-linking; (c) precipitation of a polyelectrolyte complex

size (less than 500 nm) allows it to pass through the intestinal epithelium. If necessary, its surface can be modified to increase its affinity for epithelial cells and therefore favor its passage through the intestinal epithelium.

Polyelectrolyte Complex Formation. Polyelectrolyte complexes are obtained in gentle conditions: no organic solvent, aqueous media only are used; no ultrasound, only moderate stirring at room temperature is utilized. It is therefore a technique of choice for the formulation of fragile substances such as peptides or proteins. The formation of polyelectrolyte complexes is governed by electrostatic attraction between groups of opposite charge and by the release of ions previously fixed to polyelectrolytes. Any phenomenon of macroscopic aggregation is avoided by the use of highly diluted aqueous solutions. To obtain stable particles, it is advisable to choose two polyelectrolytes of different molar masses and to mix them in nonstoichiometric proportions (Figure 14.7).

Because of electrostatic repulsion between the present charged groups along the skeleton, chains in solution are stretched. The longest serve as host to the smallest and condense them [47]. The obtained complex has a core–shell structure, the heart being hydrophobic and globally neutral, the surface being hydrophilic and charged; the charge of the surface is that of the polyelectrolyte having the higher molar mass. This charged surface confers good stability of the colloidal suspension during its storage and the centrifugation step. A polyelectrolyte complex structure is generally dynamic. At the beginning, it is a 'kinetics' amorphous complex, which is slowly reorganized in a thermodynamically more stable structure [48]. This structure can be highly ordered. This stabilization, realized by hydrogen bonds and by hydrophobic interactions, favors the ion-pair formation between polyelectrolytes.

Factors Governing Formation and Characteristics of Polyelectrolyte Complexes. Nature, distribution, and charge density of the polyelectrolytes strongly influence formation of the polyelectrolyte complexes. The interactions between strong charges are more efficient than the interaction between weak charges. Complexes stemming from these strong polyelectrolytes will be stable in a wide range of pH and in elevated concentrations in salt. The charge density is an important factor: in low density charge conditions, weak complexes will not form; in conditions of high density charge, a precipitate will be obtained instead of a coacervat. Other characteristics such as molar mass and polyelectrolyte flexibility are also critical [49]. The experimental conditions also influence the characteristics of resulting complexes [50]. The pH, by changing the ionization degree of ionic groups of weak polyelectrolytes such as CS and HA, modifies the charge density. A strong salt concentration, by creating an electrostatic screen between polyelectrolyte charges, inhibits formation of the complexes. A chosen polyanion/polycation ratio, polyelectrolyte concentration, and order of addition have an effect on charge and properties of final complexes. Perfect control of these parameters is absolutely required for the reproducible production of particles that display the chosen characteristics.

14.3.3.2 The Lipophilic Dispersant

Particles are dispersed in a pharmaceutical oil of the Miglyol® type. This oil has the role of a barrier for preventing the diffusion of hydrophilic active molecules outside particles and the diffusion of gastric medium within particles.

14.3.3.3 The Vehicle

The oily dispersion of particles is locked inside a capsule, the vehicle. The latter is designed to protect particles containing the bioactive molecule during their passage through the mouth, oesophagus, and stomach and to release them in the intestine. It is obtained by making a double emulsion (particles in oil) in aqueous ALG solution and by forming the membrane by ionotropic gelation of ALG with calcium ions.

14.4 P140 Peptide Nanostructured complex for the Treatment of Systemic Lupus Erythematosus

14.4.1 The Therapeutic Peptide P140

Systemic lupus erythematosus (SLE) represents a prototype of autoimmune rheumatic diseases. It is characterized by inflammation and damage of various tissues, including the joints, skin, kidneys, heart, lungs, blood vessels, and brain. At the immunological and biological levels, SLE is characterized by complement deficiencies, modification of cytokine secretion, hyper-gammaglobulinemia, production of autoantibodies, and formation of immune complexes, which play a crucial role in associated glomerulonephritis and cutaneous lesions. At least 60 different antigens targeted by specific antibodies have been characterized in SLE. Of these autoantibodies, anti-Sm and anti-double-stranded DNA antibodies are sufficiently discriminating to be part of the American College of Rheumatology criteria for the disease. The cause of the illness is unknown. The etiology is multigenetic, hormonal (with a female prevalence), and environmental. It has been estimated that its average prevalence in France corresponds to 75 cases per 100 000 persons (currently ~ 50 000 patients in France and 1.4 million individuals in the United States). In its worst form, lupus can be fatal: 10% of patients die from kidney disease, cardiovascular disease, or infections.

Current treatments of the disease are mainly based on immunosuppressive drugs such as corticosteroids and cyclophosphamide, which are often administered at high doses in acute exacerbation phases. Although these treatments have reduced mortality and significantly lengthened patients' life expectancies, they cause a nonspecific immune suppression leading to side effects, which are sometimes worse than the disease itself. To avoid such side effects, much effort is being put into designing potent and more specific strategies that will not affect the whole system and will preferentially target deleterious cells. In this context, the phosphorylated peptide P140/lupuzor™, which acts as an immunomodulator of the lupus response, emerges as a very promising therapeutic method. Administrated intravenously to MRL/lpr lupus-prone mice, this peptide significantly ameliorates the clinical and biological manifestations of treated animals and prolongs their survival [51, 52]. *Ex vivo* P140 peptide does not induce proliferation of human peripheral T cells but generates secretion of high levels of regulatory cytokine IL-10 [53].

The P140 peptide was included in phase I and II clinical trials conducted by ImmuPharma (Mulhouse/London). The clinical lot of P140 peptide administered in mannitol as excipient (Lupuzor) was found to be safe and well tolerated by subjects. In addition, in a phase IIa open trial [54] and then in a double-blind, multicentric, controlled phase IIb trial [55], Lupuzor administration significantly improved the clinical and biological status of lupus patients who received three subcutaneous doses of 200 µg peptide. A phase III clinical trial is currently in preparation.

The P140 mode of action is not totally understood and several pathways of intervention seem to be involved. Thus, P140 peptide might act as an altered peptide ligand (APL) of the T cell receptor (TCR) [56]. The latter can interpret subtle changes in its ligand, resulting in different phenotypic outcomes. A dramatic effect of APL stimulation with APCs is the induction of anergy as opposed to proliferation, a feature observed when patients' T cells from the peripheral blood were incubated with P140 peptide [53]. Very recently, the existence of specific $CD4^+$ T cells was demonstrated using MHCII tetramers of I-E^k containing P140 peptide [57]. A more frequent population of $CD4^+$ T cells specific for the nonphosphorylated sequence was also identified, supporting our model.

Another P140 mode of action was also proposed following our finding that the peptide selectively interacts with the constitutive heat-shock HSC70/Hsp73 protein [58]. Following P140 peptide binding to HSC70, the interaction of HSC70 to other chaperones and ligands appeared to be disrupted. This might greatly alter the endogenous (auto)antigen processing, which occurs via a mechanism involving a lysosomal degradation pathway, in which HSC70 is central.

The P140 mode of action seems to be restricted to autoreactive immune responses. Compared to naïve MRL/lpr mice, P140-treated MRL/lpr mice resist equally well to a viral challenge and B and T cells react normally to mitogens [52, 59]. Therefore P140 peptide behaves as an immunomodulator and not as an immunosuppressor.

P140 peptide, a very promising therapeutic molecule for the treatment of lupus disease, represents an excellent candidate peptide to evaluate the ability of nanovectors to improve its drugability. The purpose is to replace the subcutaneous route currently used to administrate P140 peptide into lupus patients by a noninvasive route, which is much more appreciated by patients. Below, we describe a series of studies aimed at orally administrating the P140 included in nanocapsules to MRL/lpr mice to improve the course of their disease.

14.4.2 Development of Nanoparticles Containing Hyaluronic Acid Associated to P140 peptide (HA-P140)

P140 is a 21-mer peptide corresponding to the sequence 131–151 of the U1-70 K spliceosomal protein and encompassing a phosphoserine residue at position 140 (Figure 14.8).

131 140 151

H_2N RIHMVYSKRSGKPRGYAFIEY COOH

| (Phospho attached at position 140)

Phospho

Figure 14.8 *Sequence of P140 peptide. The residues R^{131}, H^{133}, K^{138}, R^{139}, K^{142}, and R^{144} are positively charged, while phosphoserine^{140}, E^{150}, and Y^{151} are negatively charged*

In solution in Millipore® water (pH 5.5), basic residues such as arginine (R; pK_R 12.48), lysine (K; pK_R 10.54), and histidine (H; pK_R 6.04) are positively charged. The N-terminal arginine residue has two positive charges, one on its side chain (pK_R 12.48) and one on its amino group (pK_2 8.99). The glutamic acid residue (E; pK_R 4.07) is negatively charged as the C-terminal carboxylic acid group (pK_1 2.20). The phosphorylation of the serine residue confers to P140 an additional negative charge. At pH 5.5, P140 peptide presents a structure with seven positive charges and three negative charges.

14.4.2.1 Optimization of Protocol Designed for HA-P140 Nanoparticle Synthesis

A stock solution of HA was prepared by dissolving HA (5.0 mg) in Millipore® water (25 mL) and stirred during 24 h to obtain the complete dissolution of the polymer. A part of this HA solution (1.0 mL; pH 6.5) was poured into a test tube fitted with a small magnetic stirrer. In a hemolysis tube, P140 peptide (200 µg) was dissolved in Millipore® water (250 µL). With a micropipette, HA solution was added dropwise in the stirred HA solution at room temperature. The hemolysis tube was washed with Millipore® water (50 µL). This washing solution was added in the reaction medium. After P140 peptide addition in HA, the solution became cloudy. After 15 min of stirring, this suspension was transferred in a microtube (1.5 mL) and stored at 4 °C. The obtained HA-P140 nanoparticles were characterized by granulometry, zeta potential measurements, and transmission electron microscopy (TEM).

14.4.2.2 Influence of Different Synthesis Parameters on HA-P140 Nanoparticle Characteristics

Influence of the HA Molar Mass. To examine the influence of HA molar mass, HA-P140 preparations were synthesized using different HA of 1200 kDa, 400 kDa, and 27.6 kDa.

HA of high molar mass (1200 or 400 kDa) was found to form nanoparticles with an aspect of fibrillar filaments. The long chains of polymers initially stretched due to the electrostatic repulsions between the acidic groups (carboxylic acid) did not totally condense in spherical spheres. After centrifugation at 2370 *g*, an aggregation of nanoparticles occurred and the spherical morphology was destroyed. The centrifugal force reorganized the polymer chains to form fibers and the negative surface charge of nanoparticles was too weak to prevent aggregation. The use of such particles was not possible.

HA of lower molar mass (27.6 kDa) formed monodisperse, perfectly spherical nanoparticles and the polydispersity index was less than 0.1. They were isolated from the synthesis medium by centrifugation at 2370 *g* without any problem of aggregation. Their negative surface (zeta potential in the order of −30 mV) was mainly covered with HA, and P140 was condensed within the particle and protected from the outside environment. This electrical charge on the surface provides good stability to the suspension during centrifugation and storage (at least two weeks at 4 °C).

The length of HA chains influences morphology, size distribution, and HA-P140 stability. To obtain spherical nanoparticles, the polyelectrolyte has to have a chain of a size appreciably superior of that of the peptide, but this difference has to remain moderate. A factor of 10 between the molar masses of both polyelectrolytes (HA: 27.6 kDa; P140: 2.636 kDa) appears to be the correct ratio.

Influence of the HA (mHA) to P140 (mP140) Molar Weight Ratio. Another important parameter is the mass ratio in which both polyelectrolytes have to be added. This ratio influences directly the one between positive and negative charges in the solution. P140 peptide has four residual positive charges for a molar mass of 2636 g/mole and HA one negative charge for a repetitive sequence mass of 415 g/mole; a mass ratio (R_m = mHA/mP140) of 1 is equivalent to a charge ratio (R_c = number of negative charges/number of positive charges) of 1.5.

When an excess amount of P140 is added to HA (R_m = 0.75; R_c = 1.2), nanoparticles form. They are unstable and quickly coalesce. In the suspension, the number of negative charges is not enough and HA cannot cover the totality of the particle surface. The electrostatic stabilization necessary to keep them individualized is not sufficient.

When an equal amount of P140 is added to HA (R_m = 1; R_c = 1.5), spherical nanoparticles (~200 nm) are obtained. The colloidal stability is assured by a negatively charged surface (zeta potential = –30 mV). Nanoparticles are stable for several weeks at 4 °C and can be easily spin-dried.

When a faulty amount of P140 is added to HA (R_m = 2; R_c = 3), the solution remains crystal clear. Observation by MET shows the presence of monodisperse spherical particles (~50 nm). Having a diameter lower than 100 nm, they do not diffuse the light and their presence is not visible to the naked eye. These particles are stable in Millipore® water at 4 °C for some days. Their very small size makes them very difficult to isolate by centrifugation.

When a highly defective amount of P140 is added to HA (R_m = 3; R_c = 4.5), it is impossible to obtain particles. The number of positive charges brought by P140 is then defective and cannot create enough cross-linking connections with HA to precipitate the polyelectrolyte complex.

The size of HA-P140, a decisive parameter for its passage through the intestinal epithelium, is strongly dependent on the mass ratio in which P140 is added to HA. In the following, HA-P140 synthesis were made with HA of a low molar mass (27.6 kDa) using a mass ratio mHA/mP140 of 1 (1.5 negative charges for 1 positive charge).

14.4.2.3 The Efficiency of P140 Encapsulation

To estimate the encapsulation efficiency (EE), it is advisable to determine the P140 amount inside nanoparticles. This efficiency is defined by the equation:

$$\text{EE}(\%) = \frac{(\text{Encapsulated quantity of P140})}{(\text{Initial quantity of P140})} \times 100$$

Calibration Curve of P140. P140 encompasses several aromatic residues (three tyrosine residues and one phenylalanine residue) that absorb at 280 nm. The successful encapsulation of P140 in nanoparticles can be measured by UV–visible spectrophotometry. A calibration curve of the P140 absorption at 280 nm (A_{280}) is established from three P140 stock solutions and their successive dilutions. In the limit of P140 concentrations used for HA-P140 synthesis, there exists a linear relation between the P140 concentration and its A_{280}.

Determination of the P140 Encapsulation Efficiency. After their synthesis, HA-P140 are separated from free HA and free P140 by several cycles of centrifugation at 2370 *g* until nanoparticles are not detectable in the supernatant (crystal clear supernatant). The supernatant A_{280} is then measured. As HA does not absorb at this wavelength, the obtained value is directly connected to the quantity of free P140. Having determined the quantity of free P140 and knowing the initial quantity of P140 added to HA solution, the quantity of encapsulated P140 in the nanoparticles can be calculated.

Evolution with Time of the P140 Encapsulation Level. After dropwise addition of P140 (400 μg; 500 μL Millipore® water; pH 5.5) in the solution of HA (2.2 mL Millipore® water; pH 6.5), the suspension is stirred for 15 min at room temperature before being transferred in a microtube and stored in 4 °C. At regular intervals of time, a sample of 500 μL is drawn and subjected to several cycles of centrifugation. The absorbance of the supernatant is measured at 280 nm to evaluate the amount of free peptide.

It was observed that the encapsulation level increases significantly during the first 5 h and reaches a plateau at ~50% of the initial amount of P140 peptide. Later on, to optimize the encapsulation level, HA-P140 was spin-dried 24 h after synthesis.

14.4.2.4 The Evolution with Time of HA-P140 Nanoparticle Morphology

The stability of nanoparticles in the synthesis medium, as determined by MET, highlighted a progressive reorganization of chains within particles (Figure 14.9).

Some minutes after synthesis, the HA chains presented no particular organization. Some days later, the HA chains situated at the periphery began to turn in parallel to one another. This reorganization continued and, after a week, particles lost their spherical shape to become toroidal. HA chains were arranged at the same time parallel to one another and in a concentric way, P140 being junction points between each of them. This reorganization as tores was observed with HA-P140 of a size ~200 nm, while nanoparticles of lower size (50 to 100 nm) kept a morphology close to spheres.

Such morphologies were observed in the case of condensation, at very low concentrations, of double-stranded DNA with polycations. The repulsive interactions between DNA phosphate groups were neutralized and then helices could assemble at the same time to one another in a concentric way to form very orderly tores. This toroidal morphology seems to be a specific property of the semi-flexible polymers [60].

In HA-P140, two properties of HA can explain the formation of tores. First, one face of HA is hydrophilic, the other being hydrophobic. This feature confers auto-associative properties and chains can turn in parallel to one another. Helices were observed by atomic force microscopy (AFM) in HA deposits on mica [61]. Second, in aqueous solution HA behaves as a semi-flexible polymer [62]. The rotation of saccharide units is limited at the

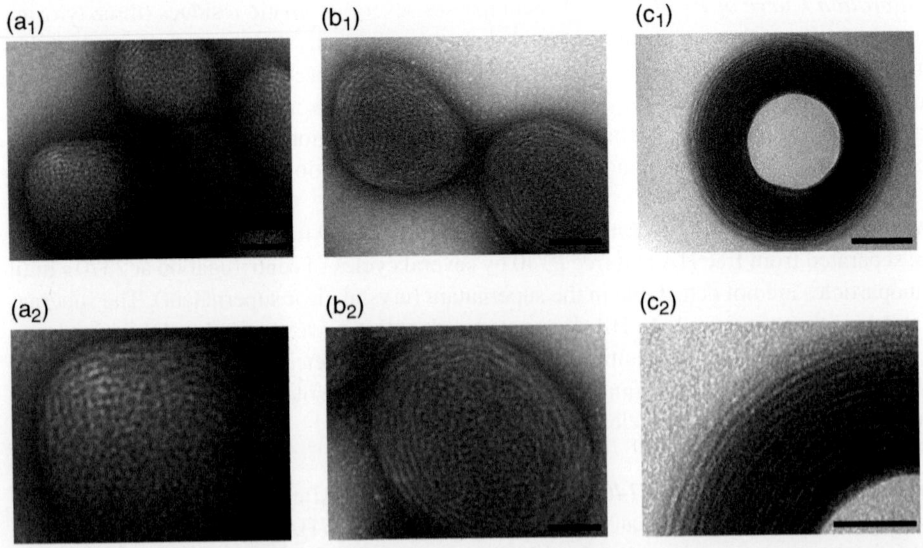

Figure 14.9 *Evolution with time of HA-P140 nanoparticle morphology observed by TEM: (a_1, a_2) images at day 1; (b_1, b_2) day 3 and (c_1, c_2) day 7. Scales: 100 nm (a_1, b_1, and c_1) and 50 nm (a_2, b_2, and c_2)*

level of the glycosidic connections: the hydrogen bonds form and break constantly between sugars and the presence of water molecules around these same sugars favors certain conformations rather than the others. This phenomenon limits HA chain mobility and increases their rigidity. In a sphere, HA is subjected to strong constraints because of the curvature of the sphere. These constraints are much less strong in a tore, which is the morphology allowing minimization of the energy of curvature and the surface energy. It then becomes the most stable thermodynamic conformation.

The 'kinetics' complexes, obtained at the end of synthesis, are reorganized in a thermodynamically favored structure in which HA is subjected to less high constraints. This reorganization involves the dissociation of certain ionic bonds, which are then reformed by optimizing the other types of interactions within the structure (hydrophobic interactions, hydrogen bond).

14.4.2.5 Study of HA-P140 Nanoparticle Stability in Simulated Media

Since HA-P140 is to be administered by the oral route within an acid-resistant vehicle, it was important to study its stability in simulated gastric medium. Once released in the intestine, HA-P140 will have to be stable enough to protect P140 peptide before passing through the intestinal epithelium. Stability in simulated intestinal medium was also measured.

Stability in Simulated Gastric Medium. The simulated gastric medium is an acidic medium (pH of between 1 and 2) containing various salts ($CaCl_2$ 0.22 g/L; KCl 2.20 g/L; NaCl 5.0 g/L; $NaHCO_3$ 1.50 g/L) and one of the most frequent enzymes, namely pepsine (500 kU of pepsinogen). HA-P140 incubation in this simulated gastric medium led to their

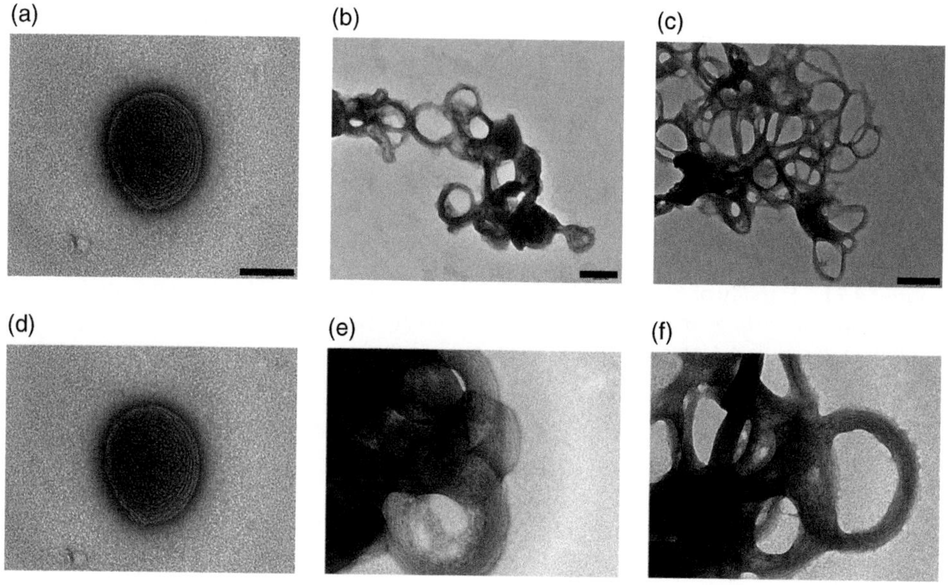

Figure 14.10 *Stability of HA-P140 nanoparticles in simulated intestinal and physiological medium observed by TEM: (a–c) simulated intestinal and (d–f) physiological medium. Images taken immediately after synthesis (a, d), after 15 min (b, e) and after 1.5 h (c, f). Scale: 200 nm*

immediate dissolution. At pH < 2, carboxylic groups of HA are not ionized (COOH form), therefore eliminating electrostatic interactions between HA and P140.

This result demonstrated the necessity of using an acid-resistant vehicle to protect the HA-P140 from the stomach medium.

Stability in Simulated Intestinal Medium. The simulated intestinal medium is a neutral medium (pH of between 6 and 8) containing various salts ($CaCl_2$ 0.22 g/L; KCl 2.20 g/L; NaCl 5.0 g/L) and 10% w/v of pancreatin, a mixture of several digestive enzymes (amylase, lipase, and trypsin) produced by the exocrine cells of the pancreas. In this medium, the dissociation of particles was very rapid, but tores were more stable (Figure 14.10a to c).

Stability in Simulated Physiological Medium. Ionic force has a very strong influence on the stability of polyelectrolyte complexes. Their cohesion is essentially due to electrostatic attractions existing between polyelectrolytes of opposite charges. Thus counterions in solution partially mask these charges and decrease attraction forces between polyelectrolytes.

To test HA-P140 stability in a simple physiological medium, particles were immersed in a salt solution (NaCl 9.0 g/L). The influence of the presence of salt on the morphology of particles of different size was followed by MET (Figure 14.10d to f). In the case of 200 nm particles, dissolution of the core was immediate whereas the more organized external chains remained condensed. This dissociation takes place during the first minutes of dumping only. Release of the P140 and HA chain leads to an aggregation of tores. In the case of

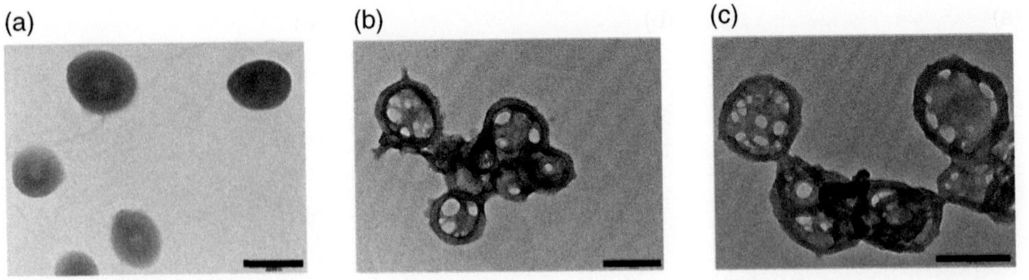

Figure 14.11 *Stability of HA-CS-P140 nanoparticles observed by TEM: (a) observation after synthesis in water; (b) after 15 min incubation in simulated intestinal medium; and (c) after 1 h incubation in simulated intestinal medium. Scale: 200 nm*

smaller particles (diameter <100 nm), dissolution of their structure was total in the first minutes of incubation in salt medium.

Stabilization of HA-P140 Nanoparticles. To stabilize HA-P140 in salt medium and simulated intestinal medium for at least 1 h, two techniques were evaluated. In the first one, CS was added to P140 solution during HA-P140 synthesis to create a denser and more complex matrix. In the second one, CS was added after HA-P140 synthesis to let chains reorganize within nanoparticles and thus optimize and reinforce the interactions between the peptide and the polyelectrolyte.

CS Embedding within HA-P140. Using the first protocol, HA (27.6 kDa) was condensed with a P140 solution containing CS (50 kDa) in different proportions (from 10 to 150% in weight), the charge ratio remaining constant (1.5 negative charges for 1 positive charge). Monodisperse nanoparticles (HA-CS-P140) of a size of ~200 nm and with a negative surface (zeta potential = –20 mV) were generated. Incubated in simulated media (intestinal and physiological) (Figure 14.11), HA-CS-P140 did not show a better stability than HA-P140.

Reorganization of Polymer Chains within HA-P140. Synthesized particles following the first protocol were spin-dried 24 h after synthesis. They were kept for one week at 4 °C before transfer to simulated media (intestinal and physiological). This period of one week allowed the chains to be reorganized within nanoparticles and HA-P140 to adopt their most stable thermodynamic conformation. HA-P140 of two different sizes (100 and 200 nm) was incubated 1 h in simulated intestinal medium and 15 min and 1.5 h in simulated physiological medium. Particles remained full and well drawn (Figure 14.12).

We could therefore conclude that reorganization of the HA-P140 makes them stable in intestinal medium for at least 1 h and in physiological medium for at least 1.5 h.

14.4.2.6 The Optimized Protocol of HA-P140 Nanoparticle Synthesis

Following these steps of optimization, HA-P140 nanoparticles tested in the *ex vivo* and *in vivo* studies were obtained using the protocol described above with HA of low

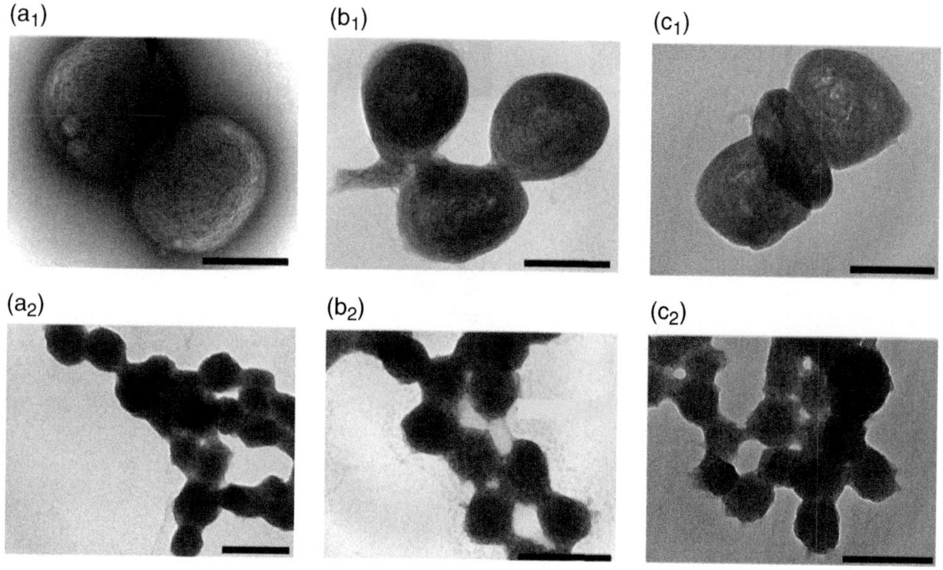

Figure 14.12 *Stability of HA-P140 nanoparticles after reorganization of polymer chains as observed by TEM: (a_1, a_2) observation after 15 min incubation in simulated physiological medium (NaCl 9.0 g/L); (b_1, b_2) after 1.5 h incubation in simulated physiological medium (NaCl 9.0 g/L); and (c_1, c_2) after 1 h incubation in simulated intestinal medium. Scales: 200 nm. Average particle sizes 200 nm (a_1, b_1, and c_1) and 100 nm (a_2, b_2, and c_2)*

molar mass (27.6 kDa) in a mass ratio of mHA/mP140 close to 1 (around 1.5 negative charges for 1 positive charge). To get good encapsulation efficiency, HA-P140 nanoparticles were isolated by centrifugation (2370 *g*) only 24 h after synthesis. To reach their thermodynamic equilibrium before use and to be stable in simulated media (intestinal and physiological), HA-P140 nanoparticles were stored in Millipore® water for a week at 4 °C.

14.4.3 The Effect of HA-P140 Nanoparticles in Healthy and Lupus Mice

14.4.3.1 HA-P140 Nanoparticle Administration to Healthy Mice

HA-P140 Administration via the Subcutaneous Route. Before administering HA-P140 to lupus mice, it was necessary to check whether P140 released by HA-P140 preserved its integrity and antigenic properties. As it is classically observed with short peptides, P140 peptide does not generate an immune response when administrated free into an animal recipient. However, P140 peptide is immunogenic when it is administrated in healthy BALB/c mice in the presence of an adjuvant. Administrated subcutaneously (s.c.) in the presence of the complete Freund's adjuvant (CFA), P140 peptide induces $CD4^+$ T cells that proliferate and secrete IL-2 and IFN-γ in the presence of P140 in the culture [51]. Four groups of 7 week-old, female, BALB/c mice (2 mice per arm) were treated as follows (s.c. injection): (1) (positive control group) P140 in the presence of CFA (P140-CFA); (2) (treated group) HA-P140 in the presence of CFA (HA-P140-CFA); (3) HA-P140 in

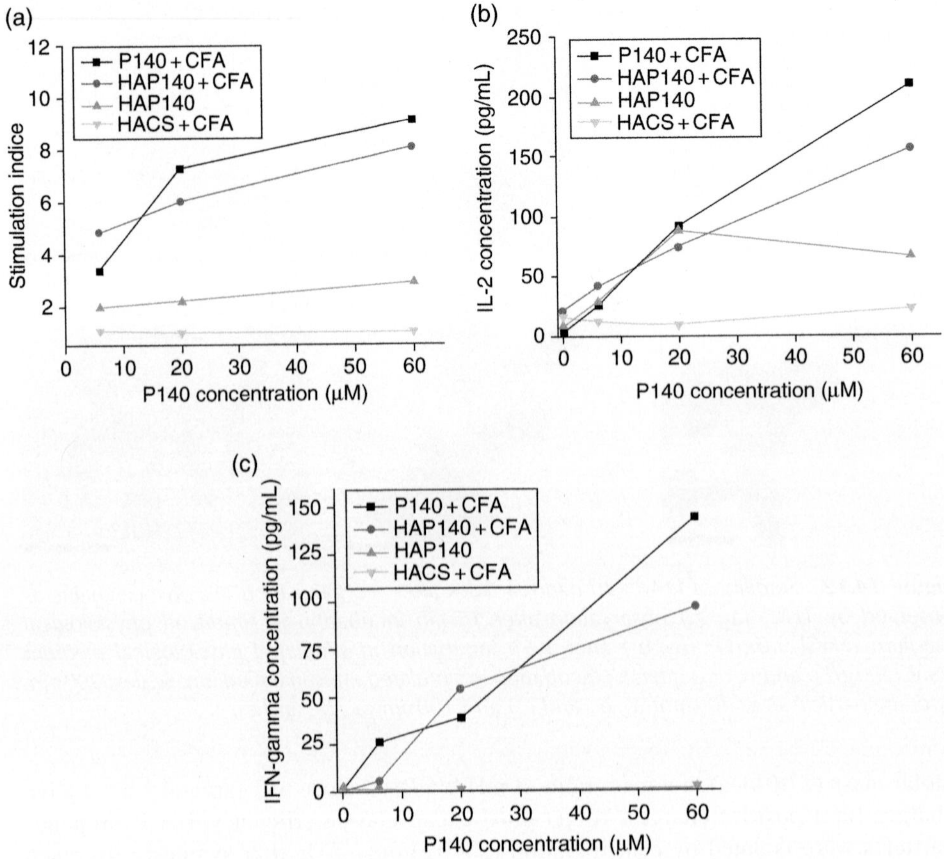

Figure 14.13 *Ex vivo response to phosphorylated peptide P140 of T lymphocytes from BALB/c mice immunized with different P140 formulations. LNCs from BALB/c mice immunized with peptide P140 peptide in CFA (■), HA-P140 nanoparticles in CFA (●), HA-P140 nanoparticles without CFA (▲), or HA-CS in CFA (▼) were recalled ex vivo after 8 days in the presence of increasing concentrations of peptide P140. A mean stimulation index (SI) ≥2.0 in the proliferation test (a) was considered as positive. IL-2 (b) and IFN-γ secretion (c) were measured by ELISA*

salt medium (to verify that HA-P140 does not display an adjuvant character); and (4) (control group) blank nanoparticles in the presence of CFA. Blank nanoparticles (HA-CS) were obtained by complex coacervation between HA and CS using the protocol described above with the difference that P140 (200 μg dissolved in 250 μL Millipore® water) was replaced by CS (100 μg dissolved in 250 μL HCl 0.002 N). Tubes were washed with HCl 0.002 N (50 μL).

Mice of subgroups 1 to 3 received the same amount of P140 (100 μg/mouse). After 8 days, mice were sacrificed, inguinal, popliteal, and periaortic lymph nodes were removed, and the proliferation assay was performed using 5×10^5 nonfractionated lymph node cells (LNC) per well and increasing P140 concentrations (0, 6, 20, and 60 μM) as recall antigen in the culture. As shown in Figure 14.13, a peptide dose-dependent proliferative response with IL-2 and

IFN-γ secretion was observed in the culture of LNCs from mice that received P140-CFA and HA-P140-CFA (groups 1 and 2). In contrast, no or very weak proliferative response without cytokine secretion was observed when LNCs from mice of groups 3 and 4 were tested.

These results indicate first that the immunogenicity of P140 released from HA-P140 is maintained and second that HA does not present a significant adjuvant effect. This result is very important since any intrinsic adjuvant effect may accelerate the disease development in lupus-prone mice [51].

HA-P140 Administration via the Intraduodenal Route. In the complex pharmaceutical vector described above, nanoparticles encapsulated into an acid-resistant vehicle are protected in the stomach and then released in the intestine before passing through the intestinal epithelium. To simulate the use of this vector, nanoparticles were directly administered in the duodenum (Figure 14.14a).

Three groups of two female BALB/c mice (7 weeks old) were immunized as follows: (1) (control group) s.c. administration P140-CFA; (2) (treated group) intraduodenal administration HA-P140-CFA; and (3) (untreated group) intraduodenal administration of HA-CS-CFA. As above, mice of groups 1 and 2 received the same amount of P140 (100 μg/mouse). Animal experiments were approved by our Regional Ethics Committee. Intraduodenal administration was made under anesthesia. A small incision was made to take out the superior segment of the intestine and realize injection in the duodenum. Then this incision was sewn and the animal was kept under a heating lamp until its wake-up. One week after the immunization, mice were sacrificed and LNCs draining the administration site were taken. In the case of intraduodenal administration mesenteric LNCs, which form a chain along the intestine and drain the intestinal mucous membrane, were collected. The protocol described above was used to evaluate the immune response *ex vivo*.

The results obtained after intraduodenal administration were compared to those obtained after s.c. administration. The proliferation of LNCs collected from mice immunized intraduodenally with the HA-P140-CFA was strong and peptide dose-dependent (Figure 14.14b). It was equivalent to that of LNCs collected from mice that received P140 peptide or HA-P140 s.c. in CFA. In contrast, very weak or no proliferation was measurable in culture wells containing LNCs from mice that received HA-CS in CFA intraduodenally. These results were confirmed when secretion of IL-2 and IFN-γ were measured in culture wells in the presence of P140 as the recall antigen (Figure 14.14c and d). All together these data show that intraduodenal administration of HA-P140-CFA leads to the activation of mesenteric LNCs with the production of type 1 cytokines and that these nanoparticles are stable enough *in vivo* in intestinal medium to protect P140 from enzyme attacks.

14.4.3.2 Administration of HA-P140 to MRL/lpr Lupus Mice

To evaluate the therapeutic potential of HA-P140 administered intraduodenally, a protocol was developed using MRL/lpr lupus mice as lupus-prone model. It was shown previously that i.v. administration to young MRL/lpr mice of P140 in saline (without adjuvant) significantly delays the appearance of disease symptoms and lengthens the expectancy life of treated animals [51, 52]. Thirteen 5-week-old female MRL/lpr lupus mice received three intraduodenal administrations (weeks 5, 7, 9) of either HA-P140 in the absence of adjuvant (8 mice; equivalent to 100 μg of P140) or HA-CS (5 mice; equivalent to 100 μg of CS). The

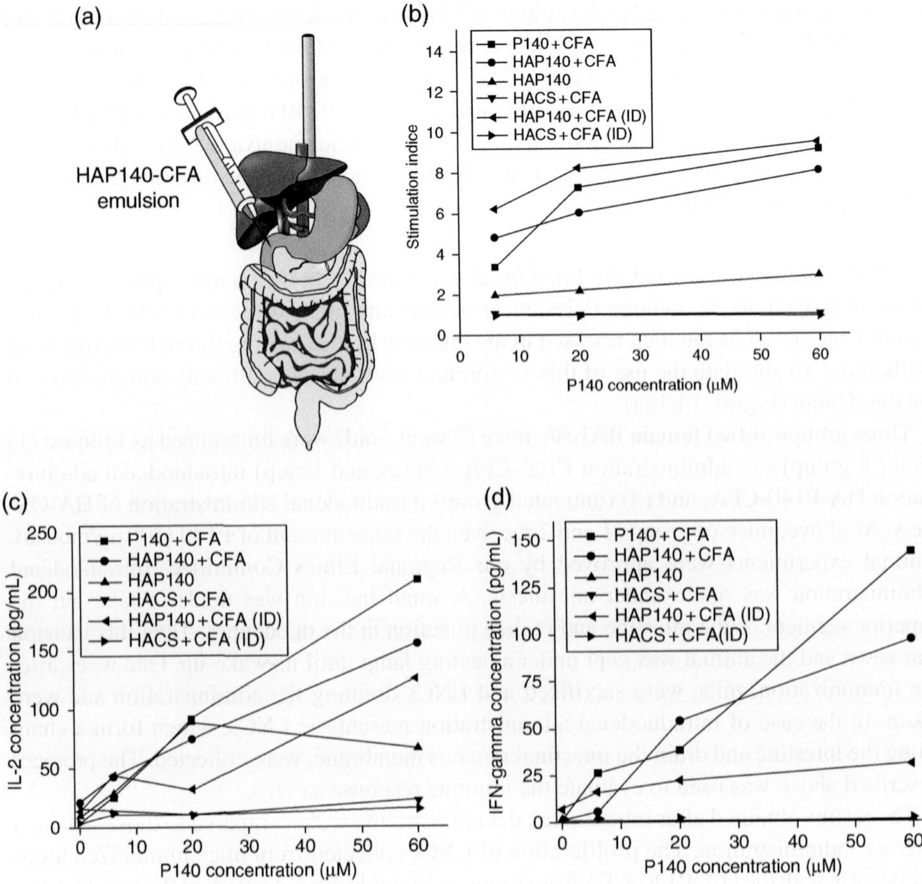

Figure 14.14 Intraduodenal administration of HA-P140 nanoparticles to BALB/c mice: (a) scheme illustrating intraduodenal administration; (b–d) ex vivo response to phosphorylated peptide P140 of T lymphocytes from BALB/c mice immunized intraduodenally with different P140 formulations (see the legend of Figure 14.13). BALB/c mice received peptide P140 peptide in CFA (■), HA-P140 nanoparticles in CFA (●), HA-P140 nanoparticles without CFA (▲), HA-CS in CFA (▼) subcutaneously, and HA-P140 nanoparticles in CFA (◀) and HA-CS in CFA (▶) intraduodenally

symptoms of lupus disease (e.g. proteinuria measured with urinary strips) were regularly followed and a profile of mortality was established.

As shown in Figure 14.15a, intraduodenal administration of HA-P140 delayed proteinuria appearance in treated MRL/lpr mice (12 weeks in nontreated MRL/lpr mice versus 15 weeks in treated animals). Furthermore in this treated group, proteinuria remained less frequent all along the course of the disease. The intraduodenal administration of HA-P140 was also shown to be beneficial on the mortality rate of treated mice. In the control group, the first mice died at 15 weeks while in the treated group the first mouse died at 22 weeks (Figure 14.15b). This delay is very important considering

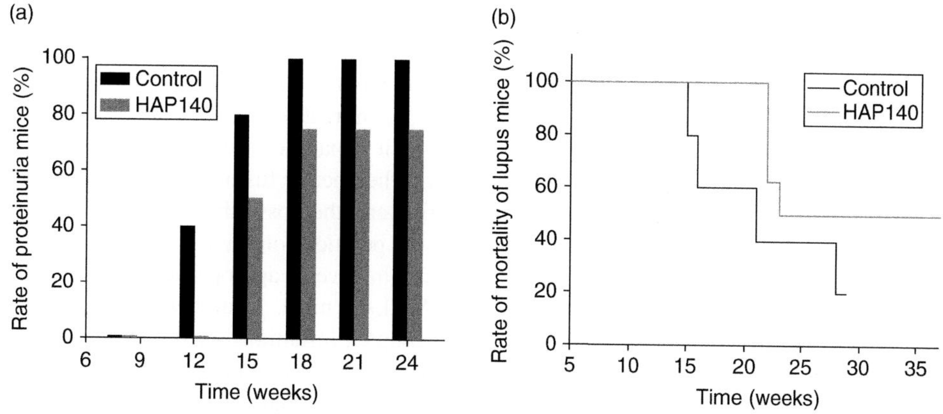

Figure 14.15 *Intraduodenal administration of HA-P140 nanoparticles to MRL/lpr lupus mice: (a) rate of proteinuria mice (proteinuria more than 3 g/L) and (b) rate of mortality. Five-week-old MRL/lpr mice received HA-P140 nanoparticles intraduodenally in saline (8 mice) or HA-CS only in saline (5 mice). They were then subjected to two further administrations in the same conditions at weeks 7 and 9. All mice were monitored regularly for different parameters until they died. The percentage of mice that had a severe proteinuria (>3 g/L; a) and that survive (b) is shown*

that the expectancy life of an MRL/lpr lupus mouse does not exceed 35 weeks. Thus, at 37 weeks, 100% of nontreated mice and only 50% of mice having received the HA-P140 were dead. These results led us to conclude that intraduodenal administration of HA-P140 to MRL/lpr lupus mice allows their life expectancy to be lengthened in a significant way.

14.4.3.3 Summary and Conclusion

The synthesis of polyelectrolyte complexes from HA and P140 led to nanoparticles having a size between 50 and 200 nm and a negative surface charge (zeta potential in the order of −30 mV). The HA-P140 nanoparticle encapsulation efficiency, estimated to be higher than 50% by UV/visible measurements, increases in an important manner within hours following the addition of P140 solution to HA. Unstable in simulated gastric medium, HA-P140 nanoparticles require the use of an acid-resistant vehicle to protect them during their passing through the stomach. In simulated intestinal medium, they are stable only when the HA chains are reorganized to optimize the interactions with P140 and when they have reached their thermodynamic equilibrium.

The s.c. or intraduodenal administration of HA-P140 in the presence of CFA in normal BALB/c mice leads to the development of an immunizing response similar to that developed by s.c. injection of free P140. Thus antigenic peptide is not affected by its incorporation into nanoparticle and can be released under a bioactive form.

To determine the therapeutic potential of HA-P140, a therapeutic protocol was engaged with female MRL/lpr mice that develop a strong form of the disease. Intraduodenal administration of HA-P140 was shown to slow down the clinical signs of the lupus disease, notably the proteinuria of animals, and significantly lengthen (at least by twice) their life expectancy. These results are extremely promising in our quest to design a procedure to administrate the P140 peptide via the oral route.

14.5 General Comments

Peptide therapeutics represents a valuable class of therapeutic agents. They possess a number of intrinsic properties that are favorable for long-term treatments. They are also versatile components that can be modified to improve their capacities without affecting their bioactivity. Peptides also offer many advantages over therapeutic fusion proteins and monoclonal antibodies, including their cost of production and the possibility to produce large amounts of drug substance under good manufacturing practice conditions.

Peptide-mediated immunotherapy has been tested in several appropriate experimental animal models, as illustrated with P140 peptide in MRL/lpr mice. A number of potent peptides are currently evaluated in clinical trials for complex inflammatory diseases such as autoimmunity, allergy, and neuropathies. Major advances have been made to improve their bioavailability and stability. A challenge is also to improve the route of peptide administration while preserving their bioactivity, their capacity to specifically target cells and tissue, and their stability. Oral delivery is particularly suitable for treating patients who are affected by chronic inflammatory diseases as well as for vaccinating a larger population. Nanostructured peptide conjugates are particularly attractive to reach this goal.

Acknowledgements

The authors thank the past and present collaborators of these studies, Anne Wawrezinieck, Caroline Eilhas, Camille Dollinger, Nicolas Schall, Monique Duval, as well as Jean-Paul Briand, for helpful discussions and carefully reading this manuscript. Research in the authors' laboratories has been financially supported by the Centre National de la Recherche Scientifique (CNRS), Région Alsace, and ImmuPharma-France.

References

[1] Zompra, A.A., Galanis, A.S., Werbitzky, O., Albericio, F. (2009) Manufacturing peptides as active pharmaceutical ingredients. Future Med. Chem. 1: 361–377.
[2] Vlieghe, P., Lisowski, V., Martinez, J., Khrestchatisky, M. (2010) Synthetic therapeutic peptides: science and market. Drug Discov. Today 15: 40–56.
[3] Briand, J.-P., Muller, S. (2005) Synthetic peptides for the analysis of B-cell epitopes. In: Autoantigens, Autoantibodies and Autoimmunity: Molecular Mechanisms in Health and Disease. Wiley-VCH, Weinheim, Germany.
[4] Briand, J.-P., Muller, S. (2010) Emerging peptide therapeutics for inflammatory autoimmune diseases. Current Pharmaceut. Des. 16: 1136–1142.
[5] Partidos, C.D., Beignon, A.-S., Semetey, V., Briand, J.-P., Muller, S. (2001) The bare skin and the nose as non-invasive routes for administering peptide vaccines. Vaccine 19: 2708–2715.
[6] Partidos, C.D., Muller, S. (2005) Decision making at the surface of the intact or barrier disrupted skin: potential applications for vaccination or therapy. Cell. Mol. Life Sci. 62: 1418–1424.
[7] Partidos, C.D., Beignon, A.-S., Briand, J.-P., Muller, S. (2004) Modulation of immune responses with transcutaneously deliverable adjuvants. Vaccine 22: 2385–2390.
[8] Glenn, G.M., Taylor, D.N., Li, X., Frankel, S., Montemarano, A., Alving, C.R. (2000) Transcutaneous immunization: a human vaccine delivery strategy using a patch. Nat. Med. 6: 1403.
[9] Itoh, T., Celis, E. (2005) Transcutaneousimmunization with cytotoxic T-cell peptide epitopes provides effective antitumor immunity in mice. J. Immunother. 28: 430–437.

[10] Patel, V.F., Liu, F., Brown, M.B. (2011) Advances in oral transmucosal drug delivery. J. Contr. Release 153: 106–116.
[11] Shojaei, A.H., Chang, R.K., Guo, X., Burnside, B.A., Couch, R.A. (2001) Systemic drug delivery via the buccal mucosal route. Pharm. Technol. 6: 70–81.
[12] Illum, L. (2002) Nasal drug delivery: new developments and strategies. Drug Discov. Today 7: 1184–1189.
[13] Urtti, A. (2006) Ocular drug delivery. Adv. Drug Deliv. Rev. 58: 1129–1130.
[14] Bayat, M., Cook, A.M. (2004) Intrapulmonary administration of medications. J. Neurosci. Nurs. 36: 231–235.
[15] Yang, L., Chu, J.S., Fix, J.A. (2002) Colon-specific drug delivery: new approaches and *in vitro*/*in vivo* evaluation. Int. J. Pharm. 235: 1–15.
[16] Pauletti, G.M., Gangwar, S., Knipp, G.T., Nerurkar, M.M., Okumu, F.W., Tamura, K., Siahaan, T.J., Borchardt, R.T. (1996) Structural requirements for intestinal absorption of peptide drugs. J. Control. Release 41: 3–17.
[17] Kompella, U.B., Lee, V.H.L. (2001) Delivery systems for penetration enhancement of peptide and protein drugs: design considerations. Adv. Drug Deliv. Rev. 46: 211–245.
[18] Woodley, J.F. (1994) Enzymatic barriers for GI peptide and protein delivery. Crit. Rev. Ther. Drug Carrier Syst. 11: 61–95.
[19] Norris, D.A., Puri, N., Sinko, P.J. (1998) The effect of physical barriers and properties on the oral absorption of particulates. Adv. Drug Deliv. Rev. 34: 135–154.
[20] Fix, J.A. (1996) Oral controlled release technology for peptides: status and future prospects. Pharm. Res. 13: 1760–1764.
[21] Langguth, P., Bohner, V., Heizmann, J., Merkle, H.P., Wolffram, S., Amidon, G.L., Yamashita, S. (1997) The challenge of proteolytic enzymes in intestinal peptide delivery. J. Control. Release 46: 39–57.
[22] Fasano, A. (1998) Innovative strategies for the oral delivery of drugs and peptides. Trends Biotechnol. 16: 152–157.
[23] Morishita, I., Morishita, M., Takayama, K., Machida, Y., Nagai, T. (1992) Hypoglycemic effect of novel oral microspheres of insulin with protease inhibitor in normal and diabetic rats. Int. J. Pharm. 78: 9–16.
[24] Mesiha, M., Plakogiannis, F., Vejosoth, S. (1994) Enhanced oral absorption of insulin from desolvated fatty acid-sodium glycocholate emulsions. Int. J. Pharm. 111: 213–216.
[25] Guichard, G., Benkirane, N., Zeder-Lutz, G., Van Rengenmortel, M.H.V., Briand, J.-P., Muller, S. (1994) Antigenic mimicry of natural L-peptides with retro-inverso-peptidomimetics. Proc. Natl Acad. Sci. 91: 9765–9769.
[26] Frère, Y., Danicher, L., Muller, S., Wawrezinieck, A. (2009) Nanoparticles containing a peptide, vectors containing the same and pharmaceutical uses of said nanoparticles and said vectors. Patent number WO2009150371.
[27] Kostarelos, K., Lacerda, L., Pastorin, G., Wu, W., Wieckowski, S., Luangsivilay, J., Godefroy, S., Pantarotto, D., Briand, J.-P., Muller, S., Prato, M., Bianco, A. (2007) Functionalised carbon nanotube cellular uptake and internalisation mechanism is independent of functional group and cell type. Nature Nanotechnol. 2: 108–113.
[28] Lian, T., Ho, R.J.Y. (2001) Trends and developments in liposome drug delivery systems. J. Pharm. Sci. 90: 667–680.
[29] Schwarz, C. (1999) Solid lipid nanoparticles (SLN) for controlled drug delivery. II. Drug incorporation and physicochemical characterization. J. Microencapsul. 16: 205–213.
[30] Silva-Cunha, A., Grossiord, J.L., Puisieux, F., Seiller, M. (1997) W/o/w multiple emulsions of insulin containing a protease inhibitor and an absorption enhancer: preparation, characterization and determination of stability towards proteases *in vitro*. Int. J. Pharm. 158: 79–89.
[31] Murakami, H., Kawashima, Y., Niwa, T., Hino, T., Takeuchi, H., Kobayashi, M. (1997) Influence of the degrees of hydrolyzation and polymerization of poly(vinylalcohol) on the preparation and properties of poly(lactide-co-glycolide) nanoparticle. Int. J. Pharm. 149: 43–49.
[32] Kang, F., Singh, J. (2005) Preparation, *in vitro* release, *in vivo* absorption and biocompatibility studies of insulin-loaded microspheres in rabbits. AAPS Pharm. Sci. Tech. 6: E487–E494.

[33] Sinha, V.R., Bansal, K., Kaushik, R., Kumria, R., Trehan, A. (2004) Poly-[epsilon]-caprolactone microspheres and nanospheres: an overview. Int. J. Pharm. 278: 1–23.
[34] Hombreiro Perez, M., Zinutti, C., Lamprecht, A., Ubrich, N., Astier, A., Hoffman, M., Bodmeier, R., Maincent, P. (2000) The preparation and evaluation of poly([epsilon]-caprolactone) microparticles containing both a lipophilic and a hydrophilic drug. J. Control. Release 65: 429–438.
[35] Barbault-Foucher, S., Gref, R., Russo, P., Guechot, J., Bochot, A. (2002) Design of poly-[epsilon]-caprolactonenanospheres coated with bioadhesive hyaluronic acid for ocular delivery. J. Control. Release 83: 365–375.
[36] Al KhouriFallouh, N., Fessi, H., RoblotTreupel, L., Devissaguet, J.P., Puisieux, F. (1986) Etude et mise au point d'un procédé original de préparation de nanocapsules polyalkylcyanoacrylates par polymérisation interfaciale. Pharm. Acta Helv. 61: 10–11.
[37] Truong-Le, V.L., Walsh, S.M., Schweibert, E., Mao, H.-Q., Guggino, W.B., August, J.T., Leong, K.W. (1999) Gene transfer by DNA–gelatin nanospheres. Arch. Biochem. Biophys. 361: 47–56.
[38] Katti, D. (1999) Preparation of albumin microspheres by an improved process. J. Microencapsul. 16: 231–242.
[39] Natsume, T., Ike, O., Okada, T., Takimoto, N., Shimizu, Y., Ikada, Y. (1993) Porous collagen sponge for esophageal replacement. J. Biomed. Mater. Res. 27: 867–875.
[40] Agnihotri, S.A., Mallikarjuna, N.N., Aminabhavi, T.M. (2004) Recent advances on chitosan-based micro- and nanoparticles in drug delivery. J. Control. Release 100: 5–28.
[41] Grant, G.T., Morris, E.R., Rees, D.A., Smith, P.J.C., Thom, D. (1973) Biological interactions between polysaccharides and divalent cations: the egg-box model. FEBS Letters 32: 195–198.
[42] Rees, D. A., Welsh, E.J. (1977) Secondary and tertiary structure of polysaccharides in solutions and gels. Angew. Chem. Int. Ed. Eng. 16: 214–224.
[43] Jeanloz, R. (1950) Structure de la chitine et de l'acide hyaluronique et oxydation des sucres aminés par l'ion periodate. Cell. Mol. Life Sci. 6: 52–53.
[44] Tian, W.M., Hou, S.P., Ma, J., Zhang, C.I., Xu, Q.Y., Lee, I.S., Li, H.D., Spector, M., Cui, F.Z. (2005) Hyaluronic acid-poly-D-lysine-based three-dimensional hydrogel for traumatic brain injury. Tissue Eng. 11: 513–525.
[45] Frère, Y., Danicher, L., Belcourt, A. (2003) Vectors for oral administration. Patent number WO2004096172.
[46] Bungenberg de Jong, H.G. (1949) Complex colloid systems. Colloid Sci. 2: 335–429.
[47] Kabanov, V.A., Zezin, A.B. (1984) Soluble interpolymeric complexes as a new class of synthetic polyelectrolytes. Pure Appl. Chem. 56: 343–354.
[48] Bakeev, K.N., Izumrudov, V.A., Kuchanov, S.I., Zezin, A.B., Kabanov, V.A. (1992) Kinetics and mechanism of interpolyelectrolyte exchange and addition reactions. Macromolecules 25: 4249–4254.
[49] Hugerth, A., Caram-Lelham, N., Sundelof, L.-O. (1997) The effect of charge density and conformation on the polyelectrolyte complex formation between carageenan and chitosan. Carbohydr. Polym. 34: 149–156.
[50] Rusu-Balaita, L., Desbrières, J., Rinaudo, M. (2003) Formation of a biocompatible polyelectrolyte complex: chitosan–hyaluronan complex stability. Polym. Bull. 50: 91–98.
[51] Monneaux, F., Lozano, J.M., Patarroyo, M.E., Briand, J.-P., Muller, S. (2003) T cell recognition and therapeutic effects of a phosphorylated synthetic peptide of the 70 K snRNP protein administered in MRL/lpr lupus mice. Eur. J. Immunol. 33: 287–296.
[52] Page, N., Gros, F., Schall, N., Décossas, M., Bagnard, D., Briand, J.-P., Muller, S. (2011) HSC70 blockade by the therapeutic peptide P140 affects autophagic processes and endogenous MHCII presentation in murine lupus. Ann. Rheum. Dis. 70: 837–843.
[53] Monneaux, F., Hoebeke, J., Sordet, C., Nonn, C., Briand, J.-P., Maillère, B., Sibillia, J., Muller, S. (2005) Selective modulation of CD4+ T cells from lupus patients by a promiscuous, protective peptide analogue. J. Immunol.175: 5839–5847.
[54] Muller, S., Monneaux, F., Schall, N., Rashkov, R.K., Oparanov, B.A., Wiesel, P., Geiger, J.M., Zimmer, R. (2008) Spliceosomal peptide P140 for immunotherapy of systemic lupus erythematosus. Results of an early phase II clinical trial. Arthritis Rheum. 58: 3873–3883.

[55] Zimmer, R., Scherbarth, H.R., Rillo, O.L., Gomez-Reino, J., Muller, S. (2013) Lupuzor/P140 peptide in patients with systemic lupus erythematosus: a randomised, double-blind, placebo-controlled phase IIb clinical trial. Ann. Rheum. Dis., in press.

[56] Monneaux, F., Parietti, V., Briand, J.P., Muller, S. (2004) Intramolecular T cell spreading in unprimed MRL/lpr mice: importance of the U1-70K protein sequence 131–151. Arthritis Rheum. 50: 3232–3238.

[57] Utz, P.J. Personal communication.

[58] Page, N., Schall, N., Strub, J.-M., Quinternet, M., Chaloin, O., Décossas, M., Cung, M.-T., Van Dorsselaer, A., Briand, J.-P., Muller, S. (2009) The splice osomalphosphopeptide P140 controls the lupus disease by interacting with the HSC70 protein and via a mechanism mediated by γδ T cells. PloS ONE 4: e5273.

[59] Monneaux, F., Parietti, V., Briand, J.-P., Muller, S. (2007) Importance of spliceosomal RNP1 motif for intermolecular T-B cell spreading and tolerance restoration in lupus. Arthritis Res. Ther. 9: R111.

[60] Cooke, I.R., Williams, D.R.M. (2004) Condensed states of a semiflexible homopolymer: ordered globules and toroids. Physica A 339: 45–52.

[61] Cowman, M.K., Li, M., Balazs, E.A. (1998) Tapping mode atomic force microscopy of hyaluronan: extended and intramolecularly interacting chains. Biophys. J. 75: 2030–2037.

[62] Spagnoli, C., Korniakov, A., Ulman, A., Balazs, E.A., Lyubchenko, Y.L., Cowman, M.K. (2005) Hyaluronan conformations on surface: effect of surface charge and hydrophobicity. Carbohydr. Polym. 340: 929–941.

15
Identification and Application of Polymer-Binding Peptides

Toshiki Sawada and Takeshi Serizawa

Department of Organic and Polymeric Materials, Tokyo Institute of Technology, Japan

Acronyms

CSD	Cell-surface display
ELISA	Enzyme-linked immunosorbent assay
PD	Phage display
PLLA	Poly(L-lactide)
PMMA	Poly(methyl methacrylate)
PPV	Poly(*p*-phenylene vinylene)
PS	Polystyrene
QCM	Quartz crystal microbalance
SPR	Surface plasmon resonance

15.1 Introduction

To understand and regulate biological phenomena such as protein adsorption and cell adhesion with a molecular level on the surface of synthetic polymers is extremely crucial for developing novel polymer materials utilized in biomedical fields [1–5]. Especially in the case of hydrophobic or charged polymers, proteins easily adsorb on to the polymer surfaces, thereby mediating various biological processes including cell adhesion, extension,

Peptide Materials: From Nanostructures to Applications, First Edition. Edited by Carlos Alemán, Alberto Bianco and Mariano Venanzi.

and proliferation. However, understanding of these interfacial phenomena is normally difficult because the interfaces are composed of multiple, complicated interactions such as electrostatic, hydrogen bonding, and/or van der Waals forces. Therefore, researchers frequently describe a vague term, 'nonspecific', when proteins adsorb on to polymer surfaces. On the other hand, nonspecific adsorption of proteins on to polymer surfaces is not stable under physiological conditions due to their noncovalent interactions, and therefore is not available for the long-term applications required in a variety of biomedical fields. Therefore, our focus has moved on to more reliable and robust interactions between the polymer surfaces and biomolecules.

It was reported that peptides could be genetically displayed on the coat proteins of filamentous phages, based on the insertion of the corresponding DNA fragment into the phage genome in 1985 [6]. The infectiousness of the resulting genetically engineered phages to a host cell such as *Escherichia coli* (*E. coli*) was maintained, followed by phage amplification. The aforementioned method to display desired peptides on the surface of phages as fused proteins is generally called as a 'phage-display' (PD) method. The most commonly utilized phage is an M13 filamentous phage, which has five coat proteins such as pIII, pVI, pII, pVIII, and pIX on its surface (Figure 15.1) [7]. When peptides with varieties of different amino acid sequences are displayed on the surface of each different phage, a peptide library displayed on the phage can be constructed in principle.

Based on the affinity-dependent selection of certain phages for a target molecule and the subsequent DNA sequencing of the phage, the peptides that specifically bind to the target molecule can be identified. Adequate genetic engineering for certain cells also enables cell-surface display (CSD) methods to be developed, in which peptides are fused into appropriate proteins on the cell-membrane surface [8–10]. Conventionally, the PD and CSD methods have been applied to identify peptides that have an affinity for biomolecules such as proteins, nucleotides, or saccharides, and the resulting peptides were potentially exploited for diagnostics and therapeutics [11–13]. The biological combinatorial methods were also applied for biomacromolecules such as collagen [14], chitin [15], and cellulose [16] to identify the specifically binding peptides.

Over the past decade, inspired by the initial researches of Brown [17] and Gaskin *et al.* [18], the PD and CSD methods have been applied to artificial materials instead of biomolecules, and unique peptides that bind specifically to the target materials surface were

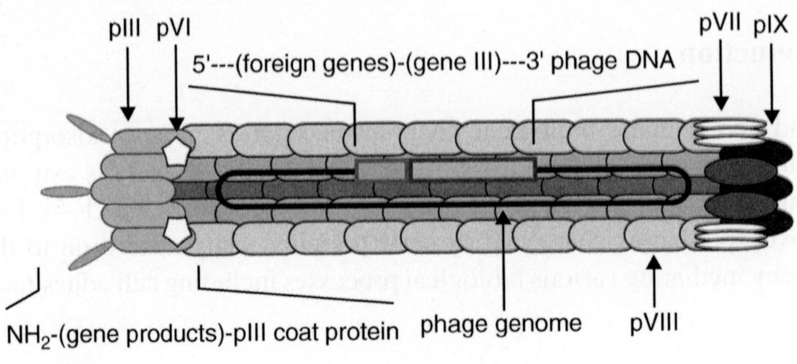

Figure 15.1 *Schematic representation of an M13 filamentous phage displaying peptides on the pIII coat proteins. Reproduced with permission from reference [7]. Copyright (2011) The Royal Society of Chemistry*

successfully identified [19–21]. The inorganic material surfaces of metal oxides [17, 18, 22–25], semiconductors [26–29], metals [30–32], and magnetics [33] were used as targets for peptides. Organic molecules such as carbon nanotubes [34, 35], carbon nanohorns [36, 37], fullerenes [38], assembled peptides [39, 40], and synthetic polymers [41–43] were also applied to these methods. The identified peptides for the target material have the potential for noncovalently constructing a stable interface between the materials and the peptides through specific binding. In fact, material-binding peptides were utilized as novel bionanomaterials such as catalysts for the preparation and dispersion stabilization of inorganic nanoparticles [44–47], nanocrystal assembly [48], adsorbents for patterning [32], surface modifiers [39, 43, 44, 49–51], and modifiers of phages [27] or proteins [52–53].

15.2 Biological Identification of Material-Binding Peptides

The protocols to construct phage-displayed peptide libraries have already been established [54, 55]; therefore, peptides with random sequences can be fused at the desired positions of phage coat proteins. In fact, M13 phages displaying 7- and 12-mer linear (or 7-mer circular) random peptide libraries on the pIII coat protein are commercially available from New England Biolabs. The enrichment process from phage libraries to target-specific phage pools is called 'biopanning' (Figure 15.2). The biopanning is essentially composed of four experimental steps as follows. In Step 1 (Interaction), an aqueous solution of the phage libraries is mounted on to the surface of target materials (film, plate, or other solid) at ambient temperature for an adequate time. In Step 2 (Wash), the target surfaces are washed with a buffer solution several times to remove unbound or weakly bound phages.

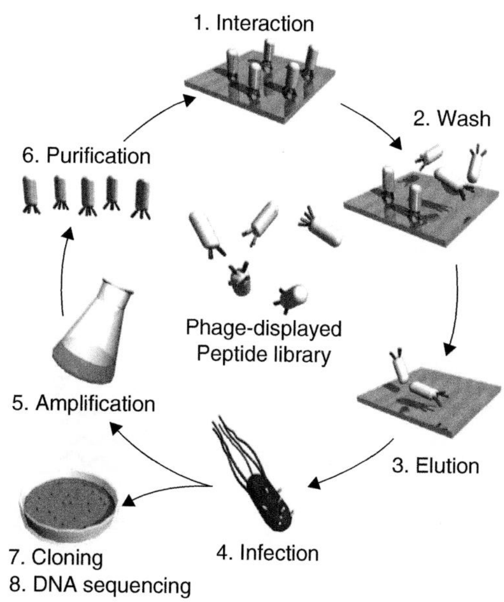

Figure 15.2 *Schematic representation of the PD method (biopanning). Reproduced with permission from reference [7]. Copyright (2011) The Royal Society of Chemistry*

The washing solution sometimes contains surfactants to more effectively wash out the unfavorable phages. In Step 3 (Elution), the strongly bound phages are eluted from the target surfaces by an elution buffer. In Step 4 (Amplification), the eluted phages are proliferated within *E. coli* to amplify the phages. These steps are repeated until the phage libraries are enriched into the desired phage clones. After adequate rounds of biopanning, the DNA sequence of displayed peptide on each phage clone, which corresponds to each amino acid sequence, is determined by conventional sequencing methods.

Ideally, a single phage clone should be isolated; however, this is not practical. In other words, at least several possible phage clones are usually isolated. There are no universal conditions to enrich phage libraries, and therefore optimization of all the parameters for the interaction, wash, elution, and biopanning rounds is needed for each target. Note that certain phage clones tend to unfavorably appear after biopanning [56]. Finally, the binding of the selected phage clones on the surface of the target or reference materials is analyzed by a semi-quantitative titer counting of the bound phages [43], enzyme-linked immunosorbent assay (ELISA) [57], or quartz crystal microbalance (QCM) measurements [23].

After the possible phage clones are identified, the corresponding peptides are chemically synthesized by a conventional solid-phase method by using the 9-fluorenylmethyloxycarbonyl (Fmoc) strategy. Analyses of the amounts of the synthetic peptides that bind to the surface of the target materials can be performed by a quantification of the unbound peptides in the supernatants [45], surface plasmon resonance (SPR) measurements [58–61], or QCM measurements [58, 62]. Binding assays with various peptide concentrations can calculate the binding constant (K_a). In general, the K_a values tend to increase with an increasing number of amino acids in the peptides. However, when the number of amino acid sequences is 7–12, the K_a values normally ranged between $10^4\,M^{-1}$ and $10^6\,M^{-1}$, regardless of the target species.

Since the aforementioned K_a values of synthetic peptides are considered to be small for the versatile application of these peptides, conformational fixation by peptide cyclization [59, 60], buildup to tandem peptides [63] and peptide clustering [61] succeeded in drastically increasing the K_a values (Figure 15.3). On the other hand, binding force measurements of peptide-modified cantilever tips against the target material surfaces have potential for quantification of the binding strength of peptides [24, 25]. After the binding assays, researchers need to determine the binding mechanism. Since it is extremely difficult to analyze the proper structure of the peptides at the solid surface of target materials, molecular

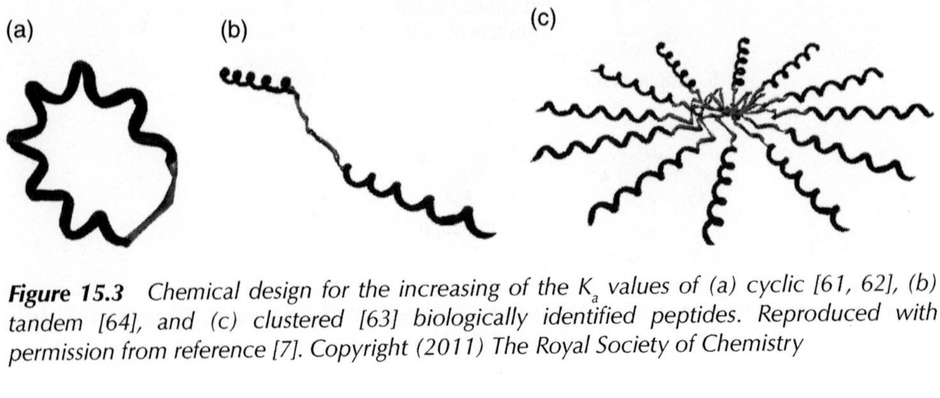

Figure 15.3 *Chemical design for the increasing of the* K_a *values of (a) cyclic [61, 62], (b) tandem [64], and (c) clustered [63] biologically identified peptides. Reproduced with permission from reference [7]. Copyright (2011) The Royal Society of Chemistry*

simulations of peptide binding are sometimes employed at present, and suggest that the conformation of the peptides is induced to fit the surface structure of the materials [48, 64].

15.3 Recognition of Polymer Stereoregularity by Peptides

We have been interested in biologically identified peptides that recognize slight differences between synthetic polymers. Although the chemical structures of synthetic polymers are considered to be too simple to be recognized by peptides, various nanostructures such as a primary sequence, stereoregularity, amphiphilicity, crystallinity, chirality, a linear/branched structure, an assembly structure, and porosity may fit into the three-dimensionally regulated nanostructures of certain peptides (Figure 15.4). In this chapter, representative methods for identification of peptides based on the PD method are outlined and subsequently our recent progress in the identification/applications of polymer-binding peptides is described.

We originally focused on the recognition of unique structural differences in polymer stereoregularity by biologically identified peptides. Among various synthetic polymers, stereoregular polymers of methacrylates, which can be synthesized by conventional living anionic polymerization [65, 66], were potential candidates for peptide targets, although these polymers have an extremely simple chemical structure composed only of alkyl and ester groups. Nonetheless, it was still attractive to clarify whether peptides discriminate between the configurational differences of methacrylate polymers with essentially the same chemical structure.

For the first target, the PD method using 7-mer random peptide libraries was applied to isotactic poly(methyl methacrylate) (it-PMMA), in which the C_α carbons of adjacent MMA units have an equal configuration (Figure 15.5a) [67]. Syndiotactic (st-)PMMA, in which the C_α carbons of adjacent MMA units have unequal configurations, was adopted as a refer-

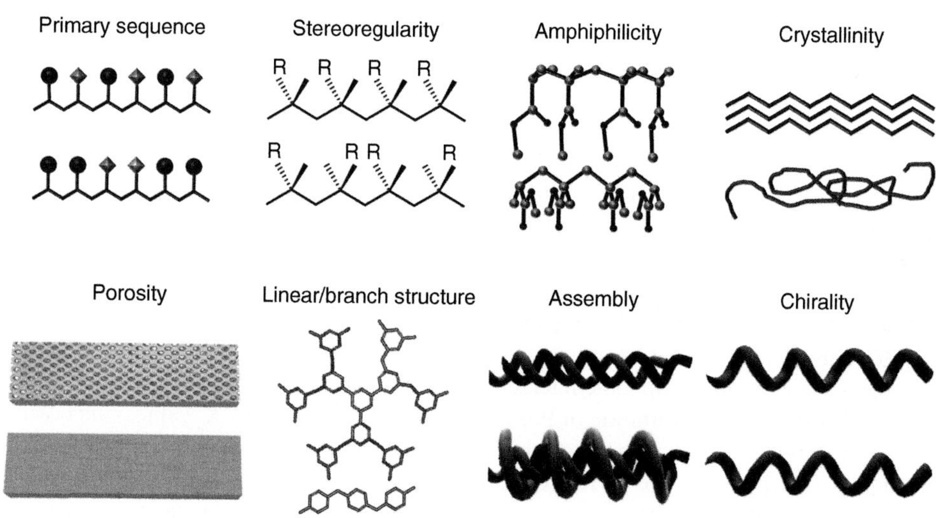

Figure 15.4 Possible polymeric nanostructures for peptide recognition. Reproduced with permission from reference [7]. Copyright (2011) The Royal Society of Chemistry

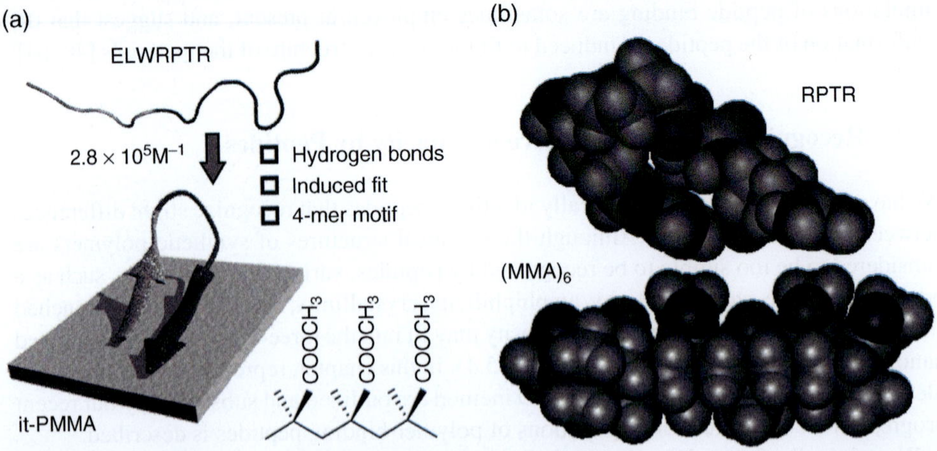

Figure 15.5 (a) Recognition of stereoregular it-PMMA by the c02 peptide (sequence: Glu–Leu–Trp–Arg–Pro–Thr–Arg), in which the C-terminal 4-mer is an essential motif for the specificity to it-PMMA. (b) Comparison of one possible structure of the 4-mer peptide (Arg–Pro–Thr–Arg) obtained by Molecular Mechanics and the hexamer of it-PMMA. It was suggested that the 4-mer motif recognizes the arrangement of methyl ester groups composed of few it-MMA units [68, 70]. Reproduced with permission from reference [7]. Copyright (2011) The Royal Society of Chemistry

ence polymer. After five rounds of biopanning for spin-cast films of it-PMMA, nine phage clones were identified. ELISA revealed that the binding amounts of all clones for it-PMMA were significantly greater than those for st-PMMA. Considering the huge molecular weight of the M13 phage (16 300 000) [68], it was surprising that the 7-mer peptides displayed on the filamentous page terminus affected the binding ability of the phage clones. Then, the binding processes of chemically synthesized 7-mer peptides for the target it-PMMA and the reference st-PMMA were monitored by SPR measurements and the K_a values were determined [69]. The maximum K_a of the c02 peptide with the Glu–Leu–Trp–Arg–Pro–Thr–Arg sequence approached $2.8 \times 10^5\ M^{-1}$ for the target, which are 40 times greater than that for the reference, st-PMMA ($6.9 \times 10^3\ M^{-1}$). The value for the target is comparable to constants obtained from 12-mer peptides that bind to titanium oxide surfaces, suggesting that peptides have potentials for recognizing subtle differences in the stereoregularity of methacrylate polymers (Figures 15.5a and 15.6a).

Each amino acid of the aforementioned c02 peptide was replaced by a neutral and hydrophobic Ala to determine the essential amino acid residues for the it-PMMA specificity [69]. Substitution of the hydrophobic amino acid resulted in a slight decrease in the K_a value, suggesting that hydrophobic interactions are not essential for the it-PMMA affinity. On the other hand, the substitution of Pro to Ala greatly decreased the K_a value, which was more than 30 times smaller than the original one, implying that the rigid peptide conformation due to the ring structure of Pro was important for the affinity. Together with other experimental results, the C-terminal 4 residues, Arg–Pro–Thr–Arg sequence, which is composed of amino acids with proton-donor hydroxy and amino lateral groups adjacent to the Pro, was an essential motif for the it-PMMA affinity. The essential sequence with a

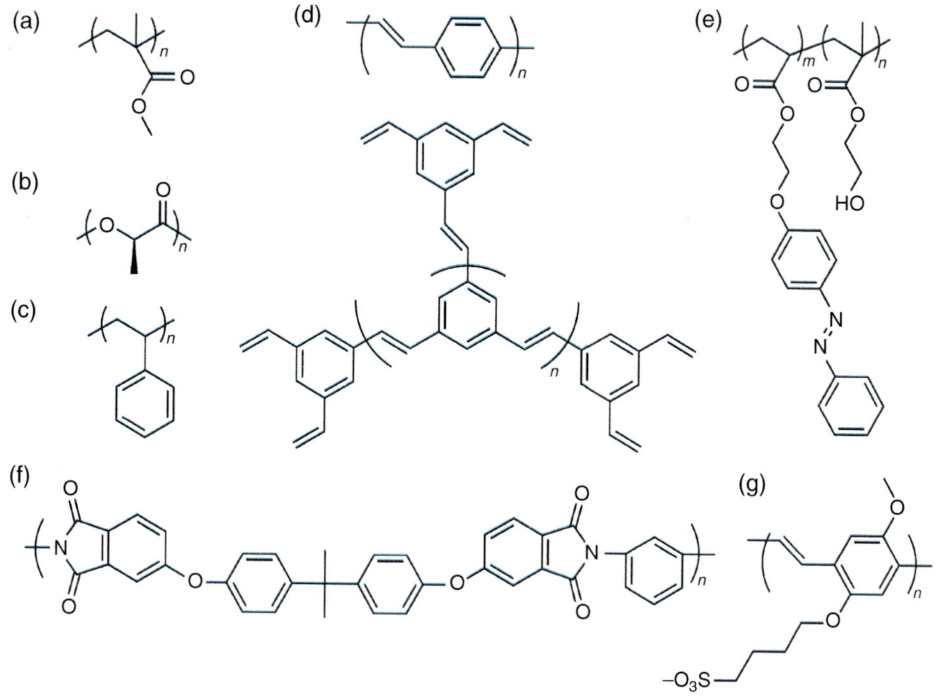

Figure 15.6 *Chemical structures of synthetic polymers applied to the PD method. (a) Poly(methyl methacrylate) [68, 72], (b) polylactide [77], (c) polystyrene [78], (d) poly (p-phenylene vinylene) [85], (e) azobenzene-containing polymer [93], (f) polyetherimide [110], and (g) poly(2-methoxy-5-propyloxysulfonate-1,4-phenylene vinylene) [111]. Reproduced with permission from reference [7]. Copyright (2011) The Royal Society of Chemistry*

rigid conformation, which would be derived from a kinked Pro, might rearrange three-dimensionally the lateral hydroxyl and/or amino groups of the Thr and two Arg as proton-donors, followed by successful recognition of the it-PMMA ester groups through hydrogen bonding interactions, even in aqueous phase (Figure 15.5b). In fact, binding force measurements using cantilever tips modified with the c02 peptide also demonstrated that the c02 peptide showed the preferential affinity of the C-terminus rather than the N-terminus for the target it-PMMA [70].

St-PMMA, which was previously utilized as in reference [67], was applied to the 7-mer PD method to determine whether peptides can generally discriminate between subtle differences in polymer stereoregularity [71]. Moreover, st-PMMA has an interesting property as a novel target for peptides described below. Conventional material surfaces for peptide targets have static characters. Here, we first investigated the recognition of variable film surfaces composed of amphiphilic polymers. Surface functional groups on suitable polymer films can be controlled by the external environment. In the case of st-PMMA films prepared on glass slides in air, it was reported that hydrophilic esters and hydrophobic alkyl groups were accumulated on the glass and air sides, respectively [72]. Therefore, st-PMMA films with variable surfaces are suitable candidates for demonstrating both the potential specificity of peptides against materials and the novel method controlling peptide–polymer interactions. After three rounds of biopanning, the possible phage clone displaying the His–Lys–Pro–Asp–Ala–Asn–Arg sequence was identified successfully. This sequence was also composed of several amino acid residues with amino groups and a Pro residue. Therefore, the presence was also anticipated of hydrogen bonding interactions between the peptide and st-PMMA. Expectedly, the amount of phage clone bound to the target st-PMMA was evidently greater than to the reference it-PMMA, suggesting that peptides can generally recognize subtle structural differences in polymers derived from stereoregularity. In fact, the K_a values of the synthetic peptide for st- and it-PMMAs were calculated to $9.1\times10^4\,M^{-1}$ and $3.0\times10^3\,M^{-1}$, respectively, thus demonstrating excellent st-PMMA specificity [69].

More interestingly, greater binding amounts of the phage clones for water-conditioned st-PMMA films, which were utilized for the peptide target, than to nonconditioned films were observed [71]. In fact, the static contact angles of the st-PMMA films decreased with increasing conditioning time, suggesting that the hydrophilic ester groups of st-PMMA rather than the hydrophobic alkyl groups were gradually exposed into the water side. Therefore, the peptides would recognize nanoscale arrangements of the ester groups due to the syndiotacticities. Similarly, peptide motifs that recognize variability in polymer film surfaces were discovered based on their affinity-dependent selection from phage-displayed peptide libraries against adequately treated st-PMMA films. Furthermore, peptides that recognize helical stereocomplexes formed by it- and st-PMMAs could be identified from phage-displayed peptide libraries [73]. Thus, assembled nanostructures prepared from stereoregular PMMAs would generally be a definite target for peptides.

15.4 Recognition of Other Polymer Nanostructures by Peptides

Poly(L-lactide) (PLLA) (Figure 15.6b) is one of the most commonly utilized synthetic polymers for biomedical materials due to its excellent properties such as thermal stability, mechanical properties, biocompatibility, and biodegradability to nontoxic products [74, 75]. However, it is difficult to introduce functionalities on the surface of PLLA because of its poor hydrophilicity and the few functionalized groups, thus resulting in a restriction for biomedical applications. Therefore, the 7-mer peptides that specifically bind to PLLA surfaces composed of the α-form crystal were selected by the PD method [76]. Kinetic analyses using SPR measurements showed that the Gln–Leu–Met–His–Asp–Tyr–Arg sequence bound to the target PLLA with a K_a value of $6.1\times10^4\,M^{-1}$, which was 10 times greater than those for the reference amorphous PLLA. Therefore, the peptide recognition for polymers could be regulated by simple annealing processes that enhance polymer crystallinity. The peptide also discriminates a slight difference in the helical pitch between the PLLA polymorphs. The K_a value for the 10_3-helical α-form was fourfold greater than that for the 3_1-helical β-form. This was the first demonstration that peptides can recognize the crystal polymorphs of synthetic polymers.

Polystyrene (PS) (Figure 15.6c) is an all-purpose vinyl polymer frequently utilized for plastic plates in biological assays such as microtiter plates. PS binding peptides were previously inadvertently obtained [41], even though it has a simple chemical structure composed of an alkyl main chain and a lateral phenyl group conjugated to every two main chain carbons. We focused on the unique nanostructure formed by syndiotactic PS (sPS), and its specific peptides were identified by the PD method using 7-mer peptide libraries [77]. sPS forms TTGG helical conformations (T and G indicate *trans* and *gauche* conformations, respectively), when the films were adequately prepared by suitable solvents [78–80]. The evaporation of free solvents fabricates δ-form films complexed with the solvent molecules. Further evaporation of the solvents fabricates empty δ-form (δ_e) porous films. Here, the δ_e-formed sPS film was utilized as the target. The phage clone displaying the peptide with the sequence of Phe–Ser–Trp–Glu–Ala–Phe–Ala, which is composed of greater amounts of aliphatic and aromatic amino acids, was successfully isolated. It was suggested that sPS and the peptides interact through hydrophobic effects and π–π interactions. Binding analyses by ELISA using the target sPS, atactic PS, isotactic PS, and sPS/toluene complex surfaces clearly demonstrated the δ_e form sPS specific affinity of the peptides. Therefore, it was indicated that peptides have the potential to recognize porous nanostructures of synthetic polymers.

Conjugated polymers are attractive materials and have been utilized in many applications, which include light emitting diodes [81], photovoltaic cells [82], and biosensing [83], because of the unique optical and electronic properties. A representative conjugated polymer, poly(*p*-phenylene vinylene) (PPV) (Figure 15.6d) and its derivatives, which have a simple chemical structure with phenyl rings and ethenyl groups, are the most extensively studied conjugated polymers. We identified 12-mer peptides that specifically recognize the linear or branched PPV based on the PD method [84]. The identified peptides for each target showed excellent specificity, indicating that the peptides recognize the slight structural difference in the PPV isomers. The aromatic amino acids were enriched in these sequences, thereby suggesting that π–π interactions between the peptides and PPV were essential. The peptide with the Trp–Arg–Leu–Gly–Thr–Trp–His sequence showed specific affinity for the branched PPV rather than the linear PPV. Molecular mechanics suggested that the two Trp residues in the peptide caused suitable interactions with the phenyl rings in the PPV. This was the first attempt to demonstrate the potential of peptides that significantly recognize linear versus branch structures in conjugated polymers.

Control of certain reactions by external stimuli facilitates the development of stimuli-responsive material systems. Stimuli such as light [85–87], temperature [88], pH [89], electric fields [90], and molecules [91] have been utilized to control material responses. Because photoirradiation with the adequate power and time conserves biomolecular structure and function, photostimuli are particularly advantageous for biological reactions. We applied the PD method using 7-mer libraries to film surfaces composed of azobenzene-containing synthetic copolymers (Figure 15.6e) [92]. Peptides with the characteristic amino acid sequence were enriched after four rounds of biopanning. ELISA using the phage clones and QCM measurements using the chemically synthesized peptide with the representative sequence (Trp–His–Thr–Leu–Pro–Asn–Ala) suggested that the peptide showed the preferential affinity to *cis* forms of azobenzene groups on the film surfaces. Binding amounts of the peptide could be controlled by the photoisomerization of the azobenzene groups, followed by stimuli-responsive switching of peptide affinity. This was the first report demonstrating that polymer-binding peptides can respond to external stimuli based on extreme specificity.

15.5 Applications of Polymer-Binding Peptides

Material-binding peptides have been utilized as effective tools for protein modification. Previously, the peptides were fused with functional proteins such as apoferritin [51, 62], cytokine [93], and green fluorescent protein [94]. The fused proteins achieved more stable immobilization on to inorganic and carbonaceous materials through noncovalent specific binding. We also reported the extremely enhanced adsorption of a model protein on to the it-PMMA film surface after fusion with the c02 peptide (Glu–Leu–Trp–Arg–Pro–Thr–Arg sequence) specific for it-PMMA (Figure 15.7a) [95]. DnaK419-607 (a blocking peptide fragment, or BPF), which is part of the substrate-binding domain of the molecular chaperon DnaK from *E. coli* [96, 97], functions as an effective blocking reagent due to its good adsorptive properties for plastic substrates through its hydrophobic domain [98]. We selected a BPF derivative (dBPF) fused with a C-terminal RGD, which would promote cell adhesion on material surfaces [99]. The K_a value of the N-terminal fused protein for it-PMMA was $4.9 \times 10^8\,M^{-1}$, and was 75 times greater than that of the original protein. Thus, the fusion protein showed the specific affinity for it-PMMA. On the other hand, genetically engineered green fluorescent protein fused with the Trp–His–Thr–Leu–Pro–Asn–Ala sequence, which specifically binds to azobenzene polymers with *cis* forms [92, 100], preferentially bound to the target polymers compared with *trans* forms [102]. These observations strongly indicate that polymer-binding peptides generally function as a specific enhancement module for immobilization of functional proteins on the polymer surface.

Au nanoparticles have potential applications in the biological and nanotechnological fields due to their unique surface plasmon properties [102–104]. Functional molecule-capping Au nanoparticles are able to direct novel and desired functions. Therefore, the surface of Au nanoparticles was capped with the Arg–Pro–Thr–Arg sequence, which was an essential motif for it-PMMA binding (Figure 15.7b) [69]. The 5-mer peptide with the Cys–Arg–Pro–Thr–Arg sequence was designed to anchor for the Au surface through thiol-Au strong bonding. $AuCl^{4-}$ was reduced in 2-[4-(2-hydroxyethyl)-1-piperazinyl]ethanesulfonic acid (HEPES) buffer containing the designed peptide under mild conditions (pH 7.2, ambient temperature), thus resulting in peptide-capped Au nanoparticles, which were stably dispersed in HEPES buffer [105]. The prepared nanoparticles preferentially adsorbed on to the it-PMMA film surface as compared to st-PMMA surfaces. Accordingly, it was concluded that polymer-binding peptides could be used as connecters for the hetereogeneous interface between inorganic compounds and synthetic polymers.

Functional modification of material surfaces plays a crucial role in recent material science and technology. In various surface modification techniques, construction of self-assembled monolayers on inorganic materials are often used for their surface modification practically and experimentally, due to the simplest methods to prepare a wide variety of functional surfaces with well-defined compositions on substrates [106–108]. If one could extend such a SAM concept to other materials, it would broaden the range of technical applications. Therefore, peptides that specifically bind to polyetherimide (PEI), a representative engineering plastic, were identified and utilized for surface modification due to its potential biomedical applications (Figures 15.6f and 15.7c) [109]. The K_a value of the identified 7-mer peptide with the sequence of Thr–Gly–Ala–Asp–Leu–Asn–Thr for PEI

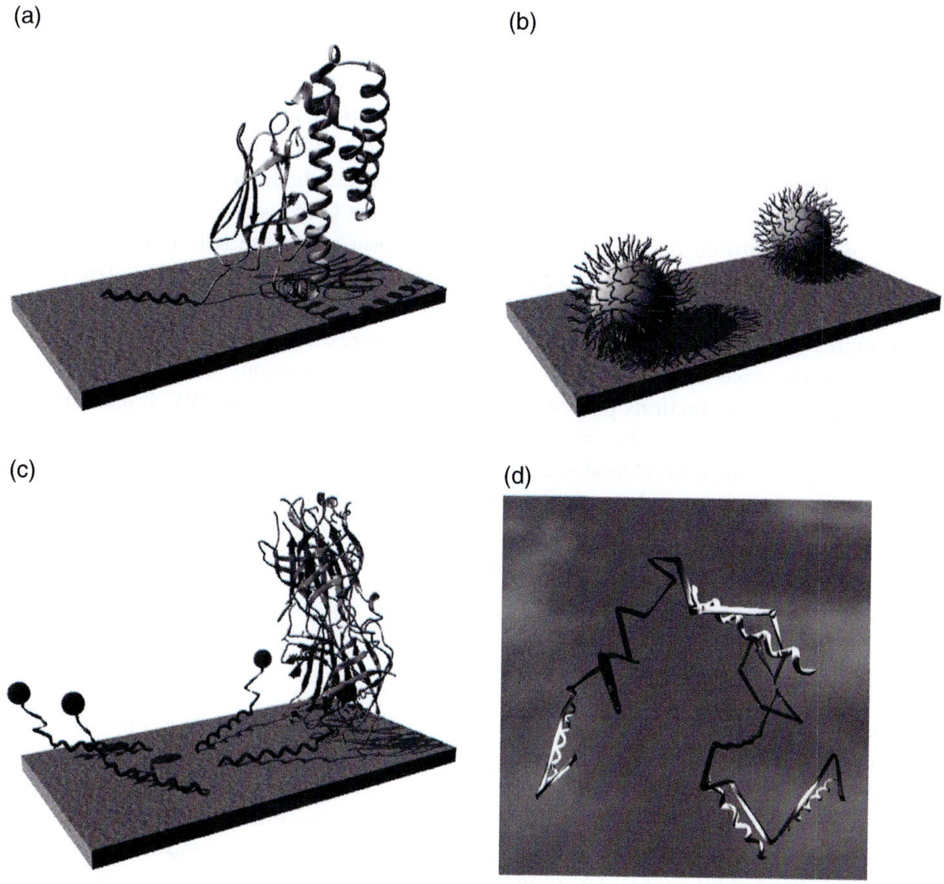

Figure 15.7 *Applications of polymer-binding peptides. (a) Fusion with proteins [96], (b) capping of nanoparticles [106], (c) surface modification of polymeric materials [110], and (d) polymer fluorescence modulation in aqueous solutions [111]. Reproduced with permission from reference [7]. Copyright (2011) The Royal Society of Chemistry*

films was $5.6 \times 10^8\,M^{-1}$, which was extremely high and more than three orders of magnitude greater than that for the reference polystyrene films. For immobilization of streptavidin (SAv), the peptide was then biotinylated and immobilized on to the PEI films. The binding amount of SAv depended on the density of immobilized peptide, resulting in a bell-shape profile, suggesting that effective immobilization of SAv requires appropriate peptide densities. The immobilized SAv via the peptide was able to further immobilize probe DNA to efficiently hybridize with its complementary DNA as compared to SAv being physically immobilized directly. Therefore, it was suggested that the SAv immobilized via the biotinylated specific peptide formed uniform monolayers with less structural denaturing on the PEI films. These results indicated that polymer-binding peptides have great potential for functional modification of polymer surfaces.

In the aforementioned paragraphs, water-insoluble synthetic polymers have been applied to the PD method. In order to extend potentials of polymer-binding peptides, we concentrated ondevoted to "water-soluble" polymers. Thus, commonly-used water-soluble PPV, poly(2-methoxy-5-propyloxysulfonate-1,4-phenylenevinylene) (mpsPPV), was selected as the peptide target (Figures. 15.6 g and 15.7d) [110]. Because of their high and sensitive fluorescence due to amplification by a collective response, water-soluble conjugated polymers have been attracting much attention as fluorescent probes in aqueous solution for biomedical applications [111–114]. To construct the target films composed of water-soluble PPV, the layer-by-layer assembly technique [115] was exploited, while the binding of the identified peptides to mpsPPV was observed in the solution phase. The identified 12-mer peptide with the sequence of His–Asn–Ala–Tyr–Trp–His–Trp–Pro–Pro–Ser–Met–Thr bound to mpsPPV, not only to films but also in aqueous solution. Furthermore, the peptide binding in aqueous solution resulted in enhancement of the polymer fluorescence at a relatively low peptide concentration. Mutation analyses suggested that π–π interactions played a major role in binding to mpsPPV. Interestingly, enzymatic degradation of the binding peptides achieved watchable fluorescence enhancement, realizing a novel biodevice that reversibly responds to bioactive components. As a consequence, it was demonstrated that biologically identified polymer-binding peptides successfully worked even in aqueous solutions based on adequate interactions with water-soluble synthetic polymers.

15.6 Summary

Combinatorial biotechnologies such as the PD and CSD methods applied to identify peptides that specifically recognize the surfaces of artificial materials are described. Then, recent progress in the identification, characterization, and applications of polymer-binding peptides that specifically discriminate the unique regular nanostructures of synthetic polymers are summarized. Observations clearly demonstrated that polymer nanostructures such as monomer structure, stereoregularity, amphiphilicity, crystallinity, porosity, linear/branch structure, and lateral functional groups could be the targets for peptides. These polymer-binding peptides are applicable to functional design of the interfaces between synthetic polymers and other contacting materials, evidenced by possible applications. Therefore, biologically identified peptides can be considered as versatile tools for manipulating the interfaces and developing material innovations.

References

[1] Elbert, D.L., Hubbell, J.A. (1996) Surface treatments of polymers for biocompatibility. Annu. Rev. Mater. Sci. 26: 365–394.
[2] Ha, C.-S., Gardella, J. (2005) Surface chemistry of biodegradable polymers for drug delivery systems. Chem. Rev. 105: 4205–4237.
[3] Vogler, E. (1998) Structure and reactivity of water at biomaterial surfaces. Adv. Colloid Interface Sci. 74: 69–186.
[4] Kasemo, B. (2002) Biological surface science. Surf. Sci. 500: 656–677.
[5] Wang, Y.-X., Robertson, J., Spillman, W., Claus, R. (2004) Effects of the chemical structure and the surface properties of polymeric biomaterials on their biocompatibility. Pharm. Res. 21: 1362–1435.
[6] Smith, G. (1985) Filamentous fusion phage: novel expression vectors that display cloned antigens on the virion surface. Science 228: 1315–1322.

[7] Serizawa, T., Matsuno, H., Sawada, T. (2011) Specific interactions between synthetic polymers and biologically identified peptides. J. Mater. Chem. 21: 10252–10260.
[8] Georgiou, G., Stathopoulos, C., Daugherty, P., Nayak, A., Iverson, B., Curtiss, R. (1997) Display of heterologous proteins on the surface of microorganisms: from the screening of combinatorial libraries to live recombinant vaccines. Nat. Biotechnol. 15: 29–63.
[9] Schreuder, M., Mooren, A., Toschka, H., Verrips, C., Klis, F. (1996) Immobilizing proteins on the surface of yeast cells. Trends Biotechnol. 14: 115–135.
[10] Maassen, C., Laman, J., den Bak-Glashouwer, M., Tielen, F., van Holten-Neelen, J., Hoogteijling, L., Antonissen, C., Leer, R., Pouwels, P., Boersma, W., Shaw, D. (1999) Instruments for oral disease-intervention strategies: recombinant *Lactobacillus casei* expressing tetanus toxin fragment C for vaccination or myelin proteins for oral tolerance induction in multiple sclerosis. Vaccine 17: 2117–2145.
[11] Brown, K. (2000) New approaches for cell-specific targeting: identification of cell-selective peptides from combinatorial libraries. Curr. Opin. Chem. Biol. 4: 16–37.
[12] Dower, W., Mattheakis, L. (2002) *In vitro* selection as a powerful tool for the applied evolution of proteins and peptides. Curr. Opin. Chem. Biol. 6: 390–398.
[13] Hoess, R. (2001) Protein design and phage display. Chem. Rev. 101: 3205–3223.
[14] Rothenfluh, D., Bermudez, H., O'Neil, C., Hubbell, J. (2008) Biofunctional polymer nanoparticles for intra-articular targeting and retention in cartilage. Nat. Mater. 7: 248–254.
[15] Fukusaki, E., Ogawa, K., Okazawa, A., Kajiyama, S., Kobayashi, A. (2004) A chitin-oligomer binding peptide obtained by screening of a phage display random peptide library and its affinity modulation corresponding to oxidation-reduction state. J. Mol. Catal. B: Enzym. 28: 181–184.
[16] Serizawa, T., Iida, K., Matsuno, H., Kurita, K. (2007) Cellulose-binding heptapeptides identified by phage display methods. Chem. Lett. 36: 988–989.
[17] Brown, S. (1992) Engineered iron oxide-adhesion mutants of the *Escherichia coli* phage lambda receptor. Proc. Natl Acad. Sci. USA 89: 8651–8656.
[18] Gaskin, D.J.H., Starck, K., Vulfson, E.N. (2000) Identification of inorganic crystal-specific sequences using phage display combinatorial library of short peptides: a feasibility study. Biotechnol. Lett. 22: 1211–2427.
[19] Sarikaya, M., Tamerler, C., Jen, A., Schulten, K., Baneyx, F. (2003) Molecular biomimetics: nanotechnology through biology. Nat. Mater. 2: 577–585.
[20] Patwardhan, S.V., Patwardhan, G., Perry, C.C. (2007) Interactions of biomolecules with inorganic materials: principles, applications and future prospects. J. Mater. Chem. 17: 2875–5759.
[21] Baneyx, F., Schwartz, D. (2007) Selection and analysis of solid-binding peptides. Curr. Opin. Biotechnol. 18: 312–319.
[22] Barbas, C., Rosenblum, J., Lerner, R. (1993) Direct selection of antibodies that coordinate metals from semisynthetic combinatorial libraries. Proc. Natl Acad. Sci. USA 90: 6385–6394.
[23] Sano, K.-I., Shiba, K. (2003) A hexapeptide motif that electrostatically binds to the surface of titanium. J. Am. Chem. Soc. 125: 14234–14239.
[24] Hayashi, T., Sano, K.-I., Shiba, K., Kumashiro, Y., Iwahori, K., Yamashita, I., Hara, M. (2006) Mechanism underlying specificity of proteins targeting inorganic materials. Nano Lett. 6: 515–524.
[25] Dickerson, M.B., Jones, S.E., Cai, Y., Ahmad, G., Naik, R.R., Kröger, N., Sandhage, K.H. (2008) Identification and design of peptides for the rapid, high-yield formation of nanoparticulate TiO_2 from aqueous solutions at room temperature. Chem. Mater. 20: 1578–1584.
[26] Whaley, S., English, D., Hu, E., Barbara, P., Belcher, A. (2000) Selection of peptides with semiconductor binding specificity for directed nanocrystal assembly. Nature 405: 665–673.
[27] Lee, S.-W., Mao, C., Flynn, C., Belcher, A. (2002) Ordering of quantum dots using genetically engineered viruses. Science 296: 892–897.
[28] Umetsu, M., Mizuta, M., Tsumoto, K., Ohara, S., Takami, S., Watanabe, H., Kumagai, I., Adschiri, T. (2005) Bioassisted room-temperature immobilization and mineralization of zinc oxide – the structural ordering of ZnO nanoparticles into a flower-type morphology. Adv. Mater. 17: 2571–2575.
[29] Ahmad, G., Dickerson, M., Cai, Y., Jones, S., Ernst, E., Vernon, J., Haluska, M., Fang, Y., Wang, J., Subramanyam, G., Naik, R., Sandhage, K. (2008) Rapid bioenabled formation of

ferroelectric $BaTiO_3$ at room temperature from an aqueous salt solution at near neutral pH. J. Am. Chem. Soc. 130: 4–9.
[30] Brown, S. (1997) Metal-recognition by repeating polypeptides. Nat. Biotechnol. 15: 269–272.
[31] Brown, S., Sarikaya, M., Johnson, E. (2000) A genetic analysis of crystal growth. J. Mol. Biol. 299: 725–760.
[32] Naik, R., Stringer, S., Agarwal, G., Jones, S., Stone, M. (2002) Biomimetic synthesis and patterning of silver nanoparticles. Nat. Mater. 1: 169–172.
[33] Mao, C., Solis, D., Reiss, B., Kottmann, S., Sweeney, R., Hayhurst, A., Georgiou, G., Iverson, B., Belcher, A. (2004) Virus-based toolkit for the directed synthesis of magnetic and semiconducting nanowires. Science 303: 213–220.
[34] Wang, S., Humphreys, E., Chung, S.-Y., Delduco, D., Lustig, S., Wang, H., Parker, K., Rizzo, N., Subramoney, S., Chiang, Y.-M., Jagota, A. (2003) Peptides with selective affinity for carbon nanotubes. Nat. Mater. 2: 196–200.
[35] Li, X., Chen, W., Zhan, Q., Dai, L., Sowards, L., Pender, M., Naik, R. (2006) Direct measurements of interactions between polypeptides and carbon nanotubes. J. Phys. Chem. B 110: 12621–12626.
[36] Zhu, J., Kase, D., Shiba, K., Kasuya, D., Yudasaka, M., Iijima, S. (2003) Binary nanomaterials based on nanocarbons: a case for probing carbon nanohorns' biorecognition properties. Nano Lett. 3: 1033–1036.
[37] Kase, D., Kulp, J., Yudasaka, M., Evans, J., Iijima, S., Shiba, K. (2004) Affinity selection of peptide phage libraries against single-wall carbon nanohorns identifies a peptide aptamer with conformational variability. Langmuir 20: 8939–8941.
[38] Morita, Y., Ohsugi, T., Iwasa, Y. (2004) A screening of phage displayed peptides for the recognition of fullerene (C60). J. Mol. Catal. B: Enzym. 28: 185–190.
[39] Sawada, T., Takahashi, T., Mihara, H. (2009) Affinity-based screening of peptides recognizing assembly states of self-assembling peptide nanomaterials. J. Am. Chem. Soc. 131: 14434–14441.
[40] Sawada, T., Mihara, H. (2012) Dense surface functionalization using peptides that recognize differences in organized structures of self-assembling nanomaterials. Mol. BioSyst. doi: 10.1039/C1032MB05435C.
[41] Adey, N., Mataragnon, A., Rider, J., Carter, J., Kay, B. (1995) Characterization of phage that bind plastic from phage-displayed random peptide libraries. Gene 156: 27–31.
[42] Berglund, J., Lindbladh, C., Mosbach, K., Nicholls, I.A. (1998) Selection of phage display combinatorial library peptides with affinity for a yohimbine imprinted methacrylate polymer. Anal. Commun. 35: 3–10.
[43] Sanghvi, A., Miller, K., Belcher, A., Schmidt, C. (2005) Biomaterials functionalization using a novel peptide that selectively binds to a conducting polymer. Nat. Mater. 4: 496–998.
[44] Slocik, J., Stone, M., Naik, R. (2005) Synthesis of gold nanoparticles using multifunctional peptides. Small 1: 1048–1100.
[45] Sano, K.-I., Sasaki, H., Shiba, K. (2005) Specificity and biomineralization activities of Ti-binding peptide-1 (TBP-1). Langmuir 21: 3090–3095.
[46] Ahmad, G., Dickerson, M.B., Church, B.C, Cai, Y., Jones, S.E., Naik, R.R., King, J.S., Summers, C.J., Kröger, N., Sandhage, K.H. (2006) Rapid, room-temperature formation of crystalline calcium molybdate phosphor microparticles via peptide-induced precipitation. Adv. Mater. 18: 1759–1763.
[47] Slocik, J.M., Naik, R.R. (2006) Biologically programmed synthesis of bimetallic nanostructures. Adv. Mater. 18: 1988–1992.
[48] Oren, E., Tamerler, C., Sarikaya, M. (2005) Metal recognition of septapeptides via polypod molecular architecture. Nano Lett. 5: 415–424.
[49] Pender, M., Sowards, L., Hartgerink, J., Stone, M., Naik, R. (2006) Peptide-mediated formation of single-wall carbon nanotube composites. Nano Lett. 6: 40–44.
[50] Sano, K.-I., Sasaki, H., Shiba, K. (2006) Utilization of the pleiotropy of a peptidic aptamer to fabricate heterogeneous nanodot-containing multilayer nanostructures. J. Am. Chem. Soc. 128: 1717–1722.
[51] Matsui, T., Matsukawa, N., Iwahori, K., Sano, K. I., Shiba, K., Yamashita, I. (2007) Realizing a two-dimensional ordered array of ferritin molecules directly on a solid surface utilizing carbonaceous material affinity peptides. Langmuir 23: 1615–1618.

[52] K. Kokubun, K. Kashiwagi, M. Yoshinari, T. Inoue, K. Shiba, (2008) Motif-programmed artificial extracellular matrix. Biomacromolecules 9: 3098–3105.

[53] K.-I. Sano, K. Ajima, K. Iwahori, M. Yudasaka, S., Iijima, I. Yamashita, K. Shiba (2005) Endowing a ferritin-like cage protein with high affinity and selectivity for certain inorganic materials. Small 1: 826–858.

[54] Smith, G., Petrenko, V. (1997) Phage display. Chem. Rev. 97: 391–801.

[55] Kehoe, J., Kay, B. (2005) Filamentous phage display in the new millennium. Chem. Rev. 105: 4056–4072.

[56] Menendez, A., Scott, J. (2005) The nature of target-unrelated peptides recovered in the screening of phage-displayed random peptide libraries with antibodies. Anal. Biochem. 336: 145–157.

[57] Huang, Y., Chiang, C.-Y., Lee, S.K., Gao, Y., Hu, E.L., Yoreo, J.D., Belcher, A.M. (2005) Programmable assembly of nanoarchitectures using genetically engineered viruses. Nano Lett. 5: 1429–1434.

[58] Tamerler, C., Oren, E., Duman, M., Venkatasubramanian, E., Sarikaya, M. (2006) Adsorption kinetics of an engineered gold binding peptide by surface plasmon resonance spectroscopy and a quartz crystal microbalance. Langmuir 22: 7712–7720.

[59] Seker, U., Wilson, B., Dincer, S., Kim, I., Oren, E., Evans, J., Tamerler, C., Sarikaya, M. (2007) Adsorption behavior of linear and cyclic genetically engineered platinum binding peptides. Langmuir 23: 7895–7900.

[60] Hnilova, M., Oren, E., Seker, U., Wilson, B., Collino, S., Evans, J., Tamerler, C., Sarikaya, M. (2008) Effect of molecular conformations on the adsorption behavior of gold-binding peptides. Langmuir 24: 12440–12445.

[61] Helms, B., Reulen, S., Nijhuis, S., de Graaf-Heuvelmans, P., Merkx, M., Meijer, E. (2009) High-affinity peptide-based collagen targeting using synthetic phage mimics: from phage display to dendrimer display. J. Am. Chem. Soc. 131: 11683–11688.

[62] Sano, K.-I., Ajima, K., Iwahori, K., Yudasaka, M., Iijima, S., Yamashita, I., Shiba, K. (2005) Endowing a ferritin-like cage protein with high affinity and selectivity for certain inorganic materials. Small 1: 826–858.

[63] Seker, U., Wilson, B., Sahin, D., Tamerler, C., Sarikaya, M. (2009) Quantitative affinity of genetically engineered repeating polypeptides to inorganic surfaces. Biomacromolecules 10: 250–257.

[64] Kantarci, N., Tamerler, C., Sarikaya, M., Haliloglu, T., Doruker, P. (2005) Molecular dynamics simulations on constraint metal binding peptides. Polymer 46: 4307–4313.

[65] Hatada, K., Ute, K., Tanaka, K., Okamoto, Y., Kitayama, T. (1986) Living and highly isotactic polymerization of methyl methacrylate by *t*-C_4H_9MgBr in toluene. Polym. J. 18: 1037–1047.

[66] Kitayama, T., Shinozaki, T., Sakamoto, T., Yamamoto, M., Hatada, K. (1989) Living and highly syndiotactic polymerization of methyl methacrylate and other methacrylates by *tert*-butyllithiumtrialkylaluminium in toluene. Makromol. Chem. 15: 167–185.

[67] Serizawa, T., Sawada, T., Matsuno, H., Matsubara, T., Sato, T. (2005) A peptide motif recognizing a polymer stereoregularity. J. Am. Chem. Soc. 127: 13780–13781.

[68] Barbas, C.F., Burton, D.R., Scott, J.K., Silverman, G.J. (2001) Phage Display: A Laboratory Manual. Cold Spring Harbor Press, New York.

[69] Serizawa, T., Sawada, T., Matsuno, H. (2007) Highly specific affinities of short peptides against synthetic polymers. Langmuir 23: 11127–11133.

[70] Date, T., Tanaka, K., Nagamura, T., Serizawa, T. (2008) Directional affinity of short peptides for synthetic polymers. Chem. Mater. 20: 4536–4538.

[71] Serizawa, T., Sawada, T., Kitayama, T. (2007) Peptide motifs that recognize differences in polymer-film surfaces. Angew. Chem., Int. Ed. 46: 723–729.

[72] Tretinnikov, O.N. (1999) Wettability and microstructure of polymer surfaces: stereochemical and conformational aspects. J. Adhes. Sci. Technol. 13: 1085–1102.

[73] Date, T., Yoshino, S., Matsuno, H., Serizawa, T. (2012) Surface modification of stereoregular and stereocomplex poly(methyl methacrylate) films with biologically identified peptides. Polym. J. 44: 366–369.

[74] Wang, S., Cui, W., Bei, J. (2005) Bulk and surface modifications of polylactide. Anal. Bioanal. Chem. 381: 547–556.

[75] Holland, T.A., Mikos, A.G. (2006) Biodegradable polymeric scaffolds. Improvements in bone tissue engineering through controlled drug delivery. Adv. Biochem. Eng./Biotechnol. 102: 161–185.
[76] Matsuno, H., Sekine, J., Yajima, H., Serizawa, T. (2008) Biological selection of peptides for poly(l-lactide) substrates. Langmuir 24: 6399–6403.
[77] Serizawa, T., Techawanitchai, P., Matsuno, H. (2007) Isolation of peptides that can recognize syndiotactic polystyrene. ChemBioChem 8: 989–993.
[78] Tsutsui, K., Tsujita, Y., Yoshimizu, H. (1998) The presence of nanopores in mesophase of syndiotactic polystyrene estimated from gas sorption behaviour. Polymer: 5177–5182.
[79] Tsutsui, K., Katsumata, T., Yamamoto, Y., Fukatsu, H., Yoshimizu, H., Kinoshita, T., Tsujita, Y. (1999) The structural organization in syndiotactic polystyrene film induced by toluene vapour sorption. Polymer 40: 3815–3819.
[80] Yamamoto, Y., Nakai, Y., Katsumata, T., Tsutsui, K., Tsujita, Y., Yoshimizu, H., Okamoto, S. (2005) Structure and properties of the mesophase of syndiotactic polystyrene. V: preferential sorption performance of the mesophase for aromatic hydrocarbons. J. Mol. Struct. 739: 13–30.
[81] Burroughes, J., Bradley, D.D.C., Brown, A., Marks, R., Mackay, K., Friend, R., Burns, P., Holmes, A. (1990) Light-emitting diodes based on conjugated polymers. Nature 347: 539–541.
[82] Yu, G., Gao, J., Hummelen, J., Wudl, F., Heeger, A. (1995) Polymer photovoltaic cells: enhanced efficiencies via a network of internal donor–acceptor heterojunctions. Science 270: 1789–1791.
[83] Feng, F., He, F., An, L., Wang, S., Li, Y., Zhu, D. (2008) Fluorescent conjugated polyelectrolytes for biomacromolecule detection. Adv. Mater. 20: 2959–2969.
[84] Ejima, H., Matsuno, H., Serizawa, T. (2010) Biological identification of peptides that specifically bind to poly(phenylene vinylene) surfaces: recognition of the branched or linear structure of the conjugated polymer. Langmuir 26: 17278–17285.
[85] Yu, Y., Nakano, M., Ikeda, T. (2003) Photomechanics: directed bending of a polymer film by light. Nature 425: 145.
[86] Liu, M., Asanuma, H., Komiyama, M. (2006) Azobenzene-tethered T7 promoter for efficient photoregulation of transcription. J. Am. Chem. Soc. 128: 1009–1015.
[87] Ferri, V., Elbing, M., Pace, G., Dickey, M., Zharnikov, M., Samorì, P., Mayor, M., Rampi, M. (2008) Light-powered electrical switch based on cargo-lifting azobenzene monolayers. Angew. Chem., Int. Ed. 47: 3407–3416.
[88] Yamato, M., Akiyama, Y., Kobayashi, J., Yang, J., Kikuchi, A., Okano, T. (2007) Temperature-responsive cell culture surfaces for regenerative medicine with cell sheet engineering. Prog. Polym. Sci. 32: 1123–1133.
[89] Bae, Y., Fukushima, S., Harada, A., Kataoka, K. (2003) Design of environment-sensitive supramolecular assemblies for intracellular drug delivery: polymeric micelles that are responsive to intracellular pH change. Angew. Chem., Int. Ed. 42: 4640–4643.
[90] Osada, Y., Okuzaki, H., Hori, H. (1992) A polymer gel with electrically driven motility. Nature 355: 242–244.
[91] Miyata, T., Asami, N., Uragami, T. (1999) A reversibly antigen-responsive hydrogel. Nature 399: 766–775.
[92] Chen, J., Serizawa, T., Komiyama, M. (2009) Peptides recognize photoresponsive targets. Angew. Chem., Int. Ed. 48: 2917–2920.
[93] Kashiwagi, K., Tsuji, T., Shiba, K. (2009) Directional BMP-2 for functionalization of titanium surfaces. Biomaterials 30: 1166–1175.
[94] Park, T., Lee, S., Lee, S., Park, J., Yang, K., Lee, K.-B., Ko, S., Park, J., Kim, T., Kim, S., Shin, Y., Chung, B., Ku, S.-J., Kim, D. H., Choi, I. (2006) Protein nanopatterns and biosensors using gold binding polypeptide as a fusion partner. Anal. Chem. 78: 7197–7205.
[95] Matsuno, H., Date, T., Kubo, Y., Yoshino, Y., Tanaka, N., Sogabe, A., Kuroita, T., Serizawa, T. (2009) Enhanced adsorption of protein fused with polymeric material-binding peptide. Chem. Lett. 38: 834–835.
[96] Zhu, X., Zhao, X., Burkholder, W., Gragerov, A., Ogata, C., Gottesman, M., Hendrickson, W. (1996) Structural analysis of substrate binding by the molecular chaperone DnaK. Science 272: 1606–1620.

[97] Tanaka, N., Nakao, S., Wadai, H., Ikeda, S., Chatellier, J., Kunugi, S. (2002) The substrate binding domain of DnaK facilitates slow protein refolding. Proc. Natl Acad. Sci. USA 99: 15398–15403.

[98] Kuroita, T., Sogabe, A., Takarada, Y., Tanaka, N. (2005) PCT Int. Appl. WO 2005/003155.

[99] Huang, Y., Chiang, C.-Y., Lee, S., Gao, Y., Hu, E., De Yoreo, J., Belcher, A. (2005) Programmable assembly of nanoarchitectures using genetically encoded viruses. Nano Lett. 5: 1429–1434.

[100] Chen, J., Serizawa, T., Komiyama, M. (2011) Binding analysis of peptides that recognize preferentially *cis*-azobenzene groups of synthetic polymers. J. Pept. Sci. 17: 163–171.

[101] Chen, J., Serizawa, T., Komiyama, M. (2011) Recognition of photoresponsive polymer targets by protein fused with *cis*-form azobenzene-binding peptide. Chem. Lett. 40: 482–483.

[102] Pasquato, L., Pengo, P., Scrimin, P. (2004) Functional gold nanoparticles for recognition and catalysis. J. Mater. Chem. 14: 3481–3487.

[103] Ghosh, S., Pal, T. (2007) Interparticle coupling effect on the surface plasmon resonance of gold nanoparticles: from theory to applications. Chem. Rev. 107: 4797–4862.

[104] De, M., Ghosh, P.S., Rotello, V.M. (2008) Applications of nanoparticles in biology. Adv. Mater. 20: 4225–4241.

[105] Serizawa, T., Hirai, Y., Aizawa, M. (2009) Novel synthetic route to peptide-capped gold nanoparticles. Langmuir 25: 12229–12234.

[106] Ulman, A. (1996) Formation and structure of self-assembled monolayers. Chem. Rev. 96: 1533–1554.

[107] Onclin, S., Ravoo, B.J., Reinhoudt, D.N. (2005) Engineering silicon oxide surfaces using self-assembled monolayers. Angew. Chem. Int. Ed. 44: 6282–6304.

[108] Love, J.C., Estroff, L.A., Kriebel, J.K., Nuzzo, R.G., Whitesides, G.M. (2005) Self-assembled monolayers of thiolates on metals as a form of nanotechnology. Chem. Rev. 105: 1103–1169.

[109] Date, T., Sekine, J., Matsuno, H., Serizawa, T. (2011) Polymer-binding peptides for the noncovalent modification of polymer surfaces: effects of peptide density on the subsequent immobilization of functional proteins. ACS Appl. Mater. Interfaces 3: 351–360.

[110] Ejima, H., Kikuchi, H., Matsuno, H., Yajima, H., Serizawa, T. (2010) Peptide-based switching of polymer fluorescence in aqueous phase. Chem. Mater. 22: 6032–6034.

[111] Zhou, Q., Swager, T.M. (1995) Method for enhancing the sensitivity of fluorescent chemosensors: energy migration in conjugated polymers. J. Am. Chem. Soc. 117: 7017–7018.

[112] Chen, L., McBranch, D., Wang, H., Helgeson, R., Wudl, F., Whitten, D. (1999) Highly sensitive biological and chemical sensors based on reversible fluorescence quenching in a conjugated polymer. Proc. Natl Acad. Sci. USA 96: 12287–12292.

[113] Thomas, S., Joly, G., Swager, T. (2007) Chemical sensors based on amplifying fluorescent conjugated polymers. Chem. Rev. 107: 1339–1386.

[114] Herland, A., Inganäs, O. (2007) Conjugated polymers as optical probes for protein interactions and protein conformations. Macromol. Rapid Commun. 28: 1703–1713.

[115] Decher, G. (1997) Fuzzy nanoassemblies: toward layered polymeric multicomposites. Science 277: 1232–1237.

Index

Peptide Materials: From Nanostructures to Applications, First Edition. Edited by Carlos Alemán, Alberto Bianco and Mariano Venanzi.

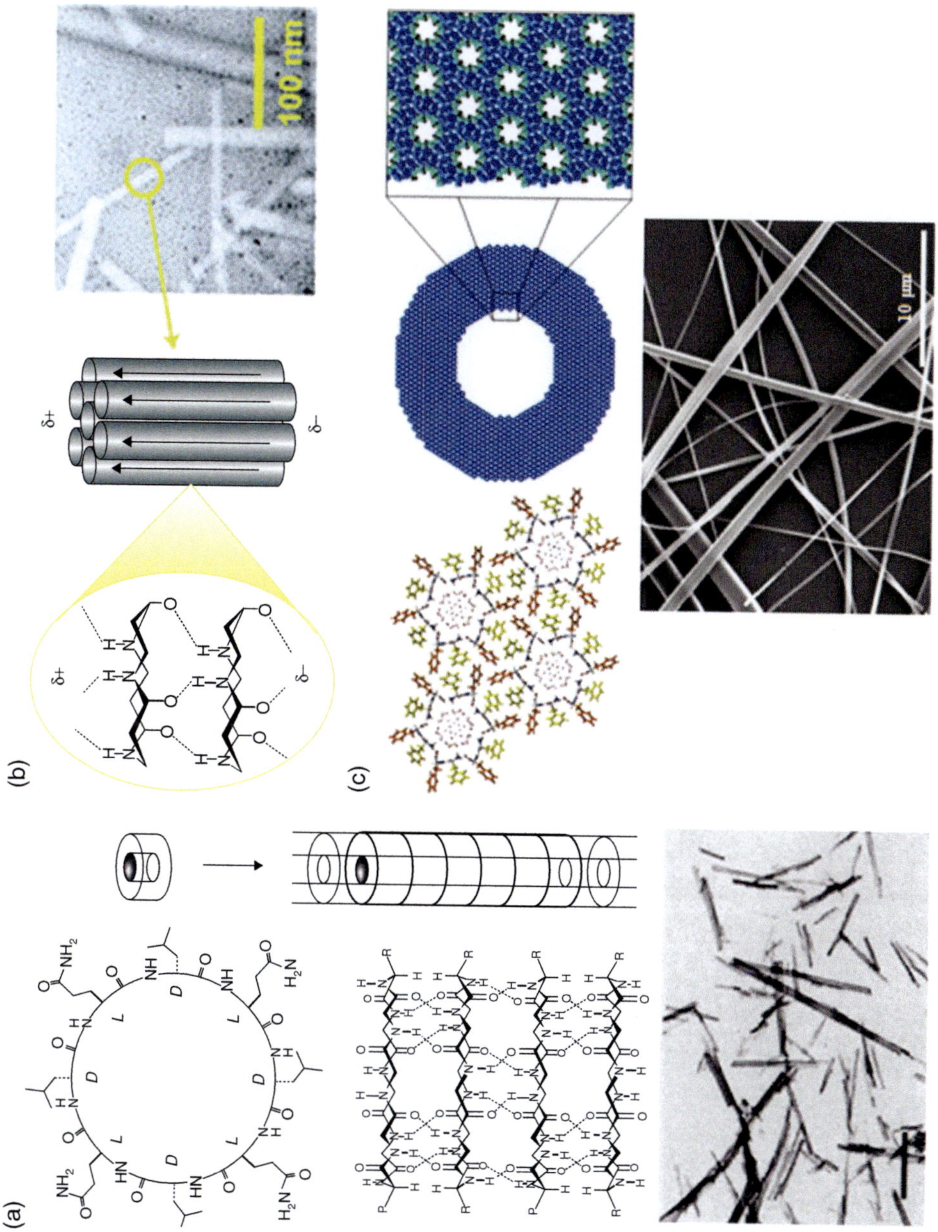

Figure 1.1 *(a) Alternating D- and L-amino acids PNT: scheme (upper part) and morphology (bottom part) (b) The concept of dipole engineering, molecular packing scheme and morphology. (c) Molecular packing (upper part) and morphology (bottom part) of FF PNTs. Reproduced with permission from references [9], [22], [40], and [45]. Copyright (1993) Nature Publishing Group, (2006) Royal Society of Chemistry, and (2010) American Chemical Society*

Peptide Materials: From Nanostructures to Applications, First Edition. Edited by Carlos Alemán, Alberto Bianco and Mariano Venanzi.

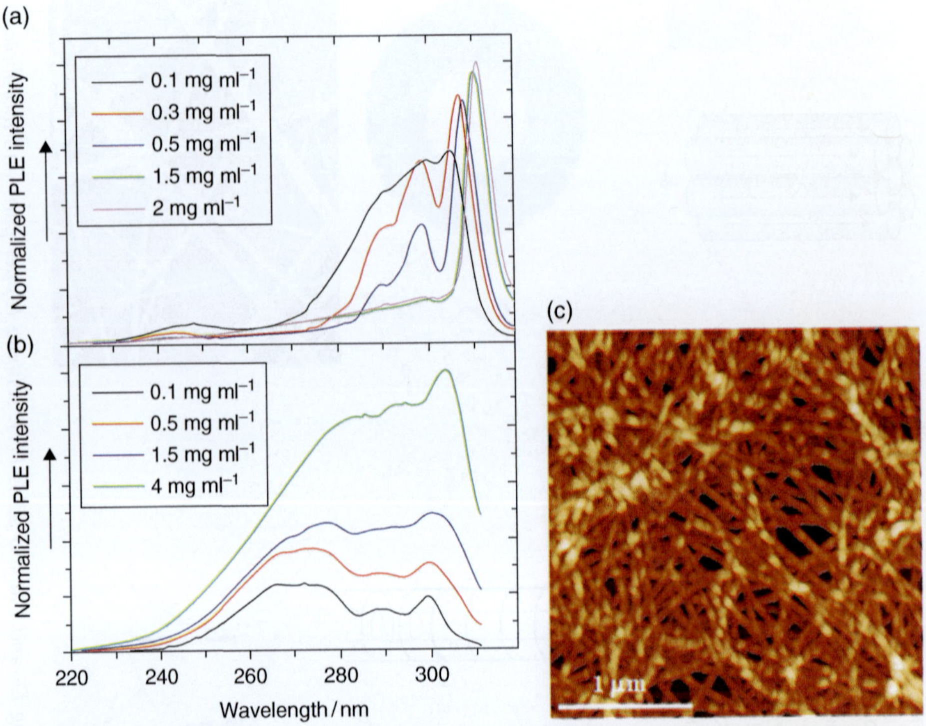

Figure 1.4 *PLE spectrum of a solution of Fmoc-FF (a) monomers and (b) aggregates at different concentrations. (c) AFM image of the Fmoc-FF fibrillar structure. Reproduced with permission from reference [42]. Copyright (2010) John Wiley & Sons, Ltd*

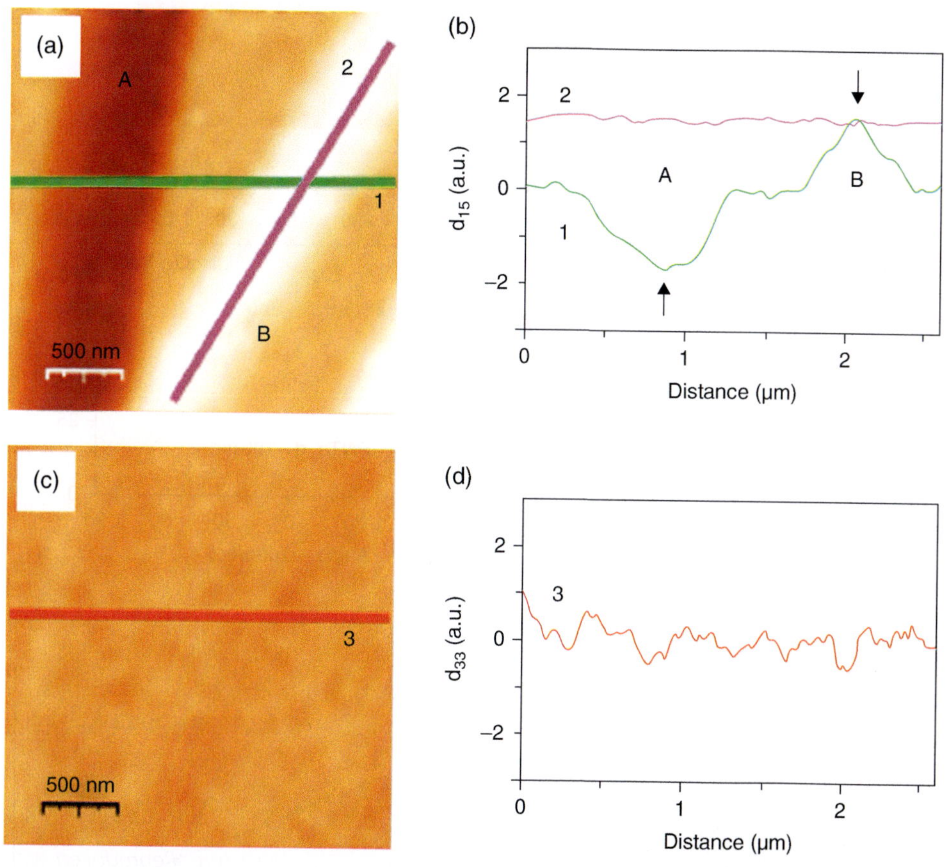

Figure 1.8 *(a) PFM IP signal of FF-PNTs. (b) Cross-section along the lines in (a). (c) PFM OOP signal of FF-PNTs. (d) Cross-section along the line in (c). Reproduced with permission from reference [25]. Copyright (2009) American Chemical Society*

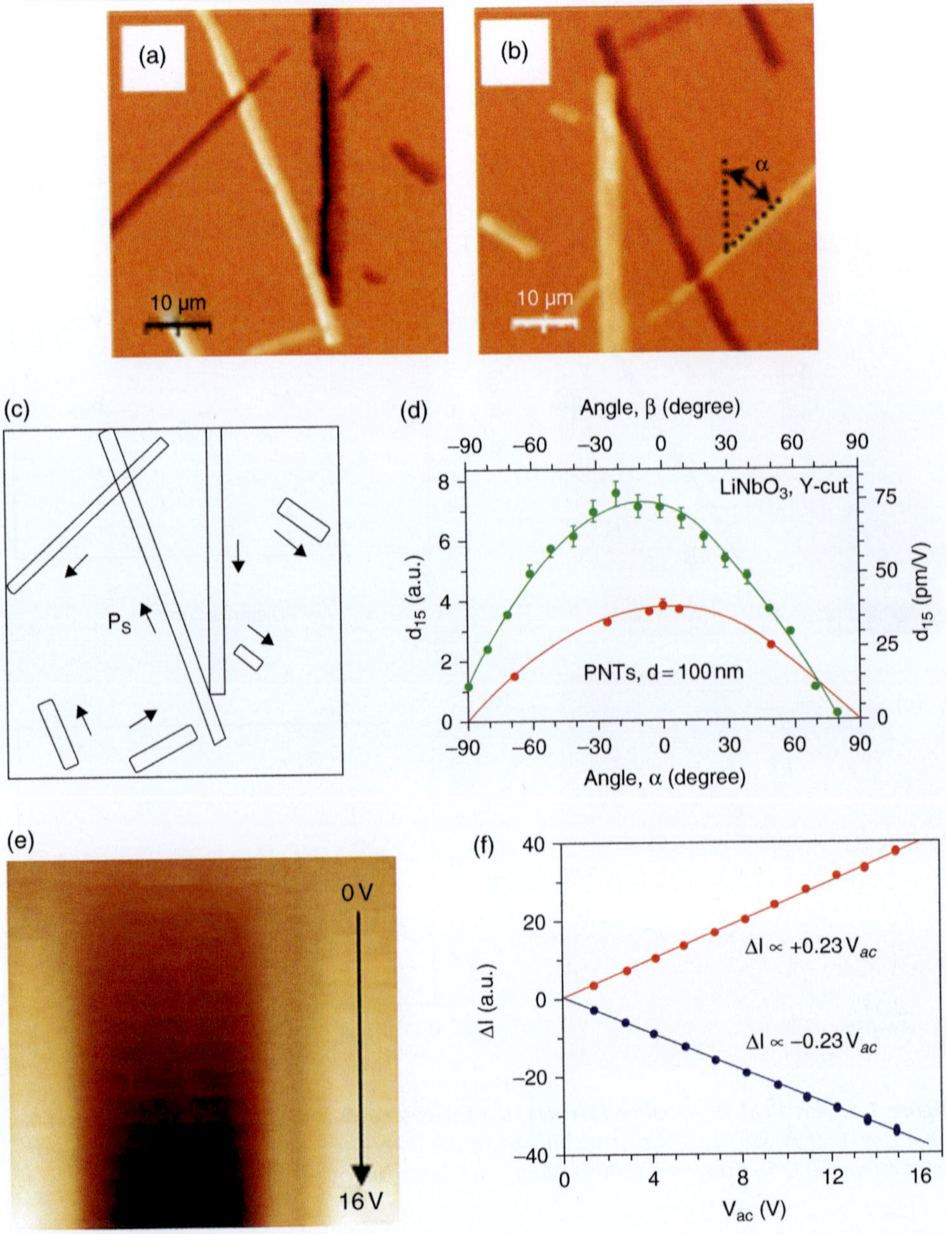

Figure 1.9 *PFM IP signal of FF-PNTs (a) before and (b) after the rotation of 180°. (c) Scheme of the spontaneous polarization vector along the tubes in (a) and (b). (d) PFM IP signal of FF-PNTs as a function of α, in comparison to an LNO crystal. (e) PFM IP signal of FF-PNTs at different AC biases, from 0 V to 16 V. (f) PFM IP signal of FF-PNTs as a function of AC bias. Reproduced with permission from reference [25]. Copyright (2010) American Chemical Society*

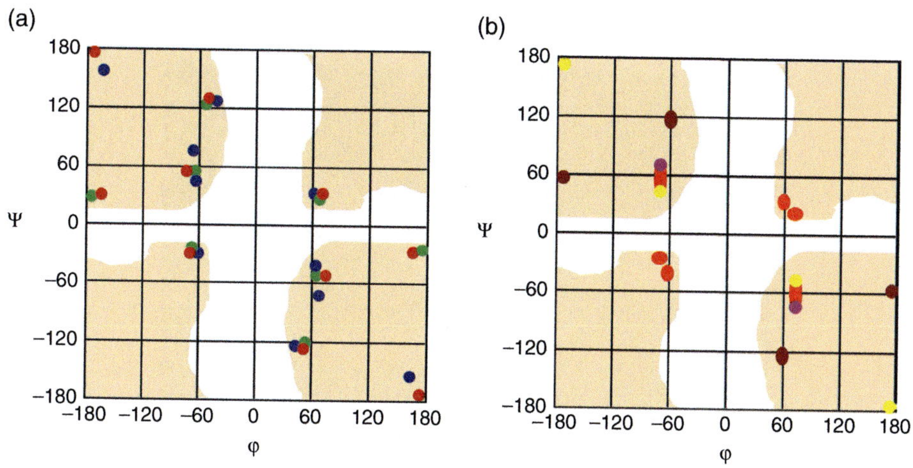

Figure 3.2 *Ramachandran maps indicating the position in the φ,ψ-map of the minimum energy conformations predicted by DFT calculations for peptides containing nc-aa. The regions with flesh color (see the color plate figure) correspond to the sterically accessible regions of the φ,ψ-map for the Ac–Gly–NHMe dipeptide. (a) Positions of the minimum energy conformations found for Ac–Aib–NHMe (red), Ac–Dφg–NHMe (green), and Ac–Db$_z$g–NHMe (dark blue). The energy gap between the minima of the lowest and highest energy is 3.4, 9.8, and 4.6 kcal/mol, respectively (taken from references [28], [29], and [30], respectively). (b) Minimum energy conformations found for Ac–Ac$_n$c–NHMe dipeptides (taken from references [31], [32], [33], and [34]): yellow indicates the positions shared by Ac$_n$c-containing dipeptides with n = 3, 4, 5, and 6; pink refers to the positions shared by Ac$_n$c-containing dipeptides with n = 4, 5, and 6; red corresponds to the positions shared by Ac–Ac$_5$c–NHMe and Ac–Ac$_6$c–NHMe; and dark red are positions exclusively found for Ac–Ac$_6$c–NHMe*

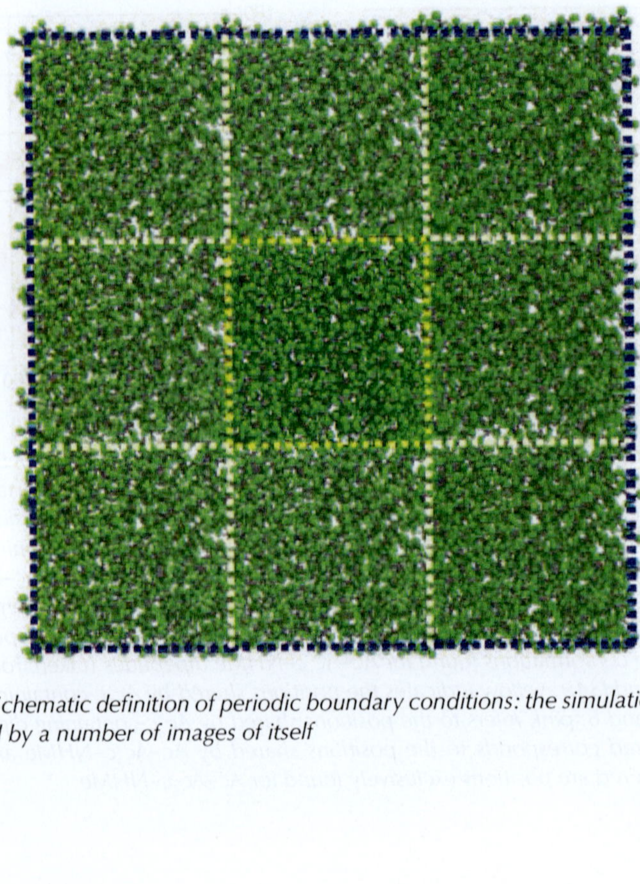

Figure 3.7 *Schematic definition of periodic boundary conditions: the simulation box (lighted) is surrounded by a number of images of itself*

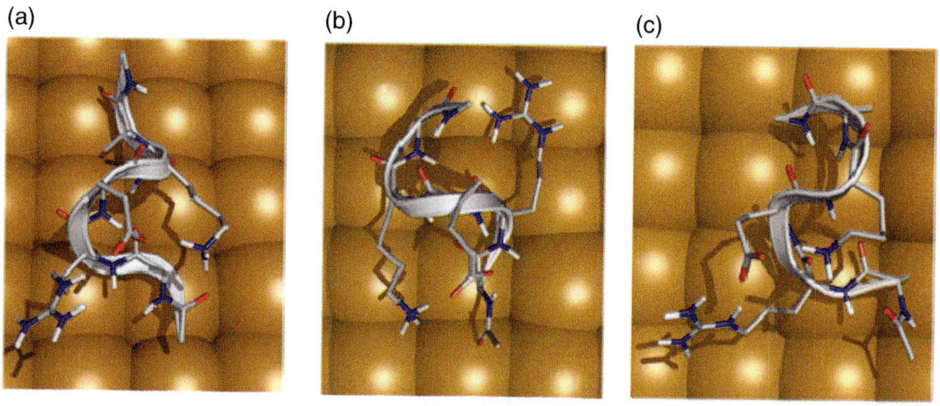

Figure 3.12 *(a) Bioactive conformation proposed for CREKA and its analogs, (b) CR-αMeE-KA, and (c) CR-NMeE-KA (αMeE and NMeE refer to α-methyl- and N-methyl-glutamic acid, respectively) attached to a nanoparticle. These conformations were determined by applying the modified SA-MD procedure*

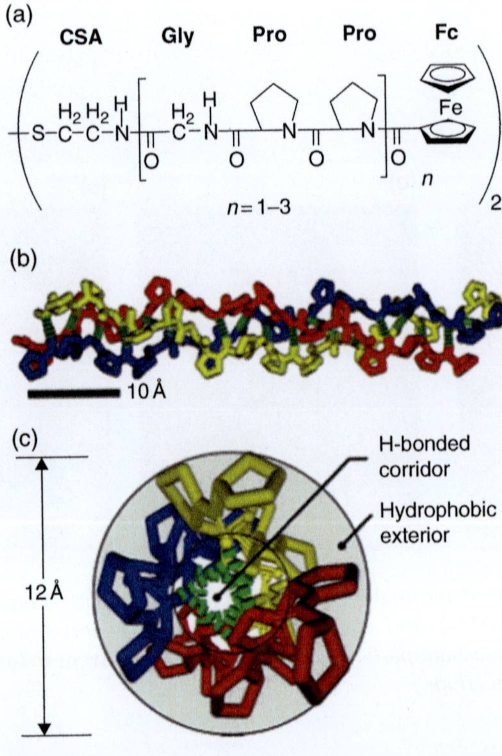

Figure 4.10 *(a) Molecular structure of the peptide investigated; (b) supramolecular coiled-coil structure adopted by a Pro–Pro–Gly repeat unit; (c) along helical axis view of the coiled-coil structure, showing the H-bonded inner corridor and the hydrophobic outer surface. Reproduced with permission from reference [57]. Copyright (2004) American Chemical Society*

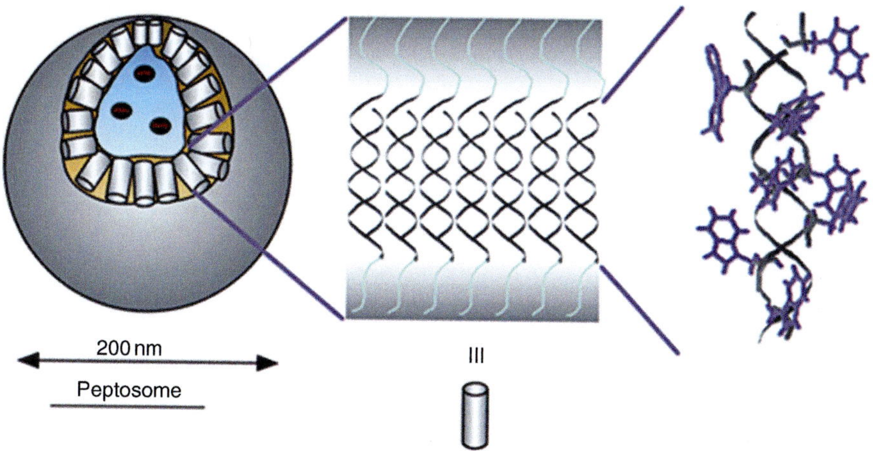

Figure 5.2 *Schematic illustration of peptosome prepared from a conjugate of gramicidin A and PEG. The gramicidin A blocks took a double helical structure to constitute the hydrophobic core region of the vesicular membrane. Reproduced with permission from reference [6]. Copyright (2004) The Royal Society of Chemistry*

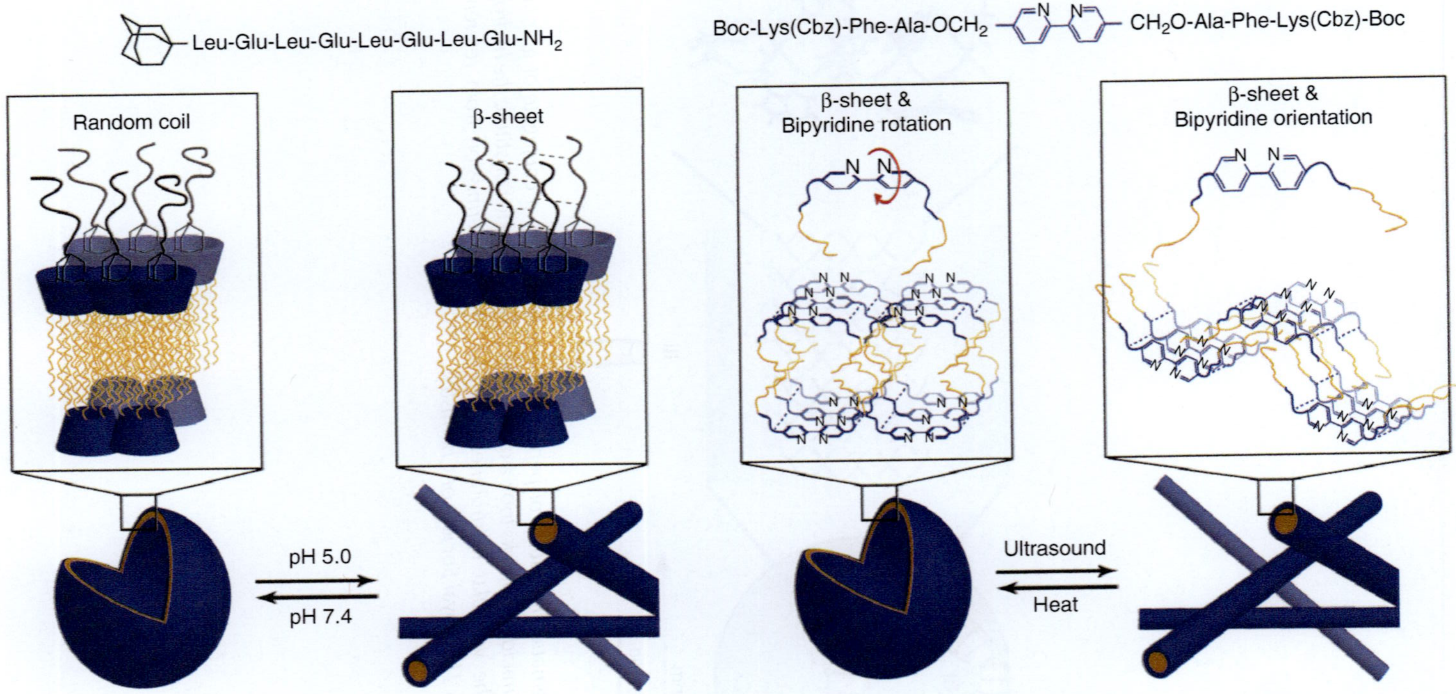

Figure 5.8 *The schematic illustrations of the transformations between vesicles and fibers by stimulations of pH change and heating. Redrawn from references [57] and [58]*

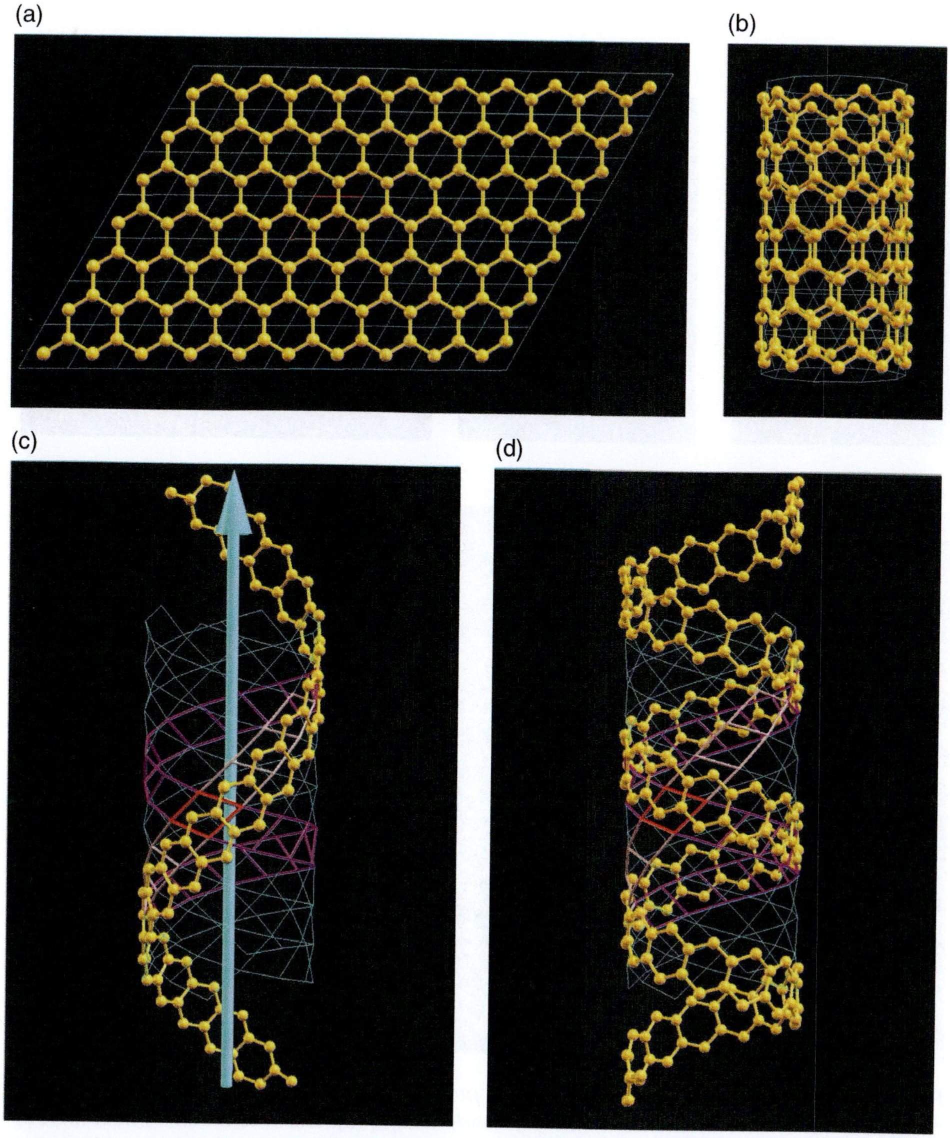

Figure 6.3 An illustration of the mechanisms of tube formation using a single-walled carbon nanotube as an example. (a) A planar sheet of carbon atoms is represented by 10 × 7 cells (in thin cyan lines) with two carbon atoms in every 2D lattice cell. (b) 10 cells are used to wrap up the tube with no helical twist or rise; therefore this is a (10,0) nanotube [18]. (c) and (d) The lattice-based tube can also be viewed as two different helical tubes when viewed along the two edges of the lattice. Both tubes shown are identified as a (10, 4) nanotube with N1 cells highlighted in magenta and N2 cells highlighted in pink. The helices of the hexagonal carbon around the tube axis are clearly displayed. In (c), the helix growth process can be described in terms of rotational duplications (four times) by the tube axis and in (d) as translational duplications (one time) along the tube axis. Note that the carbon atoms in (c) and (d) are the rotational/translational units respectively

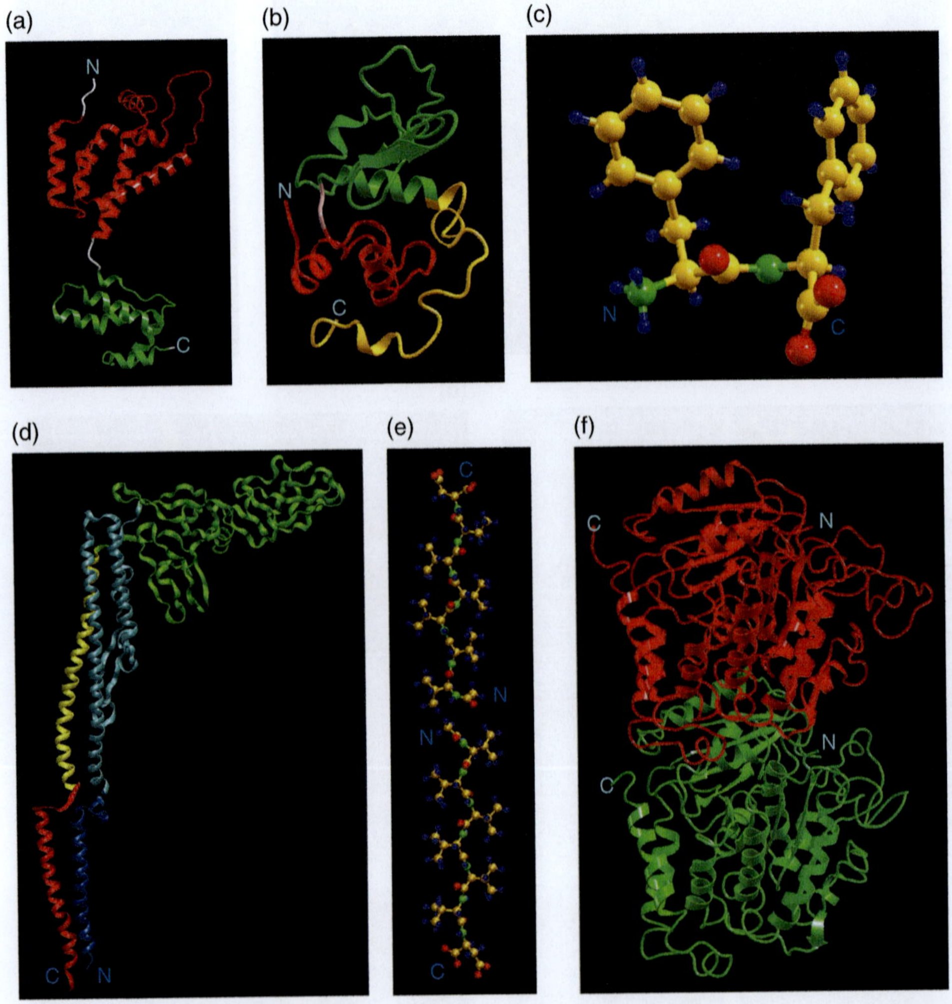

Figure 6.4 *Experimental examples of natural protein monomers that self-assemble to create nanotubes. (a) The HIV-1 CA monomer (pdb code: 1e6j) [45]. The N-terminal domain (in red) self-assembles to create a hexamer, and the C-terminal domain (in green) self-assembles into a dimer. A short linker of six residues (in white) connects the domains. (b) The full atomic structure of α-lactalbumin (pdb code: 1a4v) [47]. The colors (red, green, yellow) represent different interacting domains. (c) A model of the detailed atomic structure of diphenylalanine. The phenylalanine rings are perpendicular to each other. (d) The bacterial flagellar protein monomer (pdb code: 1ucu) [49]. Each color represents a different domain. (e) The double-layered surfactant-like peptide* AcV_6D. *The hydrophilic heads point outwards whereas the long acetylated hydrophobic tails point inward, towards each other. (f) The structure of the αβ-tubulin heterodimer (pdb code: 1tub) [52]. Each color represents a monomer. Both monomers form the asymmetric unit of the microtubule*

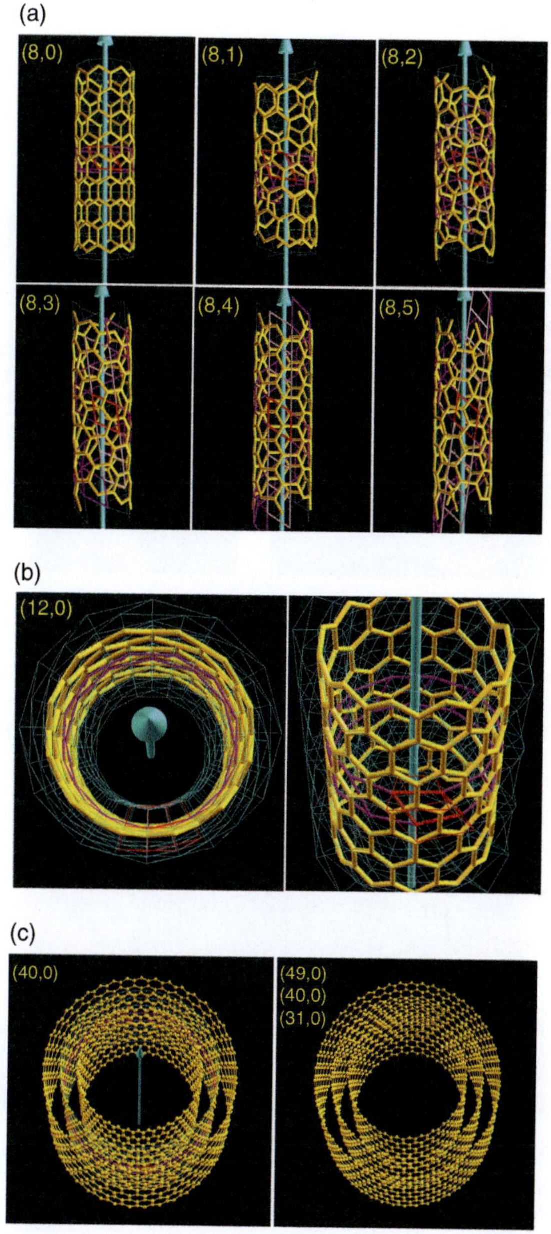

Figure 6.6 *Computational aspects of design and construction demonstrated on carbon nanotubes. (a) Six examples of single-walled carbon nanotubes having eight cells that wrap one full round and a different helicity extent [0–5]. As the helicity extent increases, the diameter decreases, and the nanotube becomes more fibrillar in shape. (b) The distortion formed as a result of the tube curvature. The inner area of the tube wall is more condensed than the outside area. (c) As a result of the distortion, in a multiwalled carbon nanotube, the number of cells needed to complete one full round decreases with the inner diameter and increases with the outer one*

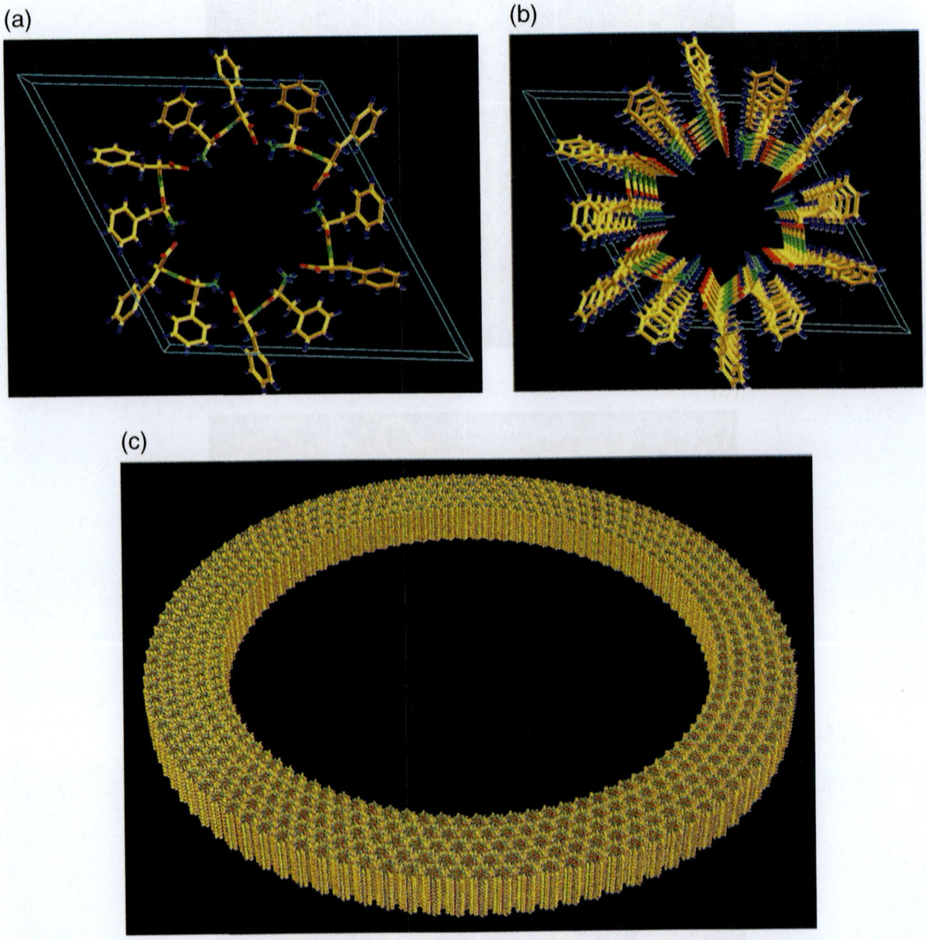

Figure 6.7 An atomic model of the diphenylalanine nanotube. (a) The unit cell consists of six dipeptides in a hexagonal arrangement such that the backbones face the tube interior while the hydrophobic side chains face its exterior. (b) The lamination of the unit cell along the hexagonal axis to create a tiny pore. (c) The FF nanotube with the tiny pores parallel to the tube axis. The width of the illustrated nanotube is 50 nm

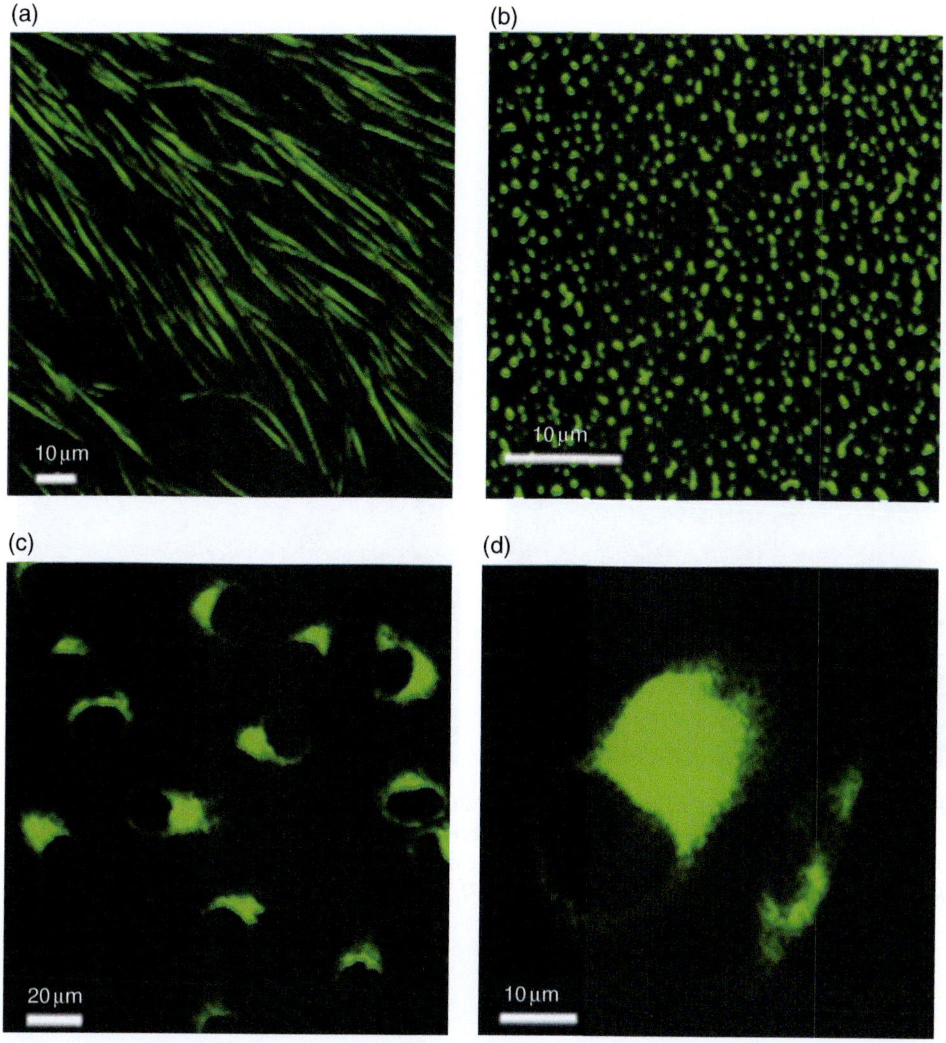

Figure 7.4 *Confocal laser scanning microsopy images of: (a) fluorescently labeled peptide nanotubes–ssDNA complex; (b) nanospheres upon dilution; (c) and (d) cellular uptake of peptide nanospheres–ssDNA complex by HeLa cells. Reproduced with permission from reference [18]. Copyright (2007) John Wiley & Sons, Ltd*

(a) (b)

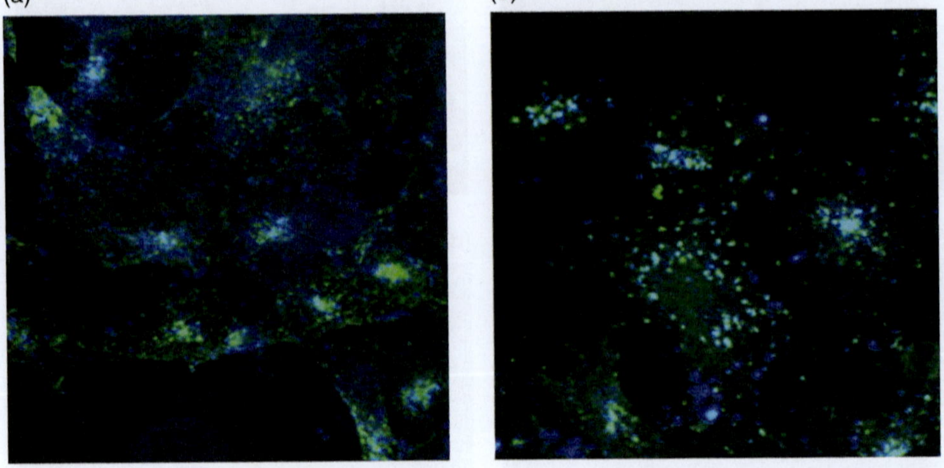

Figure 7.13 *Cell uptake of vesicle-forming SA2C3 peptides: (a) co-localization of peptides (blue) with transferrin (green) was absent; (b) co-localization of peptides (blue) with dextran 10000 (green) is visualized as multiple spots of dotted patterned co-localization (cyan). Reproduced with permission from reference [37]. Copyright (2010) Elsevier*

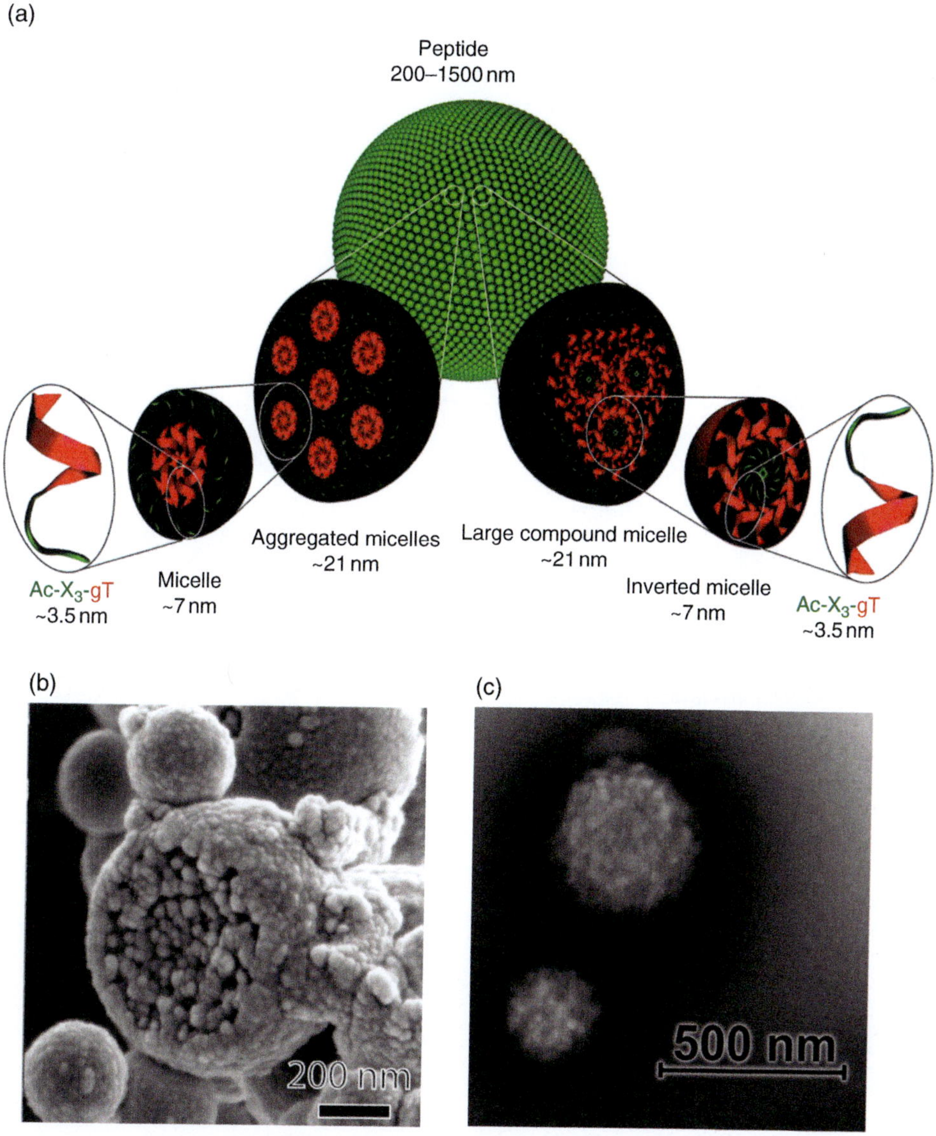

Figure 7.16 *(a) Schematic drawing shows self-assembly of smaller spherical subunits into peptide beads; (b) SEM image of a peptide bead revealing smaller spherical subunits in its core; (c) TEM image of peptide beads. Reproduced with permission from reference [63]. Copyright (2011) American Chemical Society*

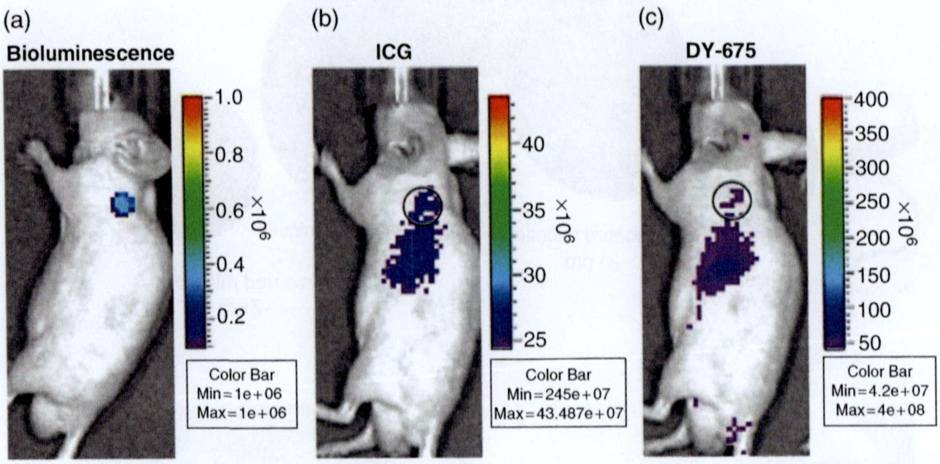

Figure 7.19 *In vivo cancer imaging using labeled peptosome: (a) bioluminescence image of SUIT2/EF-luc xenografts; (b) fluorescence image with Indocyanin green; (c) fluorescence image of DY-675. Reproduced with permission from reference [81]. Copyright (2005) American Chemical Society*

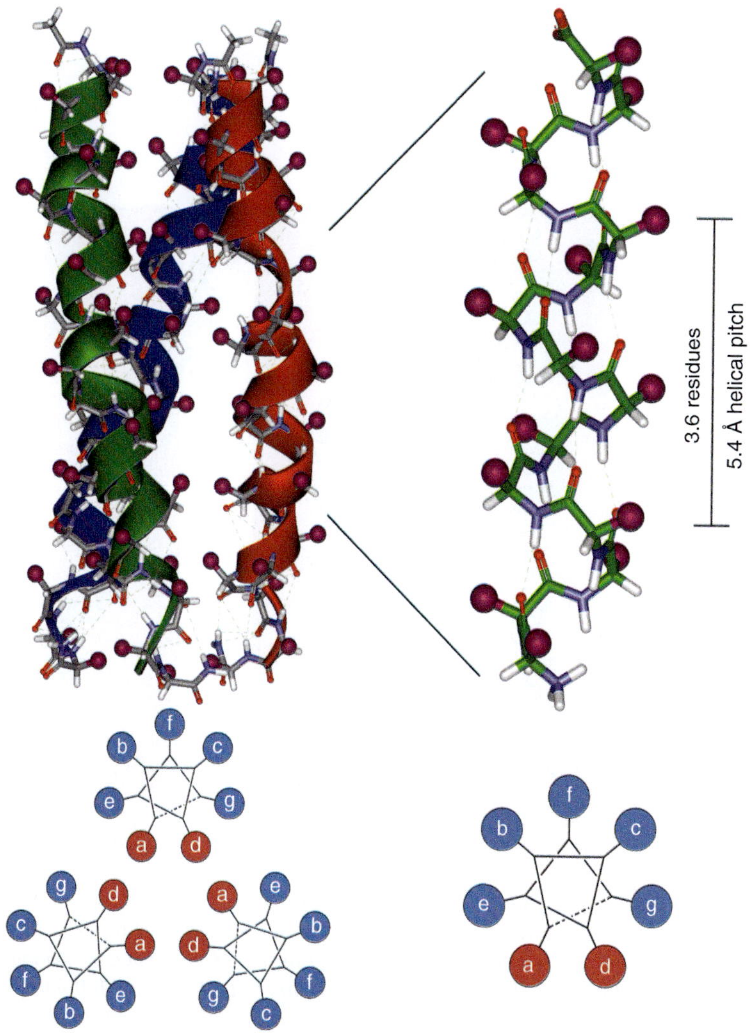

Figure 8.4 *(Top left) Molecular model of a trimeric α-helical supercoil. (Top right) Enhanced structural view of a segment in one of the supercoil's substituent α-helices. Below the models are helical wheel diagrams of the two structures summarizing apolar (red) and polar (blue) AA regions; the 'a' through 'g' labels indicate the AA positions in the heptad. The supercoil helical wheel diagram accents the positioning of the helices with apolar AAs facing the core of the supercoil*

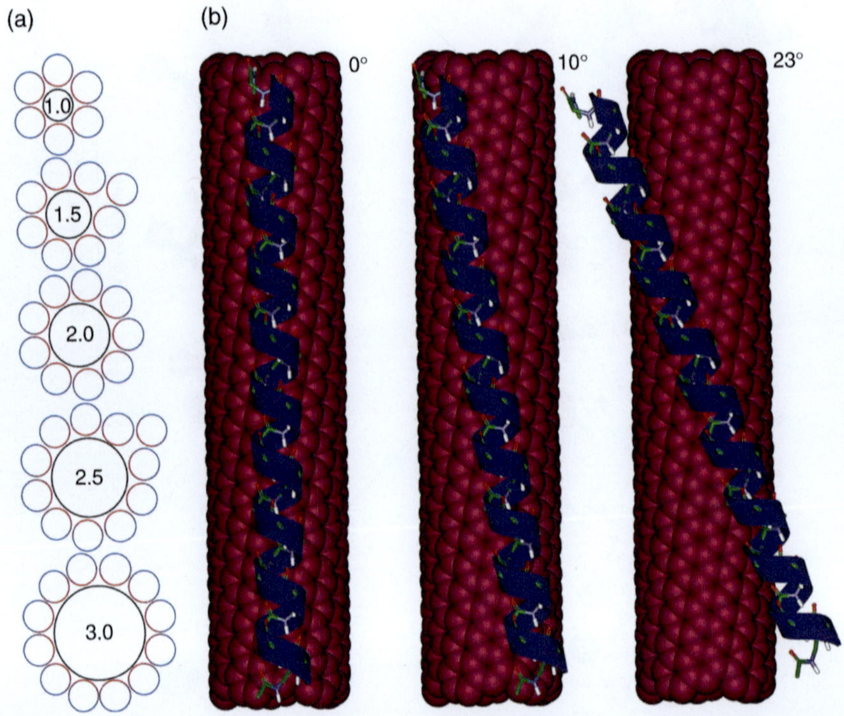

Figure 8.8 *(a) Illustration of optimal α-helical peptide packing about a CNT as a function of CNT diameter, where d is the average diameter of the helix as defined by its volume. In the intermediate stage (1.5d) there is not enough space around the CNT to accommodate another α-helix. The peptide is forced to deviate from its optimum contact angle with the CNT in order to cover a larger circumference; the contact angle is the deviation of the peptide from the CNT axis (b). The distance each set of helices covers along the CNT length is decreased and is compensated by additional helices*

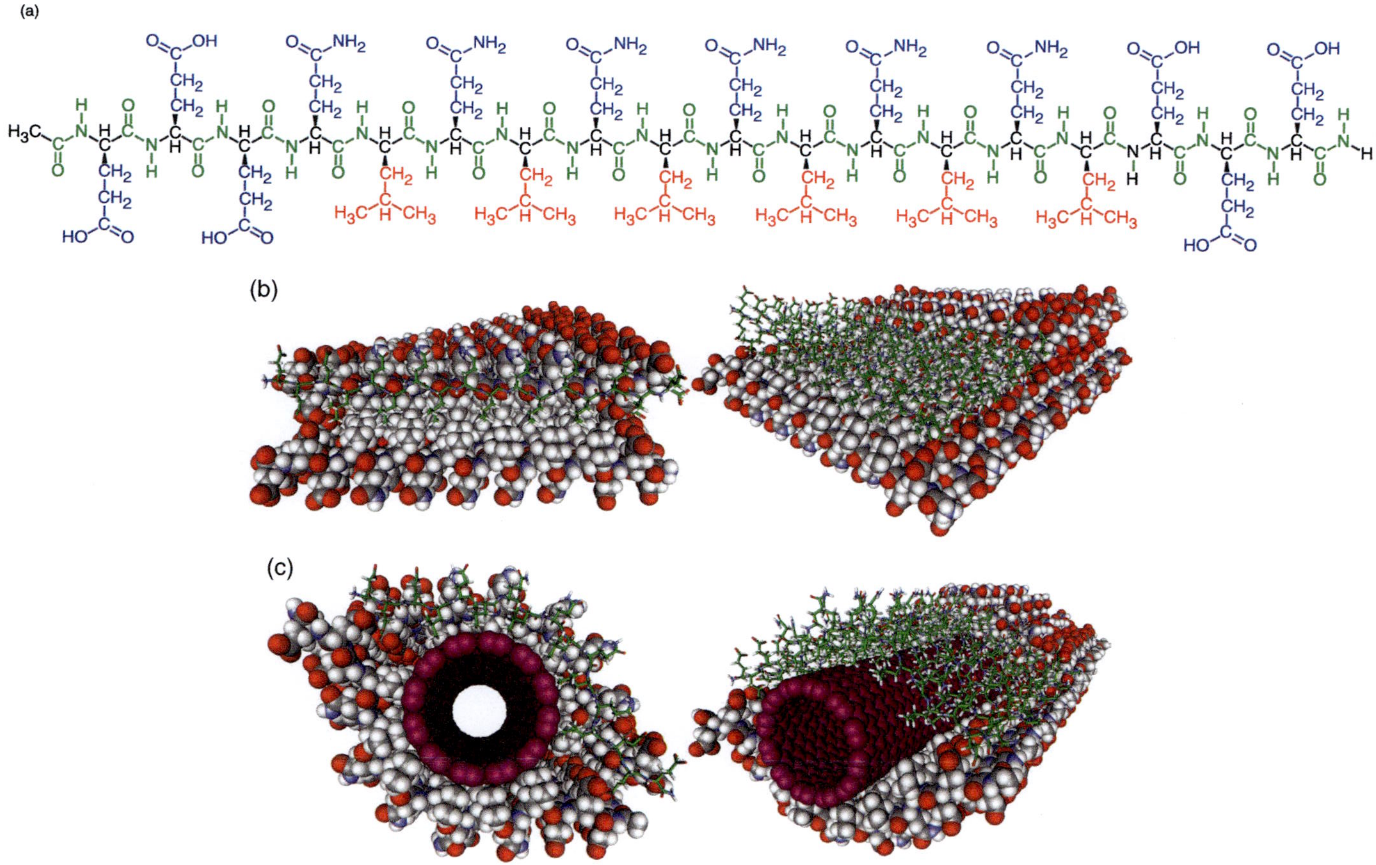

Figure 8.9 *Structure and interactions of a β-sheet surfactant peptide (Ac–$E_3(LQ)_6E_3$–NH_2). (a) Skeleton schematic of a single peptide strand, with blue polar side chains, red apolar side chains, and green peptide bonds that participate in β-sheet hydrogen bonding. (b) Space-filling and stick depiction of a β-sheet structure forming though self-aggregation in the absence of SWNTs. (c) Space-filling and stick depiction of a potential β-sheet structure formed by the peptide around a CNT*

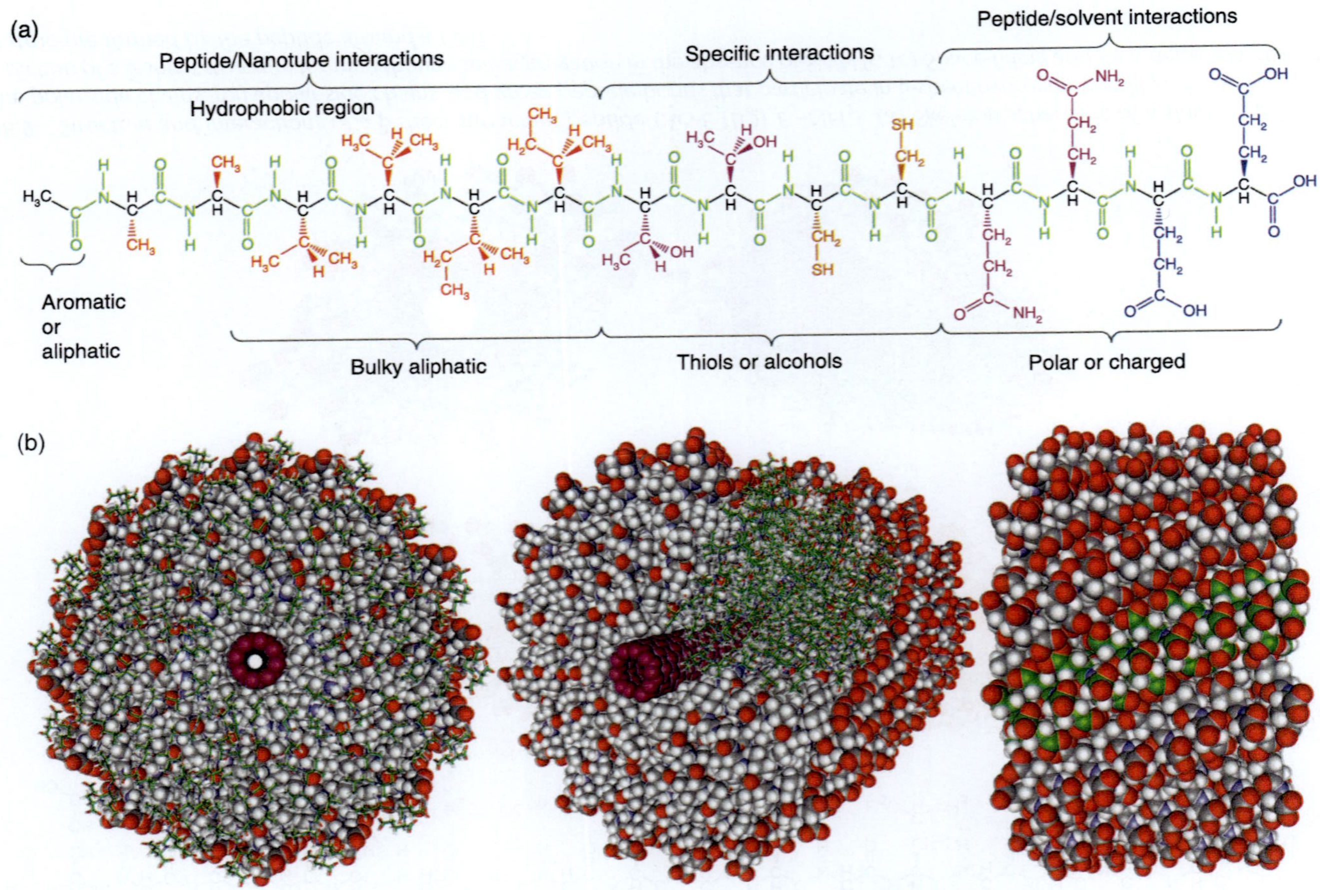

Figure 8.10 *Structure and interactions of an extended surfactant peptide. (A) Skeleton schematic of a single peptide strand, with blue polar AAs, red apolar AAs, and purple AAs integral in controlling peptide distribution and interpeptide interactions. (B) Space-filling and stick depiction of a surfactant peptide ($Ac–A_2V_2L_2E_2–NH_2$) in a β-sheet-like structure around a CNT. Extensive β-like H-bonding exists along the length of the CNT*

(a)

(b)

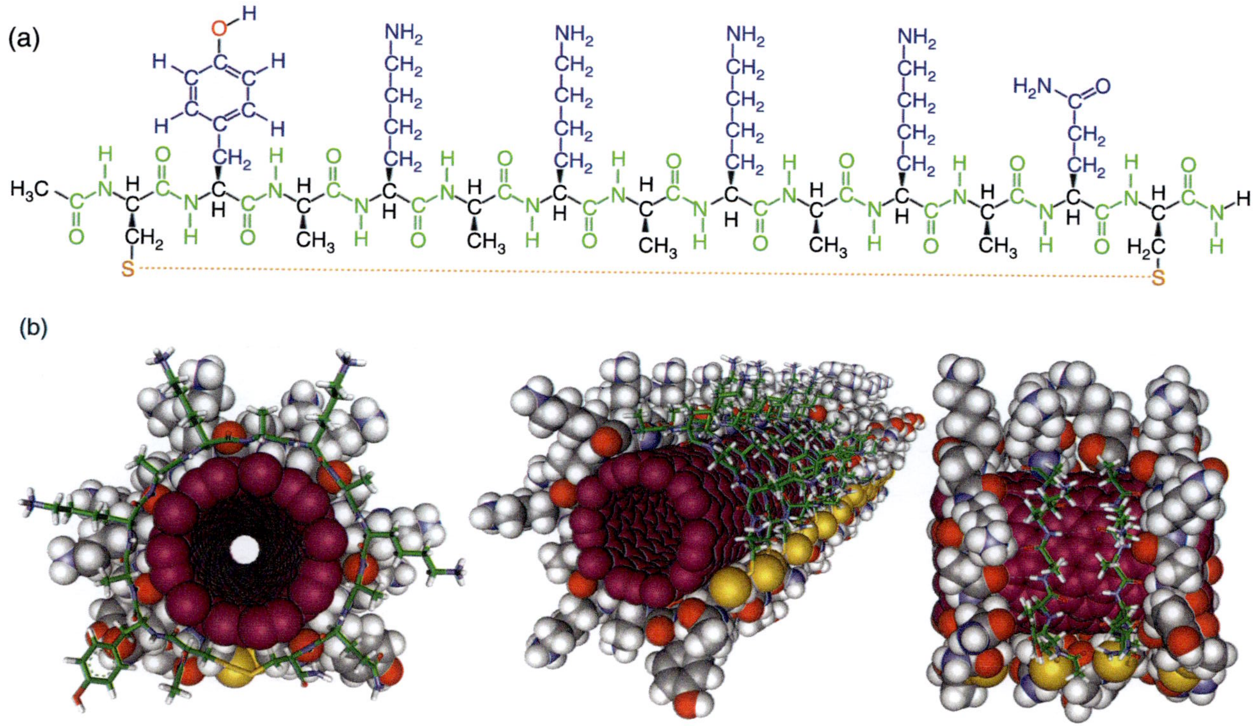

Figure 8.12 *Structure and interactions of a reversible cyclic peptide RC5–Cys. (a) Skeleton schematic of a single peptide strand, with blue polar side chains, red apolar side chains, and green peptide bonds that participate in β-sheet-like H-bonding. (b) Space-filling and stick depiction of oxidized RCP rings covering the surface of a CNT*

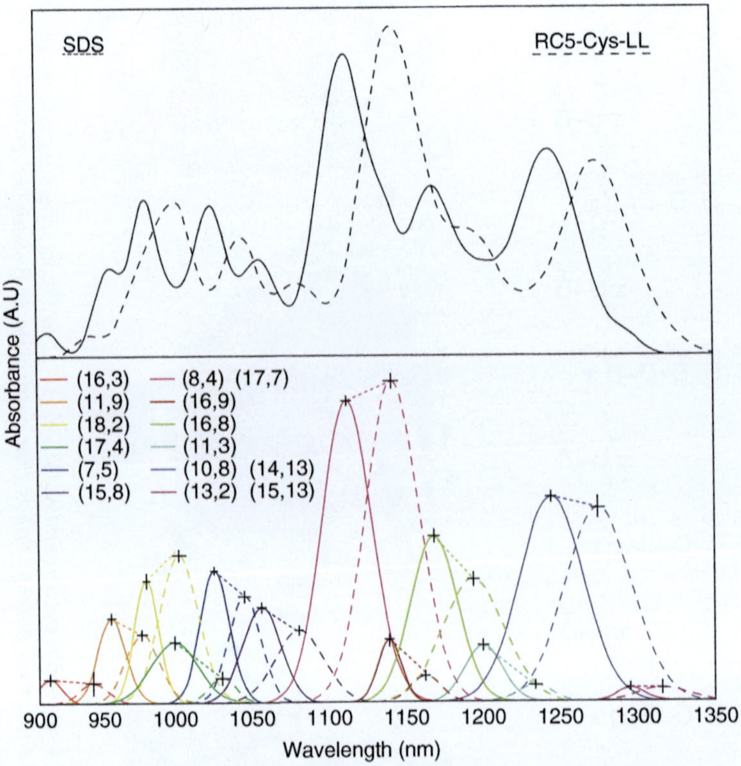

Figure 8.14 *Compilation and deconvolution of UV/Vis/NIR spectra for RC5–Cys–LL dispersed SWNTs after 100 000 g centrifugation and dialysis against 3 L of deionized water. The deconvoluted peaks attributed to van Hove singularities of various SWNT chiralities are designated by color (bottom). The summation of each set of van Hove singularities for the corresponding peptides (top). All spectra were background subtracted to remove contributions from π-plasmon, NTC, and bundled CNT absorption features*

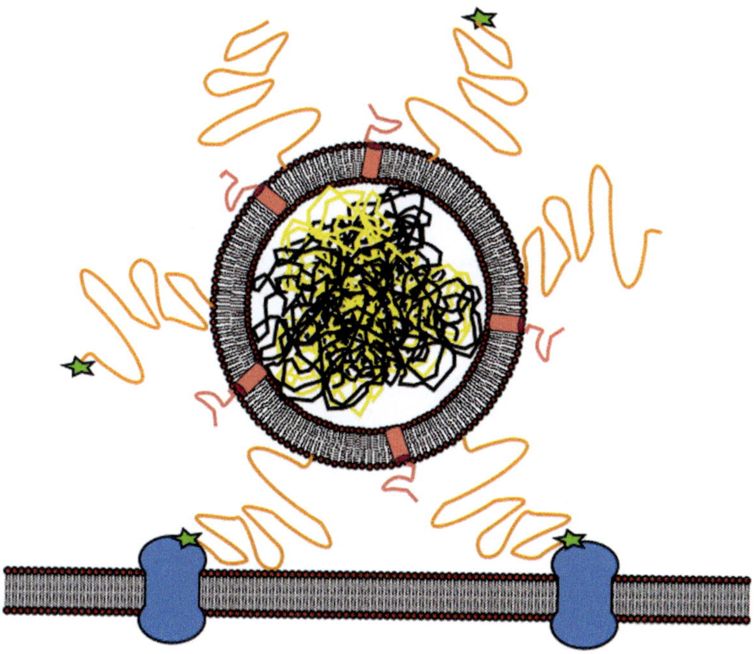

Figure 9.1 *The interaction of a multifunctional envelope-type nano device (MEND) with a specific receptor (blue) is shown. The complex of DNA (black) and a cationic polymer (yellow) is enveloped by a lipid bilayer (pink), which has been functionalised by peptides crossing the bilayer (pink), PEG tentacles (orange) and targeting ligands (green stars). The MEND particles are typically 300 nm in diameter and exhibit a negative surface charge due to the acidic lipids of the membrane*

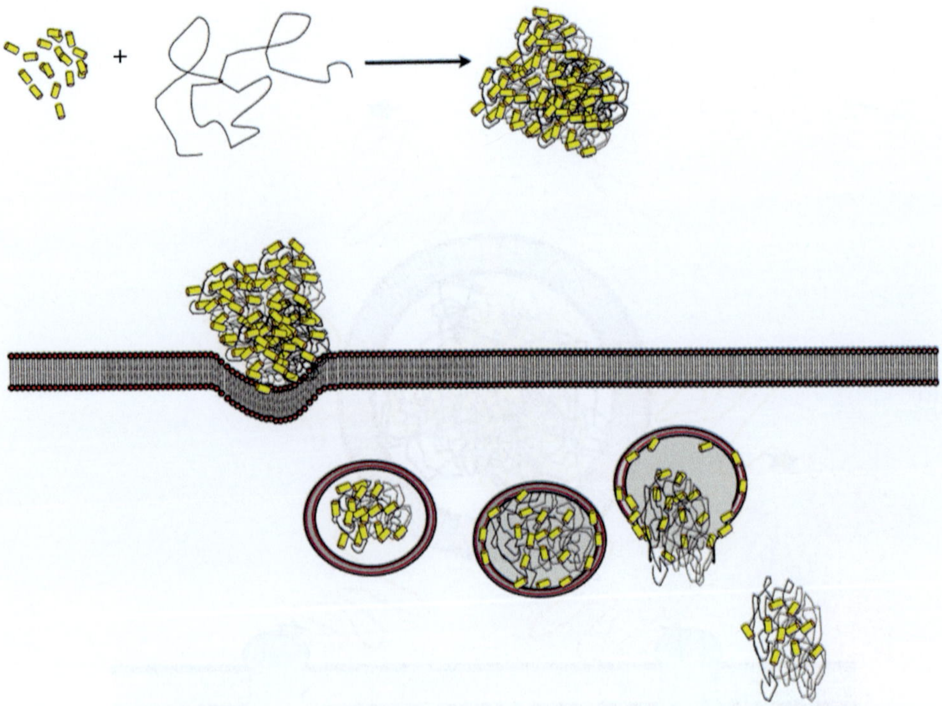

Figure 9.2 *The formation (first row), endoscytosis (second row) and endosomal escape (bottom row) of an LAH4–DNA transfection complex is shown. These complexes form spontaneously through electrostatic attraction when the two polymers are mixed. The peptide has been shown active in the transfection of plasmids of 7.6 kbp [16] or of a 21-base siRNA [17]. The resulting structures are typically 100 nm to micrometers in size and carry a slightly positive surface charge [18]. The peptides are shown in yellow and the nucleic acids in black. The resulting complexes are taken up by endocytosis. Acidification of this organelle results in the release of about half of the peptides from the DNA and at least some of them become available for interactions with the membrane. This liberation of peptide from the complex results in membrane lysis and the release of DNA (or siRNA) into the cytoplasm*

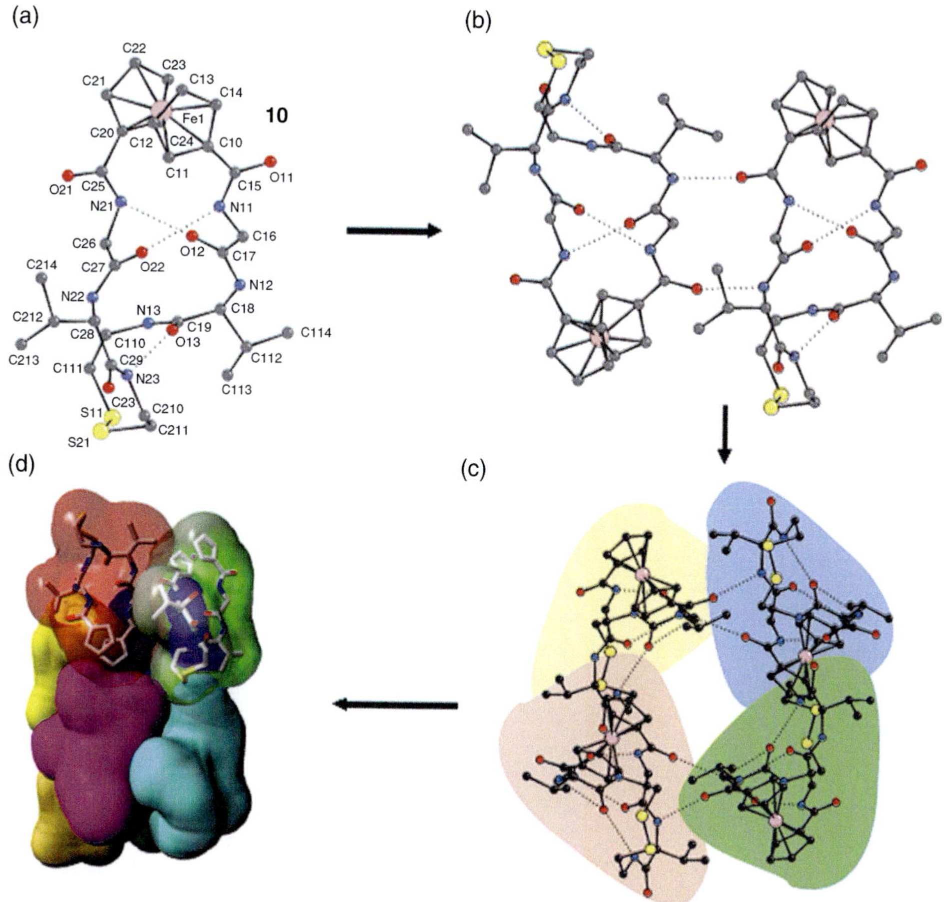

Figure 10.5 *(a) The molecular structure of Fc[CO–Gly–Val–CSA]$_2$ (**10**) showing the intramolecular H-bonding between the peptide strands on the two Cp rings. (b) The formation of an antiparallel β-sheet through intermolecular N(H)···O=C H-bonding. Molecules are arranged in a head-to-tail fashion. (c) Curvature of the molecular building blocks promotes H-bonding to form a β-barrel-like structure. (d) Tiling of cyclic ferrocene peptides is illustrated in this molecular surface representation of the barrel viewed along the c-axis. Reproduced with permission from reference [26]. Copyright (2006) Wiley-VCH Verlag GmbH & Co. KGa*

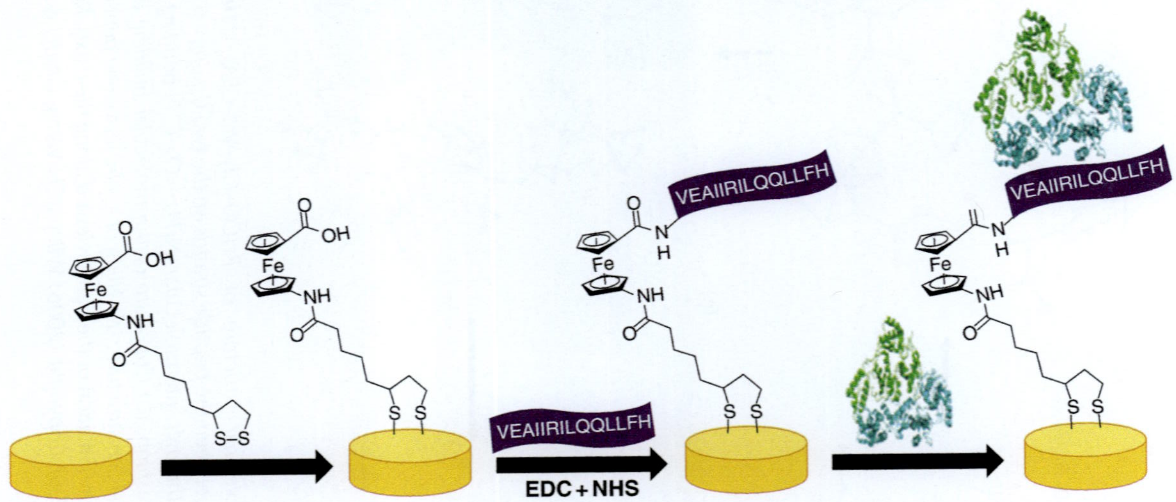

Figure 10.15 *Illustration of HIV-1 protein analysis by using disubstituted Fca conjugate on Au surfaces. Reproduced with permission from reference [75]. Copyright (2011) Elsevier*

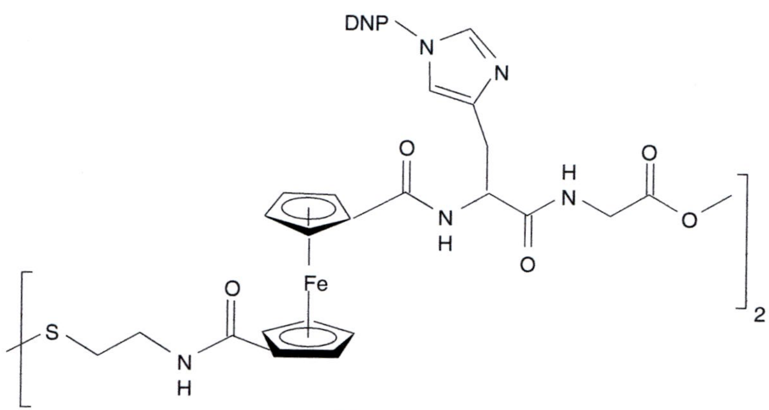

31

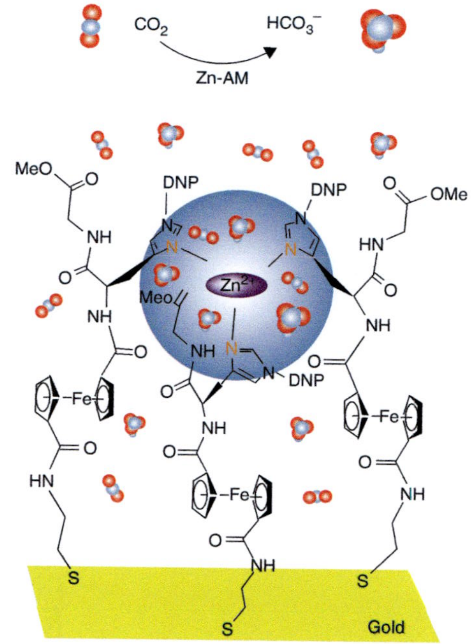

Figure 10.16 *Molecular structure of the artifical carbonic anhydrase model Fc-conjugates* ***31*** *and an illustration of the principle used for coordination of Zn(II) and reduction of CO_2 to HCO_3^-. Reproduced with permission from reference [77]. Copyright (2011) The Royal Society of Chemistry*

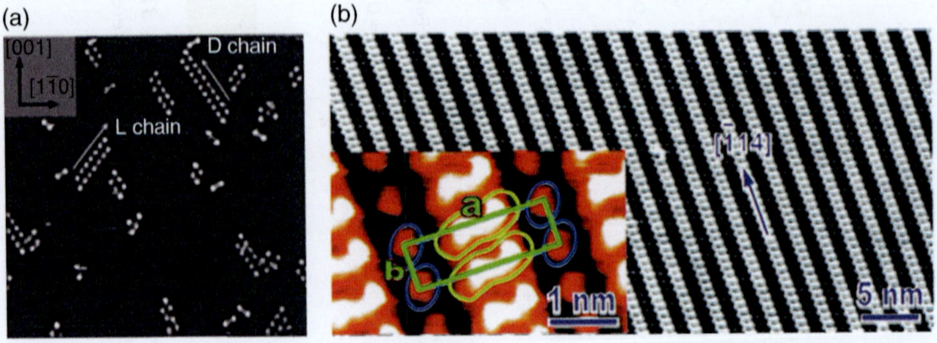

Figure 11.3 *STM images of Phe–Phe dipeptide molecules adsorbed on Cu(110). In (a) single molecules (two white blobs) and chains of molecules [35]; in (b) extended 2D arrays of Phe–Phe molecules stabilized by coadsorption of TPA molecules [36]. Reproduced with permission from references [35] and [36]. Copyright (2007) John Wiley & Sons, Ltd and (2007) American Chemical Society*

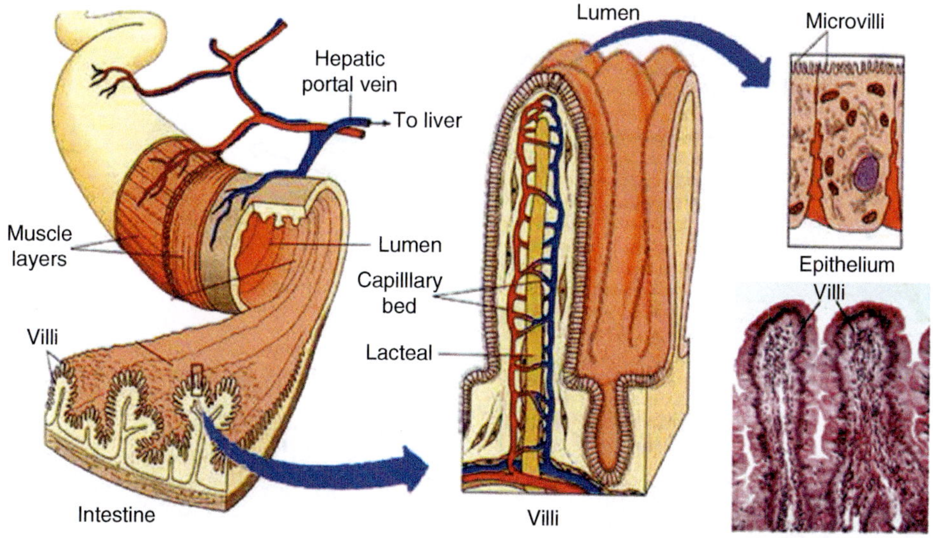

Figure 14.1 *Schema of intestinal epithelium, villi, and microvilli. From http://www.daviddarling.info/encyclopedia/S/small_intestine.html*

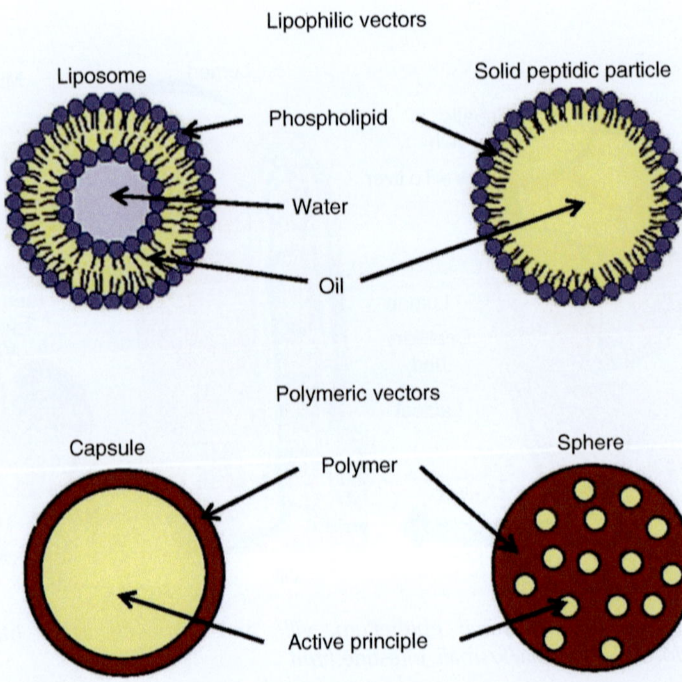

Figure 14.2 *Schematic representation of lipidic and polymeric vectors*

Printed and bound by CPI Group (UK) Ltd, Croydon, CR0 4YY
17/09/2026
14964269-0006